Understanding Modeling and Simulation of Polymerization Reactions

Esmaiel Jabbari

University of South Carolina
Columbia, SC, USA

WILEY

Published by John Wiley & Sons, Inc., Hoboken, New Jersey.

Library of Congress Cataloging-in-Publication Data has been applied for:

Print ISBN: 9781394400799
ePDF ISBN: 9781394400812
ePub ISBN: 9781394400805

Cover Design: Wiley
Cover Image: © theasis/Getty Images

To my mother and father who encouraged me to
"seek knowledge from the cradle to the grave"

Contents

Preface

Plastics, which are made from synthetic polymers, are ubiquitous in everyday life. The word "plastic" expresses the salient feature of polymers, that is, plasticity or the ability to change shape or deform without breaking. Polymers are used in everyday life because they are lightweight, deformable, unbreakable, yet durable materials. Polymers possess a wide range of forms from high-performance fibers to a glassy solid, rubber, sponge, gel, and even as a viscous liquid. Polymers can be highly stable or can be environmentally sustainable with tunable degradation. Aside from the chemical structure of the monomer, these unique properties are rooted in the polymer chain architecture, like molecular weight, weight distribution, branching, crosslinking, and sequence distribution, among others. The polymer properties depend on the method of polymerization. Considering that polymers have a wide range of properties, the study of methods of modeling polymerization to predict polymer properties will undoubtedly aid the process of designing and developing novel polymers. This book presents a textbook with a structured approach to teaching modeling and simulation of polymerization reactions for students of science and engineering in academia.

The book starts with a chapter on mechanisms of chain formation from monomers, including addition, step-growth, and their combinations. Within the addition polymerization, different methods like radical, reversible-deactivation, anionic, cationic, ring-opening, and group transfer polymerizations are described. In reversible-deactivation, different chemistries like atom-transfer (ATRP), nitroxide-mediated (NMP), and addition-fragmentation chain (RAFT) are described. Chapter 2 continues with coverage of mechanisms from the perspective of elementary reactions that are common among different polymerization methods, like chain initiation, propagation, and termination. As polymerization reactions produce a distribution of chain lengths, Chapter 3 is dedicated to the functions of distributed variables and the determination of their moments, averages, variance, and dispersity. Chapter 3 also describes the standard distributions used to model the chain length distribution of polymers. Some elementary reactions in polymerization, like polycondensation involve convolution of two chain distributions. Therefore, Chapter 4 covers the method of z-transform and its application to convoluted distributed functions, their moments in the z-domain, and the application of z-transform in addition and step-growth polymerization. At any instance of time during polymerization, propagating and terminated chains are produced and consumed by different elementary reactions, with each production or consumption reaction having its own unique distribution. Therefore, Chapter 5 first describes the mixing theory of polymer chains with separate distributions. Then, the mixing theory is used to derive the conservation equations for the molar moments of propagating, terminated, and dormant chains in a differential control volume of the reaction in Cartesian, cylindrical, and spherical coordinate systems. Next, in Chapter 5, the derived conservation equations are applied to modeling several types of reactors, like batch stirred vessel, continuous stirred tank, and plug flow tubular reactors. As the propagating species in a polymerization reaction are highly reactive and short-lived, next, the quasi-steady state of the propagating species is described and used to model the composition of propagating species in polymerization. Then, in the next section, theoretical models for predicting the composition of propagating species are presented for several types of polymerization methods, including conventional radical, reversible deactivation radical, anionic, and cationic polymerization. In a differential time dt during the reaction, chains are instantaneously produced or consumed from the distribution of existing chains at time t in the reaction. Therefore, Chapter 6 applies the method of z-transform to develop models to relate the instantaneous distribution of chains produced or consumed by different elementary reactions in differential time dt to the chain distribution at time t in the reaction. Then, in Chapter 7, the methodologies developed in previous chapters are applied to modeling of several reaction systems including conventional radical polymerization with thermal and photoinitiation, reversible-deactivation radical polymerization, and anionic and cationic polymerizations. The viscosity of a polymerizing mixture increases exponentially with monomer conversion, especially in bulk polymerization, which in turn exponentially decreases the diffusivity of propagating and terminated chains and results in a diffusion-limited reaction. Therefore, Chapter 8 discusses theoretical models to relate viscosity of the reaction medium to monomer conversion, chain diffusivity, and polymer molecular weight in different reaction regimes including dilute, semi-dilute, and concentrated below and above the onset of chain entanglement. Then, in Chapter 8, the conservation equations for the molar moment of polymer chains in diffusion-limited reactions are derived, which results in time constants for diffusivity of different chain types and their apparent reaction rate constants. Next, the derived moment conservation equations are applied to modeling bulk radical polymerization of methyl methacrylate. The events based on elementary reactions in polymerization occur stochastically with a random distribution function until the polymer chain reaches the termination state. A Markov chain defines the sequence of events in a stochastic process before reaching an absorbing state. Therefore, Chapter 9 describes a Markovian process and its elements like state space, transition probability matrix, and absorbing states. Then, the remainder of the chapter is devoted to developing stochastic models of chain distribution, based on the Markovian process, for addition and step-growth polymerization. The properties of copolymers strongly depend on sequence distribution of the different monomers in the polymer chains. The different monomers in copolymerization are added stochastically to the propagating chains, in a Markovian process, based on their composition and reactivity. Therefore, Chapter 10 starts with the representation of copolymer sequences, followed by defining monomer sequence distributions and copolymer randomness. Then, the Markov process is used to develop stochastic models for terminal copolymerization of two and three monomers for predicting copolymer composition, n-mer sequence distribution, and averages for each monomer in the copolymer. Then, based on the developed models, the conditions to produce random, ideal, alternate, or block copolymers are defined. Furthermore, the data required to assess the validity of the stochastic copolymerization model is discussed. The Monte Carlo approach that simulates a representative microscopic volume of the reaction is the preferred method to model complex polymerizations involving chains with multiple reactive species. Therefore, Chapter 11 describes the Monte Carlo simulation method and its application to modeling polymerization reactions. The stochastic method of Monte Carlo simulates the reaction at the molecular scale by assigning a probability of occurrence to each elementary reaction in polymerization. As random numbers are used to choose one of the elementary reactions at each step of the simulation, statistical distributions and methods for testing uniformity, interval, and serial correlation between the generated random numbers are described. Then, the Monte Carlo method is applied to the complex emulsion polymerization of butadiene to model the extent of tri-functional and tetra-functional branching.

Finally, polymers are among the most stable materials ever made by humankind. Due to their persistence in the environment, plastic pollution is widespread in nature from land to rivers, lakes, and oceans, which threatens the well-being of humankind as well as other species. I hope that the modeling methodologies covered in this book will lead to a rigorous understanding of polymer degradation and the development of environmentally friendly and sustainable plastics in the future.

Esmaiel Jabbari

Acknowledgment

I would like to thank my undergraduate and graduate students and post-doctoral researchers who contributed immensely to my knowledge of polymers and soft materials. I would like to thank Mr. Ahmed Iftikhar (Univ. South Carolina) for assistance in preparation of the book index.

I would like to thank my academic advisors for their mentorship and guidance throughout my academic career, for which I am very grateful. These include the late Dr. Gerhard H. Beyer (chemical engineering, Virigina Tech), the late Dr. Thomas C. Ward (Chemistry, Virginia Tech), the late Dr. Jack D. Graybeal (chemistry, Virginia Tech), Dr. Nicholas A. Peppas (Chemical Engineering, U. Texas Austin), Dr. Antonios G. Mikos (Bioengineering, Rice U.), and Dr. Michael J. Yaszemski (Orthopedic Surgery, Mayo Clinic). I especially like to thank my PhD mentor, Dr. Nicholas Peppas, for supporting my career. I would not have been able to complete this book without their guidance and mentorship.

I would like to thank the Departments of Biomedical Engineering and Chemical Engineering in the College of Engineering and Computing at the University of South Carolina for nurturing a scholarly environment for dissemination of knowledge.

Finally, I would like to thank my family, especially my sons Saman and Erfan, for inspiring me to write this book. You are the light of my life.

About the Companion Website

This book is accompanied by a companion website for BCS Standard: Instructor.

www.wiley.com/go/jabbaripolymerization

This website includes:

- Solution Manual

1

Mechanisms and Methods of Polymerization

1.1 Introduction

Polymers are chain-like molecules consisting of smaller repeat units. They are synthesized from smaller molecules that are chemically linked together. The smaller molecules are called monomers, and the repeat units on a long polymer chain are called monomer residues. Polymers are divided into natural polymers and synthetic polymers. Natural polymers or biopolymers are made by nature as part of living organisms, including microorganisms, plants, animals, and humans. They include polysaccharides from smaller saccharide or sugar monomers, proteins from smaller amino acid monomers, and polynucleotides from smaller nucleotide monomers. Biopolymers, due to having different types of repeat units in their molecular structure, like 20 different amino acids in proteins, display multiple and specific functions in living organisms. Synthetic polymers are designed and produced by mankind by linking small molecules with at least two reactive functional groups to form long chains. Unlike natural polymers, synthetic polymers are made of one, two, or at most three different types of monomers. Most synthetic polymers are homopolymers, that is, they are made of one monomer, like polyethylene and polymethylmethacrylate. Examples of copolymers include polyethylene–vinyl acetate and polystyrene–methyl methacrylate. An example of a terpolymer is polyacrylonitrile–butadiene–styrene or ABS terpolymer. There are very few, if any, synthetic polymers made up of four different monomers. Another difference between biologic and synthetic polymers is that biopolymer chains are, in general, monodisperse, whereas chains in synthetic polymers are polydisperse. For example, the chains in a sample of albumin protein have identical molecular weights, whereas chains in a polyethylene sample have a distribution of molecular weights with a specified average value and a specified polydispersity. The unique property that distinguishes polymeric materials from other man-made materials like metals and ceramics is the aspect ratio of the molecules making up the material. The aspect ratio of a molecule is defined as the ratio of its contour length to its diameter. The aspect ratio for materials made of small molecules is between 1 and 10. Conversely, the aspect ratio of polymer molecules is generally greater than one thousand. The high aspect ratio imparts unique properties to polymers. For example, unlike metals and ceramics, polymers do not exhibit a boiling point with increasing temperature because the long chains in a polymer degrade prior to gaining sufficient energy to evaporate. Instead, polymers exhibit a glassy temperature below which the chains in the polymer sample are frozen with no translational diffusion, even though the polymer is above its melting temperature. As a result of their high aspect ratio, polymers possess unique physical, mechanical, chemical, and biological properties. The aspect ratio of polymers is characterized by the number of monomeric repeat units (N) or degree of polymerization (DP), chain length (L), which is N or DP multiplied by the length of a repeat unit, or molecular weight (M), which is N or DP multiplied by molecular weight of the repeat unit. As chains in a sample have a distribution of N or DP, polymers are characterized by their average number of repeat units $(\overline{N})$, average degree of polymerization $(\overline{DP})$, average chain length $(\overline{L})$, or average molecular weight $(\overline{M})$. The $\overline{DP}$ of a polymer depends on reaction conditions for polymerization. These include temperature, pressure, pH, ionic strength, concentrations of monomers, solvent, initiator, co-initiator, catalyst, activating agent, crosslinker, chain transfer agent, chain terminating agent, retarder and inhibitor, impurities and oxygen, ratios of monomers and solvents, among others. Reaction conditions not only affect molecular weight and its distribution but also density of short and long branching, chain-end functional groups, unsaturated groups along the chain in multifunctional monomers, intra- and inter-molecular branching, and their distributions. Reaction conditions also affect chain architecture (linear, short-branch, long-branch, star-like, comb-like), distribution of monomers in copolymers (ideal, random, alternate, block), and polymer tacticity. As there is a wide range of reaction conditions affecting properties, there is a need to develop theoretical models and simulation methods to relate molecular weight and its distribution to reaction conditions. As the first step, this chapter covers mechanisms and methods of polymerization, which include addition and step-growth polymerization. Addition polymerization further includes conventional and reversible-deactivation radical polymerization (RDRP), anionic and cationic polymerization, ring-opening, and group transfer polymerization.

1.2 Mechanisms of Polymerization

1.2.1 Addition Polymerization

In the addition polymerization, monomer molecules react with active species formed by initiator dissociation to form propagating chains [1, 2]. Polymerizations in which the active species on propagating chains is a radical are called radical polymerization, whereas those with an anion or cation are called anionic polymerization and cationic polymerization, respectively [3–6].

1.2.1.1 Conventional Radical Polymerization

Monomers with unsaturated vinyl groups like styrene, methyl methacrylate, ethylene, propylene, vinyl alcohol, vinyl chloride, and alkyl vinyl ethers, as well as monomers with unsaturated triple bonds, polymerize by addition polymerization according to the following reaction:

$$\sim\!\sim\!\sim\!\sim \underset{X}{\overset{H}{C^{\bullet}}} + \underset{H}{\overset{H}{C}}{=}\underset{X}{\overset{H}{C}} \xrightarrow{k_p} \sim\!\sim\!\sim\!\sim \underset{X}{\overset{H}{C}}{-}\underset{H}{\overset{H}{C}}{-}\underset{X}{\overset{H}{C^{\bullet}}} \tag{1.1}$$

where k_p is the rate constant for the propagation reaction. In the above reaction, the radical attached to the terminal carbon atom of the propagating chain reacts with a vinyl monomer by dissociating the π-bond of the vinylic carbon. Following the reaction, the radical is transferred to the terminal carbon atom of the added monomer to increase the number of monomer residues on the propagating chain radical by one residue. The radical on the propagating chain is normally positioned on the carbon atom with a side group, shown by "X" in the above reaction, as the side group increases the stability of the radical by resonance stabilization [7]. Radical polymerization requires an initiator to form the primary reactive species. Common reactions in addition polymerization are chain initiation, chain propagation, and chain termination. In the initiation step, the initiator dissociates by the addition of a reducing or oxidizing agent, or by a change in temperature, pH, or other external factors like ultraviolet radiation to form one or more reactive species [8, 9]. The initiation reaction for a typical temperature-activated initiator like benzoyl peroxide (BPO) is as follows [10]:

$$H_5C_6—\overset{O}{\overset{\|}{C}}—O—O—\overset{O}{\overset{\|}{C}}—C_6H_5 \xrightarrow{k_d} {}^2H_5C_6^{\bullet} + {}^2CO_2 \tag{1.2}$$

In the above reaction, the weak —O—O— bond of BPO dissociates upon raising the reaction temperature to 80–100 °C, with a rate constant k_d to generate two benzoyloxy radicals. As the benzoyloxy radical is highly unstable, it further disintegrates into a more stable but transient benzyl radical with the liberation of a molecule of carbon dioxide [10]. Some initiators dissociate into two active species while others dissociate into one active species and an inactive fragment [9]. Prior to escaping the dissociation cage, the benzyl radicals may recombine to form an inactive biphenyl compound as follows [11–14]:

$$H_5C_6^{\bullet} + H_5C_6^{\bullet} \xrightarrow{k_c} H_5C_6—C_6H_5 \tag{1.3}$$

In the chain initiation step, the generated radical reacts with a vinyl monomer in the reaction mixture to form a 1-mer propagating radical as follows [11–13]:

$$H_5C_6^{\bullet} + \underset{H}{\overset{H}{C}}=\underset{X}{\overset{H}{C}} \xrightarrow{k_{ci}} H_5C_6—\underset{H}{\overset{H}{C}}—\underset{X}{\overset{H}{C}}{}^{\bullet} \tag{1.4}$$

where k_{ci} is the rate constant for initiation of a propagating radical. In the propagation step, n vinyl monomers react with a 1-mer propagating chain in n propagation reaction steps, as shown in Rxn. 1.1, to generate a propagating chain radical with $n + 1$ repeat units [15]. As the propagating chain radical is transient, it ultimately abstracts a hydrogen from a chain or combines with another propagating chain to terminate the propagating radical, and forms one or two terminated chains as follows [16, 17]:

$$\begin{aligned} &R—\left[\underset{H}{\overset{H}{C}}—\underset{X}{\overset{H}{C}}\right]_n—\underset{H}{\overset{H}{C}}—\underset{X}{\overset{H}{C}}{}^{\bullet} + R—\left[\underset{H}{\overset{H}{C}}—\underset{X}{\overset{H}{C}}\right]_m—\underset{H}{\overset{H}{C}}—\underset{X}{\overset{H}{C}}{}^{\bullet} \\ &\xrightarrow{k_t} R—\left[\underset{H}{\overset{H}{C}}—\underset{X}{\overset{H}{C}}\right]_n—\underset{H}{\overset{H}{C}}—\underset{X}{\overset{H}{C}}—\underset{H}{\overset{H}{C}}—\underset{X}{\overset{H}{C}}—\left[\underset{H}{\overset{H}{C}}—\underset{X}{\overset{H}{C}}\right]_m—R \end{aligned} \tag{1.5}$$

where k_t is the rate constant for termination. The termination reaction may take place by termination by disproportionation or termination by combination (shown above), or transfer reactions, which will be discussed in Chapter 2. The defining feature of radical polymerization is that the molecular weight of the synthesized polymer is relatively independent of the extent of polymerization or monomer conversion [18]. Conventional radical polymerization is not a living polymerization unless termination reactions are suppressed, as will be discussed in the reversible deactivation radical polymerization section [19, 20]. Initiation in radical polymerization is generally not instantaneous; that is, initiator dissociation is time-dependent, which leads to time-dependent chain initiation/termination reactions and a relatively broad chain molecular weight distribution.

1.2.1.2 Anionic Polymerization

Vinyl monomers with electrophilic or polar groups that stabilize a propagating anion undergo anionic polymerization [21–23]. Examples include styrene, butadiene, and isoprene with electrophilic groups and acrylonitrile, methyl methacrylate, and cyanoacrylate with polar groups [24]. A commonly used anionic polymerization initiator is butyl lithium [25, 26]. The carbon–lithium bond in *n*-butyl lithium is highly polarized with a significant ionic character [27]. Consequently, *n*-butyl lithium spontaneously dissociates into an ion pair in a suitable organic solvent like hexane as follows [25, 26]:

$$\underset{H}{\overset{H}{HC}}-\underset{H}{\overset{H}{C}}-\underset{H}{\overset{H}{C}}-\underset{H}{\overset{H}{C}}-Li \xrightarrow{n-hexane} \underset{H}{\overset{H}{HC}}-\underset{H}{\overset{H}{C}}-\underset{H}{\overset{H}{C}}-\underset{H}{\overset{H}{C^-}}\ Li^+ \tag{1.6}$$

The dissociation of butyl lithium in an organic solvent, akin to that of sodium chloride in water, is spontaneous as the butyl anion–lithium cation pair is far more stable in the solvent compared to the nonionized form. The propagation reaction for anionic polymerization of styrene in *n*-hexane with butyl lithium as the initiator under inert argon atmosphere is given by the following reaction [25, 26]:

$$\begin{aligned} &HC_4H_8-\left[\underset{H}{\overset{H}{C}}-\underset{C_6H_5}{\overset{H}{C}}\right]_{n-1}-\underset{H}{\overset{H}{C}}-\underset{C_6H_5}{\overset{H}{C^{\bullet}}}\ Li^+ + \underset{H}{\overset{H}{C}}=\underset{C_6H_5}{\overset{H}{C}} \\ &\xrightarrow{k_p} HC_4H_8-\left[\underset{H}{\overset{H}{C}}-\underset{C_6H_5}{\overset{H}{C}}\right]_{n}-\underset{H}{\overset{H}{C}}-\underset{C_6H_5}{\overset{H}{C^{\bullet}}}\ Li^+ \end{aligned} \tag{1.7}$$

where k_p is the rate constant for anionic chain propagation. Following the addition of styrene to the reaction, the monomer is inserted between the pair of butyl anion and lithium cation as shown above. This is attributed to resonance stabilization by delocalization of the anion in the product to the benzene ring of the styrene residue at the terminal end of the propagating anion [28]. The highly polarized nature of butyl lithium implies spontaneous initiation and dissociation of all initiator molecules to form propagating anions. Generally, anionic polymerization is carried out under inert conditions because oxygen in the atmosphere or traces of water in the reaction react with anions to deactivate the propagating chains. Under inert atmosphere, the propagating anions continue to grow until the monomer in the reaction mixture is depleted. The polymer molecular weight in anionic polymerization depends strongly on the solvent type and the initiator counterion [29]. The propagating anions are quenched deliberately by exposure to oxygen, carbon dioxide, or moisture, or unintentionally by impurities in the monomer or solvent [30, 31]. Consequently, anionic polymerization is carried out with monomers and solvents with high purity and under an inert atmosphere (nitrogen or argon) to prevent quenching of the propagating anions [32]. As the anions at the terminal end of the propagating chains repel each other and the solvated anion–cation pairs are stable, anionic polymerization is considered a living polymerization system; hence, propagating anions grow indefinitely by the addition of more monomer to the reaction mixture. Due to the absence of termination reactions, the chain distribution in anionic polymerization is narrow and described by a Poisson distribution (see Chapter 3) [33, 34].

1.2.1.3 Cationic Polymerization

Vinyl and heterocyclic monomers with nucleophilic groups that react with electrophilic initiators to stabilize a propagating cation undergo cationic polymerization [35–37]. Examples include styrene [38], 1,3-alkyl butadiene [39], tetrahydrofuran [40, 41], and 2-oxazoline [42] monomers. Lewis acids such as aluminum chloride ($AlCl_3$), boron trifluoride (BF_3), tris(pentafluorophenyl)borane [$B(C_6F_5)_3$], and titanium tetrachloride ($TiCl_4$) are commonly used as Lewis acid initiators or co-initiators in cationic polymerization [43–48]. Elementary reactions in cationic polymerization include initiation [49], propagation [50], termination [51], and chain transfer reactions [52]. The elementary initiation reactions in cationic polymerization of styrene with cumyl alcohol $[C_6H_5C\,(CH_3)_2\,OH]$ as initiator, tris(pentafluorophenyl)borane $(C_6F_5)_3\,B$ as co-initiator, and water as catalyst [53] are as follows:

$$(C_6F_5)_3\,B + {}^2H_2O \xrightarrow{k_{cpx}} (C_6F_5)_3\,\overset{OH}{B^-}\,H_3^+O \tag{1.8}$$

$$(C_6F_5)_3\,\overset{OH}{B^-}\,H_3^+O + C_6H_5C\,(CH_3)_2\,OH \xrightarrow{k_{cxi}} (C_6F_5)_3\,\overset{OH}{B^-}\,C_6H_5\underset{CH_3}{\overset{CH_3}{C^+}} + {}^2H_2O \tag{1.9}$$

In the first reaction above, $(C_6F_5)_3\,B$ as a Lewis acid accepts an electron pair from H_2O (electron donor), with a rate constant k_{cpx} to generate a tightly bound $(C_6F_5)_3\,B\,(OH)^-\ H_3^+O$ ion pair complex. The strong interaction between the negative and positive ions prevents the reaction of the hydronium ion H_3^+O with the monomer in the ion pair to form a propagating cation. Hence, in the second reaction, cumyl alcohol as a weak electron acceptor reacts with the ion pair complex with a rate constant k_{cxi} to replace the hydronium ion with $C_6H_5\overset{+}{C}\,(CH_3)_2$ cation. The styrene monomer more efficiently adds to the less tightly bound $C_6H_5\overset{+}{C}\,(CH_3)_2$ cation to form a propagating cation. Next, the styrene monomer is efficiently inserted in the solvent-separated ion pair with a rate constant k_{ci} to form a propagating cation as follows:

$$(C_6F_5)_3\,\overset{OH}{B^-}\,C_6H_5\underset{CH_3}{\overset{CH_3}{C^+}} + \underset{H}{\overset{H}{C}}{=}\underset{C_6H_5}{\overset{H}{C}} \xrightarrow{k_{ci}} C_6H_5\underset{CH_3}{\overset{CH_3}{C}}-\underset{H}{\overset{H}{C}}-\underset{C_6H_5}{\overset{H}{C^+}}\,(C_6F_5)_3\,\overset{OH}{B^-} \tag{1.10}$$

The propagating cation reacts with n monomers with a rate constant k_p to form an (n+1)-mer propagating cation as follows:

$$\begin{aligned} &C_6H_5\underset{CH_3}{\overset{CH_3}{C}}-\underset{H}{\overset{H}{C}}-\underset{C_6H_5}{\overset{H}{C^+}}\,(C_6F_5)_3\,\overset{OH}{B^-} + {}^n\underset{H}{\overset{H}{C}}{=}\underset{C_6H_5}{\overset{H}{C}} \\ &\xrightarrow{k_i} C_6H_5\underset{CH_3}{\overset{CH_3}{C}}-\left[\underset{H}{\overset{H}{C}}-\underset{C_6H_5}{\overset{H}{C}}\right]_n-\underset{H}{\overset{H}{C}}-\underset{C_6H_5}{\overset{H}{C^+}}\,(C_6F_5)_3\,\overset{OH}{B^-} \end{aligned} \tag{1.11}$$

The propagating cation is terminated by intramolecular β-hydrogen abstraction [51], anion combination with propagating cation [54], intentional addition of a nucleophile or terminating agent [55, 56], chain transfer to water [57–59] or monomer [60–63], hydride ion transfer to monomer or polymer [64], and chain transfer to additives [57, 65]. The termination reaction by β-hydrogen abstraction in $(C_6F_5)_3B$ co-initiated styrene polymerization is as follows [53]:

$$\begin{aligned} &C_6H_5\underset{CH_3}{\overset{CH_3}{C}}-\left[\underset{H}{\overset{H}{C}}-\underset{C_6H_5}{\overset{H}{C}}\right]_n-\underset{H}{\overset{H}{C_\beta}}-\underset{C_6H_5}{\overset{H}{C^+}}\,(C_6F_5)_3\,\overset{OH}{B^-}H_2O \\ &\xrightarrow{k_{t\beta H}} C_6H_5\underset{CH_3}{\overset{CH_3}{C}}-\left[\underset{H}{\overset{H}{C}}-\underset{C_6H_5}{\overset{H}{C}}\right]_n-\overset{H}{C}{=}\underset{C_6H_5}{\overset{H}{C}} + (C_6F_5)_3\,\overset{OH}{B^-}H_3^+O \end{aligned} \tag{1.12}$$

where $k_{t\beta H}$ is the rate constant for termination by β-hydrogen abstraction. In the above reaction, the $(C_6F_5)_3\,B\,(OH)^-$ anion abstracts a hydrogen from the β-carbon adjacent to the cation-carrying carbon to form a new $(C_6F_5)_3\,B(OH)^-H_3^+O$ ion pair and a polymer chain with an unsaturated allylic end-group. The termination reaction by combination with the anion in cationic polymerization of isobutene with stannic chloride ($SnCl_4$) and water as the catalyst and co-catalyst, respectively, is given by the following reaction [54, 66, 67]:

$$\begin{aligned} &(CH_3)_3\,C-\left[\underset{H}{\overset{H}{C}}-\underset{CH_3}{\overset{CH_3}{C}}\right]_n-\underset{H}{\overset{H}{C}}-\underset{CH_3}{\overset{CH_3}{C^+}}\,Cl_4\overset{OH}{Sn^-} \xrightarrow{k_{tcan}} \\ &(CH_3)_3\,C-\left[\underset{H}{\overset{H}{C}}-\underset{CH_3}{\overset{CH_3}{C}}\right]_n-\underset{H}{\overset{H}{C}}-\overset{CH_3}{C}{=}CH_2 + SnCl_4 + H_2O \end{aligned} \tag{1.13}$$

where k_{tcan} is the rate constant for termination by the combination of the propagating cation with anion. The anion abstracts a hydrogen atom from one of the methyl groups attached to the cation-carrying carbon to form a terminated polymer chain ending in an allylic group. The hydrogen abstraction by anion regenerates the catalyst and co-catalyst to initiate a new propagating cation. The termination reaction by intentional addition of methanol to $(C_6F_5)_3\,B$ initiated styrene polymerization is as follows [53]:

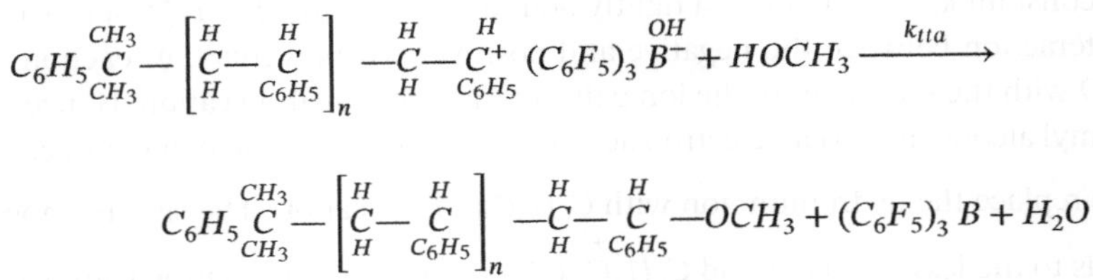

(1.14)

Where k_{tta} is the rate constant for termination by the terminating agent. In the above reaction, the methoxy group in methanol acts as a nucleophile to react and terminate the propagating cation while concurrently donating its electrophilic hydrogen to $(C_6F_5)_3\ B(OH)^-$ anion to regenerate the co-initiator and catalyst. In a reaction similar to that for methanol, termination by chain transfer to water in $(C_6F_5)_3\,B$ co-initiated styrene polymerization is as follows [53]:

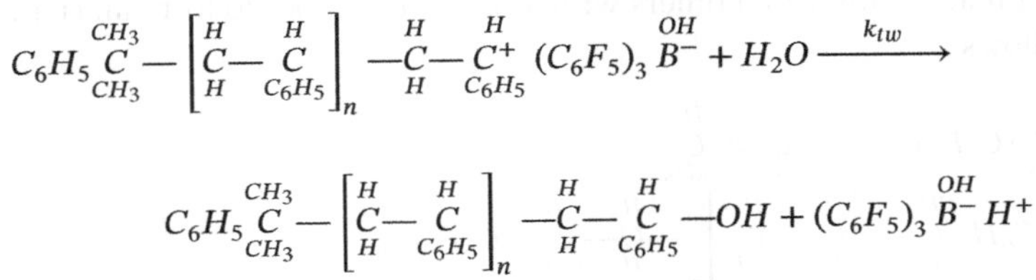

(1.15)

where k_{tw} is the rate constant for termination by chain transfer to water. The above reaction generates a tightly bound $(C_6F_5)_3\,B\,(OH)^-$ ion pair, which can initiate a new propagating chain according to Rxn. 1.9. The chain transfer to monomer in $SnCl_4$ catalyzed polymerization of isobutene is given by the following reaction [66]:

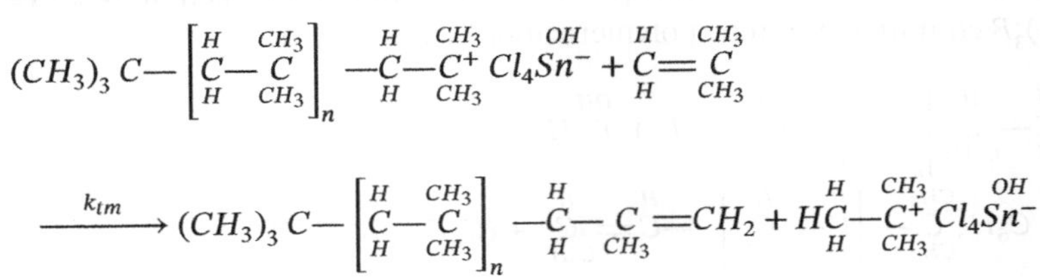

(1.16)

where k_{tm} is the rate constant for termination by chain transfer to monomer. The intermolecular hydride ion transfer to monomer and polymer in cationic polymerization of isobutene are given by the following reactions [64]:

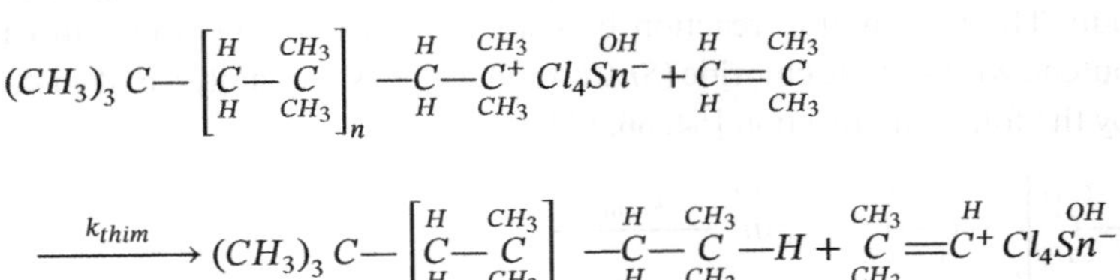

(1.17)

$$(CH_3)_3\,C-\left[\underset{H}{\overset{H}{C}}-\underset{CH_3}{\overset{CH_3}{C}}\right]_n-\underset{H}{\overset{H}{C}}-\underset{CH_3}{\overset{CH_3}{C^+}}\,\overset{OH}{Cl_4Sn^-}+\sim\sim\sim\underset{H}{\overset{H}{C}}-\underset{CH_3}{\overset{CH_3}{C}}\sim\sim\sim$$

$$\xrightarrow{k_{thip}}(CH_3)_3\,C-\left[\underset{H}{\overset{H}{C}}-\underset{CH_3}{\overset{CH_3}{C}}\right]_n-\underset{H}{\overset{H}{C}}-\underset{CH_3}{\overset{CH_3}{C}}-H+\sim\sim\sim\underset{H}{C^+}-\underset{CH_3}{\overset{CH_3}{C}}\sim\sim\sim\overset{OH}{Cl_4Sn^-} \tag{1.18}$$

where k_{thim} and k_{thip} are the rate constants for hydride ion transfer to monomer and polymer, respectively. Unlike hydride ion transfer, termination by proton transfer to monomer or polymer does not form a C=C bond at one end of the terminated chain. Proton or hydride ion transfer to monomer generates a 1-mer propagating cation, whereas hydride transfer to polymer generates an n-mer propagating cation. Chain transfer can also take place between the propagating cation and additives or monomer impurities [65, 68]. As a result of the aforementioned chain transfer and termination reactions, the molecular weight distribution in cationic polymerization is generally broader than anionic polymerization, which depends strongly on solvent type, counter ion, and initiation mechanism [37, 49]. Cationic polymerization should be carried out under inert atmosphere, as oxygen reacts with the propagating cation and reduces its reactivity. However, unlike anionic polymerization, water is more tolerated in cationic polymerization reactions and, in some cases, serves as a catalyst in the initiation reaction [69, 70].

1.2.1.4 Reversible-deactivation Radical Polymerization

Synthetic approaches to narrow the molecular weight distribution in radical polymerization by suppression of termination reactions are termed RDRP [71–73]. RDRP suppresses the termination reaction between two propagating radicals to achieve a living polymerization and a narrow molecular weight distribution [72, 74–76]. The narrow molecular weight distribution in anionic polymerization is rooted in the absence of termination reactions, resulting in stable and active anions that continue to propagate with the addition of more monomer [77, 78]. Conversely, the broad molecular weight distribution in conventional radical polymerization is rooted in time-dependent dissociation of initiator molecules and termination of propagating radicals by combination or disproportionation [79, 80], resulting in short-lived, transient radicals with a lifetime on the order of seconds [81]. RDRP uses reversible deactivation to maintain the propagating chains in a state of dormancy (dormant chains) to increase their lifetime from a few seconds to the entire reaction time (minutes or hours) [82]. The agent for establishing dormancy of the propagating radicals is selected such that the rate of deactivation is much greater than activation. This results in the propagating radicals maintaining dormancy most of the time, which is the hallmark of RDRP [74, 83]. Furthermore, the rate of chain termination between two propagating chain radicals is suppressed as compared to the rate of propagation, which narrows the chain length distribution [83, 84].

There are multiple approaches to achieve reversible deactivation of propagating radicals, including catalyzed reversible coupling [85–90], spontaneous reversible coupling [91–94], and deactivation by degenerative transfer [95–99]. Atom transfer radical polymerization (ATRP) is a method of chain deactivation by catalyzed reversible coupling [87–90]. Nitroxide-mediated polymerization [91–94] and organometallic-mediated radical polymerization [100–102] are two methods of chain deactivation by spontaneous reversible coupling. Methods of reversible deactivation by degenerative transfer include reversible addition-fragmentation chain transfer (RAFT) [103–108] and catalytic chain transfer polymerization mediated by an organo-iodine [109, 110], germanium, tin, phosphorus [111], tellurium, antimony and bismuthine [112–117], selenium [118–120], boron [121–123], copper [124–127], or cobalt [128–130]. Initiators that function both as a chain transfer agent and as a chain terminator, known as an iniferter, reversibly deactivate propagating radicals to maintain dormancy and reduce the rate of termination reactions in radical polymerization

[131–133]. These controlled, living radical polymerization methods are discussed in the following sections.

1.2.1.4.1 Atom Transfer Radical Polymerization

ATRP uses a halide initiator and a redox transition metal catalyst having two oxidation states with affinity for the halogen to reversibly deactivate the propagating radicals. A typical initiator is acetyl bromide, and a typical catalyst is copper bromide/N,N,N′,N′-tetrakis(2-pyridylmethyl)-ethylenediamine (TPEN) [134]. In this catalyst system, copper is the transition metal, bromine is the halogen, and TPEN is the ligand that controls the kinetics of activation and deactivation of propagating radicals. The initiation and reversible deactivation of propagating radicals in ATRP are given by the following reactions [135–138]:

$$R{-}X + LM^{I}X \underset{k_{daci}}{\overset{k_{aci}}{\rightleftharpoons}} LM^{II}X_2 + R^{\bullet} \xrightarrow{nM} RP_n^{\bullet} \tag{1.19}$$

$$RP_n^{\bullet} + LM^{II}X_2 \underset{k_{rac}}{\overset{k_{dac}}{\rightleftharpoons}} DOP_n{-}X + LM^{I}X \tag{1.20}$$

where $R{-}X$ is the halide initiator with X as the halide and $R^{\bullet}$ as the radical; $LM^{I}X$ and $LM^{II}X_2$ are the catalyst in first (M^{I}) and second (M^{II}) oxidation states of the transition metal, respectively, and L is the ligand; $RP_n^{\bullet}$ and $DOP_n{-}X$ are the n-mer propagating radical and n-mer dormant polymer chain, respectively; and M is the monomer. In the first reaction above, the catalyst transitions from low $(LM^{I}X)$ to high $(LM^{II}X_2)$ oxidation state by abstraction of a halide atom from the initiator to generate a radical. The k_{aci} and k_{daci} are rate constants for forward and reverse initiation reactions, respectively. The generated radical reacts with n monomers to form an n-mer propagating radical $RP^{\bullet}$. In the second reaction, the catalyst transitions from high to low oxidation state by donating a halide atom to the n-mer propagating radical to generate a dormant chain $DOP_n{-}X$. The k_{dac} and k_{rac} are rate constants for the forward deactivation and reverse reactivation reactions, respectively. For a living radical polymerization with low polydispersity, propagating chains should be in the dormant state most of the time. This condition implies that the ATRP equilibrium constant, as defined by the following equation, should be much less than one ($K_{ATRP} \ll 1$) [135–138]:

$$K_{ATRP} = \frac{k_{rac}}{k_{dac}} = \frac{[RP_n^{\bullet}]}{[DOP_n{-}X]}\frac{[LM^{II}X_2]}{[LM^{I}X]} \ll 1 \tag{1.21}$$

As $k_{aci}/k_{daci} \sim k_{rac}/k_{dac} \ll 1$, the above condition results in a slow rate of chain initiation. This indicates that termination reactions are suppressed at the expense of slow initiation or longer reaction times. ATRP is viewed as a polymerization method that not only narrows the molecular weight distribution by dormancy but it also suppresses chain transfer and termination reactions. To overcome the slow rate of initiation, a bimolecular initiation approach consisting of a conventional radical initiator, like BPO, and a halide initiator is used to increase the rate of chain initiation and maintain $k_{rac}/k_{dac} \ll 1$ according to the following reactions [135]:

$$I \xrightarrow{k_d} R^{\bullet} \xrightarrow{nM} RP_n^{\bullet} \tag{1.22}$$

$$RP_n^{\bullet} + LM^{II}X_2 \underset{k_{rac}}{\overset{k_{dac}}{\rightleftharpoons}} DOP_n{-}X + LM^{I}X \tag{1.23}$$

where I is a conventional radical initiator that dissociates with a rate constant k_d to form primary radicals. Polymers with low polydispersity in the range of 1.05–1.15 can be produced by ATRP using a wide variety of monomers. The drawbacks of ATRP are sensitivity of the reaction to oxygen and the additional steps required to separate the toxic transition metal catalyst from the polymer. Several ATRP methods have been developed to overcome these limitations [139–154].

1.2.1.4.2 Nitroxide-mediated Reversible-deactivation Radical Polymerization

To overcome the additional step of separating the toxic catalyst from the polymer, nitroxide-mediated reversible-deactivation radical polymerization (NMP) uses a highly substituted, sterically hindered, alkoxyamine as the initiator and chain transfer agent to reversibly deactivate propagating radicals [155–157]. Alkoxyamines undergo homolysis of the C—O bond, yielding a stable $\rangle NO^{\bullet}$ radical for reversible deactivation and a radical $R^{\bullet}$ to initiate polymerization. Steric hindrance around the nitroxyl group weakens the C—O bond in the structure of alkoxyamines, which increases the rate of C—O dissociation [158]. Further, steric hindrance decreases the rate of combination of a propagating radical with a nonpropagating nitroxide radical, which results in the formation of transient dormant chains as opposed to terminated polymer chains [155]. NMP is used in the controlled polymerization of most conjugated vinyl monomers, including styrene and its derivatives, acrylate, methacrylate, acrylamide, and acrylonitrile monomers, as well as some dienes like isoprene [91]. Typical nitroxide-mediated chain initiator/transfer agents include 2,2,6,6-tetramethyl-1-piperidinyloxy (TEMPO) and 2,2,5-trimethyl-4-phenyl-3-azahexane-3-nitroxide (TIPNO) [93]. The initiation and reversible deactivation of propagating radicals in NMP are given by the following reactions:

$$R_1C\text{—}ON\,(R_2R_3) \underset{k_{daci}}{\overset{k_{aci}}{\rightleftharpoons}} R_1C^{\bullet} + {}^{\bullet}ON\,(R_2R_3) \xrightarrow{nM} R_1CP_n^{\bullet} \tag{1.24}$$

$$R_1CP_n^{\bullet} + {}^{\bullet}ON\,(R_2R_3) \underset{k_{rac}}{\overset{k_{dac}}{\rightleftharpoons}} R_1C\text{—}DOP_n\text{—}ON\,(R_2R_3) \tag{1.25}$$

where $R_1C\text{—}ON\,(R_2R_3)$ is the alkoxyamine with R_1 as the alkyl group attached to the alkoxy end, and R_2 and R_3 as the hydrocarbons attached to the nitroxyl end. In the first reaction above, the alkoxyamine initiator is reversibly activated with forward and reverse rate constants k_{aci} and k_{daci}, respectively, to form a propagating radical $R_1C^{\bullet}$ and a relatively stable ${}^{\bullet}ON\,(R_2R_3)$ radical. Next, the propagating radical $R_1C^{\bullet}$ reacts with n monomers to form an n-mer propagating radical $R_1CP_n^{\bullet}$. In the second reaction above, the propagating radical $R_1CP_n^{\bullet}$ reversibly reacts with the nitroxyl radical ${}^{\bullet}ON\,(R_2R_3)$ with forward deactivation and reverse reactivation rate constants k_{dac} and k_{rac}, respectively, to form a dormant polymer chain $R_1C\text{—}DOP_n\text{—}ON\,(R_2R_3)$. Rapid activation and deactivation of the propagating chains allow all polymer chains (active and dormant) to react with monomer and propagate during the entire reaction. For a living radical polymerization with low polydispersity, propagating chains should be in the dormant state most of the time. Like ATRP, this condition implies that the NMP equilibrium constant, as defined by the following equation, should be much less than one ($K_{NMP} \ll 1$):

$$K_{NMP} = \frac{k_{rac}}{k_{dac}} = \frac{[R_1C\text{—}RP_n^{\bullet}]\,[{}^{\bullet}ON\,(R_2R_3)]}{[R_1C\text{—}DOP_n\text{—}ON\,(R_2R_3)]} \ll 1 \tag{1.26}$$

The condition $\frac{k_{aci}}{k_{daci}} \sim \frac{k_{rac}}{k_{dac}} \ll 1$ results in a slow rate of chain initiation. To overcome the slow rate of initiation, a combination of conventional and NMP radical initiators is used to increase the chain initiation rate and maintain the condition $\frac{k_{rac}}{k_{dac}} \ll 1$ [159]. Unlike ATRP, NMP does not require a purification step to remove the toxic transition metal catalyst [158], but polydispersity is adversely affected by the undesirable homolysis of $O\text{—}N$ bond in $CO\text{—}N$, as opposed to the desirable $C\text{—}O$ bond, as well as intramolecular decomposition of alkoxyamines [160, 161].

1.2.1.4.3 Reversible Addition-fragmentation Chain Transfer

RAFT makes use of thiocarbonylthio compounds $S{=}C(R_1)\,SR_2$ as a RAFT agent to reversibly deactivate propagating radicals [106, 162]. The RAFT agent, depending on monomer type and reaction temperature, is selected from dithioesters [106, 163, 164], dithiocarbamates [165], trithiocarbonates [166, 167], and xanthates [168, 169]. A typical RAFT agent is 2-phenylprop-2-yl dithiobenzoate [162] with phenyl and 2-phenylprop-2-yl groups as R_1 and R_2, respectively. The R_1 group stabilizes the polymeric RAFT adduct radical intermediate, whereas the R_2 group shifts the chain transfer reaction toward initiation of a new propagating radical [170]. The reaction mixture in RAFT polymerization consists of a radical initiator, monomer, and the RAFT agent dissolved in a suitable solvent. The initiation and chain transfer in RAFT polymerization are given by the following reactions [106, 162, 171]:

$$I \xrightarrow{k_d} R^{\bullet} \xrightarrow{nM} RP_n^{\bullet} \tag{1.27}$$

$$\begin{aligned} RP_n^{\bullet} + \underset{R_1}{S{=}C}\,SR_2 &\underset{k_{dacai}}{\overset{k_{acai}}{\rightleftharpoons}} RP_n{-}S{-}\underset{R_1}{\dot{C}}SR_2 \\ &\underset{k_{dacfi}}{\overset{k_{acfi}}{\rightleftharpoons}} RDOP_nS{-}\underset{R_1}{C}{=}S + R_2^{\bullet} \xrightarrow{nM} R_2P_m^{\bullet} \end{aligned} \tag{1.28}$$

$$\begin{aligned} R_2P_m^{\bullet} + \underset{R_1}{S{=}C}\,SP_nR &\underset{k_{daca}}{\overset{k_{aca}}{\rightleftharpoons}} R_2P_m{-}S{-}\underset{R_1}{\dot{C}}SP_nR \\ &\underset{k_{dacf}}{\overset{k_{acf}}{\rightleftharpoons}} R_2DOP_nS{-}\underset{R_1}{C}{=}S + RP_n^{\bullet} \end{aligned} \tag{1.29}$$

where I, $S{=}C\,(R_1)\,SR_2$, and M are the initiator, RAFT agent, and monomer, respectively. The first reaction generates an n-mer propagating radical $RP_n^{\bullet}$ using a conventional initiator. In the second reaction above, the propagating radical $RP_n^{\bullet}$ reversibly reacts with the RAFT agent by addition, with forward and reverse activation rate constants k_{acai} and k_{dacai}, respectively, to form the initial RAFT adduct radical intermediate $RP_n{-}S{-}\dot{C}\,(R_1)\,SR_2$. The RAFT intermediate reversibly undergoes fragmentation, with forward and reverse rate constants k_{acfi} and k_{dacfi}, respectively, to form the initial dormant, polymeric RAFT adduct $RDOP_nS{-}C\,(R_1){=}S$ and a primary radical $R_2^{\bullet}$ [162]. The new radical $R_2^{\bullet}$ reacts with m monomers to form the propagating radical $R_2P_m^{\bullet}$. In the third reaction above, the propagating radical $R_2P_m^{\bullet}$ reversibly reacts by addition with the initial dormant, polymeric RAFT adduct with forward and reverse rate constants k_{aca} and k_{daca}, respectively, to form the polymeric RAFT adduct radical intermediate $R_2P_m{-}S{-}\dot{C}\,(R_1)\,SP_nR$. The polymeric radical intermediate reversibly undergoes fragmentation, with forward and reverse rate constants k_{acf} and k_{dacf}, respectively, to form a dormant, polymeric RAFT adduct $R_2DOP_nS{-}C\,(R_1){=}S$ and a reactivated propagating radical $RP_n^{\bullet}$ [162]. The polymeric RAFT adducts $RDOP_nS{-}C\,(R_1){=}S$ and $R_2DOP_nS{-}C\,(R_1){=}S$ serve as internally generated iniferters in RAFT polymerization. The DOP_n in $RDOP_nS{-}C\,(R_1){=}S$ and $R_2DOP_nS{-}C\,(R_1){=}S$ indicates that the produced polymeric RAFT adducts are dormant chains that can be reactivated. This deactivation and reactivation of propagating radicals $RP_n^{\bullet}$ and $R_2P_m^{\bullet}$ results in a living radical polymerization with suppressed termination and low chain polydispersity. In most RAFT polymerizations, $k_{acai} \approx k_{aca}$, $k_{dacai} \approx k_{daca}$, $k_{acfi} \approx k_{acf}$, and $k_{dacfi} \approx k_{dacf}$. The polymer molecular weight distribution and polydispersity are strongly influenced by the overall RAFT equilibrium constant for reversible addition and fragmentation reactions, K_{RAFT}, as defined by the following equation [162]:

$$K_{RAFT} = K_{a,RAFT}K_{f,RAFT} = \frac{k_{aca}}{k_{daca}}\frac{k_{acf}}{k_{dacf}} = \frac{\left[R_2DOP_nS\text{—}\underset{R_1}{C}\text{=}S\right][RP_n^{\bullet}]}{[R_2P_m^{\bullet}]\left[RDOP_nS\text{—}\underset{R_1}{C}\text{=}S\right]} \tag{1.30}$$

where $K_{a,RAFT}$ and $K_{f,RAFT}$ are the equilibrium constants for addition and fragmentation reactions, respectively. The molecular weight in RAFT polymerization increases and polydispersity decreases with conversion [172]. RAFT is especially suited for controlled/living radical polymerization of water-soluble and amphiphilic monomers with judicious selection of the initiator and RAFT agent without the undesired hydrolysis of functional groups of the RAFT agent in aqueous solution [173]. RAFT is also suited for emulsion polymerization, surface functionalization, and the synthesis of self-assembled nanoparticles in aqueous solution [174–176]. As polymers formed by RAFT contain the thiocarbonylthio functional group, the polymeric product may be colored unless the RAFT agent is neutralized/removed after completion of the reaction [177–179].

1.2.1.5 Ring-opening Polymerization

Ring-opening polymerization (ROP) is used in polymerization of strained cyclic monomers including cyclic olefins (norbornenes) [180], ethers (epoxides) [181, 182], amines [183], lactams [184], lactones [185, 186], lactides [187], thioethers [188], siloxanes [189], carbonates [190], and anhydrides [191] in which the propagating species react with the monomer to reduce bond angle strain and/or reduce steric repulsion within the cyclic monomer [192]. The reactive species in the propagating chains can be a radical, anion, cation, or a metal alkylidene complex (metathesis polymerization) [192]. ROP is a reversible reaction, and the extent of polymerization is controlled by thermodynamic equilibrium between the monomer and propagating chains. The equilibrium constant is defined in terms of the enthalpy of polymerization and entropy of polymerization [193]. Due to the loss of translational degrees of freedom, entropy decreases with polymerization for most ROP monomers. Furthermore, due to the loss of ring strain energy, enthalpy also decreases with polymerization. Consequently, there exists a ceiling temperature (T_c) above which the formation of high molecular weight polymer will not occur, as given by the following equation [194]:

$$T_c = \frac{\Delta H_p}{\Delta S_p + R\ln[M_0]} \tag{1.31}$$

where ΔH_p and ΔS_p are the enthalpy and entropy of polymerization, R is the gas constant, and $[M_0]$ is the initial monomer concentration. The T_c corresponds to the temperature at which the free energy of polymerization $\Delta G_p = 0$ or the temperature at which the rates of polymerization and depolymerization are equal. The ceiling temperature increases with increasing strain energy in the ring. The ceiling temperature is not a limiting factor for ROP monomers having three or four-membered rings, as T_c is well above ambient temperature for these monomers. In radical ROP, the driving force for the reaction of an n-mer propagating radical with a monomer is the release of ring strain energy as given by the following reaction [195]:

$$RP_n^{\bullet} + \boxed{cycM} \xrightarrow{k_p} RP_{n+1}^{\bullet} \tag{1.32}$$

where $RP_n^{\bullet}$ and $RP_{n+1}^{\bullet}$ are the n-mer and (n+1)-mer propagating radicals initiated by $R^{\bullet}$ primary radical, respectively, and $\boxed{cycM}$ represents a cyclic monomer. In anionic ROP, a nucleophilic agent such as butyl lithium (Nu^-) is used as an initiator to open the cyclic monomer containing a polar functional group as follows [196]:

$$NuP_n^- X^+ + \boxed{cycM}\Big|_{P-}^{P+} \xrightarrow{k_p} NuP_{n+1}^- X^+ \tag{1.33}$$

where NuP_n^- and NuP_{n+1}^- are the n-mer and (n+1)-mer propagating anions, respectively, initiated by the nucleophile Nu^-, X^+ is the counter cation, and $\boxed{cycM}\Big|_{P-}^{P+}$ is the cyclic monomer containing a polar group represented by positive (+) and negative (−) poles. The nucleophile on the propagating anion NuP_n^- opens the cyclic monomer by reacting with the positive pole of the monomer's polar group to release the strain energy. The anionic ROP is terminated/quenched by any electrophilic impurity or traces of water in the reaction [197]. In cationic ROP, a cation generated by a Lewis acid, such as boron trifluoride (BF_3) is used to open a cyclic monomer such as tetrahydrofuran containing a nucleophilic group [198]. Cationic polymerization is especially useful for the polymerization of hydrophilic monomers in aqueous medium, as water serves as a co-initiator to generate a positive hydronium ion by the following reactions [44, 199, 200]:

$$BF_3 + H_2O \xrightarrow{k_{prei}} H_2OH^+ \overset{OH}{BF_3^-} \tag{1.34}$$

$$H_2OH^+ \overset{OH}{BF_3^-} + \boxed{cycM}\Big|^{P-} \xrightarrow{k_{ci}} H{-}P_1^+ \overset{OH}{BF_3^-} + H_2O \tag{1.35}$$

$$H{-}P_1^+ \overset{OH}{BF_3^-} + {}^n\boxed{cycM}\Big|^{P-} \xrightarrow{k_p} H{-}P_{n+1}^+ \overset{OH}{BF_3^-} \tag{1.36}$$

In the first reaction above (preinitiation), the weak, electron-deficient Lewis acid BF_3 and a water molecule combine with a rate constant k_{prei} to form an ion pair consisting of a hydronium cation (H_2OH^+, electrophile) and a $(HO)\,BF_3^-$ anion. In the second reaction above (chain initiation), the hydronium cation reacts with the nucleophilic group in the cyclic monomer with a rate constant k_{ci} to open the ring and form a 1-mer propagating $H{-}P_1^+$ cation coupled to its anion $(HO)\,BF_3{}^-$. In the third reaction above (propagation), the 1-mer propagating cation $H{-}P_n^+$ reacts with n monomers to form an (n+1)-mer propagating cation $H{-}P_{n+1}^+$. The propagating ROP cation is terminated by combination with a counterion, chain transfer to a counterion or solvent, or by intentional addition of a proton donor or quenching agent like methanol [200]. Ring-opening metathesis polymerization (ROMP) is used to produce polymers with a narrow molecular weight distribution from cycloalkenes or dicycloalkenes [201–203]. In ROMP, like olefin metathesis, the unsaturated alkene group of the monomer undergoes complexation with the organometallic metathesis catalyst to form a metal alkylidene complex as follows [201–203]:

$$\overset{Ln}{M}{=}C{-}R + \boxed{cycM}\Big\|_{C}^{C} \xrightarrow{k_{prei}} \overset{Ln}{M}{=}C{-}R \cdot \cdot \Big\|_{C}^{C}\boxed{cycM} \xrightarrow{k_{ci}} \overset{Ln}{M}{=}P_1{=}C{-}R \tag{1.37}$$

$$\overset{Ln}{M}{=}P_n{=}C{-}R + \boxed{cycM}\Big\|_{C}^{C} \xrightarrow{k_p} \overset{Ln}{M}{=}P_{n+1}{=}C{-}R \tag{1.38}$$

where $(L_n)\,M{=}C{-}R$ is the organometallic catalyst with M as the transition metal, L as the ligand, and R as the alkyl group bonded to the carbon, which is attached to the metal in the organometallic metathesis catalyst, and $\boxed{cycM}\Big\|_{C}^{C}$ is a cyclic monomer with an unsaturated alkene ($C{=}C$) group. The $(L_n)\,M{=}C{-}R$ and $(L_n)\,M{=}P_{n+1}{=}C{-}R$ are 1-mer and (n+1)-mer living propagating chains, respectively. It should be noted that the organometallic catalyst-propagating chain complex is reactive only toward the monomer and dormant toward other reagents in the reaction mixture. In the preinitiation reaction above, the catalyst and monomer react with a rate constant k_{prei} to form a metal alkylidene complex. In chain initiation reaction above, the complex rearranges to accommodate the insertion of the monomer between the transition metal and carbon atom ($M{=}C$) in the catalyst with a rate constant k_{ci} to form a 1-mer propagating chain. In the second reaction above, the 1-mer propagating chain reacts with n monomers to form an (n+1)-mer propagating chain. The $(L_n)\,M{=}P_{n+1}{=}C{-}R$ propagating chain reacts with the monomer in the reaction but not with other reagents, because only the monomer has the molecular configuration for insertion between the metal and the carbon atom in the propagating complex. The propagating complex is terminated by dissociation of the transition metal from the propagating chain, followed by deactivation using a terminating agent [201–203].

1.2.1.6 Group Transfer Polymerization

Group transfer polymerization (GTP) is used for the polymerization of monomers of alkyl methacrylate, acrylate, acrylamide, acrylonitrile, and vinyl ketone at ambient temperature to produce polymers with well-defined chain architecture and narrow molecular weight distribution [204–206]. GTP is initiated by a silyl ketene acetal like 1-methoxy-1-(trimethylsiloxy)-2-methylprop-1-ene (MTS) and a Lewis acid catalyst like hydrogen fluoride. The reaction steps in GTP between the initiator (MTS) and a nucleophile (Nu^-), which is formed by dissociation of a Lewis acid, are as follows in the polymerization of methyl methacrylate ($H_2C{=}C\,(CH_3)\,OC\,({=}O)\,CH_3$) [207, 208]:

$$\underset{CH_3}{\overset{CH_3}{C}}\,C{=}\overset{OCH_3}{C}{-}OSi\,(CH_3)_3 \underset{k_{gtdcx}}{\overset{Nu^-,\,k_{gtcx}}{\rightleftharpoons}} \underset{CH_3}{\overset{CH_3}{C}}\,C{=}\overset{OCH_3}{C}{-}O\overset{Nu^-}{Si}\,(CH_3)_3 \tag{1.39}$$

$$\underset{CH_3}{\overset{CH_3}{C}}\,C{=}\overset{OCH_3}{C}{-}O\overset{Nu^-}{Si}\,(CH_3)_3 \underset{k_{gtdf}}{\overset{k_{gtf}}{\rightleftharpoons}} \underset{CH_3}{\overset{CH_3}{C}}{=}\overset{OCH_3}{C}{-}O^- + NuSi\,(CH_3)_3 \tag{1.40}$$

$$\underset{CH_3}{\overset{CH_3}{C}}\,C{=}\overset{OCH_3}{C}{-}O^- + H_2C{=}\overset{CH_3}{C}\,\overset{\overset{O}{\|}}{C}OCH_3 \xrightarrow{k_{ci}} O{=}\overset{OCH_3}{C}{-}\underset{CH_3}{\overset{CH_3}{C}}{-}P_1^- \tag{1.41}$$

$$O{=}\overset{OCH_3}{C}{-}\underset{CH_3}{\overset{CH_3}{C}}{-}P_n^- + H_2C{=}\overset{CH_3}{C}\,\overset{\overset{O}{\|}}{C}OCH_3 \xrightarrow{k_p} O{=}\overset{OCH_3}{C}{-}\underset{CH_3}{\overset{CH_3}{C}}{-}P_{n+1}^- \tag{1.42}$$

$$\begin{aligned} O{=}\overset{OCH_3}{C}\,\underset{CH_3}{\overset{CH_3}{C}}{-}P_n^- + NuSi\,(CH_3)_3 &\underset{k_{gtac}}{\overset{k_{gtdac}}{\rightleftharpoons}} O{=}\overset{OCH_3}{C}\,\underset{CH_3}{\overset{CH_3}{C}}{-}P_n{-}Nu^-Si\,(CH_3)_3 \\ &\underset{k_{gtcx}}{\overset{k_{gtdcx}}{\rightleftharpoons}} O{=}\overset{OCH_3}{C}\,\underset{CH_3}{\overset{CH_3}{C}}{-}DOP_n{-}Si\,(CH_3)_3 + Nu^- \end{aligned} \tag{1.43}$$

In the first reaction above (preinitiation or complexation), the nucleophile (Nu^-) reversibly bonds to the electron acceptor silica in the trimethoxy silane side group of the silyl ketene acetal (initiator) with forward and reverse rate constants k_{gtcx} and k_{gtdcx}, respectively, to form an $MTS{-}Nu^-$ anion adduct intermediate. In the second reaction above (fragmentation), the $MTS{-}Nu^-$ adduct intermediate dissociates reversibly with forward and reverse rate constants k_{gtf} and k_{gtdf}, respectively, to form the reactive ester enolate anion $(CH_3)_2\,C{=}C\,(OCH_3){-}O^-$ and a stable nucleophile–trimethyl silane adduct $[NuSi\,(CH_3)_3]$. In the third reaction above (chain initiation), the ester enolate anion reacts with a methacrylate monomer and transfers its anion to the carbonyl oxygen of the monomer by bond rearrangement with a rate constant k_{ci} to form a 1-mer propagating chain $O{=}C\,(OCH_3){-}C\,(CH_3)_2{-}P_1^-$. In the fourth reaction (propagation), the n-mer propagating anion reacts with a monomer with a rate constant k_p to form an (n+1)-mer propagating anion. In the fifth reaction (chain deactivation), the $NuSi\,(CH_3)_3$ adduct produced in the second reaction reversibly reacts with an n-mer propagating anion with forward and reverse rate constants k_{gtdac} and k_{gtac}, respectively, to form an unstable polymeric anion complex which reversibly dissociates to produce an n-mer dormant polymer chain with the same rate constants as the first reaction. The last reaction also regenerates the nucleophile to initiate a new propagating chain. Upon dissociation from the propagating anion, the $NuSi\,(CH_3)_3$ adduct reversibly reacts with the carbonyl group of the last polymerized monomer to reactivate the dormant polymer chain. The advantage of GTP is that polymerization can be carried out at ambient temperatures as opposed to sub-ambient temperatures required in anionic polymerization [209]. Furthermore, GTP alleviates the need to use toxic organometallic catalysts as in RDRP [209].

1.2.2 Step-growth Polymerization

In step-growth polymerization, monomers possess complementary functional groups that react with one another to form longer chains in a stepwise manner, that is, an n-mer reacts with an m-mer to form a longer (m+n)-mer chain [210–212]. Polyesters, polyamides, polyurethanes, polyureas, polyanhydrides, and polyacetals polymerize by step polymerization [213, 214]. Monomers in step polymerization should be at least difunctional, and the functional groups should be complementary to react and form dimers, trimers, n-mers, and (n+m)-mers according to the following reactions:

$$A\text{—}R\text{—}B + A\text{—}R\text{—}B \xrightarrow{k_p} A\text{—}\,(R\text{—}BA\text{—})_1\,R\text{—}B \tag{1.44}$$

$$A\text{—}(R\text{—}BA\text{—})_1\,R\text{—}B + A\text{—}(R\text{—}BA\text{—})_1\,R\text{—}B \xrightarrow{k_p} A\text{—}(R\text{—}BA\text{—})_3\,R\text{—}B \tag{1.45}$$

$$A\text{—}(R\text{—}BA\text{—})_n\,R\text{—}B + A\text{—}(R\text{—}BA\text{—})_m\,R\text{—}B \xrightarrow{k_p} A\text{—}(R\text{—}BA\text{—})_{n+m+1}\,R\text{—}B \tag{1.46}$$

where $A\text{—}R\text{—}B$ is a difunctional monomer with A and B as the complementary functional groups attached to the monomer residue R. The reactive group A is complementary to the reactive group B and the two groups react to form a condensed BA linkage, which is a characteristic group of the produced polymer. In most step-growth polymerizations like polyesters and polyamides, the reaction results in the formation of an AB linkage and a small molecule like water or carbon dioxide. As a small molecule is produced by the reaction between two chains, the term condensation polymerization is used to describe step-growth polymerization. In the first reaction above, an $A\text{—}R\text{—}B$ monomer reacts with another $A\text{—}R\text{—}B$ monomer in the reaction mixture with a rate constant k_p to form a dimer with one BA linkage as the characteristic group. In the second reaction, a dimer with one characteristic group BA linkage reacts with another dimer with one characteristic group BA with a rate constant k_p to form a tetramer with three BA linkages. In the third reaction, an (n+1)-mer chain with n characteristic BA groups reacts with an (m+1)-mer chain with m characteristic BA groups to form an (n+m+2)-mer chain with (n+m+1) characteristic BA groups. For example, polylactic acid is synthesized by condensation polymerization of difunctional lactic acid monomer with complementary hydroxyl and carboxylic acid groups [215]. The defining feature of step-growth polymerization is that the polymer molecular weight correlates with reaction conversion (α), that is, molecular weight increases stepwise as monomers, dimers, trimers, oligomers, etc., react to form higher molecular weight polymer chains in late stages of the reaction at high conversions [216]. The complementary functional groups do not have to be on one monomer. Two bifunctional monomers, one having functional groups A ($A—R_1—A$) and the other having complementary functional groups B ($B—R_2—B$) undergo step-growth polymerization according to the following reactions:

$$A—R_1—A + B—R_2—B \xrightarrow{k_p} A—R_1—AB—R_2—B \tag{1.47}$$

$$\begin{aligned} &A—R_1—AB—R_2—B + A—R_1—AB—R_2—B \\ &\xrightarrow{k_p} A—R_1\left(—AB—R_2—BA—R_1\right)_1—AB—R_2—B \end{aligned} \tag{1.48}$$

$$\begin{aligned} &A—R_1\left(—AB—R_2—BA—R_1\right)_n—AB—R_2—B \\ &+A—R_1\left(—AB—R_2—BA—R_1\right)_m—AB—R_2—B \\ &\xrightarrow{k_p} A—R_1\left(—AB—R_2—BA—R_1\right)_{n+m+1}—AB—R_2—B \end{aligned} \tag{1.49}$$

In the first reaction above, the complementary functional groups A and B on the two monomers react to form a bifunctional dimer $A—R_1—AB—R_2—B$ having both A and B functional groups. In the second reaction, the bifunctional dimers react to form bifunctional trimers, tetramers, and longer oligomers. In the third reaction, an n-mer bifunctional chain with 2n characteristic AB groups reacts with another m-mer bifunctional chain with 2m characteristic AB groups to form an (n+m+1)-mer bifunctional chain with 2n+2m+3 characteristic AB groups. For example, poly(hexamethylene adipamide), known commercially as nylon-66, is synthesized by the reaction of hexamethylene diamine having two amine ($—NH_2$) groups with adipic acid having two complementary carboxylic acid ($—COOH$) groups, such that the two complementary groups condense to form a bifunctional monomer $—C\left(=O\right)—N\left(H\right)—$ with a characteristic amide linkage.

The is no initiator in step-growth polymerization; that is, once the monomer and the catalyst are added to the reaction, polymerization continues so long as there are unreacted monomer and the reaction conditions, like temperature, pressure, and viscosity, are favorable. Unlike addition polymerization, step-growth polymerization is very sensitive to monomer purity, as a small fraction of a monofunctional monomer or an imbalance in the ratio of complementary functional groups can terminate chain growth. A slight difference in the amounts of the two bifunctional monomers $A–R_1–A$ and $B–R_2–B$ leads to termination of the propagating chains. As a result, polymer molecular weight in step-growth polymerization is controlled by intentional addition of a small fraction of a monofunctional monomer to the reaction [212, 217, 218]. The polymer molecular weight increases with conversion in step-growth polymerization. At low conversions, the product is predominantly monomers, dimers, trimers, and other low molecular weight oligomers, whereas high molecular weight polymer is produced at conversions exceeding 99% [212, 219].

1.2.3 Step Transfer-addition and Radical Termination Polymerization

Step transfer-addition and radical termination (START) is a step-growth polymerization reaction in which the functional groups of at least one of the monomers in the polymerization reaction are activated by a radical initiator [220, 221]. This approach is used in step-growth polymerization of α,ω-unconjugated dienes ($H_2C=C\left(H\right)RC\left(H\right)=CH_2$, monomer B) with α,ω-diiodoperfluoroalkanes ($I—\left(CF_2\right)_n—I$, monomer A) under visible light irradiation [222]. In this approach, monomers A and B are polymerized using tris(bipyridine) ruthenium(II) chloride ($Cl_2\left[Ru\left(bpy\right)_3\right]$) and sodium ascorbate ($Asc$) as photocatalyst and reducing agent, respectively [222]. The chain initiation steps in START are given by the following reactions [222]:

$$Cl_2\left[Ru\,(bpy)_3\right] \xrightarrow{k_{ext},\, h\nu(BLR)} Cl_2\left[Ru\,(bpy)_3\right]^{*} \xrightarrow{k_{rq},\, Asc} \left[Ru\,(bpy)_3\right]^{+} \tag{1.50}$$

$$I\!-\!(CF_2)\!-\!I \xrightarrow{k_{rci},\left[Ru(bpy)_3\right]^{+}} {}^{\bullet}(CF_2)_n^{\bullet} + {}^{2}I^{\bullet} \tag{1.51}$$

$${}^{\bullet}(CF_2)_n^{\bullet} + H_2C{=}\overset{H}{C}R\overset{H}{C}{=}CH_2 \xrightarrow{k_{ra}} {}^{\bullet}(CF_2)_n\,\underset{H}{\overset{H}{C}}\underset{H}{\overset{\bullet}{C}}R\overset{H}{C}{=}CH_2 \tag{1.52}$$

In the first photoinitiation reaction above, the photocatalyst is elevated to an excited energy state $Cl_2[Ru\ (bpy)_3]^{*}$ by irradiation with blue light (BLR) with a rate constant k_{ext}, which in turn is reductively quenched with a rate constant k_{rq} to produce a reactive tris(bipyridine) ruthenium cation $\left[Ru\ (bpy)_3\right]^{+}$. In the second radical initiation reaction, the generated cation abstracts an iodine from monomer A by reductive coupling with a rate constant k_{rci} to form two iodine $I^{\bullet}$ and iodoperfluoroalkane $I(CF_2)_n^{\bullet}$ radicals. In the third chain initiation reaction, the generated $I(CF_2)_n^{\bullet}$ radical reacts with one of the $C{=}C$ bonds in monomer B with a rate constant k_{ci} to form an $AB^{\bullet}$ radical. The first repeat unit of the propagating chain between monomers A and B is formed by the following reactions:

$${}^{\bullet}(CF_2)_n\,\underset{H}{\overset{H}{C}}\underset{H}{\overset{\bullet}{C}}R\overset{H}{C}{=}CH_2 + I^{\bullet} \xrightarrow{k_{tia}} {}^{\bullet}(CF_2)_n\,\underset{H}{\overset{H}{C}}\underset{H}{\overset{I}{C}}R\overset{H}{C}{=}CH_2 \tag{1.53}$$

$$\begin{gathered} {}^{\bullet}(CF_2)_n\,\underset{H}{\overset{H}{C}}\underset{H}{\overset{I}{C}}R\overset{H}{C}{=}CH_2 + {}^{\bullet}(CF_2)_n\,\underset{H}{\overset{H}{C}}\underset{H}{\overset{I}{C}}R\overset{H}{C}{=}CH_2 \xrightarrow{k_{ci}} \xrightarrow{k_{tia},I^{\bullet}} \\ {}^{\bullet}(CF_2)_n\,\underset{H}{\overset{H}{C}}\!-\!\left\{\underset{H}{\overset{I}{C}}R\underset{H}{\overset{I}{C}}CH_2\,(CF_2)_n\,CH_2\right\}_1 -\!\overset{I}{C}R\underset{H}{\overset{H}{C}}{=}CH_2 \end{gathered} \tag{1.54}$$

In the first reaction above, an iodine radical $I^{\bullet}$ reacts with an $AB^{\bullet}$ radical via termination by iodine addition with a rate constant k_{tia} to form the half repeat unit ${}^{\bullet}(CF_2)_n\,CH_2C\,(IH)\,RCH{=}CH_2$ of the propagating chain, which is stable in the presence of photocatalyst and reducing agent [222]. In the second reaction, the radical on the half repeat unit reacts with a diene $C{=}C$ bond on another half repeat unit with a rate constant k_{ci} to dissociate the π-bond and form a new radical comprising the first full repeat unit of the diene-perfluoroalkane propagating chain. The new radical reacts with an iodine radical with a rate constant k_{tia} to produce the first repeat unit $-C\,(IH)\,RC\,(IH)\,CH_2-(CF_2)_n-CH_2-$ or $-AB_1-$ of the propagating chain (shown in braces in Rxn. 1.54). The general propagation step for START reaction is as follows [222]:

$$\begin{gathered} {}^{\bullet}(CF_2)_n\,\underset{H}{\overset{H}{C}}\!-\!\left\{\underset{H}{\overset{I}{C}}R\underset{H}{\overset{I}{C}}CH_2\,(CF_2)_n\,CH_2\right\}_i -\!\overset{I}{C}R\underset{H}{\overset{H}{C}}{=}CH_2 + \\ I\,(CF_2)_n\,\underset{H}{\overset{H}{C}}\!-\!\left\{\underset{H}{\overset{I}{C}}R\underset{H}{\overset{I}{C}}CH_2\,(CF_2)_n\,CH_2\right\}_j -\!\overset{I}{C}R\underset{H}{\overset{H}{C}}{=}CH_2 \xrightarrow{k_p,\left[Ru(bpy)_3\right]^{+},I^{\bullet}} \\ I\,(CF_2)_n\,\underset{H}{\overset{H}{C}}\!-\!\left\{\underset{H}{\overset{I}{C}}R\underset{H}{\overset{I}{C}}CH_2\,(CF_2)_n\,CH_2\right\}_{i+j+1} -\!\overset{I}{C}R\underset{H}{\overset{H}{C}}{=}CH_2 \end{gathered} \tag{1.55}$$

In the above reaction, a $-\dot{C}F_2$ radical on an i-mer AB_i propagating chain reacts with a $C{=}C$ bond on a j-mer AB_j propagating chain to form an AB_{i+j+1} chain with (i+j+1) repeat units. The polymerization reaction continues so long as there are $-\dot{C}F_2$ radicals and unsaturated $C{=}C$ bonds on the propagating chains.

1.3 Living Polymerization

Living polymerization refers to those reactions such that termination and chain transfer reactions are absent or sharply suppressed [6]. Polymers produced by living polymerization are nearly monodisperse or have a very narrow molecular weight distribution, and the average molecular weight is given by the following simple relation:

$$\overline{M}_{lp} = \frac{[M_0] - [M]}{[I_0]} \tag{1.56}$$

where $\overline{M}_{lp}$ is the average molecular weight of the polymer produced by living polymerization, $[M]_0$ and $[M]$ are the initial and time t concentrations of monomer in the reaction, respectively, and $[I]_0$ is the initial concentration of initiator. Upon adding more monomers to a living polymerization, the reaction will continue to generate longer chains. Anionic polymerization is one of the most effective methods to produce living polymers (see Section 1.2.1.2). In anionic polymerization, the anion and cation generated by initiator dissolution are stabilized by solvation [223, 224]. As a result, the generated anions are stable and remain reactive throughout the polymerization reaction. Furthermore, in the absence of moisture and under inert atmosphere, there are relatively few undesired side reactions in anionic polymerization to quench the propagating anions. Anionic polymerization reactions are often quenched by the intentional addition of methanol or moisture to the reaction to quench the propagating anions. Another highly effective method to produce living polymers is group transfer polymerization (GTP, see Section 1.2.1.6) [204–206]. In GTP, the propagating species is an ester enolate anion, which forms a complex with a silyl ketene acetal to reversibly form dormant species. As a result, a small fraction of the propagating chains is active at any given time during the reaction, and the reversible reactivation of dormant chains produces a living polymer with a very narrow molecular weight distribution [208].

Cationic polymerization undergoes many chain termination reactions, including intramolecular β-hydrogen abstraction [51], combination with the opposite charge [54], and hydride ion transfer to monomer/polymer (see Section 1.2.1.3) [64]. Therefore, conventional cationic polymerization is a controlled, but not a living, polymerization system. The side reactions in cationic polymerization are attributed to the formation of two distinct propagating species. These include the dissociated nonliving and associated living cation–anion species [225]. It is posited that the dissociated, nonliving species is responsible for most of the undesired termination reactions. This fact implies that cationic polymerization can be a living polymerization if the formation of dissociated nonliving propagating species is suppressed. It is reported that propagation in cationic polymerization of styrene in methylene chloride at −15 °C with 1-phenyl halide as the initiator and tin tetrachloride ($SnCl_4$) as the activator involves two cationic species as follows:

$$X{-}C_6H_5 + SnCl_4 \xrightarrow{k_{dsip}} H_5C_6^+ + Cl_4SnX^- \tag{1.57}$$

$$H_5C_6^+ + Cl_4SnX^- \xrightleftharpoons{k_{ndsip}} H_5C_6^+ ..^-XSnCl_4 \tag{1.58}$$

In the first reaction above, 1-phenyl halide and tin tetrachloride react to form a dissociated nonliving cationic $H_5C_6^+$ and anionic Cl_4SnX^- ion pair with a rate constant k_{dsip}. In the second reaction, the dissociated nonliving ionic species reversibly convert to an associated ion-pair $H_5C_6^+ ..^-XSnCl_4$ with a rate constant k_{ndsip}. The presence of both species leads to the nonliving character of polymerization with a bimodal molecular weight distribution. Upon the addition of n-butylammonium halide ($nBuN^+Y^-$ with Y as the halide) to the reaction as a catalyst, the dissociated nonliving propagating ion-pairs are irreversibly converted to associated living ion-pairs by the following reaction [225]:

$$H_5C_6^+ + Cl_4SnX^- \xrightarrow{nBu_4N^+Y^-} H_5C_6^+ ..^-XSnCl_4 \tag{1.59}$$

The conversion of dissociated nonliving species to associated living species leads to a living cationic polymerization with a narrow molecular weight distribution [225]. Therefore, some cationic polymerizations can be converted to living polymerization by suppressing the formation of undesired reactive species.

Conventional radical polymerization is not living, as the transient radical species undergo termination by combination or termination by disproportionation. However, synthetic approaches like RDRP that maintain the propagating radicals in a state of reversible dormancy to suppress termination reactions yield a practically living polymerization reaction (see Section 1.2.1.4) [226, 227]. RDRP uses a deactivating agent to reversibly deactivate and maintain the propagating chains in a state of dormancy. In the ATRP approach to RDRP, the deactivator is a combination of a transition metal catalyst LMX with two oxidation states and an alkyl halide initiator $R—X$ to maintain dormancy as follows [135–138]:

$$RP_n^{\bullet} + LM^{II}X_2 \underset{k_{rac}}{\overset{k_{dac}}{\rightleftharpoons}} DOP_n—X + LM^IX \tag{1.60}$$

The transition metal is lowered from a high to a low oxidation state ($LM^{II}X$ to LM^IX) by the transfer of a ligand X from the transition metal catalyst to the propagating chain to form a dormant polymer chain $DOP_n—X$. As the propagating chains are dormant most of the time and active a fraction of the time, the bimolecular termination reaction between the two propagating radicals is sharply reduced to form a living radical polymerization. In the NMP approach to RDRP, the deactivator is a highly substituted, sterically hindered alkoxyamine which serves both as initiator and chain transfer agent as follows [155–157]:

$$R_1P_n^{\bullet} + R_1C—ON\,(R_2R_3) \underset{k_{rac}}{\overset{k_{dac}}{\rightleftharpoons}} R_1DOP_n—ON\,(R_2R_3) + R_1C^{\bullet} \tag{1.61}$$

The alkoxyamine dissociates into a radical fragment $R_1C^{\bullet}$and a chain transfer fragment $^{\bullet}ON\,(R_2R_3)$ followed by transfer of $^{\bullet}ON\,(R_2R_3)$ to the propagating radical to form a dormant polymer chain $R_1DOP_n—ON\,(R_2R_3)$ and a reactive radical fragment $R_1C^{\bullet}$ for chain initiation. In RAFT approach to RDRP, the deactivator is a thiocarbonylthio chain transfer agent as follows [106, 162]:

$$RP_n^{\bullet} + S{=}\overset{R_1}{C}SR_2 \underset{k_{daca}}{\overset{k_{aca}}{\rightleftharpoons}} \underset{k_{dacf}}{\overset{k_{acf}}{\rightleftharpoons}} RDOP_nS—\overset{R_1}{C}{=}S + R_2^{\bullet} \tag{1.62}$$

A propagating chain $RP_n^{\bullet}$ reacts with the thiocarbonylthio activator to form an activator-propagating chain complex, which undergoes fragmentation to form a dormant $RDOP_nS—C\,(R_1)=S$ chain and a reactive radical fragment $R_2^{\bullet}$ for chain initiation. In RDRP methods, the polymerization reaction approaches a living polymerization as the rate of deactivation of propagating chains becomes much greater than their reactivation rate.

ROP can be living if the propagating species do not undergo termination [228]. As discussed in Section 1.2.1.5, ROP may proceed by radical, cationic, anionic, or ROMP mechanisms. Anionic ROP of three or four-membered ring monomers with metal alkoxide or alkoxyamine initiators is considered living as the propagating anion is highly stable and side reactions are nearly absent [229]. ROMP is also a living method so long as the organometallic metathesis catalyst is present in the reaction [230]. Complexation of the metathesis catalyst $L_nM{=}C{-}R$ with monomer, followed by its molecular rearrangement, forms a 1-mer propagating chain as follows [201–203]:

$$\overset{L_n}{M}{=}CR \xrightarrow{k_{prei}\boxed{cycM}\|^{c}_{c}} \boxed{cycM}\|^{c}_{c}\cdot\cdot\overset{L_n}{M}{=}CR \xrightarrow{k_{ci}} \overset{L_n}{M}{=}P_1{=}CR$$
$$\xrightarrow{k_p,^{n}\boxed{cycM}\|^{c}_{c}} \overset{L_n}{M}{=}P_{n+1}{=}CR \qquad (1.63)$$

The 1-mer catalyst–monomer complex $L_nM{=}P_1{=}CR$ adds only monomers by insertion in the $M{=}C$ bond to form an n-mer living propagating chain $L_nM{=}P_{n+1}{=}CR$ without termination side reaction. The living chains may be terminated intentionally by deactivation of the metathesis catalyst [201–203].

Unlike anionic ROP and ROMP, conventional cationic and radical ROP are not considered a living polymerization method because the generated cation or radical is unstable and undergoes termination or partakes in other side reactions. A common approach to inhibit side reactions in cationic or radical ROP is the use of a stabilizing agent that reversibly converts the reactive propagating species to the more stable dormant or non-dissociated species [225, 228, 231]. In the dormant approach, the cation–anion complex $(H_5C_6)_3\,C^+\,(C_6F_5)_4\,B^-$ formed by initiator/co-initiator/catalyst (see Section 1.2.1.3) is transformed to a dormant state by the addition of a weak cyclic ether Lewis base like hexamethylene-oxide $O\boxed{hexmo}$ in the polymerization of cyclohexene oxide $\left(O-\boxed{cychox}\right)$ as follows [231]:

$$(H_5C_6)_3\,C^+\,(C_6F_5)_4\,B^- + O\boxed{cychox} \xrightarrow{k_{ci}}$$
$$(H_5C_6)_3\,C{-}\overset{+}{O}{-}\boxed{cychox} + (C_6F_5)_4\,B^- \xrightarrow[O\boxed{hexmo}]{k_{dorm}}$$
$$(H_5C_6)_3\,C{-}\left(O{-}\boxed{cychox}\right)_1{-}\overset{+}{O}\boxed{hexmo} \xrightarrow[nO\boxed{cychox}]{k_p}$$
$$(H_5C_6)_3\,C{-}\left(O{-}\boxed{cychox}\right)_{n+1}{-}\overset{+}{O}{-}\boxed{hexmo} \qquad (1.64)$$

where k_{dorm} is the rate constant for cation transfer from the oxygen of monomer to the oxygen of hexamethylene oxide for the transformation of the propagating cation to a dormant state. In the above reactions, the reactive cation on the oxygen atom of cyclohexene oxide monomer is transferred to the more stable (dormant) oxygen of $O\boxed{hexmo}$ after the addition of hexamethylene oxide [231]. Monomers that undergo living ring-opening cationic polymerization using cyclic ethers include three-membered ring butylene oxide (DMBO), five-membered cyclopentene oxide (CPO), six-membered ring cyclohexene oxide (CHO), seven-membered ring cycloheptene oxide (CHPO), and their derivatives [231]; cyclic ethers used as a weak Lewis base include five-membered ring tetrahydrofuran (THF) and 1,3-dioxolane (1,3-DO), six-membered ring pyran (THP) and 1,3- and 1,4-dioxane (1,3-DO or 1,4-DO), seven-membered ring hexamethylene oxide (HMO), and their derivatives [231]. In the non-dissociated approach, a strong Bronsted acid like trifluoromethane sulfonimide (HNTf, $F_3CS\,(O_2){-}N\,(H){-}S\,(O_2)\,CF$) is used as the catalyst to form a non-dissociated propagating cationic species, as shown by the following reactions for polymerization of δ-valerolactone ($\left(\boxed{\delta val}{=}O\right)$) with 3-phenyl-1-propanol ($\left(\boxed{3ph}(CH_2)_3\,OH\right)$) as the initiator [232]:

$$\boxed{\delta val}{=}O + F_3C\overset{O}{\underset{O}{S}}{-}\overset{H}{N}{-}\overset{O}{\underset{O}{S}}CF_3 \xrightarrow{k_{ci}}$$

$$F_3C\overset{O}{\underset{O}{S}}{-}\overset{-}{N}{-}\overset{O}{\underset{O}{S}}CF_3\ {}^{+}HO{=}\boxed{\delta val} \xrightarrow[\boxed{3ph}(CH_2)_3{-}\ddot{O}{-}H]{k_{nds}}$$

$$\boxed{3ph}(CH_2)_3\,O\left(-\overset{O}{C}(CH_2)_6\,O-\right)_1 H + F_3C\overset{O}{\underset{O}{S}}{-}\overset{H}{N}{-}\overset{O}{\underset{O}{S}}CF_3 \tag{1.65}$$

$$\xrightarrow[nF_3C\overset{OHO}{\underset{O\ O}{SNS}}CF_3\ {}^{+}HO{=}\boxed{\delta val}]{k_p} \boxed{3ph}(CH_2)_3\,O\left(-\overset{O}{\overset{\|}{C}}(CH_2)_6\,O-\right)_{n+1} H$$

where k_{nds} is the rate constant for the reaction of the cation–anion complex with the lone pair electrons of oxygen on the 3-phenyl-1-propanol initiator to form a non-dissociated monomer–initiator adduct. In the above reactions, the cation–anion complex formed by the interaction of the catalyst and monomer reacts with the lone pair electrons of oxygen on the 3-phenyl-1-propanol initiator to form a non-dissociated monomer–initiator adduct [232]. The non-dissociated adduct propagates by reaction with n monomer–catalyst cation–anion complexes to form an (n+1) non-dissociated propagating chain.

In radical ROP, the atom transfer approach is used for living polymerization of 5,6-benzo-2-methylene-1,3-dioxepane (BMDO, $C_6H_5{-}\boxed{doxp}\overset{O}{\underset{O}{\rangle}}{=}$) to produce a polymer with low polydispersity and linear increase in molecular weight with conversion [233]. In this ROP polymerization, ethyl α-bromobutyrate ($C_2H_5C\,(BrH){-}OC\,(O)\,C_2H_5$) and 2,2′-bipyridine/copper bromide ($BiPy{-}Cu^{I}Br$) are used as the halide initiator and transition metal catalyst, respectively, as shown by the following reactions [233]:

$$C_2H_5\overset{Br}{\underset{H}{C}}{-}O\overset{O}{C}C_2H_5 + BiPy{-}Cu^{I}Br \underset{k_{daci}}{\overset{k_{acl}}{\rightleftharpoons}} BiPyCu^{II}Br_2 + C_2H_5\overset{\bullet}{\underset{H}{C}}{-}O\overset{O}{C}C_2H_5 \tag{1.66}$$

$$C_2H_5\overset{\bullet}{\underset{H}{C}}{-}O\overset{O}{C}C_2H_5 + C_6H_5{-}\boxed{doxp}\overset{O}{\underset{O}{\rangle}}{=} \xrightarrow{k_p}$$
$$C_2H_5\underset{CO_2C_2H_5}{\overset{H}{C}} - \left(CH_2CO\underset{H}{\overset{O\ H}{C}}{-}C_6H_4{-}\underset{H}{\overset{H}{C}}\right)_n {-}CH_2CO\overset{O\ H}{C}C_6H_4\underset{H}{\overset{H}{C}}\bullet \tag{1.67}$$

$$C_2H_5\underset{CO_2C_2H_5}{\overset{H}{C}} - \left(CH_2CO\underset{H}{\overset{O\ H}{C}}C_6H_4\underset{H}{\overset{H}{C}}\right)_n {-}CH_2CO\underset{H}{\overset{O\ H}{C}}C_6H_4\underset{H}{\overset{H}{C}}\bullet$$
$$\underset{k_{rac}}{\overset{k_{dac},BiPyCu^{II}Br_2}{\rightleftharpoons}} BPyCu^{I}Br \tag{1.68}$$
$$+C_2H_5\underset{CO_2C_2H_5}{\overset{H}{C}} - \left(CH_2CO\underset{H}{\overset{O\ H}{C}}C_6H_4\underset{H}{\overset{H}{C}}\right)_n {-}CH_2CO\underset{H}{\overset{O\ H}{C}}C_6H_4\underset{H}{\overset{H}{C}}Br$$

where $BiPyCu^{I}Br$ and $BiPyCu^{II}Br_2$ are the transition metal catalysts in low and high oxidation states, respectively. In the first reaction above, a bromine atom is reversibly transferred from the initiator to the catalyst to form an ethyl α-butyrate radical with the radical on a secondary carbon. With the transfer of the bromine atom, the transition metal catalyst shifts from low to high oxidation state. In the second reaction, the ethyl α-butyrate radical reacts with the monomer to form a propagating BMDO radical. In the third reaction, a bromine atom is reversibly transferred from the catalyst in the high oxidation state to the propagating BMDO radical to form a dormant BMDO chain with the catalyst in the low oxidation state (catalyst regeneration). As discussed previously (see Section 1.2.1.4.1 on ATRP), the polymerization is living if the catalyst is selected such that the rate of deactivation (k_{dac}, dormant chain formation) is much greater than the rate of activation (k_{rac}, propagating chain formation) or the ATRP equilibrium constant $K_{ATRP} = k_{rac}/k_{dac} \ll 1$.

1.4 Copolymerization

A homopolymer is formed from the polymerization of a single monomer type, whereas a copolymer is formed from the copolymerization of two or more monomer types [234, 235]. Most monomers used in polymerization are miscible with one another at the molecular scale, whereas most polymers, due to their low entropy of mixing, are immiscible and undergo phase separation upon physical blending [236–240]. As an example, styrene (S) and butadiene (B) are miscible as monomers, whereas polystyrene (PS) and polybutadiene (PB) are immiscible and phase separate upon blending [241]. PS and PB have complementary properties as PS is a brittle high modulus polymer and PB is a low modulus rubbery polymer that absorbs a significant amount of energy prior to rupturing [242–244]. In this respect, copolymerization of monomers with complementary properties allows molecular blending of polymers that are otherwise immiscible as homopolymers [245, 246]. In copolymerization, the reaction rate of a monomer with a propagating chain depends on the terminal repeat unit of the chain, as shown below for free radical copolymerization of styrene (S) and butadiene (B) monomers:

$$\sim\sim\sim\sim S\bullet + S \xrightarrow{k_{SS}} \sim\sim\sim\sim S\bullet \tag{1.69}$$

$$\sim\sim\sim\sim S\bullet + B \xrightarrow{k_{SB}} \sim\sim\sim\sim B\bullet \tag{1.70}$$

$$\sim\sim\sim\sim B\bullet + S \xrightarrow{k_{BS}} \sim\sim\sim\sim S\bullet \tag{1.71}$$

$$\sim\sim\sim\sim B\bullet + B \xrightarrow{k_{BB}} \sim\sim\sim\sim B\bullet \tag{1.72}$$

In the first and second reactions, two propagating chains ending in a styrene repeat unit react with a styrene or butadiene monomer, respectively, with rate constants k_{SS} and k_{SB} to produce two propagating chains ending in a styrene or butadiene repeat unit. Likewise, in the third and fourth reactions, two propagating chains ending in a butadiene repeat unit react with a styrene or butadiene monomer, respectively, with rate constants k_{BS} and k_{BB} to produce two propagating chains ending in a styrene or butadiene repeat unit. The rate of addition of monomers to the propagating chain in copolymerization depends on the fraction of monomers in the feed, their reactivity, and the terminal monomer on the propagating chain. Depending on monomer reactivity, the terminal monomer, and the order of addition of monomers to the reaction, a random copolymer, an ideal copolymer, an alternating copolymer, or a block copolymer is produced, as shown in the sequences of Figure 1.1 for two monomers (open and filled circles) with equal feed molar composition [247–251]:

In a random copolymer (sequence A in Figure 1.1), the copolymer composition is the same as the feed, as the monomers are equally reactive with the propagating chain and copolymerization is independent of the chains' terminal repeat unit [252]. In an ideal copolymer (sequence B in Figure 1.1), the copolymer composition depends on the monomer reactivity with the radical

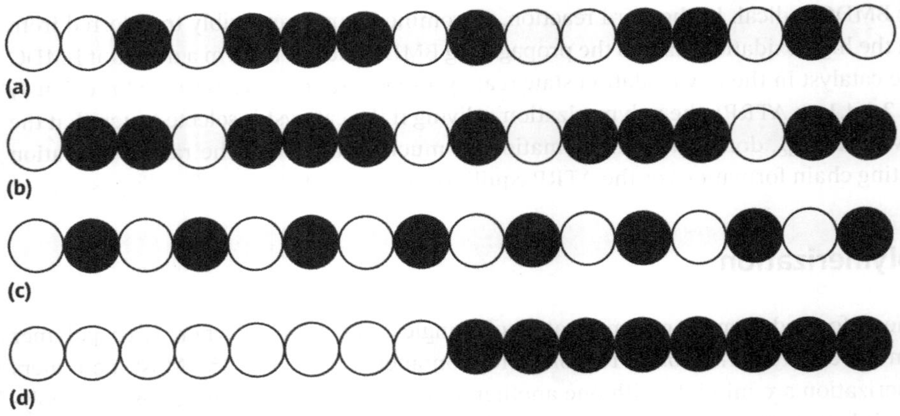

Figure 1.1 Schematic sequences of two monomers (open and filled circles) with equal feed molar composition showing (a) random copolymer, (b) ideal copolymer, (c) alternating copolymer, and (d) block copolymer.

but is independent of the type of terminal monomer repeat unit on the propagating chain [253]. In sequence B in Figure 1.1, the fraction of monomer shown by filled circles is higher in the copolymer for equal monomer feed composition because it has a higher reactivity with the radical on the propagating chain. For random and ideal copolymers, the monomers are randomly distributed on the copolymer chains. In an alternating copolymer (sequence C in Figure 1.1), the two monomers are added alternatively to the propagating chain because the reactivity of the monomer with the propagating chain having a dissimilar terminal repeat unit is much higher than that with a similar repeat unit [254]. Consequently, the composition of an alternating copolymer is independent of the feed composition. In a block copolymer (sequence D in Figure 1.1), the reactivity of the monomer with a propagating chain having a similar terminal repeat unit is much higher than that with a dissimilar repeat unit [249, 255]. As a result, the chains in a block copolymer contain long sequences of the same monomer.

1.5 Architecture of Polymer Chains

Polymer chains can be linear, branched, or crosslinked depending on the chemical nature and number of functional/unsaturated groups in the monomer [256]. Side reactions can also influence the architecture of polymer chains. The chain A in Figure 1.2 is a linear chain architecture, whereas the chains B through D show chains with nonlinear short branches, long branches, tree-like branches, and crosslinked chain architecture, respectively [257–259]:

Vinyl monomers with one unsaturated group polymerize by addition polymerization, while bifunctional monomers polymerize by step polymerization, and produce linear chains in the absence of side reactions (chain (a) in Figure 1.2). Polymerizations that undergo chain transfer to polymer produce chains with short or long branches. For example, propagating chains in the radical polymerization of ethylene undergo chain transfer to self or chain backbiting to produce short branches, as shown by chain (b) in Figure 1.2 [260]. Conversely, chain transfer to propagating or terminated polymer chains other than self produces chains with long branches, as shown by chain (c) in Figure 1.2 [261–263]. Polymerization reactions carried out in the presence of a multifunctional monomer, with a concentration below the threshold for crosslinking, produce tree-like or

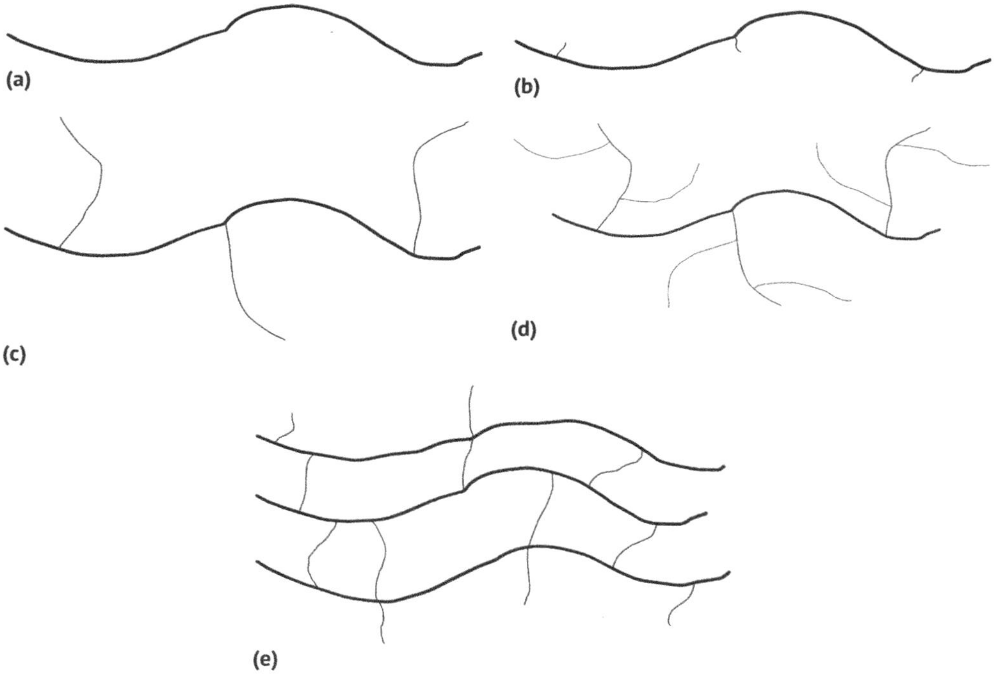

Figure 1.2 Schematic of chains with (a) linear architecture, nonlinear (b) short branches, (c) long branches, (d) tree-like branches, and (e) crosslinked chain architectures.

hyperbranched polymer chains [264, 265] as shown by chain (d) in Figure 1.2, whereas those with a concentration above the threshold for crosslinking produce a network of crosslinked chains as shown by chain (e) in Figure 1.2 [266, 267].

1.6 Summary

Mechanistically, polymerization reactions fall into addition or step-growth polymerization. In the addition polymerization, monomer molecules add successively to reactive species generated by the dissociation of initiator molecules to form reactive propagating chains. The reactive species in addition polymerization may be a radical, anion, cation, or an initiator–catalyst complex (group transfer polymerization or metathesis polymerization). The monomer in group transfer polymerization or metathesis polymerization is inserted within the complex's molecular structure. Cyclic monomers with polarized functional groups undergo ROP to reduce bond angle strain and steric repulsion within the monomer. Due to the stability of active species in anionic polymerization, metathesis polymerization, and group transfer polymerization, these reactions tend to form living polymers with little or no termination reaction and a narrow molecular weight distribution. Cationic and radical polymerization, due to the instability or transient nature of the reactive species, tend to undergo chain transfer reactions and termination, which broadens their molecular weight distribution. One approach to form a living radical polymerization is to reversibly deactivate the propagating chains to suppress chain termination reactions by forming dormant chains. The choice of deactivating agent is contingent on a much higher deactivation rate compared to the rate of activation. Radical polymerization methods that reversibly deactivate propagating chains and form dormant species include ATRP, NMP, and RAFT. In cationic polymerization, the approach of forming associated ionic species from dissociated ionic species is used to produce a living polymer. In the associated approach, a Bronsted acid is used to convert the dissociated cation–anion pair into an associated pair, which suppresses the undesired side reactions that broaden the molecular weight distribution. Depending on whether polymerization is living or nonliving, three different types of polymer chains can be defined. These are propagating polymer chains or reactive polymer species ($RP_n^{\bullet}, RP_n^-,\ or\ RP_n^+$), dormant polymer chains that are not reactive toward monomer or other molecules in the reaction mixture until they are activated (DOP_n), and terminated polymer chains (TP_n). A homopolymer is produced by the polymerization of a single monomer, whereas a copolymer is produced from the polymerization of two or more monomers. The rate of addition of monomers to a propagating chain in copolymerization and the sequence of monomers on the chains depend on the fraction of each monomer in the feed, their reactivity, and the terminal monomers on the propagating chain. Polymer chains can be linear with no branches or nonlinear with short, long, or hyper-branch architecture. Polymers can also have a crosslinked chain architecture. The monomer type, mechanism of polymerization, living or nonliving nature of the propagating species, sequence of monomers in a copolymer, and chain architecture can significantly affect physicochemical, mechanical, biochemical, and environmental properties of polymeric materials.

1.7 List of Symbols and Abbreviations

CTA: chain transfer agent deliberately added to a polymerization reaction to control the polymer molecular weight

DOP_n: n-mer dormant polymer chain (not reactive and not terminated) in controlled/living polymerization

I: initiator molecule in addition to polymerization

$[I]$: initiator concentration in a polymerization reaction (mole cm^{-3})

$[I_0]$: initial concentration of initiator in a polymerization reaction (mole m^{-3})

k_{aca}: forward rate constant for reversible addition of a propagating radical to a polymeric RAFT chain transfer agent to form a RAFT adduct bi-chain propagating radical in RAFT polymerization (m^3 mole^{-1} s^{-1})

k_{acai}: forward rate constant for reversible addition of a propagating radical to a RAFT chain transfer agent to form a RAFT adduct propagating chain radical in the initiation step in RAFT polymerization (m^3 mole^{-1} s^{-1})

k_{acf}: forward rate constant for reversible fragmentation of a RAFT adduct bi-chain propagating radical to form a new polymeric RAFT chain transfer agent and a new propagating radical in RAFT polymerization (s^{-1})

k_{acfi}: forward rate constant for reversible fragmentation of a RAFT adduct propagating chain radical to form a new polymeric RAFT chain transfer agent and a new propagating radical in the initiation step in RAFT polymerization (s^{-1})

k_{aci}: forward (activation) rate constant for reversible transfer of the leaving group (halogen) from an initiator to a transition metal catalyst to form a primary radical (activation) in ATRP (m^3 mole^{-1} s^{-1})

$K_{a,RAFT}$: equilibrium constant for reversible addition of a propagating radical to a polymeric RAFT chain transfer agent to form a RAFT adduct bi-chain propagating radical in RAFT polymerization (dimensionless)

k_{BS}: rate constant for reaction between a growing chain terminating in monomer B and a monomer S in copolymerization of monomers S and B (m^3 $mole^{-1}$ s^{-1})

k_{BB}: rate constant for reaction between a growing chain terminating in monomer B and a monomer B in copolymerization of monomers S and B (m^3 $mole^{-1}$ s^{-1})

k_c: rate constant for recombination of two radicals generated from dissociation of the initiator in radical polymerization (m^3 $mole^{-1}$ s^{-1})

k_{ci}: rate constant for chain initiation between a primary reactive species and a monomer to form a 1-mer propagating chain in radical, cationic, or metathesis polymerization (m^3 $mole^{-1}$ s^{-1})

k_{cpx}: rate constant for preinitiation complex formation between a co-initiator and catalyst in cationic polymerization (m^3 $mole^{-1}$ s^{-1})

k_{cxi}: rate constant for reaction between the complex, formed by co-initiator and catalyst, and initiator to generate a primary reactive cation in cationic polymerization (m^3 $mole^{-1}$ s^{-1})

k_d: rate constant for dissociation of the initiator in radical polymerization (s^{-1})

k_{dac}: rate constant for deactivation of an active propagating species to form dormant species in controlled/living polymerization (m^3 $mole^{-1}$ s^{-1})

k_{daca}: reverse rate constant for reversible addition of a propagating radical to a polymeric RAFT chain transfer agent to form a RAFT adduct bi-chain propagating radical in RAFT polymerization (s^{-1})

k_{dacai}: reverse rate constant for reversible addition of a propagating radical to a RAFT chain transfer agent to form a RAFT adduct propagating chain radical in the initiation step in RAFT polymerization (s^{-1})

k_{dacf}: reverse rate constant for reversible fragmentation of a RAFT adduct bi-chain propagating radical to form a new polymeric RAFT chain transfer agent and a new propagating radical in RAFT polymerization (m^3 $mole^{-1}$ s^{-1})

k_{dacfi}: reverse rate constant for the reversible fragmentation of a RAFT adduct propagating chain radical to form a new polymeric RAFT chain transfer agent and a new propagating radical in the initiation step in RAFT polymerization (m^3 $mole^{-1}$ s^{-1})

k_{daci}: reverse (deactivation) rate constant for reversible transfer of the leaving group (halogen) from a transition metal catalyst to an initiator to regenerate the initiator in ATRP (m^3 $mole^{-1}$ s^{-1})

k_{dorm}: rate constant for cation transfer from the oxygen of monomer to the oxygen of hexamethylene oxide for transformation of the propagating cation to a formant state in living cationic polymerization

k_{ext}: rate constant for photoexcitation of a photocatalyst with blue light irradiation in START polymerization (rad^{-1} s^{-1})

$K_{f,RAFT}$: equilibrium constant for reversible fragmentation of a RAFT adduct bi-chain propagating radical to form a polymeric RAFT chain transfer agent and a new propagating radical in RAFT polymerization (dimensionless)

k_{gtac}: rate constant for chain activation by dissociation of the unstable polymeric anion complex to regenerate the n-mer propagating anion and nucleophile–trimethyl silane adduct in group transfer polymerization (s^{-1})

k_{gtcx}: rate constant for complexation of a nucleophile with the electron acceptor silica in trimethoxy silane side group of the silyl ketene acetal (initiator) to form an unstable anion adduct in group transfer polymerization (m^3 $mole^{-1}$ s^{-1})

k_{gtdac}: rate constant for chain deactivation between an n-mer propagating anion and nucleophile–trimethyl silane adduct to form an unstable polymeric anion complex in group transfer polymerization (m^3 $mole^{-1}$ s^{-1})

k_{gtdcx}: rate constant for decomplexation of a nucleophile from the electron acceptor silica in trimethoxy silane side group of the silyl ketene acetal to regenerate the initiator in group transfer polymerization (s^{-1})

k_{gtdf}: rate constant for defragmentation of the reactive ester enolate anion and nucleophile–trimethyl silane adduct to form an unstable initiator-nucleophile adduct in group transfer polymerization (m^3 $mole^{-1}$ s^{-1})

k_{gtf}: rate constant for fragmentation of the unstable initiator-nucleophile adduct to form a reactive ester enolate anion and a stable nucleophile–trimethyl silane adduct in group transfer polymerization (s^{-1})

k_{gti}: rate constant for chain initiation reaction between the ester enolate anion and a monomer to form a 1-mer propagating anion in group transfer polymerization (m^3 $mole^{-1}$ s^{-1})

k_{nds}: rate constant for the reaction of cation–anion complex with the lone pair electrons of oxygen on the 3-phenyl-1-propanol initiator to form a non-dissociated monomer–initiator adduct

k_p: rate constant for reaction between an n-mer propagating chain (radical, cationic, or anionic) and a monomer to form a new (n+1)-mer propagating chain, or the reaction between an n-mer and an m-mer with complementary functional groups in step-growth polymerization (m^3 $mole^{-1}$ s^{-1})

k_{prei}: rate constant for preinitiation reaction to form a complex between the catalyst and initiator/co-initiator to activate the initiator/co-initiator for the reaction with monomer in radical, cationic, or metathesis polymerization (m^3 $mole^{-1}$ s^{-1})

k_{rac}: rate constant for reactivation of dormant species to form an active propagating species in controlled/living radical polymerization (m^3 $mole^{-1}$ s^{-1})

k_{rci}: rate constant for reductive cationic initiation between a α, ω-diiodoperfluoroalkane and a cationic catalyst to form two iodine radicals and a perfluoroalkane di-radical in START polymerization (m^3 $mole^{-1}$ s^{-1})

k_{rq}: rate constant for reductive quenching of an excited photocatalyst in the presence of reducing agent sodium ascorbate to generate a reductive cationic catalyst in START polymerization (m^3 $mole^{-1}$ s^{-1})

k_{SB}: rate constant for reaction between a growing chain terminating in monomer S and a monomer B in copolymerization of monomers S and B (m^3 $mole^{-1}$ s^{-1})

k_{SS}: rate constant for reaction between a growing chain terminating in monomer S and a monomer S in copolymerization of monomers S and B (m^3 $mole^{-1}$ s^{-1})

k_t: rate constant for termination reaction between two propagating chains to produce one or more terminated polymer chains (m^3 $mole^{-1}$ s^{-1})

k_{tcan}: rate constant for termination of a propagating cation by combination of the cation–anion pair in cationic polymerization (s^{-1})

k_{thim}: rate constant for termination of a propagating cation by hydride ion transfer from a propagating cation to a monomer in cationic polymerization (m^3 $mole^{-1}$ s^{-1})

k_{thip}: rate constant for termination of a propagating cation by hydride ion transfer from a propagating cation to a polymer chain or another propagating cation in cationic polymerization (m^3 $mole^{-1}$ s^{-1})

k_{tia}: rate constant for termination by iodine addition between a perfluoroalkane-unconjugated diene radical and an iodine radical to form the first repeat unit of the propagating chain in START polymerization (m^3 $mole^{-1}$ s^{-1})

k_{tm}: rate constant for termination of a propagating species (radical, cationic, anionic) by reaction of a propagating species with a monomer (m^3 $mole^{-1}$ s^{-1})

k_{tta}: rate constant for termination of a propagating cation by a chain-terminating agent in cationic polymerization (m^3 $mole^{-1}$ s^{-1})
k_{tw}: rate constant for termination of a propagating species (cationic or anionic) by reaction of a propagating chain with water in cationic or anionic polymerization (m^3 $mole^{-1}$ s^{-1})
$k_{t\beta H}$: rate constant for abstraction of a hydrogen from the carbon beta to the cation-carrying carbon by the anion in the propagating cation pair to form a terminated chain in cationic polymerization (s^{-1})
K_{ATRP}: equilibrium constant for activation/deactivation of dormant species in ATRP (dimensionless)
K_{NMP}: equilibrium constant for activation/deactivation of dormant species in NMP (dimensionless)
K_{RAFT}: equilibrium constant for reversible formation and fragmentation of a RAFT chain transfer agent in RAFT polymerization (dimensionless)
LM^IX: transition metal catalyst in ATRP with M as the metal, L as the ligand, and X as the leaving group in low oxidation state (I)
$LM^{II}X$: transition metal catalyst in ATRP with M as the metal, L as the ligand, and X as the leaving group in high oxidation state (II)
M: monomer molecule in a polymerization reaction
$[M]$: concentration of monomer in a polymerization reaction (mole m^{-3})
$[M_0]$: initial concentration of monomer in a polymerization reaction (mole m^{-3})
$\overline{M}_{lp}$: average molecular weight of a polymer produced by living polymerization (g $mole^{-1}$)
R: gas constant (8.314 J $mole^{-1}$ K^{-1})
$R^{\bullet}$: primary radical or a reactive species generated from the dissociation of initiator in radical polymerization
R_1, R_2: designated alky groups
$RP_1^{\bullet}$: 1-mer propagating chain or reactive polymer chain produced by the reaction of a primary radical with a monomer in radical polymerization
$RP_n^{\bullet}$: n-mer propagating chain produced by the reaction of an (n-1)-mer propagating chain with a monomer in radical polymerization
t: time in a polymerization reaction (s)
T_c: ceiling temperature in ring-opening polymerization, at which $\Delta G_p = 0$ or the temperature at which the rates of polymerization and depolymerization are equal (K)
TP_n: n-mer terminated (not reactive and not dormant) polymer chain in a polymerization reaction
α: fraction of monomer converted or conversion in a polymerization reaction
ΔG_p: Gibbs free energy of polymerization (J $mole^{-1}$)
ΔH_p: enthalpy of polymerization (J $mole^{-1}$)
ΔS_p: entropy of polymerization (J $mole^{-1}$ K^{-1})

1.8 Practice Problems

Problem 1.1 Ziegler–Natta catalysts are used to synthesize stereoregular, high molecular weight polymers of 1-alkenes like propylene. The catalyst system in Ziegler–Natta polymerization consists of a group four transition metal catalyst, like titanium chloride, and an organo-aluminum co-catalyst like triethyl-aluminum [268]. The catalyst and co-catalyst form a complex for the insertion and growth of the polymer chain. 1-alkenes like propylene are also polymerized by metallocene catalysts. Metallocene catalysts are transition metal complexes based on zirconium with sandwich-type biscyclopentadienyl ligands and methylaluminoxane ($Al_4O_3Me_6$) [269]. Write the reactions for propylene polymerization using Zeigler–Natta and Metallocene catalysts and explain the mechanistic differences between the two polymerization reactions.

Problem 1.2 Cycloalkenes can be polymerized by olefin metathesis polymerization with Grubbs catalysts. Grubbs catalysts are transition metal complexes based on ruthenium and saturated *N*-heterocyclic carbene. Write the reactions for polymerization of cycloalkenes using Metathesis catalysts and describe the mechanism of polymerization with this catalyst system.

References

1 Moad, G. and Solomon, D.H., Invited review – understanding and controlling radical polymerization. *Australian Journal of Chemistry*, 1990, **43**(2), 215.

2 Matyjaszewski, K. Introduction: towards well-defined polymeric nanostructures by better understanding of the mechanisms of controlled/living radical polymerization, in *224th ACS national meeting*. 2002, American Chemical Society.

3 Jagur-Grodzinski, J., ed. *Living and controlled polymerization: synthesis, characterization, and properties of the respective polymers and copolymers*. 2006, Nova Science Publishers, 385.

4 Bisht, H.S. and Chatterjee, A.K., Living free-radical polymerization – a review. *Journal of Macromolecular Science-Polymer Reviews*, 2001, **C41**(3), 139.

5 Webster, O.W., Living polymerization methods. *Science*, 1991, **251**(4996), 887.

6 Yagci, Y. and Tasdelen, M.A., Mechanistic transformations involving living and controlled/living polymerization methods. *Progress in Polymer Science*, 2006, **31**(12), 1133.

7 Tedder, J.M., Which factors determine the reactivity and regioselectivity of free-radical substitution and addition reactions. *Angewandte Chemie-International Edition in English*, 1982, **21**(6), 401.

8 Braun, D., Origins and development of initiation of free radical polymerization processes. *International Journal of Polymer Science*, 2009, **2009**, 893234.

9 Moad, G. et al., Initiating free radical polymerization. *Macromolecular Symposia*, 2002, **182**, 65.

10 Bevington, J.C., The nature of the initiation reaction in the polymerization of styrene sensitized by benzoyl peroxide. *Proceedings of the Royal Society of London Series A: Mathematical and Physical Sciences*, 1957, **239**(1218), 420.

11 Buback, M., Elementary steps of initiation and termination processes in radical polymerization. *Macromolecular Symposia*, 2005, **226**, 121.

12 Autrey, T., Klepzig, K., and Franz, J.A., An experimental investigation of solvent-cage induced radical termination. Recombination efficiencies of a photochemical caged pair of phenylthiyl radicals. *Abstracts of Papers of the American Chemical Society*, 1999, **217**, U287.

13 Huggenberger, C., Lipscher, J., and Fischer, H., Self-termination of benzoyl radicals to ground-state and excited-state benzil – symmetry control of a radical combination. *Journal of Physical Chemistry*, 1980, **84**(25), 3467.

14 Martin, J.C. and Hargis, J.H., Cage recombination of benzoyloxy radicals – an important pathway in decomposition of benzoyl peroxide. *Journal of the American Chemical Society*, 1969, **91**(19), 5399.

15 Yamada, B., Westmoreland, D.G., Kobatake, S., and Konosu, O., ESR spectroscopic studies of radical polymerization. *Progress in Polymer Science*, 1999, **24**(4), 565.

16 Gromov, V.F. et al., Effect of solvent on chain propagation and termination reaction-rates in radical polymerization. *European Polymer Journal*, 1980, **16**(6), 529.

17 Nakamura, Y., Lee, R., Coote, M.L., and Yamago, S., Termination mechanism of the radical polymerization of acrylates. *Macromolecular Rapid Communications*, 2016, **37**(6), 506.
18 Pojman, J.A. et al., Factors affecting propagating fronts of addition polymerization - velocity, front curvature, temperature profile, conversion, and molecular-weight distribution. *Journal of Polymer Science Part A:Polymer Chemistry*, 1995, **33**(4), 643.
19 Han, H.F. et al., Initiation and termination in styrene free-radical polymerization initiated by redox initiation. *Macromolecular Chemistry and Physics*, 2020, **221**, 2000277.
20 Li, X.P. et al., Evidence for polarity- and viscosity-controlled pathways in the termination reaction in the radical polymerization of acrylonitrile. *Macromolecules*, 2021, **54**(10), 4497.
21 Szwarc, M., Living polymers. *Nature*, 1956, **178**(4543), 1168.
22 Szwarc, M., Living polymers. Their discovery, characterization, and properties. *Journal of Polymer Science Part A: Polymer Chemistry*, 1998, **36**(1), IX–XV.
23 Vanbeylen, M. et al., Developments in anionic-polymerization – a critical review. *Advances in Polymer Science*, 1988, **86**, 87.
24 Iovu, M.C., Mapolie, S., and Britchi, A.G., Styrene-butadiene rubber synthesized by anionic polymerization. *Macromolecular Symposia*, 2001, **165**, 55.
25 Frankel, M., Ottolenghi, A., Albeck, M., and Zilkha, A., Anionic polymerisation of vinyl monomers with butyl-lithium. *Journal of the Chemical Society*, 1959, 3858.
26 Nurkeeva, Z.S., Plate, N.A., and Maltsev, V.V., Anionic polymerization and copolymerization of isoprene and styrene with butyl lithium in presence of tetraethyltin. *Vysokomolekulyarnye Soedineniya Section A*, 1972, **14**(10), 2047.
27 Margerison, D. and Newport, J.P., Degree of association of *N*-butyl lithium in hydrocarbon media. *Transactions of the Faraday Society*, 1963, **59**(489), 2058.
28 Bullock, A.T., Cameron, G.G., and Elsom, J.M., Electron-spin resonance studies of spin-labeled synthetic polymers. 12. polymerization of methyl-methacrylate initiated by butyl lithium. *European Polymer Journal*, 1977, **13**(10), 751.
29 Kawalec, M., Smiga-Matuszowicz, M., and Kurcok, P., Counterion and solvent effects on the anionic polymerization of β-butyrolactone initiated with acetic acid salts. *European Polymer Journal*, 2008, **44**(11), 3556.
30 Hirohara, H., Takaya, K., and Ise, N., Study of effect of impurities in living anionic polymerization. *Macromolecules*, 1971, **4**(3), 288.
31 Wenger, F., Effects of impurities on control of anionic batch polymerizations. *Journal of Polymer Science*, 1962, **60**(170), 99.
32 Vega, J.R., Frontini, G.L., and Meira, G.R., Optimal periodic control of a continuous living anionic-polymerization. 3. presence of impurities in the monomer feed. *Journal of Applied Polymer Science*, 1991, **42**(12), 3181.
33 Schulz, G.V., Reaction mechanism and distribution of molecular weights in anionic polymerisation of styrene. *Chemistry & Industry*, 1966, (5), 212.
34 Lee, W. et al., Molecular weight distribution of polystyrene made by anionic polymerization. *Macromolecules*, 2000, **33**(14), 5111.
35 Marvel, C.S. et al., Cationic polymerization of butadiene and copolymerization of butadiene and styrene. *Journal of Polymer Science*, 1951, **6**(4), 483.
36 Plesch, P.H., Developments in the theory of cationic polymerization. *Journal of Applied Chemistry*, 1951, **1**(6), 269.
37 Aoshima, S. and Kanaoka, S., A renaissance in living cationic polymerization. *Chemical Reviews*, 2009, **109**(11), 5245.

38 Yang, M.L., Li, K., and Stover, H.D.H., Living cationic polymerization of styrene monomers. *Macromolecular Rapid Communications*, 1994, **15**(5), 425.

39 Hasegawa, K., Asami, R., and Higashimura, T., Cationic polymerization of alkyl-1,3-butadienes. *Macromolecules*, 1977, **10**(3), 592.

40 Matyjaszewski, K., Slomkowski, S., and Penczek, S., Kinetics and mechanism of the cationic polymerization of tetrahydrofuran in solution – THF–CH_3NO_2 system. *Journal of Polymer Science Part A: Polymer Chemistry*, 1979, **17**(1), 69–80.

41 Matyjaszewski, K., Slomkowski, S., and Penczek, S., Kinetics and mechanism of the cationic polymerization of tetrahydrofuran in solution – THF–CH_2Cl_2 and THF–CH_2Cl_2/CH_3NO_2 systems. *Journal of Polymer Science Part A: Polymer Chemistry*, 1979, **17**(8), 2413.

42 Kagiya, T., Matsuda, T., and Hirata, R., Polymerization of 2-oxazolines. 3. Influence of polymerization media on propagation reaction in cationic polymerization of 2-phenyl-2-oxazoline. *Journal of Macromolecular Science A: Chemistry*, 1972, **6**(3), 451.

43 Sage, V., Clark, J.H., and Macquarrie, D.J., Cationic polymerization of styrene using mesoporous silica supported aluminum chloride. *Journal of Molecular Catalysis A-Chemical*, 2003, **198**(1–2), 349.

44 Satoh, K., Kamigaito, M., and Sawamoto, M., Direct living cationic polymerization of *p*-hydroxystyrene with boron trifluoride etherate in the presence of water. *Macromolecules*, 2000, **33**(15), 5405.

45 Yang, X.M., Stern, C.L., and Marks, T.J., Cationic zirconocene olefin polymerization catalysts based on the organo-Lewis acid tris(Pentafluorophenyl)borane – a synthetic, structural, solution dynamic, and polymerization catalytic study. *Journal of the American Chemical Society*, 1994, **116**(22), 10015.

46 Kostjuk, S.V., Radchenko, A.V., and Ganachaud, F., Controlled/living cationic polymerization of *p*-methoxystyrene in solution and aqueous dispersion using tris(pentafluorophenyl)borane as a Lewis acid: acetonitrile does the job. *Macromolecules*, 2007, **40**(3), 482.

47 Cheradame, H., Mazza, M., Hung, N.A., and Sigwalt, P., Cationic polymerization of indene initiated by titanium tetrachloride. 1. Influence of impurities of solvent, methylene chloride, and cocatalytic effects of water and hydrochloric acid. *European Polymer Journal*, 1973, **9**(5), 375.

48 Hung, N.A., Cheradam. H, and Sigwalt, P., Cationic polymerization of indene initiated by titanium tetrachloride. 2. Influence of temperature on rate of polymerization. *European Polymer Journal*, 1973, **9**(5), 385.

49 Chen, Y.N. et al., Recent advances in living cationic polymerization with emerging initiation/controlling systems. *Macromolecular Rapid Communications*, 2021, **42**(13), 2100148.

50 De, P. and Faust, R., Determination of the absolute rate constants of propagation for ion pairs and free ions in the living cationic polymerization of isobutylene. *Macromolecules*, 2005, **38**(23), 9897.

51 Tobisch, S., Mechanism of the chain termination of the allylnickel(II)-catalyzed polymerization of 1,3-butadiene. A density functional investigation for the cationic $[Ni^{II}(RC_3H_4)(cis\text{-}C_4H_6)L](+)$ active catalyst. *Macromolecules*, 2003, **36**(16), 6235.

52 Fodor, Z., Bae, Y.C., and Faust, R., Temperature effects on the living cationic polymerization of isobutylene: determination of spontaneous chain transfer constants in the presence of terminative chain transfer (vol 31, pg 4439, 1998). *Macromolecules*, 1998, **31**(15), 5170.

53 Zhang, J.H. et al., Characteristics and mechanism of styrene cationic polymerization in aqueous media initiated by cumyl alcohol/$B(C_6F_5)_{(3)}$. *Macromolecular Chemistry and Physics*, 2019, **220**(4), 1800419.

54 Cai, G.F., Yan, D.Y., and Litt, M., A kinetic theory on cationic polymerization with termination by combination with a negative fragment. *Macromolecules*, 1988, **21**(3), 578.

55 Overberger, C.G. and Newton, M.G., Ionic polmerization. 15. Further molecular terminating agents in the cationic polymerization of styrene. *Journal of the American Chemical Society*, 1960, **82**(14), 3622.

56 Overberger, C.G., Endres, G.F., and Monaci, A., Ionic polymerization. 7. Relative reactivities of mono-dialkylbenzenes and para-dialkylbenzenes as molecular terminating agents in the cationic polymerization of styrene. *Journal of the American Chemical Society*, 1956, **78**(9), 1969.

57 Sigwalt, P. et al., Inhibiting or cocatalytic effect of water and other additives on cationic polymerization of cyclodimethylsiloxanes. *Makromolekulare Chemie-Macromolecular Symposia*, 1991, **42**(3), 229.

58 Gonzalez, R. et al., Influence of water on the cationic polymerisation of 2-ethenylfuran by trifluoroacetic acid in methylene dichloride. *Macromolecular Chemistry and Physics*, 2001, **202**(5), 710.

59 Huang, R., Ficek, B.A., Glover, S.O., and Scranton, A.B., Effect of water in cationic photopolymerizations: reversible inhibition. *Redtech Report*, 2007, **21**, 30.

60 Yan, D.Y., Jiang, H., Dong, H., and Muller, A.H.E., Kinetic analysis of "living" polymerization processes exhibiting slow equilibria. 5. Effect of monomer transfer in cationic polymerization and similar living processes. *Macromolecules*, 1996, **29**(15), 5065.

61 Penczek, S., Kinetics of chain transfer to monomer and intramolecular termination in cationic polymerization of alpha-epichlorohydrin. *Bulletin De L Academie Polonaise Des Sciences-Serie Des Sciences Chimiques*, 1970, **18**(1), 39.

62 Destephen, A. and Ballard, N., On the limitations of cationic polymerization of vinyl monomers in aqueous dispersed media. *Polymer Chemistry*, 2021, **12**(44), 6444.

63 Endres, G.F. and Overberger, C.G., Ionic polymerization. 5. Molecular termination in the cationic polymerization of styrene catalyzed by stannic chloride. *Journal of the American Chemical Society*, 1955, **77**(8), 2201.

64 Utyanskaya, E.Z., Role of intermolecular hydride transfer mechanism in initiation and chain transfer in cationic polymerization of cyclic ethers. *Vysokomolekulyarnye Soedineniya Section A*, 1971, **13**(3), 523.

65 Aoshima, S. and Higashimura, T., Living cationic polymerization of vinyl monomers by organo-aluminum halides. 3. Living polymerization of isobutyl vinyl ether by $EtAlCl_2$ in the presence of ester additives. *Macromolecules*, 1989, **22**(3), 1009.

66 Norrish, R.G.W. and Russell, K.E., Polymerization of isobutene by stannic chloride. *Transactions of the Faraday Society*, 1952, **48**(1), 91.

67 Bauer, R.F. and Russell, K.E., Cationic polymerization of isobutene initiated by stannic chloride and phenols – polymer end group studies. *Journal of Polymer Science Part A-1:Polymer Chemistry*, 1971, **9**(5), 1451.

68 Aoshima, S., Fujisawa, T., and Kobayashi, E., Living cationic polymerization of isobutyl vinyl ether by $EtAlCl_2$ in the presence of ether additives – cyclic ethers, cyclic formals, and acyclic ethers with oxyethylene units. *Journal of Polymer Science Part A: Polymer Chemistry*, 1994, **32**(9), 1719.

69 Ghanem, N.A. and Marek, M., Role and fate of water as co-catalyst in cationic polymerization. *European Polymer Journal*, 1972, **8**(8), 999.

70 Vergnaud, J., Sarazin, Y., Strub, H., and Carpentier, J.F., Water-tolerant catalyst systems for the bulk cationic polymerization of para-methylstyrene and indene. *European Polymer Journal*, 2010, **46**(5), 1093.

71 Corrigan, N. et al., Reversible-deactivation radical polymerization (Controlled/living radical polymerization): from discovery to materials design and applications. *Progress in Polymer Science*, 2020, **111**, 101311.

72 Kubisa, P., Radical polymerization with reversible deactivation of active species. *Polimery*, 2000, **45**(11–12), 741.

73 Shipp, D.A., Reversible-deactivation radical polymerizations. *Polymer Reviews*, 2011, **51**(2), 99.

74 Harrisson, S., The chain length distribution of an ideal reversible deactivation radical polymerization. *Polymers*, 2018, **10**(8), 887.

75 Jenkins, A.D., Jones, R.G., and Moad, G., Terminology for reversible-deactivation radical polymerization previously called "controlled" radical or "living" radical polymerization (IUPAC Recommendations 2010). *Pure and Applied Chemistry*, 2010, **82**(2), 483.

76 Stenzel, M.H. and Barner-Kowollik, C., The living dead – common misconceptions about reversible deactivation radical polymerization. *Materials Horizons*, 2016, **3**(6), 471.

77 Kawabata, N. and Tsuruta, T., Elementary reactions of metal alkyl in anionic polymerization. I. Reaction mode of *N*-butyllithium in initiation step of methyl acrylate and methyl methacrylate polymerization. *Makromolekulare Chemie*, 1965, **86**(Aug), 231.

78 Kawabata, N. and Tsuruta, T., Elementary reactions of metal alkyl in anionic polymerization. 2. Reaction mode of *N*-butyllithium in initiation step of acrylonitrile and methacrylonitrile polymerization. *Makromolekulare Chemie*, 1966, **98**(Nov), 262.

79 Tadmor, Z. and Biesenbe, J.A., Influence of chain initiation rate on molecular weight dispersion in free radical polymerizations. *Journal of Polymer Science Part B-Polymer Letters*, 1965, **3**(9), 753.

80 Clay, P.A. and Gilbert, R.G., Molecular-weight distributions in free-radical polymerizations. 1. Model development and implications for data interpretation. *Macromolecules*, 1995, **28**(2), 552.

81 Bamford, C.H. and Dewar, M.J.S., Studies in polymerization. 1. A method for determining the velocity constants in polymerization reactions and its application to styrene. *Proceedings of the Royal Society of London Series A: Mathematical and Physical Sciences*, 1948, **192**(1030), 309.

82 Matyjaszewski, K., Lifetimes of polystyrene chains in atom transfer radical polymerization. *Macromolecules*, 1999, **32**(26), 9051.

83 Charleux, B., Nicolas, J., and Guerret, O., Theoretical expression of the average activation-deactivation equilibrium constant in controlled/living free-radical copolymerization operating via reversible termination. Application to a strongly improved control in nitroxide-mediated polymerization of methyl methacrylate. *Macromolecules*, 2005, **38**(13), 5485.

84 Wang, Y. et al., Reversible-deactivation radical polymerization in the presence of metallic copper. Comproportionation-disproportionation equilibria and kinetics. *Macromolecules*, 2013, **46**(10), 3793.

85 Goto, A., Wakada, T., Fukuda, T., and Tsujii, Y., A systematic kinetic study in reversible chain transfer catalyzed polymerizations (RTCPs) with germanium, tin, phosphorus, and nitrogen catalysts. *Macromolecular Chemistry and Physics*, 2010, **211**(5), 594.

86 Vana, P. and Goto, A., Kinetic simulations of reversible chain transfer catalyzed polymerization (RTCP): guidelines to optimum molecular weight control. *Macromolecular Theory and Simulations*, 2010, **19**(1), 24.

87 Patten, T.E., Xia, J.H., Abernathy, T., and Matyjaszewski, K., Polymers with very low polydispersities from atom transfer radical polymerization. *Science*, 1996, **272**(5263), 866.

88 Wang, J.S. and Matyjaszewski, K., Controlled living radical polymerization - halogen atom-transfer radical polymerization promoted by a Cu(I)Cu(II) redox process. *Macromolecules*, 1995, **28**(23), 7901.

89 Wang, J.S. and Matyjaszewski, K., Living controlled radical polymerization – transition-metal-catalyzed atom-transfer radical polymerization in the presence of a conventional radical initiator. *Macromolecules*, 1995, **28**(22), 7572.

90 Wang, J.S. and Matyjaszewski, K., Controlled living radical polymerization – atom-transfer radical polymerization in the presence of transition-metal complexes. *Journal of the American Chemical Society*, 1995, **117**(20), 5614.

91 Grubbs, R.B., Nitroxide-mediated radical polymerization: limitations and versatility. *Polymer Reviews*, 2011, **51**(2), 104.

92 Sciannamea, V., Jerome, R., and Detrembleur, C., In-situ nitroxide-mediated radical polymerization (NMP) processes: their understanding and optimization. *Chemical Reviews*, 2008, **108**(3), 1104.

93 Nicolas, J. et al., Nitroxide-mediated polymerization. *Progress in Polymer Science*, 2013, **38**(1), 63.

94 Studer, A. and Schulte, T., Nitroxide-mediated radical processes. *Chemical Record*, 2005, **5**(1), 27.

95 Matyjaszewski, K., Gaynor, S., and Wang, J.S., Controlled radical polymerizations – the use of alkyl iodides in degenerative transfer. *Macromolecules*, 1995, **28**(6), 2093.

96 Gaynor, S.G., Wang, J.S., and Matyjaszewski, K., Well-defined polymers obtained through the use of controlled radical polymerization – the use of alkyl iodides as degenerative transfer reagents. *Abstracts of Papers of the American Chemical Society*, 1995, **209**, 106.

97 Gaynor, S.G., Wang, J.S., and Matyjaszewski, K., Controlled radical polymerization by degenerative transfer – effect of the structure of the transfer agent. *Macromolecules*, 1995, **28**(24), 8051.

98 Lansalot, M. et al., Controlled free-radical miniemulsion polymerization of styrene using degenerative transfer. *Macromolecules*, 1999, **32**(22), 7354.

99 Coelho, J.F.J. et al., Synthesis of poly(lauryl acrylate) by single-electron transfer/degenerative chain transfer living radical polymerization catalyzed by $Na_2S_2O_4$ in water. *Macromolecular Chemistry and Physics*, 2007, **208**(11), 1218.

100 Allan, L.E.N., Perry, M.R., and Shaver, M.P., Organometallic mediated radical polymerization. *Progress in Polymer Science*, 2012, **37**(1), 127.

101 Hurtgen, M., Detrembleur, C., Jerome, C., and Debuigne, A., Insight into organometallic-mediated radical polymerization. *Polymer Reviews*, 2011, **51**(2), 188.

102 Poli, R. and Xue, Z.G., Organometallic-mediated radical polymerization of vinyl acetate with iron, in *242nd ACS National Meeting*. 2011, American Chemical Society.

103 Moad, G. et al., Living free radical polymerization with reversible addition-fragmentation chain transfer (the life of RAFT). *Polymer International*, 2000, **49**(9), 993–1001.

104 Semsarilar, M. and Perrier, S., 'Green' reversible addition-fragmentation chain-transfer (RAFT) polymerization. *Nature Chemistry*, 2010, **2**(10), 811.

105 Favier, A. and Charreyre, M.T., Experimental requirements for an efficient control of free-radical polymerizations via the reversible addition-fragmentation chain transfer (RAFT) process. *Macromolecular Rapid Communications*, 2006, **27**(9), 653.

106 Chiefari, J. et al., Living free-radical polymerization by reversible addition-fragmentation chain transfer: the RAFT process. *Macromolecules*, 1998, **31**(16), 5559.

107 Perrier, S. and Takolpuckdee, P., Macromolecular design via reversible addition-fragmentation chain transfer (RAFT)/Xanthates (MADIX) polymerization. *Journal of Polymer Science Part A: Polymer Chemistry*, 2005, **43**(22), 5347.

108 Keddie, D.J., A guide to the synthesis of block copolymers using reversible-addition fragmentation chain transfer (RAFT) polymerization. *Chemical Society Reviews*, 2014, **43**(2), 496.

109 Liu, X.D., Zhang, L.F., Cheng, Z.P., and Zhu, X.L., Straightforward catalyst/solvent-free iodine-mediated living radical polymerization of functional monomers driven by visible light irradiation. *Chemical Communications*, 2016, **52**(72), 10850.

110 Ni, Y.Y., Zhang, L.F., Cheng, Z.P., and Zhu, X.L., Iodine-mediated reversible-deactivation radical polymerization: a powerful strategy for polymer synthesis. *Polymer Chemistry*, 2019, **10**(20), 2504.

111 Goto, A. et al., Living radical polymerizations with germanium, tin, and phosphorus catalysts – reversible chain transfer catalyzed polymerizations (RTCPs). *Journal of the American Chemical Society*, 2007, **129**(43), 13347.

112 Yamago, S., Iida, K., and Yoshida, J., Organotellurium compounds as novel initiators for controlled/living radical polymerizations. Synthesis of functionalized polystyrenes and end-group modifications. *Journal of the American Chemical Society*, 2002, **124**(12), 2874.

113 Yamago, S., Ukai, Y., Matsumoto, A., and Nakamura, Y., Organotellurium-mediated controlled/living radical polymerization initiated by direct C—Te bond photolysis. *Journal of the American Chemical Society*, 2009, **131**(6), 2100.

114 Sugihara, Y., Kagawa, Y., Yamago, S., and Okubo, M., Organotellurium-mediated living radical polymerization in miniemulsion. *Macromolecules*, 2007, **40**(26), 9208.

115 Kwak, Y. et al., Kinetic study on role of ditelluride in organotellurium-mediated living radical polymerization (TERP). *Macromolecules*, 2007, **40**(6), 1881.

116 Kumar, S. et al., Synthesis of well-defined amphiphilic block copolymers by organotellurium-mediated living radical polymerization (TERP). *Macromolecular Rapid Communications*, 2011, **32**(19), 1576.

117 Kayahara, E., Kondo, N., and Yamago, S., Substituent effect on the antimony atom in organostibine-mediated living radical polymerization. *Heteroatom Chemistry*, 2011, **22**(3–4), 307.

118 Zeng, J.D. et al., Organoselenium compounds: development of a universal "living" free radical polymerization mediator. *Polymer Chemistry*, 2013, **4**(12), 3453.

119 Zeng, J.D. et al., New selenium-based iniferter agent for living free radical polymerization of styrene under UV irradiation. *Journal of Polymer Science Part A: Polymer Chemistry*, 2012, **50**(11), 2211.

120 Botha, C., Weber, W., Pfukwa, H., and Pasch, H., Controlled radical polymerization using a novel symmetrical selenium RAFT agent. *Macromolecular Chemistry and Physics*, 2014, **215**(17), 1625.

121 Dodonov, V.A. et al., Macroinitiators of controlled free-radical polymerization of methyl methacrylate that are based on the tributyl boron-naphthoquinone system. *Polymer Science Series B*, 2010, **52**(9–10), 520.

122 Kuznetsova, Y.L., Chesnokov, S.A., Zaitsev, S.D., and Ludin, D.V., The catalytic system tri-*n*-butyl boron-*p*-quinone in the free-radical polymerization of styrene. *Polymer Science Series B*, 2012, **54**(9–10), 434.

123 Renaud, P. et al., Boron: a key element in radical reactions. *Pure and Applied Chemistry*, 2007, **79**(2), 223.

124 Matyjaszewski, K., Wei, M.L., Xia, J.H., and Gaynor, S.G., Atom transfer radical polymerization of styrene catalyzed by copper carboxylate complexes. *Macromolecular Chemistry and Physics*, 1998, **199**(10), 2289.

125 Boyer, C. et al., Copper-mediated living radical polymerization (atom transfer radical polymerization and copper(0) mediated polymerization): from fundamentals to bioapplications. *Chemical Reviews*, 2016, **116**(4), 1803.

126 Patten, T.E. and Matyjaszewski, K., Copper(I)-catalyzed atom transfer radical polymerization. *Accounts of Chemical Research*, 1999, **32**(10), 895.

127 Pintauer, T. and Matyjaszewski, K., Atom transfer radical addition and polymerization reactions catalyzed by ppm amounts of copper complexes. *Chemical Society Reviews*, 2008, **37**(6), 1087.

128 Gridnev, A.A. and Ittel, S.D., Catalytic chain transfer in free-radical polymerizations. *Chemical Reviews*, 2001, **101**(12), 3611.

129 Li, Z.H. et al., Atom transfer radical polymerization of methyl methacrylate catalyzed by cobalt catalyst system. *Journal of Polymer Science Part A: Polymer Chemistry*, 2005, **43**(21), 5207.

130 Chen, Y.H. et al., Cobalt(II) phenoxy-imine complexes in radical polymerization of vinyl acetate: the interplay of catalytic chain transfer and controlled/living radical polymerization. *Journal of Polymer Science*, 2020, **58**(1), 101.

131 Chen, M., MacLeod, M.J., and Johnson, J.A., Visible-light-controlled living radical polymerization from a trithiocarbonate iniferter mediated by an organic photoredox catalyst. *ACS Macro Letters*, 2015, **4**(5), 566.

132 Otsu, T., Iniferter concept and living radical polymerization. *Journal of Polymer Science Part A: Polymer Chemistry*, 2000, **38**(12), 2121.

133 Otsu, T. and Yoshida, M., Role of initiator-transfer agent-terminator (iniferter) in radical polymerizations: polymer design by organic disulfides as iniferters. *Macromolecular Rapid Communications*, 2005, **26**(3), 137.

134 Tang, H.D. et al., Highly active copper-based catalyst for atom transfer radical polymerization. *Journal of the American Chemical Society*, 2006, **128**(50), 16277.

135 Krys, P. and Matyjaszewski, K., Kinetics of atom transfer radical polymerization. *European Polymer Journal*, 2017, **89**, 482.

136 Parker, J., Jones, R.G., and Holder, S.J., Direct evidence for the interaction of the mechanisms of thermally initiated and atom transfer radical polymerization. *Macromolecules*, 2000, **33**(25), 9166.

137 Matyjaszewski, K., Atom transfer radical polymerization (ATRP): current status and future perspectives. *Macromolecules*, 2012, **45**(10), 4015.

138 Matyjaszewski, K., Atom transfer radical polymerization: from mechanisms to applications. *Israel Journal of Chemistry*, 2012, **52**(3–4), 206.

139 Liu, X.H. et al., Copper-mediated initiators for continuous activator regeneration atom transfer radical polymerization of acrylonitrile. *Journal of Polymer Science Part A: Polymer Chemistry*, 2012, **50**(20), 4358.

140 Okada, S., Park, S., and Matyjaszewski, K., Initiators for continuous activator regeneration atom transfer radical polymerization of methyl methacrylate and styrene with *N*-heterocyclic carbene as ligands for Fe-based catalysts. *ACS Macro Letters*, 2014, **3**(9), 944.

141 Martinez, M.R. et al., Depolymerization of P($PDMS_{11}$MA) bottlebrushes via atom transfer radical polymerization with activator regeneration. *Macromolecules*, 2021, **54**(12), 5526.

142 Jakubowski, W., Min, K., and Matyjaszewski, K., Activators regenerated by electron transfer for atom transfer radical polymerization of styrene. *Macromolecules*, 2006, **39**(1), 39.

143 Li, N. et al., Universal strategy for efficient fabrication of blood compatible surfaces via polydopamine-assisted surface-initiated activators regenerated by electron transfer atom-transfer radical polymerization of zwitterions. *ACS Applied Materials & Interfaces*, 2020, **12**(10), 12337.

144 Nicol, E. and Nze, R.P., Supplemental activator and reducing agent atom transfer radical polymerization of 2-hydroxyethyl acrylate from high molar mass poly(ethylene oxide) macroinitiator in dilute solution. *Macromolecular Chemistry and Physics*, 2015, **216**(13), 1405.

145 Bortolamei, N. et al., Controlled aqueous atom transfer radical polymerization with electrochemical generation of the active catalyst. *Angewandte Chemie-International Edition*, 2011, **50**(48), 11391.

146 Dadashi-Silab, S., Tasdelen, M.A., and Yagci, Y., Photoinitiated atom transfer radical polymerization: current status and future perspectives. *Journal of Polymer Science Part A: Polymer Chemistry*, 2014, **52**(20), 2878.

147 Xia, J.H. and Matyjaszewski, K., Controlled/"living" radical polymerization. Homogeneous reverse atom transfer radical polymerization using AIBN as the initiator. *Macromolecules*, 1997, **30**(25), 7692.

148 Gromada, J. and Matyjaszewski, K., Simultaneous reverse and normal initiation in atom transfer radical polymerization. *Macromolecules*, 2001, **34**(22), 7664.

149 Jakubowski, W. and Matyjaszewski, K., Activator generated by electron transfer for atom transfer radical polymerization. *Macromolecules*, 2005, **38**(10), 4139.

150 Huang, Z.X., Zhang, Y.M., Li, H., and Liu, Y.G., A novel immobilized cobalt(II)/copper(II) bimetallic catalyst for atom transfer radical polymerization (ATRP) of methyl methacrylate. *Applied Catalysis A: General*, 2007, **332**(2), 192.

151 Huang, Z.X. et al., Atom transfer radical polymerization of methyl methacrylate catalyzed by ion exchange resin immobilized Co(II) hybrid catalyst. *Journal of Polymer Science Part A: Polymer Chemistry*, 2008, **46**(4), 1416.

152 Treat, N.J. et al., Metal-free atom transfer radical polymerization. *Journal of the American Chemical Society*, 2014, **136**(45), 16096.

153 Pan, X.C., Lamson, M., Yan, J.J., and Matyjaszewski, K., Photoinduced metal-free atom transfer radical polymerization of acrylonitrile. *Acs Macro Letters*, 2015, **4**(2), 192.

154 Degirmenci, M., Catalgil-Giz, H., and Yagci, Y., Synthesis of block copolymers by combined ultrasonic irradiation and reverse atom transfer radical polymerization processes. *Journal of Polymer Science Part A: Polymer Chemistry*, 2004, **42**(3), 534.

155 Dervan, P., Aldabbagh, F., Zetterlund, P.B., and Yamada, B., Mechanism and kinetics of the imidazolidinone nitroxide-mediated free-radical polymerization of styrene. *Journal of Polymer Science Part A: Polymer Chemistry*, 2003, **41**(2), 327.

156 Veregin, R.P.N., Odell, P.G., Michalak, L.M., and Georges, M.K., Mechanism of rate enhancement using organic acids in nitroxide-mediated living free-radical polymerizations. *Macromolecules*, 1996, **29**(11), 4161.

157 Studer, A. et al., New sterically hindered nitroxides for the living free radical polymerization: X-ray structure of an α-H-bearing nitroxide. *Macromolecules*, 2004, **37**(1), 27.

158 Lamontagne, H.R. and Lessard, B.H., Nitroxide-mediated polymerization: a versatile tool for the engineering of next generation materials. *Acs Applied Polymer Materials*, 2020, **2**(12), 5327.

159 Veregin, R.P.N., Odell, P.G., Michalak, L.M., and Georges, M.K., Molecular weight distributions in nitroxide-mediated living free radical polymerization: kinetics of the slow equilibria between growing and dormant chains. *Macromolecules*, 1996, **29**(10), 3346.

160 Hodgson, J.L. et al., Side reactions of nitroxide-mediated polymerization: N—O versus O—C cleavage of alkoxyamines. *Journal of Physical Chemistry A*, 2010, **114**(38), 10458.

161 Gryn'ova, G., Lin, C.Y., and Coote, M.L., Which side-reactions compromise nitroxide mediated polymerization? *Polymer Chemistry*, 2013, **4**(13), 3744.

162 Hawthorne, D.G., Moad, G., Rizzardo, E., and Thang, S.H., Living radical polymerization with reversible addition-fragmentation chain transfer (RAFT): direct ESR observation of intermediate radicals. *Macromolecules*, 1999, **32**(16), 5457.

163 Li, C.Z. and Benicewicz, B.C., α-Cyanobenzyl dithioester reversible addition-fragmentation chain-transfer agents for controlled radical polymerizations. *Journal of Polymer Science Part A: Polymer Chemistry*, 2005, **43**(7), 1535.

164 Zhou, N. et al., Reversible addition-fragmentation chain transfer polymerization of styrene using a novel thiophene dithioester as the reversible addition-fragmentation chain transfer agent. *Journal of Applied Polymer Science*, 2007, **105**(4), 2357.

165 Moad, G., A critical survey of dithiocarbamate reversible addition-fragmentation chain transfer (RAFT) agents in radical polymerization. *Journal of Polymer Science Part A: Polymer Chemistry*, 2019, **57**(3), 216.

166 Liu, X.H. et al., Dibenzyl trithiocarbonate mediated reversible addition-fragmentation chain transfer polymerization of acrylonitrile. *Journal of Polymer Science Part A: Polymer Chemistry*, 2006, **44**(1), 490.

167 Strover, L.T., Postma, A., Horne, M.D., and Moad, G., Anthraquinone-mediated reduction of a trithiocarbonate chain-transfer agent to initiate electrochemical reversible addition-fragmentation chain transfer polymerization. *Macromolecules*, 2020, **53**(23), 10315.

168 Jiang, B. et al., Reversible addition fragmentation chain transfer polymerization mediated by xanthate. *Progress in Chemistry*, 2008, **20**(7–8), 1128.

169 Wolpers, A. et al., Aromatic xanthates and dithiocarbamates for the polymerization of ethylene through reversible addition-fragmentation chain transfer (RAFT). *Angewandte Chemie-International Edition*, 2019, **58**(40), 14295.

170 Destarac, M., On the critical role of RAFT agent design in reversible addition-fragmentation chain transfer (RAFT) polymerization. *Polymer Reviews*, 2011, **51**(2), 163.

171 Moad, G. et al., Advances in RAFT polymerization: the synthesis of polymers with defined end-groups. *Polymer*, 2005, **46**(19), 8458.

172 Zhang, M. and Ray, W.H., Modeling of "living" free-radical polymerization with RAFT chemistry. *Industrial & Engineering Chemistry Research*, 2001, **40**(20), 4336.

173 McCormack, C.L. and Lowe, A.B., Aqueous RAFT polymerization: recent developments in synthesis of functional water-soluble (co)polymers with controlled structures. *Accounts of Chemical Research*, 2004, **37**(5), 312.

174 Kim, H.J. et al., Synthesis of low glass transition temperature worms comprising a poly(styrene-stat-*n*-butyl acrylate) core segment via polymerization-induced self-assembly in RAFT aqueous emulsion polymerization. *Polymer Chemistry*, 2022, **13**(12), 1719.

175 Zhang, W.J. et al., RAFT-mediated one-pot aqueous emulsion polymerization of methyl methacrylate in presence of poly(methacrylic acid-co-poly(ethylene oxide) methacrylate) trithiocarbonate macromolecular chain transfer agent. *Polymer*, 2013, **54**(8), 2011.

176 Hatton, F.L., Derry, M.J., and Armes, S.P., Rational synthesis of epoxy-functional spheres, worms and vesicles by RAFT aqueous emulsion polymerisation of glycidyl methacrylate. *Polymer Chemistry*, 2020, **11**(39), 6343.

177 Wu, Y. et al., Fast conversion of terminal thiocarbonylthio groups of RAFT polymers to "clickable" thiol groups via versatile sodium azide. *Polymer Chemistry*, 2014, **5**(19), 5546.

178 Perrier, S., Takolpuckdee, P., and Mars, C.A., Reversible addition-fragmentation chain transfer polymerization: end group modification for functionalized polymers and chain transfer agent recovery. *Macromolecules*, 2005, **38**(6), 2033.

179 Perrier, S. Synthesis, use and recovery of chain transfer agents for RAFT polymerization: from laboratory scale to industrial production, in *229th ACS National Meeting*. 2005, American Chemical Society.

180 Allcock, H.R., Laredo, W.R., deDenus, C.R., and Taylor, J.P., Ring-opening metathesis polymerization of phosphazene-functionalized norbornenes. *Macromolecules*, 1999, **32**(23), 7719.

181 Vandenberg, E.J., Mechanism aspects of ring-opening polymerization of episulfides compared to epoxides. *Journal of Polymer Science Part A-1-Polymer Chemistry*, 1972, **10**(2), 329.

182 Watanabe, Y., Yasuda, T., Aida, T., and Inoue, S., Stereochemistry of ring-opening reactions of epoxide with aluminum and zinc porphyrins – relation to the mechanism of ring-opening polymerization. *Macromolecules*, 1992, **25**(5), 1396.

183 Yang, R.H. et al., Carboxylic acid initiated organocatalytic ring-opening polymerization of *N*-sulfonyl aziridines: an easy access to well-controlled polyaziridine-based architectural and functionalized polymers. *Macromolecules*, 2019, **52**(22), 8793.

184 Hashimoto, K., Ring-opening polymerization of lactams. Living anionic polymerization and its applications. *Progress in Polymer Science*, 2000, **25**(10), 1411.

185 Phelan, M., Aldabbagh, F., Zetterlund, P.B., and Yamada, B., Mechanism and kinetics of the free radical ring-opening polymerization of cyclic allylic sulfide lactones. *Polymer*, 2005, **46**(26), 12046.

186 Shakaroun, R.M. et al., Organocatalyzed ring-opening polymerization (ROP) of functional β-lactones: new insights into the ROP mechanism and poly(hydroxyalkanoate)s (PHAs) macromolecular structure. *Polymer Chemistry*, 2020, **11**(15), 2640.

187 Dutta, S., Hung, W.C., Huang, B.H., and Lin, C.C., Recent developments in metal-catalyzed ring-opening polymerization of lactides and glycolides: preparation of polylactides, polyglycolide, and poly(lactide-co-glycolide). *Synthetic Biodegradable Polymers*, 2012, **245**, 219.

188 Zhang, C.J. et al., Poly(thioether)s from closed-system one-pot reaction of carbonyl sulfide and epoxides by organic bases. *Journal of the American Chemical Society*, 2019, **141**(13), 5490.

189 Wang, Q.J. et al., Cationic ring-opening polymerization of cyclosiloxanes initiated by electron-deficient organosilicon reagents. *Macromolecules*, 1996, **29**(21), 6691.

190 Yan, B.K. et al., Synthesis of main chain sulfur-containing aliphatic polycarbonates by organocatalytic ring-opening polymerization of macrocyclic carbonates. *Polymer Chemistry*, 2019, **10**(38), 5191.

191 Ropson, N., Dubois, P., Jerome, R., and Teyssie, P., Macromolecular engineering of polylactones and polylactides. 17. Cryometric and viscosimetric analysis of the species active in the ring-opening polymerization of lactones, lactides, and cyclic anhydrides as initiated by aluminum triisopropoxide. *Macromolecules*, 1994, **27**(21), 5950.

192 Nuyken, O. and Pask, S.D., Ring-opening polymerization-an introductory review. *Polymers*, 2013, **5**(2), 361.

193 Olsen, P., Odelius, K., and Albertsson, A.C., Thermodynamic presynthetic considerations for ring-opening polymerization. *Biomacromolecules*, 2016, **17**(3), 699.

194 Olsen, P. et al., Switching from controlled ring-opening polymerization (cROP) to controlled ring-closing depolymerization (cRCDP) by adjusting the reaction parameters that determine the ceiling temperature. *Biomacromolecules*, 2016, **17**(12), 3995.

195 Bailey, W.J., Free-radical ring-opening polymerization. *Polymer Journal*, 1985, **17**(1), 85.

196 Ageyeva, T., Sibikin, I., and Karger-Kocsis, J., Polymers and related composites via anionic ring-opening polymerization of lactams: recent developments and future trends. *Polymers*, 2018, **10**(4), 357.

197 Sutar, A.K. et al., Ring-opening polymerization by lithium catalysts: an overview. *Chemical Society Reviews*, 2010, **39**(5), 1724.

198 Aouissi, A., Al-Deyab, S.S., and Al-Shahri, H., The cationic ring-opening polymerization of tetrahydrofuran with 12-tungstophosphoric acid. *Molecules*, 2010, **15**(3), 1389.

199 Aida, T., Living and immortal polymerizations. *Progress in Polymer Science*, 1994, **19**(3), 469.

200 Dubreuil, M.F., Farcy, N.G., and Goethals, E.J., Influence of the alkyl group of triflate esters on their initiation ability for the cationic ring-opening polymerization of tetrahydrofuran. *Macromolecular Rapid Communications*, 1999, **20**(7), 383.

201 Ivin, K.J., Mechanism of ring-opening polymerization of cycloalkenes initiated by metathesis catalysts. *Pure and Applied Chemistry*, 1980, **52**(7), 1907.

202 Ivin, K.J., Lapienis, G., and Rooney, J.J., Double-bond-shift reactions in polymers made by ring-opening polymerization of cycloalkenes initiated by metathesis catalysts. *Polymer*, 1980, **21**(4), 367.

203 Ivin, K.J. et al., Some recent applications of C-13 NMR-spectroscopy to the study of the ring-opening polymerization of cycloalkenes and related reactions initiated by metathesis Catalysts. *Pure and Applied Chemistry*, 1982, **54**(2), 447.

204 Dicker, I.B. et al., Oxyanions catalyze group-transfer polymerization to give living polymers. *Macromolecules*, 1990, **23**(18), 4034.

205 Sogah, D.Y. et al., Catalyzed silicon-mediated living polymerizations – structure control, catalyst design and mechanisms. *Makromolekulare Chemie-Macromolecular Symposia*, 1990, **32**, 75.

206 Sogah, D.Y. and Webster, O.W., Sequential silyl aldol condensation in controlled synthesis of living poly(vinyl-alcohol) precursors. *Macromolecules*, 1986, **19**(6), 1775.

207 Zhang, Y. and Chen, E.Y.X., Controlled polymerization of methacrylates to high molecular weight polymers using oxidatively activated group transfer polymerization initiators. *Macromolecules*, 2008, **41**(1), 36.

208 Fuchise, K., Chen, Y.G., Satoh, T., and Kakuchi, T., Recent progress in organocatalytic group transfer polymerization. *Polymer Chemistry*, 2013, **4**(16), 4278.

209 Webster, O.W. et al., Group-transfer polymerization. 1. A new concept for addition polymerization with organo-silicon initiators. *Journal of the American Chemical Society*, 1983, **105**(17), 5706.

210 Rogers, M.E. and Long, T.E., *Synthetic methods in step-growth polymers*. 2003, Wiley, 605.

211 Carraher, C.E. and Swift, G.G., eds. *Functional condensation polymers*. 2002, Kluwer Academic/Plenum Publishers, 315.

212 Flory, P.J., Fundamental principles of condensation polymerization. *Chemical Reviews*, 1946, **39**(1), 137.

213 Stille, J.K., Step-growth polymerization. *Journal of Chemical Education*, 1981, **58**(11), 862.

214 Gupta, S.K. and Kumar, A., *Reaction engineering of step growth polymerization*, 1e. The Plenum Chemical Engineering Series. 2012, Springer, 444.

215 Inkinen, S., Hakkarainen, M., Albertsson, A.C., and Sodergard, A., From lactic acid to poly(lactic acid) (PLA): characterization and analysis of PLA and its precursors. *Biomacromolecules*, 2011, **12**(3), 523.

216 Gan, L.M. and Koh, L.L., A discussion of the derivation of molecular weight distribution functions of linear step-growth polymerization. *Polymer Communications*, 1985, **26**(3), 76.

217 Digiacomo, A., Size distribution and gelation in condensation polymers with one unsymmetric reactant. *Journal of Polymer Science*, 1960, **47**(149), 435.

218 Beers, K.J. and Ray, W.H., A linkage moment approach to modeling condensation polymerization with multiple monomers. I. Linear polymers. *Journal of Applied Polymer Science*, 2001, **79**(2), 246.

219 Ramakrishnan, S., Condensation polymerization. *Resonance*, 2017, **22**, 355.

220 Xu, T.C., Zhang, L.F., Cheng, Z.P., and Zhu, X.L., Insight into the polymerization mechanism of photoinduced step transfer-addition & radical-termination (START) polymerizations. *Polymer Chemistry*, 2017, **8**(26), 3910.

221 Xu, T.C., Zhang, L.F., Cheng, Z.P., and Zhu, X.L., The positive effect of water on photo-induced step transfer-addition & radical-termination (START) polymerization. *RSC Advances*, 2017, **7**(29), 17988.

222 Xu, T.C. et al., Step transfer-addition and radical-termination (START) polymerization of α,ω-unconjugated dienes under irradiation of blue LED light. *Macromolecular Rapid Communications*, 2017, **38**(13), 1600587.

223 Tsvetanov, C.B., Dimov, D.K., Petrova, E.B., and Dotcheva, D.T., Solvation effects in anionic-polymerization of polar vinyl monomers. *Makromolekulare Chemie-Macromolecular Symposia*, 1992, **60**, 297.

224 East, G.C. and Harrison, M.A., Monomer solvation of ion pairs in anionic polymerization. *Nature*, 1967, **215**(5098), 273.

225 Higashimura, T., Ishihama, Y., and Sawamoto, M., Living cationic polymerization of styrene – new initiating systems based on added halide salts and the nature of the growing species. *Macromolecules*, 1993, **26**(4), 744.

226 Matyjaszewski, K. and Poli, R., Comparison of bond dissociation energies of dormant species relevant to degenerative transfer and atom transfer radical polymerization. *Macromolecules*, 2005, **38**(19), 8093.

227 Gillies, M.B. et al., A DFT study of R—X bond dissociation enthalpies of relevance to the initiation process of atom transfer radical polymerization. *Macromolecules*, 2003, **36**(22), 8551.

228 Penczek, S., Pretula, J., and Lewinski, P., Dormant polymers and their role in living and controlled polymerizations: influence on polymer chemistry, particularly on the ring opening polymerization. *Polymers (Basel)*, 2017, **9**(12), 646.

229 Kawamoto, M., Mori, Y., Tsuge, A., and Endo, T., Anionic ring-opening polymerization behavior of *trans*-cyclohexene carbonate using metal *tert*-butoxides: construction of living anionic ring-opening polymerization by lithium *tert*-butoxide. *Journal of Polymer Science*, 2022, **60**(8), 1416.

230 Bielawski, C.W. and Grubbs, R.H., Living ring-opening metathesis polymerization. *Progress in Polymer Science*, 2007, **32**(1), 1.

231 Inoue, M., Kanazawa, A., and Aoshima, S., Living cationic ring-opening homo- and copolymerization of cyclohexene oxide by "Dormant" species generation using cyclic ethers as Lewis basic additives. *Macromolecules*, 2021, **54**(11), 5124.

232 Kakuchi, R. et al., Controlled/living ring-opening polymerization of delta-valerolactone using triflylimide as an efficient cationic organocatalyst. *Macromolecules*, 2010, **43**(17), 7090.

233 Yuan, J.Y., Pan, C.Y., and Tang, B.Z., "Living" free radical ring-opening polymerization of 5,6-benzo-2-methylene-1,3-dioxepane using the atom transfer radical polymerization method. *Macromolecules*, 2001, **34**(2), 211.

234 Mayo, F.R. and Walling, C., Copolymerization. *Chemical Reviews*, 1950, **46**(2), 191.

235 Walling, C., Briggs, E.R., Cummings, W., and Mayo, F.R., Copolymerization. 14. Copolymerization by non-radical mechanisms. *Journal of the American Chemical Society*, 1950, **72**(1), 48.

236 Krause, S., Polymer–polymer miscibility. *Pure and Applied Chemistry*, 1986, **58**(12), 1553.

237 Balazs, A.C., Karasz, F.E., and Macknight, W.J., The effect of chain microstructure on polymer–polymer miscibility. *Croatica Chemica Acta*, 1987, **60**(1), 147.

238 Belfiore, L.A., Thermodynamic miscibility in polymer–polymer and polymer–diluent blends. *Abstracts of Papers of the American Chemical Society*, 1987, **194**, 33.

239 Jiang, W.H. and Han, S.J., An improved criterion of polymer–polymer miscibility determined by viscometry. *European Polymer Journal*, 1998, **34**(11), 1579.

240 Oh, S.Y. and Bae, Y.C., Molecular thermodynamics approach for polymer–polymer miscibility. *European Polymer Journal*, 2010, **46**(6), 1328.

241 Han, C.D. et al., Phase behavior of polystyrene/polybutadiene and polystyrene/hydrogenated polybutadiene mixtures: effect of the microstructure of polybutadiene. *Macromolecules*, 1998, **31**(2), 394.

242 Roe, R.J. and Zin, W.C., Determination of the polymer–polymer interaction parameter for the polystyrene–polybutadiene pair. *Bulletin of the American Physical Society*, 1980, **25**(3), 354.

243 Tomlins, P.E. and Higgins, J.S., Molecular conformation and interactions in oligomeric mixtures of polystyrene with polybutadiene: a small-angle neutron-scattering study. *Polymer*, 1985, **26**(10), 1554.

244 Narasimhan, V., Huang, R.Y.M., and Burns, C.M., Polymer polymer interaction parameters of polystyrene and polybutadiene from studies in solutions of toluene or tetrahydrofuran. *Journal of Polymer Science Part B-Polymer Physics*, 1983, **21**(10), 1993.

245 Robeson, L., Historical perspective of advances in the science and technology of polymer blends. *Polymers*, 2014, **6**(5), 1251.

246 Robeson, L.M., *Polymer blends: a comprehensive review*. 2007, Carl Hanser Verlag, 459.

247 Harwood, H.J., Sequence distribution in copolymers. Chemical studies. *Angewandte Chemie International*, 1965, **4**, 394.

248 Herbert, I.R., Statistical analysis of copolymer sequence distribution, in *NMR spectroscopy of polymers* (ed. R.N. Ibbet). 1993, Blackie Academic & Professional, pp. 50–79.

249 Hadjichristidis, N., Pispas, S., and Floudas, G., *Block copolymers: synthetic strategies, physical properties, and applications*. 2002, John Wiley, 419.

250 Yin, L.Y., Liu, L., and Zhang, N., Brush-like polymers: design, synthesis and applications. *Chemical Communications*, 2021, **57**(81), 10484.

251 Advicula, R.C., Brittain, W.J., Caster, K.C., and Ruhe, J., *Polymer brushes: synthesis, characterization, applications*. 2004, Wiley-VCH Verlag GmbH.

252 Soteros, C.E. and Whittington, S.G., The statistical mechanics of random copolymers. *Journal of Physics A: Mathematical and General*, 2004, **37**(41), R279.

253 Gordon, M. and Taylor, J.S., Ideal copolymers and the 2nd-order transitions of synthetic rubbers. 1. Non-crystalline copolymers. *Journal of Applied Chemistry*, 1952, **2**(9), 493.

254 Taniguchi, M., Kawasaki, A., and Furukawa, J., Alternate copolymerization of acrylonitrile and isobutylene. *Bulletin of the Chemical Society of Japan*, 1969, **42**(9), 2728.

255 Noshay, A. and McGrath, J.E., *Block copolymers: overview and critical survey*. 1977, Academic Press, 516.

256 Gao, Y.S. et al., Complex polymer architectures through free-radical polymerization of multivinyl monomers. *Nature Reviews Chemistry*, 2020, **4**(4), 194.

257 Schappacher, M. and Deffieux, A., New polymer chain architecture: synthesis and characterization of star polymers with comb polystyrene branches. *Macromolecules*, 2000, **33**(20), 7371.

258 Horkay, F. and Amis, E.J., eds. *Biological and synthetic polymer networks and gels*. Macromolecular Symposium. Vol. 227. 2005, Wiley-VCH Verlag GmbH, 392.

259 Khabaz, F. and Khare, R., Effect of chain architecture on the size, shape, and intrinsic viscosity of chains in polymer solutions: a molecular simulation study. *Journal of Chemical Physics*, 2014, **141**(21), 214904.

260 Bovey, F.A., Schilling, F.C., McCrackin, F.L., and Wagner, H.L., Short-chain and long-chain branching in low-density polyethylene. *Macromolecules*, 1976, **9**(1), 76.

261 Liu, P.W. et al., A comprehensive review on controlled synthesis of long-chain branched polyolefins: part 1. Single catalyst systems. *Macromolecular Reaction Engineering*, 2016, **10**(3), 156.

262 Liu, P.W. et al., A comprehensive review on controlled synthesis of long-chain branched polyolefins: part 3. Characterization of long-chain branched polymers. *Macromolecular Reaction Engineering*, 2017, **11**(1), 1600012.

263 Liu, W.F. et al., A comprehensive review on controlled synthesis of long-chain-branched polyolefins: part 2. Multiple catalyst systems and prepolymer modification. *Macromolecular Reaction Engineering*, 2016, **10**(3), 180.

264 Frey, H. and Kautz, H., Hyperbranched polymers: from tree-like macromolecules to functional materials. *Nachrichten Aus Der Chemie*, 2002, **50**(11), 1218.

265 Azam, N.A., Shurbevski, A., and Nagamochi, H., Enumerating tree-like graphs and polymer topologies with a given cycle rank. *Entropy*, 2020, **22**(11), 1295.
266 Seiffert, S. et al., *Supramolecular polymer networks and gels*. 2015, Springer.
267 Varnava, C.K. and Patrickios, C.S., Polymer networks one hundred years after the macromolecular hypothesis: a tutorial review. *Polymer*, 2021, **215**, 123322.
268 Soga, K. and Shiono, T., Ziegler–Natta catalysts for olefin polymerizations. *Progress in Polymer Science*, 1997, **22**(7), 1503.
269 Kaminsky, W. and Laban, A., Metallocene catalysis. *Applied Catalysis A: General*, 2001, **222**(1–2), 47.

2

Elementary Reactions in Polymerization

2.1 Introduction

Polymers are synthesized from their corresponding monomers by two reaction mechanisms, namely addition and step-growth polymerization. In the addition polymerization, reactive species are formed by initiator dissociation, followed by monomer addition to the reactive species to form propagating chains, followed by termination or chain transfer to produce terminated polymer chains. For example, the organotellurium-mediated polymerization of styrene proceeds by light-activated radical formation, followed by chain initiation, propagation, and combination of propagating chains to form terminated polymer chains [1]. As a result, the molecular weight of the produced polymer chains, in the addition polymerization, depends on the ratio of reactive species formed by initiator dissociation to the number of monomers in the reaction. Conversely, in step-growth polymerization, monomers having complementary functional groups react by combination in a stepwise manner to form dimers, trimers, oligomers, macromers, and polymer chains. For example, the hydroxyl and carboxylic acid functional groups in ethylene glycol and 2,5-furandicarboxylic acid monomers, respectively, react in a stepwise manner to form a poly(ethylene-2,5-furandicarboxylate) polymer by bulk polymerization [2]. Consequently, the molecular weight in step-growth polymerization depends on the conversion of complementary functional groups of the monomers and their initial stoichiometric imbalance [3]. Although the reaction mechanisms in addition and step-growth polymerizations differ, there are common elements between the two mechanisms. In the addition polymerization of styrene, two propagating chains combine to form a terminated polymer chain, whereas two complementary functional groups in step-growth polymerization of ethylene glycol and 2,5-furandicarboxylic acid combine to form a longer chain. The common elementary reaction in these two reactions is the "combination" of two growing chains. Chapter 1 covered polymerization reactions with respect to mechanism and method. This chapter aims to describe polymerization reactions in terms of their elementary reactions. The elementary reactions covered include reversible initiation and irreversible initiation, propagation reaction, chain transfer reaction, combination reaction, termination reaction, and chain dissociation reaction.

2.2 Initiation Reaction

Initiation is used in the addition polymerization to generate reactive species for the formation of propagating chains. Initiation is not required in step-growth polymerization as the monomer molecules possess complementary functional groups that can react to form dimer, trimer, oligomer, and polymer chains. A typical initiator is a compound that dissociates into reactive species by a change in reaction conditions that, in turn, generates propagating chains. The reactive species in radical, anionic, and cationic polymerization are a radical, an anion, and a cation, respectively.

2.2.1 Radical Initiation

Radical initiators in general are divided into two main types, namely reversible initiators and irreversible initiators. In the irreversible type, initiator dissociation is controlled by the rate of dissociation and initiator concentration [4–52]. In the reversible type, initiator dissociation is controlled by the equilibrium between the concentrations of initiator and the dissociated fragments [53–71].

2.2.1.1 Irreversible Radical Initiation

Irreversible initiators include those that generate one reactive, transient species upon activation and those that undergo homolytic dissociation to form two identical reactive species [5]. Initiators require a change in reaction conditions, like an increase in temperature, exposure to ultraviolet (UV) light or visible radiation, addition of a reducing agent, catalyst, or co-initiator to cause dissociation and the formation of reactive species [6]. A common method to generate free radicals is to use azo or triazine compounds with thermally sensitive —N—N— linkages or organic and inorganic peroxides with —O—O— linkages [7–10]. The most common thermal initiators for radical polymerization are 2,2′-azobis(isobutyronitrile) (AIBN) and benzoyl peroxide (BPO) [11–14]. The homolytic dissociation of BPO is as follows:

$$(H_5C_6)\overset{O}{\overset{\|}{C}}O\text{—}O\overset{O}{\overset{\|}{C}}(C_6H_5) \xrightarrow[\Delta T]{k_d} (H_5C_6)\overset{O}{\overset{\|}{C}}O^{\bullet} + (H_5C_6)\overset{O}{\overset{\|}{C}}O^{\bullet} \tag{2.1A}$$

$$(H_5C_6)\overset{O}{\overset{\|}{C}}O^{\bullet} + (H_5C_6)\overset{O}{\overset{\|}{C}}O^{\bullet} \xrightarrow[\Delta T]{k_f} H_5C_6^{\bullet} + H_5C_6^{\bullet} + CO_2\uparrow \tag{2.1B}$$

$$H_5C_6^{\bullet} + H_5C_6^{\bullet} \xrightarrow[\Delta T]{k_c} H_5C_6\text{—}C_6H_5 \tag{2.1C}$$

In the first reaction above, the weak —O—O— bond of BPO dissociates with a rate constant k_d upon increasing the reaction temperature to 80–100 °C to generate two benzoyloxy radicals [31]. The two benzoyloxy radicals, due to their highly unstable nature, do not recombine but undergo fragmentation by the second reaction with a rate constant k_f to liberate two stable carbon dioxide gas molecules and two equally reactive, transient benzyl radicals [4]. The fragmentation occurs prior to the two benzoyloxy radicals escaping the initiator dissociation cage. The two benzyl radicals are sufficiently stable to escape the dissociation cage and react by combination with a rate constant k_c to form a relatively stable, compared to benzyl radical, biphenyl compound, as shown by the third reaction. Therefore, the dissociation of BPO to two benzyl radicals is irreversible; that is, the benzyl radicals cannot combine to reproduce BPO. Further, the two benzyl radicals formed are transient, equally reactive, and short lived, such that they react with a monomer to initiate polymerization. The homolytic dissociation of AIBN is as follows:

$$(H_3C)_2 \overset{CN}{C} N{=}N \overset{CN}{C} (CH_3)_2 \xrightarrow[\Delta T]{k_d} (H_3C)_2 \overset{CN}{C^\bullet} + (H_3C)_2 \overset{CN}{C^\bullet} + N_2 \uparrow \tag{2.2A}$$

$$(H_3C)_2 \overset{CN}{C^\bullet} + (H_3C)_2 \overset{CN}{C^\bullet} \xrightarrow[\Delta T]{k_c} (H_3C)_2 \overset{CN}{C} - \overset{CN}{C} (H_3C)_2 \tag{2.2B}$$

The dissociation of AIBN with a rate constant k_d generates two 2-cyanoprop-2-yl radicals and a molecule of nitrogen gas [31, 32]. Unlike BPO, a nitrogen molecule is liberated during the dissociation reaction concurrent with the formation of two 2-cyanoprop-2-yl radicals. The two 2-cyanoprop-2-yl radicals may escape the dissociation cage and react by combination with a rate constant k_c to form a relatively stable bi(2-cyanoprop-2-yl) compound, as shown by the second reaction. Like BPO, the dissociation of AIBN to two 2-cyanoprop-2-yl is irreversible. Like BPO, the two radicals are transient, equally reactive, and short lived; thus, they react with a monomer to initiate polymerization. The reactions of the two equally reactive radicals generated by homolytic dissociation of thermally sensitive initiators like BPO and AIBN are as follows [33, 34]:

$$R^\bullet + R^\bullet \xrightarrow{k_c} R{-}R \tag{2.3}$$

$$R^\bullet + O_2 \xrightarrow{k_{O_2}} R{-}O{-}O^\bullet \tag{2.4}$$

$$R^\bullet + M \xrightarrow{k_{ci}} RP_1^\bullet \tag{2.5}$$

where k_c, k_{O_2}, and k_{ci} are the rate constants for radical combination, radical reaction with dissolved oxygen, and radical reaction with monomer, respectively. As monomer concentration in polymerization is much higher than that of radical or oxygen, the chain initiation reaction between the monomer and primary radical is favored to form a 1-mer propagating chain (i.e., Rxn. 2.5). Reactions 2.3 and 2.4 reduce the efficiency of chain initiation as the $R{-}R$ and $R{-}O{-}O^\bullet$ adducts are much less reactive than the $RP_1^\bullet$ propagating radical [35].

A different approach to initiate polymerization, especially in aqueous solution under mild conditions, is the use of redox initiators like persulfates, peroxomonosulfates, and peroxydiphosphates in the presence of bidentate ligands [15–18]. Common redox initiation systems include ascorbic acid–peroxydisulfate [19] and ammonium persulfate–N,N,N′,N′-tetramethylethylenediamine (TMEDA) [20]. Redox initiators generate radicals at ambient conditions, which is especially useful for the delivery of proteins and cells via *in-situ* polymerizing hydrogel precursor solutions at physiological conditions [36]. Ammonium persulfate (APS), like AIBN and BPO, dissociates thermally to the persulfate radical to initiate polymerization [37]. Upon the addition of TMEDA, the redox reaction between APS and TMEDA reduces the energy of activation for APS dissociation, which in turn lowers the reaction temperature for initiation to the ambient temperature. The reactions for redox dissociation of APS–TMEDA initiator system are as follows [38]:

$$\begin{aligned} &(NH_4^+)^- O\overset{O}{\underset{O}{S}}O{-}O\overset{O}{\underset{O}{S}}O^- (NH_4^+) + (CH_3)_2 N\overset{HH}{\underset{HH}{CC}}N (CH_3)_2 \\ &\xrightarrow[a.t.]{k_{cpx}} NH_4^+ : SO_4^- + (CH_3)_2 N\overset{HH}{\underset{HH}{CC}} \quad \overset{+}{\underset{\substack{O \\ O_2SO^- NH_4^+}}{N}} \quad (CH_3)_2 \end{aligned} \tag{2.6}$$

$$(CH_3)_2\,N\underset{HH}{\overset{HH}{CC}}\;\underset{\underset{O_2SO^-NH_4^+}{O}}{\overset{+}{N}}\;(CH_3)_2 \xrightarrow[a.t.]{k_{dcpx}} \tag{2.7}$$

$$HO_3SO^{\bullet} + (CH_3)_2\,N\underset{HH}{\overset{HHCH_3}{CC\,N}}\,CH_2^{\bullet} + NH_4 : H_2O$$

$$(CH_3)_2\,N\underset{HH}{\overset{HHCH_3}{CC\,N}}\,CH_2^{\bullet} + NH_4 : H_2O \xrightarrow[a.t.]{k_{creg}} \tag{2.8}$$

$$(CH_3)_2\,N\underset{HH}{\overset{HH}{CC}}N\,(CH_3)_2 + NH_3 \uparrow + H_2O$$

where k_{cpx}, k_{dcpx}, and k_{creg} are the rate constants for APS–TMEDA complexation, dissociation of the complex, and catalyst regeneration, respectively, and "a.t." stands for ambient temperature. The first reaction to form the APS–TMEDA complex is bimolecular, requiring the diffusion of the reactants prior to complexation, whereas the dissociation of the complex in the second reaction is unimolecular. Therefore, the first reaction to form the complex is the rate-limiting step. The radical for chain initiation is $HO_3SO^{\bullet}$ in the APS–TMEDA initiation system. The third reaction above transfers a hydrogen atom from the freed NH_4 to $—CH_2^{\bullet}$ of TMEDA to regenerate the catalyst and release a molecule of ammonia. A similar reaction mechanism is proposed for the initiation system consisting of potassium persulfate (PPS) and ascorbic acid (AsA) as follows [39]:

$$(K^+)\,^-O\underset{O}{\overset{O}{S}}O—O\underset{O}{\overset{O}{S}}O^-\,(K^+) + \boxed{O—\overset{O}{C}—\overset{OH}{C}=\overset{OH}{C}—\overset{H}{C}}—\underset{H}{\overset{OH}{C}}—\overset{OH}{C}H_2 \tag{2.9}$$

$$\xrightarrow[a.t.]{k_{cpx}} K^+ : SO_4^- + \boxed{O—\overset{O}{C}—\overset{OH}{C}=\overset{OH}{C}—\overset{H}{C}}—\underset{H}{\overset{OH}{C}}—\underset{H}{\overset{OH}{C}}—O_3SO^-K^+$$

$$\boxed{O—\overset{O}{C}—\overset{OH}{C}=\overset{OH}{C}—\overset{H}{C}}—\underset{H}{\overset{OH}{C}}—\underset{H}{\overset{OH}{C}}—O_3SO^-K^+ \xrightarrow[a.t.]{k_{dcpx}} \tag{2.10}$$

$$HO_3SO^{\bullet} + \boxed{O—\overset{O}{C}—\overset{OH}{C}=\overset{OH}{C}—\overset{H}{C}}—\underset{H}{\overset{OH}{C}}—\underset{H}{\overset{OH}{C}}^{\bullet} + K : H_2O$$

$$\boxed{O—\overset{O}{C}—\overset{OH}{C}=\overset{OH}{C}—\overset{H}{C}}—\underset{H}{\overset{OH}{C}}—\underset{H}{\overset{OH}{C}}^{\bullet} + K : H_2O \xrightarrow[a.t.]{k_{creg}} HO_3SO^{\bullet} \tag{2.11}$$

$$\boxed{O—\overset{O}{C}—\overset{OH}{C}=\overset{OH}{C}—\overset{H}{C}}—\underset{H}{\overset{OH}{C}}—\overset{OH}{C}H_2 + KOH$$

where the chain inside the rectangle indicates a five-membered ascorbate ring structure. The PPS–AsA initiation system generates a $HO_3SO^{\bullet}$ radical. The advantage of persulfate initiators over peroxide initiators is that radical initiation is less sensitive to the presence of oxygen in the reaction mixture [37]. The redox-generated radical reacts with a monomer to produce a 1-mer reactive propagating chain, as shown by Rxn. 2.5.

Another approach to initiate polymerization in aqueous or organic solution is the dissociation of a photolabile compound by irradiation [21]. Photolabile compounds that are excitable by irradiation in the ultraviolet, visible, or near-infrared region of the electromagnetic spectrum are used as initiators in photopolymerization [26, 28, 43, 44]. Photoinitiation is used extensively in the polymerization of thin films, coatings, and 3D-printed plastics because photoinitiation is fast and initiation can be controlled externally from the light source at any time during the reaction [25, 40]. Photopolymerization is also used in medical applications for cell encapsulation or protein delivery because polymerization/gelation can be done rapidly and under inert conditions (ambient temperature and neutral pH) [41, 42, 47]. Water-soluble photoinitiators that dissolve in the aqueous physiological medium, like α-hydroxyketones, are used in medical applications [28]. Common photolabile compounds that undergo dissociation with UV irradiation include 2-hydroxy-1,2-diphenylethanone (benzoin), Bis(2,4,6-trimethylbenzoyl)-phenyl phosphineoxide (BAPO), dimethoxyphenylacetophenone (DMPA) [21–25]. Photosensitizer molecules that dissociate with visible radiation include eosin-Y, riboflavin, or rose Bengal in combination with a hydrogen donor (co-initiator) like 2,7-dianilinothioxanthen-9-one or other thioxanthene derivatives [26–30].

Photoinitiator molecules are divided into three types, namely cleavable (type I), bimolecular (type II), and two-photon (type III) photoinitiators [46]. Type I photoinitiator is directly dissociated upon irradiation with UV light to form two radicals that react with the monomer to form reactive propagating chains. For instance, the covalent bond between the carbonyl group and 2-hydroxy dimethyl in 2-hydroxy-4′-(2-hydroxyethoxy)-2-methylpropiophenone (commercially known as Irgacure 2959) is cleaved upon UV irradiation with a wavelength of 365 nm to form two radicals by the following reaction [45, 46]:

$$HO\,(CH_2)_2\,O-\boxed{C_6H_4}-\overset{O}{\overset{\|}{C}}-C\,\underset{OH}{\overset{CH_3}{C}}\,CH_3 \xrightarrow[a.t.]{k_d,h\nu} HO\,(CH_2)_2\,O-\boxed{C_6H_4}-\overset{O}{\overset{\|}{C^{\bullet}}} + CH_3\,\underset{OH}{\overset{CH_3}{C}}\,C^{\bullet} \tag{2.12}$$

where the chemical group within the rectangle has a ring structure, $h\nu$ represents the energy of radiated photons, and k_d is the rate constant for dissociation. The above reaction is irreversible, radicals are transient and short lived, and they equally react with monomers to form propagating chains.

In type II photoinitiation, a combination of a photosensitizer like thioxanthone and an electron donor like a tertiary amine is used to generate radicals [25, 26, 46, 49]. The photoinitiation reactions of thioxanthone as the photosensitizer with 2-ethylhexyl-4-dimethylaminobenzoate (EHA) as the tertiary amine to generate a transient free radical are as follows [49]:

$$\boxed{C_6H_4}\boxed{\overset{O}{\overset{\|}{CS}}}\boxed{C_6H_4} \xrightarrow[a.t.]{k_{ext},h\nu} \left[\boxed{C_6H_4}\boxed{\overset{O}{\overset{\|}{CS}}}\boxed{C_6H_4}\right]^{*} \tag{2.13}$$

$$\left[\boxed{C_6H_4}\boxed{\overset{O}{\overset{\|}{CS}}}\boxed{C_6H_4}\right]^{*} + \underset{CH_3}{\overset{CH_3}{N}}-\boxed{C_6H_4}-\overset{O}{\overset{\|}{C}}O\overset{H}{C}\,\underset{\underset{CH_3}{HCH_2}}{\overset{H}{C}}\,(CH_2)_2\,CH_3 \xrightarrow{k_{etr}} \left[\boxed{C_6H_4}\boxed{\underset{\bullet}{\overset{O^{\bullet-}}{\overset{|}{CS}}}}\boxed{C_6H_4}\underset{CH_3}{\overset{CH_3}{N^{\bullet+}}}-\boxed{C_6H_4}-\overset{O}{\overset{\|}{C}}OCH_2\,\underset{\underset{CH_3}{CH_2}}{\overset{H}{C}}\,(CH_2)_2\,CH_3\right] \tag{2.14}$$

$$\xrightarrow{k_{ptr}} \boxed{C_6H_4}\boxed{\underset{\bullet}{\overset{OH}{\overset{|}{CS}}}}\boxed{C_6H_4} + \underset{CH_3}{\overset{H_2C^{\bullet}}{N}}-\boxed{C_6H_4}-\overset{O}{\overset{\|}{C}}OCH_2\,\underset{\underset{CH_3}{CH_2}}{\overset{H}{C}}\,(CH_2)_2\,CH_3 \tag{2.15}$$

wherek_{ext}, k_{etr}, and k_{ptr} are the rate constants for excitation of the photosensitive molecule (TX), electron transfer from TX to initiator (EHA), and proton transfer from EHA to TX. In the first reaction, the photosensitizer TX is elevated to an excited state TX^* by absorbing a photon of energy. In the second reaction, the energy of the excited state TX^* is transferred to EHA to generate an unstable sensitizer–donor $(TX^-—EHA^+)$ ion pair. Then, in the third reaction, TX^- abstracts a proton (H^+) from EHA^+ in the ion pair to form a transient, tertiary amine $EHA^\bullet$radical and a stable $TX^\bullet$species. The above photoinitiation reaction is not reversible. The transient, short-lived $EHA^\bullet$ radical reacts with a monomer to form a 1-mer reactive propagating chain. The $TX^\bullet$ radical is stabilized by resonance structures of the phenyl groups. Thioxanthone (TX) and its derivatives are also used as a photosensitizer with visible light as the source of photons with intra- or inter-molecular hydrogen abstraction [27]. Natural water-soluble visible and fluorescent dyes used in staining biological tissues fall into type II. Examples include eosin-Y, riboflavin (vitamin B_2), camphorquinone, erythrosine, and rose Bengal, among others [27, 28, 30, 48, 49]. These natural photoinitiators are transformed into an electron acceptor pair following excitation to a triplet state and react with an electron donor, like ethanolamine, to form radical species [28].

In type III photoinitiation, a highly Π-bonded photoinitiator with strong donor–acceptor groups, like WSPI or BDEA [50, 51], elevates to an excited state by absorption of two photons of energy under visible light irradiation, followed by transfer of an electron from the donor–acceptor groups of the photoinitiator to the Π-bonded electrons [28]. Next, the negatively charged Π-bond interacts directly with a monomer in the polymerization mixture to form an exciplex, which generates a radical to initiate a propagating chain. The reactions for a highly conjugated photoinitiator with donor–acceptor groups are as follows [28, 50, 51]:

$$WSPI \xrightarrow[a.t.,2h\nu]{k_{ext}} WSPI^* \xrightleftharpoons{k_{etda}} WSPI^{\pm}_{-}\!: \tag{2.16}$$

$$WSPI^{\pm}_{-}\!: + M \xrightarrow[a.t.]{k_{ci}} WSPI^{\bullet}_{+} + M^{\bullet}_{-} \tag{2.17}$$

$$WSPI^{\bullet}_{+} \xrightarrow[a.t.]{k_{rlx}} WSPI \tag{2.18}$$

where $WSPI$ and $WSPI^*$ are the ground and excited states of the photoinitiator, respectively, $WSPI^{\pm}_{-}\!:$ and $WSPI^{\bullet}_{+}$ are the positively-negatively charged species and positively charged conjugated radical, $M^{\bullet}_{-}$ is a negatively charged monomer radical, and k_{ext}, k_{etda}, k_{ci}, and k_{rlx} are the rate constants for two-photon excitation, electron transfer by donor–acceptor group in the excited initiator, chain initiation by radical transfer to monomer, and relaxation of the excited initiator, respectively. The side reactions that reduce the efficiency of photoinitiation include low dissociation quantum yield and quenching of radicals by molecular oxygen [52]. The above reactions, due to the highly unstable nature of ionic–radical species, are irreversible, and the transient, short-lived ionic–radical exciplex transfers a radical to a monomer to form a 1-mer reactive propagating chain.

An efficiency factor f is used to quantify the fraction of primary radicals generated by initiator dissociation that react with a monomer to form a propagating chain. The initiation efficiency ranges from $0 < f < 1$. Highly efficient thermal and redox initiators and photoinitiators generate a reactive propagating radical for every primary radical that is generated by the initiation reaction.

2.2.1.2 Reversible Radical Initiation

Conventional radical initiators, as discussed in the previous section, dissociate irreversibly in a homolytic mode to generate one or more unstable, very short-lived fragments that further decompose to form transient, short-lived fragments that react with monomer to produce propagating chains. Reversible radical initiators dissociate in a heterotypic mode to generate a transient, short-lived radical and a relatively persistent, long-lived radical. The transient radical reacts with the monomer to form an n-mer propagating chain, whereas the persistent radical reacts with a transient propagating radical to deactivate it and form a dormant chain. As the dormant chains can be reactivated and deactivated many times throughout the reaction, reversible initiators lead to the formation of living radical polymerization with a narrow distribution of degree of polymerization. The reactivation and deactivation of propagating chains by the persistent radical is controlled by the equilibrium state established between the activated and dormant species [71]. The specifics of chemical reactions to cause reversible deactivation of propagating chains were discussed in Chapter 1. Here, we present the elementary reactions for reversible deactivation by atom transfer radical polymerization (ATRP), nitroxide-mediated polymerization (NMP), and reversible addition fragmentation chain transfer (RAFT).

ATRP approach uses the combination of an initiator having a labile group X ($R—X$) and a chain transfer agent (CTA) to reversibly generate radicals for propagating chain activation and deactivation (dormant chain formation) [62]. In ATRP, an alkyl halide and a redox transition metal catalyst are used as initiator and CTA, respectively, to reversibly deactivate the propagating chains and form dormant chains [63]. The elementary reactions for reversible initiation in ATRP are as follows [62]:

$$R—X + CTA \xrightarrow{k_{ltr}} R^{\bullet} + CTA—X^{\bullet} \tag{2.19}$$

$$R^{\bullet} +^{n} M \xrightarrow{k_p} RP_n^{\bullet} \tag{2.20}$$

$$RP_n^{\bullet} + CTA—X^{\bullet} \underset{k_{rac}}{\overset{k_{dac}}{\rightleftharpoons}} DOP_n—X + CTA \tag{2.21}$$

where $R—X$ is the initiator having a labile ligand X, CTA is a chain transfer agent that accepts the labile group X, $R^{\bullet}$ is the primary, transient radical that reacts with monomer M to form an n-mer reactive propagating chain $RP_n^{\bullet}$, $CTA—X^{\bullet}$ is the activated CTA or the persistent radical, and $DOP_n—X$ is an n-mer dormant chain; $k_{ltr}$is the rate constant for ligand transfer from initiator to CTA, and k_{dac} and k_{rac} are the rate constants for deactivation and activation of propagating chains, respectively. In the first reaction, the labile group X is irreversibly transferred from the initiator to CTA to form the transient, short-lived radical $R^{\bullet}$ and the persistent, long-lived radical $CTA—X^{\bullet}$. In the second reaction, $R^{\bullet}$ reacts with the monomer to form an n-mer propagating chain $RP_n^{\bullet}$. In the third reaction, $CTA—X^{\bullet}$ donates its ligand X to the carbon atom holding the radical of the n-mer propagating chain to reversibly deactivate and form an n-mer dormant chain. The third reaction also frees the CTA to react with a new initiator molecule.

The NMP approach uses a chemical agent that acts simultaneously as initiator and CTA to reversibly deactivate a propagating chain and form a dormant chain [64, 65]. This chemical agent is called an iniferter. A commonly used iniferter is 2,2,6,6-tetramethyl-1-piperidinyloxy (TEMPO) that dissociates to generate a transient radical and a persistent radical species [72–74]. The elementary reactions for initiation with an iniferter in NMP approach are as follows [64, 65]:

$$R{-}CTA \underset{k_{daci}}{\overset{k_{aci}}{\rightleftharpoons}} R^{\bullet} + {}^{\bullet}CTA \tag{2.22}$$

$$R^{\bullet} + {}^{n}M \xrightarrow{k_p} RP_n^{\bullet} \tag{2.23}$$

$$RP_n^{\bullet} + {}^{\bullet}CTA \underset{k_{rac}}{\overset{k_{dac}}{\rightleftharpoons}} DOP_n{-}CTA \tag{2.24}$$

where $R{-}CTA$ is the iniferter, $R^{\bullet}$ as the initiating radical to form a propagating chain $RP_n^{\bullet}$, $CTA^{\bullet}$ is the active chain transfer agent, and $DOP_n{-}CTA$ is a reversibly deactivated propagating chain or a dormant chain; and k_{aci} and k_{daci} are the rate constants for activation and deactivation of inifeter, respectively.

RAFT approach uses a RAFT agent that combines with a propagating chain to form a polymeric RAFT radical complex. The complex then undergoes fragmentation to generate a new m-mer propagating chain and an n-mer dormant chain iniferter by reversible deactivation of a propagating chain [66–68]. In this approach, a thiocarbonylthio compound ($S{=}C\,(Z)\,SR_1$) is used as the RAFT agent to reversibly deactivate propagating chains [67–69]. The elementary reactions for the RAFT approach using 2-phenylprop-2-yl dithiobenzoate as the RAFT agent with phenyl and 2-phenylprop-2-yl as Z and R_1 groups, respectively, are as follows [66–69]:

$$I \xrightarrow{k_d} R^{\bullet} \xrightarrow{{}^{n}M,\,k_p} RP_n^{\bullet} \tag{2.25}$$

$$\begin{aligned} RP_n^{\bullet} + S{=}\underset{Z}{C}SR_1 &\underset{k_{daca}}{\overset{k_{aca}}{\rightleftharpoons}} RP_n{-}S\underset{Z}{C}^{\bullet}{-}SR_1 \\ &\underset{k_{dacf}}{\overset{k_{acf}}{\rightleftharpoons}} RDOP_n{-}S\underset{Z}{C}{=}S + R_1^{\bullet} \end{aligned} \tag{2.26}$$

$$R_1^{\bullet} + {}^{m}M \xrightarrow{k_p} R_1P_m^{\bullet} \tag{2.27}$$

$$\begin{aligned} R_1P_m^{\bullet} + S{=}\underset{Z}{C}S{-}RDOP_n &\underset{k_{daca}}{\overset{k_{aca}}{\rightleftharpoons}} \\ R_1P_m{-}S{-}\underset{Z}{C}^{\bullet}{-}S{-}RP_n &\underset{k_{dacf}}{\overset{k_{acf}}{\rightleftharpoons}} RP_n^{\bullet} + S{=}\underset{Z}{C}S{-}DOP_m \end{aligned} \tag{2.28}$$

where I is a conventional radical polymerization initiator like benzoyl peroxide to start the polymerization; $R^{\bullet}$ is a primary radical generated by dissociation of I, $R_1^{\bullet}$ is a primary radical generated by addition/fragmentation of the RAFT agent with a propagating chain; $RP_n^{\bullet}$ and $R_1P_n^{\bullet}$ are the n-mer propagating chains generated by the addition of monomer to $R^{\bullet}$ and $R_1^{\bullet}$, respectively; $S{=}\underset{Z}{C}SR_1$ is the RAFT agent; and $RP_n{-}S\underset{Z}{C}^{\bullet}{-}SR_1$ and $R_1P_m{-}S{-}\underset{Z}{C}^{\bullet}{-}S{-}RP_n$ are the polymeric RAFT adduct radical intermediates that undergo fragmentation to form a $DOP_n{-}S\underset{Z}{C}{=}S$ polymeric RAFT adduct (dormant chain), and new polymeric radicals $RP_n^{\bullet}$ and $R_1P_m^{\bullet}$, respectively. The $DOP_n{-}S\underset{Z}{C}{=}S$ dormant chain-RAFT adduct is the internally generated iniferter in RAFT polymerization. The Z group in RAFT agent stabilizes the polymeric RAFT adduct radical intermediate, whereas R_1 group shifts the chain transfer reaction toward the formation of a new propagating chain [70]. The reversible activation allows the propagating chains to intermittently experience the entire reaction conditions to narrow the molecular weight distribution. The k_d and k_p are the rate constants for dissociation of the conventional initiator and chain propagation, respectively; k_{aca} and k_{daca} are the rate constants for reversible activation-addition and reversible deactivation-addition reactions between a propagating chain and a polymeric RAFT adduct; and the rate constants k_{acf} and k_{dacf} are for reversible activation-fragmentation and reversible deactivation-fragmentation of the polymeric RAFT adduct radical intermediate. The reactions 2.21, 2.24, and 2.28 above are the "characteristic" elementary reactions associated with reversible radical initiation in ATRP, NMP, and RAFT living radical polymerization, respectively.

2.2.2 Anionic Initiation

Initiation involves the reaction of a neutral (Nu:) or negative (Nu:⁻) nucleophile with a monomer to form a propagating chain [75]. A nucleophile is generated by solvation of a nucleophilic compound, like an organometallic compound, in an organic solvent. The most common organometallic initiator used in anionic polymerization is *n*-butyl lithium [76]. The butyl lithium initiator is instantaneously and almost completely solvated in a non-proton-donating (aprotic), but moderately polar solvent like tetrahydrofuran (THF) as shown below [77]:

$$HCH_2CH_2CH_2\overset{H}{\underset{H}{C}}Li \xrightarrow[inert\ atm.]{k_{dslv},\ aprotic\ solvent} HCH_2CH_2CH_2\overset{H}{\underset{H}{C^-}}\ Li^+ \tag{2.29}$$

where k_{dslv} is the rate constant for dissociation by solvation. The dissociation by solvation of butyl lithium to butyl anion and lithium cation in THF is both complete and instantaneous, like the dissociation of sodium chloride to sodium cation and chlorine anion in aqueous solution. Furthermore, as the outermost p orbitals of the anion-carrying carbon in butyl anion are occupied and the reaction is carried out under inert atmosphere (no oxygen) and in the absence of moisture, the produced anion is relatively stable throughout the reaction without undergoing termination. The anion then propagates by dissociating the π-bond on a vinyl monomer and transferring the anion to the terminal carbon atom of the propagating chain, which further stabilizes the propagating anion. Therefore, the initiation step in anionic polymerization generates living propagating chains that are terminated by intentional addition of a terminating agent like hydrogen or oxygen gas, water, or ethanol [78, 79]. However, for some monomers, intra-/intermolecular termination with the functional groups of the monomer is possible [80–82].

The initiation step in group transfer polymerization (GTP) generates anions that propagate to form polymer chains. Therefore, GTP is treated as a special case of anionic polymerization in which the nucleophile catalyzes the dissociation of a silyl ketene acetal initiator by complexation to form anionic initiating species according to the following reaction (see subsection 1.2.1.6 in Chapter 1) [83, 84]:

$$\begin{aligned}\overset{CH_3}{\underset{CH_3}{C}}C{=}\overset{OCH_3}{C}{-}OSi\,(CH_3)_3 &\underset{k_{gtdcx}}{\overset{Nu^-,\,k_{gtcx}}{\rightleftharpoons}} \overset{CH_3}{\underset{CH_3}{C}}C{=}\overset{OCH_3}{C}{-}O\overset{Nu^-}{Si}\,(CH_3)_3\\ &\underset{k_{gtdf}}{\overset{k_{gtf}}{\rightleftharpoons}} \overset{CH_3}{\underset{CH_3}{C}}{=}\overset{OCH_3}{C}{-}O^- + NuSi\,(CH_3)_3\end{aligned} \tag{2.30}$$

where k_{gtcx} and k_{gtdcx} are the rate constants for initiator/catalyst complexation and decomplexation, respectively, whereas k_{gtf} and k_{gtdf} are the rate constants for fragmentation and defragmentation of the complex to form the ester enolate propagating anion. In anionic polymerization, solvation of the initiator in the solvent to form propagating anions is complete and irreversible. Conversely, the ester enolate propagating anion in GTP is in equilibrium with its silyl ketene acetal form (initiator) and is reversibly deactivated in a manner akin to reversible activation/deactivation in living radical polymerization. However, both methods produce polymers with low polydispersity because the propagating anions are active throughout the reaction.

2.2.3 Cationic Initiation

Initiation involves the reaction of a strong Lewis acid electron acceptor as a co-initiator with a catalyst to form a tightly bound ion pair, followed by the reaction of the ion pair with a weak electron acceptor as the initiator to form a loosely bound reactive cation [85–90]. The initiation reaction between tris(pentafluorophenyl)borane [$B(C_6F_5)_3$] as the co-initiator (strong Lewis acid), cumyl alcohol $\left[C_6H_5C\left(CH_3\right)_2OH\right]$ as the initiator (weak acid), and water as the catalyst to form a reactive propagating cation is as follows [91]:

$$(C_6F_5)_3\,B + H_2O \xrightarrow{k_{coic}} (C_6F_5)_3\,\overset{OH}{B^-}H^+ \xrightarrow{k_{cpxi},C_6H_5C(CH_3)_2OH} (C_6F_5)_3\,\overset{OH}{B^-}C_6H_5\overset{CH_3}{\underset{CH_3}{C^+}} + H_2O \tag{2.31}$$

where k_{coic} and k_{cpxi} are the rate constants for the reactions between co-initiator and the catalyst to form an ion pair complex and between the complex and initiator, respectively. Unlike GTP, the produced cation is not reversibly deactivated/activated, and it is not as stable as the anion in anionic polymerization because the atom carrying the cation has an unfilled, empty orbital in its outermost electron shell. As a result, propagating cations are more likely to be terminated by side reactions, and the molecular weight distribution is somewhat wider as compared with anionic polymerization.

2.3 Propagation Reaction

The propagation elementary step adds a monomer to the n-mer reactive propagating chain to form an (n+1)-mer propagating chain according to the following reaction [92]:

$$RP_n^{\bullet} + M \xrightarrow{k_p} RP_{n+1}^{\bullet} \tag{2.32}$$

where $RP_n^{\bullet}$ and $RP_{n+1}^{\bullet}$ are reactive propagating chains with n and (n+1) monomers, respectively, and k_p is the rate constant for propagation. Propagation increases the kinetic chain length of the growing chains by one monomer residue [93, 94].

2.4 Termination Reactions

Termination reactions in radical polymerization include termination by disproportionation and combination [95–100]. In termination by disproportionation, two propagating chains with n and m repeat units react by disproportionation to terminate the two propagating chains and produce two terminated polymer chains TP_n and TP_m, respectively, and one unsaturated double bond at the terminus of one of the chains. Termination by disproportionation for two propagating methyl methacrylate chains is shown by the following reaction [1, 101–103]:

$$P_{n-1}\!-\!\overset{H}{\underset{H}{C_\beta}}\!-\!\overset{CH_3}{\underset{CO_2CH_3}{C_\alpha^{\bullet}}} + \overset{CH_3}{\underset{CO_2CH_3}{C_\alpha^{\bullet}}}\!-\!\overset{H}{\underset{H}{C_\beta}}\!-\!P_{m-1} \xrightarrow{k_{td}} P_{n-1}\!-\!\underset{H}{C_\beta}\!=\!\overset{CH_3}{\underset{CO_2CH_3}{C_\alpha}} + H\!-\!\overset{CH_3}{\underset{CO_2CH_3}{C_\alpha}}\!-\!\overset{H}{\underset{H}{C_\beta}}\!-\!P_{m-1} \tag{2.33}$$

where k_{td} is the rate constant for termination by disproportionation. In the above reaction, the radical on the α carbon of the m-mer propagating chain abstracts a hydrogen atom from the β carbon of the n-mer propagating chain to form a $C{-}H$ bond that terminates the m-mer propagating chain. Following hydrogen abstraction, the two lone electrons on the α and β carbons of the n-mer propagating chain form a terminal $C{=}C$ bond, which terminates the n-mer propagating chain.

In termination by combination, two propagating chains with m and n repeat units react by combination to terminate the two propagating chains and produce one terminated polymer chain TP_{n+m} with (n+m) repeat units. Termination by the combination of two propagating chains in radical polymerization of styrene monomer is shown by the following reaction [1, 104, 105]:

$$P_{n-1}-\underset{H}{\overset{H}{C_\beta}}-\underset{C_6H_5}{\overset{H}{C_\alpha^{\bullet}}} + \underset{C_6H_5}{\overset{H}{C_\alpha^{\bullet}}}-\underset{H}{\overset{H}{C_\beta}}-P_{m-1} \xrightarrow{k_{tc}} P_{n-1}-\underset{H}{\overset{H}{C_\beta}}-\underset{C_6H_5}{\overset{H}{C_\alpha}}-\underset{C_6H_5}{\overset{H}{C_\alpha}}-\underset{H}{\overset{H}{C_\beta}}-P_{m-1} \tag{2.34}$$

where k_{tc} is the rate constant for termination by combination. The average degree of polymerization of polymer chains produced by combination is twice that of disproportionation. As this is a combination reaction, it will also be discussed in the section on "combination reactions." It is reported that the dominant mechanism of termination in polystyrene is a combination, likely due to the attractive π–π interaction between the phenyl rings attached to the terminal carbon atom on the two propagating chains, even though the combination reaction is sterically hindered [1, 102, 106, 107]. Conversely, the dominant mechanism of termination in polymethyl methacrylate is disproportionation due to the steric hindrance of the bulky methacrylate groups [1]. Both mechanisms contribute equally to chain termination in polymerization of monomers with less bulky side groups like ethylene or propylene [108–111].

A propagating chain can also terminate by itself without the involvement of other propagating or terminated chains, especially in ionic polymerization, by group transfer to the counterion [112–114]. For example, in propylene polymerization with a Ziegler–Natta catalyst system comprising titanium tetrachloride, magnesium chloride, and triethyl aluminum $[TiCl_4/MgCl_2/Al\,(C_2H_5)_3]$, the monomer is inserted in the transient bond between the terminal carbon of the propagating chain and the transition metal. In this polymerization, the transition metal in the propagating chain–metal complex abstracts a hydrogen atom or a methyl group from the β-carbon of the propagating chain to terminate the chain according to the following reactions [113, 114]:

$$P_{n-1}-\underset{H}{\overset{H}{C_\beta}}-\underset{CH_3}{\overset{H}{C_\alpha^{+}}}\ldots{}^{-}\underset{Cl_2}{\overset{Cl_2}{Ti}}C_2H_5 \xrightarrow{k_{t\beta H}} TP_{n-1}-\overset{H}{C_\beta}{=}\underset{CH_3}{\overset{H}{C_\alpha}} + H-\underset{Cl_2}{\overset{Cl_2}{Ti}}C_2H_5 \tag{2.35}$$

$$P_{n-1}-\underset{CH_3}{\overset{H}{C_\beta}}-\underset{H}{\overset{H}{C_\alpha^{+}}}\ldots{}^{-}\underset{Cl_2}{\overset{Cl_2}{Ti}}C_2H_5 \xrightarrow{k_{t\beta CH3}} TP_{n-1}-\overset{H}{C_\beta}{=}\underset{H}{\overset{H}{C_\alpha}} + CH_3-\underset{Cl_2}{\overset{Cl_2}{Ti}}C_2H_5 \tag{2.36}$$

where $k_{t\beta H}$ and $k_{t\beta CH3}$ are the rate constants for termination by hydride ion and termination by methyl group abstraction, respectively, from the β carbon to the transition metal catalyst. Both termination reactions produce a terminated polymer chain with a terminal C=C bond. Reactions like hydride/methyl abstraction can also take place in reversible deactivation radical polymerization [71, 112, 115–119]. However, these terminations are transient and generate dormant chains, which were covered in the section on reversible initiation. In general form, reactions for termination are written as follows:

$$RP_n^{\bullet} + RP_m^{\bullet} \xrightarrow{k_{td}} TP_n + TP_m + DB \tag{2.37}$$

$$RP_n^{\bullet} + RP_m^{\bullet} \xrightarrow{k_{tc}} TP_{n+m} \tag{2.38}$$

$$RP_n^{\bullet} \xrightarrow{k_{t\beta}} TP_n + DB \tag{2.39}$$

Where $k_{t\beta}$ is the rate constant for termination by group transfer from the β carbon of the propagating chain to the counterion, and DB is a C═C bond formed at one end of a terminated chain. Chain termination can also occur by the transfer of active species on propagating chains to molecules other than propagating chains in the reaction mixture, which is covered in the section on chain transfer reactions.

2.5 Chain Transfer Reactions to Small Molecules

The reactive species on propagating chains, due to their instability, are transferred to other molecules in the reaction mixture by chain transfer reactions [120–124]. In some polymerizations, a chain transfer agent (CTA) is added to the reaction to control the degree of polymerization [125–127]. Chain transfer can take place in radical, anionic, or cationic polymerization [120, 128, 129]. The reactive species on a propagating chain may be transferred to any small molecule in the reaction mixture, including monomer, initiator or co-initiator, catalyst, crosslinking agent, diluent, additive, or solvent [130–137]. Due to their low concentration, the rate of chain transfer to initiator, co-initiator, catalyst, or crosslinking agent is orders of magnitude lower than the rate of transfer to monomer or solvent. The extent of chain transfer to monomer depends on monomer type [113, 114, 124, 131, 138]. In methyl methacrylate (MMA) polymerization, chain transfer to monomer takes place by hydrogen abstraction and donation according to the following reactions [124]:

$$P_{n-1}\text{—}\underset{H}{\overset{H}{C_\beta}}\text{—}\underset{CO_2CH_3}{\overset{CH_3}{C_\alpha^{\bullet}}} + \underset{CO_2CH_3}{\overset{CH_3}{C}}{=}\underset{H}{\overset{H}{C}} \xrightarrow{k_{ctmab}} TP_{n-1}\text{—}\underset{H}{\overset{H}{C_\beta}}\text{—}\underset{CO_2CH_3}{\overset{CH_3}{C_\alpha}}\text{—}H + CH_2{=}C\,(CO_2CH_3)\text{—}\underset{H}{\overset{H}{C^{\bullet}}} \tag{2.40}$$

$$P_{n-1}\text{—}\underset{H}{\overset{H}{C_\beta}}\text{—}\underset{CO_2CH_3}{\overset{CH_3}{C_\alpha^{\bullet}}} + \underset{CO_2CH_3}{\overset{CH_3}{C}}{=}\underset{H}{\overset{H}{C}} \xrightarrow{k_{ctmdn}} P_{n-1}\text{—}\overset{H}{C_\beta}{=}\underset{CO_2CH_3}{\overset{CH_3}{C_\alpha}} + H\text{—}\underset{H}{\overset{H}{C}}\text{—}\underset{CO_2CH_3}{\overset{CH_3}{C^{\bullet}}} \tag{2.41}$$

where k_{ctmab} and k_{ctmdn} are the rate constants for chain transfer to monomer by hydrogen abstraction and hydrogen donation, respectively. In the first reaction above, the α-carbon of the terminal monomer residue on a propagating chain abstracts a hydrogen atom from the methyl side group of MMA monomer to form an n-mer terminated polymer chain and a 1-mer reactive propagating monomer. In the second reaction, the β-carbon of the terminal monomer residue on a propagating chain donates a hydrogen atom to the CH_2 group of the MMA monomer to form an n-mer terminated polymer chain with a terminal C═C bond and a 1-mer reactive propagating monomer. In the polymerization of vinyl chloride (VC), the high rate of chain transfer to monomer is attributed to the radical on secondary carbons of polymer chains, which are formed by chain transfer to polymer [138]. The reactive species on a secondary carbon of a propagating chain in VC polymerization is transferred to monomer by hydrogen or chlorine atom donation by the following reactions [138]:

$$P_{n-m}\text{—}\underset{H}{\overset{\bullet}{C_\alpha}}\text{—}\underset{Cl}{\overset{H}{C_\beta}}\text{—}P_{m-1} + \underset{H}{\overset{H}{C}}{=}\underset{Cl}{\overset{H}{C}} \xrightarrow{k_{ctmdncl}} P_{n-m}\text{—}\underset{H}{\overset{H}{C_\alpha}}{=}C_\beta\text{—}P_{m-1} + Cl\text{—}\underset{H}{\overset{H}{C}}\text{—}\underset{Cl}{\overset{H}{C^{\bullet}}} \tag{2.42}$$

$$P_{n-m}-\underset{H}{\overset{\bullet}{C_\alpha}}-\underset{Cl}{\overset{H}{C_\beta}}-P_{m-1}+\underset{H}{\overset{H}{C}}=\underset{Cl}{\overset{H}{C}}\xrightarrow{k_{ctmdnh}}P_{n-m}-\underset{H}{C_\alpha}=\overset{Cl}{C_\beta}-P_{m-1}+H-\underset{H}{\overset{H}{C}}-\underset{Cl}{\overset{H}{C^\bullet}} \tag{2.43}$$

where $k_{ctmdncl}$ and k_{ctmdnh} are the rate constants for chain transfer to monomer by donation of a chlorine or hydrogen atom, respectively. In the above reactions, the β-carbon of an n-mer propagating chain donates a chlorine or a hydrogen atom to a VC monomer to form an n-mer terminated polymer chain with a C=C terminal bond and a 1-mer reactive propagating chain. In Zeigler–Natta polymerization of propylene (PP) with $TiCl_4/MgCl_2/Al\,(C_2H_5)_3$ catalyst system, the transition metal catalyst in the propagating chain-catalyst ion pair is transferred to a monomer by the following reaction steps [113]:

$$\begin{aligned}&P_{n-1}-\underset{H}{\overset{H}{C_\beta}}-\underset{CH_3}{\overset{H}{C_\alpha^+}}\ldots{}^-\underset{Cl_2}{\overset{Cl_2}{Ti}}C_2H_5+\underset{H}{\overset{H}{C}}=\underset{CH_3}{\overset{H}{C}}\xrightarrow{k_{ctmc}}\\&P_{n-1}-\underset{H}{\overset{H}{C_\beta}}-\underset{CH_3}{\overset{H}{C_\alpha^+}}\ldots\underset{H}{\overset{H}{C}}=\underset{CH_3}{\overset{H}{C}}\ldots{}^-\underset{Cl_2}{\overset{Cl_2}{Ti}}C_2H_5\xrightarrow{k_{ctmd}}\\&TP_{n-1}-\overset{H}{C_\beta}=\underset{CH_3}{\overset{H}{C_\alpha}}+H-\underset{H}{\overset{H}{C}}-\underset{CH_3}{\overset{H}{C^+}}\ldots{}^-\underset{Cl_2}{\overset{Cl_2}{Ti}}C_2H_5\end{aligned} \tag{2.44}$$

where k_{ctmc} and k_{ctmd} are the rate constants for complexation and dissociation of the propagating chain/monomer/catalyst complex, respectively. In the first reaction above, the propagating chain-catalyst ion pair and PP monomer form a complex. In the next reaction, the complex dissociates by concurrent hydride atom transfer from the β-carbon of the propagating chain to the monomer, cleavage of the transient electrostatic bond in the propagating chain–catalyst ion pair, and the formation of a new transient ion pair between the monomer and the catalyst. The dissociation reaction produces an n-mer terminated polymer chain with a terminal C=C bond and a monomer-catalyst ion pair for propagation.

If polymerization is carried out in a solvent, the active species may be transferred from the propagating chain to the solvent. Chain transfer to the solvent depends on the ability of the solvent to donate a hydrogen atom. The chain transfer constant to solvent in styrene polymerization increases as the solvent is replaced from benzene to chlorobenzene, toluene, ethylbenzene, and dichloroethane [134, 139]. Chain transfer to solvent in polymerization of *N-tert*-butyl acrylamide increases in linear alcohols with the number of methylene units in the solvent, that is, propanol and nonanol have the lowest and highest transfer constants, respectively [132]. Aprotic solvents like dichloromethane, chloroform, tetrahydrofuran, dimethylformamide, and dimethyl sulfoxide, which are not hydrogen donors, are commonly used as solvents in polymerization to reduce chain transfer to solvent [132, 134–137, 140]. As a typical reaction, chain transfer to toluene in the polymerization of styrene is as follows [134]:

$$P_{n-1}-\underset{H}{\overset{H}{C_\beta}}-\underset{C_6H_5}{\overset{H}{C_\alpha^\bullet}}+H\underset{H}{\overset{H}{C}}-C_6H_5\xrightarrow{k_{cts}}TP_{n-1}-\underset{H}{\overset{H}{C_\beta}}-\underset{C_6H_5}{\overset{H}{C_\alpha}}-H+C_6H_5-\underset{H}{\overset{H}{C^\bullet}} \tag{2.45}$$

where k_{cts} is the rate constant for chain transfer to solvent. In the above reaction, the radical on the α-carbon of the n-mer propagating chain abstracts a hydrogen from the methyl group of toluene to form an n-mer terminated polymer chain and a methyl benzene primary radical. For polymerization of ethylene in chloroform, the propagating chain can abstract a hydrogen or chlorine atom from the solvent as illustrated by the following reactions [141]:

$$P_{n-1}-\underset{H}{\overset{H}{C_\beta}}-\underset{H}{\overset{H}{C_\alpha^\bullet}}+H-\underset{Cl}{\overset{Cl}{C}}Cl\xrightarrow{k_{ctsH}}TP_{n-1}-\underset{H}{\overset{H}{C_\beta}}-\underset{H}{\overset{H}{C_\alpha}}-H+ClC\underset{Cl}{\overset{Cl}{{}^\bullet}} \tag{2.46}$$

$$P_{n-1}\text{—}\underset{H}{\overset{H}{C_\beta}}\text{—}\underset{H}{\overset{H}{C_\alpha^{\bullet}}} + Cl\text{—}\underset{Cl}{\overset{Cl}{CH}} \xrightarrow{k_{ctsCl}} TP_{n-1}\text{—}\underset{H}{\overset{H}{C_\beta}}\text{—}\underset{H}{\overset{H}{C_\alpha}}\text{—}H + H\underset{Cl}{\overset{Cl}{C^{\bullet}}} \tag{2.47}$$

where k_{ctsH} and k_{ctsCl} are the rate constants for chain transfer to solvent by hydrogen abstraction and chain transfer to solvent by chlorine abstraction, respectively. In the first reaction, a trichloromethyl radical is formed by hydrogen abstraction, whereas a dichloromethyl radical is formed by chlorine abstraction in the second reaction.

In some polymerizations, a chain transfer agent (CTA) is intentionally added to the reaction to control the average degree of polymerization or its distribution [139, 141]. CTAs are also used in reversible chain transfer catalyzed polymerization to reversibly deactivate propagating chains and form a living polymerization reaction [125–127], which is covered in a separate section on reversible chain initiation. CTAs are used not only in radical polymerization but also in ring-opening, cationic, or anionic polymerization to control the degree of polymerization [142–144]. Water-soluble thiols like glutathione and *N*-acetyl-L-cysteine are used as CTA in polymerizations carried out in aqueous solution [145]. CTAs are compounds that contain a donating atom that is abstracted by a propagating chain to terminate the chain and form a new radical for chain initiation. The two most common classes of CTAs are halocarbons like chloroform (—H or —Cl leaving atoms) [141] and thiols (mercaptans) like *n*-dodecyl mercaptan (—H leaving atom) [145]. The chain transfer reaction for benzenethiol, as a model CTA, is as follows [124, 141]:

$$\begin{aligned} &P_{n-1}\text{—}\underset{H}{\overset{H}{C_\beta}}\text{—}\underset{H}{\overset{H}{C_\alpha^{\bullet}}} + H\text{—}\underset{\bullet\bullet}{\overset{\bullet\bullet}{S}}\text{—}\boxed{phenyl} \xrightarrow{k_{cta}} \\ &\qquad TP_{n-1}\text{—}\underset{H}{\overset{H}{C_\beta}}\text{—}\underset{H}{\overset{H}{C_\alpha}}\text{—}H + \boxed{phenyl}\text{—}\underset{\bullet\bullet}{\overset{\bullet\bullet}{S^{\bullet}}} \end{aligned} \tag{2.48}$$

where $\boxed{phenyl}$ is a phenyl ring and k_{cta} is the rate constant for chain transfer to CTA. In the above reaction, the α-carbon of the terminal monomer repeat unit of an n-mer propagating chain abstracts the hydrogen attached to the sulfur atom of the CTA to form a C—H bond and a terminated polymer chain. The thiobenzyl radical formed following hydrogen atom abstraction is stabilized by the two lone pair electrons of the sulfur atom as well as the resonance structures of the phenyl ring [141]. The resonance-stabilized thiobenzyl radical reacts with a monomer to generate a 1-mer propagating chain. The addition of a CTA affects the degree of polymerization but not the number of propagating chains. To control the polymer molecular weight, the reactivity of CTA with the propagating chain in the reaction mixture should be much higher than that of other small molecules, like monomer or solvent.

For some highly reactive monomers, a polymerization inhibitor is added to the pure liquid monomer to prevent premature polymerization during shipping and storage [146–148]. An inhibitor is a compound that triggers an induction period after which the polymerization starts [149, 150]. For example, the addition of a small quantity of monomethyl ether hydroquinone (MEHQ) or *tert*-butylcatechol to acrylic acid monomer prevents its premature polymerization during storage [151]. The reaction of MEHQ with the acrylic acid radical is as follows:

$$\begin{aligned} &H_3C\text{—}\underset{CO_2H}{\overset{H}{C^{\bullet}}} + H\text{—}\underset{\bullet\bullet}{\overset{\bullet\bullet}{O}}\text{—}\boxed{phenyl}\text{—}\underset{\bullet\bullet}{\overset{\bullet\bullet}{O}}\text{—}CH_3 \xrightarrow{k_{ctih}} \\ &\qquad H_3C\text{—}\underset{CO_2H}{\overset{H}{C}}\text{—}H + {}^{\bullet}\underset{\bullet\bullet}{\overset{\bullet\bullet}{O}}\text{—}\boxed{phenyl}\text{—}\underset{\bullet\bullet}{\overset{\bullet\bullet}{O}}\text{—}CH_3 \end{aligned} \tag{2.49}$$

where k_{ctih} is the rate constant for the inhibition of acrylic acid polymerization by MEHQ. In the above reaction, MEHQ readily donates its hydroxyl hydrogen to quench the acrylic acid radical.

The MEHQ radical formed after hydrogen donation is not reactive as it is stabilized by resonance with oxygen electron pairs of the hydroxyl and methyl ether groups as well as those of the phenyl group. Prior to polymerization, the monomer should be passed through a silica or alumina gel to adsorb and remove the inhibitor [152]. Alternatively, the monomer can be polymerized without removing the inhibitor by adding a slightly higher concentration of initiator to the reaction to neutralize the inhibitor [153]. Oxygen from the atmosphere (O_2) and sulfur impurities (S_8) in the monomer also inhibit polymerization according to the following reactions for acrylic acid and vinyl acetate, respectively [151, 154–156]:

$$P_{n-1}-\underset{H}{\overset{H}{C}}-\underset{CO_2H}{\overset{H}{C^{\bullet}}} + \ddot{O}=\ddot{O} \xrightarrow{k_{ctO2}} P_{n-1}-\underset{H}{\overset{H}{C}}-\underset{CO_2H}{\overset{H}{C}}-\ddot{O}-\ddot{O}^{\bullet} \tag{2.50}$$

$$P_{n-1}-\underset{H}{\overset{H}{C}}-\underset{CO_2CH_3}{\overset{H}{C^{\bullet}}} + \boxed{S}_8 \xrightarrow{k_{ctS8}} P_{n-1}-\underset{H}{\overset{H}{C}}-\underset{CO_2CH_3}{\overset{H}{C}}-\boxed{S}_8^{\bullet} \tag{2.51}$$

where $\boxed{S}_8$ is an octa-sulfur ring, and k_{ctO2} and k_{ctS8} are the rate constants for inhibition of polymerization by oxygen and sulfur, respectively. The resulting peroxide and persulfide radicals are stabilized by resonance with oxygen and sulfur electron pairs, respectively. Polymerization reactions are generally purged with an inert gas like nitrogen or argon prior to adding the initiator to remove dissolved oxygen from the reaction mixture [35, 157–160]. As polymerization reactions are exothermic, a polymerization retarder is used in some reactions to control the overall rate of polymerization and reaction temperature as a preventive measure to prevent the formation of a runaway reaction [148–150, 161, 162]. Retarders are also used to control the polymer molecular weight and its distribution [163].

In general form, chain transfer reactions to monomer, solvent, CTA, inhibitor, and retarder are described by the following elementary reactions:

$$RP_n^{\bullet} + M \xrightarrow{k_{ctm}} TP_n + RP_1^{\bullet} \xrightarrow{k_p, mM} RP_m^{\bullet} \tag{2.52}$$

$$RP_n^{\bullet} + S \xrightarrow{k_{cts}} TP_n + S^{\bullet} \xrightarrow{k_{ps}, mM} RP_m^{\bullet} \tag{2.53}$$

$$RP_n^{\bullet} + CTA \xrightarrow{k_{cta}} TP_n + CTA^{\bullet} \xrightarrow{k_p, mM} RP_m^{\bullet} \tag{2.54}$$

$$RP_n^{\bullet} + IH \xrightarrow{k_{ctih}} TP_n + IIH^{\bullet} \tag{2.55}$$

$$RP_n^{\bullet} + RT \xrightarrow{k_{ctrt}} TP_n + RT^{\bullet} \tag{2.56}$$

where k_{ctm}, k_{cts}, k_{cta}, k_{ctih}, and k_{ctrt} are the rate constants for chain transfer to monomer, solvent, CTA, inhibitor, and retarder, respectively; IH and RT are inhibitor and retarder, respectively; $IIH^{\bullet}$ and $RT^{\bullet}$ are inactive inhibitor and retarder radicals, respectively.

2.6 Chain Backbiting Reaction

A reactive propagating chain may react intramolecularly and transfer its radical to other carbon atoms on the chain. This intramolecular chain transfer reaction is known as chain backbiting [164]. Propagating chains formed by monomers with no side groups, like ethylene, with repeating C—C single bonds, are less sterically hindered. The radical at the terminal end of these chains can turn on itself and undergo a backbiting reaction [164, 165]. Backbiting is also reported in the polymerization of other monomers, including propylene, methyl acrylate, α-butyl acrylate, ethylhexyl acrylate, arylamide, acrylonitrile, and conjugated dienes [166–173]. The backbiting reaction for a propagating polyethylene radical is as follows [174]:

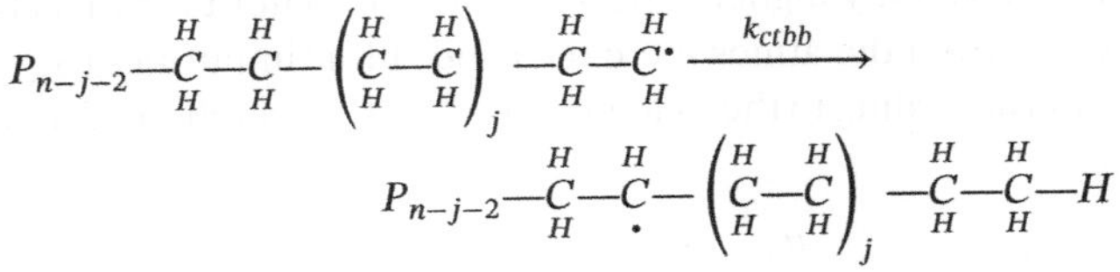
(2.57)

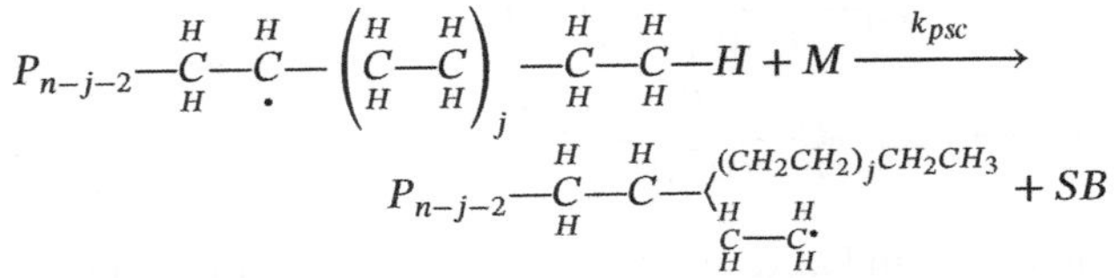
(2.58)

where k_{ctbb} and k_{psc} are the rate constants for radical transfer by backbiting and the reaction of a radical on a secondary carbon with a monomer, respectively [175, 176]. In the first reaction above, the radical on the primary carbon of the terminal monomer of an n-mer propagating chain abstracts a hydrogen intramolecularly from the carbon atom (j+1) monomer units away to form a C—H bond and transfer its radical to the secondary carbon atom on the chain. In the second reaction, the radical on the secondary carbon of the n-mer propagating chain reacts with a monomer to form a radical on a primary carbon and a short side chain *SB*. Chain backbiting generally occurs with monomer residues 2–9 units away from the carbon with the radical ($j = 2$–9 in the above reaction) on the propagating chain. Assuming k_p and k_{psc} are of the same order of magnitude and the backbiting reaction is the limiting step [175, 176], the two reactions above can be combined to form the following elementary reaction for backbiting:

$$RP_n^{\bullet} + M \xrightarrow{k_{ctbb}} TP_{n+1}^{\bullet} + SB \tag{2.59}$$

The backbiting reaction generates a short branch on the propagating chain, and the radical is restored to a primary carbon by the reaction of a monomer with the secondary radical. Backbiting reaction does not change the concentration of propagating chains nor their molecular weight distribution. There are three types of polyethylene, namely low-density (LDPE), linear low-density (LLDPE), and high-density (HDPE). LDPE has both short and long branches, whereas LLDPE and HDPE have short branches. The density of short branches on HDPE is much lower than LLDPE. Stiffness, yield point, melting point, and solvent sorption properties of HDPE are superior to those of LLDPE, which is attributed to the lower short-chain branching density and higher crystallinity [177]. Chain transfer reactions to a polymer other than backbiting are covered in the next section on chain combination reactions.

2.7 Chain Combination Reactions

Reactions like termination by combination, intermolecular chain transfer to polymer, reaction of a propagating chain with an unsaturated carbon–carbon bond at the terminal end or along the main chain of a terminated polymer chain are considered combination reactions [95, 178]. Intramolecular chain transfer in propagating chains forms short branches [164], whereas intermolecular chain transfer to terminated polymer chains results in the formation of long branches [178–184].

In theory, chain transfer to polymer can occur in any polymerization, including anionic and cationic polymerization [185, 186], but due to the higher instability of radical species, it is more likely to happen in radical polymerization. In the polymerization of vinyl monomers, a chain transfer monomer, which is a monomer that increases the rate of chain transfer to polymer, or a latent chain transfer monomer, is added to the reaction to increase long-chain branching, chain crosslinking, or gelation. For example, the addition of a small quantity of 3-mercaptohexyl methacrylate (MHM, a monomer with a hydrogen-donating thiol group) to methyl methacrylate or styrene polymerization increases long-chain branching [187]. A latent chain transfer to monomer like sodium 4-vinylbenzyl thiosulfate (VBTS) with a latent sulfonyl group, which can be activated during polymerization, is also used to increase long-chain branching [188]. In some polymerizations, chain transfer to polymer results in the formation of polyradicals that affect the chain length distribution and gelation [189]. As an example of combination reactions, two radicals in the polymerization of styrene or acrylates with n and m repeat units, respectively, react and combine to form a terminated polymer chain which is on average twice the length of the propagating chains (m+n repeat units) [1, 102, 190–193]. As another example of combination reactions in polymerization of n-butyl acrylate or vinyl acetate, a propagating radical with n repeat units reacts and combines with a terminated polymer chain with m repeat units to form a new longer propagating chain with (m+n) repeat units [179, 181, 183, 184, 186, 189, 194–199]. Combination reactions affect the degree of branching, average molecular weight, and chain length distribution of the produced polymer [178, 182, 184, 195]. The following sections cover termination by combination, radical transfer to polymer, chain transfer to the unsaturated carbon bond (C=C) in the terminal repeat unit of the polymer chains, and chain transfer to the unsaturated carbon bond (C=C) along the main chain.

2.7.1 Termination by Combination

It is covered here because termination is a combination reaction. In combination, two radicals on two propagating chains $RP_n^{\bullet}$ and $RP_m^{\bullet}$ with n and m repeat units, respectively, combine at time t in the reaction to produce a polymer chain $TP_{n+m}^{\bullet}$ by the following reaction [1, 104, 105]:

$$RP_n^{\bullet} + RP_m^{\bullet} \xrightarrow{k_{tc}} TP_{n+m} \tag{2.60}$$

The combination of two propagating radicals results in the formation of a carbon–carbon single bond. Termination by combination is favored in polymerization of vinyl monomers with attractive interaction between the side groups, like polystyrene and polyacrylonitrile [1, 200], and not favored in monomers with relatively bulky side groups like methyl methacrylate, vinyl acetate, and vinyl chloride [1, 101, 201].

2.7.2 Radical Transfer to Polymer Reaction

Another type of combination reaction is the transfer of a reactive species from a propagating chain to a α-carbon of a monomer residue along the main chain of a polymer chain in the reaction mixture. Radical transfer to polymer is a common side reaction in polymerization of vinyl monomers, which has been used to synthesize macromonomers [202]. Vinyl monomers that favor termination by disproportionation via hydrogen abstraction from one propagating chain to another, like vinyl acetate and n-butyl methacrylate, tend to have a higher rate of chain transfer to polymer [183, 184, 194, 203]. In vinyl acetate polymerization, a radical on a propagating chain is transferred to an α-carbon of a monomer residue along the main chain of a terminated polymer chain via hydrogen abstraction by the following reaction [194]:

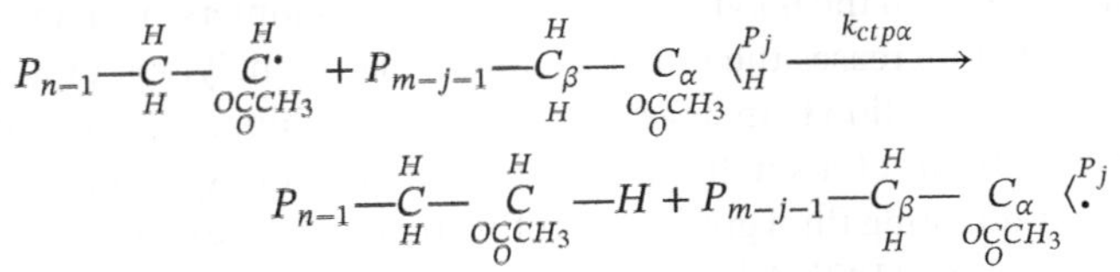

(2.61)

where $k_{ctp\alpha}$ is the rate constant for radical transfer to polymer by α-hydrogen abstraction. The radical may also be transferred to the methyl moiety of the acetate side group on the α-carbon of a monomer residue along the main chain by hydrogen abstraction as follows [194]:

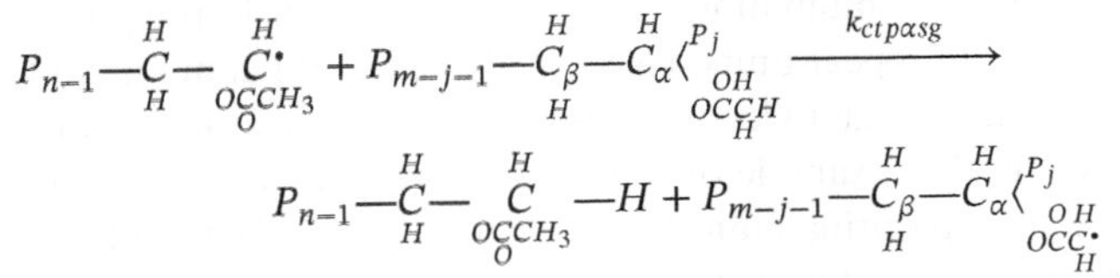

(2.62)

where $k_{ctp\alpha sg}$ is the rate constant for radical transfer to polymer by side group α-hydrogen abstraction. Either reaction terminates the propagating chain and generates a new radical on an α-carbon or a methyl radical on the side group of an α-carbon along the main chain of a terminated chain. Following hydrogen abstraction, the transferred radical will react with the monomer to form a long branch as follows [194]:

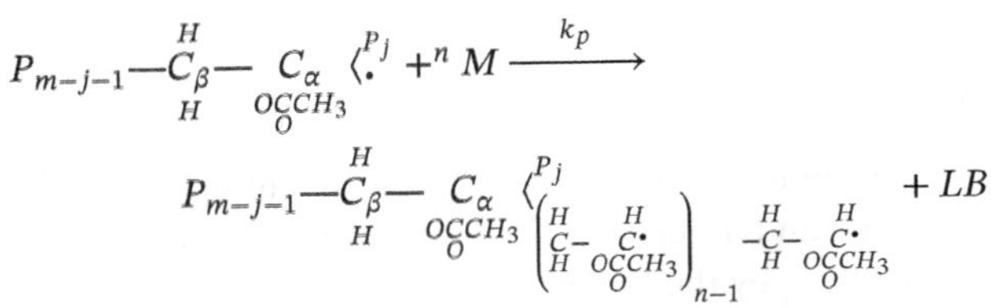

(2.63)

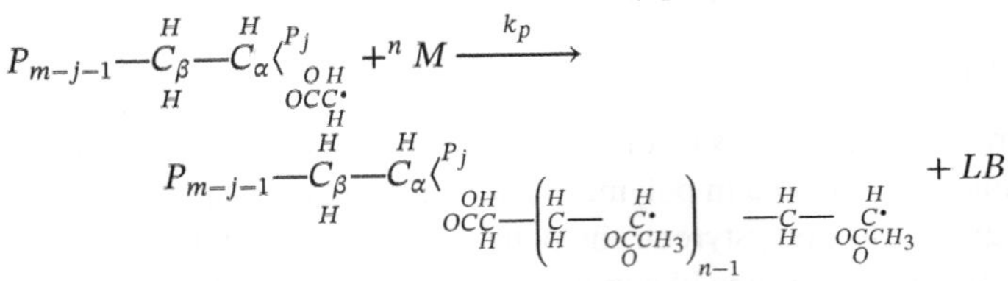

(2.64)

The above reactions produce a long branch, *LB*, in the polymer chain. Hydrogen abstraction from the α-carbon results in a backbone branch, whereas hydrogen abstraction from the methyl of the acetate side group results in a side group branch [182]. A similar chain transfer reaction occurs in polymerization of *n*-butyl methacrylate [184]. Long branches also form in polymerization of ethylene by chain transfer to polymer [204–207]. However, radical transfer in polyethylene occurs only by hydrogen abstraction from the α-carbon on the main chain. Based on the above reactions, the elementary reactions for chain transfer to polymer chains are as follows:

$$RP_n^{\bullet} + TP_m \xrightarrow{k_{ctp}} TP_n + RP_m^{\bullet} \tag{2.65}$$

$$RP_m^{\bullet} + {}^{n}M \xrightarrow{k_p} RP_{n+m}^{\bullet} + LB \tag{2.66}$$

Chain transfer to polymer results in termination of an n-mer propagating chain and reactivation of a terminated chain to a propagating chain with a growing long branch.

2.7.3 Radical Transfer to Terminal Double Bond of Polymer Reaction

A terminal double bond is formed on one of the chains in polymerization of vinyl monomers when two propagating chains are terminated by disproportionation [104, 105]. In methyl methacrylate polymerization, a radical on an n-mer propagating chain reacts by combination with the β-carbon of the terminal double bond of an m-mer terminated chain to form a longer (n+m)-mer propagating radical as follows [182]:

$$P_{n-1}-\underset{H}{\overset{H}{C_\beta}}-\underset{CO_2CH_3}{\overset{CH_3}{C_\alpha^{\bullet}}} + \underset{H}{\overset{P_{m-1}}{C_\beta}} = \underset{CO_2CH_3}{\overset{CH_3}{C_\alpha}} \xrightarrow{k_{ctptdb}} P_{n-1}-\underset{H}{\overset{H}{C_\beta}}-\underset{CO_2CH_3}{\overset{CH_3}{C_\alpha}}-\underset{H}{\overset{P_{m-1}}{C_\beta}}-\underset{CO_2CH_3}{\overset{CH_3}{C_\alpha^{\bullet}}} \quad (2.67)$$

where k_{ctptdb} is the rate constant for radical transfer to the terminal double bond. The chain transfer reaction by combination generates a new (n+m)-mer propagating chain with a radical attached to the terminal α-carbon and the m-mer polymer chain attached to the terminal β-carbon. Following radical transfer, a long branch is formed by reaction of q monomers with the radical on the new (n+m)-mer propagating chain to form a propagating chain with (n+m+q) repeat units and a long branch LB by the following reaction:

$$P_n-\underset{H}{\overset{P_{m-1}}{C_\beta}}-\underset{CO_2CH_3}{\overset{CH_3}{C_\alpha^{\bullet}}} +^q M \xrightarrow{k_p} P_n-\underset{H}{\overset{P_{m-1}}{C_\beta}}-\underset{CO_2CH_3}{\overset{CH_3}{C_\alpha}}-\left(\underset{H}{\overset{H}{C}}-\underset{CO_2CH_3}{\overset{CH_3}{C}}\right)_{q-1}-\underset{H}{\overset{H}{C}}-\underset{H}{\overset{P_{m-1}}{C^{\bullet}}} \quad (2.68)$$

The elementary reactions for chain transfer to the terminal double bond of a polymer chain are as follows:

$$RP_n^{\bullet} + TDBTP_m \xrightarrow{k_{ctptdb}} RP_{n+m}^{\bullet} \quad (2.69)$$

$$RP_{n+m}^{\bullet} +^q M \xrightarrow{k_p} RP_{n+m+q}^{\bullet} + LB \quad (2.70)$$

where $TDBTP_m$ is a terminal double bond on a terminated chain. Based on the above elementary reactions, chain transfer to terminal double bond is a combination reaction between an n-mer propagating chain, an m-mer terminated polymer chain, and q monomers to produce an (n+m+q)-mer propagating chain with a long branch.

2.7.4 Radical Transfer to Pendant Double Bond of Polymer Reaction

In polymerization of vinyl monomers, a small fraction of divinyl monomer is intentionally added to the reaction to generate long-chain branches because rheological properties and processability of polymers improve with branching [208–210]. In polymerization of 2-hydroxypropyl methacrylate (HPMA), a small fraction of ethylene glycol dimethacrylate (EGDMA) is added to the polymerization to increase long-chain branching [210]. In this polymerization, a radical on an n-mer propagating chain reacts by combination with the α-carbon of the pendant double bond of an EGDMA residual monomer on a terminated chain to form a longer (n+m)-mer propagating radical and a long branch by the following reaction [210]:

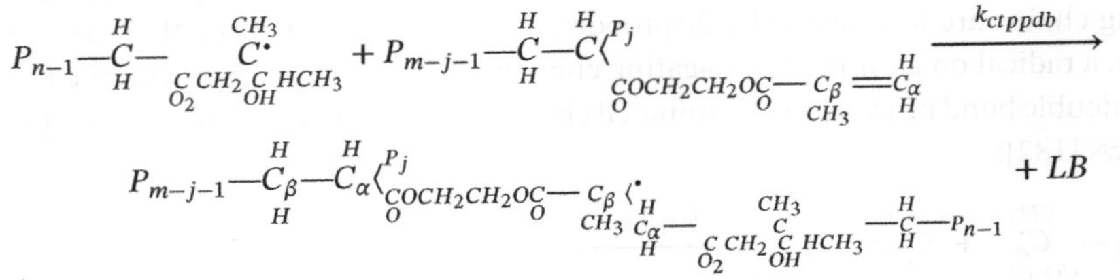

(2.71)

where k_{ctppdb} is the rate constant for radical transfer to a pendant double bond. Next, the newly generated radical on the β-carbon of the pendant double bond reacts with q monomers to form a propagating chain with (n+m+q) repeat units and a second long branch with q repeat units by the following reaction:

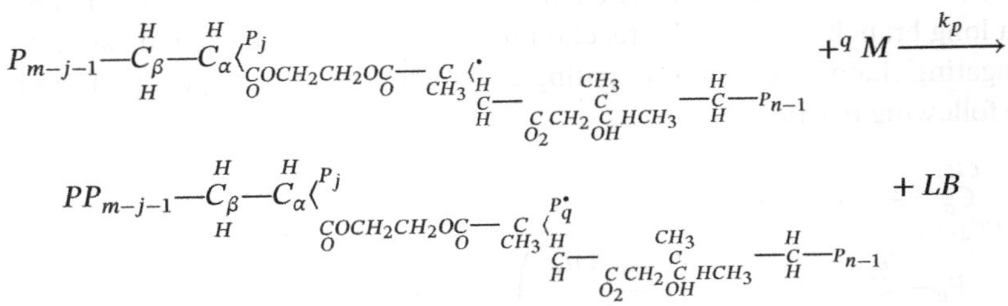

(2.72)

The elementary reactions for chain transfer to a pendant double bond of a chain are as follows:

$$RP_n^{\bullet} + PDBTP_m \xrightarrow{k_{ctppdb}} RP_{n+m}^{\bullet} + LB \tag{2.73}$$

$$RP_{n+m}^{\bullet} + {}^{q}M \xrightarrow{k_p} RP_{n+m+q}^{\bullet} + LB \tag{2.74}$$

where $PDBTP_m$ is a pendant double bond on a terminated polymer chain. Chain transfer to pendant double bond is a combination reaction between an n-mer propagating chain, an m-mer terminated chain, and q monomers in the reaction mixture to produce an (n+m+q)-mer propagating chain with two long branches.

2.7.5 Condensation Reaction in Step-growth Polymerization

Longer polymer chains are formed in step-growth polymerization by condensation of two complementary reactive groups on any two chains, like two monomers, a monomer and an oligomer, two oligomers, an oligomer and a polymer chain, or two polymer chains. Consider the condensation polymerization of a monomer with two reactive isocyanate groups (a diisocyanate) like 1,6-hexamethylene diisocyanate (HMDI) and another monomer with two reactive complementary hydroxyl groups (a diol) like diethylene glycol (DEG) [211]. First, HMDI monomer undergoes condensation with DEG monomer to form a urethane monomer with complementary isocyanate and hydroxyl groups at each end by the following reaction [211]:

$$\begin{aligned} &OCN{-}(CH_2)_6{-}NCO + HO{-}(C_2H_4O)_2{-}H \\ &\xrightarrow{k_{pc}} OCN{-}(CH_2)_6{-}\overset{\overset{H\ O}{|\ \ \|}}{NCO}{-}(C_2H_4O)_2{-}H \end{aligned} \tag{2.75}$$

where k_{pc} is the rate constant for propagation by condensation of two shorter chains and $—N\,(H)\,C\,(=O)\,O—$ is a urethane linkage. Next, a urethane monomer undergoes condensation with another urethane monomer by the following reaction to form a urethane dimer with one urethane repeat unit:

$$\begin{aligned}&OCN—(CH_2)_6—\overset{H}{\overset{|}{N}}\overset{O}{\overset{\|}{C}}O—(C_2H_4O)_2—H+\\&OCN—(CH_2)_6—\overset{H}{\overset{|}{N}}\overset{O}{\overset{\|}{C}}O—(C_2H_4O)_2—H\xrightarrow{k_{pc}}\\&OCN\,(CH_2)_6—\left[\overset{H}{\overset{|}{N}}\overset{O}{\overset{\|}{C}}O\,(C_2H_4O)_2\,\overset{H}{\overset{|}{N}}\overset{O}{\overset{\|}{C}}O\,(CH_2)_6\right]_1—\overset{H}{\overset{|}{N}}\overset{O}{\overset{\|}{C}}O—(C_2H_4O)_2\,H\end{aligned}\tag{2.76}$$

where the bracketed term is the first urethane repeat unit. As the step-growth reaction proceeds, a urethane oligomer with n repeat units condenses with another urethane oligomer with m repeat units to form a longer urethane oligomer with (n+m+1) repeat units by the following reaction:

$$\begin{aligned}&OCN\,(CH_2)_6—\left[\overset{H}{\overset{|}{N}}\overset{O}{\overset{\|}{C}}O\,(C_2H_4O)_2\,\overset{H}{\overset{|}{N}}\overset{O}{\overset{\|}{C}}O\,(CH_2)_6\right]_n—\overset{H}{\overset{|}{N}}\overset{O}{\overset{\|}{C}}O—(C_2H_4O)_2\,H+\\&OCN\,(CH_2)_6—\left[\overset{H}{\overset{|}{N}}\overset{O}{\overset{\|}{C}}O\,(C_2H_4O)_2\,\overset{H}{\overset{|}{N}}\overset{O}{\overset{\|}{C}}O\,(CH_2)_6\right]_m—\overset{H}{\overset{|}{N}}\overset{O}{\overset{\|}{C}}O—(C_2H_4O)_2\,H\xrightarrow{k_{pc}}\\&OCN\,(CH_2)_6—\left[\overset{H}{\overset{|}{N}}\overset{O}{\overset{\|}{C}}O\,(C_2H_4O)_2\,\overset{H}{\overset{|}{N}}\overset{O}{\overset{\|}{C}}O\,(CH_2)_6\right]_{n+m+1}—\overset{H}{\overset{|}{N}}\overset{O}{\overset{\|}{C}}O—(C_2H_4O)_2\,H\end{aligned}\tag{2.77}$$

The elementary reaction for condensation between reactive group A in $P_n—A$ and complementary reactive group B in $P_n—B$ is as follows:

$$B—P_n—A + B—P_m—A \xrightarrow{k_{pc}} B—P_{n+m+1}—A \tag{2.78}$$

where A and B are the complementary reactive or functional groups, and $—P_{n+m+1}—$ represents the remainder of the polymer chain with (n+m+1) repeat units. In step-growth polymerization, one or both complementary groups are depleted as the reaction progresses to completion. The rate constant k_{pc} can vary with chain length, especially at high conversions when the reaction is limited by chain diffusion [212–214].

2.8 Chain Dissociation Reactions

A secondary $_{H}\rangle C\langle^{\bullet}$ or tertiary $\rangle C\langle^{\bullet}$ radical is unstable and undergoes chain scission to form a relatively more stable primary radical $^{H}_{H}\rangle C\langle^{\bullet}$ [215–218]. A chain scission reaction results in the formation of a shorter propagating chain with a secondary or primary radical and a terminated polymer chain with a terminal double bond. There are two types of chain scission reactions, namely β_2 chain scission and β_3 chain scission [219–227]. In β_2 chain scission, an n-mer propagating chain, with a radical on a secondary carbon atom, splits by cleavage of α and β carbons to form a shorter propagating chain with a primary radical and a terminated chain with a terminal double bond.

As an example, a secondary radical in polyethylene, created by intramolecular chain backbiting (see Section 2.2.6) or intermolecular chain transfer to polymer (see Section 2.2.7), splits by β_2 scission into two chains according to the following reaction [228]:

$$P_{n-j-2}-\overset{H}{\underset{H}{C_\beta}}-\overset{H}{\underset{\bullet}{C_\alpha}}-\overset{H}{\underset{H}{C_\beta}}\cdots\overset{H}{\underset{H}{C_\alpha}}-P_{j-1}\xrightarrow{k_{cs\beta2}} TP_{n-j-2}-\overset{H}{\underset{H}{C_\beta}}-\overset{H}{C_\alpha}=\overset{H}{\underset{H}{C_\beta}}+P_{j-1}-\overset{H}{\underset{H}{C^\bullet}} \tag{2.79}$$

where $k_{cs\beta2}$ is the rate constant for β_2 chain scission. In the above reaction, the bond between the α and β carbons, shown with a dotted line, adjacent to the α carbon bearing the secondary radical of an n-mer propagating chain, splits to form two chains. One of the split chains is a terminated polyethylene chain with (n-j-1) repeat units and a terminal double bond. The other split chain is a (j-1) polyethylene propagating chain with a primary radical on the terminal carbon of the chain. Therefore, the elementary reaction for β_2 chain scission is given by the following reaction:

$$RP_n^\bullet \xrightarrow{k_{cs\beta2}} TP_{n-j-1} + DB + RP_{j-1}^\bullet \tag{2.80}$$

where $RP_n^\bullet$ is an n-mer propagating chain with a secondary radical, and $RP_{j-1}^\bullet$ is a (j-1)-mer propagating chain with a primary radical such that n>j; TP_{n-j-1} is a terminated polymer chain with (n-j-1) repeat units, and DB is a double bond.

In β_3 chain scission, an n-mer propagating chain, with a tertiary radical, splits by cleavage of the α and β carbons and forms a shorter propagating chain with a secondary radical and a terminated chain with a terminal double bond. For example, a tertiary radical in poly(2-ethlyhexyl acrylate) (PEHA), created by intermolecular chain transfer (see Section 2.2.7), splits by β_3 scission into two chains according to the following reaction [222]:

$$P_{n-j-2}-\overset{H}{\underset{H}{C_\beta}}-\overset{\bullet}{\underset{CO_2C_8H_{17}}{C_\alpha}}-\overset{H}{\underset{H}{C_\beta}}\cdots\overset{H}{\underset{CO_2C_8H_{17}}{C_\alpha}}-P_{j-1}\xrightarrow{k_{cs\beta3}} TP_{n-j-2}-\overset{H}{\underset{H}{C_\beta}}-\underset{CO_2C_8H_{17}}{C_\alpha}=\overset{H}{\underset{H}{C_\beta}}+P_{j-1}-\overset{H}{\underset{CO_2C_8H_{17}}{C^\bullet}} \tag{2.81}$$

where $k_{cs\beta3}$ is the rate constant for β_3 chain scission. Like β_2 scission, the bond between the α and β carbons, shown with a dotted line, adjacent to the α carbon bearing the tertiary radical of an n-mer propagating chain, splits to form two chains. One of the split chains is a terminated PEHA chain with (n-j-1) repeat units and a terminal double bond. The other chain is a (j-1) PEHA propagating chain with a secondary radical on the terminal carbon of the main chain. Therefore, the elementary reaction for β_3 chain scission is given by the following reaction:

$$RP_n^\bullet \xrightarrow{k_{cs\beta3}} TP_{n-j-1} + DB + RP_{j-1}^\bullet \tag{2.82}$$

where $RP_n^\bullet$ is an n-mer propagating chain with a tertiary radical, and $RP_{j-1}^\bullet$ is a (j-1)-mer propagating chain with a secondary radical with n>j.

2.9 Summary

Polymerization reactions, regardless of mechanism or method, can be divided into elementary reactions for the purpose of producing terminated polymer chains. These elementary reactions include irreversible and reversible chain initiation, propagation, chain transfer to small molecules, chain backbiting, chain combination, and chain scission. The initiation reaction can be further subdivided into activation, dissociation, recombination, chain initiation, and reversible deactivation. Reversible deactivation of propagating chains includes an initiator with a labile group activated by a chain transfer agent, iniferter, or RAFT agent. Reversible deactivation deactivates a propagating chain to generate a dormant chain that can be reactivated. Propagation is a necessary step in all polymerization reactions to generate long chains. Chain transfer to small molecules includes transfer to monomer, solvent, chain transfer agent, inhibitor, and retarder. Inhibition of chain propagation by the inhibitor is fast, whereas inhibition by retarder is relatively slow. Termination includes combination and disproportionation of two propagating chains. Chain combination reactions include termination by combination, transfer to polymer, transfer to terminal double bond, transfer to double bonds along the chain, and condensation of polymer chains in step-growth polymerization. A polymerization reaction can be modeled if its elementary reactions are known. Each elementary reaction in a polymerization produces propagating, dormant, or terminated polymer chains with a degree of polymerization unique to that elementary reaction. The overall distribution of the degree of polymerization of chains in a differential unit of time during polymerization is the product of the degree of polymerization of the individual elementary reactions, which will be covered in later chapters.

2.10 List of Symbols and Abbreviations

$B—P_n—A$: a growing polymer chain with n repeat units in step-growth polymerization with complementary reactive groups A and B at chain ends

CTA: chain transfer agent deliberately added to reaction in controlled/living radical polymerization

$CTA^{\bullet}$: activated chain transfer agent formed by radical transfer to a chain transfer agent

$CTA—X^{\bullet}$: activated chain transfer agent formed by transfer of a labile ligand X from initiator to chain transfer agent in controlled/living radical polymerization

DB: double bond on a polymer chain

$DOP_n—CTA$: dormant polymer chain with n repeat units formed by transfer of a chain transfer agent from an iniferter in nitroxide-mediated radical polymerization or reversible addition-fragmentation chain-transfer polymerization

$DOP_n—X$: dormant polymer chain with n repeat unit formed by transfer of a labile ligand X from activated chain transfer agent to propagating chain in atom transfer radical polymerization

f: efficiency of chain initiation, which is the ratio of number of generated propagating chains to the total number of generated active species in a differential time interval during a polymerization reaction ($0 < f < 1$)

I: initiator molecule

IH: inhibitor used to inhibit premature polymerization of monomer during storage

$IIH^{\bullet}$: inactive inhibitor radical formed by chain transfer to inhibitor, which is unable to initiate a propagating chain

k_{aca}: rate constant for reversible activation, addition of propagating chain to RAFT agent to form a complex in RAFT method of reversible deactivation radical polymerization (m^3 $mole^{-1}$ s^{-1})

k_{acf}: rate constant for reversible fragmentation of polymeric RAFT adduct radical intermediate to form a dormant chain and a new primary radical in RAFT method of reversible deactivation radical polymerization (s^{-1})

k_{aci}: rate constant for iniferter activation to generate a radical and an activated chain transfer agent in nitroxide-mediated method of reversible deactivation radical polymerization (s^{-1})

k_c: rate constant for recombination of two radicals formed by dissociation of initiator (m^3 $mole^{-1}$ s^{-1})

k_{ci}: rate constant for reaction of a primary radical with monomer to form a 1-mer propagating chain (m^3 $mole^{-1}$ s^{-1})

k_{coic}: rate constant for the reaction between co-initiator and catalyst to form an ion pair complex in cationic polymerization (m^3 $mole^{-1}$ s^{-1})

k_{cpx}: rate constant for complexation of oxidizing and reducing agents in redox initiation (m^3 $mole^{-1}$ s^{-1})

k_{cpxi}: rate constant for the reaction between the ion pair complex formed by co-initiator and catalyst and initiator to form a propagating cation–anion pair in cationic polymerization (m^3 $mole^{-1}$ s^{-1})

k_{creg}: rate constant for catalyst regeneration in APS–TMEDA redox initiation (m^3 $mole^{-1}$ s^{-1})

$k_{cs\beta 2}$: rate constant for β_2 chain scission of an n-mer propagating chain by splitting the bond between α and β carbons, where α is the carbon carrying a secondary to form two shorter chains in addition to polymerization (s^{-1})

$k_{cs\beta 3}$: rate constant for β_2 chain scission of an n-mer propagating chain by splitting the bond between α and β carbons, where α is the carbon carrying a tertiary radical to form two shorter chains in addition to polymerization (s^{-1})

k_{cta}: rate constant for transfer of a propagating species to a chain transfer agent in addition polymerization (m^3 $mole^{-1}$ s^{-1})

k_{ctbb}: rate constant for chain transfer by intramolecular backbiting in addition polymerization (m^3 $mole^{-1}$ s^{-1})

k_{ctih}: rate constant for chain transfer to inhibitor in addition polymerization (m^3 $mole^{-1}$ s^{-1})

k_{ctm}: rate constant for chain transfer to monomer (m^3 $mole^{-1}$ s^{-1})

k_{ctmab}: rate constant for chain transfer to monomer by α carbon of terminal monomer residue of a propagating chain abstracting a hydrogen from the side group in polymerization of acrylates (m^3 $mole^{-1}$ s^{-1})

k_{ctmc}: rate constant for chain transfer to monomer by complexation of the propagating cation and its negative ion pair with monomer to form a 1-mer propagating chain in cationic polymerization (m^3 $mole^{-1}$ s^{-1})

k_{ctmd}: rate constant for chain transfer to monomer by dissociation of the complex formed by the propagating cation–anion ion pair with monomer to form a 1-mer propagating chain in cationic polymerization (s^{-1})

k_{ctmdn}: rate constant for chain transfer to monomer by β carbon of terminal monomer residue of a propagating chain donating a hydrogen to —CH_2 group in polymerization of acrylates (m^3 $mole^{-1}$ s^{-1})

$k_{ctmdncl}$: rate constant for chain transfer to monomer by donation of a chlorine atom to monomer from a secondary carbon with a chlorine group to form a double bond to terminate the propagating chain in polymerization of vinyl chloride (m^3 $mole^{-1}$ s^{-1})

k_{ctO2}: rate constant for chain transfer to oxygen as an inhibitor in addition polymerization (m^3 $mole^{-1}$ s^{-1})

k_{ctp}: rate constant for radical transfer to polymer (m^3 $mole^{-1}$ s^{-1})

k_{ctppdb}: rate constant for chain transfer to a pendant double bond along the main chain of a terminated polymer chain to form two long branches (m^3 $mole^{-1}$ s^{-1})

k_{ctptdb}: rate constant for chain transfer to the terminal double bond on a terminated polymer chain (m^3 $mole^{-1}$ s^{-1}) to form a long branch (m^3 $mole^{-1}$ s^{-1})

$k_{ctp\alpha}$: rate constant for radical transfer to polymer by hydrogen abstraction from an α carbon of a terminated polymer chain (m^3 $mole^{-1}$ s^{-1})

$k_{ctp\beta}$: rate constant for radical transfer to polymer by hydrogen abstraction from a β carbon of a terminated polymer chain (m^3 $mole^{-1}$ s^{-1})

k_{ctrt}: rate constant for chain transfer to retarding agent in addition polymerization (m^3 $mole^{-1}$ s^{-1})

k_{cts}: rate constant for chain transfer to solvent in addition polymerization (s^{-1})

k_{ctsCl}: rate constant for chain transfer to solvent methylene chloride by chlorine abstraction from solvent to form a dichloromethyl radical and a terminated polymer chain (m^3 $mole^{-1}$ s^{-1})

k_{ctsH}: rate constant for chain transfer to solvent methylene chloride by hydrogen abstraction from solvent to form a trichloromethyl radical and a terminated polymer chain (m^3 $mole^{-1}$ s^{-1})

k_{ctS8}: rate constant for chain transfer to sulfur as an inhibitor in addition polymerization (m^3 $mole^{-1}$ s^{-1})

k_d: rate constant for dissociation of the initiator molecule (s^{-1})

k_{dac}: rate constant for deactivation of propagating chains to form dormant chains in reversible deactivation radical polymerization (m^3 $mole^{-1}$ s^{-1})

k_{daca}: rate constant for reversible deactivation of propagating chain-RAFT agent complex in RAFT method of reversible deactivation radical polymerization (s^{-1})

k_{dacf}: rate constant for reversible deactivation-fragmentation of a dormant chain and primary radical to form a polymeric RAFT adduct radical intermediate in RAFT method of reversible deactivation radical polymerization (m^3 $mole^{-1}$ s^{-1})

k_{daci}: rate constant for iniferter deactivation by recombination of generated radical with activated chain transfer agent in nitroxide-mediated method of reversible deactivation radical polymerization (m^3 $mole^{-1}$ s^{-1})

k_{dcpx}: rate constant for dissociation of the complex formed by oxidizing and reducing agents in redox initiation (s^{-1})

k_{dslv}: rate constant for initiator dissociation by solvation in anionic polymerization (s^{-1})

k_{etda}: rate constant for electron transfer by donor–acceptor groups in the excited initiator in photoinitiated polymerization (s^{-1})

k_{etr}: rate constant for electron transfer from excited photosensitive compound to electron donor compound (2-ethylhexyl-4-dimethylaminobenzoate, EHA) to form sensitizer–donor ion pair in photoinitiated polymerization (m^3 $mole^{-1}$ s^{-1})

k_{ext}: rate constant for two-photon excitation of photosensitive compound (thioxanthone, TX) in photoinitiated polymerization

k_f: rate constant for fragmentation of benzoyloxy radical to liberate a molecule of carbon dioxide and a reactive benzyl radical (s^{-1})

k_{gtcx}: rate constant for initiator–catalyst complexation in group transfer polymerization (m^3 $mole^{-1}$ s^{-1})

k_{gtdcx}: rate constant for initiator–catalyst decomplexation in group transfer polymerization (s^{-1})

k_{gtdf}: rate constant for defragmentation of fragmented initiator–catalyst complex in group transfer polymerization (m^3 $mole^{-1}$ s^{-1})

k_{gtf}: rate constant for fragmentation of the initiator–catalyst complex in group transfer polymerization (s^{-1})

k_{ltr}: rate constant for ligand transfer from initiator with a labile ligand to chain transfer agent to form an activated chain transfer agent in ATRP method of reversible deactivation radical polymerization (m^3 $mole^{-1}$ s^{-1})

k_{o_2}: rate constant for the reaction of a primary radical with oxygen dissolved in the reaction initiator (m^3 $mole^{-1}$ s^{-1})

k_p: rate constant for propagation reaction between a growing chain and a monomer (m^3 $mole^{-1}$ s^{-1})

k_{pc}: rate constant for chain propagation in step-growth polymerization by condensation of complementary reactive groups between an n-mer and m-mer to form an (n+m)-mer chain (m^3 $mole^{-1}$ s^{-1})

k_{pcta}: rate constant for addition of a monomer to activated chain transfer agent following transfer to chain transfer agent to form a 1-mer propagating chain in addition polymerization (m^3 $mole^{-1}$ s^{-1})

k_{prt}: rate constant for addition of a monomer to activated retarding agent following transfer to retarding agent to form a 1-mer propagating chain in addition polymerization (m^3 $mole^{-1}$ s^{-1})

k_{ps}: rate constant for addition of a monomer to the activated solvent molecule following chain transfer to solvent to form a 1-mer propagating chain in addition polymerization (m^3 $mole^{-1}$ s^{-1})

k_{ptr}: rate constant for proton transfer in the sensitizer–donor ion pair to form a transient tertiary amine radical (EHA*) and a stable sensitizer radical (TX*) in photoinitiated polymerization (s^{-1})

k_{rac}: rate constant for reactivation of dormant chains to form an active propagating chain in reversible deactivation radical polymerization (m^3 $mole^{-1}$ s^{-1})

k_{rlx}: rate constant for relaxation of the excited initiator in photoinitiated polymerization (s^{-1})

k_{tc}: rate constant for termination of two propagating chains by combination (m^3 $mole^{-1}$ s^{-1})

k_{td}: rate constant for termination of two propagating chains by disproportionation (m^3 $mole^{-1}$ s^{-1})

$k_{t\beta}$: rate constant for termination by group transfer from β carbon of the propagating chain to the counter ion in cationic polymerization to form a terminated polymer chain with a terminal double bond (s^{-1})

$k_{\beta H}$: rate constant for termination by hydride atom abstraction from β carbon to the transition metal of the counterion in cationic polymerization to form a terminated polymer chain with a terminal double bond (s^{-1})

$k_{\beta CH_3}$: rate constant for termination by methyl group transfer from β carbon to the transition metal of the counterion in cationic polymerization to form a terminated polymer chain with a terminal double bond (s^{-1})

LB: long branch formed by transfer of active species from propagating chain to a terminated polymer chain

M: monomer

$M_-^{\bullet}$: negatively charged monomer radical generated in photoinitiated polymerization

$R^{\bullet}$: radical formed by dissociation of an initiator

$R—R$: inactive compound formed by the recombination of two primary radicals

$RP_1^{\bullet}$: reactive propagating chain with one repeat unit

$RP_n^{\bullet}$: reactive propagating chain with n repeat units

RT: retarder added to monomer to retard/slow polymerization of monomer during storage

$RT^{\bullet}$: active retarder radical formed by chain transfer to the retarder, which can slowly initiate a propagating chain

$R—X$: initiator with a labile ligand X that is transferable to a chain transfer agent to generate a radical and activate the chain transfer agent in atom transfer radical polymerization

S: solvent

$S^{\bullet}$: active solvent radical formed by chain transfer to solvent

SB: short branch formed on a propagating chain by chain backbiting in polymerization

TP_n: terminated polymer chain with n repeat units or an n-mer terminated polymer chain

$WSPI$: highly conjugated photoinitiator in the ground state

$WSPI^*$: highly conjugated photoinitiator in the excited state

$WSPI_-^{+\bullet}$: positively and negatively charged, radicalized, highly conjugated photoinitiator

2.11 Practice Problems

Problem 2.1 In the melt polymerization of styrene with *t*-butyl benzoyl peroxide initiator, chain transfer reaction takes place between the propagating chain and a butoxy radical formed by initiator dissociation according to the following reaction to produce a tert-butyl methyl ketone molecule $(CH_3)_3\overset{H}{C}{=}O$ and a terminated chain with a terminal double bond. The tert-butyl methyl ketone molecule is stable and does not react with the monomer to initiate a new chain.

$$(CH_3)_3\overset{O}{\overset{\|}{C^{\bullet}}} + P_{n-1}{-}\underset{H}{\overset{H}{C}}{-}\underset{C_6H_5}{\overset{H}{C^{\bullet}}} \xrightarrow{k_{cti}} P_{n-1}\underset{H}{C}{=}\underset{C_6H_5}{\overset{H}{C}} + (CH_3)_3\overset{H}{C}{=}O \tag{2.83}$$

where k_{cti} is the rate constant for chain transfer to the initiator fragment. Write all the elementary reactions including initiation, propagation, and termination reactions for this styrene polymerization, assuming chain transfer to initiator fragment and chain termination by disproportionation.

Problem 2.2 The ceiling temperature in polymerization of α-methyl styrene is 61 °C. This means that polymerization of α-methyl styrene and its copolymerization with other monomers like styrene at temperatures between 25 and 100 °C is reversible [229]. Write all the elementary reactions for the melt polymerization of styrene and α-methyl styrene at 60 °C, assuming no chain transfer reactions, termination by combination, and reversibility of α-methyl styrene polymerization.

Problem 2.3 Polymerization of N-isopropylacrylamide (NIPAAm) is carried out on the surface of hydroxyl-functionalized super-activated carbon nanoparticles (NPs) in a batch reactor with 1, 4-dioxane as the solvent, AIBN as the initiator, and 2-[(dodecylsulfanyl)carbonothioyl]- sulfanylpropanoic acid (DSCTSP) as the chain transfer agent [230]. The purpose of this polymerization is to produce temperature-sensitive carbon NPs. First, DSCTSP is grafted to the surface of NPs to produce DSCTSP-CB NPs. Next, NIPAAm is polymerized in 1,4-dioxane in the presence of AIBN and DSCTSP-CB NPs. Assuming no chain transfer to solvent, monomer, or initiator, and termination by disproportionation, write all the elementary reactions for the polymerization of NIPAAm on the surface of CB NPs.

Problem 2.4 A hydrogel is a crosslinked hydrophilic polymer that interacts with water molecules by hydrogen bonding, polar, and ionic interactions to absorb large quantities of water and swell in physiological medium. Hydrogels are used as environmentally sensitive excipients in drug delivery and tissue regeneration [231]. They are synthesized by copolymerization of a bifunctional monomer with a multifunctional crosslinker. In the synthesis of acrylic acid hydrogel, acrylic acid monomer is copolymerized by free radical polymerization with methylene bisacrylamide (MBIS) crosslinker in the presence of ammonium persulfate (APS) and sodium disulfite (SDS) as the redox initiation system. Write all elementary reactions for crosslinking of acrylic acid to a hydrogel, assuming no chain transfer reactions and termination by disproportionation.

Problem 2.5 *p*-methoxy styrene (pMOS) is polymerized by cationic polymerization in polar methylene chloride (CH_2Cl_2) solvent with hydrogen iodide (HI) as initiator and zinc iodide (ZnI_2) as the Lewis acid [232]. The initial ZnI_2 concentration is half of HI. The initiator HI is added first to the solution of pMOS in CH_2Cl_2, followed by ZnI_2. Upon addition, the initiator reacts with the monomer to form H-pMOS-I compound. Following the addition of Lewis acid, a dissociated $(HpMOS^+ + I^-ZnI_2)$ ion pair and an associated $(HpMOS^+..I^-ZnI_2)$ ionic pair are formed. Both dissociated and associated ion pairs propagate by adding pMOS monomer. As the $HpMOS^+ + I^-ZnI_2$ ion pair is dissociated, the cation $HpMOS^+$ can react with I^- from excess HI to form a terminated chain. Conversely, the associated $HpMOS^+..I^-ZnI_2$ is a living ion pair and terminates slowly.

a. Write all the elementary reactions for polymerization of pMOS in CH_2Cl_2 solvent with HI/ZnI_2 initiating system.
b. A small amount (10% of initiator) of a quaternary ammonium salt of chloride (*tert*-butyl ammonium chloride) is added to the above reaction. Addition of the salt causes termination of the dissociated ($HpMOS^+ + I^-ZnI_2$) ion-pair. Write all the elementary reactions for polymerization of pMOS in CH_2Cl_2 solvent with HI/ZnI_2 initiating system in the presence of nBu_4NCl.

Problem 2.6 A metallocene catalyst system is used in the polymerization of propylene with toluene as the solvent. A metallocene catalyst system consists of a transition metal like titanium (Ti) between two substituted cyclopentadienyl (Cp) groups, which is activated by a co-catalyst like methyl aluminoxane (MAO) to initiate the polymerization. Upon activation, the MAO reacts with the Ti catalyst to form a reactive, cationic Ti species for propylene initiation as follows:

$$Cp_1Cp_2Ti\,(III) + MAO + H_2C{=}\underset{CH_3}{C}H \xrightarrow{k_{ci}} Cp_1Cp_2Ti\,(IV){-}\underset{H}{\overset{H}{C}}{-}\underset{CH_3}{\overset{H}{C^{\bullet}}}{}^{+}\ MAO^- \tag{2.84}$$

Multiple side reactions have been reported in this polymerization system [233]. First, the catalyst-propagating chain complex can undergo homolysis of Ti—C bond to form a $Cp_1Cp_2Ti\,(IV)^{\bullet+}\ MAO^-$ and a reactive propagating radical $RP_n^{\bullet}$. In one radical transfer reaction, $RP_n^{\bullet}$ abstracts a hydrogen atom from the solvent to form a terminated TP_n polymer chain and a solvent radical $H_5C_6CH_2^{\bullet}$. Furthermore, the solvent radical and $Cp_1Cp_2Ti\,(IV)^{\bullet+}\ MAO^-$ may react reversibly to form a catalyst-solvent complex for monomer insertion and polymerization. In a second radical transfer reaction, $RP_n^{\bullet}$ abstracts a hydrogen from the monomer to form a terminated TP_n polymer chain and a monomer radical $H_2C{=}\underset{H}{C}{-}CH_2^{\bullet}$. The monomer radical and $Cp_1Cp_2Ti\,(IV)^{\bullet+}\ MAO^-$ can react reversibly to form a catalyst-monomer complex for monomer insertion and polymerization. In a third radical transfer reaction, the solvent radical $H_5C_6CH_2^{\bullet}$ produced in the first transfer reaction can abstract a hydrogen from a monomer to form a solvent molecule and a monomer radical. The monomer radical $H_2C{=}\underset{H}{C}{-}CH_2^{\bullet}$ produced in the second transfer reaction can abstract a hydrogen from the solvent to form a monomer and a solvent radical. Write all the elementary reactions for polymerization of propylene in toluene using a metallocene catalyst system consisting of Ti sandwiched between two Cp groups activated by MAO.

References

1 Nakamura, Y. and Yamago, S., Termination mechanism in the radical polymerization of methyl methacrylate and styrene determined by the reaction of structurally well-defined polymer end radicals. *Macromolecules*, 2015, **48**(18), 6450–6456.

2 Lima, R.C. et al., Modeling and parameter estimation of step-growth polymerization of ethylene glycol and 2,5-furandicarboxylic acid (FDCA) using the method of moments. *Macromolecular Symposia*, 2020, **394**(1), 2000105.

3 Novitsky, T. et al., Effect of stoichiometric imbalances on the melt condensation polymerization of poly(dodecamethylene terephthalamide) studied by intrinsic viscosity and C-13-NMR. *Journal of Applied Polymer Science*, 2010, **116**(6), 3388–3395.

4 Bevington, J.C., The nature of the initiation reaction in the polymerization of styrene sensitized by benzoyl peroxide. *Proceedings of the Royal Society of London Series A-Mathematical and Physical Sciences*, 1957, **239**(1218), 420–432.

5 Moad, G. et al., Initiating free radical polymerization. *Macromolecular Symposia*, 2002, **182**, 65–80.

6 Braun, D., Origins and development of initiation of free radical polymerization processes. *International Journal of Polymer Science*, 2009, **2009**, 893234.

7 Bamford, C.H. and Mullik, S.U., New thermal initiators of free-radical polymerization and block copolymerization. *Polymer*, 1976, **17**(1), 94–95.

8 Kim, B. et al., Thermal radical initiator derivatives based on O-imino-isourea: synthesis, polymerization, and characterization. *Journal of Polymer Science Part A-Polymer Chemistry*, 2016, **54**(22), 3593–3600.

9 Kang, S. et al., New optimized triazene radical initiators for thermal polymerization. *Journal of Industrial and Engineering Chemistry*, 2019, **79**, 274–278.

10 Noguchi, N. et al., 2,2′-azobis (4-methoxy-2,4-dimethylvaleronitrile), a new lipid-soluble azo initiator: application to oxidations of lipids and low-density lipoprotein in solution and in aqueous dispersions. *Free Radical Biology and Medicine*, 1998, **24**(2), 259–268.

11 Huang, N.J. and Sundberg, D.C., Fundamental studies of grafting reactions in free-radical copolymerization. 2. Grafting of styrene, acrylate, and methacrylate monomers onto *cis*-polybutadiene using AIBN initiator in solution polymerization. *Journal of Polymer Science Part A-Polymer Chemistry*, 1995, **33**(15), 2551–2570.

12 Huang, N.J. and Sundberg, D.C., Fundamental studies of grafting reactions in free-radical copolymerization. 4. Grafting of styrene, acrylate, and methacrylate monomers onto vinyl-polybutadiene using benzoyl peroxide and AIBN initiators in solution polymerization. *Journal of Polymer Science Part A-Polymer Chemistry*, 1995, **33**(15), 2587–2603.

13 Yamago, S. et al., Practical protocols for organotellurium-mediated living radical polymerization by in situ generated initiators from AIBN and ditellurides. *Macromolecules*, 2003, **36**(11), 3793–3796.

14 Berger, K.C., Deb, P.C., and Meyerhoff, G., Radical vinyl polymerization - reactions of benzoyl peroxide during bulk-polymerization of styrene with labeled initiator. *Macromolecules*, 1977, **10**(5), 1075–1080.

15 Mitra, S.C. et al., Effect of temperature on the rates of aqueous polymerization of vinyl-acetate initiated by SO_4 free-radicals produced from potassium persulfate and the redox reaction of ferrous ammonium-sulfate and potassium persulfate (Vol 34a, Pg 602, 1995). *Indian Journal of Chemistry Section A-Inorganic Bio-Inorganic Physical Theoretical & Analytical Chemistry*, 1995, **34**(12), 1014.

16 Mitra, S.C. et al., Effect of temperature on the rates of aqueous polymerization of vinyl-acetate initiated by SO_4-center-dot-free-radicals produced from potassium persulfate and the redox reaction of ferrous Ammonium-sulfate and potassium persulfate. *Indian Journal of Chemistry Section A-Inorganic Bio-Inorganic Physical Theoretical & Analytical Chemistry*, 1995, **34**(8), 602–610.

17 Du, X.Y. et al., Generation of a carbon dots/ammonium persulfate redox initiator couple for free radical frontal polymerization. *Polymer Chemistry*, 2018, **9**(4), 420–427.

18 Sarac, A.S., Redox polymerization. *Progress in Polymer Science*, 1999, **24**(8), 1149–1204.

19 Pattnaik, S. et al., Aqueous polymerization of methyl-methacrylate initiated by potassium peroxydisulfate-ascorbic acid redox system. *Journal of Macromolecular Science-Chemistry*, 1979, **A13**(6), 797–805.

20 Temenoff, J.S. et al., In vitro osteogenic differentiation of marrow stromal cells encapsulated in biodegradable hydrogels. *Journal of Biomedical Materials Research Part A*, 2004, **70**(2), 235–244.

21 Yoshida, H. et al., Initiation mechanism of radiation-induced polymerization. *Advances in Chemistry Series*, 1968, (82), 548.

22 Gunzler, F. et al., Quantifying the efficiency of photoinitiation processes in methyl methacrylate free radical polymerization via electrospray ionization mass spectrometry. *Macromolecules*, 2009, **42**(5), 1488–1493.

23 Voll, D. et al., Quantifying photoinitiation efficiencies in a multiphotoinitiated free-radical polymerization. *Macromolecular Rapid Communications*, 2012, **33**(1), 47–53.

24 Bao, Y.Y., Recent trends in advanced photoinitiators for vat photopolymerization 3D printing. *Macromolecular Rapid Communications*, 2022, **43**(14), 2200202.

25 Bagheri, A. and Jin, J.Y., Photopolymerization in 3D printing. *ACS Applied Polymer Materials*, 2019, **1**(4), 593–611.

26 Hola, E. et al., Visible-light amine thioxanthone derivatives as photoredox catalysts for photopolymerization processes. *ACS Applied Polymer Materials*, 2021, **3**(11), 5547–5558.

27 Shao, J.Z., Huang, Y., Fan, Q.U., Visible light initiating systems for photopolymerization: status, development and challenges. *Polymer Chemistry*, 2014, **5**(14), 4195–4210.

28 Tomal, W. and Ortyl, J., Water-soluble photoinitiators in biomedical applications. *Polymers*, 2020, **12**(5), 1073.

29 Grotzinger, C. et al., Visible light-induced photopolymerization: speeding up the rate of polymerization by using co-initiators in dye/amine photoinitiating systems. *Polymer*, 2003, **44**(13), 3671–3677.

30 Valdesaguilera, O. et al., Photopolymerization studies using visible-light photoinitiators. *Macromolecules*, 1992, **25**(2), 541–547.

31 Ng, S.C. and Chee, K.K., Determination of rate constants of benzoyl peroxide and azobisisobutyronitrile dissociation via polymerization kinetics. *Journal of Polymer Science Part A-Polymer Chemistry*, 1982, **20**(2), 409–415.

32 Arutyunyan, R.S., Nalbandyan, D.M., and Beileryan, N.M., Mechanism of dissociation of 2,2′-azobisisobutyronitrile in aqueous emulsions. *Kinetics and Catalysis*, 1985, **26**(5), 906–910.

33 Ito, K., A theory generalized to cage effects in radical formation for determining initiator efficiency. *Polymer Journal*, 1985, **17**(3), 421–426.

34 Vogt, W., Radical efficiency at thermal decomposition of azobisisobutyronitrile in presence of oxygen - cage reaction and crossed termination. *Makromolekulare Chemie*, 1971, **145**(Jul), 309.

35 Yeh, C.J., Hu, M., and Shull, K.R., Oxygen inhibition of radical polymerizations investigated with the rheometric quartz crystal microbalance. *Macromolecules*, 2018, **51**(15), 5511–5518.

36 Johnson, L.M. et al., Enzyme-mediated redox initiation for hydrogel generation and cellular encapsulation. *Biomacromolecules*, 2009, **10**(11), 3114–3121.

37 Okamura, S. and Motoyama, T., Ammonium persulfate, benzoyl peroxide and azobisisobutyronitrile as polymerization initiators in homogeneous phase. *Journal of Polymer Science*, 1955, **17**(85), 428–431.

38 Huang, C. et al., Enhancing the self-recovery and mechanical property of hydrogels by macromolecular microspheres with thermal and redox initiation systems. *RSC Advances*, 2017, **7**(26), 16015–16021.

39 Nagaraja, G.K., Demappa, T., and Mahadevaiah, Polymerization kinetics of acrylonitrile by oxidation: reduction system using potassium persulfate/ascorbic acid in an aqueous medium. *Journal of Applied Polymer Science*, 2011, **121**(3), 1299–1303.

40 Corrigan, N. et al., Seeing the light: advancing materials chemistry through photopolymerization. *Angewandte Chemie-International Edition*, 2019, **58**(16), 5170–5189.

41 Bao, Y.Y., Paunovic, N., and Leroux, J.C., Challenges and opportunities in 3D printing of biodegradable medical devices by emerging photopolymerization techniques. *Advanced Functional Materials*, 2022, **32**(15), 2109864.

42 Xu, X.Y. et al., Vat photopolymerization 3D printing for advanced drug delivery and medical device applications. *Journal of Controlled Release*, 2021, **329**, 743–757.

43 Neidinger, P. et al., Near infrared light induced radical polymerization in water. *Angewandte Chemie-International Edition*, 2022, **61**(42), e202209177.

44 Colley, C.S. et al., Probing the reactivity of photoinitiators for free radical polymerization: time-resolved infrared spectroscopic study of benzoyl radicals. *Journal of the American Chemical Society*, 2002, **124**(50), 14952–14958.

45 Muller, S.M. et al., Recent Advances in Type I photoinitiators for visible light induced photopolymerization. *Chemphotochem*, 2022, **6**(11), e202200091.

46 Ligon, S.C. et al., Polymers for 3D printing and customized additive manufacturing. *Chemical Reviews*, 2017, **117**(15), 10212–10290.

47 Wang, Z.J. et al., Visible light photoinitiation of cell-adhesive gelatin methacryloyl hydrogels for stereolithography 3D bioprinting. *ACS Applied Materials & Interfaces*, 2018, **10**(32), 26859–26869.

48 Avens, H.J., Randle, T.J., and Bowman, C.N., Polymerization behavior and polymer properties of eosin-mediated surface modification reactions. *Polymer*, 2008, **49**(22), 4762–4768.

49 Allushi, A. et al., Conventional Type II photoinitiators as activators for photoinduced metal-free atom transfer radical polymerization. *Polymer Chemistry*, 2017, **8**(12), 1972–1977.

50 Qin, X.-H. et al., Additive manufacturing of photosensitive hydrogels for tissue engineering application. *BioNanoMaterials*, 2014, **15**(3–4), 49–70.

51 Wloka, T., Gottschaldt, M., and Schubert, U.S., From light to structure: photo initiators for radical two-photon polymerization. *Chemistry-A European Journal*, 2022, **28**(32), e202104191.

52 Eibel, A., Fast, D.E., and Gescheidt, G., Choosing the ideal photoinitiator for free radical photopolymerizations: predictions based on simulations using established data. *Polymer Chemistry*, 2018, **9**(41), 5107–5115.

53 Fischer, H., The persistent radical effect: a principle for selective radical reactions and living radical polymerizations. *Chemical Reviews*, 2001, **101**(12), 3581–3610.

54 Goto, A. and Fukuda, T., Kinetics of living radical polymerization. *Progress in Polymer Science*, 2004, **29**(4), 329–385.

55 Lorandi, F. et al., New protocol to determine the equilibrium constant of atom transfer radical polymerization. *Electrochimica Acta*, 2018, **260**, 648–655.

56 Zhang, H.Q. et al., Effect of Cu(II) on the kinetics of the homogeneous atom transfer radical polymerization of methyl methacrylate. *Macromolecules*, 2001, **34**(18), 6169–6173.

57 Keating, J.J. and Plawsky, J.L., Radical lifetimes in atom transfer radical polymerization: a Monte Carlo and deterministic investigation. *Macromolecules*, 2020, **53**(17), 7224–7238.

58 Greszta, D. and Matyjaszewski, K., Mechanism of controlled/"living" radical polymerization of styrene in the presence of nitroxyl radicals. Kinetics and simulations. *Macromolecules*, 1996, **29**(24), 7661–7670.

59 Matyjaszewski, K., Patten, T.E., and Xia, J.H., Controlled/"living" radical polymerization. Kinetics of the homogeneous atom transfer radical polymerization of styrene. *Journal of the American Chemical Society*, 1997, **119**(4), 674–680.

60 Tang, W., Tsarevsky, N.V., and Matyjaszewski, K., Determination of equilibrium constants for atom transfer radical polymerization. *Journal of the American Chemical Society*, 2006, **128**(5), 1598–1604.

61 Tang, W. et al., Understanding atom transfer radical polymerization: effect of ligand and initiator structures on the equilibrium constants. *Journal of the American Chemical Society*, 2008, **130**(32), 10702–10713.

62 Charleux, B., Nicolas, J., and Guerret, O., Theoretical expression of the average activation-deactivation equilibrium constant in controlled/living free-radical copolymerization operating via reversible termination. Application to a strongly improved control in nitroxide-mediated polymerization of methyl methacrylate. *Macromolecules*, 2005, **38**(13), 5485–5492.

63 Tang, H.D. et al., Highly active copper-based catalyst for atom transfer radical polymerization. *Journal of the American Chemical Society*, 2006, **128**(50), 16277–16285.

64 Otsu, T., Iniferter concept and living radical polymerization. *Journal of Polymer Science Part A-Polymer Chemistry*, 2000, **38**(12), 2121–2136.

65 Otsu, T. and Yoshida, M., Role of initiator-transfer agent-terminator (iniferter) in radical polymerizations: polymer design by organic disulfides as iniferters. *Macromolecular Rapid Communications*, 2005, **26**(3), 137–142.

66 Moad, G. et al., Advances in RAFT polymerization: the synthesis of polymers with defined end-groups. *Polymer*, 2005, **46**(19), 8458–8468.

67 Chiefari, J. et al., Living free-radical polymerization by reversible addition-fragmentation chain transfer: the RAFT process. *Macromolecules*, 1998, **31**(16), 5559–5562.

68 Hawthorne, D.G. et al., Living radical polymerization with reversible addition-fragmentation chain transfer (RAFT): direct ESR observation of intermediate radicals. *Macromolecules*, 1999, **32**(16), 5457–5459.

69 Yin, H. et al., Selection principle of RAFT chain transfer agents and universal RAFT chain transfer agents. *Progress in Chemistry*, 2022, **34**(6), 1298–1307.

70 Destarac, M., On the critical role of RAFT agent design in reversible Addition-fragmentation chain transfer (RAFT) polymerization. *Polymer Reviews*, 2011, **51**(2), 163–187.

71 Harrisson, S., The chain length distribution of an ideal reversible deactivation radical polymerization. *Polymers*, 2018, **10**(8), 887.

72 Dervan, P. et al., Mechanism and kinetics of the imidazolidinone nitroxide-mediated free-radical polymerization of styrene. *Journal of Polymer Science Part A-Polymer Chemistry*, 2003, **41**(2), 327–334.

73 Veregin, R.P.N. et al., Mechanism of rate enhancement using organic acids in nitroxide-mediated living free-radical polymerizations. *Macromolecules*, 1996, **29**(11), 4161–4163.

74 Studer, A. et al., New sterically hindered nitroxides for the living free radical polymerization: X-ray structure of an α-H-bearing nitroxide. *Macromolecules*, 2004, **37**(1), 27–34.

75 Baskaran, D. and Muller, A.H.E., Anionic vinyl polymerization - 50 years after Michael Szwarc. *Progress in Polymer Science*, 2007, **32**(2), 173–219.

76 Frankel, M. et al., Anionic polymerisation of vinyl monomers with butyl-lithium. *Journal of the Chemical Society*, 1959, 778, 3858–3864.

77 Bywater, S. and Worsfold, D.J., Anionic polymerization of styrene - effect of tetrahydrofuran. *Canadian Journal of Chemistry-Revue Canadienne De Chimie*, 1962, **40**(8), 1564–1570.

78 Hirohara, H., Takaya, K., and Ise, N., Study of effect of impurities in living anionic polymerization. *Macromolecules*, 1971, **4**(3), 288–290.

79 Wenger, F., Effects of impurities on control of anionic batch polymerizations. *Journal of Polymer Science*, 1962, **60**(170), 99–111.

80 Fathi, H., Hamielec, A.E., and Davison, E.J., Modelling the anionic solution polymerization of butadiene - the effects of chain termination and long chain branching on molecular weight distribution development. *Polymer Reaction Engineering*, 1996, **4**(4), 243–277.

81 Gerner, F.J. et al., On the termination reaction in the anionic polymerization of methyl-methacrylate in polar solvents. 1. Kinetic studies. *European Polymer Journal*, 1984, **20**(4), 349–355.

82 Chaplin, R.P. and Yaddehige, S., Termination mechanism in the anionic-polymerization of methyl-methacrylate. *Journal of Macromolecular Science-Chemistry*, 1980, **A14**(1), 23–31.

83 Zhang, Y. and Chen, E.Y.X., Controlled polymerization of methacrylates to high molecular weight polymers using oxidatively activated group transfer polymerization initiators. *Macromolecules*, 2008, **41**(1), 36–42.

84 Fuchise, K. et al., Recent progress in organocatalytic group transfer polymerization. *Polymer Chemistry*, 2013, **4**(16), 4278–4291.

85 Sage, V., Clark, J.H., and Macquarrie, D.J., Cationic polymerization of styrene using mesoporous silica supported aluminum chloride. *Journal of Molecular Catalysis A-Chemical*, 2003, **198**(1–2), 349–358.

86 Satoh, K., Kamigaito, M., and Sawamoto, M., Direct living cationic polymerization of *p*-hydroxystyrene with boron trifluoride etherate in the presence of water. *Macromolecules*, 2000, **33**(15), 5405–5410.

87 Yang, X.M., Stern, C.L., and Marks, T.J., Cationic zirconocene olefin polymerization catalysts based on the organo-Lewis acid tris(pentafluorophenyl)borane - a synthetic, structural, solution dynamic, and polymerization catalytic study. *Journal of the American Chemical Society*, 1994, **116**(22), 10015–10031.

88 Kostjuk, S.V., Radchenko, A.V., and Ganachaud, F., Controlled/living cationic polymerization of *p*-methoxystyrene in solution and aqueous dispersion using tris(pentafluorophenyl)borane as a Lewis acid: acetonitrile does the job. *Macromolecules*, 2007, **40**(3), 482–490.

89 Cheradame, H. et al., Cationic polymerization of indene initiated by titanium tetrachloride. 1. Influence of impurities of solvent, methylene chloride, and cocatalytic effects of water and hydrochloric acid. *European Polymer Journal*, 1973, **9**(5), 375–384.

90 Hung, N.A., Cheradam, H., and Sigwalt, P., Cationic polymerization of indene initiated by titanium tetrachloride. 2. Influence of temperature on rate of polymerization. *European Polymer Journal*, 1973, **9**(5), 385–397.

91 Zhang, J.H. et al., Characteristics and mechanism of styrene cationic polymerization in aqueous media initiated by cumyl alcohol/$B(C_6F_5)_3$. *Macromolecular Chemistry and Physics*, 2019, **220**(4), 1800419.

92 Nikitin, A.N. et al., Determination of propagation rate coefficient of acrylates by pulsed-laser polymerization in the presence of intramolecular chain transfer to polymer. *Macromolecular Rapid Communications*, 2003, **24**(13), 778–782.

93 Ito, K., Relationship between kinetic chain-length and radical polymerization rate. *Journal of Polymer Science Part A-Polymer Chemistry*, 1976, **14**(7), 1761–1771.

94 Williams, G. and Coe, J.S., The kinetic chain length in the polymerization of styrene by stannic chloride in carbon tetrachloride. *Chemistry & Industry*, 1954(2), 43–44.

95 Berger, K.C., Disproportionation and combination as termination mechanisms in radical polymerization of styrene. 2. Analysis of temperature dependencies. *Makromolekulare Chemie-Macromolecular Chemistry and Physics*, 1975, **176**(12), 3575–3592.

96 Berger, K.C. and G. Meyerhoff, Disproportionation and combination as termination mechanisms in radical polymerization of styrene. 1. Experiments with C-14-labeled 2,2′-azoisobutyronitrile. *Makromolekulare Chemie-Macromolecular Chemistry and Physics*, 1975, **176**(7), 1983–2003.

97 Hatada, K., Kitayama, T., and Masuda, E., Evidence for disproportionation in termination reaction of styrene polymerization by α,α′-azobisisobutyronitrile. *Polymer Journal*, 1985, **17**(8), 985–989.

98 Ribelli, T.G. et al., Disproportionation or combination? The termination of acrylate radicals in ATRP. *Macromolecules*, 2017, **50**(20), 7920–7929.

99 Yan, D.Y. and Cai, G.F., Influence of termination and transfer on molecular-weight distribution of polymers. 1. Disproportionation and combination. *Journal of Macromolecular Science-Chemistry*, 1986, **A23**(1), 37–57.

100 Cai, G.F., Yan, D.Y., and Litt, M., A kinetic theory on cationic polymerization with termination by combination with a negative fragment. *Macromolecules*, 1988, **21**(3), 578–584.

101 Ballard, N. et al., On the termination mechanism in the radical polymerization of acrylates. *Macromolecular Rapid Communications*, 2016, **37**(16), 1364–1368.

102 Nakamura, Y. et al., Termination mechanism of the radical polymerization of acrylates. *Macromolecular Rapid Communications*, 2016, **37**(6), 506–513.

103 Barth, J. et al., Chain-length-dependent termination in radical polymerization of acrylates. *Macromolecular Chemistry and Physics*, 2011, **212**(13), 1366–1378.

104 Shen, J.C. et al., Termination process and the termination parameter-Kt of methyl-methacrylate bulk radical polymerization. *Science in China Series B-Chemistry*, 1990, **33**(9), 1040–1045.

105 Zhu, S. et al., Radical trapping and termination in free-radical polymerization of methyl methacrylate. *Macromolecules*, 1990, **23**(4), 1144–1150.

106 Sato, E., Zetterlund, P.B., and Yamada, B., Macromonomer synthesis using oligomers of ω-unsaturated methacrylate as addition-fragmentation chain transfer agents: increased efficiency by manipulation of steric hindrance. *Macromolecules*, 2004, **37**(7), 2363–2370.

107 Bevington, J.C., Melville, H.W., and Taylor, R.P., The termination reaction in radical polymerizations. 2. Polymerizations of styrene at 60° and of methyl methacrylate at 0 and 60°, and the copolymerization of these monomers at 60°. *Journal of Polymer Science*, 1954, **14**(77), 463–476.

108 Laita, Z., The Kinetics of radical polymerization of ethylene. 1. General kinetics. *Journal of Polymer Science*, 1959, **38**(133), 247–258.

109 Laita, Z. and Machacek, Z., The kinetics of radical polymerization of ethylene. 2. Determination of the kinetic coefficients. *Journal of Polymer Science*, 1959, **38**(134), 459–469.

110 Mortimer, G.A. and Arnold, L.C., Free-radical polymerization of olefins. *Journal of Polymer Science Part A-General Papers*, 1964, **2**(9), 4247–4253.

111 Boghetich, L., Daues, G.W., and Mortimer, G.A., Free-radical copolymerization of α-olefins with ethylene. *Journal of Polymer Science*, 1962, **61**(171), 3–8.

112 Giammarino, G. and Villani, V., Termination mechanisms via H-transfer for polyolefin living catalysis at DFT level. *Macromolecular Theory and Simulations*, 2014, **23**(6), 365–368.

113 Bukatov, G.D., Zakharov, V.A., and Barabanov, A.A., Mechanism of olefin polymerization on supported Ziegler–Natta catalysts based on data on the number of active centers and propagation rate constants. *Kinetics and Catalysis*, 2005, **46**(2), 166–176.

114 Yu, Y., Fu, Z.S., and Fan, Z.Q., Chain transfer reactions of propylene polymerization catalyzed by $AlEt_3$ activated $TiCl_4/MgCl_2$ catalyst under very low monomer addition rate. *Journal of Molecular Catalysis A-Chemical*, 2012, **363**, 134–139.

115 Corrigan, N. et al., Reversible-deactivation radical polymerization (Controlled/living radical polymerization): from discovery to materials design and applications. *Progress in Polymer Science*, 2020, **111**, 101311.

116 Kubisa, P., Radical polymerization with reversible deactivation of active species. *Polimery*, 2000, **45**(11–12), 741–750.

117 Shipp, D.A., Reversible-deactivation radical polymerizations. *Polymer Reviews*, 2011, **51**(2), 99–103.

118 Jenkins, A.D., Jones, R.G., and Moad, G., Terminology for reversible-deactivation radical polymerization previously called “controlled” radical or “living” radical polymerization (IUPAC recommendations 2010). *Pure and Applied Chemistry*, 2010, **82**(2), 483–491.

119 Stenzel, M.H. and Barner-Kowollik, C., The living dead - common misconceptions about reversible deactivation radical polymerization. *Materials Horizons*, 2016, **3**(6), 471–477.

120 Yamada, B. and Kobatake, S., Radical polymerization, co-polymerization, and chain transfer of α-substituted acrylic esters. *Progress in Polymer Science*, 1994, **19**(6), 1089–1152.

121 Ayrey, G. and Haynes, A.C., Chain transfer-reaction of azoisobutyronitrile during polymerization of methyl-methacrylate at 60 degrees C. *Makromolekulare Chemie-Macromolecular Chemistry and Physics*, 1974, **175**(5), 1463–1470.

122 Basu, S., Sen, J.N., and Palit, S.R., Degree of polymerization and chain transfer in methyl methacrylate. *Proceedings of the Royal Society of London Series A-Mathematical and Physical Sciences*, 1950, **202**(1071), 485–498.

123 Ghosh, P. and Mukhopadhyay, G., Chain Transfer with pyridine in the polymerization of methyl-methacrylate, styrene, and vinyl-acetate using benzoyl peroxide as initiator at 60 °C. *Makromolekulare Chemie-Macromolecular Chemistry and Physics*, 1979, **180**(9), 2253–2256.

124 Obrien, J.L. and Gornick, F., Chain transfer in the polymerization of methyl methacrylate. 1. Transfer with monomer and thiols - the mechanism of the termination reaction at 60°. *Journal of the American Chemical Society*, 1955, **77**(18), 4757–4763.

125 Prescott, S.W., Chain-length dependence in living/controlled free-radical polymerizations: physical manifestation and Monte Carlo simulation of reversible transfer agents. *Macromolecules*, 2003, **36**(25), 9608–9621.

126 Moad, G. and Solomon, D.H., Invited review - understanding and controlling radical polymerization. *Australian Journal of Chemistry*, 1990, **43**(2), 215–239.

127 Moad, G., A critical survey of dithiocarbamate reversible addition-fragmentation chain transfer (RAFT) agents in radical polymerization. *Journal of Polymer Science Part A-Polymer Chemistry*, 2019, **57**(3), 216–227.

128 Arest-Yakubovich, A.A. and Litvinenko, G.I., Chain transfer reactions in anionic polymerization of nonpolar monomers. *Progress in Polymer Science*, 1996, **21**(2), 335–398.

129 Penczek, S., Some kinetic and mechanistic consequences of chain transfer to polymer in cationic polymerization of cyclic ethers and acetals. *Makromolekulare Chemie*, 1970, **134**, 299–304.

130 Fazekas, T.J. et al., Diversification of aliphatic C–H bonds in small molecules and polyolefins through radical chain transfer. *Science*, 2022, **375**(6580), 545–550.

131 Kukulj, D., Davis, T.P., and Gilbert, R.G., Chain transfer to monomer in the free-radical polymerizations of methyl methacrylate, styrene, and α-methylstyrene. *Macromolecules*, 1998, **31**(4), 994–999.

132 Magee, C. et al., Chain transfer to solvent in the radical polymerization of structurally diverse acrylamide monomers using straight-chain and branched alcohols as solvents. *Polymer Chemistry*, 2014, **5**(7), 2259–2265.

133 Balic, R., Fellows, C.M., and van Herk, A.M., Chain transfer to monomer and polymer in the radical polymerization of vinyl neo-decanoate. *Macromolecular Research*, 2004, **12**(4), 325–335.

134 Mayo, F.R., Chain transfer in the polymerization of styrene: the reaction of solvents with free radicals. *Journal of the American Chemical Society*, 1943, **65**, 2324–2329.

135 Tan, S.B., Li, J.J., and Zhang, Z.C., Study of chain transfer reaction to solvents in the initiation stage of atom transfer radical polymerization. *Macromolecules*, 2011, **44**(20), 7911–7916.

136 Sugihara, Y. et al., Chain transfer to solvent in the radical polymerization of *N*-isopropylacrylamide. *Journal of Polymer Science Part A-Polymer Chemistry*, 2011, **49**(8), 1856–1864.

137 Ji, Y.Y., and Klausen, R.S., Chain transfer to solvent in BN 2-vinylnaphthalene radical polymerization. *Journal of Polymer Science*, 2021, **59**(21), 2521–2529.

138 Starnes, W.H. et al., Branch Structures in Poly(vinyl-chloride) and the mechanism of chain transfer to monomer during vinyl-chloride polymerization. *Macromolecules*, 1983, **16**(5), 790–807.

139 Gregg, R.A. and Mayo, F.R., Chain transfer in the polymerization of styrene. 7. Compounds containing halogens, oxygen and nitrogen. *Journal of the American Chemical Society*, 1953, **75**(14), 3530–3533.

140 Yamamoto, T. et al., The effects of solvents on chain-transfer reaction in radical polymerizations of styrene, methyl-methacrylate and vinyl-acetate. *Nippon Kagaku Kaishi*, 1985, **4**, 767–770.

141 Furuncuoglu, T. et al., Role of chain transfer agents in free radical polymerization kinetics. *Macromolecules*, 2010, **43**(4), 1823–1835.

142 Gatzke, A.L., Chain transfer in anionic polymerization. Determination of chain-transfer constants by using carbon-14-labeled chain transfer agents. *Journal of Polymer Science Part A-1-Polymer Chemistry*, 1969, **7**(8), 2281–2292.

143 Peng, L., Haucourt, N., and Goethals, E.J., Acetals as chain transfer agents in the cationic polymerization of vinyl ethers. *Macromolecular Rapid Communications*, 1994, **15**(3), 197–202.

144 Benedicto, A.D., Claverie, J.P., and Grubbs, R.H., Molecular-weight distribution of living polymerization involving chain-transfer agents - computational results, analytical solutions, and experimental investigations using ring-opening metathesis polymerization. *Macromolecules*, 1995, **28**(2), 500–511.

145 Henriquez, C. et al., Thiols as chain transfer agents in free radical polymerization in aqueous solution. *Polymer*, 2003, **44**(19), 5559–5561.

146 Stolzenberg, K. and Kirchner, K., Continuous free-radical polymerization of styrene in the stirred tank reactor - effect of inhibitors. *Angewandte Makromolekulare Chemie*, 1981, **95**, 185–197.

147 Tanczos, I., Foldesberezsnich, T., and Tudos, F., Kinetics of radical Polymerization. 40. Investigation of the effect of molecule inhibitors in polymerization in solution. *European Polymer Journal*, 1983, **19**(3), 225–229.

148 Li, L. and Lee, L.J., Effects of inhibitors and retarders on low temperature free radical crosslinking polymerization between styrene and vinyl ester resin. *Polymer Engineering and Science*, 2001, **41**(1), 53–65.

149 Bovey, F.A. and Kolthoff, I.M., Studies of retarders and inhibitors in the emulsion polymerization of styrene. 2. Inhibitors. *Journal of Polymer Science*, 1950, **5**(5), 569–586.

150 Kolthoff, I.M. and Bovey, F.A., Studies of retarders and inhibitors in the emulsion polymerization of styrene. 1. Retarders. *Journal of the American Chemical Society*, 1948, **70**(2), 791–799.

151 Li, R.J. and Schork, F.J., Modeling of the inhibition mechanism of acrylic acid polymerization. *Industrial & Engineering Chemistry Research*, 2006, **45**(9), 3001–3008.

152 Nedez, C. and Ray, J.L., Efficient removal of polymerization inhibitors by adsorption on the surface of an optimized alumina. *Langmuir*, 1999, **15**(18), 5932–5936.

153 Cutie, S.S. et al., The effects of MEHQ on the polymerization of acrylic acid in the preparation of superabsorbent gels. *Journal of Applied Polymer Science*, 1997, **64**(3), 577–589.

154 Allen, T.L., Oxygen inhibition of the polymerization of styrene. *Journal of Applied Chemistry*, 1954, **4**(6), 289–290.

155 Bartlett, P.D. and Trifan, D.S., Some experiments on sulfur as inhibitor for the polymerization of styrene. *Journal of Polymer Science*, 1956, **20**(96), 457–476.

156 Bartlett, P.D. and Kwart, H., Some inhibitors and retarders in the polymerization of liquid vinyl acetate. 2. 1,3,5-trinitrobenzene and sulfur. *Journal of the American Chemical Society*, 1952, **74**(16), 3969–3973.

157 Batch, G.L. and Macosko, C.W., Oxygen inhibition in differential scanning calorimetry of free-radical polymerization. *Thermochimica Acta*, 1990, **166**, 185–198.

158 Li, N. and Pan, X.C., Controlled radical polymerization: from oxygen inhibition and tolerance to oxygen initiation. *Chinese Journal of Polymer Science*, 2021, **39**(9), 1084–1092.

159 Simic, R. et al., Oxygen inhibition of free-radical polymerization is the dominant mechanism behind the "mold effect" on hydrogels. *Soft Matter*, 2021, **17**(26), 6394–6403.

160 Aromaa, M.K. and Vallittu, P.K., Delayed post-curing stage and oxygen inhibition of free-radical polymerization of dimethacrylate resin. *Dental Materials*, 2018, **34**(9), 1247–1252.

161 Kirchner, K. and Rintelen, T., Continuous free-radical polymerization of styrene in the stirred tank reactor - effect of retarders. *Angewandte Makromolekulare Chemie*, 1986, **141**, 85–93.

162 Bovey, F.A. and Kolthoff, I.M., Inhibition and retardation of vinyl polymerization. *Chemical Reviews*, 1948, **42**(3), 491–525.

163 Jia, R. et al., Fine control of the molecular weight and polymer dispersity via a latent monomeric retarder. *Polymer Chemistry*, 2021, **12**(3), 349–355.

164 Elliott, J.R., Myers, R.L., and Roedel, G.F., Suspension polymerization of chlorotrifluoroethylene. *Industrial and Engineering Chemistry*, 1953, **45**(8), 1786–1790.

165 Ishihara, Y. et al., Back biting reactions during the catalytic decomposition of polyethylene. *Bulletin of the Chemical Society of Japan*, 1991, **64**(12), 3585–3592.

166 Liu, S. et al., Backbiting and β-scission reactions in free-radical polymerization of methyl acrylate. *International Journal of Quantum Chemistry*, 2014, **114**(5), 345–360.

167 Hamzehlou, S. et al., Analyzing the discrepancies in the activation energies of the backbiting and β-scission reactions in the radical polymerization of *n*-butyl acrylate. *Polymer Chemistry*, 2016, **7**(11), 2069–2077.

168 Vir, A.B. et al., From *n*-butyl acrylate Arrhenius parameters for backbiting and tertiary propagation to β-scission via stepwise pulsed laser polymerization. *Polymer Chemistry*, 2019, **10**(30), 4116–4125.

169 Quintens, G. and T. Junkers, Pulsed laser polymerization-size exclusion chromatography investigations into backbiting in ethylhexyl acrylate polymerization. *Polymer Chemistry*, 2022, **13**(14), 2019–2025.

170 Dossi, M., Storti, G., and Moscatelli, D., Relevance of backbiting and β-scission reactions in the free radical polymerization of acrylonitrile. *Macromolecular Symposia*, 2010, **289**, 119–123.

171 Kattner, H. and Buback, M., EPR study of backbiting in the aqueous-solution polymerization of acrylamide. *Macromolecular Rapid Communications*, 2015, **36**(24), 2186–2191.

172 Improta, R. and Peluso, A., The influence of back-biting interaction oil the polymerization of conjugated dienes in the presence of Ziegler–Natta catalysts. *Macromolecules*, 1999, **32**(20), 6852–6855.

173 Palza, H., Quijada, R., and Wilhelm, M., Effect of short-chain branching on the melt behavior of polypropylene under small-amplitude oscillatory shear conditions. *Macromolecular Chemistry and Physics*, 2013, **214**(1), 107–116.

174 McCord, E.F., Shaw, W.H., and Hutchinson, R.A., Short-chain branching structures in ethylene copolymers prepared by high-pressure free-radical polymerization: an NMR analysis. *Macromolecules*, 1997, **30**(2), 246–256.

175 Chatgilialoglu, C., Ingold, K.U., and Scaiano, J.C., Rate constants and Arrhenius parameters for the reactions of primary, secondary, and tertiary alkyl radicals with tri-*n*-butyltin hydride. *Journal of the American Chemical Society*, 1981, **103**(26), 7739–7742.

176 Franz, J.A., Bushaw, B.A., and Alnajjar, M.S., Absolute rate expressions for the abstraction of hydrogen by primary, secondary, and tertiary alkyl Radicals from thiophenol. *Journal of the American Chemical Society*, 1989, **111**(1), 268–275.

177 Sperati, C.A., Franta, W.A., and Starkweather, H.W., The molecular structure of polyethylene. 5. The effect of chain branching and molecular weight on physical properties. *Journal of the American Chemical Society*, 1953, **75**(24), 6127–6133.

178 Iedema, P.D. and Hoefsloot, H.C.J., Computing the trivariate chain length/degree of branching/number of combination points distribution for radical polymerization with transfer to polymer and recombination termination. *Macromolecular Theory and Simulations*, 2005, **14**(8), 505–518.

179 Tobita, H. and Hatanaka, K., Long-chain branching in free-radical polymerization due to chain transfer to polymer. *Journal of Polymer Science Part B-Polymer Physics*, 1995, **33**(5), 841–853.

180 Tobita, H. and Hamashima, N., Monte Carlo simulation of size exclusion chromatography for branched polymers formed through free-radical polymerization with chain transfer to polymer. *Macromolecular Theory and Simulations*, 2000, **9**(8), 453–462.

181 Dias, R.C.S. and Costa, M.R.P.F.N., A new look at kinetic modeling of nonlinear free radical polymerizations with terminal branching and chain transfer to polymer. *Macromolecules*, 2003, **36**(23), 8853–8863.

182 Iedema, P.D., Grcev, S., and Hoefsloot, H.C.J., Molecular weight distribution modeling of radical polymerization in a CSTR with long chain branching through transfer to polymer and terminal double bond (TDB) propagation. *Macromolecules*, 2003, **36**(2), 458–476.

183 Ahmad, N.M., Heatley, F., and Lovell, P.A., Chain transfer to polymer in free-radical solution polymerization of *n*-butyl acrylate studied by NMR spectroscopy. *Macromolecules*, 1998, **31**(9), 2822–2827.

184 Ahmad, N.M. et al., Chain transfer to polymer and branching in controlled radical polymerizations of *n*-butyl acrylate. *Macromolecular Rapid Communications*, 2009, **30**(23), 2002–2021.

185 Rozenberg, B.A. et al., Effect of chain transfer to polymer in cationic polymerization of oxygen-containing cyclic compounds (Reprinted from Vysokomolekulyarnye Soedineniya, Vol 7, Pg 188, 1965). *Vysokomolekulyarnye Soedineniya Seriya A & Seriya B*, 1999, **41**(12), 1929–1929.

186 Litvinenko, G.I. and Arest-Yakubovich, A.A., Effect of chain transfer to polymer on molecular mass characteristics of polymers in anionic polymerization - comparison with radical one. *Vysokomolekulyarnye Soedineniya Seriya B*, 1988, **30**(6), 1218–1225.

187 Jiang, L. et al., Radical polymerization in the presence of chain transfer monomer: an approach to branched vinyl polymers. *Macromolecules*, 2012, **45**(10), 4092–4100.

188 Jiang, J.Q. et al., A novel approach to branched polymers via latent thiol chain-transfer polymerization. *Chemistry Letters*, 2011, **40**(12), 1378–1380.

189 Iedema, P.D. and Hoefsloot, H.C.J., Role of polyradicals in predicting chain length distribution and gelation for radical polymerization with transfer to polymer. *Macromolecules*, 2004, **37**(26), 10155–10164.

190 Lalevee, J., Allonas, X., and Fouassier, J.P., Direct measurements of the addition and recombination of acrylate radicals: access to propagation and termination rate constants? *Journal of Polymer Science Part A-Polymer Chemistry*, 2006, **44**(11), 3577–3587.

191 Funt, B.L. and Pasika, W., Radical termination mechanisms in bulk polymerization. *Canadian Journal of Chemistry-Revue Canadienne De Chimie*, 1960, **38**(10), 1865–1870.

192 Hensley, D.R. et al., 2d-Inadequate NMR evidence for the termination mechanism of styrene free-radical polymerization. *Macromolecules*, 1995, **28**(5), 1586–1591.

193 Andrzejewska, E. et al., Termination mechanism during the photo-induced radical cross-linking polymerization in the presence and absence of oxygen. *Physical Chemistry Chemical Physics*, 2003, **5**(12), 2635–2642.

194 Britton, D., Heatley, F., and Lovell, P.A., Chain transfer to polymer in free-radical bulk and emulsion polymerization of vinyl acetate studied by NMR spectroscopy. *Macromolecules*, 1998, **31**(9), 2828–2837.

195 Heatley, F., Lovell, P.A., and Yamashita, T., Chain transfer to polymer in free-radical solution polymerization of 2-ethylhexyl acrylate studied by NMR spectroscopy. *Macromolecules*, 2001, **34**(22), 7636–7641.

196 Irzhak, T.F. and Irzhak, V.I., Kinetics of formation of the topological structure during radical polymerization accompanied by the reaction of chain transfer to the polymer. *Polymer Science Series B*, 2013, **55**(5–6), 391–399.

197 Pugh, C., Fan, G.Y., and Kasko, A.M., Measurement of chain transfer constants to polymer using oligomers and model compounds: chain transfer to poly [11-(4′-cyanophenyl-4“-phenoxy)undecyl acrylate] in radical polymerization. *Macromolecules*, 2005, **38**(19), 8071–8077.

198 Reyes, Y. and Asua, J.M., Revisiting chain transfer to polymer and branching in controlled radical polymerization of butyl acrylate. *Macromolecular Rapid Communications*, 2011, **32**(1), 63–67.

199 Thomas, S., Hamielec, A., and Soares, J., Free radical polymerization - an elegant method of solving the population balance equations with chain transfer to polymer. *Polymer Reaction Engineering*, 1997, **5**(4), 183–203.

200 Bailey, B.E. and Jenkins, A.D., The mechanism of termination in the polymerization of acrylonitrile. *Transactions of the Faraday Society*, 1960, **56**, 903–906.

201 Park, S.G. and Smith, D.G., Vinyl chloride studies. II. Initiation and termination in the homogeneous polymerization of vinyl chloride. *Die Makromolekulare Chemie*, 1970, **131**(1), 1–6.

202 Chiefari, J. et al., Chain transfer to polymer: a convenient route to macromonomers. *Macromolecules*, 1999, **32**(22), 7700–7702.

203 Britton, D., Heatley, F., and Lovell, P.A., Effect of monomer feed rate on chain transfer to polymer in semibatch emulsion polymerization of vinyl acetate studied by NMR spectroscopy. *Macromolecules*, 2000, **33**(14), 5048–5052.

204 Buback, M. and Schweer, J., Chain-length dependence of free-radical rate coefficients. 1. Termination rate coefficient in pure ethylene polymerization. *Makromolekulare Chemie-Rapid Communications*, 1988, **9**(10), 699–704.

205 Jensen, M.D. et al., Kinetics of long-chain branch formation in polyethylene. *ACS Catalysis*, 2018, **8**(1), 725–737.

206 Reinking, M.K., Orf, G., and McFaddin, D., Novel mechanism for the formation of long-chain branching in polyethylene. *Journal of Polymer Science Part A-Polymer Chemistry*, 1998, **36**(16), 2889–2898.

207 Simon, L.C. and Soares, J.B.P., Polyethylene made with combinations of single-site-type catalysts: Monte Carlo simulation of long-chain branch formation. *Macromolecular Theory and Simulations*, 2002, **11**(2), 222–232.

208 Huang, W.Y. et al., Studies on the branching atom transfer radical polymerization using different divinyl monomers as the branching agent. *Acta Polymerica Sinica*, 2011(11), 1253–1257.

209 Huang, W.Y. et al., Branching copolymerization of styrene and methyl methacrylate with divinylbenzene. *Iranian Polymer Journal*, 2010, **19**(8), 589–598.

210 Bannister, I. et al., Development of branching in living radical copolymerization of vinyl and divinyl monomers. *Macromolecules*, 2006, **39**(22), 7483–7492.

211 Nishio, A. et al., Synthesis and characterization of ordered polyurethanes from a pair of diisocyanate monomers and ethylene glycol. *High Performance Polymers*, 2001, **13**(2), S233–S246.

212 Flory, P.J., Fundamental principles of condensation polymerization. *Chemical Reviews*, 1946, **39**(1), 137–197.

213 Nanda, V.S. and Jain, S.C., Statistical study of effect of variation of bimolecular rate constant in condensation polymerization. *Journal of Polymer Science Part A-1-Polymer Chemistry*, 1970, **8**(7), 1871–1885.

214 Park, O.O., Molecular-weight distribution and moments for condensation polymerization with variant reaction-rate constant depending on chain lengths. *Macromolecules*, 1988, **21**(3), 732–735.

215 Mayer, P.M. et al., The effects of protonation on the structure, stability, and thermochemistry of carbon-centered organic radicals. *Journal of the American Chemical Society*, 1997, **119**(52), 12889–12895.

216 Menon, A.S. et al., Effect of substituents on the stabilities of multiply-substituted carbon-centered radicals. *Organic & Biomolecular Chemistry*, 2011, **9**(10), 3636–3657.

217 Theilacker, W. and M.L. Ewald, On the theory of the stability of carbon radicals. *Naturwissenschaften*, 1943, **31**, 302–302.

218 Wen, Z. et al., On the direction and magnitude of radical substituent effects: the role of polar interaction on thermodynamic stabilities of benzylic C—H bonds and related carbon radicals. *Journal of Organic Chemistry*, 2001, **66**(4), 1466–1472.

219 Triacca, V.J. et al., Free-radical degradation of polypropylene - random chain scission. *Polymer Engineering and Science*, 1993, **33**(8), 445–454.

220 Pinheiro, L.A., Chinelatto, M.A., and Canevarolo, S.V., The role of chain scission and chain branching in high-density polyethylene during thermo-mechanical degradation. *Polymer Degradation and Stability*, 2004, **86**(3), 445–453.

221 Kumagai, T. et al., Specific polymerization mechanism involving β-scission of mid-chain radical yielding oligomers in the free-radical polymerization of vinyl ethers. *Macromolecules*, 2008, **41**(20), 7347–7351.

222 Lena, J.B. et al., Effects of chain transfer agent and temperature on branching and β-scission in radical polymerization of 2-ethylhexyl acrylate. *Macromolecular Chemistry and Physics*, 2018, **219**(9), 1700579.

223 Tobita, H., Markovian approach to free-radical polymerization with simultaneous long-chain branching and scission: effect of branching and scission kinetics. *Macromolecular Reaction Engineering*, 2015, **9**(3), 245–258.

224 Tobita, H., Free-radical polymerization with long-chain branching and scission: Markovian solution of the weight-average molecular weight. *Macromolecular Theory and Simulations*, 2014, **23**(8), 477–489.

225 Junkers, T. and Barner-Kowollik, C., The role of mid-chain radicals in acrylate free radical polymerization: branching and scission. *Journal of Polymer Science Part A-Polymer Chemistry*, 2008, **46**(23), 7585–7605.

226 Szymanski, R. and Baran, J., Molecular weight distribution in living polymerization proceeding with reshuffling of polymer segments due to chain transfer to polymer with chain scission, 2(a) - Monte Carlo simulation of polymer reshuffling proceeding with disproportionation of chain functionalities. *Macromolecular Theory and Simulations*, 2002, **11**(8), 836–844.

227 Iedema, P.D. and Hoefsloot, H.C.J., The influence of branching architecture on scission in modeling free-radical polymerization with long chain branching and scission. *Chimia*, 2001, **55**(3), 256–258.

228 Ueno, T., Nakashima, E., and Takeda, K., Quantitative analysis of random scission and chain-end scission in the thermal degradation of polyethylene. *Polymer Degradation and Stability*, 2010, **95**(9), 1862–1869.

229 Rudin, A. and Chiang, S.S.M., Kinetics of free-radical copolymerization of α-methylstyrene and styrene. *Journal of Polymer Science Part A-Polymer Chemistry*, 1974, **12**(10), 2235–2254.

230 Yang, Q. et al., A temperature-responsive carbon black nanoparticle prepared by surface-induced reversible addition-fragmentation chain transfer polymerization. *Polymer*, 2007, **48**(12), 3444–3451.

231 Jabbari, E., Bioconjugation of hydrogels for tissue engineering. *Current Opinion in Biotechnology*, 2011, **22**(5), 655–660.

232 Kojima, K., Sawamoto, M., and Higashimura, T., Living cationic polymerization of P-methoxystyrene by the Hi/Zni_2 and Hi/I_2 initiating systems - effects of tetrabutylammonium halides in a polar-solvent. *Macromolecules*, 1990, **23**(4), 948–953.

233 Zaccaria, F. et al., Chain transfer to solvent and monomer in early transition metal catalyzed olefin polymerization: mechanisms and implications for catalysis. *Catalysts*, 2021, **11**(2), 215.

3

Functions with Distributed Variables

3.1 Introduction

Molecular weight distribution of chains in a polymer sample is measured by gel permeation chromatography (GPC), matrix-assisted laser desorption ionization-time of flight mass spectrometry (MALDI-TOF MS), or dynamic rheometry [1–6]. A macroscopic characteristic like polymer viscosity or melt temperature depends on the average property (chain length) of the ensemble of polymer chains. These average chain length dependent properties are measured by a variety of techniques, including viscometry, osmometry, freezing-point depression (cryoscopy), vapor pressure depression, boiling-point elevation (ebulliometry), and light scattering. Viscometry is widely used in industry to determine the molecular weight of polymers, which is a measure of viscosity-average molecular weight, $\overline{M}_\eta$ [7–10]. If a colligative property like freezing-point depression of a polymer solution is used to characterize the polymer, then the number-average molecular weight, $\overline{M}_n$, is measured [11]. The colligative property of a polymer solution depends on the number of chains in solution, not their size. If light scattering from a polymer solution is used, the weight-average molecular weight, $\overline{M}_w$, of the polymer is measured [12, 13]. If the polymer is dissolved in a good solvent in which the chains are solvated, the intensity of the scattered light would be proportional to the volume of the polymer plus that of the solvent in the solvated polymer, which is proportional to the z-average molecular weight, $\overline{M}_z$, of the polymer distribution [14]. In other words, longer chains in the ensemble of polymer chains take up more solvent and make a greater contribution to the average molecular weight of the polymer. In extreme cases like rod-like chains in solution or in gelation [15], the longest chains in the ensemble tend to dominate the polymer properties, thus higher-order averages, like the (z+1)-average molecular weight $\left(\overline{M}_{z+1}\right)$, are used to define the average molecular weight of the polymer.

Knowing the average molecular weight, the average degree of polymerization, $\overline{DP}$, is calculated by dividing $\overline{M}$ by the molecular weight of repeat unit, M_0, as follows:

$$\overline{DP} = \frac{\overline{M}}{M_0} \tag{3.1}$$

The average contour length of the polymer chains, $\overline{L}_c$, is calculated by multiplying $\overline{DP}$ by the length of repeat unit, l_r, as follows:

$$\overline{L}_c = \overline{DP}\, l_r = \frac{\overline{M}}{M_0} l_r \tag{3.2}$$

In some polymerizations, like Ziegler–Natta, group transfer, atom transfer, or anionic polymerization, the molecular weight distribution is relatively narrow, whereas in others, like radical or cationic polymerization, the distribution is relatively broad. The physical and mechanical properties of a polymer, like viscosity, glass transition temperature (T_g), melting point (T_m), elastic modulus, fracture toughness, tensile strength, and strain to break, depend strongly on the average molecular weight as well as the breadth of chain distribution [16–19]. There is a need to measure and quantify both averages and the breadth of the polymer molecular weight distribution for practical applications. This chapter covers the moments of chain distribution and their relation to different average molecular weights of the polymer distribution.

3.2 Moments of a Distribution

Distributed variables like polymer chain length, degree of polymerization, or molecular weight are defined by their "mathematical expectation," "expected value," or "expectation." If x is a discontinuous random variable in the real number space i with a statistical probability distribution of $P_x(i)$, the expectation of property $f(x)$, represented by $\langle f(x)\rangle$, is given by the following expression [20]:

$$\langle f(x)\rangle = \sum_{i=-\infty}^{+\infty} f(x=i)\,P_x(i) \tag{3.3}$$

where $f(x=i)$ is a function of the variable x. In polymerization reactions, the discontinuous random variable x is the chain length or the degree of polymerization, DP, while $P_x(i)$ is the statistical distribution function of the polymer chains or $P_{DP}(i)$. Therefore, the expectation of the function $f(DP)$ for an ensemble of polymer chains, represented by $\langle f(Dp)\rangle$, is given by the following expression:

$$\langle f(DP)\rangle = \sum_{i=1}^{+\infty} f(DP=i)\,P_{DP}(i) \tag{3.4}$$

The above equation is subject to the following condition that the sum of probabilities for all chains in the ensemble should be equal to unity:

$$\sum_{i=1}^{+\infty} P_{DP}(i) = 1 \tag{3.5}$$

Consider the following $f(DP)$ function for the purpose of defining the moments of a distribution:

$$f(DP) = DP^k \tag{3.6}$$

where the exponent k indicates the order of dependence of the function f on DP. Based on the defining Eq. (3.6), the kth moment of the chain distribution in a polymer sample, expressed as M_k or $\langle DP^k\rangle$, is defined by the following equation:

$$M_k = \langle DP^k\rangle = \sum_{i=1}^{\infty} i^k P_{DP}(i) \tag{3.7}$$

If the distribution function is based on the number of chains with degree of polymerization i, the function is represented by $P_{DPn}(i)$ and the corresponding k^{th} moment represented by $M_{n,k}$. Conversely, if the distribution is based on the weight of the chains with degree of polymerization i, the function is represented by $P_{DPw}(i)$ and the corresponding k^{th} moment by $M_{w,k}$ as defined by the following equations:

$$M_{n,k} = \langle DP_n^k \rangle = \sum_{i=1}^{\infty} i^k P_{DPn}(i) \tag{3.8}$$

$$M_{w,k} = \langle DP_w^k \rangle = \sum_{i=1}^{\infty} i^k P_{DPw}(i) \tag{3.9}$$

The kth moment of a distribution can also be defined based on concentration or density of the ith fraction C_i and ρ_i of chains with degree of polymerization i by replacing $P_{DPn}(i)$ with C_i in Eq. (3.8) or $P_{DPw}(i)$ with ρ_i in Eq. (3.9) as follows:

$$Q_{n,k} = C\langle DP_n^k \rangle = \sum_{i=1}^{\infty} i^k C_i \tag{3.10}$$

$$Q_{w,k} = \rho\langle DP_w^k \rangle = \sum_{i=1}^{\infty} i^k \rho_i \tag{3.11}$$

where $Q_{n,K}$ and $Q_{w,K}$ are the kth concentration moment and the kth density moment of the chain distribution, respectively. The concentration distribution and density distribution are related to the number and weight probability distributions, respectively, by the following relations:

$$P_{DPn}(i) = C_i / C \tag{3.12}$$

$$P_{DPw}(i) = \rho_i / \rho \tag{3.13}$$

where C and ρ are the total molar concentration and density of chains in the polymer sample, respectively. The zero-order number $(M_{n,0})$, weight $(M_{w,0})$, concentration $(Q_{n,0})$, and density $(Q_{w,0})$ moments of the chains in a polymer sample are derived by setting the exponent k in the defining Eqs. (3.8–3.11) equal to zero as follows:

$$M_{n,0} = \langle DP_n{}^0 \rangle = \sum_{i=1}^{+\infty} i^0 P_{DPn}(i) = \sum_{i=0}^{+\infty} P_{DPn}(i) = 1 \tag{3.14}$$

$$M_{w,0} = \langle DP_w{}^0 \rangle = \sum_{i=1}^{+\infty} i^0 P_{DPw}(i) = \sum_{i=0}^{+\infty} P_{DPw}(i) = 1 \tag{3.15}$$

$$Q_{n,0} = C\langle DP_n{}^0 \rangle = \sum_{i=1}^{+\infty} i^0 C_i = \sum_{i=1}^{+\infty} C_i = C \tag{3.16}$$

$$Q_{w,0} = C\langle DP_w{}^0 \rangle = \sum_{i=1}^{+\infty} i^0 \rho_i = \sum_{i=1}^{+\infty} \rho_i = \rho \tag{3.17}$$

The zero-order number and weight distribution moments are equal to one, whereas the zero-order concentration and density moments are equal to the total concentration and total density of the polymer, respectively. The first-order number $(M_{n,1})$, weight $(M_{w,1})$, concentration $(Q_{n,1})$, and density $(Q_{w,1})$ moments of the chains in a polymer sample are obtained by setting the exponent k in Eqs. (3.8–3.11) equal to one as follows:

$$M_{n,1} = \langle DP_n{}^1 \rangle = \sum_{i=1}^{+\infty} i^1 P_{DPn}(i) = \sum_{i=1}^{+\infty} i P_{DPn}(i) \tag{3.18}$$

$$M_{w,1} = \langle DP_w{}^1 \rangle = \sum_{i=1}^{+\infty} i^1 P_{DPw}(i) = \sum_{i=1}^{+\infty} i P_{DPw}(i) \tag{3.19}$$

$$\begin{aligned} Q_{n,1} &= C\langle DP_n{}^1 \rangle = \sum_{i=1}^{+\infty} i^1 C_i = \sum_{i=1}^{+\infty} i C P_{DPn}(i) = C \sum_{i=1}^{+\infty} i P_{DPn}(i) \\ &= C\, M_{n,1} = Q_{n,0}\, M_{n,1} \end{aligned} \tag{3.20}$$

$$\begin{aligned} Q_{w,1} &= C\langle DP_w{}^1 \rangle = \sum_{i=1}^{+\infty} i^1 \rho_i = \sum_{i=1}^{+\infty} i \rho P_{DPw}(i) = \rho \sum_{i=1}^{+\infty} i P_{DPw}(i) \\ &= \rho\, M_{w,1} = Q_{w,0}\, M_{w,1} \end{aligned} \tag{3.21}$$

The first-order concentration moments $Q_{n,1}$and $Q_{w,1}$ are related to $M_{n,1}$ and $M_{w,1}$ by zero-order moments $Q_{n,0}$ and $Q_{w,0}$, respectively. The second-order number $(M_{n,2})$, weight $(M_{w,2})$, concentration $(Q_{n,2})$, and density $(Q_{w,2})$ moments of the chains in a polymer sample are derived by setting the exponent k in Eqs. (3.8–3.11) equal to two as follows:

$$M_{n,2} = \langle DP_n{}^2 \rangle = \sum_{i=1}^{+\infty} i^2 P_{DPn}(i) \tag{3.22}$$

$$M_{w,2} = \langle DP_w{}^2 \rangle = \sum_{i=1}^{+\infty} i^2 P_{DPw}(i) \tag{3.23}$$

$$\begin{aligned} Q_{n,2} &= C\langle DP_n^2 \rangle = \sum_{i=1}^{+\infty} i^2 C_i = \sum_{i=1}^{+\infty} i^2 C P_{DPn}(i) \\ &= C \sum_{i=1}^{+\infty} i^2 P_{DPn}(i) = C M_{n,2} = Q_{n,0} M_{n,2} \end{aligned} \tag{3.24}$$

$$\begin{aligned} Q_{w,2} &= C\langle DP_w^2 \rangle = \sum_{i=1}^{+\infty} i^2 \rho_i = \sum_{i=1}^{+\infty} i^2 \rho P_{DPw}(i) \\ &= \rho \sum_{i=1}^{+\infty} i^2 P_{DPw}(i) = \rho M_{w,2} = Q_{w,0} M_{w,2} \end{aligned} \tag{3.25}$$

The number-average and weight-average degree of polymerization of the chains in a polymer sample are derived from the moments of the chain distribution, as will be shown in the following sections.

3.3 Relating Molar-based Distribution of Polymer Chains to Weight-based Distribution

The weight probability distribution $P_{DPw}(i)$ is related to the number probability distribution $P_{DPn}(i)$. If $P_{DPn}(i)$ is known, the molar concentration of chains with i degree of polymerization is equal to $CP_{DPn}(i)$ where C is the molar concentration of chains in the polymer sample. The density of chains with i repeat units, ρ_i, is equal to the molar concentration of chains with i repeat units times the mass of a chain with i units, M_i, as follows:

$$\rho_i = M_i\, CP_{DPn}(i) = M_0 i\, CP_{DPn}(i) \tag{3.26}$$

where M_0 is the molecular mass of a repeat unit in the polymer chain, and i is the number of repeat units. The weight probability distribution is derived by dividing the mass of chains with i repeat units, ρ_i, to the total mass of the polymer chains, ρ, as follows:

$$P_{DPw}(i) = \frac{\rho_i}{\rho} = \frac{M_0 i C P_{DPn}(i)}{\sum_{i=1}^{\infty} M_0\, i\, C\, P_{DPn}(i)} = \frac{i\, P_{DPn}(i)}{\sum_{i=1}^{\infty} i\, P_{DPn}(i)} = \frac{i\, P_{DPn}(i)}{M_{n,1}} \tag{3.27}$$

Using the same approach, $P_{DPn}(i)$ which is equal to the ratio of concentration of chains with i repeat units in the polymer sample divided by the total concentration of chains is expressed in terms of $P_{DPw}(i)$ as follows:

$$P_{DPn}(i) = \frac{C_i}{C} = \frac{\rho_i / M_i}{\sum_{i=1}^{\infty} \rho_i / M_i} = \frac{\rho / M_0 i^{-1} P_{DPw}(i)}{\rho / M_0 \sum_{i=1}^{\infty} i^{-1} P_{DPw}(i)} = \frac{i^{-1} P_{DPw}(i)}{M_{w,-1}} \tag{3.28}$$

where the moment $M_{w,-1}$ in the above equation is the negative one-order weight moment of the chains in the polymer sample defined by the following equation:

$$M_{w,-1} = \sum_{i=1}^{+\infty} i^{-1} P_{DPw}(i) \tag{3.29}$$

Above equations demonstrate that $P_{DPn}(i)$ and $P_{DPw}(i)$ of the chains are independent of C, ρ, or M_0 of the polymer. Using a similar approach, other polymer distributions like z or $z+1$ distributions are derived in terms of the number or weight probability distribution of the chains. Consider the z distribution $P_{Dpz}(i)$ and the kth order z moment of the distribution $M_{z,k}$ defined as follows:

$$M_{z,k} = \langle DP_z^k \rangle = \sum_{i=1}^{\infty} i^k\, P_{DPz}(i) \tag{3.30}$$

where $P_{DPz}(i) = K_z i^2 P_{DPn}(i)$ (3.31)

subject to the following condition for the proportionality constant K_z:

$$\sum_{i=1}^{+\infty} P_{DPz}(i) = 1 \tag{3.32}$$

The z distribution $P_{DPz}(i)$ of chains in the polymer sample is expressed in terms of $P_{DPn}(i)$ and $P_{DPw}(i)$ using the following relations:

$$P_{DPz}(i) = \frac{i^2 P_{DPn}(i)}{\sum_{i=1}^{+\infty} i^2 P_{DPn}(i)} = \frac{i^2 P_{DPn}(i)}{M_{n,2}} = \frac{i P_{DPw}(i)}{\sum_{i=1}^{+\infty} i P_{DPw}(i)} = \frac{i P_{DPw}(i)}{M_{w,1}} \tag{3.33}$$

The above relations demonstrate that $P_{DPw}(i)$ and $P_{DPz}(i)$ as well as other higher-order distributions are derived from $P_{DPn}(i)$ without the knowledge of the concentration C, density ρ, or molecular weight of the repeat unit M_0 of the polymer chains.

3.4 Relating Molar-based Moments of Polymer Chains to Weight-based Moments

The kth order weight-based moment of the chain distribution, $M_{w,k}$, can be derived from the kth molar-based moment, $M_{n,k}$, of the polymer sample and vice versa. The $M_{w,k}$ is expressed in terms of $M_{n,k}$ by substituting for $P_{DPw}(i)$ in Eq. (3.9) from Eq. (3.27) as follows:

$$M_{w,k} = \langle DP_w^k \rangle = \sum_{i=1}^{\infty} i^k P_{DPw}(i) = \sum_{i=1}^{\infty} i^k \frac{i P_{DPn}(i)}{M_{n,1}} = \frac{M_{n,k+1}}{M_{n,1}} = \frac{Q_{n,k+1}}{Q_{n,1}} \quad (3.34)$$

In the above equation, $M_{w,k}$ is related to $M_{n,k+1}$ or $Q_{n,k+1}$. Using the same approach, $M_{n,k}$ is expressed in terms of $M_{w,k}$ by substituting for $P_{DPn}(i)$ in Eq. (3.8) from Eq. (3.28) as follows:

$$M_{n,k} = \langle DP_n^k \rangle = \sum_{i=1}^{\infty} i^k P_{DPn}(i) = \sum_{i=1}^{\infty} i^k \frac{i^{-1} P_{DPw}(i)}{M_{w,-1}} = \frac{M_{w,k-1}}{M_{w,-1}} = \frac{Q_{w,k-1}}{Q_{w,-1}} \quad (3.35)$$

In the above equation, $M_{n,k}$ is related to $M_{w,k}$ or $Q_{w,k}$. It can be shown that higher moments like the kth order z or $z+1$ moment are expressed in terms of the kth order molar-based moments or weight-based moments using the following relations:

$$M_{z,k} = \sum_{i=1}^{\infty} i^k P_{DPz}(i) = \frac{M_{n,k+2}}{M_{n,2}} = \frac{Q_{n,k+2}}{Q_{n,2}} = \frac{M_{w,k+1}}{M_{w,1}} = \frac{Q_{w,k+1}}{Q_{w,1}} \quad (3.36)$$

$$M_{z+1,k} = \sum_{i=1}^{\infty} i^k P_{DPz+1}(i) = \frac{M_{n,k+3}}{M_{n,3}} = \frac{Q_{n,k+3}}{Q_{n,3}} = \frac{M_{w,k+2}}{M_{w,2}} = \frac{Q_{w,k+2}}{Q_{w,2}} \quad (3.37)$$

3.4.1 Example Problem 1

The distribution of chains in two polymer samples S1 (white bars) and S2 (gray bars) in terms of chain concentration C_i in moles L^{-1} with degree of polymerization DP_i is plotted in Figure 3.1. Sample S1 has a unimodal distribution while sample S2 has a bimodal distribution with a small fraction of high DP chains. Determine $M_{n,k}$, $Q_{n,k}$, $M_{w,k}$, and $M_{z,k}$ for $k = 0$, 1 *and* 2 and compare $P_{DPn}(i)$, $P_{DPw}(i)$ and $P_{DPz}(i)$ of the two polymer samples.

Solution: The chain distributions in Figure 3.1 are tabulated in Table 3.1. The total concentrations of the two samples are given by the following equations:

$$C(S1) = \sum C_i(S1) = 39.1\ mole/L \qquad C(S2) = 53.6\ mole/L \quad (3.38)$$

The mole-fraction chain distribution $P_{DPn}(i)$ of the two samples, calculated using Eqs. (3.12) and (3.38), and the data in Table 3.1 are listed in the second and fifth columns of Table 3.2.

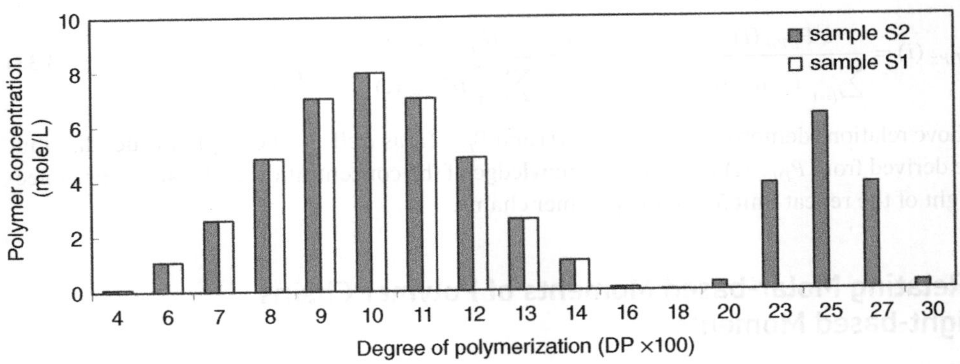

Figure 3.1 Distribution of chains in polymer samples S1 and S2.

Table 3.1 Molar concentration of chains as a function of DP_i for polymer samples S1 and S2 of Example 3.1.

	Sample S1	Sample S2		Sample S1	Sample S2
DP(i)	Ci(mol lit^{-1})	Ci(mol lit^{-1})	DP(i)	Ci(mol lit^{-1})	Ci(mol lit^{-1})
400	0.088	0.088	1400	1.078	1.078
600	1.078	1.078	1600	0.088	0.089
700	2.585	2.585	1800	0.003	0.017
800	4.829	4.829	2000	0.000	0.280
900	7.026	7.026	2300	0.000	3.863
1000	7.962	7.962	2500	0.000	6.369
1100	7.026	7.026	2700	0.000	3.863
1200	4.829	4.829	3000	0.000	0.280
1300	2.585	2.585			

Table 3.2 The P_{DPn} (i), P_{DPw} (i), and P_{DPz} (i) chain distributions for polymer samples S1 and S2 of Example 3.1.

	S1	S1	S1	S2	S2	S2
DP(i)	P_{DPn}(i)	P_{DPw}(i)	P_{DPz}(i)	P_{DPn}(i)	P_{DPw}(i)	P_{DPz}(i)
600	0.028	0.017	0.010	0.002	0.001	0.000
700	0.066	0.046	0.031	0.020	0.009	0.003
800	0.124	0.099	0.076	0.048	0.025	0.011
900	0.180	0.162	0.140	0.090	0.053	0.025
1000	0.204	0.203	0.196	0.131	0.085	0.045
1100	0.180	0.197	0.210	0.149	0.106	0.062
1200	0.124	0.148	0.171	0.131	0.102	0.065
1300	0.066	0.086	0.108	0.090	0.076	0.052
1400	0.028	0.039	0.052	0.048	0.044	0.032
1600	0.002	0.004	0.006	0.020	0.021	0.018
1800	0.000	0.000	0.000	0.002	0.002	0.002
2000	0.000	0.000	0.000	0.000	0.000	0.000
2300	0.000	0.000	0.000	0.005	0.008	0.009
2500	0.000	0.000	0.000	0.072	0.117	0.155
2700	0.000	0.000	0.000	0.119	0.209	0.298
3000	0.000	0.000	0.000	0.072	0.141	0.223
sum	1.000	1.000	1.000	1.000	1.000	1.000

The weight fraction chain distribution P_{DPw} (i) of the two samples, calculated using Eqs. (3.18) and (3.27), and the data in Table 3.2 are listed in the third and sixth columns of Table 3.2, respectively. The z-based P_{DPz} (i) chain distribution of the two samples, calculated by Eqs. (3.19) and (3.33), and the data in Table 3.2 are listed in the fourth and seventh columns of Table 3.2, respectively.

The $Q_{n,0}$, given by Eq. (3.16), is equal to the total polymer concentration for each of the two samples (see Eq. 3.38). The molar-based moments $M_{n,1}$, $Q_{n,1}$, $M_{n,2}$, and $Q_{n,2}$ are given by Eqs. (3.18),

(3.20), (3.22), and (3.24), respectively. The $M_{n,1}$, $Q_{n,1}$, $M_{n,2}$, and $Q_{n,2}$ moments of the two polymers are determined as follows:

$$\begin{cases} M_{n,1}(S1) = \sum_{i=1}^{+\infty} iP_{Dpn}(S1, i) = 1000 & M_{n,1}(S2) = 1537 \\ Q_{n,1}(S1) = C(S1)\, M_{n,1}(S1) = 3.91 \times 10^4 & Q_{n,1}(S2) = 8.23 \times 10^4\ mole/L \\ M_{n,2}(S1) = \sum_{i=1}^{+\infty} i^2 P_{Dpn}(S1, i) = 1.03 \times 10^6 & M_{n,2}(S2) = 2.91 \times 10^6 \\ Q_{n,2}(S1) = C(S1)\, M_{n,2}(S1) = 4.06 \times 10^7 & Q_{n,2}(S2) = 1.56 \times 10^8\ mole/L \end{cases} \quad (3.39)$$

The weight-based $M_{w,1}$ and $M_{w,2}$ moments for the two polymer samples are determined using Eqs. (3.19) and (3.23), respectively, as follows:

$$\begin{cases} M_{w,1}(S1) = \sum_{i=1}^{+\infty} iP_{DPw}(S1, i) = 1036 & M_{w,1}(S2) = 1895 \\ M_{w,2}(S1) = \sum_{i=1}^{+\infty} i^2 P_{DPw}(S1, i) = 1.11 \times 10^6 & M_{w,2}(S2) = 4.26 \times 10^6 \end{cases} \quad (3.40)$$

The z-based $M_{z,1}$ and $M_{z,2}$ moments of the two polymer samples are determined using Eq. (3.30) as follows:

$$\begin{cases} M_{z,1}(S1) = \sum_{i=1}^{+\infty} iP_{DPz}(S1, i) = 1070 & M_{z,1}(S2) = 2250 \\ M_{z,2}(S1) = \sum_{i=1}^{+\infty} i^2 P_{DPz}(S1, i) = 1.18 \times 10^6 & M_{z,2}(S2) = 5.64 \times 10^6 \end{cases} \quad (3.41)$$

The $P_{DPn}(i)$, $P_{DPw}(i)$, and $P_{DPz}(i)$ chain distributions for polymer samples S1 and S2 are compared in Figure 3.2. The distribution for the unimodal polymer sample S1 shifts slightly to higher DP values from molar to weight to z (top figure). The distribution for the bimodal polymer sample S2 not only shifts slightly to the right but also the intensity of the second (higher) peak of degree of polymerization increases as compared to the first (lower) peak (bottom figure).

3.5 Relating Number-average, Weight-average, and z-average Degree of Polymerization of Polymer Chains to Molar-based Moments of Chain Distribution

The moments of a polymer distribution are used to determine the different average molecular weights or the degree of polymerization. The number-average degree of polymerization $\left(\overline{DP}_n\right)$ of a polymer chain distribution is defined by the following relations:

$$\overline{DP}_n = \sum_{i=1}^{\infty} i\, P_{DPn}(i) = M_{n,1} = \frac{M_{n,1}}{M_{n,0}} = \frac{C\, M_{n,1}}{C\, M_{n,0}} = \frac{Q_{n,1}}{Q_{n,0}} \quad (3.42)$$

According to the above equation, moments based on molar probability distribution $(M_{n,1})$ as well as molar concentration distribution ($Q_{n,1}$ and $Q_{n,0}$) can be used to calculate $\overline{DP}_n$ of a polymer chain

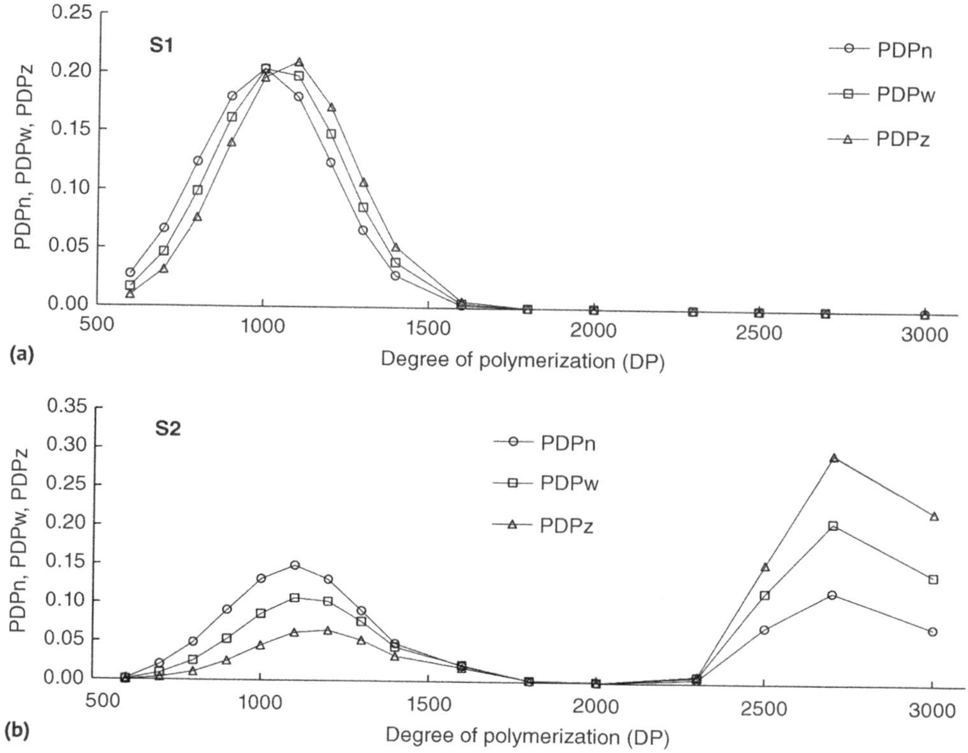

Figure 3.2 The $P_{DPn}(i)$, $P_{DPw}(i)$, and $P_{DPz}(i)$ distributions for polymer samples S1 and S2 of Example 3.1.

distribution. Moments based on weight probability distribution may also be used to calculate $\overline{DP}_n$ by substituting for $P_{DPn}(i)$ in Eq. (3.18) from Eq. (3.28) as follows:

$$\overline{DP}_n = \sum_{i=1}^{\infty} i\,P_{DPn}(i) = \sum_{i=1}^{\infty} i\frac{i^{-1}P_{DPn}(i)}{M_{w,-1}} = \frac{1}{M_{w,-1}} = \frac{C}{CM_{w,-1}} = \frac{Q_{w,0}}{Q_{w,-1}} \tag{3.43}$$

The weight-average degree of polymerization $\left(\overline{DP}_w\right)$ of a polymer chain distribution is defined by the following relations:

$$\overline{DP}_w = \sum_{i=1}^{\infty} i\,P_{DPw}(i) = M_{w,1} = \frac{M_{w,1}}{M_{w,0}} = \frac{\rho M_{w,1}}{\rho M_{w,0}} = \frac{Q_{w,1}}{Q_{w,0}} \tag{3.44}$$

According to the above equation, moments based on weight probability distribution $(M_{w,1})$ as well as mass density distribution ($Q_{w,1}$ and $Q_{w,0}$) can be used to calculate $\overline{DP}_w$ of a polymer chain distribution. $\overline{DP}_w$ can also be expressed in terms of moments based on molar probability distribution by substituting for $P_{DPw}(i)$ in Eq. (3.44) from Eq. (3.27) as follows:

$$\overline{DP}_w = \sum_{i=1}^{\infty} i\,P_{DPw}(i) = \sum_{i=1}^{\infty} i\frac{i\,P_{DPn}(i)}{M_{n,1}} = \frac{M_{n,2}}{M_{n,1}} = \frac{CM_{n,2}}{CM_{n,1}} = \frac{Q_{n,2}}{Q_{n,1}} \tag{3.45}$$

In a similar approach, it can be shown that the z-average degree of polymerization $\left(\overline{DP}_z\right)$ is defined in terms of z moments or the moments based on molar or weight probability distribution of the polymer chains by the following relations:

$$\overline{DP}_z = \sum_{i=1}^{\infty} i\, P_{DPz}\,(i) = M_{z,1} = \frac{M_{w,2}}{M_{w,1}} = \frac{Q_{w,2}}{Q_{w,1}} = \frac{M_{n,3}}{M_{n,2}} = \frac{Q_{n,3}}{Q_{n,2}} \tag{3.46}$$

Analysis of Eqs. (3.43–3.46) shows that the different DP averages of a polymer chain distribution can be expressed in terms of $M_{n,k}$, $M_{w,k}$, $M_{z,k}$ or in terms of $Q_{n,k}$, $Q_{w,k}$, $Q_{z,k}$ using the following generalized equations:

$$\overline{DP}_j = \frac{M_{n,j}}{M_{n,j-1}} = \frac{M_{w,j-1}}{M_{w,j-2}} = \frac{M_{z,j-2}}{M_{z,j-3}} = \frac{M_{z+1,j-3}}{M_{z+1,j-4}} \tag{3.47}$$

$$\overline{DP}_j = \frac{Q_{n,j}}{Q_{n,j-1}} = \frac{Q_{w,j-1}}{Q_{w,j-2}} = \frac{Q_{z,j-2}}{Q_{z,j-3}} = \frac{Q_{z+1,j-3}}{Q_{z+1,j-4}} \tag{3.48}$$

where M and Q represent the fraction-based and concentration-based moments of a polymer chain distribution, respectively; n, w, z, and $z+1$ stand for molar (number), weight, z, and z+1 distributions, j stands for the jth moment of the distribution, and $\overline{DP}_{j=1} = \overline{DP}_n$, $\overline{DP}_{j=2} = \overline{DP}_w$, $\overline{DP}_{j=3} = \overline{DP}_z$, and $\overline{DP}_{j=4} = \overline{DP}_{z+1}$.

3.5.1 Example Problem 2

Consider the two polymer samples S1 and S2 in Example 3.1. Determine $\overline{DP}_n$, $\overline{DP}_w$, and $\overline{DP}_z$ for the unimodal and bimodal polymers S1 and S2, respectively.

Solution: $\overline{DP}_n$, $\overline{DP}_w$, and $\overline{DP}_z$ are calculated using the data in Table 3.2 and Eqs. (3.42), (3.44), and (3.46), respectively, as follows:

$$\begin{cases} \overline{DP}_n\,(S1) = \sum_{i=1}^{+\infty} iP_{DPn}\,(S1, i) = M_{n,1}\,(S1) = 1001 \quad \overline{DP}_n\,(S2) = 1537 \\ \overline{DP}_w\,(S1) = \sum_{i=1}^{+\infty} iP_{DPw}\,(S1, i) = M_{w,1}\,(S1) = 1036 \quad \overline{DP}_w\,(S2) = 1895 \\ \overline{DP}_z\,(S1) = \sum_{i=1}^{+\infty} iP_{DPz}\,(S1, i) = M_{z,1}\,(S1) = 1069 \quad \overline{DP}_z\,(S2) = 2250 \end{cases} \tag{3.49}$$

Analysis of the different $\overline{DP}$ values for the two polymer samples demonstrate that the higher order averages, like $\overline{DP}_z$ are markedly affected by the higher DP fractions in bimodal distributions or distributions with a high DP tail. The $\overline{DP}_n$, $\overline{DP}_w$, and $\overline{DP}_z$ for the unimodal sample S1 increased slightly from 1001 to 1036 and 1069, respectively, whereas those for sample S2 markedly increased from 1537 to 1895 and 2250. Further, higher-order DP averages amplify the contribution of longer chains in a polymer chain distribution.

3.6 Relating Viscosity-average Degree of Polymerization of Polymer Chains to Molar-based Moments of Chain Distribution

The viscosity-average degree of polymerization, $\overline{DP}_\eta$, is an average obtained from measurements of the viscosity of a polymer sample dissolved in a good solvent [21–25]. The viscosity of a polymer solution depends on the degree of polymerization as well as chain architecture, polymer concentration, solvent type, shear rate, and temperature [26]. As the polymer concentration in solution is reduced to approach zero, the measured viscosity becomes the intrinsic viscosity of the polymer, represented by $[\eta]$, which is dependent on polymer molecular weight, solvent type, and temperature but independent of concentration [27–30]. Intrinsic viscosity is determined by extrapolating the data for viscosity versus polymer concentration in the low concentration range to zero polymer concentration [31–33]. According to Mark–Houwink–Sakurada (MHS) equation (also known as Khun–Mark–Houwink–Sakurada equation) [34–36], the intrinsic viscosity of a monodisperse polymer, $[\eta]_i$, is related to molecular weight, M_i, or the degree of polymerization, DP_i, by the following relations:

$$[\eta]_i = K_\eta M_i^a = K_\eta (M_0 DP_i)^a = K_\eta M_0^a DP_i^a \tag{3.50}$$

where M_0 is molecular weight of the polymer repeat unit and the viscosity constant K_η and exponent a depend on solvent type and temperature. Accordingly, the viscosity of a polydisperse polymer solution is defined by replacing DP_i in Eq. (3.50) with the viscosity-average degree of polymerization, $\overline{DP}_\eta$, as follows [37]:

$$[\eta] = K_\eta \overline{M}_\eta^{\,a} = K_\eta \left(M_0 \overline{DP}_\eta\right)^a = K_\eta M_0^a \overline{DP}_\eta^{\,a} \tag{3.51}$$

where $\overline{M}_\eta$ is the viscosity-average molecular weight. Since the viscosity of a polymer solution depends strongly on the size of polymer chains in solution, the weight distribution, $P_{DPw}(i)$, is used to relate intrinsic viscosity to the degree of polymerization of a polydisperse polymer solution as follows:

$$[\eta] = \sum_{i=1}^{\infty} [\eta]_i \, P_{DPw}(i) = K_\eta M_0^a \sum_{i=1}^{\infty} i^a P_{DPw}(i) \tag{3.52}$$

Upon equating the right side of Eqs. (3.51) and (3.52), $\overline{DP}_\eta$ is defined by the following equation:

$$\overline{DP}_\eta^{\,a} = \sum_{i=1}^{\infty} i^a P_{DPw}(i) \tag{3.53}$$

Upon substitution for $P_{Dpw}(i)$ in the above equation from Eq. (3.27), the following relation is obtained between $\overline{DP}_\eta$ and moments of the chain distribution of a polymer:

$$\overline{DP}_\eta = \left[\sum_{i=1}^{\infty} i^a P_{DPw}(i)\right]^{1/a} = \left[\frac{\sum_{i=1}^{\infty} i^{1+a} P_{DPn}(i)}{M_{n,1}}\right]^{1/a} = \left[\frac{M_{n,1+a}}{M_{n,1}}\right]^{1/a} = \left[\frac{Q_{n,1+a}}{Q_{n,1}}\right]^{1/a} \tag{3.54}$$

where $M_{n,1+a}$ and $Q_{n,1+a}$ are the $1 + a$ order moments based on mole-fraction and concentration distribution of the polymer chains, respectively. As the exponent $0 < a < 1$ for most polymers, $\overline{DP}_\eta$ falls between that of $\overline{DP}_n$ and $\overline{DP}_w$. Therefore, the order of different averages of $\overline{DP}$ is as follows:

$$\overline{DP}_n < \overline{DP}_\eta < \overline{DP}_w < \overline{DP}_z < \overline{DP}_{z+1} \tag{3.55}$$

3.6.1 Example Problem 3

The viscosity exponent a in MHS equation for unimodal and bimodal polymer samples S1 and S2 in Example 3.1 is $a = 0.77$. Determine $M_{n,1+a}$, $Q_{n,1+a}$ and $\overline{DP}_\eta$ for the two polymer samples S1 and S2.

Solution: As exponent $a = 0.77$, the moment $M_{n,1.77}$ is given by the following equation:

$$M_{n,1.77} = \sum_{i=1}^{\infty} i^{1+0.77} P_{DPn}(i) = \sum_{i=1}^{\infty} i^{1.77} P_{DPn}(i) \tag{3.56}$$

Using the data in Table 3.2, the values of $i^{1.77} P_{DPn}(i)$ versus DP_i for samples S1 and S2 are given in Table 3.3. Using the values of $i^{1.77} P_{DPn}(i)$ and Eq. (3.56), the viscosity weight fraction probability distribution of the chains is also given in Table 3.3.

The calculations for $M_{n,1.77}$, $Q_{n,1.77}$ and $\overline{DP}_\eta$ for the two polymer samples S1 and S2 are as follows:

$$M_{n,1.77}(S1) = \sum_{i=1}^{+\infty} i^{1+0.77} P_{DPn,S1}(i) = 2.1 \times 10^5 \tag{3.57}$$
$$M_{n,1.77}(S2) = 5.0 \times 10^5$$

$$Q_{n,1.77} = C(S1)\, M_{n,1.77}(S1) = 39.1 \times 2.1 \times 10^5 = 8.2 \times 10^6 mol/L \tag{3.58}$$
$$Q_{n,1.77} = 2.7 \times 10^7 mol/L$$

$$\overline{DP}_\eta(S1) = \left[\frac{M_{n,1.77}(S1)}{M_{n,1}(S1)}\right]^{1/2} = \left[\frac{2.1 \times 10^5}{1001}\right]^{1/2} = 1030 \tag{3.59}$$
$$\overline{DP}_\eta(S2) = 1850$$

Table 3.3 The DP_i, $i^{1.77} P_{DPn}(i)$, and $P_{DP\eta}(i)$ probability chain distribution for polymer samples S1 and S2 of Example 3.1.

	S1	S1	S1	S2	S2	S2
DP(i)	Pdpn(i)	$i^{1.77}$*Pdpn(i)	Pdpη(i)	Pdpn(i)	$i^{1.77}$*Pdn(i)	Pdpη(i)
600.00	0.03	2278.80	0.01	0.00	136.49	0.00
700.00	0.07	7181.40	0.03	0.02	2184.48	0.00
800.00	0.12	16993.74	0.08	0.05	6637.45	0.01
900.00	0.18	30457.16	0.15	0.09	15274.78	0.03
1000.00	0.20	41587.89	0.20	0.13	26780.97	0.05
1100.00	0.18	43445.54	0.21	0.15	35923.45	0.07
1200.00	0.12	34831.36	0.17	0.13	36980.86	0.07
1300.00	0.07	21481.54	0.10	0.09	29285.02	0.06
1400.00	0.03	10209.97	0.05	0.05	17872.22	0.04
1600.00	0.00	1061.53	0.01	0.02	9436.63	0.02
1800.00	0.00	39.49	0.00	0.00	956.91	0.00
2000.00	0.00	0.53	0.00	0.00	215.84	0.00
2300.00	0.00	0.00	0.00	0.01	4659.48	0.01
2500.00	0.00	0.00	0.00	0.07	74543.47	0.15
2700.00	0.00	0.00	0.00	0.12	140837.04	0.28
3000.00	0.00	0.00	0.00	0.07	102934.37	0.20
sum		209568.94	1.00	1.00	504659.46	1.00

3.7 Variance and Polydispersity of a Polymer Chain Distribution

Polymer properties depend not only on the average values of the chain distribution but also on breadth of the distribution. In probability theory, the breadth of a distribution is defined by variance. The variance of a random variable X, represented by V_X, is defined by the following equation [38]:

$$V_X = \left\langle [\langle X \rangle - X]^2 \right\rangle = \langle X^2 \rangle - \langle X \rangle^2 \tag{3.60}$$

where X is a randomly distributed variable, $\langle X \rangle$ is the average value or the first distribution moment of the variable X, $\langle X^2 \rangle$ is the second distribution moment of the variable X, and $\left\langle [\langle X \rangle - X]^2 \right\rangle$ is the average of the square of the differences between each value of X and its average. Based on the defining Eq. (3.60), the variance of the degree of polymerization of a polymer chain distribution, represented by V_{DP}, is defined by the following equation:

$$V_{DP} = \langle DP^2 \rangle - \langle DP \rangle^2 = \sum_{i=1}^{\infty} i^2 P_{DPn}(i) - \left(\sum_{i=1}^{\infty} i P_{DPn}(i) \right)^2 \tag{3.61}$$

where $\langle DP \rangle = \overline{DP}_n$ and $\langle DP^2 \rangle$ are the average and the squared average of the degree of polymerization, respectively. Upon substitution for the summations in the above equation using Eqs. (3.18) and (3.22), V_{DP} is given by the following equation in terms of moments of the distribution:

$$V_{DP} = M_{n,2} - (M_{n,1})^2 \tag{3.62}$$

Upon substitution for $M_{n,1}$ and $M_{n,2}$ in the above equation, using Eqs. (3.42) and (3.45), V_{DP} is expressed in terms of $\overline{DP}_n$ and $\overline{DP}_w$ as follows:

$$V_{DP} = \overline{DP}_w \overline{DP}_n - \left(\overline{DP}_n\right)^2 = \overline{DP}_n^{\,2} \left(\frac{\overline{DP}_w}{\overline{DP}_n} - 1 \right) \tag{3.63}$$

The standard deviation from the mean for the degree of polymerization of a polymer sample, represented by σ_{DP}, is equal to the square root of the variance, which is given by the following equation:

$$\sigma_{DP} = \sqrt{V_{DP}} = \overline{DP}_n \left(\frac{\overline{DP}_w}{\overline{DP}_n} - 1 \right)^{1/2} \tag{3.64}$$

The standard deviation is a measure of the breadth of a distribution in combination with the mean value. In other words, the standard deviation as defined by the above equation depends on the mean, $\overline{DP}_n$. To make it independent of the mean value, the normalized standard deviation, represented by Δ, is defined by dividing σ_{DP} by $\overline{DP}_n$ as follows:

$$\Delta = \frac{\sigma_{DP}}{\overline{DP}_n} = \sqrt{\frac{\overline{DP}_w}{\overline{DP}_n} - 1} = \sqrt{\frac{\overline{M}_w}{\overline{M}_n} - 1} \tag{3.65}$$

The parameter Δ, which is a function of $\overline{M}_n$ and $\overline{M}_w$, is a statistical measure of the breadth or dispersity of the polymer molecular weight distribution. In practice, the polydispersity index, represented by PDI, is used to describe the breadth of a polymer chain distribution, which is the ratio of weight-average to number-average molecular weights as follows:

$$PDI = \frac{\overline{DP}_w}{\overline{DP}_n} = \frac{\overline{M}_w}{\overline{M}_n} \tag{3.66}$$

Upon comparing Eqs. (3.65) and (3.66), PDI is related to Δ by the following equation:

$$PDI = \frac{\overline{DP}_w}{\overline{DP}_n} = \frac{\overline{M}_w}{\overline{M}_n} = \Delta^2 + 1 \tag{3.67}$$

According to Eq. (3.67), for an ideal monodisperse polymer $\Delta = 0$ and $PDI = 1$. It should be noted that Δ is a true statistical measure of the breadth of a distribution because it is derived from variance.

3.7.1 Example Problem 4

Determine V_{DP}, σ_{DP}, Δ, and PDI for unimodal S1 and bimodal S2 polymer samples of Example 3.1.

Solution: V_{DP}, σ_{DP}, Δ and PDI are defined by Eqs. (3.63–3.66), respectively. From Example 3.2, $\overline{DP}_w$ and $\overline{DP}_n$ for samples S1 are 1036 and 1001, respectively, and for sample S2 are 1895 and 1537. The values of V_{DP}, σ_{DP}, Δ, and PDI for polymer samples S1 and S2 are calculated by the following equations:

$$V_{DP}(S1) = M_{n,2}(S1) - \left[M_{n,1}(S1)\right]^2 = 1.03 \times 10^6 - (1001)^2 = 3.5 \times 10^4$$
$$V_{DP}(S2) = 5.5 \times 10^5 \tag{3.68}$$

$$\sigma_{DP}(S1) = \sqrt{V_{DP}(S1)} = \sqrt{3.5 \times 10^4} = 187 \quad \sigma_{DP}(S2) = 742 \tag{3.69}$$

$$\Delta(S1) = \sqrt{\frac{\overline{DP}_w}{\overline{DP}_n} - 1} = \sqrt{\frac{1036}{1001} - 1} = 0.19 \qquad \Delta(S2) = 0.48 \tag{3.70}$$

$$PDI(S1) = \Delta^2 + 1 = 0.19^2 + 1 = 1.03 \qquad PDI(S2) = 1.23 \tag{3.71}$$

The calculated Δ values show that breadth of the distribution for bimodal sample S2 is significantly larger than the unimodal sample S1.

3.8 Skewness and Kurtosis of a Polymer Chain Distribution

Side reactions, blending, copolymerization, or the use of multifunctional initiators can lead to asymmetric distribution of degree of polymerization of polymer chains [39–42]. Polymer properties are controlled not only by the average molecular weight and polydispersity but also by the shape of the molecular weight distribution [16, 43]. The asymmetry in the shape of a polymer chain distribution is based on deviation from the normal or Gaussian distribution as defined by skewness and kurtosis [16, 44, 45]. Skewness is a measure of shape asymmetry, whereas kurtosis is a measure of the skewness of a molecular weight distribution as compared to a normal distribution [46, 47]. A normal or Gaussian distribution is perfectly symmetric with respect to the mean with expectation μ and standard deviation σ as defined by the following equation for the distribution of degree of polymerization [48]:

$$P_{gaus}(i) = \frac{1}{\sigma\sqrt{2\pi}} e^{-\frac{1}{2}\left(\frac{i-\mu}{\sigma}\right)^2} \qquad -\infty < i < +\infty \tag{3.72}$$

where $P_{gaus}(i)$ is the probability function of a normal distribution of $DP = i$ with i ranging from negative infinity to positive infinity. For a normal distribution, 68.2% of the DP values fall within one σ, 95.4% fall within 2σ, and 99.7% fall within 3σ. Further, for a normal distribution, the expectation value $\overline{DP}_n$ is equal to the mode, or peak value (DP_P) and the mean value (DP_M). The distribution can be negatively or positively skewed depending on whether the left or right tail is longer with respect to DP_P of the distribution. If the left tail is longer, the distribution is negatively skewed. Conversely, if the right tail is longer, the distribution is positively skewed. In mathematical form, the skewness index (SI) is defined by the expectation of the third power of DP with respect to DP_P using the following equation:

$$SI = E\left[\left(\frac{i - DP_P}{\sigma_{DP}}\right)^3\right] = \sum_{i=1}^{\infty}\left(\frac{i - DP_P}{\sigma_{DP}}\right)^3 P_{DPn}(i) = \sum_{i=1}^{\infty}\left(\frac{i - DP_P}{\overline{DP}_n\Delta}\right)^3 P_{DPn}(i) \tag{3.73}$$

where Eq. (3.65) is used to express σ_{DP} in terms of Δ. Upon removing Δ from the summation, expanding the term $(i - DP_p)^3$, and using the defining Eq. (3.8) for k^{th} molar-based moments of a distribution, SI is given by the following equation:

$$SI = \left(\frac{DP_P}{\overline{DP}_n\Delta}\right)^3 \left[\frac{M_{n,3}}{DP_P{}^3} - 3\frac{M_{n,2}}{DP_P{}^2} + 3\frac{M_{n,1}}{DP_P} - 1\right] \tag{3.74}$$

where $M_{n,k}$ is the kth order molar-based moment of the distribution. Upon expressing $M_{n,k}$ in terms of $\overline{DP}_z$, $\overline{DP}_w$, and $\overline{DP}_n$ using Eq. (3.47), SI is expressed in terms of jth average degree of polymerization $\left(\overline{DP}_j\right)$ by the following equation:

$$SI = \left(\frac{DP_P}{\overline{DP}_n\Delta}\right)^3 \left[\frac{\overline{DP}_n{}^3}{DP_P{}^3}\frac{\overline{DP}_z}{\overline{DP}_w}PDI^2 - 3\frac{\overline{DP}_n{}^2}{DP_P{}^2}PDI + 3\frac{\overline{DP}_n}{DP_P} - 1\right] \tag{3.75}$$

Depending on the magnitude of DP_P in relation to $\overline{DP}_n$, SI can be negative or positive. If $DP_P > \overline{DP}_n$, the left tail is longer, the distribution is negatively skewed, and $SI < 0$. Conversely, if $DP_P < \overline{DP}_n$, the right tail is longer, the distribution is positively skewed, and $SI > 0$. If $DP_P = \overline{DP}_n$, the distribution is symmetric and $SI = 0$.

Kurtosis is a measure of skewness (flat versus peaked) of a polymer distribution with respect to the normal distribution [46, 47]. As kurtosis is a measure of flatness, as opposed to asymmetry, it should be defined with respect to the number-average degree of polymerization $\left(\overline{DP}_n\right)$. In mathematical form, the kurtosis index (KI) is defined by the expectation of the fourth power of DP around DP_p using the following expression [46, 47]:

$$KI = E\left[\left(\frac{i - DP_P}{\sigma_{DP}}\right)^4\right] = \sum_{i=1}^{\infty}\left(\frac{i - DP_P}{\sigma_{DP}}\right)^4 P_{DPn}(i) = \sum_{i=1}^{\infty}\left(\frac{i - DP_P}{\overline{DP}_n\Delta}\right)^4 P_{DPn}(i) \tag{3.76}$$

Upon removing Δ from the summation, expanding the term $(i - DP_P)^4$, and using the defining Eq. (3.8) for k^{th} molar-based moments of a distribution, KI is expressed by the following equation:

$$KI = \left(\frac{DP_P}{\overline{DP}_n\Delta}\right)^4 \left[\frac{M_{n,4}}{DP_P{}^4} - 4\frac{M_{n,3}}{DP_P{}^3} + 6\frac{M_{n,2}}{DP_P{}^2} - 4\frac{M_{n,1}}{DP_P} + 1\right] \tag{3.77}$$

Upon expressing $M_{n,k}$ in terms of $\overline{DP}_{z+1}$, $\overline{DP}_z$, $\overline{DP}_w$ and $\overline{DP}_n$ using Eq. (3.47), SI is expressed in terms of $\overline{DP}_j$ by the following equation:

$$KI = \left(\frac{DP_P}{\overline{DP}_n \Delta}\right)^4 \left[\frac{\overline{DP}_n^{\,4}}{DP_P^{\,4}} \frac{\overline{DP}_{z+1}\overline{DP}_z}{\overline{DP}_w^{\,2}} PDI^3 - 4\frac{\overline{DP}_n^{\,3}}{DP_P^{\,3}} \frac{\overline{DP}_z}{\overline{DP}_w} PDI^2 + 6\frac{\overline{DP}_n^{\,2}}{DP_P^{\,2}} PDI - 4\frac{\overline{DP}_n}{DP_P} + 1\right] \quad (3.78)$$

Kurtosis index can be negative, zero, or positive. If the polymer distribution is flatter than the normal distribution, the kurtosis index is negative (mesokurtic). If the distribution is more peaked/elevated than the normal, the kurtosis index is positive (leptokurtic). If $KI = 0$, the distribution is neither flat nor peaked.

3.8.1 Example Problem 5

The distribution S1 in Example 3.1 is unimodal with a peak at $DP = 1000$, and it can be treated like a normal distribution with or without asymmetry. The distribution S2 in Example 3.1 is bimodal with a relatively larger peak at $DP = 1000$ and a smaller peak at $DP = 2500$. If the S2 distribution can be treated as an asymmetric normal distribution with a main peak at $DP = 1000$, and a tail to the right centered at $DP = 2500$ determine the skewness and kurtosis indexes of S1 and S2 polymer samples.

Solution: Skewness index and Kurtosis index are defined by Eqs. (3.75) and (3.78), respectively. Based on previous examples, we have the following data for distributions S1 and S2:

$$\begin{cases} \overline{DP}_n(S1) = 1001 & \overline{DP}_w(S1) = 1036 & \overline{DP}_z(S1) = 1069 \\ DP_P(S1) = 1000 & PDI(S1) = 1.03 & \Delta(S1) = 0.19 \end{cases} \quad (3.79)$$

$$\begin{cases} \overline{DP}_n(S1) = 1537 & \overline{DP}_w(S1) = 1895 & \overline{DP}_z(S1) = 2250 \\ DP_P(S1) = 1000 & PDI(S1) = 1.23 & \Delta(S1) = 0.48 \end{cases} \quad (3.80)$$

$\overline{DP}_{z+1}$ can be calculated from the number distributions of S1 and S2 in Table 3.3 using the defining Eq. (3.37) as follows:

$$\begin{aligned} \overline{DP}_{z+1}(S1) &= \frac{M_{n,4}(S1)}{M_{n,3}(S1)} = \frac{1.221\times10^{12}}{1.110\times10^{9}} = 1100 \\ \overline{DP}_{z+1}(S2) &= \frac{M_{n,4}(S2)}{M_{n,3}(S2)} = \frac{1.643\times10^{13}}{6.556\times10^{9}} = 2505 \end{aligned} \quad (3.81)$$

Based on the defining Eq. (3.75), SI for distributions S1 and S2 is equal to the following values:

$$SI(S1) = \left(\frac{1000}{1001\,(0.19)}\right)^3 \left[\frac{1001^3}{1000^3}\frac{1069}{1036}1.03^2 - 3\frac{1001^2}{1000^2}1.03 + 3\frac{1001}{1000} - 1\right] = +0.70$$

$$SI(S2) = \left(\frac{1000}{1537\,(0.48)}\right)^3 \left[\frac{1537^3}{1000^3}\frac{2250}{1895}1.23^2 - 3\frac{1537^2}{1000^2}1.23 + 3\frac{1537}{1000} - 1\right] = +3.53$$

Based on Eq. (3.78), KI for distributions S1 and S2 are equal to the following values:

$$KI\,(S1) = \left(\frac{1000}{1001\,(0.19)}\right)^4 \left[\frac{1001^4}{1000^4}\frac{1100\,(1069)}{1036^2}1.03^3 - 4\frac{1001^3}{1000^3}\frac{1069}{1036}1.03^2 + 6\frac{1001^2}{1000^2}1.03 - 4\frac{1001}{1000} + 1\right] = -1.20$$

$$KI\,(S2) = \left(\frac{1000}{1537\,(0.48)}\right)^4 \left[\frac{1537^4}{1000^4}\frac{2505\,(2250)}{1895^2}1.23^3 - 4\frac{1537^3}{1000^3}\frac{2250}{1895}1.23^2 + 6\frac{1537^2}{1000^2}1.23 - 4\frac{1537}{1000} + 1\right] = -2.58$$

Based on the calculations, polymer sample S1 has a slightly longer tail to the right of the peak DP (SI > 0) and its distribution is flatter (KI < 0) as compared to a normal distribution. The sample S2, due to having a second peak centered at DP = 2500, has a longer tail to the right of the peak DP (SI > 0) and its distribution is more flat (KI < 0) with respect to the normal distribution as compared to sample S1.

3.9 Standard Distribution Functions Used in Polymerization Reactions

The chains in a polymer sample have a distribution of molecular weights or degrees of polymerization ranging from one monomer to countably infinite number of monomer residues. The likelihood of finding a molecule with a specified degree of polymerization is given by the probability distribution function for the polymer sample. Standard distribution functions are used in statistics to mathematically define the distribution of random variables, fit the experimental data, and find moments, averages, and variance of distributions. A good example is the normal distribution function, also called the Gaussian distribution function, which is used to determine the expected value, variance, probability of occurrence, and risk of failure of a natural phenomenon, an event, or a system property [49]. Standard distribution functions are also used in polymerization reactions to fit experimental data, model, and elucidate the mechanism of polymerization [16, 50]. This section describes common distribution functions used to evaluate the variance and dispersity of the chain distribution in different polymerization methods.

3.9.1 Poisson Distribution Function

A Poisson distribution is a discrete, countable probability distribution with one adjustable parameter λ. A Poisson distribution gives the probability of an event happening i times randomly and independently within a given interval of time or space. A given interval of time or space is defined by the mean number of events, λ, occurring in that time or space interval [51, 52]. A Poisson probability distribution is given by the following:

$$P_{n,psn}\,(i) = \frac{1}{i!}\lambda^i e^{-\lambda} \qquad 0 \leq i < \infty \tag{3.82}$$

where $P_{n,psn}\,(i)$ is the number Poisson probability of an event happening i times. The Poisson distribution is a one-parameter model (λ) characterized by ideal events occurring randomly and independently, with the number of events ranging from zero to countable infinity. According to

Eq. (3.35), the first $(M_{n,1})$ and second $(M_{n,2})$ moments of a Poisson distribution are given by the following equations:

$$M_{n,1,psn} = \sum_{i=0}^{\infty} iP_{n,psn}(i) = \sum_{i=0}^{\infty} i\frac{1}{i!}\lambda^i e^{-\lambda} = e^{-\lambda}\sum_{i=0}^{\infty}\frac{i}{i!}\lambda^i = \lambda \tag{3.83}$$

$$M_{n,2,psn} = \sum_{i=0}^{\infty} i^2 P_{n,psn}(i) = \sum_{i=0}^{\infty} i^2\frac{1}{i!}\lambda^i e^{-\lambda} = e^{-\lambda}\sum_{i=0}^{\infty}\frac{i^2}{i!}\lambda^i = \lambda^2 + \lambda \tag{3.84}$$

where the equalities $\sum_{i=0}^{\infty} i\lambda^i/i! = \lambda e^{\lambda}$ and $\sum_{i=0}^{\infty} i^2\lambda^i/i! = (\lambda^2+\lambda)\,e^{\lambda}$ for infinite series of exponential functions [53, 54] are used in the above derivation. According to Eqs. (3.42), (3.63), and (3.65), the mean $\left(\overline{DP}_{n,psn}\right)$, variance (V_{psn}), and normalized standard deviation (Δ_{psn}) of a Poisson distribution are given by the following equations:

$$\overline{DP}_{n,psn} = M_{n,1,psn} = \lambda \tag{3.85}$$

$$V_{psn} = M_{n,2,psn} - \left(M_{n,1,psn}\right)^2 = \lambda^2 + \lambda - \lambda^2 = \lambda \tag{3.86}$$

$$\Delta_{psn} = \frac{V_{psn}^{1/2}}{\overline{DP}_{n,psn}} = \frac{\lambda^{1/2}}{\lambda} = \lambda^{-1/2} \tag{3.87}$$

The breadth of a Poisson distribution, as defined by Δ_{psn}, is narrower than the Gaussian distribution. The Δ_{psn} of a Poisson distribution approaches zero (monodisperse polymer) as the number-average degree of polymerization, λ, takes increasingly larger values. In this respect, Poisson distribution is used to represent probability distributions in polymerization reactions that are nearly ideal, like living anionic polymerization, especially in the late stages of the reaction. In living anionic polymerization, propagating species are formed instantaneously at the start of the reaction, propagation is random, and termination or chain transfer reactions are absent [55, 56].

3.9.2 Flory–Schulz Distribution Function

The Flory–Schulz distribution is used to model the distribution of degree of polymerization in step-growth polymerization, in which longer chains are formed by the stepwise reaction of shorter chains, assuming reactivity is independent of the polymer chain length [57]. Like the Poisson distribution, it is characterized by ideal events occurring randomly and independently with events ranging from zero to countable infinity with one adjustable parameter p or α. The Flory–Schulz distribution is defined in terms of the probability that a reactive group in the polymerization volume has reacted (p), which is equal to conversion of the reacting groups (α). A propagating chain with i repeat units must have undergone $(i-1)$ "reaction" events and one "no reaction" event to account for the reactive group at the end of the polymer chain. Assuming random and independent reaction events and reactions independent of chain length, the probability of $(i-1)$ reaction events is α^{i-1} and the probability of one no reaction event is $(\alpha - 1)$. Therefore, the number probability distribution function for step-growth polymerization, $P_{n,sg}(i)$, is given by the following equation:

$$P_{n,sg}(i) = (1-\alpha)\,\alpha^{i-1} \qquad 1 \le i < \infty \tag{3.88}$$

There is one adjustable parameter α in the step-growth distribution. According to Eq. (3.35), the first and second moments of a step-growth distribution function are given by the following equations:

$$M_{n,1,sg} = \sum_{i=1}^{\infty} i\,P_{sg}(i) = (1-\alpha)\sum_{i=1}^{\infty} i\,\alpha^{i-1} = \frac{(1-\alpha)}{\alpha}\sum_{i=1}^{\infty} i\,\alpha^{i} = \frac{1}{(1-\alpha)} \tag{3.89}$$

$$M_{n,2,sg} = \sum_{i=1}^{\infty} i^2 P_{sg}(i) = (1-\alpha)\sum_{i=1}^{\infty} i^2\alpha^{i-1} = \frac{(1-\alpha)}{\alpha}\sum_{i=1}^{\infty} i^2\alpha^{i} = \frac{(1+\alpha)}{(1-\alpha)^2} \tag{3.90}$$

where the equalities $\sum_{i=1}^{\infty} i\,\alpha^i = \alpha/(1-\alpha)^2$ and $\sum_{i=1}^{\infty} i^2\alpha^i = \alpha(1+\alpha)/(1-\alpha)^3$ for infinite series of power functions [53] were used in the above derivation. The mean $\left(\overline{DP}_{n,sg}\right)$, variance (V_{sg}), and normalized standard deviation (Δ_{sg}) of a step-growth distribution are as follows:

$$\overline{DP}_{n,sg} = M_{n,1,sg} = 1/(1-\alpha) \tag{3.91}$$

$$V_{sg} = M_{n,2,sg} - \left(M_{n,1,sg}\right)^2 = \frac{(1+\alpha)}{(1-\alpha)^2} - \frac{1}{(1-\alpha)^2} = \frac{\alpha}{(1-\alpha)^2} \tag{3.92}$$

$$\Delta_{sg} = \frac{V_{sg}^{1/2}}{\overline{DP}_{n,sg}} = \frac{\alpha^{1/2}/(1-\alpha)}{1/(1-\alpha)} = \alpha^{1/2} \tag{3.93}$$

As each reaction event reduces the number of propagating chains by one, the total number of chains, N_{sg}, and the number of chains with i repeat units, $N_{sg}(i)$, are given as a function of conversion by the following equations:

$$N_{sg} = N_0(1-\alpha) \tag{3.94}$$

$$N_{sg}(i) = N_{sg}P_{n,sg}(i) = N_0(1-\alpha)^2\alpha^{i-1} \tag{3.95}$$

The weight probability distribution, obtained from N_{sg} and $N_{sg}(i)$, is given by the following equation:

$$P_{w,sg}(i) = \frac{i\,N_{sg}(i)}{\sum_{i=1}^{\infty} i\,N_{sg}(i)} = \frac{i\,N_0(1-\alpha)^2\alpha^{i-1}}{N_0} = i(1-\alpha)^2\alpha^{i-1} \qquad 1 \le i < \infty \tag{3.96}$$

The above weight distribution, which is the most probable distribution of selecting an i-mer from the ensemble of polymer chains, is known as the Flory–Schulz distribution of polymer chains $[P_{FS}(i)]$ as follows:

$$P_{FS}(i) = i(1-\alpha)^2\alpha^{i-1} \qquad 1 \le i < \infty \tag{3.97}$$

According to Eq. (3.35), the first and second moments of a Flory–Schulz distribution are given by the following equations:

$$M_{n,1,FS} = \sum_{i=1}^{\infty} i^1 P_{FS}(i) = (1-\alpha)^2\sum_{i=1}^{\infty} i^2\alpha^{i-1} = \frac{(1-\alpha)^2}{\alpha}\sum_{i=1}^{\infty} i\,\alpha^{i} = \frac{1+\alpha}{1-\alpha} \tag{3.98}$$

$$M_{n,2,FS} = \sum_{i=1}^{\infty} i^2 P_{FS}(i) = (1-\alpha)^2 \sum_{i=1}^{\infty} i^3 \alpha^{i-1} = \frac{(1-\alpha)^2}{\alpha} \sum_{i=1}^{\infty} i^2 \alpha^i = \frac{(1+4\alpha+\alpha^2)}{(1-\alpha)^2} \tag{3.99}$$

where the equalities $\sum_{i=1}^{\infty} i^2\alpha^i = \alpha(1+\alpha)/(1-\alpha)^3$ and $\sum_{i=1}^{\infty} i^3\alpha^i = \alpha(1+4\alpha+\alpha^2)/(1-\alpha)^4$ for infinite series of power functions [53] are used in the above derivation. The mean, variance, and normalized standard deviation of a Flory–Schulz distribution are given by the following equations:

$$\overline{DP}_{n,FS} = M_{n,1,FS} = \frac{(1+\alpha)}{(1-\alpha)} \tag{3.100}$$

$$V_{FS} = M_{n,2,FS} - (M_{n,1,FS})^2 = \frac{(1+4\alpha+\alpha^2)}{(1-\alpha)^2} - \frac{(1+\alpha)^2}{(1-\alpha)^2} = \frac{2\alpha}{(1-\alpha)^2} \tag{3.101}$$

$$\Delta_{FS} = \frac{V_{FS}^{1/2}}{\overline{DP}_{n,FS}} = \frac{\sqrt{2\alpha}/(1-\alpha)}{(1+\alpha)/(1-\alpha)} = \sqrt{2\alpha}/(1+\alpha) \tag{3.102}$$

The normalized standard deviation for a Flory–Schulz distribution, Δ_{FS}, increases with α and approaches $\sqrt{2}/2$ whereas Δ_{psn} approaches zero for large values of λ for a Poisson distribution. Therefore, the Flory–Schulz distribution is broader than the Poisson distribution.

3.9.3 Exponential Distribution Function

The step-growth distribution function (Eq. 3.88) can be rewritten in the following form using the equality $a^x = e^{x \ln a}$:

$$P_{n,sg}(i) = (1-\alpha)\alpha^{i-1} = (1-\alpha)e^{(i-1)\ln\alpha} \tag{3.103}$$

For large $\overline{DP}_n$ or $\alpha \to 1$, the ln α term can be expanded in a Taylor series around $\alpha = 1$ [58]. Taking the first two terms of the series and setting $i - 1 \approx i$ for large $\overline{DP}_n$, the distribution function reduces to the following form:

$$P_{n,sg}(i)\,|_{\alpha\to 1} = (1-\alpha)e^{i(\alpha-1)} = (1-\alpha)e^{-i(1-\alpha)} = P_{\exp}(i) \tag{3.104}$$

The above distribution, which applies to polymerizations with large $\overline{DP}_n$ or $\alpha \to 1$, is called the exponential distribution function [54, 59]. Upon substitution for α from Eq. (3.91), the exponential distribution is expressed in terms of $\overline{DP}_n$ by the following equation:

$$P_{\exp}(i) = (1-\alpha)e^{-i(1-\alpha)} = \frac{1}{\overline{DP}_n} e^{-\frac{i}{\overline{DP}_n}} \tag{3.105}$$

The exponential distribution function has one adjustable parameter α or $\overline{DP}_n$. The adjustable parameter $\overline{DP}_n$ is called the scale parameter of the exponential distribution. The spread of a distribution increases with the scale parameter [60]. According to Eq. (3.35), the first and second moments of the exponential distribution are given by the following equations:

$$\begin{aligned} M_{n,1,\exp} &= \sum_{i=1}^{\infty} i^1 P_{\exp}(i) = \frac{1}{\overline{DP}_n} \sum_{i=1}^{\infty} i \left(e^{-\overline{DP}_n^{-1}}\right)^i \\ &= \frac{1}{\overline{DP}_n} \frac{e^{-\overline{DP}_n^{-1}}}{\left(1-e^{-\overline{DP}_n^{-1}}\right)^2} = \frac{1}{\overline{DP}_n} \frac{1-\overline{DP}_n^{-1}}{\left(1-\left(1-\overline{DP}_n^{-1}\right)\right)^2} \approx \overline{DP}_n \end{aligned} \tag{3.106}$$

$$M_{n,2,\exp} = \sum_{i=1}^{\infty} i^2 P_{\exp}(i) = \frac{1}{\overline{DP}_n} \sum_{i=1}^{\infty} i^2 \left(e^{-\overline{DP}_n^{-1}}\right)^i$$

$$\frac{1}{\overline{DP}_n} \frac{e^{-\overline{DP}_n^{-1}}\left(1+e^{-\overline{DP}_n^{-1}}\right)}{\left(1-e^{-\overline{DP}_n^{-1}}\right)^3} = \frac{1}{\overline{DP}_n} \frac{\left(1-\overline{DP}_n^{-1}\right)\left(1+\left(1-\overline{DP}_n^{-1}\right)\right)}{\left(1-\left(1-\overline{DP}_n^{-1}\right)\right)^3} \approx 2\overline{DP}_n^2 \tag{3.107}$$

where the equalities $e^{ax} = (e^a)^x$ and exponential expansion series $e^a = \sum_{n=0}^{\infty} a^n/n!$ were used in the above derivation. It was also assumed that $\overline{DP}_n \gg 1$. The mean $\left(\overline{DP}_{n,\exp}\right)$, variance $(V_{\exp})$, and normalized standard deviation $(\Delta_{\exp})$ of the exponential distribution are given by the following equations:

$$\overline{DP}_{n,\exp} = M_{n,1,\exp} = \overline{DP}_n \tag{3.108}$$

$$V_{\exp} = M_{n,2,\exp} - \left(M_{n,1,\exp}\right)^2 = 2\overline{DP}_n^2 - \overline{DP}_n^2 = \overline{DP}_n^2 \tag{3.109}$$

$$\Delta_{\exp} = \frac{V_{\exp}^{1/2}}{\overline{DP}_{n,\exp}} = \frac{\overline{DP}_n}{\overline{DP}_n} = 1 \tag{3.110}$$

Due to its simplicity, the exponential distribution is used to model radical polymerization under conditions of low radical concentration and high chain molecular weight, such that at least one very long chain is involved in the termination reaction [61]. It is also used to model the molecular weight distribution of living radical chains during auto-acceleration, in which living chains are immobilized by entanglements and terminated only with short, unentangled, mobile chains [62–64]. In the same approach, it can be shown that the Flory–Schulz distribution can be written in exponential form for $\alpha \to 1$ or large $\overline{DP}_n$ as given by the following equations:

$$P_{FS}(i)\,|_{\alpha\to 1} = P_{FS,\exp} = (1-\alpha)^2\, i\, e^{-i(1-\alpha)} = \frac{i}{\overline{DP}_n^2} e^{-\frac{i}{\overline{DP}_n}} \tag{3.111}$$

In some reports, the exponential form of the Flory–Schulz distribution is referred to as the exponential distribution [4, 65]. The exponential distribution is a special case of the gamma distribution, which will be discussed in the next section.

3.9.4 Schulz–Zimm Distribution Function

The exponential distribution function is a special case of gamma distribution function [66]. The gamma distribution function is a two-parameter model given by the following equation [66]:

$$P_{gam}(i) = \frac{i^{h-1} e^{-\frac{i}{\lambda}}}{\lambda^h \Gamma(h)} \qquad 0 < i < \infty \qquad h, \lambda > 0 \tag{3.112}$$

where h and λ are adjustable parameters of the distribution, and $\Gamma(h) = (h-1)!$ if h is a positive integer. The adjustable parameters λ and h are called the scale and shape parameters of the gamma distribution, respectively [67]. If h is set equal to one, the above equation reduces to the exponential distribution function (Eq. 3.105) with $\lambda = 1/\overline{DP}_n$ as the only adjustable parameter. If h is set equal to two, the above equation reduces to the exponential form of the Flory–Schulz distribution function (Eq. 3.111) with $\lambda = 1/\overline{DP}_n$ as the only adjustable parameter. The above gamma distribution, when applied to the distribution of polymer chains, is known as the Schulz–Zimm distribution function with two adjustable parameters [68, 69] as follows:

$$P_{SZ}(i) = P_{gam}(i) = \frac{i^{h-1}e^{-\frac{i}{\lambda}}}{\lambda^h \Gamma(h)} \qquad 0 < i < \infty \qquad h, \lambda > 0 \tag{3.113}$$

The mean value and breadth of the distribution can be adjusted independently using λ and h, respectively. The first $(M_{n,1,gam})$ and second $(M_{n,2,gam})$ moments of the gamma distribution are given by the following equations [70]:

$$M_{n,1,gam} = \sum_{i=1}^{\infty} i^1 P_{gam}(i) = h\lambda \tag{3.114}$$

$$M_{n,2,gam} = \sum_{i=1}^{\infty} i^2 P_{gam}(i) = h(h+1)\lambda^2 \tag{3.115}$$

The mean $\left(\overline{DP}_{n,gam}\right)$, variance (V_{gam}), and normalized standard deviation (Δ_{gam}) of gamma distribution are given by the following equations:

$$\overline{DP}_{n,gam} = M_{n,1,gam} = h\lambda \tag{3.116}$$

$$V_{gam} = M_{n,2,gam} - (M_{n,1,gam})^2 = h(h+1)\lambda^2 - h^2\lambda^2 = h\lambda^2 \tag{3.117}$$

$$\Delta_{gam} = \frac{V_{gam}^{1/2}}{\overline{DP}_{n,gam}} = \frac{h^{1/2}\lambda}{h\lambda} = h^{-1/2} \tag{3.118}$$

As a two-parameter distribution function with independently adjustable mean and variance, gamma distribution function is used to fit molecular weight distribution data of polymers measured experimentally by pulsed gradient spin echo nuclear magnetic resonance (PGSE-NMR) data [71, 72]. In another work, gamma distribution function is used to model molecular weight distribution of degradation products of a polymer in solution [73]. In another study, gamma distribution function is used to model the molecular weight distribution of block and graft copolymers [74].

3.9.5 Lognormal Distribution Function

The normal or Gaussian distribution is widely used to express the probability distribution of events in engineering systems mathematically and to predict the expected value and variance [1]. The two-parameter normal or Gaussian distribution, $P_{gaus}(i)$, is given by Eq 3.72 in Section 3.8, where μ and σ are the expected value (mean) and the standard deviation from the mean (68.2% of the data within $\pm 1\sigma$). Gaussian distribution is seldom used to represent polymer chain distribution as *DP* cannot take negative values. It is sometimes used to represent polymers with a very narrow breadth of distribution, such that the distribution is not affected by disregarding the negative values. It is also used to fit distribution data in polymer chain scission and degradation [75]. If the random variable $X = i$ in the Gaussian distribution is replaced with $\ln(X = i)$, the new function, known as the lognormal distribution function, is given by the following equation [48]:

$$P_{\log n}(i) = \frac{1}{i\,\sigma_{\log n}\sqrt{2\pi}} e^{-\frac{1}{2}\left(\frac{\ln(i)-\mu_{\log n}}{\sigma_{\log n}}\right)^2} \qquad 0 < i < \infty \qquad -\infty < \ln i < \infty \tag{3.119}$$

where $\mu_{\log n}$ and $\sigma_{\log n}$ are the expected value (mean) and standard deviation of a lognormal distribution, respectively. The lognormal distribution naturally curtails the probability of occurrence of extreme values of the independent variable $X = i$ (both positive and negative large numbers). The adjustable parameters μ and σ in the Gaussian distribution and $\mu_{\log n}$ and $\sigma_{\log n}$ in lognormal distribution are called the location parameter and the scale parameter, respectively [66]. The first $(M_{n,1,\log n})$ and second $(M_{n,2,\log n})$ moments of lognormal distribution are given by the following equations [76]:

$$M_{n,1,\log n} = \sum_{i=1}^{\infty} i^1 P_{\log n}(i) = e^{\mu_{\log n} + \frac{\sigma_{\log n}^2}{2}} \tag{3.120}$$

$$M_{n,2,\log n} = \sum_{i=1}^{\infty} i^2 P_{\log n}(i) = e^{2\mu_{\log n} + 2\sigma_{\log n}^2} \tag{3.121}$$

The mean $(\bar{i}_{\log n})$, variance $(V_{\log n})$, and normalized standard deviation $(\Delta_{\log n})$ of lognormal distribution are given by the following equations:

$$M_{n,1,\log n} = \bar{i}_{\log n} = e^{\mu_{\log n} + \frac{\sigma_{\log n}^2}{2}} \tag{3.122}$$

$$V_{\log n} = M_{n,2,\log n} - \left(M_{n,1,\log n}\right)^2 = \left(e^{\mu_{\log n} + \frac{\sigma_{\log n}^2}{2}}\right)^2 \left(e^{\sigma_{\log n}^2} - 1\right) \tag{3.123}$$

$$\Delta_{\log n} = \frac{V_{\log n}^{1/2}}{\bar{i}_{\log n}} = \left[\left(e^{\mu_{\log n} + \frac{\sigma_{\log n}^2}{2}}\right)^2 \left(e^{\sigma_{\log n}^2} - 1\right)\right]^{1/2} \Bigg/ e^{\mu_{\log n} + \frac{\sigma_{\log n}^2}{2}} = \sqrt{e^{\sigma_{\log n}^2} - 1} \tag{3.124}$$

The lognormal distribution is especially useful in modeling the distribution of polymer chains with a high molecular weight tail [77–81]. In one study, a lognormal distribution was used to fit molecular weight data of polymers synthesized by living radical polymerization, like atom transfer radical polymerization (ATRP) [82, 83]. In another study, lognormal distribution was used to mathematically model and quantitatively predict the experimental molecular weight distribution of precisely designed polymers synthesized in tubular flow reactors [84].

3.9.6 Weibull Distribution Function

In the two-parameter gamma distribution function, the spread and shape of the distribution are adjusted independently through the scale parameter and shape parameter, respectively [67]. In the two-parameter Gaussian and lognormal distributions, the spread and mean of the distribution are adjusted independently through the scale and location parameters, respectively [66]. In the three-parameter Weibull distribution function, there are three adjustable parameters, namely the scale parameter σ_{wb}, the shape parameter h, and the location parameter μ_{wb}. The three-parameter Weibull distribution function is given by the following equation [85]:

$$P_{wb}(i) = \frac{h}{\sigma_{wb}} \left(\frac{i - \mu_{wb}}{\sigma_{wb}}\right)^{h-1} e^{-\left(\frac{i-\mu_{wb}}{\sigma_{wb}}\right)^h} \quad i \geq 0 \quad h, \lambda_{wb} > 0 \tag{3.125}$$

In the Weibull distribution function, the mean, spread, and shape of the distribution are adjusted independently through μ_{wb}, σ_{wb}, and h, respectively. The mean ($\bar{i}_{wb}$), variance (V_{wb}), and normalized standard deviation (Δ_{wb}) of the three-parameter Weibull distribution are given by the following equations:

$$M_{n,1,wb} = \bar{i}_{wb} = \mu_{wb} + \sigma_{wb}\,\Gamma\left(1+\frac{1}{h}\right) \tag{3.126}$$

$$V_{\log n} = M_{n,2,\log n} - \left(M_{n,1,\log n}\right)^2 = {\sigma_{wb}}^2\,\Gamma\left(1+\frac{2}{h}\right) - {\mu_{wb}}^2 \tag{3.127}$$

$$\Delta_{wb} = \frac{V_{wb}^{1/2}}{\bar{i}_{wb}} = \left[\sigma_{wb}^2\,\Gamma\left(1+\frac{2}{h}\right) - \mu_{wb}^2\right]^{1/2} \Big/ \left[\mu_{wb} + \sigma_{wb}\,\Gamma\left(1+\frac{1}{h}\right)\right] \tag{3.128}$$

The Weibull distribution is used to fit the data and model the properties of a mixed population of polymer samples, like polymer blends and composites [86]. In one study, the Weibull distribution was used to model the molecular weight distribution of a mixed population of a cellulose nitrate polymer because of partial degradation by pyrolysis [87].

3.9.7 Example Problem 6

The calculated first $(M_{n,1})$ and second $(M_{n,2})$ moments of distribution of polymer samples S1 and S2 of Example 3.1 are given in Eq. (3.39). Use the $M_{n,1}$ and $M_{n,2}$ of the two samples to mathematically represent the *DP* probability distribution data of the two samples, given in Figure 3.1, by the following distribution functions: exponential, Schulz–Zimm (gamma), Gaussian, and lognormal distribution functions. Plot the actual distributions on the same graph with exponential, Schulz–Zimm, Gaussian, and lognormal distributions.

Solution: The $M_{n,1}$ and $M_{n,2}$ of the two polymer samples, as given by Eq. (3.39), are as follows:

$$M_{n,1}\,(S1) = 1.001 \times 10^3 \qquad M_{n,2}\,(S1) = 1.038 \times 10^6$$
$$M_{n,1}\,(S2) = 1.537 \times 10^3 \qquad M_{n,2}\,(S2) = 2.914 \times 10^6$$

Exponential distribution: This distribution function is given by Eq. (3.105). As it is a one-parameter model, only $M_{n,1}$ of each polymer sample is needed for fitting the data to the model as follows:

$$M_{n,1}\,(S1) = 1.001 \times 10^3 = \overline{DP}_n\,(S1) \;\Rightarrow\; P_{\exp}\,(S1) = \frac{1}{1.001 \times 10^3} e^{-\frac{i}{1.001\times10^3}}$$

$$M_{n,1}\,(S2) = 1.537 \times 10^3 = \overline{DP}_n\,(S1) \;\Rightarrow\; P_{\exp}\,(S1) = \frac{1}{1.537 \times 10^3} e^{-\frac{i}{1.537\times10^3}}$$

Schulz–Zimm distribution: This distribution function is given by Eq. (3.113). As it is a two-parameter model, both $M_{n,1}$ and $M_{n,2}$ of each polymer sample are needed to compute λ and h using Eqs. (3.114) and (3.115) and fit the data to the model as follows:

$$\begin{cases} M_{n,1}\,(S1) = 1.001 \times 10^3 = h\,(S1)\,\lambda\,(S1) \\ M_{n,2}\,(S1) = 1.038 \times 10^6 = h\,(S1)\,[h\,(S1)+1]\,\lambda\,(S1)^2 \end{cases} \Rightarrow \begin{cases} h\,(S1) = 0.634 \\ \lambda\,(S1) = 1579 \end{cases}$$

$$\begin{cases} M_{n,1}(S2) = 1.537\times10^3 = h(S2)\,\lambda(S2) \\ M_{n,2}(S2) = 2.914\times10^6 = h(S2)\,[h(S2)+1]\,\lambda(S2)^2 \end{cases} \Rightarrow \begin{cases} h(S2) = 0.718 \\ \lambda(S2) = 2141 \end{cases}$$

$$P_{SZ}(S1) = \frac{i^{-0.366}e^{-i/1579}}{(1579)^{0.634}\,\Gamma(0.634)} = 6.625\times10^{-3} i^{-0.366}e^{-i/1579}$$

$$P_{SZ}(S2) = \frac{i^{-0.282}e^{-i/2141}}{(2141)^{0.718}\,\Gamma(0.718)} = 3.197\times10^{-3} i^{-0.282}e^{-i/2141}$$

Gaussian distribution: This distribution function is given by Eq. (3.72). As it is a two-parameter model, both $M_{n,1}$ and $M_{n,2}$ of each polymer sample are needed to compute μ and σ. The fit of the data to the model for each with $\mu = M_{n,1}$ and $\sigma^2 = V_{gaus} = M_{n,2} - (M_{n,1})^2$ is as follows:

$$\begin{cases} \mu_n(S1) = M_{n,1}(S1) = 1.001\times10^3 \\ \sigma_n(S1) = \sqrt{M_{n,2}(S1) - [M_{n,1}(S1)]^2} = \sqrt{1.038\times10^6 - (1.001\times10^3)^2} = 189.7 \end{cases}$$

$$\begin{cases} \mu_n(S2) = M_{n,1}(S2) = 1.537\times10^3 \\ \sigma_n(S2) = \sqrt{M_{n,2}(S2) - [M_{n,1}(S2)]^2} = \sqrt{2.914\times10^6 - (1.537\times10^3)^2} = 742.7 \end{cases}$$

$$P_{gaus}(S1) = 2.103\times10^{-3} e^{-\frac{1}{2}\left(\frac{i-1001}{189.7}\right)^2}$$

$$P_{gaus}(S2) = 5.370\times10^{-4} e^{-\frac{1}{2}\left(\frac{i-1537}{742.7}\right)^2}$$

Lognormal distribution: This distribution function is given by Eq. (3.119). As it is a two-parameter model, both $M_{n,1}$ and $M_{n,2}$ of each polymer sample are needed to compute $\mu_{\log n}$ and $\sigma_{\log n}$ using Eqs. (3.120) and (3.121). For lognormal distribution, $\mu_{\log n}$ and $\sigma_{\log n}$ are expressed in terms of the moments as follows:

$$\begin{aligned} (\Delta_{\log n})^2 &= \frac{V_{\log n}}{(M_{n,1,\log n})^2} = \frac{M_{n,2,\log n} - (M_{n,1,\log n})^2}{(M_{n,1,\log n})^2} = \frac{M_{n,2,\log n}}{(M_{n,1,\log n})^2} - 1 = e^{\sigma^2_{\log n}} - 1 \\ &\Rightarrow \sigma^2 = \ln\left(\frac{M_{n,2,\log n}}{(M_{n,1,\log n})^2}\right) \Rightarrow \sigma = SQRT\ \ln\left(\frac{M_{n,2,\log n}}{(M_{n,1,\log n})^2}\right) \end{aligned} \tag{3.129}$$

$$\begin{aligned} M_{n,1,\log n} &= e^{\mu_{\log n} + \frac{\sigma^2_{\log n}}{2}} \quad \Rightarrow \quad \mu_{\log n} = \ln(M_{n,1,\log n}) - \frac{\sigma^2_{\log n}}{2} \\ &\Rightarrow \mu_{\log n} = \ln(M_{n,1,\log n}) - \ln\left(\frac{M_{n,2,\log n}}{(M_{n,1,\log n})^2}\right)^{1/2} = \ln\left(\frac{(M_{n,1,\log n})^2}{(M_{n,2,\log n})^{1/2}}\right) \end{aligned} \tag{3.130}$$

Using the above equations, $\mu_{\log n}$ and $\sigma_{\log n}$ for polymer samples S1 and S2 are calculated as follows:

$$\begin{cases} \mu_{\log n} = \ln\left(\dfrac{(M_{n,1,\log n})^2}{(M_{n,2,\log n})^{1/2}}\right) = \ln\left(\dfrac{(1.001\times 10^3)^2}{(1.038\times 10^6)^{1/2}}\right) = 6.891 \\ \sigma = SQRT\ \ln\left(\dfrac{M_{n,2,\log n}}{(M_{n,1,\log n})^2}\right) = SQRT\ \ln\left(\dfrac{1.038\times 10^6}{(1.001\times 10^3)^2}\right) = 0.188 \end{cases}$$

$$\begin{cases} \mu_{\log n} = \ln\left(\dfrac{(M_{n,1,\log n})^2}{(M_{n,2,\log n})^{1/2}}\right) = \ln\left(\dfrac{(1.537\times 10^3)^2}{(2.914\times 10^6)^{1/2}}\right) = 7.233 \\ \sigma = SQRT\ \ln\left(\dfrac{M_{n,2,\log n}}{(M_{n,1,\log n})^2}\right) = SQRT\ \ln\left(\dfrac{2.914\times 10^6}{(1.537\times 10^3)^2}\right) = 0.458 \end{cases}$$

$$P_{\log n}\,(S1) = \frac{2.122}{i} e^{-\frac{1}{2}\left(\frac{\ln(i)-6.891}{0.188}\right)^2}$$

$$P_{\log n}\,(S2) = \frac{0.871}{i} e^{-\frac{1}{2}\left(\frac{\ln(i)-7.233}{0.458}\right)^2}$$

Fit of the data for samples S1 and S2 to exponential, Schulz–Zimm, Gaussian, and Lognormal distributions is shown in Figures 3.3 and 3.4, respectively.

For both samples, the exponential and Schulz–Zimm distributions did not provide a satisfactory fit to the data. For the unimodal sample S1, the Gaussian and lognormal distribution functions provided a satisfactory fit, with the Gaussian function giving a slightly better fit than the lognormal. For the bimodal sample S2, the Gaussian distribution overpredicted probabilities at low DPs, which was attributed to the negative values of i in the Gaussian distribution. Further, the Gaussian function predicted an average value between the two peaks for sample S2. Conversely, the lognormal function did not overpredict probabilities at low DPs, predicted a mean value close to the larger peak, and a tail at DPs larger than the mean, which was attributed to the second smaller peak in the distribution.

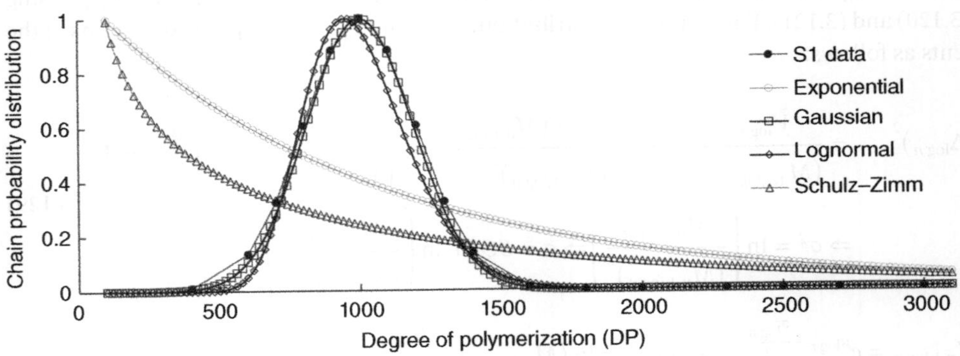

Figure 3.3 Probability distribution data for DP of polymer sample S1 (filled circle) and the fit to the data using exponential (open circle), Schulz–Zimm (open triangle), Gaussian (open square), and lognormal (closed square) distribution functions. The probabilities on the y-axis are normalized to the highest probability in the distribution.

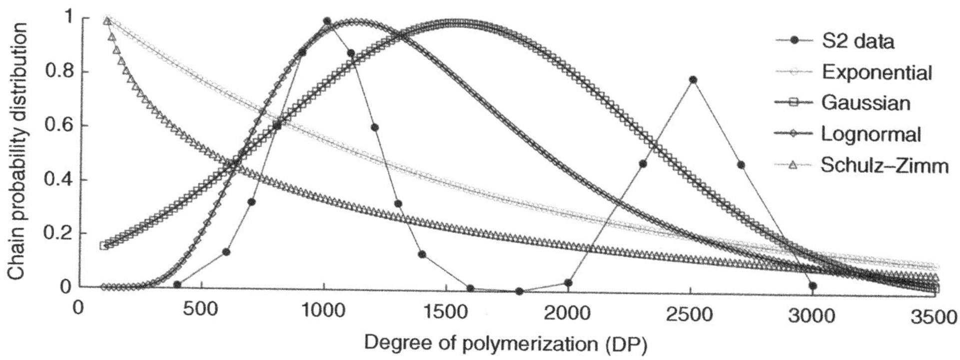

Figure 3.4 Probability distribution data for DP of polymer sample S2 (filled circle) and the fit to the data using exponential (open circle), Schulz–Zimm (open triangle), Gaussian (open square), and lognormal (closed square) distribution functions. The probabilities on the y-axis are normalized to the highest probability in the distribution.

3.10 Summary

The degree of polymerization (DP) or molecular weight of polymer chains follows a probability distribution. Distributed variables like DP are defined by their expected value or mean of the distribution. The probability distribution can be determined based on the number fraction, weight fraction, or even higher order z and $z + 1$ fraction of chains in a polymer sample with $DP = i$. Distributed variables like DP are characterized by their expected value or mean $\left(\overline{DP}_n\right)$ and variance (V_{DP}). The distribution function of the degree of polymerization is used to find the kth molar-based moments of the distribution $\left(M_{n,k}\right)$. The kth moments in turn are used to determine the different averages like number-average $\left(\overline{DP}_n\right)$, weight-average $\left(\overline{DP}_w\right)$, z-average $\left(\overline{DP}_z\right)$, z+1-average $\left(\overline{DP}_{z+1}\right)$ or viscosity-average $\left(\overline{DP}_\eta\right)$ degree of polymerization of the distribution. Higher moments of distribution place greater significance on chains in the distribution with higher degrees of polymerization. The variance of a polymer distribution is used to determine the standard deviation from the mean (σ_{DP}) and normalized standard deviation (Δ). The ratio of weight-to-number-average degree of polymerization $\left(\overline{DP}_w/\overline{DP}_n\right)$, which is related to the statistically based Δ is used to quantify the breadth or spread of a polymer chain distribution. Asymmetry in a polymer chain distribution is quantified by skewness and kurtosis of the distribution with respect to a normal or Gaussian distribution as the reference. Skewness and kurtosis, which are measures of shape asymmetry and flatness of a distribution, are quantified from the first four moments of the distribution, namely $M_{n,1}$ to $M_{n,4}$. Standard distribution functions are often used to fit data from polymerization reactions to model the molecular weight distribution. These include uniparameter Poisson, Flory–Schulz, and exponential distribution functions, two-parameter Schulz–Zimm and lognormal distribution functions, and three-parameter Weibull distribution function.

3.11 List of Symbols and Abbreviations

a: viscosity exponent in Mark–Houwink–Sakurada (MHS) equation
C_i: concentration of polymer chains with i repeat units, mole m^{-3}
C: concentration of all polymer chains, mole m^{-3}

DP: degree of polymerization
DP_M: median degree of polymerization of polymer chains
DP_P: peak degree of polymerization of polymer chains
$\overline{DP}$: average degree of polymerization
$\overline{DP}_n$: number-average degree of polymerization
$\overline{DP}_w$: weight-average degree of polymerization
$\overline{DP}_z$: z-average degree of polymerization
$\overline{DP}_{z+1}$: z+1-average degree of polymerization
$\overline{DP}_\eta$: viscosity-average degree of polymerization
$\langle DP_n{}^k \rangle$: kth moment of the degree of polymerization based on number probability distribution
$\langle DP_w{}^k \rangle$: kth moment of the degree of polymerization based on weight probability distribution
$\langle DP_z{}^k \rangle$: kth moment of the degree of polymerization based on z probability distribution
$\langle DP_{z+1}{}^k \rangle$: kth moment of the degree of polymerization based on $z+1$ probability distribution
$f\,(DP)$: distribution function of degree of polymerization of polymer chains
$\langle f\,(DP) \rangle$: expectation of a distribution function of degree of polymerization of polymer chains
h: shape parameter in gamma and Weibull distributions
K_w: constant relating number and weight probability distributions
K_z: constant relating number and z probability distributions
K_{z+1}: constant relating number and $z+1$ probability distributions
K_{z+1}: constant relating number and viscosity probability distributions
K_η: viscosity constant in Mark–Houwink–Sakurada (MHS) equation
KI: Kurtosis index of polymer chains
M_0: molecular weight of the polymer repeat unit, Da
M_i: molecular weight of the ith fraction of chains in a polymer sample, Da
$\overline{L}_c$: average contour length of chains in a polymer sample, m
l_r: length of a polymer repeat unit, m
N_{sg}: total number of polymer chains in a step-growth polymerization reaction
$N_{sg}\,(i)$: number of polymer chains in a step-growth polymerization reaction with $DP = i$
$M_{n,k}$: kth moment of the mole-fraction probability distribution
$M_{w,k}$: kth moment of the weight-fraction probability distribution
$M_{z,k}$: kth moment of the z fraction probability distribution
$M_{z+1,k}$: kth moment of the $z+1$ fraction probability distribution
$\overline{M}$: average molecular weight of a polymer sample, Da
$\overline{M}_n$: number average molecular weight of a polymer sample, Da
$\overline{M}_w$: weight average molecular weight of a polymer sample, Da
$\overline{M}_z$: z average molecular weight of a polymer sample, Da
$\overline{M}_{z+1}$: z+1 average molecular weight of a polymer sample, Da
$\overline{M}_\eta$: viscosity average molecular weight of a polymer sample, Da
p: probability that a reacting group has reacted in a step-growth polymerization
$P_{DPn}(i)$: fraction probability distribution of polymer chains with $DP = i$
$P_{DPw}(i)$: weight fraction probability distribution of polymer chains with $DP = i$
$P_{DPz}(i)$: z fraction probability distribution of polymer chains with $DP = i$
$P_{DPz+1}(i)$: $z+1$ fraction probability distribution of polymer chains with $DP = i$
PDI: polydispersity index of polymer chains
$P_X\,(i)$: probability distribution of a random variable X in the real number space i
$P_{\exp}\,(i)$: exponential probability distribution function of polymer chains with $DP = i$

$P_{\log n}$ (i): lognormal probability distribution function of polymer chains with $DP = i$
P_{gaus} (i): Gaussian or normal probability distribution function of polymer chains with $DP = i$
$P_{n,psn}$ (i): number fraction Poisson probability distribution function of polymer chains with $DP = i$
$P_{n,sg}$ (i): number fraction probability distribution function of polymer chains with $DP = i$
$P_{w,sg}$ (i): weight fraction probability distribution function of polymer chains with $DP = i$
P_{FS} (i): Flory–Schulz probability distribution function of polymer chains with $DP = i$
P_{SZ} (i): Schulz–Zimm probability distribution function of polymer chains with $DP = i$
$Q_{n,k}$: kth moment of molar density distribution of polymer chains
$Q_{w,k}$: kth moment of mass density distribution of polymer chains
$Q_{z,k}$: kth moment of z density distribution of polymer chains
$Q_{z+1,k}$: kth moment of z+1 density distribution of polymer chains
SI: skewness index of polymer chains
V_X: variance of a random variable X in the real number space
V_{DP}: variance of the degree of polymerization of polymer chains
X: a random variable X in the real number space
$\langle X \rangle$: average value of a random variable X in the real number space
$\langle X^k \rangle$: kth moment of the distribution of variable X
α: conversion of monomers in step-growth polymerization
$[\eta]$: intrinsic viscosity, kg m^{-1} s^{-1}
$[\eta_i]$: intrinsic viscosity of polymer chains with i degree of polymerization, kg m^{-1} s^{-1}
ρ_i : density of chains with i degree of polymerization, kg m^{-3}
ρ: density of a polymer sample, kg m^{-3}
σ: standard deviation of a Gaussian distribution
σ_{DP} : standard deviation from the mean of the degree of polymerization of polymer chains
$\sigma_{\log n}$: standard deviation of a lognormal distribution
σ_{wb}: standard deviation of a Weibull distribution
μ: mean value of a Gaussian distribution
$\mu_{\log n}$: mean value of a lognormal distribution
μ_{wb}: mean value of a Weibull distribution
λ: scale parameter in gamma distribution
Δ: statistical polydispersity index of polymer chains

3.12 Practice Problems

Problem 3.1 It was demonstrated in Section 3.3 that the weight distribution of the degree of polymerization P_{DPw} (i) can be expressed in terms of the number distribution P_{DPn} (i). Similarly, the z-distribution P_{DPz} (i) can be expressed in terms of P_{DPw} (i) and P_{DPn} (i). Derive the following relations between P_{DPz} (i), P_{DPw} (i), and P_{DPn} (i):

$$P_{DPz}(i) = \frac{i^2 P_{DPn}(i)}{M_{n,2}} = \frac{i P_{DPw}(i)}{M_{w,1}}$$

Problem 3.2 It was demonstrated in Section 3.4 that $M_{n,k}$ can be expressed in terms of $M_{w,k}$ or $Q_{w,k}$. Derive the following relations between $M_{n,k}$, $M_{w,k}$, and $M_{z,k}$:

$$M_{z,k} = \frac{M_{n,k+2}}{M_{n,2}} = \frac{Q_{n,k+2}}{Q_{n,2}} = \frac{M_{w,k+1}}{M_{w,1}} = \frac{Q_{w,k+1}}{Q_{w,1}}$$

Problem 3.3 The molecular weight distribution (MWD) of a polymethyl methacrylate (PMMA) sample is measured by gel permeation chromatography (GPC). The GPC results are presented in Table 3.4 as the PMMA molecular weight (g mole^{-1}) versus the concentration of the ith fraction in units of g L^{-1}. The MHS viscosity exponent "a" for PMMA is 0.63. Determine the following parameters for the PMMA sample.

Table 3.4 Molecular weight versus weight concentration of PMMA.

M_i (g mole^{-1})	C_i(g L^{-1})
100	0.01
1000	0.1
10 000	1
1 00 000	10
10 00 000	100

a. chain distributions $P_{DPn}(i)$, $P_{DPw}(i)$, $P_{DP\eta}(i)$, and $P_{DPz}(i)$.
b. molar moments $M_{n,1}$, $M_{n,2}$, $M_{n,1+a}$, $M_{n,z}$, and $M_{n,z+1}$.
c. molar concentration moments $Q_{n,0}$, $Q_{n,1}$, $Q_{n,1+a}$, $Q_{n,2}$, $Q_{n,z}$, and $Q_{n,z+1}$.
d. averages of the degree of polymerization $\overline{DP}_n$, $\overline{DP}_w$, $\overline{DP}_\eta$, $\overline{DP}_z$, and $\overline{DP}_{z+1}$.
e. averages of the molecular weight $\overline{M}_n$, $\overline{M}_w$, $\overline{M}_\eta$, $\overline{M}_z$, and $\overline{M}_{z+1}$.
f. parameters for spread of the distribution V_{DP}, σ_{DP}, Δ, and *PDI*.

Problem 3.4 The distribution of degree of polymerization (*DP*) of a polyacrylic acid (PAA) sample is presented in Table 3.5 as DP_i versus molar concentration C_i in units of mole mL^{-1}. The molecular weight of PAA repeat unit in the chains is 72 g mole^{-1}, and the MHS viscosity component is $a = 0.65$. Determine the following for the PAA sample.

Table 3.5 Degree of polymerization versus molar concentration of PAA.

DP_i	C_i (mole mL^{-1})
2	5
25	20
130	50
1200	10
15 000	2

a. chain distributions $P_{DPn}(i)$, $P_{DPw}(i)$, $P_{DP\eta}(i)$, and $P_{DPz}(i)$.
b. molar moments $M_{n,1}$, $M_{n,2}$, $M_{n,1+a}$, $M_{n,z}$, and $M_{n,z+1}$.
c. molar concentration moments $Q_{n,0}$, $Q_{n,1}$, $Q_{n,1+a}$, $Q_{n,2}$, $Q_{n,z}$, and $Q_{n,z+1}$.
d. averages of the degree of polymerization $\overline{DP}_n$, $\overline{DP}_w$, $\overline{DP}_\eta$, $\overline{DP}_z$, and $\overline{DP}_{z+1}$.
e. averages of the molecular weight $\overline{M}_n$, $\overline{M}_w$, $\overline{M}_\eta$, $\overline{M}_z$, and $\overline{M}_{z+1}$.
f. parameters for spread of the distribution V_{DP}, σ_{DP}, Δ, and *PDI*.

Problem 3.5 The molecular weight distribution of a polymer sample is given in Table 3.6 as molecular weight versus weight percent. If the MHS viscosity exponent of the sample is $a = 0.68$, determine molecular weight averages $\overline{M}_n$, $\overline{M}_w$, $\overline{M}_\eta$, $\overline{M}_z$, and $\overline{M}_{z+1}$.

Table 3.6 Molecular weight of the polymer versus weight fraction.

M_i(g mole^{-1})	w_i(%)
100	10
10 000	70
10 00 000	20

Problem 3.6 A polystyrene (PS) sample is prepared by mixing eight monodisperse PS polymers. The average molecular weight and weight fraction of the mixed eight monodisperse polymers are presented in Table 3.7. If the MHS viscosity exponent of the sample is $a = 0.6$, determine the molecular weight averages $\overline{M}_n$, $\overline{M}_w$, and $\overline{M}_\eta$ of the mixed PS sample.

Table 3.7 Molecular weight versus weight fraction for monodisperse PS.

DP_i (kmol)	15	27	39	56	78	104	120	153
W_i (g)	0.1	0.18	0.25	0.17	0.12	0.08	0.06	0.04

Problem 3.7 In a free-radical polymerization reaction, polymer chains are produced by termination via combination as well as termination by disproportionation. In the time interval Δt_i, 2.0 g of polymer is produced via termination by combination with $\overline{M}_n$ and PDI of 2.0×10^5 g mol^{-1} and 1.5, respectively. In the same time interval, 1.0 g of polymer is produced via termination by disproportionation with $\overline{M}_n$ and PDI of 1.0×10^5 g mol^{-1} and 2.0, respectively. If the viscosity exponent of the polymer is $a = 0.7$, determine the molecular weight averages $\overline{M}_n$, $\overline{M}_w$, and $\overline{M}_\eta$ of the polymer chains produced in the time interval Δt_i.

Problem 3.8 The distribution of degree of polymerization of a polymethyl methacrylate (PMMA) sample is given in Table 3.8 as DP_i versus C_i in mol L^{-1} and DP_i versus P_{DPn} (i). If the MHS viscosity exponent of the sample is equal to $a = 0.75$, determine the following parameters for the sample:

Table 3.8 Distribution of the degree of polymerization of PMMA.

DP_i	1	10	100	1000	10 000
C_i(mole L^{-1})	10	8	5	1	0.1
P_{DPn} (i)	0.415	0.332	0.207	0.042	0.004

a. chain distributions P_{DPn} (i), P_{DPw} (i), $P_{DP\eta}$ (i), and P_{DPz} (i).
d. averages of the degree of polymerization $\overline{DP}_n$, $\overline{DP}_w$, $\overline{DP}_\eta$, $\overline{DP}_z$, and $\overline{DP}_{z+1}$.
e. average of the molecular weight $\overline{M}_n$, $\overline{M}_w$, $\overline{M}_\eta$, $\overline{M}_z$, and $\overline{M}_{z+1}$.
f. parameters for spread of the distribution V_{DP}, σ_{DP}, Δ, and PDI.

Problem 3.9 One gram of polyethylene produced by the elementary reaction of chain transfer to polymer with $\overline{M}_n = 20$ kg mole^{-1} and $PDI = 1$ is mixed with one gram of polyethylene produced by the elementary reaction of chain transfer to polymer with $\overline{M}_n = 10$ kg mole^{-1} and $PDI = 1$. If the MHS viscosity exponent of the sample is equal to $a = 0.8$, determine the molecular weights $\overline{M}_n$, $\overline{M}_w$, $\overline{M}_\eta$, $\overline{M}_z$, and $\overline{M}_{z+1}$ for the mixed sample.

Problem 3.10 Two grams of polystyrene produced by termination via combination with $\overline{M}_n = 200\,\text{kg mole}^{-1}$ and $PDI = 1.5$ is mixed with one gram of polystyrene produced by termination via disproportionation with $\overline{M}_n = 100\,\text{kg mole}^{-1}$ and $PDI = 2$. If the MHS viscosity exponent of the sample is equal to $a = 0.7$, determine the molecular weights $\overline{M}_n$, $\overline{M}_w$, $\overline{M}_\eta$, $\overline{M}_z$, and $\overline{M}_{z+1}$ for the mixed sample.

References

1 Moore, J.C., Gel permeation chromatography. I. New method for molecular weight distribution of high polymers. *Journal of Polymer Science Part A-General Papers*, 1964, **2**(2), 835.

2 Dey, M., Castoro, J.A., and Wilkins, C.L., Determination of molecular-weight distributions of polymers by MALDI-FTMS. *Analytical Chemistry*, 1995, **67**(9), 1575.

3 Schnoll-Bitai, I., Hrebicek, T., and Rizzi, A., Towards a quantitative interpretation of polymer distributions from MALDI-TOF spectra. *Macromolecular Chemistry and Physics*, 2007, **208**(5), 485.

4 Schnoll-Bitai, I., A comparative study of the properties of different distribution curves. *Macromolecular Chemistry and Physics*, 2002, **203**(12), 1754.

5 Deiber, J.A., Peirotti, M.B., and Bortolozzi, R.A., Estimation of molecular-weight distribution of elastomers and polymer melts through dynamic rheometry. *Journal of Elastomers and Plastics*, 1993, **25**(1), 22.

6 Billmeyer, F.W., Characterization of molecular weight distributions in high polymers. *Journal of Polymer Science Part C-Polymer Symposium*, 1965, **8**(1), 161.

7 Bohdanecky, M., Use of viscosity-average molecular weights for characterization of distribution functions of polymers. *Collection of Czechoslovak Chemical Communications*, 1966, **31**(10), 3979.

8 Kuo, C.Y., Provder, T., Koehler, M.E., and Kah, A.F., Use of a viscometric detector for size exclusion chromatography - characterization of molecular-weight distribution and branching in polymers. *ACS Symposium Series*, 1987, **352**, 130.

9 Yau, W.W. and Rementer, S.W., Polymer characterization by SEC-viscometry - molecular-weight (Mw), size (Rg) and intrinsic-viscosity (Iv) distribution. *Journal of Liquid Chromatography*, 1990, **13**(4), 627.

10 Malihi, F.B. et al., Development of a continuous gel-permeation chromatography viscosity detector - for the characterization of absolute molecular-weight distribution of polymers. *ACS Symposium Series*, 1984, **245**, 281.

11 Ashby, C.E., Reitenour, J.S., and Hammer, C.F., The molecular structure of polyethylene. 8. Number-average molecular weight by cryoscopy. *Journal of the American Chemical Society*, 1957, **79**(19), 5086.

12 Li, P.Y. and Rollings, J.E., Development of combined low-angle laser-light scattering detection with aqueous size exclusion chromatography for water-soluble bio-polymer characterization - molecular-weight distribution and polymer branching determination, in *191st ACS National Meeting*. 1986, American Chemical Society. p. 44.

13 Wu, C., Light-scattering characterization of the molecular weight distributions of some intractable polymers. *Polymers for Advanced Technologies*, 1997, **8**(4), 177.

14 Miyaki, Y., Einaga, Y., and Fujita, H., Excluded-volume effects in dilute polymer solutions. 7. Very high molecular-weight polystyrene in benzene and cyclohexane. *Macromolecules*, 1978, **11**(6), 1180–1186.

15 Charlesby, A., Gel formation and molecular weight distribution in long-chain polymers. *Proceedings of the Royal Society of London Series A-Mathematical and Physical Sciences*, 1954, **222**(1151), 542.

16 Gentekos, D.T., Sifri, R.J., and Fors, B.P., Controlling polymer properties through the shape of the molecular-weight distribution. *Nature Reviews Materials*, 2019, **4**(12), 761.

17 Nunes, R.W., Martin, J.R., and Johnson, J.F., Influence of molecular-weight and molecular-weight distribution on mechanical properties of polymers. *Polymer Engineering and Science*, 1982, **22**(4), 205.

18 Munstedt, H., Rheological properties and molecular structure of polymer melts. *Soft Matter*, 2011, **7**(6), 2273.

19 Snetkov, P. et al., Hyaluronic acid: the influence of molecular weight on structural, physical, physico-chemical, and degradable properties of biopolymer. *Polymers (Basel)*, 2020, **12**(8), 1800.

20 Orlova, T.M., Pavlova, S.S.A., and Dubrovina, L.V., Statistical approach to the molecular-weight distribution of condensation polymers. *Journal of Polymer Science Part A-Polymer Chemistry*, 1979, **17**(7), 2209.

21 Frisch, H.L. and Yeh, S.J., Measurement of viscosity-average molecular weight of polymer solutions of unknown concentration. *Journal of Polymer Science*, 1956, **20**(96), 431.

22 Ibrahim, F., Remark on viscosity average molecular weights. *Journal of Polymer Science Part B-Polymer Letters*, 1964, **2**(4), 441.

23 McKinney, P.V., Calculation of number-average and weight-average molecular weight of poly(vinyl chloride) polymers from intrinsic viscosity measurements. *Journal of Applied Polymer Science*, 1965, **9**(2), 583.

24 Kramer, O. and Good, W.R., Correlating Mooney viscosity to average molecular-weight. *Journal of Applied Polymer Science*, 1972, **16**(10), 2677.

25 Sokolov, N.V., Correlation of the viscosity-average, number-average, and weight-average molecular weights with the distribution of fractions of macromolecules. *Journal of Applied Chemistry of the USSR*, 1990, **63**(11), 2409.

26 Frith, E.M., The effect of solvent type on the viscosity of very dilute solutions of long-chain polymers. *Transactions of the Faraday Society*, 1945, **41**(1), 17.

27 Alfrey, T., Bartovics, A., and Mark, H., The effect of temperature and solvent type on the intrinsic viscosity of high polymer solutions. *Journal of the American Chemical Society*, 1942, **64**, 1557.

28 Peterlin, A., Effect of the velocity gradient on intrinsic viscosity of polymers in solution. *Journal of Polymer Science*, 1952, **8**(6), 621.

29 Kallistov, O.V., On the effect of velocity gradient of the intrinsic viscosity of solutions of high-molecular-weight polymer fractions. *Soviet Physics-Technical Physics*, 1959, **4**(1), 61.

30 Khabaz, F. and Khare, R., Effect of chain architecture on the size, shape, and intrinsic viscosity of chains in polymer solutions: a molecular simulation study. *Journal of Chemical Physics*, 2014, **141**(21), 214904.

31 Govaerts, R. and Smets, G., Determination of intrinsic viscosity of linear polymers in concentrated solution. *Journal of Polymer Science*, 1947, **2**(6), 612.

32 Rao, V.S.R., Determination of intrinsic viscosity at theta point and molecular weight of a linear polymer from viscosity measurements in a non-ideal solvent. *Journal of Polymer Science*, 1962, **62**(174), S157.

33 Khan, H.U., Gupta, V.K., and Bhargava, G.S., Determination of intrinsic viscosity of high polymers through a single specific viscosity measurement. *Polymer Communications*, 1983, **24**(6), 191.

34 Kamide, K. and Kataoka, A., Theoretical relationships between parameters in KUHN–MARK–HOUWINK–SAKURADA equation. *Makromolekulare Chemie*, 1969, **128**, 217.

35 Dobbin, C.J.B., Rudin, A., and Tchir, M.F., Evaluation of GPC methods for estimation of Mark–Houwink–Sakurada constants. *Journal of Applied Polymer Science*, 1980, **25**(12), 2985.

36 Zhongde, X., Mingshi, S., Hadjichristidis, N., and Fetters, L.J., Method for gel-permeation chromatography calibration and the evaluation of Mark–Houwink–Sakurada constants. *Macromolecules*, 1981, **14**(5), 1591.

37 Horta, A., Saiz, E., Barrales-Rienda, J.M., and Gomez, P.A.G., Direct estimation of the true Mark–Houwink–Sakurada parameters from gel-permeation chromatography, intrinsic viscosity and one average of molecular-weight. *Polymer*, 1986, **27**(1), 139.

38 Flory, P.J. and Yoon, D.Y., Moments and distribution functions for polymer chains of finite length. 1. Theory. *Journal of Chemical Physics*, 1974, **61**(12), 5358.

39 Uchiyama, M. et al., Asymmetric cationic polymerization of benzofuran through a reversible chain-transfer mechanism: optically active polybenzofuran with controlled molecular weights. *Journal of the American Chemical Society*, 2022, **144**(23), 10429.

40 Fueno, T. and Furukawa, J., Probability theory of asymmetric chain growth in polymerization. *Journal of Polymer Science Part A-General Papers*, 1964, **2**(8), 3681.

41 Asteasuain, M., Brandolin, A., and Sarmoria, C., Molecular weight distributions in styrene polymerization with asymmetric bifunctional initiators. *Polymer*, 2004, **45**(1), 321.

42 Corrigan, N. et al., Copolymers with controlled molecular weight distributions and compositional gradients through flow polymerization. *Macromolecules*, 2018, **51**(12), 4553.

43 Kottisch, V., Gentekos, D.T., and Fors, B.P., "Shaping" the future of molecular weight distributions in anionic polymerization. *ACS Macro Letters*, 2016, **5**(7), 796.

44 Gong, X.L., Hansen, E.W., and Chen, Q., Molecular weight distribution characteristics (of a polymer) derived from a stretched-exponential PGSTE NMR response function–simulation. *Macromolecular Chemistry and Physics*, 2012, **213**(3), 278.

45 Gentekos, D.T., Dupuis, L.N., and Fors, B.P., Beyond dispersity: deterministic control of polymer molecular weight distribution. *Journal of the American Chemical Society*, 2016, **138**(6), 1848.

46 Groeneveld, R.A. and Meeden, G., Measuring skewness and kurtosis. *Statistician*, 1984, **33**(4), 391.

47 Mardia, K.V., Measures of Multivariate Skewness and Kurtosis with Applications. *Biometrika*, 1970, **57**(3), 519.

48 Heyde, C.C., Property of lognormal distribution. *Journal of the Royal Statistical Society Series B-Statistical Methodology*, 1963, **25**(2), 392.

49 Nadarajah, S., A generalized normal distribution. *Journal of Applied Statistics*, 2005, **32**(7), 685.

50 Liu, K. et al., A comprehensive platform for the design and synthesis of polymer molecular weight distributions. *Macromolecules*, 2020, **53**(20), 8867.

51 Maurya, S.K. and Nadarajah, S., Poisson generated family of distributions: a review. *Sankhya-Series B-Applied and Interdisciplinary Statistics*, 2021, **83**(Suppl 2), 484.

52 Sahai, H. and Khurshid, A., Confidence intervals for the mean of a Poisson distribution - a review. *Biometrical Journal*, 1993, **35**(7), 857.

53 Sylvester, J.J., On arithmetical series. *Messenger of Math*, 1892, **21**(1–19), 87.

54 Kubin, M., A generalized exponential function as a model of polymer distribution curves. *Collection of Czechoslovak Chemical Communications*, 1967, **32**(4), 1505.

55 Lee, W. et al., Molecular weight distribution of polystyrene made by anionic polymerization. *Macromolecules*, 2000, **33**(14), 5111.

56 Fathi, H., Hamielec, A.E., and Davison, E.J., Modelling the anionic solution polymerization of butadiene - the effects of chain termination and long chain branching on molecular weight distribution development. *Polymer Reaction Engineering*, 1996, **4**(4), 243.

57 Flory, P.J., Molecular size distribution in linear condensation polymers. *Journal of the American Chemical Society*, 1936, **58**, 1877.

58 Goodwillie, T.G., Calculus III: Taylor series. *Geometry & Topology*, 2003, **7**(2), 645.

59 Green, J.H.S., A modified exponential molecular weight distribution function for high polymers. *Chemistry & Industry*, 1959, (29), 924.

60 Ferguson, T.S., Characterization of exponential distribution. *Annals of Mathematical Statistics*, 1964, **35**(3), 1199.

61 Clay, P.A. and Gilbert, R.G., Molecular-weight distributions in free-radical polymerizations. 1. Model development and implications for data interpretation. *Macromolecules*, 1995, **28**(2), 552.

62 Oshaughnessy, B. and Yu, J., Autoacceleration in free-radical polymerization. 1. Conversion. *Macromolecules*, 1994, **27**(18), 5067.

63 Oshaughnessy, B. and Yu, J., Autoacceleration in free-radical polymerization. 2. Molecular-weight distributions. *Macromolecules*, 1994, **27**(18), 5079.

64 Oshaughnessy, B. and Yu, J., Autoacceleration in free-radical polymerization. *Physical Review Letters*, 1994, **73**(12), 1723.

65 Meira, G.R. and Oliva, H.M., Molecular weight distributions in ideal polymerization reactors. An introductory review. *Latin American Applied Research*, 2011, **41**(4), 389.

66 Lukacs, E., A characterization of the gamma distribution. *Annals of Mathematical Statistics*, 1955, **26**(2), 319.

67 Ferguson, T.S., Characterization of gamma distribution. *Annals of Mathematical Statistics*, 1962, **33**(1), 300.

68 Zimm, B.H., The scattering of light and the radial distribution function of high polymer solutions. *Journal of Chemical Physics*, 1948, **16**(12), 1093.

69 Schulz, G.V., Experimental contributions to the thermodynamics of highly molecular solutions. The solubility of highly molecular materials VII. *Zeitschrift Fur Elektrochemie Und Angewandte Physikalische Chemie*, 1939, **45**, 652.

70 Stacy, E.W., Generalization of gamma distribution. *Annals of Mathematical Statistics*, 1962, **33**(3), 1187.

71 Williamson, N.H., Nyden, M., and Roding, M., The lognormal and gamma distribution models for estimating molecular weight distributions of polymers using PGSE NMR. *Journal of Magnetic Resonance*, 2016, **267**, 54.

72 Roding, M. et al., The gamma distribution model for pulsed-field gradient NMR studies of molecular-weight distributions of polymers. *Journal of Magnetic Resonance*, 2012, **222**, 105.

73 Wang, M., Smith, J.M., and McCoy, B.J., Continuous kinetics for thermal degradation of polymer in solution. *AIChE Journal*, 1995, **41**(6), 1521.

74 Stejskal, J. and Kratochvil, P., Distribution of the chemical composition of block and graft-copolymers assuming gamma-distribution of molecular weights of the precursor blocks. *Polymer Journal*, 1982, **14**(8), 603.

75 Tayal, A. and Khan, S.A., Degradation of a water-soluble polymer: molecular weight changes and chain scission characteristics. *Macromolecules*, 2000, **33**(26), 9488.

76 Gaddum, J.H., Lognormal distributions. *Nature*, 1945, **156**(3964), 463.

77 Conti, W. and Gigli, I., Non-Newtonian behavior of polymers with log-normal molecular weight distribution. *Journal of Polymer Science Part A-1-Polymer Chemistry*, 1966, **4**(5), 1093.

78 Solc, K., Cloud-point curves of polymers with logarithmic-normal distribution of molecular weight. *Macromolecules*, 1975, **8**(6), 819.

79 Netopilik, M., Analytical solution of the effect of axial-dispersion in the analysis of a polymer with the log-normal distribution by gel-permeation chromatography with a molecular-weight detector. *Polymer Bulletin*, 1983, **10**(9–10), 478.

80 Tan, L.Y. and Weber, T.W., A general logarithmic normal distribution for simulation of molecular-weight distribution for non-degradable polymers. *Chemical Engineering Communications*, 1994, **128**, 119.

81 Gong, X.L., Hansen, E.W., and Chen, Q., A simple access to the (log/normal) molecular weight distribution parameters of polymers using PGSTE NMR. *Macromolecular Chemistry and Physics*, 2011, **212**(10), 1007.

82 Monteiro, M.J., Williamson, N.H., Nyden, M., and Roding M., The lognormal and gamma distribution models for estimating molecular weight distributions of polymers using PGSE NMR. *Journal Magnetic Resonance*, 2016, 267, 54.

83 Monteiro, M.J., Fitting molecular weight distributions using a log-normal distribution model. *European Polymer Journal*, 2015, **65**, 197.

84 Walsh, D.J., Schinski, D.A., Schneider, R.A., and Guironnet, D., General route to design polymer molecular weight distributions through flow chemistry. *Nature Communications*, 2020, **11**(1), 3094.

85 Scholz, F.W., Characterization of the Weibull distribution. *Computational Statistics & Data Analysis*, 1990, **10**(3), 289.

86 Sundaram, R.K., Kumaran, S., Samson, E., and Pandiyarajan, R., Experimental and theoretical analysis of polymer blend composites using Weibull distribution. *Journal of the Chinese Institute of Engineers*, 2022, **45**(7), 588.

87 Broido, A. and Yow, H., Resolution of molecular-weight distributions in slightly pyrolyzed cellulose using Weibull function. *Journal of Applied Polymer Science*, 1977, **21**(6), 1677.

4

Z-transform in Polymerization Reactions

4.1 Introduction

Chain growth in polymerization occurs by coupling reactions between smaller chains, irrespective of the mechanism of polymerization. In the addition polymerization, a monomer (1-mer) reacts with the active species (radical, cation or anion) on an n-mer propagating chain to form a longer (n+1)-mer chain [1–3]. This implies that propagation in addition polymerization is by coupling of a 1-mer with an n-mer chain. In step-growth polymerization, an A-type functional group on an n-mer chain reacts with a complementary B-type functional group on an m-mer chain to produce a longer (n+m)-mer chain [4]. This implies that propagation in step-growth polymerization is by coupling of two shorter chains. Some chain termination reactions, like termination by combination, are a coupling reaction. In radical polymerization, the radical on an n-mer propagating chain and the radical on an m-mer chain combine to form a covalent bond and a longer (n+m)-mer terminated chain [5–7]. Therefore, termination by the combination of two propagating chains is a coupling reaction. Conversely, degradation of polymer chains by hydrolysis, reactive radical or ionic species, enzymes, and other environmental factors takes place by decoupling reactions to form shorter chains. In reversible addition fragmentation chain transfer (RAFT) polymerization, an m-mer propagating radical and an n-mer dormant chain combine (coupling reaction) to produce an (n+m)-mer internally generated iniferter which undergoes fragmentation (decoupling reaction) to form a new m-mer dormant chain and an n-mer propagating radical [8–11]. Therefore, deactivation of propagating radicals and activation of dormant chains in RAFT polymerization involve coupling and decoupling reactions. In β3 type chain scission reactions, the bond between the tertiary carbon carrying the radical and its neighboring carbon in an n-mer propagating chain splits to form a shorter (n-j)-mer terminated polymer chain and a shorter propagating chain with the radical on a more stable secondary carbon [12].

In radical polymerization, polymer chains are produced by the combination or disproportionation of two propagating chains. In termination by combination, a y-mer propagating chain (degree of polymerization $DP = y$) and distribution $f_R(x)$ and an (x-y)-mer propagating chain ($DP = x - y$) and distribution $f_R(x)$ undergo coupling to produce a polymer molecule with $DP = y + (x - y) = x$ and distribution $g_C(x)$ such that the following equality holds [13]:

$$g_C(x) = \sum_{y=1}^{x-1} f_R(y)\, f_R(x-y) \tag{4.1}$$

The above equation states that the probability of producing a polymer chain by combination having $DP = x$ is equal to the product of the probabilities of finding two propagating chains having DP values of y and $(x - y)$, respectively, summed over all possible values of y. In other words, the distribution of the produced polymer chain is the convolution of the distribution of the two propagating chains. In the DP domain, the distribution of produced polymer chains is the convolution of the distribution of propagating chains. If the distributions are transformed from the DP to the z domain, the distribution of produced polymer chains is just the product of the distribution of propagating chains as follows [13–15]:

$$G_c(z) = F_R(z)\, F_R(z) = F_R^2(z) \tag{4.2}$$

Algebraic operations on convoluted functions are more straightforward in the z domain as compared to the DP domain. This chapter covers the methodology for applying the z-transform to mathematical modeling of the molecular weight distribution and moments of the distribution of chains in polymerization reactions.

4.2 Definition of Z-transform

Consider the general function $f(x)$ defined with respect to an independent discrete variable in the x domain. The z-transform of the function $f(x)$ in the z domain is defined by the following relation [15, 16]:

$$F(z) = Z\{f(x), z\} = \sum_{x=-\infty}^{+\infty} z^{-x} f(x) \tag{4.3}$$

As applied to polymerization, the independent variable x is DP or the number of repeat units on a polymer chain, which is a positive integer, and the function $f(x)$ is the distribution function of DP. Therefore, the z-transform of the DP distribution function of a polymer is defined by the following relation [13]:

$$F_{DP}(z) = Z\{P_{DP}(i), z\} = \sum_{i=1}^{\infty} z^{-i} P_{DP}(i) \tag{4.4}$$

where $P_{Dp}(i)$ is the probability distribution function for DP of the polymer chains. The z-transform for a number (molar) distribution function of DP, $P_{Dpn}(i)$, is given by the following relation:

$$F_{DPn}(z) = Z\{P_{DPn}(i), z\} = \sum_{i=1}^{\infty} z^{-i} P_{DPn}(i) \tag{4.5}$$

The above equation can be slightly simplified by defining the variable $s = 1/z$ as follows:

$$F_{DP}(s) = Z\{P_{DP}(i), s\} = \sum_{i=1}^{\infty} s^{i} P_{DP}(i) \tag{4.6}$$

4.3 Moments of a Distributed Function in Z Domain

The number-average $\left(\overline{DP}_n\right)$, weight-average $\left(\overline{DP}_w\right)$, viscosity-average $\left(\overline{DP}_\eta\right)$, and other DP averages are computed from the kth order molar-based moments of the chain distribution function of a polymer, as was shown in Chapter 3. The z-transform of a polymer DP distribution function and its derivatives are used to find the kth molar-based moments of the polymer chain distribution in the z domain using the following approach. The first derivative of the z-transform of the mole-fraction distribution function, $F_{DPn}(s)$, with respect to s and evaluated at $s = 1$ is derived as follows:

$$\left.\frac{\partial F_{DPn}(s)}{\partial s}\right|_{s=1} = \left.\frac{\partial\left\{\sum_{i=1}^{\infty} s^i P_{DPn}(i)\right\}}{\partial s}\right|_{s=1} = \left.\sum_{i=1}^{\infty} P_{DPn}(i)\frac{\partial}{\partial s}\{s^i\}\right|_{s=1}$$
$$= \left.\sum_{i=1}^{\infty} i s^{i-1} P_{DPn}(i)\right|_{s=1} = \sum_{i=1}^{\infty} i P_{DPn}(i) \tag{4.7}$$

The last summation in the above equation, according to Eqs. (3.18) and (3.42) in Chapter 3, is equal to the first-order mole-fraction moment of DP distribution or $\overline{DP}_n$. Therefore, the first derivative of $F_{DPn}(s)$ evaluated at $s = 1$ is equal to $M_{n,1}$ as follows:

$$M_{n,1} = \sum_{i=1}^{\infty} i P_{DPn}(i) = \overline{DP}_n = \left.\frac{\partial F_{DPn}(s)}{\partial s}\right|_{s=1} \tag{4.8}$$

The second derivative of $F_{DPn}(s)$ with respect to s evaluated at $s = 1$ is derived as follows:

$$\left.\frac{\partial^2 F_{DPn}(s)}{\partial s^2}\right|_{s=1} = \left.\frac{\partial^2\left\{\sum_{i=1}^{\infty} s^i P_{DPn}(i)\right\}}{\partial s^2}\right|_{s=1} = \left.\sum_{i=1}^{\infty} P_{DPn}(i)\frac{\partial^2}{\partial s^2}\{s^i\}\right|_{s=1}$$
$$= \left.\sum_{i=1}^{\infty} P_{DPn}(i)\frac{\partial}{\partial s}\{i s^{i-1}\}\right|_{s=1} = \left.\sum_{i=1}^{\infty} i(i-1) s^{i-2} P_{DPn}(i)\right|_{s=1} \tag{4.9}$$
$$= \sum_{i=1}^{\infty}(i^2 - i) P_{DPn}(i) = \sum_{i=1}^{\infty} i^2 P_{DPn}(i) - \sum_{i=1}^{\infty} i P_{DPn}(i)$$

The first summation in the last expression in Eq. (4.9), according to Eq. (3.22) in Chapter 3, is equal to the second-order mole-fraction moment of the DP distribution, whereas the second summation is the first-order mole-fraction moment of the distribution. Therefore, the second derivative of $F_{DPn}(s)$ evaluated at $s = 1$ is related to $M_{n,2}$ and $M_{n,1}$ by the following relation:

$$\left.\frac{\partial^2 F_{DPn}(s)}{\partial s^2}\right|_{s=1} = \sum_{i=1}^{\infty} i^2 P_{DPn}(i) - \sum_{i=1}^{\infty} i P_{DPn}(i) = M_{n,2} - M_{n,1} \tag{4.10}$$

Upon substitution for $M_{n,1}$ from Eq. (4.8) above, $M_{n,2}$ is expressed in terms of derivatives of $F_{DPn}(s)$ by the following equation:

$$M_{n,2} = \left.\frac{\partial^2 F_{DPn}(s)}{\partial s^2}\right|_{s=1} + \left.\frac{\partial F_{DPn}(s)}{\partial s}\right|_{s=1} \tag{4.11}$$

According to the above equation, $M_{n,2}$ of the DP distribution is equal to the sum of the second and first derivatives of $F_{DPn}(s)$ evaluated at $s = 1$. Using the defining Eq. (3.45) in Chapter 3, $\overline{DP}_w$ is related to the derivatives of $F_{DPn}(s)$ by the following equation:

$$\overline{DP}_w = \frac{M_{n,2}}{M_{n,1}} = \frac{\left.\frac{\partial^2 F_{DPn}(s)}{\partial s^2}\right|_{s=1} + \left.\frac{\partial F_{DPn}(s)}{\partial s}\right|_{s=1}}{\left.\frac{\partial F_{DPn}(s)}{\partial s}\right|_{s=1}} \tag{4.12}$$

Using a similar approach, the third-order mole-fraction moment, $M_{n,3}$, and the z-average degree of polymerization, $\overline{DP}_z$, of a polymer chain distribution is given in terms of derivatives of $F_{DPn}(s)$ by the following expressions:

$$M_{n,3} = \langle DP_n^3 \rangle = \left.\frac{\partial^3 F_{DPn}(s)}{\partial s^3}\right|_{s=1} + 3\left.\frac{\partial^2 F_{DPn}(s)}{\partial s^2}\right|_{s=1} + \left.\frac{\partial^1 F_{DPn}(s)}{\partial s^1}\right|_{s=1} \tag{4.13}$$

$$\overline{DP}_z = \frac{M_{n,3}}{M_{n,2}} = \frac{\left.\frac{\partial^3 F_{DPn}(s)}{\partial s^3}\right|_{s=1} + 3\left.\frac{\partial^2 F_{DPn}(s)}{\partial s^2}\right|_{s=1} + \left.\frac{\partial^1 F_{DPn}(s)}{\partial s^1}\right|_{s=1}}{\left.\frac{\partial^2 F_{DPn}(s)}{\partial s^2}\right|_{s=1} + \left.\frac{\partial F_{DPn}(s)}{\partial s}\right|_{s=1}} \tag{4.14}$$

In general, the kth-order mole-fraction moment of a polymer distribution, $M_{n,k}$, can be expressed in terms of kth and lower order derivatives of the z-transform distribution function $F_{DPn}(s)$ evaluated at $s = 1$.

4.4 Properties of a Function in the Z Domain

Algebraic operations on distributed functions are simpler to perform in the z domain as compared to the DP domain. A useful property of functions in the z domain is scaling [17, 18]. The property of scaling states that the z-transform of a function $g(x) = A\,f(x)$ in the x domain, represented by $G(s)$, is equal to the z-transform of $f(x)$ in the z domain, $F(s)$, multiplied by the constant A, as demonstrated by the following derivation:

$$\begin{aligned} G(s) &= Z\{Af(x), s\} = \sum_{x=1}^{\infty} s^x A\, f(x) \\ &= A \sum_{x=1}^{\infty} s^x f(x) = A\, Z\{f(x), s\} = A\, F(s) \end{aligned} \tag{4.15}$$

The scaling property in the z domain is useful for converting between mole-fraction and concentration distribution function. In such cases, the z-transform concentration distribution function, $C(s)$, is obtained by multiplying the z-transform mole-fraction distribution function, $F(s)$, by the total concentration of polymer C, which is a constant, that is, $C(s) = CF(s)$. Another useful property of functions in the z domain is linearity [17, 18]. The linearity property states that the z-transform of the sum of several distribution functions $f_i(x)$ in the x domain is equal to the sum of their respective z-transform functions in the z-domain as follows. If $g(x) = \sum_{j=1}^{m} A_i f_i(x)$, then the z-transform of the function $g(x)$, namely $G(s)$, is obtained from the following derivation:

$$\begin{aligned} G(s) &= Z\left\{\sum_{j=1}^{m} A_j \{f_j(x), s\}\right\} = \sum_{x=1}^{+\infty} s^x \left\{\sum_{j=1}^{m} A_j f_j(x)\right\} \\ &= \sum_{j=1}^{m} A_j \sum_{x=1}^{\infty} s^x f_j(x) = \sum_{j=1}^{m} A_j Z\{f_j(x), s\} = \sum_{j=1}^{m} A_j F_j(s) \end{aligned} \tag{4.16}$$

The above linearity function of z-transforms is used in polymer reactions where chains are produced or consumed by multiple elementary reactions having different chain length distributions. For example, if the concentrations of polymer chains in radical polymerization produced by termination via combination and disproportionation in a time interval Δt are C_{tc} and C_{td}, respectively, and their respective distributions in the z-domain are $F_{tc}(s)$ and $F_{td}(s)$, the distribution function of the produced polymer chains in z domain, $G(s)$, in the time interval Δt is $G(s) = C_{tc}F_{tc}(s) + C_{td}F_{td}(s)$. The distribution of the produced polymer chains in the x domain can then be determined by taking the inverse z-transform of the function $G(s)$. Another property of the z-transform is the first backward difference [17, 18], which states that the z-transform of the difference between distribution functions $f(x = i)$ and $f(x = i - 1)$ in the x domain is equal to the z-transform of the distribution function $f(x)$, namely $F(s)$, multiplied by the factor $(1 - s)$. If $g(x) = f(x = i) - f(x = i - 1)$, its z transform $G(s)$ is derived as follows:

$$\begin{aligned} G(s) &= \sum_{i=1}^{\infty} s^i \left[f(x=i) - f(x=i-1)\right] = \sum_{i=1}^{\infty} s^i f(x=i) - \sum_{i=1}^{\infty} s^i f(x=i-1) \\ &= F(s) - \sum_{i=1}^{\infty} s^i f(x=i-1) = F(s) - \sum_{j=0}^{\infty} s^{j+1} f(x=j) \\ &= F(s) - s\, f(x)\big|_{x=0} - s\sum_{j=1}^{\infty} s^j f(x=j) = (1-s)\,F(s) \end{aligned} \tag{4.17}$$

In the above derivation, a new variable $j = i - 1$ was defined and $f(x = 0)$ was set equal to zero to reach the final expression for $G(s)$. Using a similar approach, the first forward difference of the function $g(x) = f(x = i + 1) - f(x = i)$ in the z domain is given by the following expression:

$$G(s) = \sum_{i=1}^{\infty} s^i \left[f(x=i) - f(x=i-1)\right] = (s-1)\,F(s) \tag{4.18}$$

The property of the first forward difference in the z domain is used in modeling the propagation reaction in addition polymerization [1–3]. In propagation, a propagating chain with $DP = x = i$ and distribution $f(x)$ reacts with a monomer M and produces a longer propagating chain with the same distribution function $f(x)$ but with $DP = x = i + 1$. Therefore, the change in the DP distribution of polymer chains due to the propagation reaction is given by the first forward difference function. Another useful property of the z-transform is convolution [17, 18]. According to this property, the convolution of two functions $f_1(x)$ and $f_2(x)$ in the x domain is the product of the two distribution functions in the z domain. If $g(x) = \sum_{y=1}^{x-1} f_1(x)\, f_2(x - y)$, then the z-transform of $g(x)$, namely $G(s)$, is given by the following derivation:

$$G(s) = Z\{g(x), z\} = \sum_{x=1}^{\infty} s^x \left\{\sum_{y=1}^{x-1} f_1(x)\, f_2(x-y)\right\} = F_1(s)\, F_2(s) \tag{4.19}$$

The convolution function is used in modeling combination or scission reactions in polymerization [19–29]. In termination by combination [5, 30–32], a propagating chain with $DP = y$ and distribution $f_R(x)$ combines with a second propagating chain with $DP = x - y$ and the same distribution $f_R(x)$ to produce a terminated polymer molecule with $DP = y + x - y = x$ and the following distribution:

$$g_C(x) = \sum_{y=1}^{x-1} f_R(y)\, f_R(x-y) \tag{4.20}$$

The distribution function of $g_c(x)$ in the z domain, $G_c(s)$, is given by the product of the distribution of propagating chains in the z domain, $F_R(s)$, as follows:

$$G_c(s) = F_R(s)\, F_R(s) = F_R^2(s) \tag{4.21}$$

The scission reaction is the reverse of the combination reaction [33–36]. In a scission reaction, a polymer chain breaks into two or more propagating chains with the same chain DP distribution function. Therefore, if the z-transform distribution function of polymer chains is $F_P(s)$, the z-transform distribution function of two chains produced by scission, $F_{SC}(s)$, is given by the following equation:

$$F_{SC}^2(s) = F_P(s) \tag{4.22}$$

4.5 Inverse Z-transform

Once algebraic operations are performed on the distribution function in the z-domain, the inverse z-transform operation is carried out to transfer the function back to the real x or *DP* domain. There are two methods to transfer the z-transform function back to the real x domain. The first method expands the function $F(s)$ in the z domain in a power series [37] in terms of the function $f(x = i)$ in the x domain using the following equation [38–41]:

$$F(s) = \sum_{x=1}^{\infty} s^x f(x) = s^1 f(x = 1) + s^2 f(x = 2) + s^3 f(x = 3) + \ldots + s^n f(x = n) \tag{4.23}$$

where $f(x = i)$ is the function $f(x)$ evaluated at $x = i$ with i taking integer values from one to infinity. Next, the above series expansion for $F(s)$ is used to derive the first, second, and third derivatives of $F(s)$ with respect to s evaluated at $s = 0$ as follows:

$$\begin{aligned}\left.\frac{\partial F(s)}{\partial s}\right|_{s=0} &= \left.\{f(x = 1) + 2s\, f(x = 2) + 3s^2 f(x = 3) + \ldots\, ns^{n-1} f(x = n)\}\right|_{s=0} \\ &= 1 \times f(x = 1)\end{aligned} \tag{4.24}$$

$$\begin{aligned}\left.\frac{\partial F^2(s)}{\partial s^2}\right|_{s=0} &= \left.\{0 + 2f(x = 2) + 3 \times 2s\, f(x = 3) + \ldots\, n \times (n-1)\, s^{n-2} f(x = n)\}\right|_{s=0} \\ &= 2 \times 1 \times f(x = 2)\end{aligned} \tag{4.25}$$

$$\begin{aligned}\left.\frac{\partial F^3(s)}{\partial s^3}\right|_{s=0} &= \left.\{0 + 0 + 3 \times 2s\, f(x = 3) + \ldots\, n \times (n-1) \times (n-2)\, s^{n-3} f(x = n)\}\right|_{s=0} \\ &= 3 \times 2 \times 1 \times f(x = 3)\end{aligned} \tag{4.26}$$

Based on the above formulas for the first, second, and third derivatives, the nth derivative of the function $F(s)$ with respect to s around $s = 0$ is given by the following equation [42]:

$$\begin{aligned}\left.\frac{\partial F^n(s)}{\partial s^n}\right|_{s=0} &= \{0 + 0 + 0 + \ldots + n \times (n-1) \times (n-2) + \ldots + 2 \times 1 f(x = n)\}|_{s=0} \\ &= n \times (n-1) \times (n-2) + \ldots + 2 \times 1 f(x = n) = n! \times f(x = n)\end{aligned} \tag{4.27}$$

Upon rearrangement of the above equation for the nth derivative, $f(x = n)$ is defined by the following expression:

$$f(x = n) = \frac{1}{n!} \left.\frac{\partial F^n(s)}{\partial s^n}\right|_{s=0} \tag{4.28}$$

The series expansion method described above does not provide a closed-form solution to the inverse z-transform of the function $F(s)$. To find a closed-form solution, the method of inverse z-transform by the sum of the remainder functions of the poles is used. In this method, if the function $G(z)$ where $z = 1/s$ is defined by the following expression in terms of the function $F(s)$:

$$G(z) = z^{n-1} F(z = 1/s) \tag{4.29}$$

Then, the inverse z-transform function of $F(s)$, namely $f(x)$, is given by the following expression:

$$f(x) = \sum_{i=1}^{\text{all poles}} R_{z=p_i} \tag{4.30}$$

where the remainder function $R_{z=p_i}$ is defined by the following expression:

$$R_{z=p_i} = \left\{ \frac{d^{m_i-1}}{dz^{m_i-1}} \left[\frac{(z-p_i)^{m_i} G(z)}{(m_i-1)!} \right] \right\}_{z=p_i} \tag{4.31}$$

The parameters p_i is the ith pole of the function $G(z)$ and m_i is the corresponding exponent. The poles of the function $G(z)$ are determined by the method of partial fraction expansion as previously described [43]. Briefly, the poles are evaluated by setting the denominator of the function $G(z)$ equal to zero and solving for z values. The corresponding exponent of each pole is the number of times that pole is repeated in the denominator.

4.5.1 Example Problem 1

The z-transform distribution function of propagating chains, $F_R(z)$, in a radical polymerization reaction is given by the following equation:

$$F_R(z) = \frac{p(1-p)}{(z-p)} \tag{4.32}$$

where p is the probability of propagation reaction, which is the ratio of the propagation rate to the rates of propagation and termination reactions. Find the inverse z-transform of the function $F_R(z)$ in closed-form using the method of the sum of the remainder functions of the poles.

Solution: The function $G_R(z)$ in terms of $F_R(z)$, according to Eq. (4.29), is given by the following expression:

$$G_R(z) = z^{n-1} F_R(z) = \frac{p(1-p) z^{n-1}}{(z-p)} \tag{4.33}$$

As the denominator is first order with respect to z, the above function has only one pole, which is evaluated by setting the denominator in the above equation equal to zero and solving for z as follows:

$$denominator\ of\ G_R(z) = 0 \implies (z-p) = 0 \implies \begin{cases} p_1 = p \\ m_1 = 1 \end{cases} \tag{4.34}$$

Knowing the pole and its corresponding exponent, the remainder function of $G_R(z)$ is determined by substituting $p_1 = p$ and $m_1 = 1$ in Eq. (4.31) as follows:

$$\begin{aligned} R_{z=p_1} = R_{z=p} &= \left\{ \frac{d^{m_1-1}}{dz^{m_1-1}} \left[\frac{(z-p)^{m_1} G(z)}{(m_1-1)!} \right] \right\}_{z=p} = \left\{ \frac{d^0}{dz^0} \left[p(1-p) z^{n-1} \right] \right\}_{z=p} \\ &= \left[p(1-p) z^{n-1} \right]_{z=p} = p(1-p) p^{n-1} \end{aligned} \tag{4.35}$$

Knowing the remainder function of $G_R(z)$, the distribution function $f(x)$ in the x domain in terms of propagation probability p and degree of polymerization n is obtained by substituting for $R_{z=p}$ from Eq. (4.35) in Eq. (4.30) as follows:

$$f(x=n) = \sum_{i=1}^{n} R_{z=p_i} = R_{z=p} = p(1-p)\,p^{n-1} \tag{4.36}$$

The above equation expresses the distribution of the number of repeat units (n) of the polymer chains as a function of propagation probability (p).

4.6 Application of Z-transform in Addition Polymerization

4.6.1 Z-transform in Termination by Combination

In radical polymerization, in the absence of chain transfer reactions, the propagating chains undergo either propagation or termination reactions. The distribution function of propagating chains, $f(x=n)$, in radical polymerization is expressed in terms of probability of propagation, p, by the following expression:

$$f_R(x=n) = (1-p)\,p^n \qquad n \geq 1 \qquad 0 \leq p \leq 1 \tag{4.37}$$

where $f_R(x=n)$ is the probability of finding a propagating chain with $DP = n$ and p is the rate of propagation divided by the sum of the rates of propagation and termination. According to Eq. (4.3), the z-transform of the above function, $F_R(z)$, is defined by the following expression:

$$F_R(z) = \sum_{n=1}^{\infty} z^{-n} f_R(x=n) = \sum_{n=1}^{\infty} z^{-n}(1-p)\,p^n = (1-p)\sum_{n=1}^{\infty}\left(\frac{p}{z}\right)^n \tag{4.38}$$

Expansion of the function $\sum_{n=1}^{\infty} x^n$ in a Taylor series, results in the following expression [44]:

$$\sum_{n=1}^{\infty} x^n = \sum_{n=0}^{\infty} x^n - x^0 = \sum_{n=0}^{\infty} x^n - 1 = \frac{1}{1-x} - 1 = \frac{x}{1-x} \tag{4.39}$$

Upon substituting for x in the above expansion series p/z and using the resulting series in Eq. (4.38), the function $F_R(z)$ converts to the following expression:

$$\begin{aligned} F_R(z) &= (1-p)\sum_{n=1}^{\infty}\left(\frac{p}{z}\right)^n \\ &= (1-p)\frac{p/z}{1-p/z} = \frac{p(1-p)}{z-p} \end{aligned} \tag{4.40}$$

After rewriting the above equation in terms of the variable $s = 1/z$, the function $F_R(s)$ is given by the following equation:

$$F_R(s) = \frac{p(1-p)}{1/s - p} = \frac{p(1-p)s}{(1-ps)} \tag{4.41}$$

According to Eq. (4.8), the first moment of the distribution of propagating radical chains, $M_{n,1,R}$, is derived by taking the first derivative of the function $F_R(s)$ with respect to s and evaluating the derivative at $s = 1$. The first moment $M_{n,1,R}$, which is equal to the number-average degree of polymerization of the propagating chains, $\overline{DP}_{n,R}$, is derived as follows:

$$M_{n,1,R} = \overline{DP}_{n,R} = \left.\frac{\partial F_R(s)}{\partial s}\right|_{s=1} = \frac{\partial}{\partial s}\left\{\frac{p(1-p)s}{(1-ps)}\right\}_{s=1} = \frac{p}{1-p} \tag{4.42}$$

Based on the above equation, the $\overline{DP}_{n,R}$ of the propagating chains increases with increasing probability of propagation. In chain termination by combination, a propagating chain with $DP = y$ and distribution $f_R(x)$ in the x domain with the corresponding distribution $F_R(s)$ in the s domain combines with another propagating chain with $DP = x - y$ from the same distribution $f_R(x)$ to produce a terminated polymer chain with $DP = y + x - y = x$ and distribution $F_c(s)$ in the s domain (see Eq. 4.20). The distribution of terminated chains produced by combination in the s domain, $F_c(s)$, is given by Eq. (4.21). The distribution function $F_c(s)$ is derived by substituting for $F_R(s)$ from Eq. (4.41) in Eq. (4.21) as follows:

$$F_c(s) = [F_R(s)]^2 = \frac{p^2(1-p)^2 s^2}{(1-ps)^2} \tag{4.43}$$

The first moment of polymer chains produced by combination, $M_{n,1,c}$, is derived by taking the first derivative of the function $F_c(s)$ with respect to s and evaluating the derivative at $s = 1$ (Eq. 4.8). The first moment $M_{n,1,c}$, which is equal to $\overline{DP}_{n,C}$, is derived as follows:

$$\begin{aligned} M_{n,1,c} = \overline{DP}_{n,C} &= \left.\frac{\partial F_c(s)}{\partial s}\right|_{s=1} = p^2(1-p)^2 \frac{\partial}{\partial s}\left[\frac{s^2}{(1-ps)^2}\right]_{s=1} \\ &= \left.\frac{2p^2(1-p)^2 s}{(1-ps)^3}\right|_{s=1} = \frac{2p^2}{(1-p)} \end{aligned} \tag{4.44}$$

The second moment of the distribution of terminated chains produced by combination, $M_{n,2,c}$, is derived using the defining Eq. (4.11) as follows:

$$\begin{aligned} M_{n,2,c} &= \left.\frac{\partial^2 F_c(s)}{\partial s^2}\right|_{s=1} + \left.\frac{\partial F_c(s)}{\partial s}\right|_{s=1} \\ &= \left.\frac{p^2(1-p)^2(2+4ps)}{(1-ps)^4}\right|_{s=1} + \left.\frac{2p^2(1-p)^2 s}{(1-ps)^3}\right|_{s=1} \\ &= \frac{2p^2(1+2p)}{(1-p)^2} + \frac{2p^2}{(1-p)} = \frac{2p^2(2+p)}{(1-p)^2} \end{aligned} \tag{4.45}$$

The weight-average degree of polymerization of chains produced by combination, $\overline{DP}_{w,c}$, is equal to the ratio of $M_{n,2,C}$ over $M_{n,1,C}$ (see Eq. 4.12). The $\overline{DP}_{w,c}$ is obtained by substituting for $M_{n,1,C}$ and $M_{n,2,C}$ in Eq. (4.12) from Eqs. (4.44) and (4.45), respectively, as follows:

$$\overline{DP}_{w,C} = \frac{M_{n,2,C}}{M_{n,1,C}} = \frac{2p^2(2+p)/(1-p)^2}{2p^2/(1-p)} = \frac{(2+p)}{(1-p)} \tag{4.46}$$

The closed-form distribution function of the terminated chains produced by combination is derived by the method of inverse z-transform by the sum of the remainder functions of the poles (Eqs. 4.29–4.31). First, the function $F_C(s)$ given by Eq. (4.43) is written in terms of $s = 1/z$ as follows:

$$F_c(z) = F_c(s)|_{s=1/z} = \left.\frac{p^2(1-p)^2 s^2}{(1-ps)^2}\right|_{s=1/z} = \frac{p^2(1-p)^2}{(z-p)^2} \tag{4.47}$$

Next, the defining Eq. (4.29) for $G(z)$ is used to find the corresponding expression for $G_c(z)$ from $F_c(z)$ as follows:

$$G_c(z) = z^{n-1} F_c(z) = \frac{p^2 (1-p)^2 z^{n-1}}{(z-p)^2} \tag{4.48}$$

Next, the poles of the above $G_C(z)$ function are determined by setting the denominator equal to zero and solving for z as follows:

$$Poles \Rightarrow (z-p)^2 = 0 \Rightarrow p_1 = p \Rightarrow m_1 = 2 \tag{4.49}$$

Then, knowing the pole and its corresponding exponent, the defining Eq. (4.31) is used to find the reminder function of $G_C(z)$ by substituting $p_1 = p$ and $m_1 = 2$ in Eq. (4.31) as follows:

$$\begin{aligned} R_{z=p} &= \left\{ \frac{d^{m_1-1}}{dz^{m_1-1}} \left[\frac{(z-p)^{m_1} G(z)}{(m_1-1)!} \right] \right\} \Bigg|_{z=p} = \left\{ \frac{d}{dz} \left[\frac{(z-p)^2 G(z)}{1!} \right] \right\} \Bigg|_{z=p} \\ &= \left[p^2 (1-p)^2 (n-1) z^{n-2} \right] \Big|_{z=p} = (n-1)(1-p)^2 p^n \end{aligned} \tag{4.50}$$

The distribution function of the polymer chains produced by combination of two radicals in the x domain, $f_c(x)$, is found by substituting for the remainder function of $G_C(z)$ from Eq. (4.50) in Eq. (4.30) as follows:

$$f_c(x = n) = \sum_{i=1}^{n} R_{z=p_i} = R_{z=p} = (n-1)(1-p)^2 p^n \tag{4.51}$$

This example illustrates the application of the method of z-transform in the determination of the distribution function, moments, and averages of the degree of polymerization of polymer chains produced by the elementary termination reaction by combination. The above approach can be used to determine the distribution function, moments, and averages of polymer chains produced or consumed by other elementary reactions in polymerization.

4.7 Application of Z-transform in Step-growth Polymerization

4.7.1 Z-transform in Condensation of Bifunctional Monomers *A—R—B*

In step-growth polymerization, a reactive functional group A on one chain in the reaction mixture condenses with a complementary reactive group B on another chain to produce a longer polymer chain [45–47]. For example, a terminal hydroxyl or amine functional group on a polyester or polyamide chain, respectively, reacts with a complementary terminal carboxylic acid group on another chain to form a longer polyester or polyamide chain [48–50]; as another example, a terminal hydroxyl or amine functional group on a polyurethane or polyurea chain reacts with an isocyanate functional group to produce a longer polyurethane or polyurea chain [51–54]. Step-growth polymerization reactions include those with complementary functional groups on separate monomers like A—R_1—A and B—R_2—B, complementary groups on one monomer like A—R—B, and complementary groups on mixed monomers like A—R_1—B and A_n—R_2, where n is the number of A groups per A_n—R_2 monomer [54]. The R, R_1, and R_2 represent the residual part of the monomers that are not involved in the condensation reaction. In this section, the chain DP distribution in condensation polymerization of the bifunctional monomer A—R—B is modeled by the method of z-transform.

In polymerization of $A{-}R{-}B$, two monomers react with a rate constant k to form a dimer by the following reaction [45, 55, 56]:

$$A{-}R{-}B + A{-}R{-}B \xrightarrow{k} A{-}R{-}(BA{-}R)_1{-}B \tag{4.52}$$

The residual functional group AB, produced by condensation of reactive group A with B, is a characteristic repeating group of the produced polymer. The reaction of monomers with hydroxyl and carboxylic acid complementary functional groups generates the characteristic ester residual group ($-COO-$) and the produced polymer is a polyester [48–50]. The reaction of monomers with amine and isocyanate groups generates the characteristic urea residual group ($-HNCONH-$) and the produced polymer is a polyurea [51–54]. Next, the produced dimer reacts with a monomer or a dimer with a reaction rate constant k, assuming the rate constant is independent of the degree of polymerization, to form a trimer or tetramer, respectively. In a stepwise process, pentamers, hexamers, heptamers, and longer polymer chains are formed by condensation of reactive groups of the shorter chains. The conversion of $A{-}R{-}B$ monomers to longer chains in the reaction, α, is defined by the following equation [57]:

$$\alpha = \frac{N_0 - N_\alpha}{N_0} \tag{4.53}$$

where N_0 and N_α are the initial number of molecules/monomers in the reaction and the number of molecules/chains at conversion α, respectively. Rearrangement and differentiation of the above equation yield the following expressions for N_α and dN_α as a function of α:

$$N_\alpha = N_0\,(1-\alpha) \tag{4.54}$$

$$dN_\alpha = N_0\, d\alpha \tag{4.55}$$

The molar probability distribution of n-mers at conversion α, $f\,(\alpha, n)$, is defined by the following equation:

$$f\,(\alpha, n) = \frac{N_n}{N_\alpha} \quad s.t. \quad \sum_{n=1}^{\infty} f\,(\alpha, n) = 1 \tag{4.56}$$

where N_n is the number of n-mer chains at conversion α. The abbreviation "s.t." in the above equation represents "subject to condition." Substitution for N_α from Eq. (4.54) in the above equation, followed by rearrangement and differentiation, yields the following expressions for N_n and dN_n as a function of α and n:

$$N_n = N_\alpha f\,(\alpha, n) = N_0\,(1-\alpha)\, f\,(\alpha, n) \tag{4.57}$$

$$dN_n = d\,[N_0\,(1-\alpha)\, f\,(\alpha, n)] = N_0\,[(1-\alpha)\, df\,(\alpha, n) - f\,(\alpha, n)\, d\alpha] \tag{4.58}$$

4.7.1.1 Conservation Equation for n-mer Chains

At any given reaction conversion, n-mer polymer chains are produced by condensation of $A{-}R{-}(BA{-}R)_{i-1}{-}B$ chains ($DP = i$) with $A{-}R{-}(BA{-}R)_{n-i-1}{-}B$ chains ($DP = n - i$) as shown by the following reaction:

$$A{-}R{-}(BA{-}R)_{i-1}{-}B + A{-}R{-}(BA{-}R)_{n-i-1}{-}B \xrightarrow{k} A{-}R{-}(BA{-}R)_{n-1}{-}B \tag{4.59}$$

where i satisfies the condition $1 \le i \le n-1$. In counting the number of repeat units in a chain, one monomer unit is assigned to the groups outside of the repeating unit. For example, the i-mer $A—R—(BA—R)_{i-1}—B$ consists of $(i-1)$ repeat units and one monomer unit of the groups $A—R—$ and $—B$ on the left and right sides of the repeat units. Conversely, n-mer polymer chains are consumed by the reaction of n-mers with any j-mer chain, subject to the condition $j \geq 1$, as shown by the following reaction:

$$A—R—(BA—R)_{n-1}—B + A—R—(BA—R)_{j-1}—B \xrightarrow{k} A—R—(BA—R)_{n+j-1}—B \tag{4.60}$$

The general conservation equation for n-mer chains is defined by the following expression:

$$\left\{\begin{array}{c}\text{rate of change}\\ \text{in the number of}\\ \text{moles of } n-\text{mers}\\ \text{with time}\end{array}\right\} = \left\{\begin{array}{c}\text{rate of molar}\\ \text{production of}\\ n-\text{mers}\end{array}\right\} - \left\{\begin{array}{c}\text{rate of molar}\\ \text{consumption}\\ \text{of } n-\text{mers}\end{array}\right\} \tag{4.61}$$

The above conservation expression in mathematical form is given by the following equation:

$$\frac{dN_n}{dt} = \frac{k}{2}\left\{\sum_{i=1}^{n-1} N_i N_{n-i}\right\} - kN_n \sum_{j=1}^{\infty} N_j \tag{4.62}$$

where N_i is the number of chains with $DP = i$, t is the reaction time, and k is the reaction rate constant. As the summation of N_j for $1 \leq j < \infty$ is equal to the total number of chains at conversion α, Eq. (4.62) reduces to the following:

$$\frac{dN_n}{dt} = \frac{k}{2}\left\{\sum_{i=1}^{n-1} N_i N_{n-i}\right\} - kN_n N_\alpha \tag{4.63}$$

where N_α is the total number of chains at conversion α. In the above equation, the term on the left side accounts for the accumulation of n-mer chains in the polymerization with reaction time, whereas the first and second terms on the right side account for the production and consumption of n-mer chains, respectively. The factor 1/2 is added to the first term on the right side of the equation to compensate for counting each combination of i and $n-i$ twice in the summation. For example, for $n = 7$, the terms of the summation are N_1N_6, N_2N_5, N_3N_4, N_4N_3, N_5N_2, N_6N_1, which is equal to $2N_1N_6$, $2N_2N_5$, and $2N_3N_4$.

4.7.1.2 Overall Conservation Equation

The overall conservation equation for polymer chains in the reaction at any conversion is derived by summing the left and right sides of Eq. (4.63) over all chains as follows:

$$\left\{\begin{array}{c}\text{rate of change}\\ \text{in the number}\\ \text{of moles of}\\ \text{all chains with time}\end{array}\right\} = \left\{\begin{array}{c}\text{rate of molar}\\ \text{production}\\ \text{of all chains}\end{array}\right\} - \left\{\begin{array}{c}\text{rate of molar}\\ \text{consumption}\\ \text{of all chains}\end{array}\right\} \tag{4.64}$$

Which in mathematical form is Eq. (4.63) summed over all values of $1 \leq n < \infty$ as follows:

$$\sum_{n=1}^{\infty} \frac{dN_n}{dt} = \sum_{n=1}^{\infty}\left\{\frac{k}{2}\sum_{i=1}^{n-1} N_i N_{n-i}\right\} - \sum_{n=1}^{\infty}\{kN_n N_\alpha\} \tag{4.65}$$

The following relationships apply to all chains in the reaction mixture:

$$\sum_{n=1}^{\infty} N_n = N_\alpha \tag{4.66}$$

$$\sum_{n=1}^{\infty}\sum_{i=1}^{n-1} N_i N_{n-i} = \sum_{n=1}^{\infty} N_n \sum_{i=1}^{\infty} N_i = N_\alpha^{\ 2} \tag{4.67}$$

$$\sum_{n=1}^{\infty} \frac{dN_n}{dt} = \frac{d}{dt}\sum_{n=1}^{\infty} N_n = \frac{dN_\alpha}{dt} \tag{4.68}$$

Upon substitution of the above relations in Eq. (4.65), the overall conservation equation reduces to the following equation:

$$\frac{dN_\alpha}{dt} = \frac{k}{2}N_\alpha^{\ 2} - kN_\alpha^{\ 2} = -\frac{k}{2}N_\alpha^{\ 2} \tag{4.69}$$

Upon substitution in the above equation for N_α and dN_α from the defining Eqs. (4.54) and (4.55), respectively, the overall conservation equation reduces to the following equation relating conversion to reaction time:

$$\frac{d\alpha}{dt} = \frac{K}{2}N_0 (1-\alpha)^2 \tag{4.70}$$

4.7.1.3 Conservation Equation for n-mer Chains in Terms of Conversion

The conservation Eq. (4.63) can be expressed in terms of fractional distribution of n-mer chains at a given conversion, $f(\alpha, n)$, using the following approach. First, N_n as defined by Eq. (4.57) in terms of $f(\alpha, n)$ is differentiated with respect to time to obtain the following relation:

$$\frac{dN_n}{dt} = N_0 \left[(1-\alpha)\frac{df(\alpha, n)}{dt} - f(\alpha, n)\frac{d\alpha}{dt}\right] \tag{4.71}$$

Using the defining Eqs. (4.54) and (4.57) for N_α and N_n, respectively, the right side of the conservation Eq. (4.63) is expressed in terms of $f(\alpha, n)$ as follows:

$$\begin{aligned}
&\frac{k}{2}\left\{\sum_{i=1}^{n-1} N_i N_{n-i}\right\} - kN_n N_\alpha \\
&\frac{k}{2}\left\{\sum_{i=1}^{n-1} N_0^{\ 2}(1-\alpha)^2 f(\alpha, i) f(\alpha, n-i)\right\} - kN_0^{\ 2}(1-\alpha)^2 f(\alpha, n) \\
&= \frac{kN_0^{\ 2}(1-\alpha)^2}{2}\left\{\left(\sum_{i=1}^{n-1} f(\alpha, i) f(\alpha, n-i)\right) - 2f(\alpha, n)\right\}
\end{aligned} \tag{4.72}$$

Upon substitution of Eq. (4.71) in the left side of Eq. (4.63) and Eq. (4.72) in the right side, the following equation is obtained for the conservation equation of the n-mers in terms of $f(\alpha, n)$, α, and t:

$$(1-\alpha)\frac{df(\alpha, n)}{dt} - f(\alpha, n)\frac{d\alpha}{dt} = \frac{kN_0^{\ 2}(1-\alpha)^2}{2}\left\{\left(\sum_{i=1}^{n-1} f(\alpha, i) f(\alpha, n-i)\right) - 2f(\alpha, n)\right\} \tag{4.73}$$

Dividing the above conservation equation for n-mers by the overall conservation equation, given by Eq. (4.70), yields the following dimensionless differential equation relating fractional n-mer distribution, $f(\alpha, n)$, to conversion α in condensation polymerization of difunctional monomer $A—R—B$:

$$(1-\alpha)\frac{df(\alpha,n)}{d\alpha}+f(\alpha,n)-\sum_{i=1}^{n-1}f(\alpha,i)\,f(\alpha,n-i)=0 \tag{4.74}$$

4.7.1.4 Z-transform of Conservation Equation for n-mers

Based on the defining Eq. (4.3), the z-transform of the conservation Eq. (4.75) is given by the following expression:

$$(1-\alpha)\frac{dZ\{f(\alpha,n)\}}{d\alpha}+Z\{f(\alpha,n)\}-Z\left\{\sum_{i=1}^{n-1}f(\alpha,i)\,f(\alpha,n-i)\right\}=0 \tag{4.75}$$

Upon representing the z-transform of the function $f(\alpha, z)$ by $F(\alpha, z)$ and using the defining Eq. (4.19) for the z transform of the convolution of two functions $f_1(x)$ and $f_2(x)$ having the same distribution, Eq. (4.75) is reduced to the following form:

$$(1-\alpha)\frac{dF(\alpha,z)}{d\alpha}+F(\alpha,z)-F(\alpha,z)^2=0 \tag{4.76}$$

The above equation can be solved for $F(\alpha, z)$ by the method of separation of variables [58–60]. The function $F(\alpha, z)$ and the variable α can be separated by rearrangement of the above equation and using the relation $1/(x^2-x)=-1/x+1/(x-1)$ to yield the following equation:

$$\frac{-dF(\alpha,z)}{F(\alpha,z)}+\frac{dF(\alpha,z)}{[F(\alpha,z)-1]}=\frac{d\alpha}{(1-\alpha)} \tag{4.77}$$

Integration of each term in Eq. (4.77) with respect to α yields the following equation relating $F(\alpha, z)$ to α:

$$-\ell nF(\alpha,z)+\ell n[F(\alpha,z)-1]=-\ell n(1-\alpha)+\ell nC \tag{4.78}$$

where the term ℓnC is a constant of integration. It can be shown, upon rearrangement of Eq. (4.78), that $F(\alpha, z)$ is given in terms of α by the following equation:

$$F(\alpha,z)=\frac{1-\alpha}{1-\alpha-C} \tag{4.79}$$

The constant C in the above equation is evaluated using the initial condition for the polymerization reaction as described in the next section.

4.7.1.5 Z-transform of Initial Reaction Condition

A common initial condition in A—R—B polymerization is that the reaction mixture contains only monomer A—R—B ($DP = 1$) at $t = 0$. This initial condition is expressed mathematically using the delta function $\delta(x)$ as follows [61, 62]:

$$f(\alpha,n)|_{\alpha=0}=f(0,n)=\begin{cases}1 & \in n=1\\ 0 & \in n>1\end{cases}=\delta(n-1) \tag{4.80}$$

The function $\delta(n-1)$ is equal to unity for $n = 1$ and zero for all other values of n. Using the defining Eq. (4.3), the z-transform of the initial condition given by Eq. (4.80) is as follows:

$$F(0,z)=Z\{f(0,n),z\}=\sum_{n=1}^{\infty}z^{-n}\delta(n-1)=z^{-1}\times 1=z^{-1} \tag{4.81}$$

The constant C in the equation for $F(\alpha, z)$ (Eq. 4.79) is evaluated by substitution of the initial condition $F(0, z)$ from the above equation in Eq. (4.79) as follows:

$$F(0, z) = F(\alpha, z)\,|_{\alpha=0} = \left[\frac{1-\alpha}{1-\alpha-C}\right]_{\alpha=0} = \frac{1}{1-C} = z^{-1} \Rightarrow C = 1 - z \tag{4.82}$$

Upon substitution for C from the above relation in Eq. (4.79), the distribution of n-mers in the z-domain, $F(\alpha, z)$, is expressed in terms of α and z by the following equation:

$$F(\alpha, z) = \frac{1-\alpha}{1-\alpha-C} = \frac{1-\alpha}{1-\alpha-(1-z)} = \frac{1-\alpha}{z-\alpha} \tag{4.83}$$

4.7.1.6 Inverse Z-transform of Chain Distribution Function

The closed-form distribution function of the polymer chains in the DP domain in closed-form is derived from the distribution function in the z domain (Eq. 4.83) using the method of sum of the remainder functions of the poles as described in Section 4.5. Based on Eqs. (4.29) to (4.31), the $G(z)$ function, poles, and remainder function of $F(\alpha, z)$ are given by the following equations in terms of conversion α:

$$G(\alpha, z) = z^{n-1} F(\alpha, z) = \frac{(1-\alpha)\, z^{n-1}}{(z-\alpha)} \tag{4.84}$$

$$poles \Rightarrow \quad (z-\alpha) = 0 \quad \Rightarrow p_1 = \alpha \quad and \quad m_1 = 1 \tag{4.85}$$

$$R_{z=\alpha} = \left\{\frac{d^0}{dz^0}\left[\frac{(z-\alpha)(1-\alpha)\, z^{n-1}}{(z-\alpha)(1-1)!}\right]\right\}\Bigg|_{z=\alpha} = (1-\alpha)\,\alpha^{n-1} \tag{4.86}$$

$$f(\alpha, n) = R_{z=\alpha} = (1-\alpha)\,\alpha^{n-1} \quad \in n \geq 1 \tag{4.87}$$

Figure 4.1 shows the solution to $f(\alpha, n)$ as a function of n and α for $1 \leq n \leq 500$ and $\alpha =$ 0.95, 0.99, 0.995.

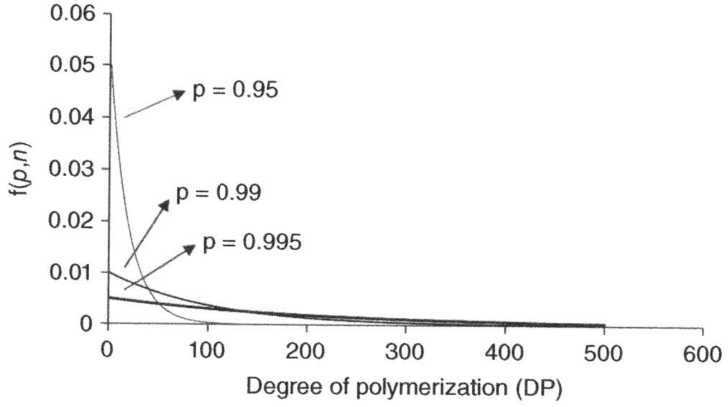

Figure 4.1 The number distribution of polymer chains produced by condensation polymerization of A—R—B monomers as a function of n and α.

The equations for number- $\left(\overline{DP}_{n,\alpha}\right)$ and weight-average $\left(\overline{DP}_{w,\alpha}\right)$ degree of polymerization of the chains produced by condensation polymerization of A—R—B monomers as a function of α are derived using the defining Eqs. (4.8) and (4.12), respectively, by the following approach.

First, the distribution function of the chains in the z domain, as given by Eq. (4.83), is rewritten in terms of $s = 1/z$ as follows:

$$F(\alpha, s) = F(\alpha, z)\big|_{z=1/s} = \left[\frac{1-\alpha}{z-\alpha}\right]_{z=1/s} = \frac{(1-\alpha)\,s}{(1-\alpha s)} \tag{4.88}$$

Next, the relation for $M_{n,1}$ or $\overline{DP}_{n,\alpha}$ (Eq. 4.8) is derived by taking the first derivative of $F(\alpha, s)$ with respect to s and evaluating it at $s = 1$ as follows:

$$\overline{DP}_{n,\alpha} = M_{n,1} = \left.\frac{\partial F(\alpha, s)}{\partial s}\right|_{s=1} = \frac{\partial}{\partial s}\left[\frac{(1-\alpha)\,s}{(1-\alpha s)}\right]\Bigg|_{s=1} = \frac{1}{1-\alpha} \tag{4.89}$$

Then, the relation for $M_{n,2}$ (Eqs. 4.10 and 4.11) is derived by taking the second derivative of $F(\alpha, s)$ with respect to s and evaluate it at $s = 1$ as follows:

$$M_{n,2} - M_{n,1} = \left.\frac{\partial^2 F(\alpha, s)}{\partial s^2}\right|_{s=1} = \frac{\partial^2}{\partial s^2}\left[\frac{(1-\alpha)\,s}{(1-\alpha s)}\right]\Bigg|_{s=1} = \frac{2\alpha}{(1-\alpha)^2} \tag{4.90}$$

$$M_{n,2} = \left.\frac{\partial^2 F(\alpha, s)}{\partial s^2}\right|_{s=1} + M_{n,1} = \frac{2\alpha}{(1-\alpha)^2} + \frac{1}{1-\alpha} = \frac{1+\alpha}{(1-\alpha)^2} \tag{4.91}$$

According to Eq. (4.12), the $\overline{DP}_{w,\alpha}$ of $F(\alpha, s)$ is given by the following equation:

$$\overline{DP}_{w,\alpha} = \frac{M_{n,2}}{M_{n,1}} = \frac{1+\alpha}{1-\alpha} \tag{4.92}$$

The above approach can be used to determine the distribution function, moments, and averages of the distribution of polymer chains produced by condensation polymerization of A—R—B monomers with any initial reaction condition.

4.7.2 Z-transform in Condensation Polymerization of Bifunctional Monomer *A—R—B* and Monofunctional Monomer *A—x*

A more common step-growth polymerization system is the condensation polymerization of bifunctional monomer *A—R—B* with monofunctional *A—x* [47, 55]. The absence of *A—x* can lead to uncontrolled reaction viscosity and temperature, runaway reaction, and uncontrolled molecular weight. In practice, the bifunctional monomer *A—R—B* contains small impurities, thus *DP* is naturally limited to a finite value. The addition of a small amount of *A—x* to the reaction places an upper limit on *DP* and consequently on the viscosity [63]. At 100% conversion, the degree of polymerization is limited by the initial ratio of the moles of monofunctional $N_{A,0}$ and difunctional $N_{AB,0}$ monomers or the imbalance ratio, r, as defined by the following equation [64, 65]:

$$r = \frac{N_{AB,0}}{N_{AB,0} + N_{Ax,0}} \tag{4.93}$$

The reactive groups are balanced when $r = 1$ or $N_{Ax} = 0$. At any point in time, there are two types of chains in the reaction, namely monofunctional and bifunctional chains. In one reaction, a bifunctional chain with $i - 1$ repeat units ($DP = i$) reacts with another bifunctional chain with $j - 1$ repeat units ($DP = j$) to form a longer bifunctional chain with $i + j - 1$ repeat units ($DP = i + j$) as follows:

$$A{-}R{-}(BA{-}R)_{i-1}{-}B + A{-}R{-}(BA{-}R)_{j-1}{-}B \xrightarrow{k} A{-}R{-}(BA{-}R)_{i+j-1}{-}B \tag{4.94}$$

In another reaction, a bifunctional chain with $i-1$ repeat units ($DP=i$) reacts with a monofunctional chain with $j-2$ repeat units ($DP=j$) to form a longer monofunctional chain with $i+j-2$ repeat units ($DP=i+j$) as follows:

$$A—R—(BA—R)_{i-1}—B + A—R—(BA—R)_{j-2}—A—x \xrightarrow{k} A—R—(BA—R)_{i+j-2}—A—x \tag{4.95}$$

However, a monofunctional chain cannot react with another monofunctional chain, as both chains are terminated in the A functional group. Therefore, the number of moles of bifunctional chains, N_{AB}, is reduced with increasing conversion, whereas the number of monofunctional chains, N_{Ax}, remains constant and equal to the initial number of monofunctional monomers throughout the reaction as follows:

$$N_{Ax} = N_{Ax,0} \tag{4.96}$$

As the functional group A is in excess, the reaction conversion is defined in terms of the limiting functional group B by the following relation:

$$\alpha_B = \frac{N_{B,0} - N_B}{N_{B,0}} = \frac{N_{AB,0} - N_{AB}}{N_{AB,0}} \tag{4.97}$$

where α_B is the conversion of B functional groups, $N_{B,0}$ and N_B are the number of moles of B functional groups initially and at conversion α_B, respectively; and $N_{AB,0}$ and N_{AB} are the number of moles of bifunctional chains in the reaction mixture initially and at conversion α_B. N_B and N_{AB} are equal as each bifunctional AB chain has one B group. Rearrangement and differentiation of the above equation yield the following relations for N_{AB} and $d\,N_{AB}$ in terms of α_B:

$$N_{AB} = N_{AB,0}\,(1-\alpha_B) \tag{4.98}$$

$$dN_{AB} = -N_{AB,0}d\alpha_B \tag{4.99}$$

The distribution of degree of polymerization of AB and Ax n-mers, $f_{AB}\,(\alpha_B, n)$ and $f_{Ax}\,(\alpha_B, n)$, respectively, are defined by the following relations:

$$f_{AB}\,(\alpha_B, n) = \frac{N_{AB,n}}{N_{AB}} \Rightarrow \sum_{n=1}^{\infty} f_{AB}\,(\alpha_B, n) = 1.0 \tag{4.100}$$

$$f_{Ax}\,(\alpha_B, n) = \frac{N_{Ax,n}}{N_{Ax}} \Rightarrow \sum_{n=1}^{\infty} f_{Ax}\,(\alpha_B, n) = 1.0 \tag{4.101}$$

Upon substitution for N_{AB} and N_{Ax} from Eqs. (4.98) and (4.96), respectively, in the above equations and rearrangement, $N_{AB,n}$ and $N_{Ax,n}$ are given by the following equations:

$$N_{AB,n} = N_{AB}f_{AB}\,(\alpha_B, n) = N_{AB,0}\,(1-\alpha_n)\,f_{AB}\,(\alpha_B, n) \tag{4.102}$$

$$N_{Ax,n} = N_{Ax,0}f_{AB}\,(\alpha_B, n) \tag{4.103}$$

The derivatives of $N_{AB,n}$ and $N_{AB,x}$ with respect to their distribution functions and α_B are given by the following relations:

$$dN_{AB,n} = N_{AB.0}\,[(1-\alpha_n)\,df_{AB}\,(\alpha_B, n) - f_{AB}\,(\alpha_B, n)\,\alpha_n] \tag{4.104}$$

$$dN_{Ax,n} = N_{Ax,0}df_{AB}\,(\alpha_B, n) \tag{4.105}$$

4.7.2.1 Conservation Equation for Bifunctional n-mer Chains

At any conversion α_B, bifunctional n-mers with $DP = n$ are produced by the reaction of $A—R—(BA—R)_{i-1}—B$ chains ($DP = i$) with $A—R—(BA—R)_{n-i-1}—B$ ($DP = n - i$) chains. The bifunctional n-mers are consumed by two reactions. The first reaction is that of a bifunctional n-mer reacting with any chain with $DP = j$ (see reaction 4.94). The second reaction is that of a bifunctional n-mer reacting with any monofunctional chain with $DP = j$ as follows:

$$\begin{aligned} &A—R—(BA—R)_{n-1}—B + A—R—(BA—R)_{j-2}—BA—x \xrightarrow{k} \\ &A—R—(BA—R)_{n+j-2}—BA—x \end{aligned} \tag{4.106}$$

Based on the defining Eq. (4.61) for the conservation of n-mers, the conservation equation for the bifunctional n-mers is given by the following equation for polymerization of $A—R—B$ and $A—x$ monomers:

$$\begin{aligned} \frac{dN_{AB,n}}{dt} &= \frac{k}{2}\left\{\sum_{j=1}^{n-1} N_{AB,j}N_{AB,n-j}\right\} - kN_{AB,n}\sum_{j=1}^{\infty} N_{AB,j} - kN_{AB,n}\sum_{j=1}^{\infty} N_{Ax,j} \\ &= \frac{k}{2}\left\{\sum_{j=1}^{n-1} N_{AB,j}N_{AB,n-j}\right\} - kN_{AB,n}N_{AB} - kN_{AB,n}N_{Ax,0} \end{aligned} \tag{4.107}$$

The term on the left side of the above equation accounts for the accumulation of bifunctional n-mers in the reaction, whereas the first, second, and third terms on the right side are the rates of production of n-mers, consumption of bifunctional n-mers, and consumption of monofunctional n-mers, respectively. The factor ½ in the production term (first term on the right side) accounts for the repetition of each combination of j and $n - j$ in the summation. The conservation equation for all polymer chains is derived by summation of the left and right sides of the above equation over all n-mer chains, leading to the following equation for the rate of change of N_{AB} with reaction time:

$$\frac{dN_{AB}}{dt} = \frac{k}{2}N_{AB}\left(N_{AB} - 2N_{Ax,0}\right) \tag{4.108}$$

Upon substitution in Eq. (4.107) for $N_{AB,n}$ and $dN_{AB,n}$ from Eqs. (4.102) and (4.104), respectively, for N_{AB} and dN_{AB} from Eqs. (4.98) and (4.99), for imbalance ratio r from Eq. (4.93) to relate $N_{AB,0}$ to $N_{Ax,0}$, and dividing Eq. (4.107) by Eq. (4.108), the following differential equation is derived for the rate of change in the distribution of bifunctional n-mers, namely $f_{AB}(\alpha_B, n)$ in terms of α_B:

$$\begin{aligned} &\left[(1-\alpha_B) + 2\left(\frac{1}{r} - 1\right)\right]\frac{df_{AB}(\alpha_B, n)}{d\alpha_B} + f(\alpha_B, n) \\ &- \sum_{j=1}^{n-1} f_{AB}(\alpha_B, n)\, f_{AB}(\alpha_B, n-j) = 0 \end{aligned} \tag{4.109}$$

The above differential equation is subject to the following initial condition:

$$f_{AB}(\alpha_B, n)\big|_{\alpha_B=0} = f_{AB}(0, n) = \begin{cases} 1 & \in n = 1 \\ 0 & \in n > 1 \end{cases} = \delta(n-1) \tag{4.110}$$

The above initial condition states that the reaction mixture initially consists only of monomers $A—R—B$ and $A—x$. Based on the defining Eq. (4.3), the z-transform of the conservation Eq. (4.109) and its initial condition are given by the following relations:

$$\left[(1-\alpha_B) + 2\left(\frac{1}{r} - 1\right)\right]\frac{\partial F_{AB}(\alpha_B, z)}{\partial \alpha_B} + F_{AB}(\alpha_B, z) - \left[F_{AB}(\alpha_B, z)\right]^2 = 0 \tag{4.111}$$

$$F_{AB}(0, z) = z^{-1} \tag{4.112}$$

The above differentiation equation, subject to its initial condition, can be solved using the method of separation of variables as described in Sections 4.7.1.4 and 4.7.1.5. Upon taking the inverse z-transform of the function $F_{AB}(\alpha_B, z)$ using the method of sum of the remainder functions of the poles as described in section 4.7.1.6, the number distribution function of the bifunctional n-mers, namely $f_{AB}(\alpha_B, n)$, as a function of α_B and imbalance ratio r is given by the following equation:

$$f_{AB}(\alpha_B, n) = \left[1 - \frac{\alpha_B}{(2/r) - 1}\right]\left[\frac{\alpha_B}{(2/r) - 1}\right]^{n-1} \quad \in n \geq 1 \tag{4.113}$$

The solution to $f_{AB}(\alpha_B, n)$ as a function of n for α values of 0.95, 0.99, and 0.995, and $r = 0.99$ is presented in Figure 4.2; the solution to $f_{AB}(\alpha_B, n)$ as a function of n and α for $r = 0.995$ is presented in Figure 4.3. These figures demonstrate that the distribution of bifunctional n-mers shifts to higher n values as α and r approach one.

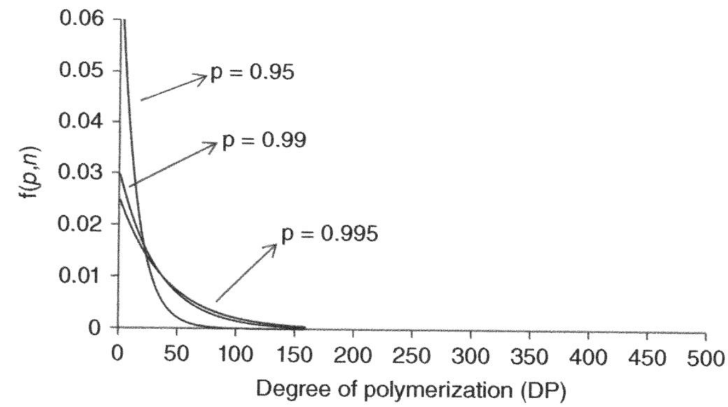

Figure 4.2 The distribution of bifunctional polymer chains produced by condensation polymerization of bifunctional A—R—B and monofunctional A—x monomers as a function of degree of polymerization n and conversion α for $r = 0.99$.

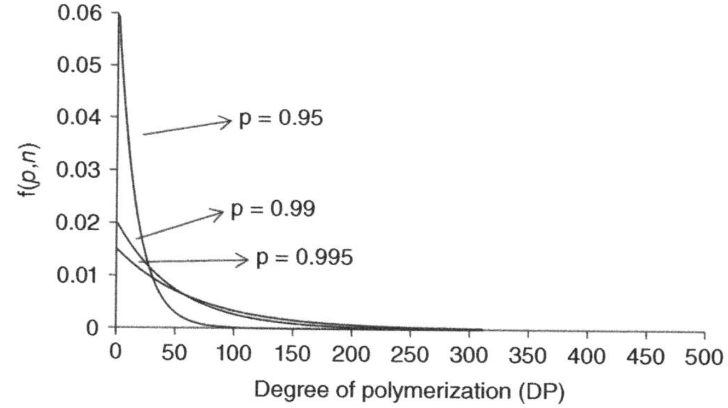

Figure 4.3 The distribution of bifunctional polymer chains produced by condensation polymerization of bifunctional A—R—B and monofunctional A—x as a function of degree of polymerization n and conversion α for $r = 0.995$.

4.7.2.2 Conservation Equation for Monofunctional n-mer Chains

At any conversion α_B, n-mer monofunctional chains with $DP = n$ are produced by the reaction of bifunctional $A—R—(BA—R)_{j-1}—B$ chains $(DP = j)$ with monofunctional $A—R—(BA—R)_{n-j-2}—BA—x$ chains $(DP = n - j)$, as shown by the following Rxn. 4.114. Similarly, n-mer monofunctional chains are consumed by the reaction of bifunctional $A—R—(BA—R)_{n-1}—B$ chains $(DP = n)$ with monofunctional $A—R—(BA—R)_{j-2}—BA—x$ chains $(DP = j)$ as shown by the following Rxn. 4.115.

$$A—R—(BA—R)_{j-1}—B + A—R—(BA—R)_{n-j-2}—BA—x \xrightarrow{k} A—R—(BA—R)_{n+j-2}—BA—x \tag{4.114}$$

$$A—R—(BA—R)_{n-1}—B + A—R—(BA—R)_{j-2}—BA—x \xrightarrow{k} A—R—(BA—R)_{n+j-2}—BA—x \tag{4.115}$$

Based on the defining Eq. (4.61) for the conservation of n-mers, the conservation equation for monofunctional n-mers is given by the following relation:

$$\begin{aligned}\frac{dN_{Ax,n}}{dt} &= k\sum_{j=1}^{n-1} N_{AB,j}N_{Ax,n-j} - kN_{Ax,n}\sum_{j=1}^{\infty} N_{AB,j}\\ &= k\sum_{j=1}^{n-1} N_{AB,j}N_{Ax,n-j} - kN_{Ax,n}N_{AB}\end{aligned} \tag{4.116}$$

The first term on the left side of the above equation accounts for the accumulation of monofunctional n-mers in the polymerization reaction. The first and second terms on the right side are the rates of production and consumption of monofunctional n-mers, respectively. Unlike Eq. (4.107) for the conservation of bifunctional n-mers, the factor ½ does not appear in the production term (the first term on the right side) of Eq. (4.116) because the combinations of j and $n - j$ in the summation correspond to bifunctional and monofunctional chains, respectively. The overall conservation equation for monofunctional polymers is derived by summing the left and right sides of Eq. (4.116) over all chains, leading to the following differential equation for the rate of change of N_{Ax} with reaction time:

$$\frac{dN_{Ax,n}}{dt} = 0 \Rightarrow N_{Ax} = N_{Ax,0} \tag{4.117}$$

Upon substitution for $N_{Ax,n}$ and $dN_{Ax,n}$ from Eqs. (4.103) and (4.105) and for the imbalance ratio r from Eq. (4.93), the following differential equation is derived for the rate of change in the distribution of monofunctional n-mers, namely $f_{Ax}(\alpha_B, n)$ in terms of α_B:

$$\begin{aligned}&\left[\left(\frac{2}{r} - 1\right) - \alpha_B\right]\frac{df_{Ax}(\alpha_B, n)}{d\alpha_B} + f_{Ax}(\alpha_B, n)\\ &-\sum_{j=1}^{n-1} f_{AB}(\alpha_B, j)\, f_{Ax}(\alpha_B, n - j) = 0\end{aligned} \tag{4.118}$$

The above differential equation is subject to the following initial condition:

$$f_{Ax}(\alpha_B, n)\big|_{\alpha_B=0} = f_{Ax}(0, n) = \begin{cases}1 & \in n = 1\\ 0 & \in n > 1\end{cases} = \delta(n - 1) \tag{4.119}$$

The above initial condition states that the reaction mixture initially consists only of monomers $A—x$ and $A—R—B$. Based on the defining Eq. (4.3), the z-transform of the conservation Eq. (4.118) and its initial condition are given by the following relations:

$$\left[\left(\frac{2}{r}-1\right)-\alpha_B\right]\frac{\partial F_{Ax}(\alpha_B,z)}{\partial \alpha_B}+2F_{Ax}(\alpha_B,z)-F_{Ax}(\alpha_B,z)\,F_{AB}(\alpha_B,z)=0 \tag{4.120}$$

$$F_{Ax}(0,z)=z^{-1} \tag{4.121}$$

Upon substitution for $F_{AB}(\alpha_B,z)$ from Eq. (4.113), the above differentiation equation, subject to its initial condition, is solved using the method of separation of variables as described in Sections 4.7.1.4 and 4.7.1.5. Upon taking the inverse z-transform of the function $F_{Ax}(\alpha_B,z)$ using the method of sum of the remainder functions of the poles as described in Section 4.7.1.6, the number distribution function of the monofunctional n-mers, namely $f_{Ax}(\alpha_B,n)$, as a function of α_B and imbalance ratio r is given by the following equation:

$$f_{Ax}(\alpha_B,n)=n\left[1-\frac{\alpha_B}{(2/r)-1}\right]^2\left[\frac{\alpha_B}{(2/r)-1}\right]^{n-1}\quad \in n\geq 1 \tag{4.122}$$

The solution to $f_{AB}(\alpha_B,n)$ as a function of n for α values of 0.95, 0.99, and 0.995, and $r=0.99$ is presented in Figure 4.4; the solution to $f_{AB}(\alpha_B,n)$ as a function of n and α for $r=0.995$ is presented in Figure 4.5. These figures demonstrate that the distribution of monofunctional n-mers shifts to higher n values as α and r approach one.

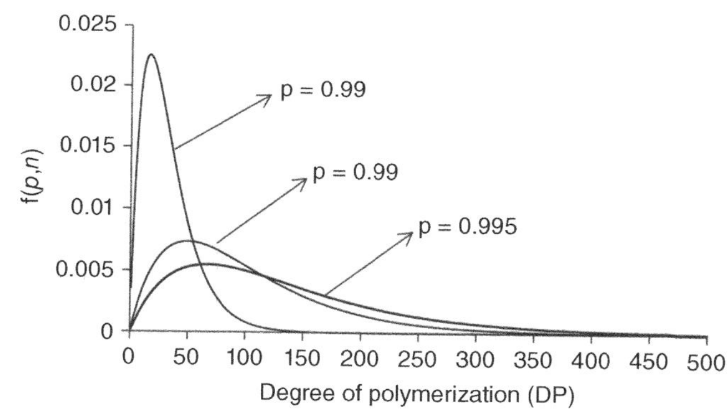

Figure 4.4 The distribution of monofunctional polymer chains produced by condensation polymerization of bifunctional $A—R—B$ and monofunctional $A—x$ monomers as a function of degree of polymerization n and conversion α for $r=0.99$.

4.8 Summary

Some elementary reactions in polymerization are combination reactions between multiple chains, like termination by combination or polymerization by step growth. Some other elementary reactions involve scission of a chain into two smaller chains. The differential conservation equations describing the distribution of the degree of polymerization of polymer chains produced/consumed by these combination/scission reactions involve the convolution of chain distributions. These differential conservation equations with convoluted distributions can be conveniently solved by transforming the equations to the z-domain, finding the solution in the z-domain, and transferring

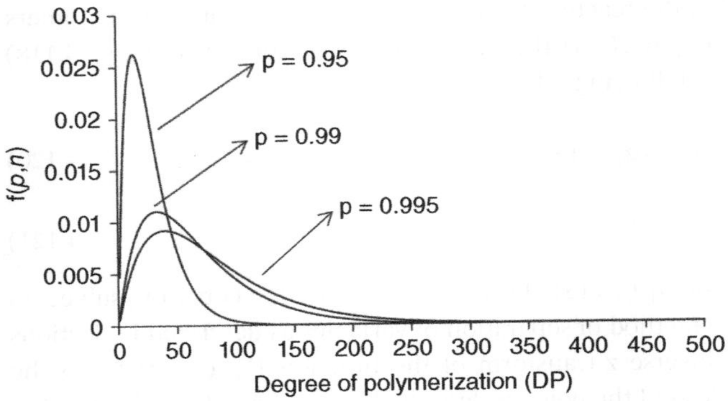

Figure 4.5 The distribution of monofunctional polymer chains produced by condensation polymerization of bifunctional A—R—B and monofunctional A—x monomers as a function of degree of polymerization n and conversion α for $r = 0.995$.

the solution back to the degree of polymerization domain. If $f(x)$ is a distributed, discrete function in the DP domain describing the probability distribution of the degree of polymerization (DP) of polymer chains, its z-transform function in the z-domain, denoted by $F(z)$, is defined as $\sum_{x=-\infty}^{x=+\infty} z^{-x} f(x)$. The following approach is used to find the moments and averages of the DP distribution of polymer chains as a function of reaction conversion. First, the differential conservation equation describing the rate of change of the DP distribution function, namely, $f(x)$ where $x = DP$, as a function of conversion and the corresponding relation for initial condition, are derived. Next, the differential conservation equation for $f(x)$ and its equation for the initial condition are transformed to the z-domain. Then, the z-transformed differential equation for $f(x)$, namely $F(z)$, is integrated and solved for $F(z)$ subject to the initial condition. The moments $M_{n,k}$ and averages $\overline{DP}$ of the distribution of polymer chains are determined from $F(z)$ and its derivatives with respect to z. The distribution function $F(z)$ is inverse transformed back to the $x = DP$ domain by defining the function $G(z) = z^{n-1}F(z)$ where $1 \le n = DP < \infty$. Next, the poles (p_i is the ith pole) of the function $G(z)$ are determined. Finally, the DP distribution function $f(x)$ of the polymer chains in closed form is determined by the method of the sum of the remainder functions $(R_{z=p_i})$ of the poles of the function $G(z)$ using the formula $f(x = n) = \sum_{i=1}^{n} R_{z=p_i}$.

4.9 List of Symbols and Abbreviations

C: concentration of polymer chains
C_{td}: concentration of polymer chains produced by termination via disproportionation
C_{tc}: concentration of polymer chains produced by termination via combination
DP: degree of polymerization
$\overline{DP}$: average degree of polymerization
$\overline{DP}_n$: number-average degree of polymerization
$\overline{DP}_w$: weight-average degree of polymerization
$\overline{DP}_z$: z-average degree of polymerization
$\overline{DP}_\eta$: viscosity-average degree of polymerization
$\overline{DP}_{n,c}$: number-average degree of polymerization of chains produced by termination via combination

$\overline{DP}_{w,c}$: weight-average degree of polymerization of chains produced by termination via combination
$\overline{DP}_{n,R}$: number-average degree of polymerization of the propagating chains
$\overline{DP}_{n,\alpha}$: number-average degree of polymerization of polymer chains at conversion α
$\overline{DP}_{w,\alpha}$:weight-average degree of polymerization of polymer chains at conversion α
$f(x)$: distribution function of a random discrete variable in the x domain, where x is the degree of polymerization
$f_C(x)$: distribution function for the degree of polymerization of chains produced by termination via combination
$f_R(x)$: distribution function $f_R(x)$ for the degree of polymerization of propagating chains
$f_{sc}(x)$: distribution function for the degree of polymerization of chains produced by scission of a polymer chain
$f(\alpha, n)$: distribution function of the polymer chains as a function of conversion α and DP
$f_{AB}(\alpha_B, n)$: distribution function of bifunctional A—R—B polymer chains in step-growth polymerization as a function of conversion of the limiting B functional group α_B and DP
$f_{Ax}(\alpha_B, n)$: distribution function of monofunctional A—x polymer chains in step-growth polymerization as a function of conversion of the limiting B functional group α_B and DP
$F(s)$: s-transform of the distribution function $f(x)$ in the s domain where $s = 1/z$
$F(z)$: z-transform of the distribution function $f(x)$ in the z domain
$F_{DP}(z)$: z-transform of the distribution function $f_{DP}(x)$ in the z domain where x is the degree of polymerization
$F_{DPn}(s)$: s-transform of the distribution function $f_{DPn}(x)$ in the s domain where x is the number distribution of DP
$F_{DPn}(z)$: z-transform of the distribution function $f_{DPn}(x)$ in the z domain where x is the number distribution of DP
$F_c(s)$: s-transform of the distribution function $f_C(x)$ in the s domain for the degree of polymerization of chains produced by termination via combination
$F_c(z)$: z-transform of the distribution function $f_C(x)$ in the z domain for the degree of polymerization of chains produced by termination via combination
$F_{sc}(s)$: s-transform of the distribution function $f_{sc}(x)$ in the s domain for the degree of polymerization of chains produced by scission of a polymer chain
$F_p(s)$: s-transform of the distribution function $f_p(x)$ in the s domain for terminated polymer chains
$F(\alpha, z)$: z-transform of the distribution function of polymer chains in step-growth polymerization in the z domain as a function of conversion α and DP
$F_R(s)$: s-transform of the distribution function $f_R(x)$ of the propagating chains in the s domain
$F_R(z)$: z-transform of the distribution function $f_R(x)$ of the propagating chains in the z domain
$F_{tc}(s)$: s-transform of the distribution function $f_{tc}(x)$ in the s domain for the polymer chains produced by termination via combination
$F_{td}(s)$: s-transform of the distribution function $f_{td}(x)$ in the s domain for the polymer chains produced by termination via disproportionation
$F_{AB}(\alpha_B, z)$: z-transform of the distribution function $f_{AB}(\alpha_B, n)$ in the z domain for bifunctional A—R—B polymer chains in step-growth polymerization as a function of conversion of the limiting B functional group α_B and DP
$F_{Ax}(\alpha_B, z)$: z-transform of the distribution function $f_{Ax}(\alpha_B, n)$ in the z domain for monofunctional A—x polymer chains in step-growth polymerization as a function of conversion of the limiting B functional group α_B and DP

$g_c\,(x)$: distribution function of the polymer chains produced by a combination reaction as the convolution of the distribution functions of two propagating chains
$G\,(z)$: distribution function $F\,(z)$ in the z-domain multiplied by z^{n-1} generated to find the inverse z-transform of $F\,(z)$
$G_C\,(z)$: distribution function in the z-domain of the polymer chains produced by the combination reaction as the convolution of the distribution functions of two propagating chains
$G_R\,(z)$: distribution function in the z-domain of the propagating chains in addition polymerization
$G\,(\alpha, z)$: distribution function $F\,(\alpha, z)$ in the z-domain multiplied by z^{n-1} generated to find the inverse z-transform of $F\,(\alpha, z)$
k: reaction rate constant independent of the degree of polymerization for step-growth polymerization
m_i exponent of the pole $z = z_{pi}$ of the z-transform function $G\,(z)$
$M_{n,k}$: kth moment of the mole-fraction probability distribution of polymer chains
$M_{n,k,R}$: kth moment of the mole-fraction probability distribution of the propagating chains in addition polymerization
$\overline{M}_n$: number-average molecular weight of the chains in a polymer sample
$\overline{M}_w$: weight average molecular weight of the chains in a polymer sample
$\overline{M}_z$: z average molecular weight of the chains in a polymer sample
$\overline{M}_{z+1}$: z+1 average molecular weight of the chains in a polymer sample
$\overline{M}_\eta$: viscosity-average molecular weight of the chains in a polymer sample
N_0: initial number of moles of the molecules in the reaction corresponding to zero conversion
$N_{AB,0}$: initial number of moles of the bifunctional $A—R—B$ chains in the reaction in step-growth polymerization corresponding to zero conversion
N_{AB}: number of moles of the bifunctional $A—R—B$ chains in the reaction in step-growth polymerization corresponding to conversion α
$N_{AB,j}$: number of moles of the bifunctional $A—R—B$ chains with $DP = j$ in the reaction in step-growth polymerization corresponding to conversion α
$N_{Ax,0}$: initial number of moles of the monofunctional $A—x$ chains in the reaction in step-growth polymerization corresponding to zero conversion
N_{Ax}: number of moles of the monofunctional chains in the reaction in step-growth polymerization, corresponding to conversion α
$N_{Ax,j}$: number of moles of the monofunctional $A—x$ chains with $DP = j$ in the reaction in step-growth polymerization corresponding to conversion α
$N_{B,0}$: initial number of moles of the B functional groups in $A—R—B$ chains in the reaction in step-growth polymerization at zero conversion
N_B: number of moles of the B functional groups in chains in the reaction in step-growth polymerization at conversion α
N_j number of moles of the chains with $DP = j$ in the reaction in step-growth polymerization at any time t or any conversion α
N_α number of moles of the chains in the reaction in step-growth polymerization at conversion α
p: probability of propagation in addition polymerization, or the probability of a functional group reacting in step-growth polymerization
p_i: ith pole of the distribution function $G(z)$ which is equal to the function $F\,(z)$ in the z-domain multiplied by z^{n-1} generated to find the inverse z-transform of $F\,(z)$
$p_{DP}\,(i)$: DP probability distribution function of polymer chains for any $DP = i$
$P_{DPn}\,(i)$: number DP probability distribution function of polymer chains for any $DP = i$
r: imbalance ratio

$R_{z=p_i}$: remainder function of the pole $z = z_{pi}$ of the function $G(z)$ in the z-domain for the determination of the inverse z-transform of the function $f(x)$

s: s-domain dependent variable where $s = 1/z$

t: reaction time

x: independent variable in the domain of the degree of polymerization (DP)

z: z-domain dependent variable

z_{pi}: value of the ith pole of the z-transform function $G(z)$

α: conversion of a polymerization reaction

α_B: conversion of the limiting functional group B in step-growth polymerization reaction

$\delta(x)$ delta function for a discontinuous random positive integer x

4.10 Practice Problems

Problem 4.1 The function $F_{DPn}(s)$ is defined as the s-transform of the distribution function $f_{DPn}(x)$ in the s domain, where x is the number distribution of DP.

a. Using the same approach used to derive Eqs. (4.8–4.11), show that the third order number-based moment of the DP distribution of the polymer chains, namely $M_{n,3}$, is given by the following equation in terms of the second and third derivatives of $F_{DPn}(s)$ with respect to s and evaluated at s = 1.

$$M_{n,3} = \left.\frac{\partial^3 F_{DPn}(s)}{\partial s^3}\right|_{s=1} + 3\left.\frac{\partial^2 F_{DPn}(s)}{\partial s^2}\right|_{s=1} + \left.\frac{\partial F_{DPn}(s)}{\partial s}\right|_{s=1}$$

b. Show that the z-average DP of the polymer chains, namely $\overline{DP}_z$, is given by the following equation in terms of the first, second, and third derivatives of $F_{DPn}(s)$ with respect to s evaluated at s = 1.

$$\overline{DP}_z = \frac{M_{n,3}}{M_{n,2}} = \frac{\left.\frac{\partial^3 F_{DPn}(s)}{\partial s^3}\right|_{s=1} + 3\left.\frac{\partial^2 F_{DPn}(s)}{\partial s^2}\right|_{s=1} + \left.\frac{\partial F_{DPn}(s)}{\partial s}\right|_{s=1}}{\left.\frac{\partial^2 F_{DPn}(s)}{\partial s^2}\right|_{s=1} + \left.\frac{\partial F_{DPn}(s)}{\partial s}\right|_{s=1}}$$

Problem 4.2 The method of inverse z-transform by the sum of the remainder functions of the poles is used to find a closed-form solution to the distribution function $f(x = n)$ in the x domain. Use the method of the sum of the remainder functions of the poles to find the probability distribution function $f(x = n)$ in the x-domain for the following functions:

a. $F(z) = \dfrac{z^2}{(z+3)^2}$

b. $F(z) = \dfrac{Kz(z+1)}{(z-1)^3}$ where K is a constant

c. $F(z)\ \dfrac{(1-\alpha)^{\beta} z}{(z-\alpha)^2}$ where α and β are constants

d. $F(z) = \dfrac{(1-\alpha)}{(z-\alpha)}$ where α is a constant

e. $F(z) = \dfrac{(1-\alpha)^{\beta} z^2}{(z-\alpha)^3}$ where α and β are constants

Problem 4.3 In a radical polymerization reaction, $\overline{DP}_n$, $\overline{DP}_w$, $\overline{DP}_z$, and $\overline{DP}_{z+1}$ averages of the propagating chains are 2L, 5L, 7L, and 9L, respectively, where L is a constant. The propagating chains terminate only by combination. Determine the following:

a. If $F_R(s)$ is the z-transform of the *DP* distribution function of the propagating chains, evaluate the following expressions:

$$\left.\frac{\partial F_R(s)}{\partial s}\right|_{s=1} \qquad \left.\frac{\partial^2 F_R(s)}{\partial s^2}\right|_{s=1} \qquad \left.\frac{\partial^3 F_R(s)}{\partial s^3}\right|_{s=1}$$

b. Determine the molar-based first, second, and third moments of the *DP* of propagating chains $M_{n,1}$, $M_{n,2}$, and $M_{n,3}$, respectively, from $F_R(s)$ and its derivatives.

c. Determine degree of polymerization averages $\overline{DP}_n$, $\overline{DP}_w$, $\overline{DP}_z$, and $\overline{DP}_{z+1}$ of the polymer chains produced by termination of propagating chains via combination.

Problem 4.4 In a radical polymerization reaction, $\overline{DP}_n$ and $\overline{DP}_w$ of the propagating chains are 1500 and 4000, respectively. If termination of the propagating chains is by combination, determine $\overline{DP}_n$ and $\overline{DP}_w$ of the produced polymer chains.

Problem 4.5 In a radical polymerization reaction, the kinetic chain length is *L*, the weight-average DP of the propagating chains $M_{n,2,R}/M_{n,1,R} = 2L$, $M_{n,3,R}/M_{n,2,R} = 3L$, and $M_{n,4,R}/M_{n,3,R} = 4L$. Determine the following for the above distribution of propagating chains:

a. find first order mole-fraction moment of the propagating chains $M_{n,1,R}$.

b. find $\partial F_R/\partial s\,|_{s=1}$ in z domain for the propagating chains.

c. find $\partial^2 F_R/\partial s^2\,|_{s=1}$ in z domain for the propagating chains.

d. find second order mole-fraction moment of the propagating chains $M_{n,2,R}$.

e. find third order mole-fraction moment of the propagating chains $M_{n,3,R}$.

f. find $\partial^3 F_R/\partial s^3\,|_{s=1}$ in z domain for the propagating chains.

g. find the first order mole-fraction moment of the terminated polymer chains via combination, $M_{n,1,p}$.

h. find the number-average DP of the terminated polymer chains produced by combination of the propagating chains, $\overline{DP}_{n,p}$.

i. find $\partial F_p/\partial s\,|_{s=1}$ in z domain for the terminated polymer chain by combination.

j. find second order mole-fraction moment of the terminated polymer chains via combination, $M_{n,2,p}$.

k. find $\partial^2 F_p/\partial s^2\,|_{s=1}$ in z domain for the terminated polymer chains by combination.

l. find the weight-average DP of the terminated polymer chains produced by combination of the propagating chains, $\overline{DP}_{w,p}$.

m. find third order mole-fraction moment of the terminated polymer chains via combination, $M_{n,3,p}$.

n. find $\partial^3 F_p/\partial s^3\,|_{s=1}$ in z domain for the terminated polymer chains by combination.

o. Find the z-average DP of the terminated polymer chains produced by combination of the propagating chains, $\overline{DP}_{z,p}$.

Problem 4.6 The probability distribution function for *DP* of the propagating chains in a radical polymerization reaction is given by the following expression in terms of the number of repeat units *i* and conversion α:

$$f_R(i) = (1 - \alpha)\,\alpha^i$$

a. determine the z-transform function of the distribution $f_R(i)$.

b. if the propagating chains terminate via combination to produce terminated polymer chains, determine the z-transform of the probability distribution function of the produced polymer chains.

c. if the viscosity exponent of the produced polymer chains is independent of molecular weight and equal to $a = 0.75$, determine $\overline{DP}_n, \overline{DP}_w, \overline{DP}_z, \overline{DP}_{a+1} = \overline{DP}_\eta$, and PDI of the produced polymer chains.

d. determine $M_{n,1}, M_{n,2}, M_{n,1+a}$ and $M_{n,z}$ of the produced polymer chains (Hint: use the first three terms of the Taylor series to determine $M_{n,1+a}$).

Problem 4.7 A step-growth polymerization reaction consists of bifunctional monomer $A—R—B$ and tetra-functional monomer $R'A_4$. The functional group A can only react with the complementary group B. The number of moles of the bifunctional monomer $A—R—B$ at the start of the reaction is $N_{AB,0}$ and the mole fraction of the functional group B is equal to r.

a. derive a differential equation for the number distribution function of $A—R—(BA—R)_{n-1}—B$ chains, namely $f_{AB,n}(\alpha, r)$, in terms of the number of repeat units n, conversion α, and r.

b. derive a differential equation for $F_{AB}(z)$, the z-transform of the distribution function $f_{AB,n}(\alpha, r)$ and solve the differential equation by the method of separation of variables.

c. derive the equation for the inverse z-transform of $F_{AB}(z)$ by the method of sum of the remainder functions of the poles to find an equation for $f_{AB,n}(\alpha, r)$ in terms of n, α, and r.

d. derive a differential equation for the mole-fraction distribution of $R'A_4$ chains, namely $f_{A4,n}(\alpha, r)$, in terms of n, α, and r.

e. derive a differential equation for $F_{A4}(z)$, the z-transform of the distribution function $f_{A4,n}(\alpha, r)$ and solve the differential equation by the method of separation of variables.

f. derive the equation for the inverse z-transform of $F_{A4}(z)$ by the method of sum of the remainder functions of the poles to find an equation for $f_{A4,n}(\alpha, r)$ in terms of n, α, and r.

g. plot the mole-fraction distribution of $A—R—(BA—R)_{n-1}—B$ and $R'A_4$ chains as a function of n, for $\alpha = 1$ and $r = 0.49$, 0.495, and 0.499.

Problem 4.8 Consider modeling the condensation polymerization of bifunctional monomers $A—R—B$ in step-growth polymerization in Section 4.7.1.

a. starting from Eq. (4.76) in subsection 4.7.1.4 for the z-transform of the distribution function of $A—R—(BA—R)_{i-1}—B$ propagating chains $F(\alpha, z)$, separate the function $F(\alpha, z)$ from the dependent variable α using the relation $1/(x^2 - x) = -1/x + 1/(x-1)$ to yield the following expression:

$$\frac{-dF(\alpha, z)}{F(\alpha, z)} + \frac{dF(\alpha, z)}{[F(\alpha, z) - 1]} = \frac{d\alpha}{(1-\alpha)}$$

b. integrate the above expression and rearrange the integrated equation to derive the following equation relating $F(\alpha, z)$ to α with C as the constant of integration:

$$F(\alpha, z) = \frac{1-\alpha}{1-\alpha-C}$$

c. use the above equation and the function for the initial reaction condition in condensation polymerization of $A—R—B$ monomers given by Eq. (4.80) to determine the constant C and use it to derive the following equation for $F(\alpha, z)$:

$$F(\alpha, z) = \frac{1-\alpha}{z-\alpha}$$

d. take the inverse z-transform of the above equation using the method of sum of the remainder functions of the poles to derive the following closed-form solution to the distribution function of $A—R—(BA—R)_{i-1}—B$ propagating chains $f(\alpha, n)$ in terms of n and α:

$$f(\alpha, n) = R_{z=\alpha} = (1-\alpha)\,\alpha^{n-1} \quad \in n \geq 1$$

References

1 Nikitin, A.N., Castignolles, P., Charleux, B., and Vairon, J.P., Determination of propagation rate coefficient of acrylates by pulsed-laser polymerization in the presence of intramolecular chain transfer to polymer. *Macromolecular Rapid Communications*, 2003, **24**(13), 778.

2 Ito, K., Relationship between kinetic chain-length and radical polymerization rate. *Journal of Polymer Science Part A-Polymer Chemistry*, 1976, **14**(7), 1761.

3 Williams, G. and Coe, J.S., The kinetic chain length in the polymerization of styrene by stannic chloride in carbon tetrachloride. *Chemistry & Industry (London)*, 1954, (2), 43.

4 Nishio, A. et al., Synthesis and characterization of ordered polyurethanes from a pair of diisocyanate monomers and ethylene glycol. *High Performance Polymers*, 2001, **13**(2), S233.

5 Nakamura, Y. and Yamago, S., Termination mechanism in the radical polymerization of methyl methacrylate and styrene determined by the reaction of structurally well-defined polymer end radicals. *Macromolecules*, 2015, **48**(18), 6450.

6 Shen, J.C., Tian, Y., Wang, G.B., and Yang, M.L., Termination process and the termination parameter-Kt of methyl-methacrylate bulk radical polymerization. *Science in China Series B-Chemistry*, 1990, **33**(9), 1040.

7 Zhu, S., Tian, Y., Hamielec, A.E., and Eaton, D.R., Radical trapping and termination in free-radical polymerization of methyl methacrylate. *Macromolecules*, 1990, **23**(4), 1144.

8 Moad, G. et al., Advances in RAFT polymerization: the synthesis of polymers with defined end-groups. *Polymer*, 2005, **46**(19), 8458.

9 Chiefari, J. et al., Living free-radical polymerization by reversible addition-fragmentation chain transfer: the RAFT process. *Macromolecules*, 1998, **31**(16), 5559.

10 Hawthorne, D.G., Moad, G., Rizzardo, E., and Thang, S.H., Living radical polymerization with reversible addition-fragmentation chain transfer (RAFT): direct ESR observation of intermediate radicals. *Macromolecules*, 1999, **32**(16), 5457.

11 Yin, H. et al., Selection principle of RAFT chain transfer agents and universal RAFT chain transfer agents. *Progress in Chemistry*, 2022, **34**(6), 1298.

12 Lena, J.B. et al., Effects of chain transfer agent and temperature on branching and β-scission in radical polymerization of 2-ethylhexyl acrylate. *Macromolecular Chemistry and Physics*, 2018, **219**(9), 1700579.

13 Mills, P.L., Determination of polymer-chain length distributions by numerical inversion of z-transforms. *Computers & Chemical Engineering*, 1986, **10**(4), 399.

14 Kliger, I. and Lipinski, W.C., Z transform of a product of 2 functions. *IEEE Transactions on Automatic Control*, 1964, **9**(4), 582.

15 Beutler, F.J., Theory and application of z-transform method. *American Mathematical Monthly*, 1965, **72**(10), 1149.

16 *Z-transform*, in *Wikipedia: the free encyclopedia*. 2025, Wikimedia Foundation.

17 El Attar, R., *Lecture notes on z-transform*. Mathematical series. 2004, Lula Press, 57.

18 Bohner, M. and Peterson, A., Laplace transform and z-transform: unification and extension. *Methods & Applications of Analysis*, 2002, **9**(1), 151.

19 Berger, K.C., Disproportionation and combination as termination mechanisms in radical polymerization of styrene. 2. Analysis of temperature dependencies. *Makromolekulare Chemie-Macromolecular Chemistry and Physics*, 1975, **176**(12), 3575.

20 Iedema, P.D. and Hoefsloot, H.C.J., Computing the trivariate chain length/degree of branching/number of combination points distribution for radical polymerization with transfer to polymer and recombination termination. *Macromolecular Theory and Simulations*, 2005, **14**(8), 505.

21 Elliott, J.R., Myers, R.L., and Roedel, G.F., Suspension polymerization of chlorotrifluoroethylene. *Industrial and Engineering Chemistry*, 1953, **45**(8), 1786.

22 Tobita, H. and Hatanaka, K., Long-chain branching in free-radical polymerization due to chain transfer to polymer. *Journal of Polymer Science Part B-Polymer Physics*, 1995, **33**(5), 841.

23 Tobita, H. and Hamashima, N., Monte Carlo simulation of size exclusion chromatography for branched polymers formed through free-radical polymerization with chain transfer to polymer. *Macromolecular Theory and Simulations*, 2000, **9**(8), 453.

24 Dias, R.C.S. and Costa, M.R.P.F.N., A new look at kinetic modeling of nonlinear free radical polymerizations with terminal branching and chain transfer to polymer. *Macromolecules*, 2003, **36**(23), 8853.

25 Iedema, P.D., Grcev, S., and Hoefsloot, H.C.J., Molecular weight distribution modeling of radical polymerization in a CSTR with long chain branching through transfer to polymer and terminal double bond (TDB) propagation. *Macromolecules*, 2003, **36**(2), 458.

26 Ahmad, N.M., Heatley, F., and Lovell, P.A., Chain transfer to polymer in free-radical solution polymerization of n-butyl acrylate studied by NMR spectroscopy. *Macromolecules*, 1998, **31**(9), 2822.

27 Ahmad, N.M. et al., Chain transfer to polymer and branching in controlled radical polymerizations of n-butyl acrylate. *Macromolecular Rapid Communications*, 2009, **30**(23), 2002.

28 Rozenberg, B.A., Ludvig, E.B., Gantmakher, A.R., and Medvedev, S.S., Effect of chain transfer to polymer in cationic polymerization of oxygen-containing cyclic compounds. *Vysokomolekulyarnye Soedineniya Seriya A & Seriya B*, 1999, **41**(12), 1929.

29 Litvinenko, G.I. and Arestyakubovich, A.A., Effect of chain transfer to polymer on molecular mass characteristics of polymers in anionic polymerization - comparison with radical one. *Vysokomolekulyarnye Soedineniya Seriya B*, 1988, **30**(6), 1218.

30 Ballard, N., Hamzehlou, S., Ruiperez, F., and Asua, J.M., On the termination mechanism in the radical polymerization of acrylates. *Macromolecular Rapid Communications*, 2016, **37**(16), 1364.

31 Nakamura, Y., Lee, R., Coote, M.L., and Yamago, S., Termination mechanism of the radical polymerization of acrylates. *Macromolecular Rapid Communications*, 2016, **37**(6), 506.

32 Barth, J., Buback, M., Russell, G.T., and Smolne, S., Chain-length-dependent termination in radical polymerization of acrylates. *Macromolecular Chemistry and Physics*, 2011, **212**(13), 1366.

33 Mayer, P.M., Glukhovtsev, M.N., Gauld, J.W., and Radom, L., The effects of protonation on the structure, stability, and thermochemistry of carbon-centered organic radicals. *Journal of the American Chemical Society*, 1997, **119**(52), 12889.

34 Menon, A.S., Henry, D.J., Bally, T., and Radom, L., Effect of substituents on the stabilities of multiply-substituted carbon-centered radicals. *Organic & Biomolecular Chemistry*, 2011, **9**(10), 3636.

35 Theilacker, W. and Ewald, M.L., On the theory of the stability of carbon radicals. *Naturwissenschaften*, 1943, **31**, 302.

36 Wen, Z., Li, Z.C., Shang, Z.F., and Cheng, J.P., On the direction and magnitude of radical substituent effects: the role of polar interaction on thermodynamic stabilities of benzylic C—H bonds and related carbon radicals. *Journal of Organic Chemistry*, 2001, **66**(4), 1466.

37 Snow, D.R., Spreadsheets, power series, generating functions, and integers. *The College Mathematics Journal*, 1989, **20**(2), 143.
38 Chen, C.F. and Shieh, L.S., A single matrix formula for closed form solution of inverse z-transform. *International Journal of Electronics*, 1969, **27**(6), 535.
39 Fletcher, R.H., Simplified inverse z-transform expression. *IEEE Transactions on Audio and Electroacoustics*, 1973, **21**(6), 552.
40 Kraus, K., Method of z transform inverse. *Electronic Engineering*, 1977, **49**(590), 29.
41 Sayood, K., *The inverse z-transform*, in *Signals and Systems: A One-Semester Course*. 2021, Springer, pp. 343–354.
42 Asher, K., Z transform. *International Journal of Applied Mathematics & Statistical Sciences*, 2013, **2**(2), 35.
43 Gosh, S., *Signals and systems – partial fraction expansion method for inverse z-transform*. 2005, Pearson India, 475.
44 Goodwillie, T.G., Calculus III: Taylor series. *Geometry & Topology*, 2003, **7**(2), 645.
45 Rogers, M.E. and Long, T.E., *Synthetic methods in step-growth polymers*. 2003, Hoboken, NJ: Wiley, 605.
46 Carraher, C.E. and Swift, G.G., eds. *Functional condensation polymers*. 2002, Kluwer Academic/Plenum Publishers, 315.
47 Flory, P.J., Fundamental principles of condensation polymerization. *Chemical Reviews*, 1946, **39**(1), 137.
48 Long, T.E. and Scheirs, J., *Modern polyesters: chemistry and technology of polyesters and copolyesters*. 2005, Wiley, 750.
49 Deopura, B.L., Alagirusamy, R., Joshi, M., and Gupta, B., eds. *Polyesters and polyamides*. 2008, Elsevier, 608.
50 Rulkens, R. and Koning, C.E., Chemistry and technology of polyamides, in *Polymer science: a comprehensive reference*. 2012, Elsevier, pp. 431–467.
51 Fahmina, Z. and Sharmin, E., eds. *Polyurethane*. 2012, IntechOpen, 470.
52 Thomson, T., *Polyurethanes as specialty chemicals: principles and applications*. 2004, CRC Press, 208.
53 Gupta, R.K. and Kahol, P.K., *Polyurethane chemistry: renewable polyols and isocyanates*. Scaling industrial green polyurethanes and recycling for sustainable manufacturing. 2021, American Chemical Society, 447.
54 Flory, P.J., *Principles of polymer chemistry*. 1953, Cornell University Press, 688.
55 Stille, J.K., Step-growth polymerization. *Journal of Chemical Education*, 1981, **58**(11), 862.
56 Rogers, M.E., Long, T.E., and R., T.S., Introduction to synthetic methods in step-growth polymers, in *Synthetic methods in step-growth polymers*. 2003, Wiley, pp. 1–16.
57 Gupta, S.K. and Kumar, A., *Reaction engineering of step growth polymerization*. 1 ed. The plenum chemical engineering series. 2012, Springer, 444.
58 Polyanin, A.D. and Zhurov, A.I., *Separation of variables and exact solutions to nonlinear PDEs*. 2021, CRC Press, 402.
59 Shearer, M. and Levy, R., *Partial differential equations: an introduction to theory and applications*. 2015, Princeton University Press, 288.
60 Olver, P., *Introduction to partial differential equations*. 2014, Berlin: Springer, 621.
61 Nedeljkov, M. and Oberguggenberger, M., Ordinary differential equations with delta function terms. *Publications De L Institut Mathematique-Beograd*, 2012, **91**(105), 125.
62 Lehnigk, S.H., Class of conservative diffusion processes with delta function initial conditions. *Journal of Mathematical Physics*, 1976, **17**(6), 973.

63 Itzinger, R., Schwarzinger, C., and Paulik, C., Investigation of the influence of impurities on the ring-opening polymerisation of L-lactide from biogenous feedstock. *Journal of Polymer Research*, 2020, **27**, 383.

64 Novitsky, T. et al., Effect of stoichiometric imbalances on the melt condensation polymerization of poly(dodecamethylene terephthalamide) studied by intrinsic viscosity and C-13-NMR. *Journal of Applied Polymer Science*, 2010, **116**(6), 3388.

65 Gao, M. et al., Stoichiometric imbalance-promoted synthesis of polymers containing highly substituted naphthalenes: rhodium-catalyzed oxidative polycoupling of arylboronic acids and internal diynes. *Polymer Chemistry*, 2013, **4**(5), 1372.

5

Conservation of Moments of Distribution of Chains in Polymerization

5.1 Introduction

The polymer chains at any time t during the reaction have a degree of polymerization (DP) distribution function $f_t(x)$, kth mole-fraction moment $M_{k,n,t}$, and kth concentration moment $Q_{k,n,t}$. In a differential time dt, some chains are concurrently consumed and produced by different elementary reactions. The elementary reactions that consume chains remove a fraction of the existing chains from the reaction with a DP distribution function $f_t(x)$, which reduces the kth order concentration moment of the existing chains. Each chain producing elementary reaction l adds new chains to the reaction mixture with a unique DP distribution function $f_{dt,l}(x)$, kth order mole-fraction moment $M_{k,n,l,dt}$, and kth order concentration moment $Q_{k,n,l,dt}$. These chain consumption and production reactions lead to a new chain distribution function $f_{t+dt}(x)$, kth order mole-fraction moment $M_{k,n,t+dt}$, and kth order concentration moment $Q_{k,n,t+dt}$ for the chains in the reaction at time $t+dt$. The $M_{k,n,t+dt}$ and $Q_{k,n,t+dt}$ of the chains in a polymerization reaction at time $t+dt$ are related to $M_{k,n,t}$, and $Q_{k,n,t}$ at time t by the conservation equation of the kth order moment of the chains and the knowledge of the DP distribution function of the chains produced/consumed in the time interval dt by each elementary reaction. This chapter covers the mixing theory for distributed variables and applies the theory to polymerization reactions to derive the conservation equation for the kth order moment of the chains, including propagating chains, dormant chains, and terminated chains, and the conservation equation for small molecule reactants, like monomer and initiator, in the reaction.

5.2 Elementary Reactions in Polymerization in a Differential Time Interval

Consider a radical polymerization reaction at time t shown in Figure 5.1. The polymer chains have a distribution function $f_t(x)$ at time t with kth order mole-fraction moment $M_{k,n,t}$ and kth order concentration moment $Q_{k,n,t}$ (see Chapter 3). In a differential time dt following time t, the propagating and terminated polymer chains as well as dormant chains are consumed and produced by elementary reactions. As the desired outcome is predicting the distribution and moments of the terminated polymer chains, the diagram in Figure 5.1 shows only those elementary reactions that produce or consume terminated polymer chains. However, it should be emphasized that the moments of the terminated chains at time t are related to the instantaneous distribution of the propagating as well

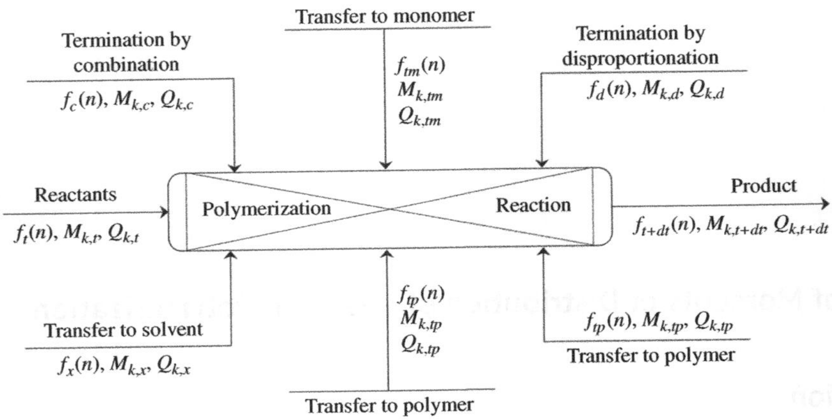

Figure 5.1 Elementary reactions for the production and consumption of polymer chains in the time interval dt.

as dormant chains in the polymerization reaction [1–13]. Propagating chains are produced by radical initiation [14–47] and propagation [48–50] while they are consumed by termination reactions [51–77]. Propagating chains are also produced and consumed by chain transfer to small molecules like monomer [69, 70, 78–80], solvent, inhibitor, retarder, and oxygen [79–99], chain transfer agent [100–102], and iniferter [103, 104], transfer to dormant [103–109] and terminated polymer chains [110–121], and chain dissociation reactions [122–130]. Dormant chains are produced and consumed by reversible deactivation of propagating chains [74] which includes, among others [131–138], deactivation by transition metal mediated atom transfer radical addition (ATRP) [105, 139], chain transfer to iniferter [103, 104], or chain transfer by reversible addition-fragmentation (RAFT) [106–109, 140]. As shown in Figure 5.1, in the differential time dt, terminated polymer chains are produced by combination or disproportionation of the propagating chains with distributions $f_c(x)$ and $f_d(x)$, respectively, and kth order mole-fraction moments $M_{k,c}$ and $M_{k,d}$ [51–56] where $x = n$ is the DP of the produced polymer chains. In that differential time dt, polymer chains are also produced by radical transfer from the propagating chains to monomer and solvent with distributions $f_{tm}(x)$ and $f_{ts}(x)$, respectively, and kth order mole-fraction moments $M_{k,tm}$ and $M_{k,ts}$ [80, 85–91]. In that differential time dt, terminated polymer chains are also produced and consumed by radical transfer from the propagating chains to the unsaturated bonds (terminal, main chain, or pendant) of polymer chains with a distribution function $f_{tp}(x)$ and kth order mole-fraction moment $M_{k,tp}$ [110–117]. These elementary reactions produce terminated polymer chains with unique distributions. The mixing of these unique chain distributions with the overall chain distribution at time t generates the overall distribution of polymer chains at time $t + dt$ in the reaction.

5.3 Mixing Theory of Polymer Chains

The mixing theory states that the kth order molar moment of a mixture of n polymer samples is equal to the sum of the kth order molar moments of the individual polymer samples. Note that "molar" in the above kth order molar moment is the "number of moles" as opposed to mole-fraction or molar concentration moment. According to the mixing theory, if $N_{k,i}$ is the kth order molar moment of the polymer sample i, the kth order molar moment of the polymer mixture, $N_{k,mix}$, is given by the following equation:

$$N_{k,mix} = \sum_{i=1}^{n} N_{k,i} \tag{5.1}$$

where, according to the defining Eq. (3.6) in Chapter 3, the kth order molar moment of a distribution, $N_{k,i}$, is defined as follows:

$$N_{k,i} = \sum_{j=1}^{+\infty} j^k\, n_{i,j} = \sum_{j=1}^{+\infty} j^k\, x_{i,j}\, n_i = n_i M_{k,i} \tag{5.2A}$$

$$n_{i,j} = C_{i,j}\, V_i \tag{5.2B}$$

where $n_{i,j}$ and $x_{i,j}$ are the number of moles and mole fraction of the chains with $DP = j$ in the sample i, respectively; n_i is the total number of moles of sample i; and $M_{k,i}$ is the kth order mole-fraction moment of the sample i. The parameter $C_{i,j}$ is the molar concentration of the chains with $DP = j$ in sample i and V_i is the volume of the sample i. Substitution of Eq. (5.2B) in Eq. (5.2A) results in the following equation relating $N_{k,i}$ to the kth order concentration moment of the sample i ($Q_{k,i}$):

$$N_{k,i} = n_i M_{k,i} = V_i\, C_i\, M_{k,i} = V_i\, Q_{0,i}\, M_{k,i} = V_i\, Q_{k,i} \tag{5.2C}$$

Equation (5.2C) applies to any polymeric sample i or a mixture of polymer samples. Substitution for $N_{k,i}$ and $N_{k,mix}$ from Eq. (5.2C) in Eq. (5.1) results in the following equation relating the kth order concentration moment of the mixture, $Q_{k,mix}$, to that of the sample i, $Q_{k,i}$:

$$Q_{k,mix} = \frac{\sum_{i=1}^{n} V_i\, Q_{k,i}}{V_{mix}} = \sum_{i=1}^{n} \varphi_i\, Q_{k,i} \tag{5.2D}$$

where V_{mix} is the volume of the polymer mixture and ϕ_i is the volume fraction of the ith sample in the mixture. Equation (5.2D) states that the kth order concentration moment of the polymer mixture is equal to the product of the kth order concentration moment of each polymer sample i and its volume fraction, summed over all samples in the mixture. As $Q_{k,i} = Q_{0,i}\, M_{k,i}$ and $Q_{0,i}$ is the molar concentration of the ith sample in the mixture (see Eq. 3.16 in Chapter 3), the kth order mole-fraction moment of the polymer mixture is given as follows:

$$M_{k,mix} = \frac{\sum_{i=1}^{n} V_i\, Q_{0,i}\, M_{k,i}}{V_{mix}\, Q_{0,mix}} = \sum_{i=1}^{n} \frac{V_i\, Q_{0,i}}{V_{mix}\, Q_{0,mix}} M_{k,i} = \sum_{i=1}^{n} \frac{V_i\, C_i}{V_{mix}\, C_{mix}} M_{k,i} \tag{5.3A}$$

As $V_i C_i$ is the number of mole of the ith sample (n_i) and $V_{mix} C_{mix}$ is the number of moles of the polymer mixture (n_{mix}), Eq. (5.3A) simplifies to the following equation:

$$M_{k,mix} = \sum_{i=1}^{n} \frac{n_i}{n_{mix}} M_{k,i} = \sum_{i=1}^{n} x_i\, M_{k,i} \tag{5.3B}$$

where x_i and $M_{k,i}$ are the mole fraction and kth order mole-fraction moment of the ith polymer sample in the mixture, respectively. Equations (5.1–5.3B) show the different forms of the mixing theory for polymers. These equations are used to find the kth order moment and molecular weight averages of the propagating and terminated polymer chains produced by elementary reactions in the differential time dt during the reaction. Mixing theory is also used in deriving the conservation equation for the kth order moment of the propagating and terminated chains in polymerization reactions.

The mixing theory is used to determine the kth order mole-fraction moment of the mixture, including the non-integer $(1 + a)^{\text{th}}$ order moment where parameter $0 < a \leq 1$ is the viscosity exponent, as defined by Eq. (3.50) in Section 3.6 of Chapter 3. The $(1 + a)^{\text{th}}$ order mole-fraction moment of a polymer mixture, $M_{1+a,mix}$, is defined by the following equation:

$$M_{1+a,mix} = \sum_{i=1}^{n} x_i M_{1+a,i} \tag{5.4}$$

where $M_{1+a,i}$ is the $(1 + a)^{\text{th}}$ order mole-fraction moment of the ith component of the mixture, as defined by Eq. (3.54) in Section 3.6 of Chapter 3. Equations (5.1–5.4) are used to find the distribution, mole-fraction moments, and the averages of the degree of polymerization of polymer blends, as shown schematically in Figure 5.2. The polymer sample A in Figure 5.2 with the mole-fraction distribution $f_A(x)$, kth order mole-fraction moment $M_{k,A}$, kth order concentration moment $Q_{k,A}$, and volume V_A is mixed with the polymer sample B with mole-fraction distribution $f_B(x)$, kth order mole-fraction moment $M_{k,B}$, kth order concentration moment $Q_{k,B}$, and volume V_B to produce the blend AB:

According to Eqs. (5.3B) and (5.2D), and Eq. (5.1), the kth order mole-fraction moment, $M_{k,AB}$, kth order concentration moment, $Q_{k,AB}$, and kth order molar moment, $N_{k,AB}$, for the polymer blend AB are defined as follows:

$$M_{k,AB} = x_A M_{k,A} + x_B M_{k,B} \tag{5.5A}$$

$$Q_{k,AB} = \frac{V_A Q_{k,A} + V_B Q_{k,B}}{V_A + V_B} = \phi_A Q_{k,A} + \phi_B Q_{k,B} \tag{5.5B}$$

$$N_{k,AB} = V_A Q_{k,A} + V_B Q_{k,B} = n_A M_{k,A} + n_B M_{k,B} \tag{5.5C}$$

where x_A and x_B are the mole fractions of samples A and B in the blend, respectively; ϕ_A and ϕ_B are the volume fractions of A and B; and V_A and V_B are the volumes of A and B. The number-average, $\overline{DP}_{n,AB}$, weight-average, $\overline{DP}_{w,AB}$, and viscosity-average, $\overline{DP}_{\eta,AB}$, of the AB polymer blend are defined as follows:

$$\overline{DP}_{n,AB} = Q_{1,AB}/Q_{0,AB} \tag{5.6A}$$

$$\overline{DP}_{w,AB} = Q_{2,AB}/Q_{1,AB} \tag{5.6B}$$

$$\overline{DP}_{\eta,AB} = Q_{1+a,AB}/Q_{1,AB}]^{(1/a)} \tag{5.6C}$$

The following example illustrates the use of the above equations to determine the DP averages of a polymer blend from the DP distribution of the individual components.

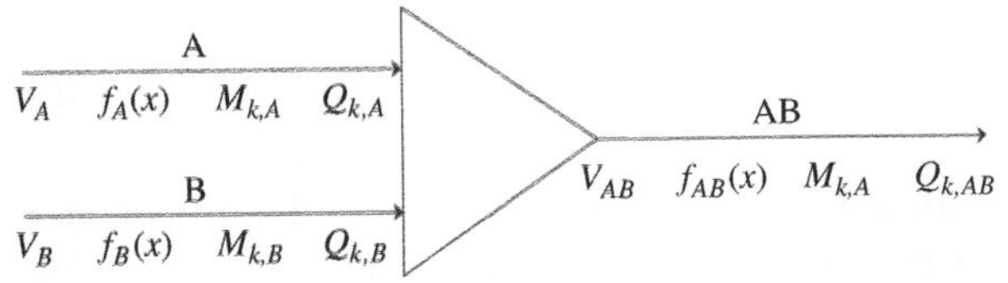

Figure 5.2 Schematic diagram for the mixing of polymer samples A and B with mole-fraction distributions $f_A(x)$ and $f_B(x)$, respectively, to produce the polymer blend AB with mole-fraction distribution $f_{AB}(x)$.

5.3.1 Example Problem 1

The mole fraction distribution, that is $DP = i$ versus $P_{DPn}(i)$, and molar concentration distribution, which is $DP = i$ versus $C_p(i)$, of DP of polymer samples A and B are given in Table 5.1. If 1 L of sample A is mixed with 4 L of sample B, determine $M_{1,AB}$, $M_{2,AB}$, $Q_{0,AB}$, $Q_{1,AB}$, $Q_{2,AB}$, $\overline{DP}_{n,AB}$, and $\overline{DP}_{w,AB}$ for the blend.

Solution: The zero, first, and second order molar concentration moments of samples A and B are calculated using Eq. (3.10) in Chapter 3 as follows:

$$Q_{0,A} = \sum_{i=1}^{+\infty} i^0 C_{PA,i} = 6\ \text{moles/L}$$

$$Q_{0,B} = \sum_{i=1}^{+\infty} i^0\, C_{PB,i} = 7\ \text{moles/L}$$

$$Q_{1,A} = \sum_{i=1}^{+\infty} i^1\, C_{PA,i} = 1320\ \text{moles/L}$$

$$Q_{1,B} = \sum_{i=1}^{+\infty} i^1\, C_{PB,i} = 855\ \text{moles/L}$$

$$Q_{2,A} = \sum_{i=1}^{+\infty} i^2\, C_{PA,i} = 1030200\ \text{moles/L}$$

$$Q_{2,B} = \sum_{i=1}^{+\infty} i^2\, C_{PB,i} = 27752\ \text{moles/L}$$

The zero, first, and second order molar concentration moments for the AB blend are calculated using the above kth order Q values of the samples A and B, their respective volumes, and the defining Eq. (5.5B) as follows:

$$Q_{0,AB} = \frac{V_A Q_{0,A} + V_B Q_{0,B}}{V_A + V_B} = 6.8\ \text{moles/L}$$

$$Q_{1,AB} = \frac{V_A Q_{1,A} + V_B Q_{1,B}}{V_A + V_B} = 948\ \text{moles/L}$$

$$Q_{2,AB} = \frac{V_A Q_{2,A} + V_B Q_{2,B}}{V_A + V_B} = 428060\ \text{moles/L}$$

Table 5.1 Mole fraction distribution, $P_{DPn}(i)$, and molar concentration distribution, $C_P(i)$, of the DP of polymer samples A and B.

Sample A			**Sample B**		
i	$P_{DPn}(i)$	$C_p(i)$, moles L^{-1}	i	$P_{DPn}(i)$	$C_p(i)$, moles L^{-1}
10	0.333	2	5	0.143	1
100	0.500	3	50	0.429	3
1000	0.167	1	100	0.286	2
			500	0.143	1

The first and second order mole-fraction moments of the blend are calculated using the kth order blend Q values and the defining Eqs. (3.20) and (3.24) in Chapter 3 as follows:

$$M_{1,AB} = \frac{Q_{1,AB}}{Q_{0,AB}} = 139$$

$$M_{2,AB} = \frac{Q_{2,AB}}{Q_{0,AB}} = 62950$$

The $\overline{DP}_{n,AB}$, and $\overline{DP}_{w,AB}$ of the blend are calculated using the above kth order blend Q values and Eqs. (3.42) and (3.45) in Chapter 3 as follows:

$$\overline{DP}_{n,AB} = \frac{Q_{1,AB}}{Q_{0,AB}} = 139$$

$$\overline{DP}_{w,AB} = \frac{Q_{2,AB}}{Q_{1,AB}} = 451$$

The $\overline{DP}_{n,AB} = 139$ for the blend is between those of $\overline{DP}_{n,A} = 220$ and $\overline{DP}_{n,B} = 122$; and $\overline{DP}_{w,AB} = 451$ for the blend is between those of $\overline{DP}_{w,A} = 780$ and $\overline{DP}_{w,A} = 325$.

5.4 Conservation of *k*th Order Moment of Polymer Chains in a Reaction

Consider the polymerization reaction in the time interval Δt between t and $t + \Delta t$, as shown schematically in Figure 5.3 [141–149]. At time t, there are n_t moles of polymeric chains in the reaction mixture with mole-fraction moment $M_{k,t}$ and concentration moments $Q_{k,t}$. During time interval Δt, new polymer chains are produced, whereas some existing chains are consumed and removed from the reaction mixture by the elementary reactions. Each elementary reaction j produces/consumes Δn_j moles of chains with instantaneous kth order mole-fraction moment $M'_{k,j}$ and concentration moment $\Delta Q_{0,j}$. The prime in $M'_{k,j}$ denotes instantaneous moment for the production or consumption of the polymer chains by reaction j and the "Δ" in Δn_j and $\Delta Q_{0,j}$ denotes a change in the number of moles or concentration of polymer chains as a result of production/consumption by the reaction j. As a result of these elementary reactions in the time interval Δt, the number of moles of polymer chains in the reaction at time $t + \Delta t$ is equal to $n_t + \Delta n_t$ with kth order mole-fraction moment $M_{k,t} + \Delta M_{k,t}$ and kth order concentration moment $Q_{k,t} + \Delta Q_{k,t}$.

As the polymer content in the reaction at time $t + \Delta t$ is the sum of the polymer content at time t and the polymer chains produced/consumed instantaneously in the time interval Δt, the mixing theory can be applied to determine the kth order molar moment of the polymer chains at time

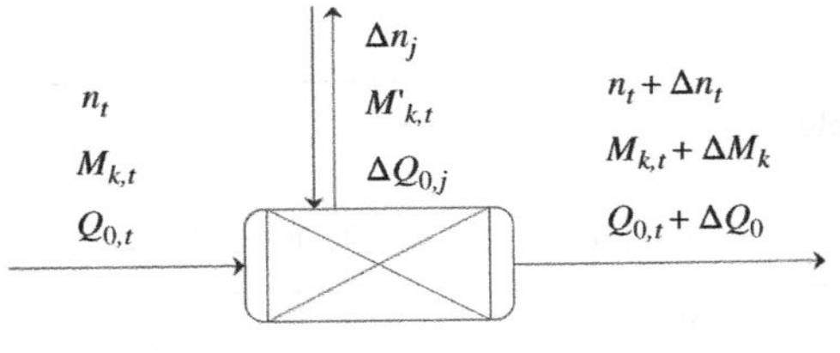

Figure 5.3 Schematic diagram of the polymerization reaction in the time interval between t and $t + \Delta t$ showing the number of moles, kth order mole-fraction moment, and concentration moment of the polymer chains.

$t + \Delta t$. Based on the defining Eq. (5.1) of the mixing theory, the kth order molar moment of the chains at time $t + \Delta t$ is expressed in terms of that at time t by the following equation:

$$N_{k,t+\Delta t} = N_{k,t} + \sum_j N'_{k,j} \tag{5.7}$$

where $N_{k,t+\Delta t}$ and $N_{k,t}$ are the kth order molar moments of the polymer chains at time $t + \Delta t$ and t, respectively; and $N'_{k,j}$ is the instantaneous molar moment of the chains produced/consumed by the jth elementary reaction. Using the defining Eq. (5.3B) for the mole-fraction moment, the conservation equation for the polymer chains in the time interval between t and $t + \Delta t$ can be written in terms of the mole-fraction moments as follows:

$$M_{k,t+\Delta t}\, n_{t+\Delta t} = M_{k,t}\, n_t + \sum_j M'_{k,j}\, \Delta n_{j,t} \tag{5.8A}$$

where $M_{k,t+\Delta t}$ and $M_{k,t}$ are the kth order mole-fraction moments of the chains at time $t + \Delta t$ and t, respectively; $M'_{k,j}$ is the instantaneous kth order mole-fraction moment of the chains produced/consumed by jth elementary reaction; $n_{t+\Delta t}$ and n_t are the moles of chains at times $t + \Delta t$ and t; and $\Delta n_{j,t}$ is the change in the number of moles of the chains produced/consumed by jth elementary reaction at time t. Substitution for $M_{k,t+\Delta t}$ and $n_{t+\Delta t}$ in terms of $M_{k,t} + \Delta M_{k,t}$ and $n_t + \Delta n_t$, respectively, results in the following relation for the conservation equation:

$$(M_{k,t} + \Delta M_{k,t})(n_t + \Delta n_t) = M_{k,t}\, n_t + \sum_j M'_{k,j}\, \Delta n_{j,t} \tag{5.8B}$$

which, after multiplication of the two terms in parentheses and deleting identical terms on both sides, results in the following equation:

$$\Delta M_{k,t}\, n_t + M_{k,t}\, \Delta n_t + \Delta M_{k,t}\, \Delta n_t = \sum_j M'_{k,j}\, \Delta n_{j,t} \tag{5.8C}$$

As Δn_t becomes infinitesimally small, the term $\Delta M_{k,t}\, \Delta n_t$ becomes much smaller compared to $\Delta M_{k,t}\, n_t + M_{k,t}\, \Delta n_t$. Upon neglecting the term $\Delta M_{k,t}\, \Delta n_t$, the above equation simplifies as follows:

$$\Delta M_{k,t}\, n_t + M_{k,t}\, \Delta n_t = \sum_j M'_{k,j}\, \Delta n_{j,t} \tag{5.8D}$$

The left side of the above equation can be further simplified by using the product rule of derivatives as follows [150]:

$$\Delta M_{k,t}\, n_t + M_{k,t}\, \Delta n_t = \Delta \left(M_{k,t}\, n_t\right) \tag{5.8E}$$

Upon substitution of the above equality in Eq. (5.8D), the conservation equation for the kth order mole-fraction moment of the polymer chains in the reaction time interval Δt simplifies as follows:

$$\Delta \left(M_{k,t}\, n_t\right) = \sum_j M'_{k,j}\, \Delta n_{j,t} \tag{5.8F}$$

The change in the number of moles of polymer chains due to the elementary reaction j can be expressed as the product of the volumetric rate of the reaction j, reaction volume, and time interval Δt as follows:

$$\Delta n_j = R_j V_R \Delta t \tag{5.8G}$$

where R_j and V_R are the volumetric reaction rate and reaction volume, respectively. The number of moles of chains in the reaction mixture, n_t, can be expressed as the product of the reaction volume and molar concentration or the zero-order molar concentration moment of the polymer chains, $Q_{0,t}$, as follows:

$$n_t = Q_{0,t} V_R \tag{5.8H}$$

Substitution for Δn_j and n_t from the above Eqs. (5.8G) and (5.8H), respectively, in Eq. (5.8F) and dividing both sides by Δt results in the following relation for the conservation equation:

$$\frac{\Delta \left(M_{k,t}\, Q_{0,t}\, V_R\right)}{\Delta t} = \sum_j M'_{k,j}\, R_j\, V_R \tag{5.8I}$$

Taking the limit as Δt approaches zero, the conservation equation in the differential form is given as follows:

$$\lim_{\Delta t \to 0} \left[\frac{\Delta \left(M_{k,t}\, Q_{0,t}\, V_R\right)}{\Delta t}\right] = \lim_{\Delta t \to 0} \left[\sum_j M'_{k,j}\, R_j\, V_R\right] \tag{5.8J}$$

$$\Rightarrow \quad \frac{d \left(M_{k,t}\, Q_{0,t}\, V_R\right)}{dt} = \sum_j M'_{k,j}\, R_j\, V_R \tag{5.9A}$$

Equation (5.8K) is the differential form of the conservation equation for kth order mole-fraction moment of the polymer chains in a reaction due to production or consumption by elementary reactions in the absence of mass transfer by convection or molecular diffusion. This equation applies to the propagating as well as dormant and terminated chains. If the reaction volume is constant, V_R can be removed to the outside of the differential and summation terms, which simplifies the above equation as follows:

$$\frac{d \left(M_{k,t}\, Q_{0,t}\right)}{dt} = \sum_j M'_{k,j}\, R_j \quad \in dV_R/dt = 0 \tag{5.9B}$$

In the above equation, R_j is positive for elementary reactions that produce chains and negative for those that consume chains.

5.4.1 Conservation of kth Order Moment of Terminated Polymer Chains in the Absence of Convective or Diffusive Mass Transfer

To distinguish between the different types of polymer chains in a reaction, we use the symbol "μ" in place of the general symbol "M" for the mole-fraction moment of terminated polymer chains. For terminated chains, the right side of the conservation Eq. (5.9B) for constant volume reaction can be expanded to include separate terms for production and consumption of terminated chains by elementary reactions as follows:

$$\frac{d \left(\mu_{k,t}\, Q_{0,tp}\right)}{dt} = \sum_i \mu'_{k,i}\, R_{tp,prd,i} - \sum_l \mu'_{k,l}\, R_{tp,con,l} \tag{5.10}$$

where $\mu_{k,t}$ and $QP_{0,tp}$ are the kth order mole-fraction moment and zero-order molar concentration moment of the terminated polymer chains at time t, respectively; $\mu'_{k,i}$ and $R_{tp,prd,i}$ are the instantaneous kth order mole-fraction moment of the produced terminated polymer chains by the ith elementary reaction and its corresponding production rate; and $\mu'_{k,l}$ and $R_{tp,con,l}$ are the instantaneous kth order mole-fraction moment of the consumed terminated polymer chains by the lth elementary reaction and its corresponding consumption rate. The symbol "tp" is used to denote terminated polymer chains. Equation (5.10) states that the rate of accumulation of the kth order mole-fraction moment of the terminated chains (the left side term) is equal to the instantaneous rate of production (the first term on right) minus the instantaneous rate of consumption (the second term on right) of their kth order moment, respectively. The mole-fraction and concentration moments of terminated polymer chains can be monitored during the reaction by measuring polymer concentration and its molecular weight distribution by gel permeation chromatography (GPC) [151], osmometry [152], light scattering [153–155], rheometry [156], capillary electrophoresis [157], ultracentrifugation [158, 159], ultraviolet–visible spectroscopy [160], acoustic spectroscopy [161], flow-modulated birefringence [162], infrared spectroscopy [163], mass spectrometry [164, 165], fluorescence lifetime imaging microscopy [166], diffusion-ordered nuclear magnetic resonance spectroscopy [167–169], protein nanopore translocation [170] among other methods [171, 172].

5.4.2 Conservation of kth Order Moment of Dormant Polymer Chains in the Absence of Convective or Diffusive Mass Transfer

We use the symbol "γ" in place of the general symbol "M" for the mole-fraction moment of dormant polymer chains in the reaction. For dormant chains, the right side of the conservation Eq. (5.9B) for constant volume reaction can be expanded to include separate terms for production and consumption of the dormant chains by elementary reactions as follows:

$$\frac{d\left(\gamma_{k,t}\, Q_{0,dop}\right)}{dt} = \sum_i \gamma'_{k,i}\, R_{dop,prd,i} - \sum_l \gamma'_{k,l}\, R_{dop,con,l} \tag{5.11}$$

where $\gamma_{k,t}$ and $Q_{0,dop}$ are the kth order mole-fraction moment and zero-order molar concentration moment of the dormant chains at time t, respectively; $\gamma'_{k,i}$ and $R_{dop,prd,i}$ are the instantaneous kth order mole-fraction moment of the produced dormant chains by the ith elementary reaction and its corresponding production rate; and $\gamma'_{k,l}$ and $R_{dop,con,l}$ are the instantaneous kth order mole-fraction moment of the consumed dormant chains by the lth elementary reaction and its corresponding consumption rate. The symbol "dop" denotes dormant polymer chains. The concentration of dormant species in reversible deactivation radical polymerization (RDRP) is related to the equilibrium constant between activation and deactivation reactions, which can be measured using the persistent radical effect [173–175].

5.4.3 Conservation of kth Order Moment of Propagating Polymer Chains in the Absence of Convective or Diffusive Mass Transfer

We use the symbol "λ" in place of the general symbol "M" to denote the mole-fraction moment of the propagating chains in the reaction. For propagating chains, the right side of the conservation Eq. (5.9B) for constant volume reaction can be expanded to include separate terms for production and consumption of the propagating chains by elementary reactions as follows:

$$\frac{d\left(\lambda_{k,t}\, Q_{0,rp}\right)}{dt} = \sum_i \lambda'_{k,i}\, R_{rp,prd,i} - \sum_l \lambda'_{k,l}\, R_{rp,con,l} \tag{5.12}$$

where $\lambda_{k,t}$ and $Q_{0,rp}$ are the kth order mole-fraction moment and zero-order molar concentration moment of the reactive propagating chains at time t, respectively; $\lambda'_{k,i}$ and $R_{rp,prd,i}$ are the instantaneous kth order mole-fraction moment of the produced propagating chains by the ith elementary reaction and its corresponding production rate; and $\lambda'_{k,l}$ and $R_{rp,con,l}$ are the instantaneous kth order mole-fraction moment of the consumed propagating chains by the lth elementary reaction and its corresponding consumption rate. The symbol "rp" is used to denote reactive propagating chains. Equation (5.12) states that the rate of accumulation of the kth order mole-fraction moment of the propagating chains (the term on left side) is equal to their instantaneous production rate (the first term on right) minus their instantaneous consumption rate (the second term on right). The concentration of active propagating species, like radicals and ions, can be measured by techniques like time-resolved electron spin resonance spectroscopy [176] and optical absorption spectrometry [177] and the experimental measurements can be used to test the accuracy of the mathematical models for the rate of production of active species or determine the dissociation constant of polymerization initiators as well as the kinetics of chain termination between the active species [178–180]. In the following section, we will derive the general conservation equation for the kth order moment of the polymeric chains and apply the general conservation equation to the modeling of different polymerization reactors.

5.5 Derivation of the General Conservation Equation for kth Order Moment of Polymer Chains in a Reaction

It was assumed in the above sections that the change in the concentration of different polymeric species (propagating, dormant, or terminated chains) within a microscopic control volume in the reactor was due only to the elementary reactions. In general, the concentration of polymeric species in a control volume changes with time not only by chemical reactions but also by convection via pressure gradients, as well as molecular diffusion via concentration gradients of the polymeric species in and out of the control volume in different directions.

Using the reactive propagating chain as an example of polymeric species, a typical delta control volume for the reactive propagating chains in the Cartesian coordinate system with dimensions Δx, Δy, and Δz in the x, y, and z directions, respectively, is shown in Figure 5.4 [181]. The production and consumption of polymeric species occur within the control volume, whereas polymeric species enter and exit through the faces of the control volume in different directions.

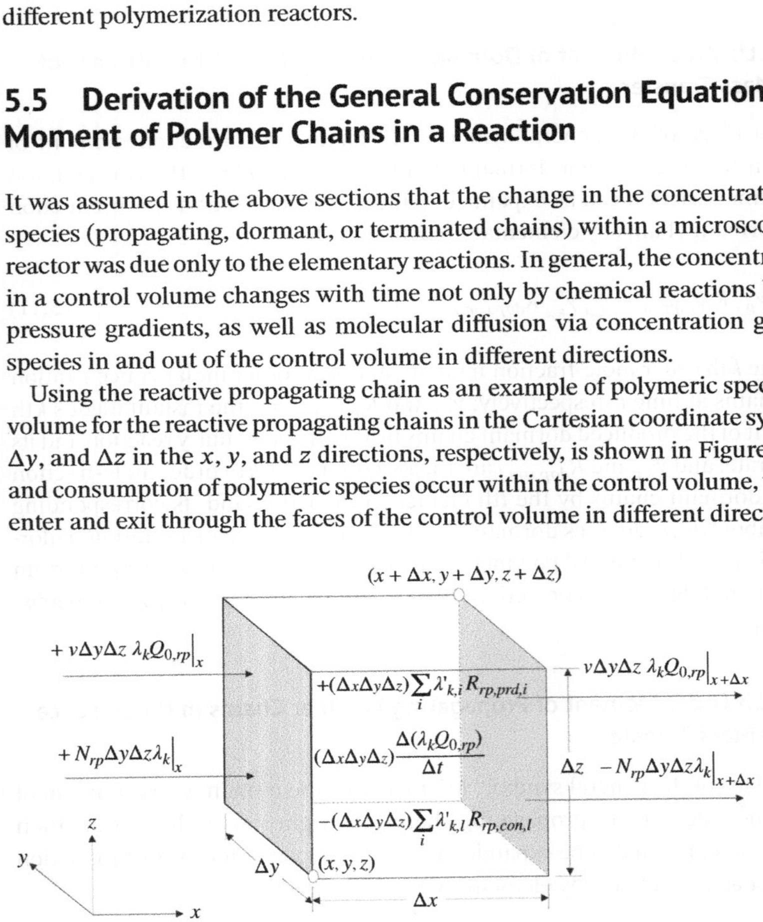

Figure 5.4 Schematic diagram of the delta control volume in the Cartesian coordinate system showing the transport of propagating chains by convection and molecular diffusion entering at x and existing at $x + \Delta x$ faces of the cube and production and consumption by the elementary reactions in the control volume. The transport of propagating chains in the y and z directions is not shown.

The general equation for the conservation of kth order molar moment of the reactive propagating chains in the control volume of the reaction is written as follows:

$$\left\{\begin{array}{l}\text{rate of change}\\ \text{in } k\text{th molar}\\ \text{moment of}\\ \text{the reactive}\\ \text{propagating}\\ \text{chains with}\\ \text{time}\end{array}\right\} = \left\{\begin{array}{c}\text{instantaneous}\\ \text{rate of}\\ \text{convective}\\ \text{transfer of}\\ k\text{th molar}\\ \text{moment of}\\ \text{the reactive}\\ \text{propagating}\\ \text{chains in}\\ x-\text{direction}\end{array}\right\} + \left\{\begin{array}{c}\text{instantaneous}\\ \text{rate of}\\ \text{convective}\\ \text{transfer of}\\ k\text{th molar}\\ \text{moment of}\\ \text{the reactive}\\ \text{propagating}\\ \text{chains in}\\ y-\text{direction}\end{array}\right\} + \left\{\begin{array}{c}\text{instantaneous}\\ \text{rate of}\\ \text{convective}\\ \text{transfer of}\\ k\text{th molar}\\ \text{moment of}\\ \text{the reactive}\\ \text{propagating}\\ \text{chains in}\\ z-\text{direction}\end{array}\right\}$$

$$+ \left\{\begin{array}{c}\text{instantaneous}\\ \text{rate of}\\ \text{diffusive}\\ \text{transfer of}\\ k\text{th molar}\\ \text{moment of}\\ \text{the reactive}\\ \text{propagating}\\ \text{chains in}\\ x-\text{direction}\end{array}\right\} + \left\{\begin{array}{c}\text{instantaneous}\\ \text{rate of}\\ \text{diffusive}\\ \text{transfer of}\\ k\text{th molar}\\ \text{moment of}\\ \text{the reactive}\\ \text{propagating}\\ \text{chains in}\\ y-\text{direction}\end{array}\right\} + \left\{\begin{array}{c}\text{instantaneous}\\ \text{rate of}\\ \text{diffusive}\\ \text{transfer of}\\ k\text{th molar}\\ \text{moment of}\\ \text{the reactive}\\ \text{propagating}\\ \text{chains in}\\ z-\text{direction}\end{array}\right\} \tag{5.13A}$$

$$+ \left\{\begin{array}{c}\text{rate of}\\ \text{instantaneous}\\ \text{production of}\\ k\text{th molar}\\ \text{moment of}\\ \text{the reactive}\\ \text{propagating}\\ \text{chains}\end{array}\right\} - \left\{\begin{array}{c}\text{rate of}\\ \text{intantaneous}\\ \text{consumption of}\\ k\text{th molar}\\ \text{moment of}\\ \text{the reactive}\\ \text{propagating}\\ \text{chains}\end{array}\right\}$$

The rate of mass transfer by convection and mass transfer by molecular diffusion of the kth order molar moment of the propagating chains entering and exiting the control volume at faces x and $x + \Delta x$ through the cross-sectional area $\Delta y\, \Delta z$ are given by the following relations:

$$+ \dot{V}Q_{k,rp}\Big|_x - \dot{V}Q_{k,rp}\Big|_{x+\Delta x} = + \nu\, \Delta y \Delta z\, \lambda_k Q_{0,rp}\big|_x - \nu\, \Delta y \Delta z\, \lambda_k Q_{0,rp}\big|_{x+\Delta x}\ \ convection \tag{5.13B}$$

$$+ N_{rp}\Delta y \Delta z \lambda_k\big|_x - N_{rp}\Delta y \Delta z \lambda_k\big|_{x+\Delta x}$$
$$= + \left(-D_{rp}\partial Q_{0,rp}/\partial x\right) \Delta y \Delta z\, \lambda_k\big|_x - \left(-D_{rp}\partial Q_{0,rp}/\partial x\right) \Delta y \Delta z\, \lambda_k\big|_{x+\Delta x}\ \ diffusion \tag{5.13C}$$

where $\dot{V}$ is the volumetric flow rate, $Q_{k,rp}$ and $Q_{0,rp}$ are the kth and zero-order concentration moment of the propagating chains, respectively, ν is the fluid velocity, and N_{rp}, λ_k, and D_{rp} are the molar flux, kth order mole-fraction moment, and diffusion coefficient of the propagating chains, respectively. In the above Eq. (5.13C), the flux of the reactive propagating chains is related to its concentration gradient in the x direction using Fick's law of diffusion [182–184]. Similarly, the transport of the kth order concentration moment of the propagating chains by convection and molecular diffusion is expressed mathematically in the y and z directions. Upon applying Eq. (5.13B) and (5.13C) to the mass balance in 5.13A, the conservation equation for the propagating chains is expressed mathematically as follows:

$$\begin{aligned}(\Delta x \Delta y \Delta z)\frac{\Delta\left(\lambda_k Q_{0,rp}\right)}{\Delta t} = &\left(\nu\, \Delta y \Delta z\, \lambda_k Q_{0,rp}\big|_x - \nu\, \Delta y \Delta z\, \lambda_k Q_{0,rp}\big|_{x+\Delta x}\right)\\ &\left(\nu\, \Delta x \Delta z\, \lambda_k Q_{0,rp}\big|_y - \nu\, \Delta x \Delta z\, \lambda_k Q_{0,rp}\big|_{y+\Delta y}\right)\\ &\left(\nu\, \Delta y \Delta x\, \lambda_k Q_{0,rp}\big|_z - \nu\, \Delta y \Delta x\, \lambda_k Q_{0,rp}\big|_{z+\Delta z}\right)\\ &-D_{rp}\partial Q_{0,rp}/\partial x\ \Delta y \Delta z\, \lambda_k\big|_x + D_{rp}\partial Q_{0,rp}/\partial x\ \Delta y \Delta z\, \lambda_k\big|_{x+\Delta x}\\ &-D_{rp}\partial Q_{0,rp}/\partial x\ \Delta x \Delta z\, \lambda_k\big|_y + D_{rp}\partial Q_{0,rp}/\partial x\ \Delta x \Delta z\, \lambda_k\big|_{y+\Delta y}\\ &-D_{rp}\partial Q_{0,rp}/\partial x\ \Delta y \Delta x\, \lambda_k\big|_z + D_{rp}\partial Q_{0,rp}/\partial x\ \Delta y \Delta x\, \lambda_k\big|_{z+\Delta z}\\ &+(\Delta x \Delta y \Delta z)\sum_i \lambda'_{k,i}\, R_{rp,prd,i} - (\Delta x \Delta y \Delta z)\sum_l \lambda'_{k,l}\, R_{rp,con,l}\end{aligned} \tag{5.13D}$$

In the above equation, it is assumed that the change in control volume $\Delta V = \Delta x \Delta y \Delta z$ is negligible with conversion of monomer to polymer, so that the term $\Delta x \Delta y \Delta z$ can be moved out of the delta terms. Upon dividing all terms by $\Delta x \Delta y \Delta z$ and taking the limit as Δx, Δy, Δz, and Δt approach zero, the above conservation equation converts to the following equation:

$$\begin{aligned}\lim_{\Delta t \to 0}\frac{\Delta\left(\lambda_k Q_{0,rp}\right)}{\Delta t} = &-\lim_{\Delta x \to 0}\left[\frac{\nu\, \lambda_k Q_{0,rp}\big|_{x+\Delta x} - \nu\, \Delta y \Delta z\, \lambda_k Q_{0,rp}\big|_x}{\Delta x}\right]\\ &-\lim_{\Delta y \to 0}\left[\frac{\nu\, \lambda_k Q_{0,rp}\big|_{y+\Delta y} - \nu\, \Delta y \Delta z\, \lambda_k Q_{0,rp}\big|_y}{\Delta y}\right]\\ &-\lim_{\Delta z \to 0}\left[\frac{\nu\, \lambda_k Q_{0,rp}\big|_{z+\Delta z} - \nu\, \Delta y \Delta z\, \lambda_k Q_{0,rp}\big|_z}{\Delta z}\right]\\ &+\lim_{\Delta x \to 0}\left[\frac{D_{rp}\lambda_k \partial Q_{0,rp}/\partial x\big|_{x+\Delta x} - D_{rp}\lambda_k \partial Q_{0,rp}/\partial x\big|_x}{\Delta x}\right]\\ &+\lim_{\Delta y \to 0}\left[\frac{D_{rp}\lambda_k \partial Q_{0,rp}/\partial x\big|_{y+\Delta y} - D_{rp}\lambda_k \partial Q_{0,rp}/\partial x\big|_y}{\Delta y}\right]\\ &+\lim_{\Delta z \to 0}\left[\frac{D_{rp}\lambda_k \partial Q_{0,rp}/\partial x\big|_{z+\Delta z} - D_{rp}\lambda_k \partial Q_{0,rp}/\partial x\big|_z}{\Delta z}\right]\\ &+\sum_i \lambda'_{k,i}\, R_{rp,prd,i} - \sum_l \lambda'_{k,l}\, R_{rp,con,l}\end{aligned} \tag{5.13E}$$

Upon taking the limit as Δx, Δy, Δz, and Δt go to zero, the general conservation equation for the propagating chains in differential form is given by the following equation:

$$\begin{aligned}\frac{d\left(\lambda_k Q_{0,rp}\right)}{dt} &= -v_x\frac{\partial\left(\lambda_k Q_{0,rp}\right)}{\partial x} - v_y\frac{\partial\left(\lambda_k Q_{0,rp}\right)}{\partial y} - v_z\frac{\partial\left(\lambda_k Q_{0,rp}\right)}{\partial z}\\ &-\lambda_k Q_{0,rp}\left(\frac{\partial v_x}{\partial x}+\frac{\partial v_y}{\partial y}+\frac{\partial v_z}{\partial z}\right)+\frac{\partial}{\partial x}\left[D_{rp}\lambda_k\frac{\partial\left(Q_{0,rp}\right)}{\partial x}\right]\\ &+\frac{\partial}{\partial y}\left[D_{rp}\lambda_k\frac{\partial\left(Q_{0,rp}\right)}{\partial y}\right]+\frac{\partial}{\partial z}\left[D_{rp}\lambda_k\frac{\partial\left(Q_{0,rp}\right)}{\partial z}\right]\\ &+\sum_i \lambda'_{k,i} R_{rp,prd,i}-\sum_l \lambda'_{k,l} R_{rp,con,l}\end{aligned} \tag{5.13F}$$

If the change in volume with conversion is negligible, it can be shown that the equation of continuity simplifies to the following equation [181]:

$$\left(\frac{\partial v_x}{\partial x}+\frac{\partial v_y}{\partial y}+\frac{\partial v_z}{\partial z}\right)=0 \quad \text{constant volume} \tag{5.13G}$$

Upon substitution of Eq. (5.13G) in Eq. (5.13F), the equation for the conservation of kth order concentration moment of the propagating chains is given by the following expression:

$$\begin{aligned}\frac{d\left(\lambda_k Q_{0,rp}\right)}{dt} &= -v_x\frac{\partial\left(\lambda_k Q_{0,rp}\right)}{\partial x} - v_y\frac{\partial\left(\lambda_k Q_{0,rp}\right)}{\partial y} - v_z\frac{\partial\left(\lambda_k Q_{0,rp}\right)}{\partial z}\\ &+\frac{\partial}{\partial x}\left[D_{rp}\lambda_k\frac{\partial\left(Q_{0,rp}\right)}{\partial x}\right]+\frac{\partial}{\partial y}\left[D_{rp}\lambda_k\frac{\partial\left(Q_{0,rp}\right)}{\partial y}\right]+\frac{\partial}{\partial z}\left[D_{rp}\lambda_k\frac{\partial\left(Q_{0,rp}\right)}{\partial z}\right]\\ &+\sum_i \lambda'_{k,i} R_{rp,prd,i}-\sum_l \lambda'_{k,l} R_{rp,con,l}\end{aligned} \tag{5.14A}$$

The term on the left side of the equation is the rate of change of the kth order concentration moment of the propagating chains with time in the control volume; the first three terms on the right side are the rates of transport of the kth order concentration moment of the propagating chains by convection in the x, y, and z directions; the next three terms on the right side are the rates of transport of the kth order concentration moment of the propagating chains by molecular diffusion in the x, y, and z directions; and the last two terms on the right side are the rates of production and consumption of the kth order concentration moment of the propagating chains by the elementary reactions within the control volume. The above differential equation is valid for any order k of the concentration moment of the propagating chains. The above coupled moment conservation equations for different k values are solved analytically or numerically, starting with $k = 0$, followed by $k = 1$ and higher order moments, subject to the initial and boundary conditions, to find the concentration or mole-fraction moments of the propagating chains. The mole-fraction moments λ_k are then used to find the different $\overline{DP}_{n,rp}$ and higher DP averages of the propagating chains. For $k = 0$, $\lambda_0 = 1$, $Q_{0,rp} = C_{rp}$ and the conservation equation simplifies to the molar mass conservation equation of the propagating chains in terms of concentration and time as follows:

$$\begin{aligned}\frac{d\left(C_{rp}\right)}{dt} &= -v_x\frac{\partial C_{rp}}{\partial x} - v_y\frac{\partial C_{rp}}{\partial y} - v_z\frac{\partial C_{rp}}{\partial z}+\frac{\partial}{\partial x}\left(D_{rp}\frac{\partial C_{rp}}{\partial x}\right)\\ &+\frac{\partial}{\partial y}\left(D_{rp}\frac{\partial C_{rp}}{\partial y}\right)+\frac{\partial}{\partial z}\left(D_{rp}\frac{\partial C_{rp}}{\partial z}\right)+\sum_i \lambda'_{k,i} R_{rp,prd,i}-\sum_l \lambda'_{k,l} R_{rp,con,l}\end{aligned} \tag{5.14B}$$

where C_{rp} is the molar concentration of propagating chains. The kth order moment of the propagating chains given by Eq. (5.13A) is the conservation equation in the Cartesian coordinate system.

It can be shown that the general equation for conservation equation in cylindrical coordinate system and conservation equation in spherical coordinate system for kth order moment of the propagating chains are given by the following equations:

$$\begin{aligned}\frac{d(\lambda_k Q_{0,rp})}{dt} &= -v_r\frac{\partial(\lambda_k Q_{0,rp})}{\partial r} - \frac{v_\theta}{r}\frac{\partial(\lambda_k Q_{0,rp})}{\partial\theta} - v_z\frac{\partial(\lambda_k Q_{0,rp})}{\partial z}\\ &+\frac{1}{r^2}\frac{\partial}{\partial r}\left[r^2 D_{rp}\lambda_k\frac{\partial(Q_{0,rp})}{\partial r}\right] + \frac{1}{r^2}\frac{\partial}{\partial\theta}\left[D_{rp}\lambda_k\frac{\partial(Q_{0,rp})}{\partial\theta}\right]\\ &+\frac{\partial}{\partial z}\left[D_{rp}\lambda_k\frac{\partial(Q_{0,rp})}{\partial z}\right] + \sum_i \lambda'_{k,i} R_{rp,prd,i} - \sum_l \lambda'_{k,l} R_{rp,con,l}\end{aligned} \tag{5.15}$$

$$\begin{aligned}\frac{d(\lambda_k Q_{0,rp})}{dt} &= -v_r\frac{\partial(\lambda_k Q_{0,rp})}{\partial r} - \frac{v_\theta}{r}\frac{\partial(\lambda_k Q_{0,rp})}{\partial\theta} - \frac{v_\phi}{r\sin(\theta)}\frac{\partial(\lambda_k Q_{0,rp})}{\partial\phi}\\ &+\frac{1}{r^2}\frac{\partial}{\partial r}\left[r^2 D_{rp}\lambda_k\frac{\partial(Q_{0,rp})}{\partial r}\right] + \frac{1}{r^2\sin(\theta)}\frac{\partial}{\partial\theta}\left[\sin(\theta) D_{rp}\lambda_k\frac{\partial(Q_{0,rp})}{\partial\theta}\right]\\ &+\frac{1}{r^2\sin^2(\theta)}\frac{\partial}{\partial\phi}\left[D_{rp}\lambda_k\frac{\partial(Q_{0,rp})}{\partial\phi}\right] + \sum_i \lambda'_{k,i} R_{rp,prd,i} - \sum_l \lambda'_{k,l} R_{rp,con,l}\end{aligned} \tag{5.16}$$

where r, θ, and z are the axes in the cylindrical coordinate and r, θ, and ϕ are the axes in the spherical coordinate system. Using a similar approach, it can be shown that the general conservation equations for the kth order moment of the terminated and dormant polymer chains in the Cartesian coordinate system are given by the following equations:

$$\begin{aligned}\frac{d(\mu_k Q_{0,tp})}{dt} &= -v_x\frac{\partial(\mu_k Q_{0,tp})}{\partial x} - v_y\frac{\partial(\mu_k Q_{0,tp})}{\partial y} - v_z\frac{\partial(\mu_k Q_{0,tp})}{\partial z}\\ &+\frac{\partial}{\partial x}\left[D_{tp}\mu_k\frac{\partial(Q_{0,tp})}{\partial x}\right] + \frac{\partial}{\partial y}\left[D_{tp}\mu_k\frac{\partial(Q_{0,tp})}{\partial y}\right] + \frac{\partial}{\partial z}\left[D_{tp}\mu_k\frac{\partial(Q_{0,tp})}{\partial z}\right]\\ &+\sum_i \mu'_{k,i} R_{tp,prd,i} - \sum_l \mu'_{k,l} R_{tp,con,l}\end{aligned} \tag{5.17}$$

$$\begin{aligned}\frac{d(\gamma_k Q_{0,dop})}{dt} &= -v_x\frac{\partial(\gamma_k Q_{0,dop})}{\partial x} - v_y\frac{\partial(\gamma_k Q_{0,dop})}{\partial y} - v_z\frac{\partial(\gamma_k Q_{0,dop})}{\partial z}\\ &+\frac{\partial}{\partial x}\left[D_{dop}\gamma_k\frac{\partial(Q_{0,dop})}{\partial x}\right] + \frac{\partial}{\partial y}\left[D_{dop}\gamma_k\frac{\partial(Q_{0,dop})}{\partial y}\right] + \frac{\partial}{\partial z}\left[D_{dop}\gamma_k\frac{\partial(Q_{0,dop})}{\partial z}\right]\\ &+\sum_i \gamma'_{k,i} R_{dop,prd,i} - \sum_l \gamma'_{k,l} R_{dop,con,l}\end{aligned} \tag{5.18}$$

In the above equations, μ_k and γ_k are the kth order mole-fraction moment of the terminated and dormant chains, respectively, $Q_{0,tp}$ and $Q_{0,dop}$ are the zero-order molar concentration moment or the molar concentration of terminated and dormant chains, and D_{tp} and D_{dop} are the diffusion coefficient of the terminated and dormant chains. For small-molecule reactants, like monomer, initiator, or chain transfer agent, that are not polymeric, the kth order mole-fraction moment term is equal to unity, and the Q_0 term is the molar concentration of the small molecule. The diffusion coefficient of small-molecule reactants is not a strong function of concentration. Therefore, the diffusion coefficient term can be removed from the differential in the mass transfer by diffusion terms. After applying the above changes, the molar conservation equation for a small-molecule reactant in the Cartesian coordinate system simplifies to the following expression:

$$\begin{aligned}\frac{dC_{sm}}{dt} &= -\left(v_x \frac{\partial C_{sm}}{\partial x} + v_y \frac{\partial C_{sm}}{\partial y} + v_z \frac{\partial C_{sm}}{\partial z}\right) \\ &+ D_{sm}\left(\frac{\partial^2 C_{sm}}{\partial x^2} + \frac{\partial^2 C_{sm}}{\partial y^2} + \frac{\partial^2 C_{sm}}{\partial z^2}\right) + \sum_i R_{sm,prd,i} - \sum_l R_{sm,con,l}\end{aligned} \tag{5.19}$$

where C_{sm} and D_{sm} are the molar concentration and diffusion coefficient of the small molecule reactant. In most practical polymerization systems, the conservation equations for the kth order moments of the propagating, dormant, and terminated chains and the small-molecule reactants would have to be solved simultaneously using numerical methods to predict the concentration and different DP averages of the produced polymer. In the following subsections, the general moment conservation equation is applied to different polymerization reactors and processes.

5.5.1 Moment Conservation Equations Applied to Batch Stirred Vessel Reactor

In batch stirred vessel reactor (BSVR), the reactants are charged in the reactor under stirring to eliminate spatial gradients, the reactor is closed, pressure and temperature are set to the desired values, and the reaction is allowed to proceed for a specified time, as shown in Figure 5.5 [185–191]. After completion of the reaction, the reactor is opened, and the product is recovered from the reactor and purified to separate the polymer from the unreacted monomer.

As the reactants are mixed during the reaction, the transport of polymeric species by convection and molecular diffusion is negligible. Hence, the conservation equations for the kth order moment of the propagating and terminated polymer chains and the concentration of small-molecule reactants for a batch stirred reactor are as follows:

$$\frac{d\left(\lambda_k Q_{0,rp}\right)}{dt} = \sum_i \lambda'_{k,i} R_{rp,prd,i} - \sum_l \lambda'_{k,l} R_{rp,con,l} \tag{5.20A}$$

$$\frac{d\left(\mu_k Q_{0,tp}\right)}{dt} = \sum_i \mu'_{k,i} R_{tp,prd,i} - \sum_l \mu'_{k,l} R_{tp,con,l} \tag{5.20B}$$

$$\frac{dC_{sm}}{dt} = -\sum_l R_{sm,con,l} \tag{5.20C}$$

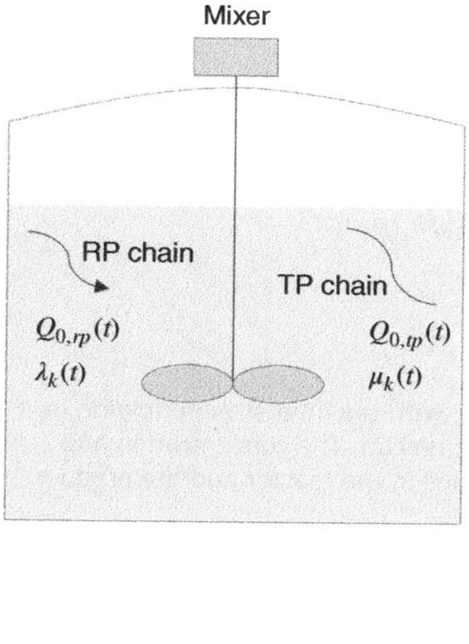

Figure 5.5 Schematic diagram of a BSVR. The reactor is closed during the reaction with no inlet or outlet streams. The concentration and kth order mole-fraction moments of the propagating (RP) and terminated (TP) polymer chains vary with time in the reactor.

In Eq. (5.20C), it is assumed that small molecules like initiator or monomer are only consumed in the reaction.

5.5.2 Moment Conservation Equations Applied to Continuous Stirred Tank Reactor

In a continuous stirred tank reactor (CSTR), the reactants continuously flow into the reactor, and the products continuously flow out, as shown in Figure 5.6. As the reaction is at steady state with no accumulation of mass in the reactor, the flow of mass entering the reactor is equal to the flow of mass exiting the reactor [190, 192–194]. Assuming the change in density with reaction conversion is negligible, the volumetric flow rates of the reactant and product streams are equal, but the two streams differ in composition. As the reaction is operating at steady state, the compositions of the two streams do not change with time. As the reactor is stirred, there is no spatial gradient of the reactants and products in the reactor, and the reactor content is homogenous. This implies that the composition of the product stream is identical to that of the reactor content.

Under these conditions and in the absence of spatiotemporal gradients in the reactor, the kth order concentration moment of the reactive propagating chains in the product stream can be related to the kth order moment in the reactant stream using a macroscopic balance as follows:

$$\left\{\begin{array}{l}\text{rate of}\\\text{addition of}\\k\text{th molar}\\\text{moment of}\\\text{the reactive}\\\text{propagating}\\\text{chains by}\\\text{entering}\\\text{reactant}\\\text{stream}\end{array}\right\} - \left\{\begin{array}{l}\text{rate of}\\\text{removal of}\\k\text{th molar}\\\text{moment of}\\\text{the reactive}\\\text{propagating}\\\text{chains by}\\\text{exiting}\\\text{product}\\\text{stream}\end{array}\right\} + \left\{\begin{array}{l}\text{rate of}\\\text{production}\\\text{of the }k\text{th}\\\text{molar}\\\text{moment of}\\\text{the reactive}\\\text{propagating}\\\text{chains in}\\\text{the reactor}\end{array}\right\} - \left\{\begin{array}{l}\text{rate of}\\\text{consumption}\\\text{of the }k\text{th}\\\text{molar}\\\text{moment of}\\\text{the reactive}\\\text{propagating}\\\text{chains in}\\\text{the reactor}\end{array}\right\} = 0 \quad (5.21\text{A})$$

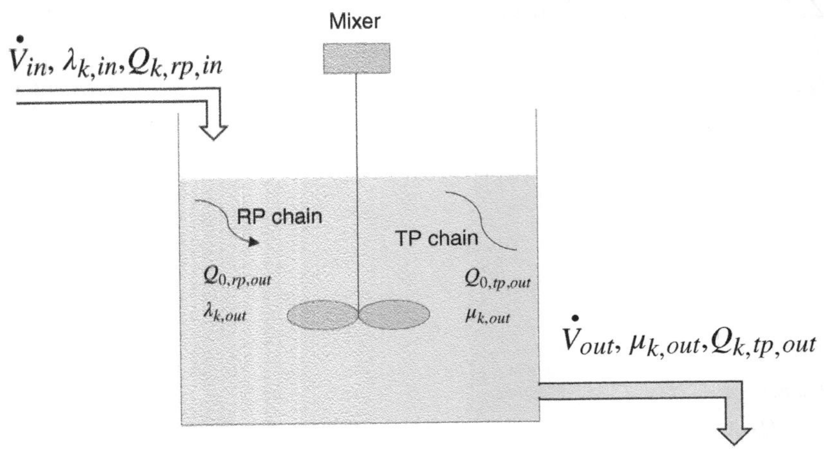

Figure 5.6 Schematic diagram of a CSTR. The reaction is at steady state with the inlet stream flowing in and the product stream flowing out with no accumulation of mass in the reactor. The concentration and mole-fraction moments of the propagating (RP) and terminated (TP) chains in the reactor and the product streams are identical.

The above macroscopic conservation equation is written in mathematical form as follows:

$$\lambda_{k,in}\, Q_{0,rp,in}\dot{V}_{in} - \lambda_{k,out}\, Q_{0,rp,out}\dot{V}_{out} + \sum_i \lambda'_{k,i}\, R_{rp,prd,i} V_r - \sum_l \lambda'_{k,l} R_{rp,con,l} V_r = 0 \tag{5.21B}$$

In the above equation, $\lambda_{k,in}$ and $\lambda_{k,out}$ are the mole-fraction moment of the propagating chains in the entering reactant and exiting product streams, respectively, $Q_{0,rp,in}$ and $Q_{0,rp,out}$ are the zero-order molar concentration moment of the propagating chains in the entering and existing streams, $\dot{V}_{in}$ and $\dot{V}_{out}$ are the volumetric flow rate of the entering and existing streams, and V_r is the volume of the CSTR. Upon rearrangement, and assuming the change in density due to the reaction is negligible $\left(\dot{V}_{in} = \dot{V}_{out} = \dot{V}\right)$, the kth order concentration moment of the propagating chains in the exiting product stream $(Q_{k,rp,out} = \lambda_{k,out}\, Q_{0,rp,out})$ is given in terms of the entering reactant stream and the reaction rates by the following expression:

$$Q_{k,rp,out} = \lambda_{k,out}\, Q_{0,rp,out} = \lambda_{k,in}\, Q_{0,rp,in} + \tau_{CSTR}\left(\sum_i \lambda'_{k,i}\, R_{rp,prd,i} - \sum_l \lambda'_{k,l} R_{rp,con,l}\right) \tag{5.21C}$$

where $\tau_{CSTR} = V_r/\dot{V}$ in the above equation, which is the ratio of the reactor volume to the volumetric flow rate of the reactant/product stream, is the mean residence time of the reactants/products in the CSTR. Using a similar approach, the kth order concentration moment of the terminated polymer chains and the concentration of small-molecule reactants for a CSTR are given as follows:

$$Q_{k,tp,out} = \mu_{k,out}\, Q_{0,tp,out} = \mu_{k,in}\, Q_{0,tp,in} + \tau_{CSTR}\left(\sum_i \mu'_{k,i}\, R_{tp,prd,i} - \sum_l \mu'_{k,l} R_{tp,con,l}\right) \tag{5.21D}$$

$$C_{sm,out} = C_{sm,in} - \tau_{CSTR}\left(\sum_i R_{sm,prd,i} - \sum_l R_{sm,con,l}\right) \tag{5.21E}$$

5.5.3 Moment Conservation Equations Applied to Plug-flow Tubular Reactor

In a plug-flow tubular reactor (PFTR), the polymerization reaction proceeds as the reactants flow in a tubular reactor. A PFTR operates at steady state, that is, the extent of reaction is independent of time but it increases in the direction of flow as shown in Figure 5.7 [195–200]. A PFTR resembles many small CSTR reactors in series with $\Delta V = A\Delta z$ where ΔV is the volume of each small CSTR of width Δz and A is the cross-sectional area of the tubular reactor.

Using the cylindrical coordinate system, assuming the flow is well mixed in the r and θ directions, and flow is only in the z-direction (i.e., plug flow, $v_r = v_\theta = 0$, no diffusion in r and θ directions), the general conservation equation for the kth order mole-fraction moment of the propagating chains (see Eq. 5.14A) simplifies as follows for a PFTR:

$$-v_z\frac{\partial\left(\lambda_k Q_{0,rp}\right)}{\partial z} + \frac{\partial}{\partial z}\left[D_{rp}\lambda_k\frac{\partial\left(Q_{0,rp}\right)}{\partial z}\right] + \sum_i \lambda'_{k,i}\, R_{rp,prd,i} - \sum_l \lambda'_{k,l}\, R_{rp,con,l} = 0 \tag{5.22A}$$

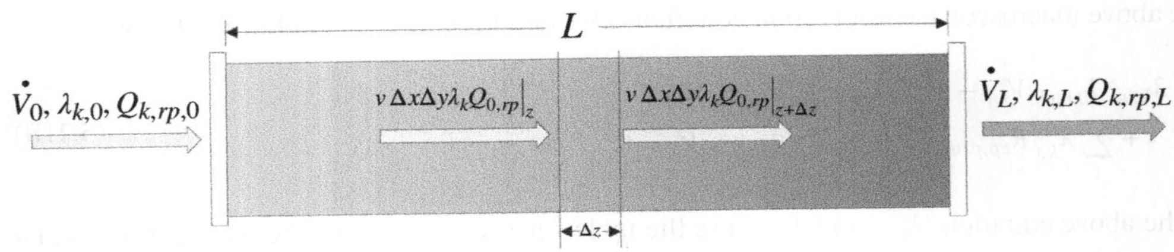

Figure 5.7 Schematic diagram of a PFTR. The inlet stream flows in at the tube inlet, and the product stream flows out at tube outlet. The reaction is at steady state with respect to time with no accumulation of mass in the reactor, but the concentration and kth order mole-fraction moments of the propagating (RP) and terminated (TP) chains vary with length of the tubular reactor in the z-direction.

In general, in a PFTR, the transport of the propagating chains by convection (the first term in the above equation) is much greater than the transport by molecular diffusion (the second term). Therefore, the second term in the above equation can be neglected, which further simplifies the conservation equation to the following:

$$-\nu_z \frac{\partial(\lambda_k Q_{0,rp})}{\partial z} + \sum_i \lambda'_{k,i} R_{rp,prd,i} - \sum_l \lambda'_{k,l} R_{rp,con,l} = 0 \tag{5.22B}$$

where ν_z is the flow velocity in the z-direction. Upon normalizing the z-direction by $z' = z/L_{PFTR}$ where L_{PFTR} is the length of the reactor, and using $\tau_{PFTR} = V_r/\dot{V}$, the above conservation equation is written as follows:

$$\frac{\partial(\lambda_k Q_{0,rp})}{\partial z'} = \tau_{PFTR}\left(\sum_i \lambda'_{k,i} R_{rp,prd,i} - \sum_l \lambda'_{k,l} R_{rp,con,l}\right) \tag{5.22C}$$

where τ_{PFTR} is the mean residence time of the reactants/products in the PFTR, $\dot{V} = \nu_z A$ and $V_r = A\,L_{PFTR}$ where $\dot{V}$, V_r, and A are the volumetric flow rate, reactor volume, and cross-sectional area of the tubular reactor, respectively. Upon integration of the above equation from $z' = 0$ to any z', the kth order concentration moment of the propagating chains $(Q_{k,rp} = \lambda_k\, Q_{0,rp})$ is given by the following relation:

$$Q_{k,rp} = \lambda_k\, Q_{0,rp} = \tau_{PFTR} \int_0^{z'} \left(\sum_i \lambda'_{k,i} R_{rp,prd,i} - \sum_l \lambda'_{k,l} R_{rp,con,l}\right) dz' \tag{5.22D}$$

Using a similar approach, it can be shown that the kth order concentration moment of the terminated polymer chains and the concentration of small-molecule reactants (assuming only consumed) for a PFTR are given as follows:

$$Q_{k,tp} = \mu_k\, Q_{0,tp} = \tau_{PFTR} \int_0^{z'} \left(\sum_i \mu'_{k,i} R_{rp,prd,i} - \sum_l \mu'_{k,l} R_{rp,con,l}\right) dz' \tag{5.22E}$$

$$C_{sm} = -\tau_{PFTR} \int_0^{z'} \left(\sum_l R_{sm,con,l}\right) dz' \tag{5.22F}$$

5.5.4 Moment Conservation Equations Applied to Batch Bulk, Solution, and Suspension Polymerizations

In bulk polymerization, the reaction is carried out with pure monomer without the addition of a solvent or other organic diluents because the solvent removal and its purification for recycling in the polymerization process can be costly. However, in bulk polymerization, the reaction viscosity increases exponentially by many orders of magnitude, which changes the kinetics of the reaction from rate-controlled to diffusion-controlled [201–205]. This change can lead to higher rates of chain propagation as compared to termination, unpredictable changes in polymer molecular weight and its distribution, and a rapid rise in the rate of polymerization. As polymerization reactions are exothermic, the rapid rise in the rate of polymerization can lead to temperature runaways (see Chapter 8 for diffusion-controlled polymerization) [206–210]. Consequently, often a solvent is added to the polymerization (i.e., solution polymerization) reaction to control the sharp rise in viscosity at moderate to high conversions [211, 212]. Another approach to control the reaction temperature and reduce medium viscosity is to disperse the polymerization reaction phase in a continuous phase, which is a nonsolvent for the monomer, as shown in Figure 5.8. As most monomers, like styrene and methyl methacrylate, are hydrophobic or sparingly soluble in aqueous solution, water is often used as the continuous phase to form a suspension. In suspension polymerization, the dispersed monomer phase (monomer, initiator, and other reactants) is added to the continuous water phase under stirring to form microdroplets with a size in the range of 10–1000 μm [213].

After forming the suspension, polymerization starts in the dispersed phase, whereas the continuous phase provides a relatively large volume for dissipation of the exothermic heat of the reaction to prevent a thermal runaway. The viscosity in the microdroplets continues to increase by orders of magnitude with monomer conversion to polymer, while the viscosity of the overall suspension remains relatively unaffected. As the temperature of the dispersed phase is controlled by the aqueous phase, the reaction continues until the polymerization mixture in the microdroplets solidifies into microparticles. As the size of the microdroplets is orders of magnitude larger than the size of the produced polymer chains, the polymer molecular weight and its distribution are not affected by

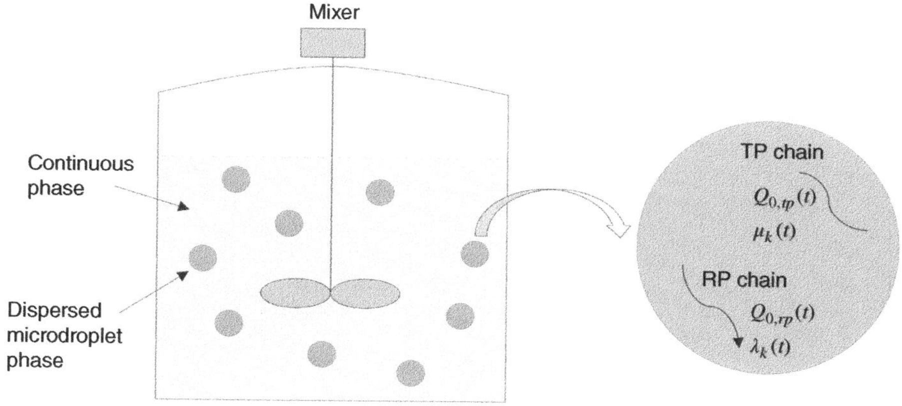

Figure 5.8 Schematic diagram of a suspension polymerization reactor. The reactants (monomer and initiator) phase is suspended as microdroplets in a continuous phase. The initiation, propagation, and termination reactions to form polymer chains occur in the dispersed microdroplet phase, whereas the continuous phase serves as a large volume for dissipating the heat of reaction. The concentration and kth order mole-fraction moments of the propagating (RP) and terminated (TP) polymer chains vary with time in the microdroplet phase.

the microscopic size of the droplets [214]. Therefore, the equations for the conservation of the kth order moment of the propagating and terminated chains derived for BSVR in Section 5.5.1 apply to the batch bulk and solution polymerization (see Eq. 5.20A–C). The conservation equations for the kth order moment of the propagating and terminated polymer chains applied to the microdroplets in a batch stirred suspension polymerization reactor are given as follows:

$$\frac{d\,Q_{k,rp,susp}}{d\,t} = \sum_{i} \lambda'_{k,i,susp}\, R_{rp,prd,i,susp} - \sum_{l} \lambda'_{k,l,susp}\, R_{rp,con,l,susp} \tag{5.23A}$$

$$\frac{d\,Q_{k,tp,susp}}{d\,t} = \sum_{i} \mu'_{k,i,susp}\, R_{tp,prd,i,susp} - \sum_{l} \mu'_{k,l,susp}\, R_{tp,con,l,susp} \tag{5.23B}$$

where the subscript "susp" is used to denote the suspended or microdroplet phase of the suspension; and $Q_{k,rp,susp}$ and $Q_{k,tp,susp}$ are the kth order molar concentration moment of the propagating and terminated polymer chains in the suspended phase of the reaction, respectively.

5.5.5 Moment Conservation Equations Applied to Batch Emulsion Polymerization

In emulsion polymerization, an emulsifier or surface-active agent, like anionic sodium lauryl sulfate or nonionic polyoxyethylene nonyl phenyl ether, is used to emulsify the monomer in an aqueous solution [215]. If the emulsifier concentration in the aqueous solution is greater than a critical concentration (i.e., critical micelle concentration or CMC), thermodynamically stable nanostructures, called micelles or colloidal particles, are formed in the aqueous solution, as shown in Figure 5.9 [216].

Upon its addition to the aqueous solution, the monomer is taken up by the micelles to form monomer-filled micelles in the size range of 10–100 nm [217, 218]. The monomer-filled micelles are thermodynamically stable because the monomer uptake reduces the free energy of the emulsion. Not unlike suspension polymerization, microdroplets of monomer with a size in the range of 10–1000 μm are also formed in the emulsion. The extent of microdroplet formation depends on the

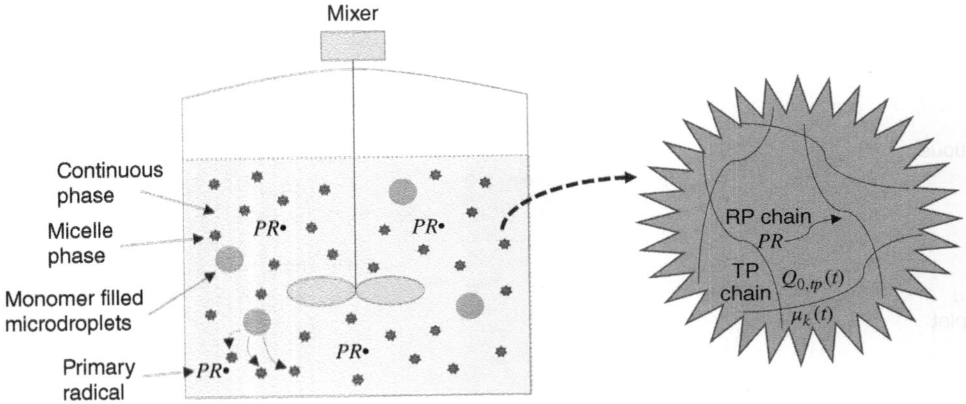

Figure 5.9 Schematic diagram of a batch emulsion polymerization reactor. The reaction consists of the colloidal (micelle) phase, stabilized by emulsifier and filled with monomer, monomer microdroplet phase stabilized by mixing, and the continuous phase. The initiator is dissolved in the continuous phase, but polymerization reaction occurs mainly in the colloidal phase. The monomer is transferred from the microdroplets to the micelles as it is depleted by the reaction. The concentration and kth order mole-fraction moment of the terminated (TP) chains in the colloidal phase increase with reaction time.

amount of added monomer and emulsifier concentration. After addition and dissociation of the initiator in the aqueous phase, the reactive fragments are absorbed by the monomer-filled micelles to initiate chain propagation [218, 219]. As the surface area of the micelles is orders of magnitude larger than that of microdroplets, almost all initiator fragments are taken up by the micelles. Therefore, unlike suspension polymerization, polymerization almost exclusively takes place inside the monomer-filled micelles, not the monomer microdroplets. As chain propagation proceeds and monomer concentration decreases in the micelles, monomer is transferred from the microdroplets to the micelles to replenish consumed monomer and maintain its equilibrium concentration with the produced polymer in the micelles. Therefore, the microdroplets in emulsion polymerization serve as a depot for the monomer during the reaction. With many growing micelles in the reaction system, there is, on average, only one propagating chain per micelle [220, 221]. As propagation is favored over termination, with on average only one reactive chain per micelle, ultra-high molecular weight polymers can be produced by emulsion polymerization [215, 222]. The molecular weight of the polymer can be controlled by the intentional addition of a chain transfer agent to the reaction. As polymerization occurs in the dispersed colloidal phase and there is, on average, only one propagating chain per micelle, the simulation method of Monte Carlo is best suited to model emulsion polymerization (see Chapter 11) [223–228]. The conservation equation for the kth order moment of the polymer chains can also be used to model emulsion polymerization by applying it to the quadrillions of thermodynamically stable micelles in the emulsion. It can be assumed that the initiator fragments, after dissociation in the aqueous phase, are rapidly absorbed and homogenously divided between the micelles; thus, the retention of fragments by the aqueous phase or absorption by monomer microdroplets is negligible. Due to their small size, it can be assumed that there is no concentration gradient within the micelles; thus, the transport of the reactants by convection or diffusion is negligible. As there is, on average, only one propagating chain in each micelle, it can be assumed that termination by disproportionation and combination is almost absent; thus, chain termination is controlled by the addition of a terminating agent or chain transfer agent. It can also be assumed that propagating chains in the micelles, due to their large size, cannot desorb; thus, they are confined to the colloidal phase. As microdroplets serve as a depot for the monomer throughout the reaction, the monomer concentration in the micelle phase is not dictated by the feed concentration, but it is controlled by its equilibrium concentration with the produced polymer chains within each micelle. Based on the above assumptions, the conservation equation for the kth order concentration moment of the terminated chains in the spherical coordinate system (Eq. 5.16) applied to the colloidal phase simplifies to the following:

$$\frac{d\,Q_{k,tp,mic}}{d\,t} = \frac{d\left(\lambda_{k,mic}\,Q_{0,tp,mic}\right)}{d\,t} = \sum_{i} \lambda'_{k,i,mic}\,R_{tp,prd,i,mic} - \sum_{l} \lambda'_{k,l}\,R_{tp,con,l,mic} \tag{5.24A}$$

where the subscript "mic" in the above equations is used to denote the micelle or colloidal phase of the emulsion; $Q_{k,tp,mic}$ is the kth order concentration moment of the terminated chains in the colloidal phase. As the monomer is in equilibrium with the polymer in the colloidal phase, the monomer concentration can be estimated from the equilibrium thermodynamics of multicomponent solutions. Using the Flory–Huggins solution theory, the equilibrium monomer concentration $(Q_{0,m,mic})$ in terms of the polymer concentration $(Q_{0,tp,mic})$ is given by the following relation:

$$\begin{aligned} &ln\left(\frac{Q_{0,m.mic}}{Q_{0,m,mic}+\mu_{1,mic}Q_{0,tp,mic}}\right)+\frac{Q_{0,tp,mic}}{Q_{0,m,mic}}ln\left(\frac{\mu_{1,mic}Q_{0,tp.mic}}{Q_{0,m,mic}+\mu_{1,mic}Q_{0,tp,mic}}\right)\\ &+\chi_{m-p}\left(\frac{\mu_{1,mic}Q_{0,tp.mic}}{Q_{0,m,mic}+\mu_{1,mic}Q_{0,tp,mic}}\right)=0 \end{aligned} \tag{5.24B}$$

In the above equation, $\mu_{1,mic}$ is the first order mole-fraction moment of the terminated polymer chains in the colloidal phase, and χ_{m-p} is the Flory–Huggins interaction parameter between the monomer and its polymer. As propagating species have a short lifetime, next sections delve into the methodology for theoretical modeling of the kth order moments of the reactive propagating species in different polymerization systems.

5.6 Formation of Reactive Species in Conventional Radical Polymerization and the Quasi-steady State Assumption

Reactants like monomer and initiator and products like terminated and dormant polymer chains are considered long-lived molecules such that they can be separated from the mixture, and their quantity measured at time t during the reaction. Conversely, species like reactive propagating radicals are in an excited, reactive, or unstable state that readily react with any molecule in the reaction mixture to return to the stable ground state. Hence, radical species cannot be separated from the polymerization mixture and quantified like stable molecules. The concentration and lifetime of radicals can be measured by *in situ* techniques [229]. The lifetime of the primary radical species is on the order of microseconds [230], whereas the lifetime of reactive propagating chains in conventional radical polymerization with initiators like benzoyl peroxide (BPO) or azobisisobutyronitrile (AIBN) that generate two equally reactive benzyl or cyanopropyl radicals is on the order of 100 microseconds to 1 milliseconds [231, 232]. Hence, for a polymerization reaction on the order of many minutes, it is appropriate to assume that the radicals formed in a differential time interval are consumed in the same interval [233, 234]. However, there are situations in which initiator dissociation gives rise to two radical species with unequal reactivity and stability [235]. The two fragments may have equal reactivity, different reactivity with initiation far from thermodynamic equilibrium, and different reactivity with initiation near equilibrium. This section is concerned with initiators that generate reactive species with equal reactivity.

If the two primary radicals formed by initiator dissociation in radical polymerization are equally reactive (i.e., unstable with short half-life), the produced fragments react instantaneously with monomer to form propagating chains or react with other fragments to form more stable species. This is the basis of the quasi-steady state assumption (QSSA) in polymerization for predicting the concentration of reactive propagating species. QSSA states that the reactive species in polymerization do not accumulate in the reaction if the reaction is not diffusion controlled [233, 234]. The reactive species produced by initiator dissociation in a differential time interval dt at time t in the reaction are consumed in the same differential time dt, that is, the produced species do not carry over to the next differential time interval. In terms of the kth moment of the reactive propagating species, QSSA states that, in a differential time dt, as compared to the reaction time, the rate of production of the kth order concentration moment of the reactive species (primary radicals or propagating chains) is equal to their consumption rate. Therefore, Eq. (5.14A) for the conservation of the kth order moment of the reactive propagating species in the absence of convective and diffusive mass transfer simplifies as follows:

$$\frac{d\left(\lambda_{k,t}\, Q_{0,rp}\right)}{d\,t} = 0 \tag{5.25A}$$

$$\Rightarrow \sum_i \lambda'_{k,i}\, R_{rp,prd,i} = \sum_l \lambda'_{k,l}\, R_{rp,con,l} \tag{5.25B}$$

Equation (5.25A) states that the kth order mole-fraction moment of the reactive propagating species does not accumulate in the reaction. Equation (5.25B) states that the production and consumption rates of the kth order mole-fraction moment of the reactive propagating species are equal in a time interval dt at time t in the reaction. If QSSA is valid, Eq. (5.25A) and (5.25B) apply to the kth order mole-fraction moment of the reactive propagating species/chains at any time t for any order k. For $k = 0$, $\lambda'_{0,i} = \lambda'_{0,l} = 1$, and Eq. (5.25B) is simplified as follows:

$$k = 0 \quad \Rightarrow \sum_i R_{rp,prd,i} = \sum_l R_{rp,con,l} \quad \in QSSA \tag{5.25C}$$

Equation (5.25C) states that reactive species (primary radical or reactive propagating chain) produced in the time interval dt between the time $t - dt$ to t is consumed in that same time interval and is not carried over to the next interval between t to $t + dt$. Therefore, if QSSA is valid, the kth order mole-fraction moment of the propagating chains at time t, $\lambda_{k,t}$, is equal to its instantaneous kth order mole-fraction moment, $\lambda'_{k,t}$, which is stated mathematically as follows:

$$\lambda_{k,t} = \lambda'_{k,t} = \frac{\sum_i \lambda'_{k,i} R_{rp,prd,i}}{\sum_i R_{rp,prd,i}} = \frac{\sum_l \lambda'_{k,l} R_{rp,con,l}}{\sum_l R_{rp,con,l}} \tag{5.25D}$$

Depending on reaction conditions (conversion and viscosity), QSSA may be valid for all reactive species in the reaction like in dilute, low viscosity, reaction-controlled regime of polymerization; QSSA may be valid for primary, small molecule reactive species but not for long, reactive propagating chains like in concentrated, viscous, diffusion controlled regime of polymerization; and QSSA may not be valid for any reactive species in the reaction like in solid state regime of polymerization.

5.6.1 Quasi-steady State in Dilute Polymerization Regime

If the polymerization reaction is not limited by diffusion of reactive species, QSSA is valid for primary radicals R as well as the short and long propagating chains RP_n as expressed by the following equations:

$$\begin{cases} d\,[R]\,/dt = 0 \\ d\left(\lambda_{k,t}\, Q_{0,rp}\right)/dt = 0 \end{cases} \quad \in \text{dilute solution} \tag{5.26A}$$

5.6.2 Quasi-steady State in Concentrated Polymerization Regime with Reaction Temperature Above Glass Transition Temperature

If the polymerization reaction is carried out in bulk or to high conversions, but above the glass transition temperature (T_g) of the polymer, such that the reaction medium is a highly viscous liquid, the reaction becomes diffusion controlled for the long propagating chains [236–238]. Under this condition, the diffusivity of the propagating chains decreases by many orders of magnitude while the diffusivity of small-molecule reactive species (i.e., primary radicals) is relatively unaffected by the increased viscosity of the medium. Therefore, QSSA applies to the primary radicals but not to the propagating chains as follows:

$$\begin{cases} d\,[R]\,/dt = 0 \\ d\left(\lambda_{k,t}\, Q_{0,rp}\right)/dt \neq 0 \end{cases} \quad \text{concentrated, } T \text{ above } T_g \tag{5.26B}$$

5.6.3 Quasi-steady State in Concentrated Polymerization Regime with Reaction Temperature Below Glass Transition Temperature

If the reaction is carried out in the glassy state at or below T_g of the polymer, all reactive species, including small-molecule primary radicals, become trapped in the polymerizing mixture, and their reactions become diffusion controlled. Under these conditions, all reactive species could accumulate in the reaction, and QSSA is invalid for all reactive species as follows:

$$\begin{cases} d\,[R]\,/dt = 0 \\ d\left(\lambda_{k,t}\, Q_{0,rp}\right)/dt \neq 0 \end{cases} \quad \text{concentrated, } T \text{ below } T_g \tag{5.26C}$$

5.7 Initiation and Formation of Reactive Species in Polymerization

5.7.1 Conventional Radical Polymerization with Equal Reactivity of Radical Species

In conventional radical polymerization, the two or more primary radicals R generated by the dissociation of initiator molecule I have equal reactivity. The dissociation reaction is initiated by temperature, a redox reaction, or ultraviolet or gamma radiation, producing, in general, r primary radicals with equal reactivity [239]. Following initiation, the produced radicals may recombine prior to diffusing away from the dissociation cage with a rate constant k_c to form an inactive molecular fragment IA [240]. Otherwise, the produced radical diffuses to the bulk of the reaction, where it reacts with a monomer with a rate constant k_{ci} to form a propagating chain with one monomer unit RP_1. The reactions for initiation, recombination, and propagating chain formation for a thermally dissociated initiator in the presence of monomer are as follows:

$$I \xrightarrow{k_d} {}^rR \tag{5.27A}$$

$$R + R \xrightarrow{k_c} IA \tag{5.27B}$$

$$R + M \xrightarrow{k_{ci}} RP_1 \tag{5.27C}$$

where k_d is the rate constant for initiator dissociation. The molar mass balance for the initiator I in the absence of transport by convection or molecular diffusion is given by the following expression (see Eq. 5.13A for general mass balance):

$$\left\{\begin{array}{l}\text{rate of change in the number}\\ \text{of moles of I with time}\\ \text{in the reaction volume}\end{array}\right\} = \left\{\begin{array}{l}\text{rate of molar}\\ \text{consumption}\\ \text{of I}\end{array}\right\} \tag{5.28A}$$

The molar mass balance above in mathematical form is written as follows:

$$-\frac{d\left([I]\,V_R\right)}{dt} = k_d\,[I]\,V_R \tag{5.28B}$$

Where $[I]$ and V_R are the initiator concentration and reaction volume, respectively. Assuming constant reaction volume, the above equation can be integrated with respect to time using the initial condition $[I] = [I]_0$ at time $t = 0$ to find the following equation relating $[I]$ to reaction time:

$$[I] = [I]_0\; e^{-k_d\, t} \tag{5.28C}$$

According to the above equation, the initiator concentration decreases exponentially with dissociation. Using a similar approach, the molar mass balance for the primary radical R is written as follows:

$$\left\{\begin{array}{l}\text{rate of change}\\ \text{in the number}\\ \text{of moles of } R\\ \text{with time}\end{array}\right\} = \left\{\begin{array}{l}\text{rate of}\\ \text{production}\\ \text{of } R \text{ by I}\\ \text{dissociation}\end{array}\right\} - \left\{\begin{array}{l}\text{rate of}\\ \text{consumption}\\ \text{of } R \text{ by}\\ \text{recombination}\end{array}\right\} - \left\{\begin{array}{l}\text{rate of}\\ \text{consumption}\\ \text{of } R \text{ by}\\ \text{chain initiation}\end{array}\right\} \quad (5.29A)$$

The above mass balance in mathematical form is given by the following equation:

$$\begin{aligned}\frac{d\,([R]\,V_R)}{d\,t} &= rR_d - R_c - R_{ci}\\ &= +r\,k_d\,V_R\,[I] - k_c V_R\,[R]^2 - k_i\,V_R\,[R]\,[M]\end{aligned} \quad (5.29B)$$

where r is the number of radicals generated per initiator dissociation, R_d, R_c, and R_{ci} are the rates of initiator dissociation, recombination, and initiation of a propagating chain, respectively; and $[R]$ and $[M]$ are the molar concentrations of primary radical and monomer. For constant reaction volume, the above equation simplifies to the following:

$$\frac{d\,[R]}{d\,t} = +r\,k_d\,[I] - k_c\,[R]^2 - k_i\,[R]\,[M] \quad (5.29C)$$

The extent of radical recombination reaction versus chain initiation depends on the type of initiator. Assuming QSSA is valid; the above equation can be solved for the primary radical concentration as follows:

$$\begin{aligned}&d\,[R]\,/dt = 0 \quad \in QSSA\\ &\Rightarrow r\,k_d\,[I] - k_c\,[R]^2 - k_i\,[R]\,[M] = 0\end{aligned} \quad (5.29D)$$

The initiator molecule is wasted if it dissociates, and the two fragments of its dissociation recombine to form an inactive species IA. Therefore, the ratio of the rate of propagating chain initiation to the rate of primary radical formation can be defined as the initiator efficiency by the following relation:

$$f_0 = \frac{R_i}{rR_d} = \frac{k_i\,[R]\,[M]}{r\,k_d\,[I]} = \frac{1}{1 + \left(rk_c\,[R]^2\,/k_i\,[R]\,[M]\right)} \quad (5.29E)$$

Upon substitution of the above equation for initiation efficiency in Eq. (5.29D), $[R]$ is expressed in terms of $[I]$ and $[M]$ by the following equation:

$$[R] = \left[\frac{r\,k_d\,(1 - f_0)\,[I]}{k_c}\right]^{1/2} = \frac{r\,k_d f_0\,[I]}{k_{ci}\,[M]} \quad (5.29F)$$

The above equation expresses $[R]$ in terms of measurable parameters. If $f_0 = 1$, then $k_c = 0$, and the second equation on the right side of Eq. (5.29F) is used to determine $[R]$. Aside from recombination and chain initiation, there are other reactions in polymerization that can deactivate reactive species, like reaction with oxygen, inhibitor, retarder, and impurities in the reaction mixture [93–99, 241–244]. As a result, f_0 is rarely equal to unity. Substitution for $[I]$ from Eq. (5.28C) in the above equation results in the following expression for $[R]$ as a function of time:

$$[R] = \left(\frac{r\,k_d\,(1 - f_0)\,[I]_0}{k_c}\right)^{1/2} e^{-k_d\,t/2} = \frac{r\,k_d f_0\,[I]_0}{k_{ci}\,[M]} e^{-k_d\,t} \quad (5.29G)$$

Based on the above equation, QSSA does not imply that $[R]$ is independent of time; it implies a state in which production and consumption of the primary reactive species are equal in a differential interval of time. The above equation relates $[R]$ to measurable quantities like $[I]$, $[M]$, and the reaction rate constants. During the early stages of the reaction when $[I] \approx [I]_0$, the radical concentration is nearly independent of time and given by the following equation:

$$[I] \approx [I]_0 \quad \Rightarrow [R] \approx \left(r\, k_d\,(1 - f_0)\,[I]_0\,/k_c\right)^{1/2} \tag{5.30A}$$

and the rate of polymerization, R_p, is linearly dependent on monomer concentration as follows:

$$\begin{aligned}[I] \approx [I]_0 \quad &\Rightarrow \quad R_p = -d\,[M]\,/dt = -k_p\,[R]\,[M] \\ &\approx \quad -k_p\left(r\, k_d\,(1 - f_0)\,[I]_0\,/k_c\right)^{1/2}\,[M]\end{aligned} \tag{5.30B}$$

where k_p is the rate constant for propagation reaction. Upon integration of the above equation, for constant initiator concentration, ln $\left([M]_0\,/\,[M]\right)$ increases linearly with reaction time during the early stage of the reaction.

5.7.2 Reversible-deactivation Radical Polymerization with Unequal Reactivity of Reactive Species

In reversible-deactivation radical polymerization (RDRP), the dissociation of initiator produces two reactive species, with one species very reactive, known as the transient radical, and the other species relatively unreactive, known as the persistent radical [235, 245, 246]. RDRP includes, among others, atom transfer radical polymerization (ATRP) and nitroxide-mediated radical polymerization (NMP) [247–251]. In ATRP, an alkyl halide as the initiator and a transition metal catalyst (TMC) at low oxidation state react to form the transient and persistent reactive species, respectively, as follows:

$$I + C \underset{k_{dac}}{\overset{k_{ac}}{\rightleftharpoons}} R + Y \tag{5.31A}$$

The above ATRP initiation reaction is bimolecular with I as the alkyl halide initiator, C as the TMC in low oxidation state, R as the reactive transient, and Y as the persistent reactive species. In NMP, an alkoxyamine initiator dissociates to form two species with unequal reactivity as follows [252–254]:

$$R_1C{-}ON\,(R_2R_3) \underset{k_{dac}}{\overset{k_{ac}}{\rightleftharpoons}} R_1C^{\bullet} + {}^{\bullet}ON\,(R_2R_3) \tag{5.31B}$$

where $R_1C{-}ON\,(R_2R_3)$ is the alkoxyamine initiator. The radical on the fragment $(R_3R_2)\,NO^{\bullet}$ has low reactivity due to resonance stabilization by relocation of the radical from the oxygen atom to the nitrogen and vice versa, whereas the radical on the carbon atom of the alkyl fragment $R_1C^{\bullet}$ is highly reactive. As a result, the more reactive fragment $R_1C^{\bullet}$ in this radical pair is termed the transient radical, whereas the relatively stable fragment $(R_3R_2)\,NO^{\bullet}$ is labeled the persistent radical. The above initiation reaction can be written as follows:

$$I \underset{k_{dac}}{\overset{k_{ac}}{\rightleftharpoons}} R + Y \tag{5.31C}$$

Both initiation reactions in ATRP and NMP are reversible because of the persistence of the less reactive Y fragment, which reacts with R to regenerate the initiator. In general, the reactions resulting in chain initiation in RDRP for the purpose of deriving an expression for the concentration of R are as follows:

$$I + C \underset{k_{dac}}{\overset{k_{ac}}{\rightleftharpoons}} R + Y \tag{5.32A}$$

$$R + R \xrightarrow{k_{tR}} R\text{—}R \tag{5.32B}$$

$$R + M \xrightarrow{k_{ci}} RP_1 \tag{5.32C}$$

where k_{tR} is the rate constant for self-termination of the reactive transient species R. In RDRP, the equilibrium constant for reversible deactivation $K = k_{ac}/k_{dac} \ll 1$ to maintain dormancy of the propagating chains most of the time, thus minimizing chain termination reactions that broaden the chain DP distribution. For example, in ATRP, $1 \times 10^{-9} < K_{ATRP} < 1 \times 10^{-4}$[255]. Due to the low value of K_{ATRP} and the positively charged nature of the Y fragment in most cases, the persistent radical Y exclusively reacts with R or RP (i.e., the reactive propagating species) to generate dormant chains [256]. As a result, the rate of Y self-termination $R_{tY} \approx 0$. The rate constants in the above reactions are $k_{ci} \approx 1 \times 10^4\, M^{-1}s^{-1}$ [257], $k_{tR} \approx 1 \times 10^9\, M^{-1}s^{-1}$, and $1 \times 10^4 M^{-1}s^{-1} < k_{dac} < 1 \times 10^6\, M^{-1}s^{-1}$ [255, 256]. These values of the rate constants imply that $[R]$ as well as $[I]$, $[C]$ and $[Y]$ are regulated by the equilibrium Rxn. 5.32A. Therefore, unlike conventional radical polymerization, QSSA does not hold for the transient radicals in RDRP, and the production of $[R]$ in a differential time interval is not inevitably equal to its consumption rate. Instead, in RDRP, it can be assumed that the equilibration time of the equilibrium Rxn. 5.32A, which is in the order of seconds, is much faster than the rate of chain propagation (order of hours in RDRP due to chain dormancy). If the concentrations of I, C, R, and Y are regulated by the equilibrium Rxn. 5.32A and self-termination Rxn. 5.32B, the molar mass conservation equation for R is written as follows:

$$\begin{Bmatrix} \text{rate of change} \\ \text{in the number} \\ \text{of moles of } R \\ \text{with time} \end{Bmatrix} = \begin{Bmatrix} \text{rate of} \\ \text{production} \\ \text{of } R \text{ by I} \\ \text{dissociation} \end{Bmatrix} - \begin{Bmatrix} \text{rate of} \\ \text{consumption} \\ \text{of } R \text{ by} \\ \text{reversible} \\ \text{deactivation} \end{Bmatrix} - \begin{Bmatrix} \text{rate of} \\ \text{consumption} \\ \text{of } R \text{ by} \\ \text{self–} \\ \text{termination} \end{Bmatrix} \tag{5.33A}$$

The above molar balance is given by the following equation in mathematical form for a constant volume reaction [235, 245, 255–259]:

$$\frac{d\,[R]}{dt} = +k_{ac}\,[I]\,[C] - k_{dac}\,[R]\,[Y] - 2k_{tR}\,[R]^2 \tag{5.33B}$$

The factor "2" in the last term of the above equation is attributed to the involvement of two transient radicals in the self-termination reaction. Similarly, the molar mass conservation equation for Y is given as follows [235, 245, 255–259]:

$$\frac{d\,[Y]}{dt} = +k_{ac}\,[I]\,[C] - k_{dac}\,[R]\,[Y] \tag{5.33C}$$

In the above two equations, $[I]$ and $[C]$ are functions of time and related to their initial values by the following relations:

$$\begin{cases} [Y'] = [Y] - [Y]_0 = [I]_0 - [I] \\ [Y'] = [Y] - [Y]_0 = [C]_0 - [C] \end{cases} \tag{5.33D}$$

where $[I]_0$, $[C]_0$, and $[Y]_0$ are the initial concentrations of initiator, catalyst, and persistent species, respectively. The above differential Eq. (5.33B) and (5.33C) are subject to the following equilibrium relation between I, C, R, and Y:

$$K_{RDRP} = \frac{k_{ac}}{k_{dac}} = \frac{[R]\,[Y]}{[I]\,[C]} \tag{5.33E}$$

where K_{RDRP} is the equilibrium constant for RDRP. Upon solving differential Eq. (5.33B) and (5.33C) subject to the initial condition 5.33D and equilibrium condition 5.33D, $[Y]'$ is given by the following expression as a function of time [235, 245, 255–259]:

$$F\left([Y]'\right) = 2k_{tR}K_{RDRP}^2\, t + F\left([I]_0, [C]_0, [Y]_0\right) \tag{5.33F}$$

where the functions $F([Y'])$ and $F\left([I]_0, [C]_0, [Y]_0\right)$ are given by the following expressions:

$$F\left([Y]'\right) = \left(\frac{[I]_0\,[C]_0}{[C]_0 - [I]_0}\right)^2 \times \left(\frac{1/[C]_0^2}{([I]_0 - [Y'])} + \frac{2/\left([I]_0\,[C]_0\right)}{([C]_0 - [I]_0)} \ln\left(\frac{[I]_0 - [Y']}{[C]_0 - [Y']}\right) + \frac{1/[I]_0^2}{([C]_0 - [Y'])}\right) \tag{5.33G}$$

$$F\left([I]_0, [C]_0, [Y]_0\right) = \left(\frac{[I]_0\,[C]_0}{[C]_0 - [I]_0}\right)^2 \times \left(\frac{1/[C]_0^2}{([I]_0 + [Y]_0)} + \frac{2/\left([I]_0\,[C]_0\right)}{([C]_0 - [I]_0)} \ln\left(\frac{[I]_0 + [Y]_0}{[C]_0 + [Y]_0}\right) + \frac{1/[I]_0^2}{([C]_0 + [Y]_0)}\right) \tag{5.33H}$$

The equilibrium constant for RDRP, defined by Eq. (5.33E), is determined from the slope of the linear line relating $F([Y'])$ as the Y-axis to time t as the X-axis. If K_{RDRP} is known, Eq. (5.33F) in conjunction with Eq. (5.33G) and (5.33H) are used to solve for $[Y']$, which in turn is used to find $[Y]$, $[I]$, and $[C]$ using Eq. (5.33D). Knowing $[Y]$, $[I]$, and $[C]$, the equilibrium Eq. (5.33E) is used to find $[R]$. If $[I]_0 = [C]_0$ and $[Y]_0 = 0$, Eq. (5.33F) simplifies to the following expression [235, 245, 255–259]:

$$\frac{[C]_0^2/3}{([C]_0 - [Y])^3} - \frac{[C]_0}{([C]_0 - [Y])^2} + \frac{1}{([C]_0 - [Y])} = 2k_{tR}K_{RDRP}^2 t + \frac{1}{3\,[C]_0} \tag{5.34}$$

Based on the above equations, if $[Y]_0 < 0.1\,[C]_0$, then $[Y]$ and $[R]$ are related to time by the following relations:

$$\frac{[Y]_0}{[C]_0} < 0.1 \Rightarrow \begin{cases} [R] = \left(K_{RDRP}\,[I]_0\,[C]_0\,/6k_{tR}\right)^{1/3}\, t^{-1/3} \\ [Y] = \left(6k_{tR}K_{RDRP}^2[I]_0^2[C]_0^2\right)^{1/3}\, t^{1/3} \end{cases} \tag{5.35A}$$

The above relation for $[R]$ results in the scaling relation $\ln\left([M]_0 / [M]\right) \propto t^{2/3}$ which is referred to as the Fischer equation [235, 245, 246, 255]. Conversely, if $[Y]_0 > 0.1\,[C]_0$, then $[R]$ is approximately independent of time and given by the following expression:

$$\frac{[Y]_0}{[C]_0} > 0.1 \Rightarrow [R] \approx K_{RDRP}\,[I]_0\,[C]_0\,/\,[Y]_0 \tag{5.35B}$$

The above relation for $[R]$ results in the scaling relation $\ln\left([M]_0 / [M]\right) \propto t$ which is referred to as the Matyjaszewski relation [246].

5.7.3 Anionic Polymerization with Free, Stable, and Reactive Anionic Species

The reactive species in anionic polymerization is an anion paired with an oppositely charged cation for electroneutrality. A typical initiator for anionic polymerization is butyl lithium, as well as lithium-related organometallic compounds [260–263]. In a typical anionic polymerization, the butyl lithium initiator is added to distilled, dried polymerization solvent (diethyl ether or tetrahydrofuran) in a reaction flask in an ice bath under stirring and a nitrogen atmosphere. After dissolution in the solvent, the organolithium initiator spontaneously dissociates into butyl anion and lithium cation according to the following reaction:

$$R\text{—}Li \xrightleftharpoons{k_{d,ani}} R^- \cdot \cdot Li^+ \tag{5.36}$$

Unlike RDRP, which shifts the equilibrium in the initiation reaction almost entirely toward deactivation (see Eq. 5.31A), the initiation reaction in anionic polymerization is shifted practically completely toward the formation of reactive anionic species. This is attributed to the high susceptibility of the lithium atom to give up its lone electron in the 2s orbital to form a stable lithium cation, with a full first shell and a vacant 2s orbital, paired with butyl anion stabilized by solvent molecules in solution. Furthermore, unlike RDRP, there is no self-termination of reactive anionic species (see Rxn. 5.32B). Following the dissociation of initiator into a stable ion pair, the monomer is added to the reaction for polymerization. Therefore, the formation of reactive species in anionic polymerization happens prior to the addition of monomer. As a result, the molar mass balance for the initiator I in anionic polymerization in the absence of transport by convection or molecular diffusion is written as follows (see Eq. 5.13A for general mass balance):

$$\left\{\begin{array}{l}\text{rate of change in the number}\\ \text{of moles of } R-Li \text{ with time}\\ \text{in the reaction volume}\end{array}\right\} = \left\{\begin{array}{l}\text{rate of molar}\\ \text{consumption of}\\ R-Li \text{ by dissociation}\end{array}\right\} \tag{5.37A}$$

The above molar mass balance in mathematical form is given by the following equation:

$$\frac{d\,([RLi]\,V_R)}{d\,t} = -k_{d,ani}\,[RLi]\,V_R \tag{5.37B}$$

where $k_{d,ani}$ is the rate constant for dissociation of anionic initiator RLi and $[RLi]$ is the initiator concentration. Integration of the above differential equation at constant reaction volume, subject to the initial condition $[RLi] = [RLi]_0$ at $t = 0$ results in the following equation for $[RLi]$ as a function of time:

$$[RLi] = [RLi]_0\; e^{-k_{d,ani}t} \tag{5.37C}$$

Similarly, the molar mass conservation equation for the produced anionic species is as follows:

$$\frac{d\,[R^-]}{dt} = -\frac{d\,([RLi])}{d\,t} \Rightarrow [R^-] = [RLi]_0\left(1 - e^{-k_{d,ani}t}\right) \tag{5.38A}$$

where $[R^-]$ is the concentration of anion in the reaction that reacts with the monomer to initiate chain propagation. As RLi dissociation is complete, the produced anions are stable until a terminating agent is intentionally added to the reaction. Therefore, after initial formation of the reactive species, the concentration of anions in the propagation step is practically constant, and monomers can be added at any time to the reaction to increase the polymer chain length. As a result, the concentration of the propagating anion, $[R^-]$, in anionic polymerization is given by the following equation:

$$k_{d,ani}t \rightarrow \infty \;\Rightarrow (1 - e^{-k_{d,ani}t}) \rightarrow 1 \;\Rightarrow [R^-] \approx [RLi]_0 \tag{5.38B}$$

The above equation implies that, due to the complete dissociation of the initiator, the concentration of reactive anionic species remains nearly constant during anionic polymerization.

5.7.4 Cationic Polymerization with Free and Ion-pair, Reactive Cationic Species

The reactive species in cationic polymerization is a cation paired with an oppositely charged anion for electroneutrality. Typically, an electrophilic Lewis acid is used to form the cationic species for chain initiation and the anionic species for electroneutrality [264, 265]. A typical initiation system for cationic polymerization is hydrogen iodide (HI) as the initiator with a metal halide like zinc halide (ZnX_2) or tin halide (SnX_2), boron trifluoride diethyl etherate (BF_3OEt_2), as well as molecular iodine as the Lewis acid to form an ion pair [264]. The reactions for initiation of cationic polymerization with hydrogen iodide (HI) and metal halide (MX_2) are as follows [264, 266–270]:

$$HI + H_2C{=}\underset{R}{\overset{H}{C}} \xrightarrow{k_{preI}} H_3C{-}\underset{R}{\overset{H}{C}}X \tag{5.39A}$$

$$H_3C{-}\underset{R}{\overset{H}{C}}X + MX_2 \underset{k_{dac,cat}}{\overset{k_{ac,cat}}{\rightleftharpoons}} H_3C{-}\underset{R}{\overset{H}{C}}{}^{+}\cdot\cdot MX_3^{-} \tag{5.39B}$$

$$H_3C{-}\underset{R}{\overset{H}{C}}{}^{+}\cdot\cdot MX_3^{-} \underset{k_{ip,cat}}{\overset{k_{fi,cat}}{\rightleftharpoons}} H_3C{-}\underset{R}{\overset{H}{C}}{}^{+} + MX_3^{-} \tag{5.39C}$$

where $H_2C{=}CHR$ is the monomer. In the pre-initiation Rxn. 5.39A, hydrogen iodide reacts with the unsaturated monomer with rate constant k_{preI} to form an alkyl halide. As the reaction is exothermic and fast, it is not the rate-limiting step of the overall initiation reaction. In the second reaction above, the electrophilic metal halide abstracts a halide from the alkyl halide to form an ion pair consisting of an alkyl cation and a metal halide anion. The second reaction is reversible with forward and reverse rate constants $k_{ac,cat}$ and $k_{dac,cat}$, respectively. In the third reaction above (observed in the presence of tetrabutylammonium salts), the cation and anion are solvated to form free ions with the cation capable of initiating chain propagation, with forward and reverse rate constants $k_{fi,cat}$ and $k_{ip,cat}$, respectively, where "fi" and "ip" stand for free ion and ion pair [268–270]. The equilibrium constants for the formation of reactive ion pair ($K_{ip,cat}$) and free ion ($K_{fi,cat}$) are defined by the following equations [269]:

$$K_{ip,cat} = [R^+MX_3^-] \,/\, [HI]\,[MX_2] \tag{5.40A}$$

$$K_{fi,cat} = [R^+]\,[MX_3^-] \,/\, [R^+MX_3^-] \tag{5.40B}$$

The values of $K_{ip,cat}$ and $K_{fi,cat}$ in the polymerization of styrene initiated by 1-phenylethyl chloride and tin tetrachloride in the presence of tetrabutyl ammonium salts are in the order of $K_{ip,cat} \approx 1 \times 10^{-5}\,M^{-1}$ and $K_{fi,cat} \approx 10^{-5} - 10^{-7}\,M$, respectively [269]. These values imply that the reactive free cations are a small fraction of the reactive species. Furthermore, the values indicate that the reaction for the formation of reactive ion pairs (Rxn. 5.39B) is shifted strongly toward the deactivated dormant species. Another initiation system for cationic polymerization consists of cumyl alcohol as the initiator, tris(pentafluorophenyl)-borane, $(C_6F_5)_3B$, as the Lewis acid, and water as the catalyst with the following reactions for initiation [265, 271–273]:

$$^2H_2O + B\,(C_6F_5)_3 \xrightarrow{k_{preI}} H_2OH^+ - \underset{OH}{\bar{B}}\,(C_6F_5)_3 \quad (5.41A)$$

$$ROH + H_2OH^+ - \underset{OH}{\bar{B}}\,(C_6F_5)_3 \underset{k_{dac,cat}}{\overset{k_{ac,cat}}{\rightleftharpoons}} R^+ \cdot\cdot \underset{OH}{\bar{B}}\,(C_6F_5)_3 + {}^2H_2O \quad (5.41B)$$

In the pre-initiation Rxn. 5.41A, $(C_6F_5)_3\,B$ reacts with two water molecules to form a tight H_3O^+ and $HO\bar{B}\,(C_6F_5)_3$ ion pair, which hinders the addition of monomer to the cation. The first reaction is fast and not the rate-limiting step in the overall initiation reaction. However, in the second reaction, cumyl alcohol initiator reacts with the tight ion pair and replaces the hydronium cation with a cumyl cation, which is reactive and adds monomer to form a propagating cation. Like Rxn. 5.39B for HI/MX_2 initiation, the reaction of cumyl alcohol with the tight $H_2OH^+ - \underset{OH}{\bar{B}}\,(C_6F_5)_3$ ion pair (Rxn 5.41B) is reversible. If the pre-initiation reaction (Rxn. 5.41A) is fast and the ion-pair formation reaction (Rxn. 5.41B) has reached equilibrium, it can be shown by mass conservation balance that the concentration of initiating ion pairs for HI/MX_2 and $ROH/B\,(C_6F_5)_3$ initiation systems is given by the following equations [274]:

$$[R^+MX_3^-] = K_{eq,cat}\,[HI]\,[MX_2] = K'_{eq,cat}\,[HI] \quad (5.42A)$$

$$\left[R^+HO\bar{B}\,(C_6F_5)_3\right] = K_{eq,cat}\,[ROH]\left[B\,(C_6F_5)_3\right] = K'_{eq,cat}\,[ROH] \quad (5.42B)$$

where $K_{eq,cat}$ is the equilibrium constant for the reversible Rxn. 5.39B or Rxn. 5.41B; $[HI]$ and $[ROH]$ are the concentration of initiators; and $[MX_2]$ and $\left[B\,(C_6F_5)_3\right]$ are the Lewis acids used to form the ion pair. In the above equations, $K'_{eq,cat}$ is the product of $K_{eq,cat}$ and the Lewis acid concentration, as the Lewis acid provides the counter ion for the propagating cation; hence, it is not used during the reaction. It can be shown by mass conservation balance that the rates of chain initiation for HI/MX_2 and $ROH/B\,(C_6F_5)_3$ initiation systems corresponding to the above reactive cation concentrations are as follows:

$$R_{ci} = k_{ci}\,[R^+MX_3^-]\,[M] = k_{ci}K'_{eq,cat}\,[HI]\,[M] \quad (5.43A)$$

$$R_{ci} = k_{ci}\left[R^+HO\bar{B}\,(C_6F_5)_3\right][M] = k_{ci}K'_{eq,cat}\,[ROH]\,[M] \quad (5.43B)$$

where R_{ci} is the rate of chain initiation. Analogous to RDRP, the $H_3C{-}RHCX$ and ROH act as dormant species that are reactivated by the equilibrium Rxns. 5.39B and 5.41B, respectively [268–270]. Furthermore, analogous to RDRP, the equilibrium reaction is shifted strongly toward the dormant species. As a result, cationic polymerizations catalyzed by a Lewis acid are considered living polymerization reactions as propagating chains shift between the dormant and reactive states. However, unlike RDRP, the reactive cationic species do not self-terminate by combination (see Eq. 5.32B). Unlike anionic polymerization, cationic polymerization can tolerate moisture, as demonstrated by the activation of tris(pentafluorophenyl)-borane with water in pre-initiation Rxn. 5.41A. Unlike anionic polymerization, in which the initiation reaction is strongly shifted toward the formation of a reactive anion, the initiation in cationic polymerization is strongly shifted toward the dormant nonreactive species [269]. In addition, unlike anionic polymerization, the propagating cation is less stable and may terminate by losing an H^+ cation, transferring the cation to a monomer, or reacting with the counterion [56, 275].

5.8 Summary

This chapter introduced the mixing theory which is used in the derivation of the conservation equation for the kth order mole-fraction moment of the propagating chains in a polymerization reaction. The mixing theory states that the kth order molar moment of a mixture of n polymer samples is equal to sum of the kth order molar moment of the individual samples. Using the mixing theory, the conservation equation for the kth order molar moment of the polymer chains in the reaction at time $t + \Delta t$ was related to the kth order molar moment of the polymer chains at time t and the kth order molar moment of the polymer chains produced/consumed in the time interval Δt. The conservation equation can be applied to the kth order molar moment of terminated polymer chains (TP_n), reactive propagating chains (RP_n), dormant chains (DOP_n), as well as any small molecule in the reaction mixture, like monomer, initiator, or the primary radical formed by initiator dissociation. Next, the general conservation equation for the kth order molar moment of the propagating chains was derived using a differential control volume approach in a Cartesian coordinate system, accounting for transfer of the propagating chains by convection and molecular diffusion as well as production and consumption of the propagating chains. The derived general conservation equation for the kth order molar moment of propagating chains was extended to include terminated and dormant chains as well as small molecule reactants. The general equation was also extended to cylindrical and spherical coordinate systems. The general conservation equation for the kth order molar moment of the polymer chains was applied to the modeling of different types of polymerization reactors and processes, including BSVR, continuous stirred tank reactor, plug flow tubular reactor, suspension polymerization, and emulsion polymerization. As propagating species have a short lifetime, we discussed methodologies for theoretical modeling of the kth order molar moment of the reactive propagating species in different polymerization systems. In conventional radical polymerization, QSSA is used to relate the concentration of reactive species to measurable parameters like the initiator concentration, rate constant for initiator dissociation, and the rate constant for chain initiation. In dilute solution polymerization or bulk polymerization at low conversions, QSSA is assumed to be valid for both primary reactive species (R) and the propagating chains (RP_n). In bulk polymerization at temperatures above T_g of the polymer, chain growth is limited by the diffusion of propagating chains; thus, QSSA is valid only for the primary radicals. In polymerizations below the T_g of the polymer, chain initiation, propagation, and termination all may be limited by diffusion; thus, QSSA does not apply for the primary reactive species as well as the propagating chains. In RDRP, the equilibrium constant for chain propagation reaction can be assumed to be shifted almost entirely toward deactivation to maintain dormancy of the propagating chains most of the time, thus minimizing chain termination reactions that broaden the chain DP distribution. Unlike RDRP, which shifts the equilibrium in the initiation reaction almost entirely toward deactivation, in anionic polymerization, it can be assumed that the initiation reaction is almost completely shifted toward the formation of reactive anionic species. Unlike anionic polymerization, the initiation in cationic polymerization is strongly shifted toward the dormant nonreactive species, and the propagating cation is less stable and may undergo termination. But unlike RDRP, the cationic propagating chains do not self-terminate by combination.

5.9 List of Symbols and Abbreviations

a: viscosity exponent in Mark–Houwink–Sakurada (MHS) equation
C_i: concentration of sample i in a polymer mixture
$C_{i,j}$: concentration of sample i with $DP = j$ in a polymer mixture
C_{mix}: concentration of polymer in a polymer mixture
$C_P(i)$: molar concentration distribution of polymer chains as a function of DP
DP: degree of polymerization
$\overline{DP}_{n,AB}$: number-average degree of polymerization of a blend of polymers A and B
$\overline{DP}_{w,AB}$: weight-average degree of polymerization of a blend of polymers A and B
$\overline{DP}_{\eta,AB}$: viscosity-average degree of polymerization of a blend of polymers A and B
DOP_n: dormant polymer chain in the reaction with $DP = n$
f_0: initiator efficiency
$f_A(x)$, $f_B(x)$: mole-fraction distribution of polymer chains for samples A and B, respectively
$f_{AB}(x)$: mole-fraction distribution of polymer chains for a blend of A and B polymers
$f_c(x)$: mole-fraction distribution of polymer chains produced by termination via combination
$f_d(x)$: mole-fraction distribution of polymer chains produced by termination via disproportionation
$f_t(x)$, $f_{t+\Delta t}(x)$: mole-fraction distribution of polymer chains at time t and $t + \Delta t$, respectively, in the reaction
$f_{tm}(x)$: mole-fraction distribution of the polymer chains produced by chain transfer to monomer
$f_{tp}(x)$: mole-fraction distribution of polymer chains produced/consumed by chain transfer to polymer
$f_{ts}(n)$: mole-fraction distribution of polymer chains produced by chain transfer to solvent
I: polymerization initiator
IA: inactive fragment produced by combination of two primary radicals
$[I]$: initiator concentration in polymerization reaction
k_d: reaction rate constant for initiator dissociation
k_c: reaction rate constant for the combination of two primary radicals produced by initiator dissociation
k_i: rate constant for the reaction of a primary radical with a monomer to produce a propagating chain with $DP = 1$
M: Polymerization monomer
$[M]$: concentration of monomer in polymerization reaction
$M_{1+a,i}$: $(1+a)^{th}$ order mole-fraction moment of chains for polymer sample i
$M_{1+a,mix}$: $(1+a)^{th}$ order mole-fraction moment of chains for a polymer mixture
$M_{k,A}$, $M_{k,B}$: kth order mole-fraction moment of polymer chains for samples A and B, respectively
$M_{k,AB}$: kth order mole-fraction moment of polymer chains for a blend of A and B polymers
$M_{k,c}$: kth order mole-fraction moment of polymer chains produced by termination via combination
$M_{k,d}$: kth order mole-fraction moment of polymer chains produced by termination via disproportionation

$M_{k,tm}$: kth order mole-fraction moment of polymer chains produced by chain transfer to monomer
$M_{k,ts}$: kth order mole-fraction moment of polymer chains produced by chain transfer to solvent
$M_{k,i}$: kth order mole-fraction moment of polymer chains produced/consumed by the elementary reaction i
$M_{k,tp}$: kth order mole-fraction moment of polymer chains produced/consumed by chain transfer to polymer
$M'_{k,j}$: kth order mole-fraction moment of polymer chains produced/consumed instantaneously by elementary reaction j
$M_{k,n,t}$: kth order mole-fraction moment of polymer chains at time t in the reaction
$M_{k,n,l,t}$: kth order mole-fraction moment of polymer chains produced and consumed by elementary reaction l at time t in the reaction
$M_{k,n,t+\Delta t}$: kth order mole-fraction moment of polymer chains at time $t + \Delta t$ in the reaction
n_i: number of moles of polymer i in a polymer mixture
n_A, n_B: number of moles of polymer A and B in a blend of polymers A and B, respectively
$n_{i,j}$: number of moles of polymer i with degree of polymerization j
n_t, $n_{t+\Delta t}$: number of moles of polymer chains at time t and $t + \Delta t$, respectively, in the reaction
Δn_t: change in the number of moles of polymer chains at time t in the reaction
$\Delta n_{j,t}$: change in the number of moles of polymer chains produced/consumed by elementary reaction j at time t in the reaction
$N_{k,A}$, $N_{k,B}$: kth order molar moment of polymer chains for samples A and B, respectively
$N_{k,AB}$: kth order molar moment of polymer AB produced by blending polymers A and B
$N_{k,mix}$: kth order molar moment of a polymer mixture
$N_{k,i}$: kth order mole-based moment of polymer chains produced/consumed by elementary reaction j at time t in the reaction
$N'_{k,j}$: instantaneous kth order mole-based moment of polymer chains produced/consumed by elementary reaction j at time t in the reaction
$N_{k,t}$,$N_{k,t+\Delta t}$: kth order moles-based moment of polymer chains at time t and $t + \Delta t$ in the reaction, respectively
$P_{DPn}(i)$: mole-fraction distribution of polymer chains as a function of DP
$Q_{k,i}$: kth order concentration moment of polymer chains produced/consumed by elementary reaction i
$Q_{0,i}$: zero-order concentration moment of polymer i in a polymer mixture
$Q_{k,mix}$: zero-order concentration moment of a polymer mixture
$Q_{k,A}$, $Q_{k,B}$: kth order concentration moment of polymers A and B, respectively
$Q_{k,AB}$: kth order concentration moment of polymer AB produced by blending polymers A and B
$Q_{k,tp}$: kth order concentration moment of terminated polymer chains in polymerization reaction
$Q_{k,dop}$: kth order concentration moment of dormant polymer chains in polymerization reaction
$Q_{k,rp}$: kth order concentration moment of reactive propagating chains in polymerization reaction
$Q_{k,n,t}$: kth order molar concentration moment of polymer chains at time t in the reaction
$Q_{k,n,l,t}$: kth order molar concentration moment of polymer chains produced, consumed by elementary reaction l at time t in the reaction
$Q_{k,n,t+\Delta t}$: kth order molar concentration moment of polymer chains at time $t + \Delta t$ in the reaction
r: number of primary radicals produced by dissociation of initiator
R: primary radical produced by dissociation of initiator
$[R]$: molar concentration of primary radical
R_d: reaction rate for initiator dissociation
R_c: reaction rate for combination of two primary radicals produced dissociation of initiator

R_i: rate of reaction of a primary radical with a monomer to generate a propagating chain
R_j: volumetric rate of polymerization for reaction j
$R_{tp,prd,i}$: rate of production of terminated polymer chains by elementary reaction i
$R_{tp,con,i}$: rate of consumption of terminated polymer chains by elementary reaction i
$R_{rp,prd,i}$: rate of production of reactive propagating chains by elementary reaction i
$R_{rp,con,i}$: rate of consumption of reactive propagating chains by elementary reaction i
$R_{dop,prd,i}$: rate of production of dormant polymer chains by elementary reaction i
$R_{dop,con,i}$: rate of consumption of dormant polymer chains by elementary reaction i
RP_1: reactive propagating chain with one monomer unit
RP_n: reactive propagating chain in the reaction with $DP = n$
$t, t + \Delta t, dt$: reaction times t and $t + \Delta t$ and differential time dt, respectively, in the reaction
TP_n: terminated polymer chain in the reaction with $DP = n$
V_A, V_B: volume of polymer samples A and B, respectively
V_{AB}: volume of polymer AB produced by blending polymers A and B
V_{mix}: volume of a mixture of polymers
V_R: volume of polymerization reaction
V_i: volume of sample i in a polymer mixture
V_R: reaction volume
x_i: mole fraction of sample i in a polymer mixture
x_A, x_B: mole fraction of polymers A and B in a polymer blend
$x_{i,j}$: mole fraction of the chains with $DP = j$ in polymer sample i in a polymer mixture
ϕ_i: volume fraction of polymer sample i in a polymer mixture
$\mu_{k,t}$: kth order mole-fraction moment of terminated polymer chains at time t in polymerization reaction
$\mu'_{k,i}$: instantaneous kth order mole-fraction moment of terminated polymer chains produced by elementary reaction i
$\gamma_{k,t}$: kth order mole-fraction moment of dormant polymer chains at time t in polymerization reaction
$\gamma'_{k,i}$: instantaneous kth order mole-fraction moment of dormant polymer chains produced by elementary reaction i
$\lambda_{k,t}$: kth order mole-fraction moment of reactive propagating chains at time t in polymerization reaction
$\lambda'_{k,i}$: instantaneous kth order mole-fraction moment of reactive propagating chains produced by elementary reaction i

5.10 Practice Problems

Problem 5.1 In a polymerization reaction at time t, the mole-fraction first and second order moments of the propagating chains (λ_1 and λ_2), the mole-fraction first and second order moments of the produced polymer chains (μ_1 and μ_2), and the weight-fraction first and second order moments of the produced polymer chains (ν_1 and ν_2) have the following values:

$$\lambda_1 = 200 \quad \lambda_2 = 50,000$$

$$\mu_1 = 500 \quad \mu_1 = 400,000$$

$$\nu_1 = 800 \quad \nu_2 = 800,000$$

If the average kinetic chain length of the propagating chains is 150, find $\overline{DP}_n$ and $\overline{DP}_w$ of the propagating chains produced by chain transfer to the terminal double bond of the polymer chains (assume that all polymer chains have one terminal double bond per chain).

Problem 5.2 In radical polymerization, termination of propagating chains via disproportionation results in the formation of a terminal double bond on one of the polymer chains. One mechanism for chain transfer is transfer to the terminal double bond of the polymer chains produced by termination via disproportionation. In this mechanism, the radical on the alpha carbon of the propagating chain reacts with the terminal double bond on a terminated polymer chain with a rate constant k_{ctptdb}, as shown below, to form a new longer propagating chain with the radical on a secondary carbon:

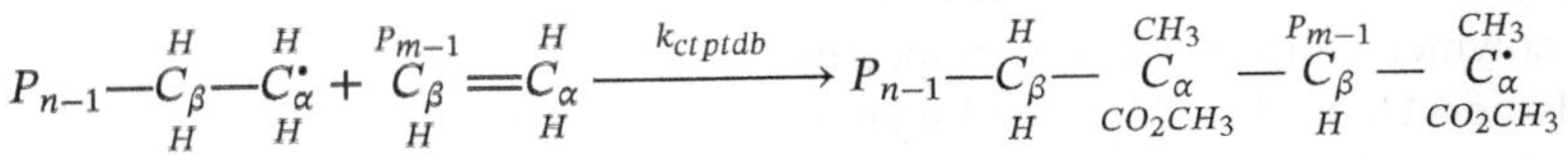

The longer propagating chain adds monomers by propagation to form a long branch. The kth order mole- and weight-fraction moments of the polymer chains in the reaction are μ_{kp} and α_{kp}, respectively, the instantaneous kth order mole-fraction moment of the propagating chains is λ'_k, and the average kinetic chain length of the propagating chains is L.

a. derive equations for $\overline{DP}_n$ and $\overline{DP}_w$ of the propagating chains with a long branch in terms of μ_{kp}, α_{kp}, λ'_k, and L using the method of z-transform.
b. derive equations for polydispersity of the propagating chains with a long branch in terms of μ_{kp}, α_{kp}, λ'_k, and L.

Problem 5.3 In the first step of chain transfer to double bonds along a polymer chain, the radical on the propagating chain reacts with a double bond on a polymer chain with a rate constant k_{ctppdb}, as shown below, to form a new propagating chain with a long branch LB:

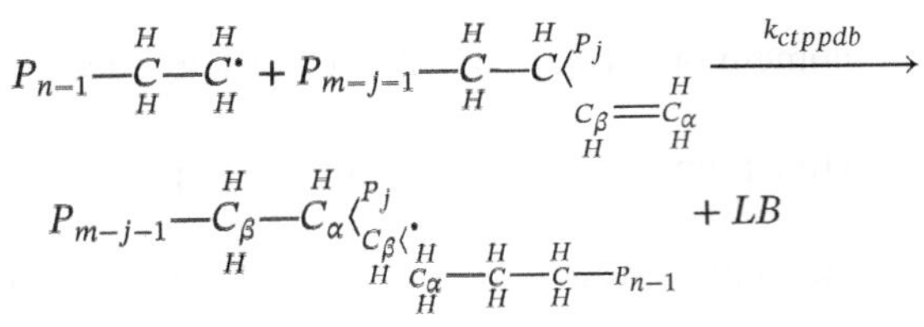

The new branched propagating chain adds monomers by propagation and produces a propagating chain with two branches, with both branches originating from the same site along the main polymer chain, as shown above. The kth order mole- and weight-fraction moments of the polymer chains in the reaction are μ_{kp} and α_{kp}, respectively, the instantaneous kth order mole-fraction moment of the propagating chains is λ'_k, and average kinetic chain length of the propagating chains is L.

a. derive equations for $\overline{DP}_n$ and $\overline{DP}_w$ of the new propagating chains with two branches in terms of μ_{kp}, α_{kp}, λ'_k, and L using the method of z-transform.
b. derive an equation for polydispersity of the new propagating chains with two branches in terms of μ_{kp}, α_{kp}, λ'_k, and L.

References

1 Veregin, R.P.N., Odell, P.G., Michalak, L.M., and Georges, M.K., Molecular weight distributions in nitroxide-mediated living free radical polymerization: kinetics of the slow equilibria between growing and dormant chains. *Macromolecules*, 1996, **29**(10), 3346.

2 Busico, V. et al., Reactivity of secondary metal-alkyls in catalytic propene polymerization: how dormant are "dormant chains"? *Journal of the American Chemical Society*, 2005, **127**(6), 1608.

3 Borisova, O.V. et al., New efficient method for activation of dormant chains in pseudoliving radical polymerization. *Polymer Science Series B*, 2011, **53**(7–8), 409.

4 Poli, R., Stoffelbach, F., Maria, S., and Mata, J., An experimental and computational study on the effect of $Al(OiPr)_{(3)}$ on atom-transfer radical polymerization and on the catalyst-dormant-chain halogen exchange. *Chemistry-a European Journal*, 2005, **11**(8), 2537.

5 Zaremski, M. et al., Influence of media polarity on the rate of activation of "dormant" chains in nitroxide-mediated radical polymerization. *Journal of Polymer Science Part a-Polymer Chemistry*, 2012, **50**(16), 3437.

6 Yang, P.J. et al., Determination and tracing of active and dormant propagation chains in 1-hexene polymerization with supported Ziegler–Natta catalyst. *Applied Catalysis A: General*, 2020, **595**, 117469.

7 Svidchenko, E.A., Surin, N.M., Sigan, A.L., and Safonov, A.A., Photochemical synthesis and formation mechanisms of polyvinyl acetate-Co-III(salen) "dormant" chains during "living" radical polymerization. *Polymer Science Series B*, 2016, **58**(6), 665.

8 Penczek, S., Pretula, J., and Lewinski, P., Dormant polymers and their role in living and controlled polymerizations: influence on polymer chemistry, particularly on the ring opening polymerization. *Polymers (Basel)*, 2017, **9**(12), 646.

9 Muller, A.H.E. et al., Kinetic analysis of living polymerization processes exhibiting slow equilibria. 2. Molecular-weight distribution for degenerative transfer (direct activity exchange between active and dormant species) at constant monomer concentration. *Macromolecules*, 1995, **28**(22), 7335.

10 Szwarc, M. and Hermans, J.J., Molecular weight distribution in non-terminated polymerization involving living + dormant polymers. *Journal of Polymer Science Part B-Polymer Letters*, 1964, **2**(8), 815.

11 Nakamura, S., Yamada, M., and Matsuzaki, K., Graft frequency of graft copolymers by chain transfer of growing polymer radicals to trunk polymers containing nitro-groups. *Journal of Applied Polymer Science*, 1978, **22**(7), 2011.

12 Bolshakov, A.I., Mikhailov, A.I., and Barkalov, I.M., Investigation of polymerization ability of acrylic and methacrylic monomers at low temperatures and activity of growing polymer-chain radicals. *Vysokomolekulyarnye Soedineniya Seriya A*, 1978, **20**(8), 1820.

13 Shamanin, V.V., Melenevskaya, E.Y., and Sgonnik, V.N., Influence of the concentration of growing chains on the polymerization rate of butadiene and the microstructure of the polymer formed in the polybutadienyllithium-butadiene-aliphatic hydrocarbon system. *Acta Polymerica*, 1982, **33**(3), 175.

14 Huang, N.J. and Sundberg, D.C., Fundamental studies of grafting reactions in free-radical copolymerization. 2. Grafting of styrene, acrylate, and methacrylate monomers onto *cis*-polybutadiene using AIBN initiator in solution polymerization. *Journal of Polymer Science Part A-Polymer Chemistry*, 1995, **33**(15), 2551.

15 Huang, N.J. and Sundberg, D.C., Fundamental studies of grafting reactions in free-radical copolymerization. 4. Grafting of styrene, acrylate, and methacrylate monomers onto

vinyl-polybutadiene using benzoyl peroxide and AIBN initiators in solution polymerization. *Journal of Polymer Science Part A-Polymer Chemistry*, 1995, **33**(15), 2587.

16 Yamago, S., Iida, K., Nakajima, M., and Yoshida, J., Practical protocols for organotellurium-mediated living radical polymerization by in situ generated initiators from AIBN and ditellurides. *Macromolecules*, 2003, **36**(11), 3793.

17 Berger, K.C., Deb, P.C., and Meyerhoff, G., Radical vinyl polymerization - reactions of benzoyl peroxide during bulk-polymerization of styrene with labeled initiator. *Macromolecules*, 1977, **10**(5), 1075.

18 Mitra, S.C., De, N., Guchhait, S., and Konar, R.S., Effect of temperature on the rates of aqueous polymerization of vinyl-acetate initiated by SO_4 free-radicals produced from potassium persulfate and the redox reaction of ferrous ammonium-sulfate and potassium persulfate. *Indian Journal of Chemistry Section A-Inorganic Bio-Inorganic Physical Theoretical & Analytical Chemistry*, 1995, **34**(12), 1014.

19 Mitra, S.C., De, N., Guchhait, S., and Konar, R.S., Effect of temperature on the rates of aqueous polymerization of vinyl-acetate initiated by SO_4-center-dot- free-radicals produced from potassium persulfate and the redox reaction of ferrous ammonium-sulfate and potassium persulfate. *Indian Journal of Chemistry Section A-Inorganic Bio-Inorganic Physical Theoretical & Analytical Chemistry*, 1995, **34**(8), 602.

20 Du, X.Y. et al., Generation of a carbon dots/ammonium persulfate redox initiator couple for free radical frontal polymerization. *Polymer Chemistry*, 2018, **9**(4), 420.

21 Sarac, A.S., Redox polymerization. *Progress in polymer science*, 1999, **24**(8), 1149.

22 Pattnaik, S., Roy, A.K., Baral, N., and Nayak, P.L., Aqueous polymerization of methyl-methacrylate initiated by potassium peroxydisulfate-ascorbic acid redox system. *Journal of Macromolecular Science A-Chemistry*, 1979, **13**(6), 797.

23 Hola, E., Fiedor, P., Dzienia, A., and Ortyl, J., Visible-light amine thioxanthone derivatives as photoredox catalysts for photopolymerization processes. *ACS Applied Polymer Materials*, 2021, **3**(11), 5547.

24 Shao, J.Z., Huang, Y., and Fan, Q.U., Visible light initiating systems for photopolymerization: status, development and challenges. *Polymer Chemistry*, 2014, **5**(14), 4195.

25 Tomal, W. and Ortyl, J., Water-soluble photoinitiators in biomedical applications. *Polymers (Basel)*, 2020, **12**(5), 1073.

26 Grotzinger, C., Burget, D., Jacques, P., and Fouassier, J.P., Visible light induced photopolymerization: speeding up the rate of polymerization by using co-initiators in dye/amine photoinitiating systems. *Polymer*, 2003, **44**(13), 3671.

27 Valdesaguilera, O. et al., Photopolymerization studies using visible-light photoinitiators. *Macromolecules*, 1992, **25**(2), 541.

28 Ng, S.C. and Chee, K.K., Determination of rate constants of benzoyl peroxide and azobisisobutyronitrile dissociation via polymerization kinetics. *Journal of Polymer Science Part A-Polymer Chemistry*, 1982, **20**(2), 409.

29 Arutyunyan, R.S., Nalbandyan, D.M., and Beileryan, N.M., Mechanism of dissociation of 2,2′-azobisisobutyronitrile in aqueous emulsions. *Kinetics and Catalysis*, 1985, **26**(5), 906.

30 Vogt, W., Radical efficiency at thermal decomposition of azobisisobutyronitrile in presence of oxygen - cage reaction and crossed termination. *Makromolekulare Chemie*, 1971, **145**(Jul), 309.

31 Yeh, C.J., Hu, M., and Shull, K.R., Oxygen inhibition of radical polymerizations investigated with the rheometric quartz crystal microbalance. *Macromolecules*, 2018, **51**(15), 5511.

32 Huang, C. et al., Enhancing the self-recovery and mechanical property of hydrogels by macromolecular microspheres with thermal and redox initiation systems. *RSC Advances*, 2017, **7**(26), 16015.

33 Nagaraja, G.K., Demappa, T., and Mahadevaiah, Polymerization kinetics of acrylonitrile by oxidation: reduction system using potassium persulfate/ascorbic acid in an aqueous medium. *Journal of Applied Polymer Science*, 2011, **121**(3), 1299.

34 Corrigan, N. et al., Seeing the light: advancing materials chemistry through photopolymerization. *Angewandte Chemie-International Edition*, 2019, **58**(16), 5170.

35 Bagheri, A. and Jin, J.Y., Photopolymerization in 3D printing. *ACS Applied Polymer Materials*, 2019, **1**(4), 593.

36 Bao, Y.Y., Paunovic, N., and Leroux, J.C., Challenges and opportunities in 3D printing of biodegradable medical devices by emerging photopolymerization techniques. *Advanced Functional Materials*, 2022, **32**(15), 2109864.

37 Xu, X.Y. et al., Vat photopolymerization 3D printing for advanced drug delivery and medical device applications. *Journal of Controlled Release*, 2021, **329**, 743.

38 Neidinger, P. et al., Near infrared light induced radical polymerization in water. *Angewandte Chemie-International Edition*, 2022, **61**(42), e202209177.

39 Colley, C.S. et al., Probing the reactivity of photoinitiators for free radical polymerization: time-resolved infrared spectroscopic study of benzoyl radicals. *Journal of the American Chemical Society*, 2002, **124**(50), 14952.

40 Ligon, S.C. et al., Polymers for 3D printing and customized additive manufacturing. *Chemical Reviews*, 2017, **117**(15), 10212.

41 Muller, S.M. et al., Recent advances in type I photoinitiators for visible light induced photopolymerization. *Chemphotochem*, 2022, **6**(11), 2200091.

42 Allushi, A. et al., Conventional type II photoinitiators as activators for photoinduced metal-free atom transfer radical polymerization. *Polymer Chemistry*, 2017, **8**(12), 1972.

43 Wang, Z.J. et al., Visible light photoinitiation of cell-adhesive gelatin methacryloyl hydrogels for stereolithography 3D bioprinting. *ACS Applied Materials & Interfaces*, 2018, **10**(32), 26859.

44 Avens, H.J., Randle, T.J., and Bowman, C.N., Polymerization behavior and polymer properties of eosin-mediated surface modification reactions. *Polymer*, 2008, **49**(22), 4762.

45 Qin, X.-H., Ovsianikov, A., Stampfl, J., and Liska, R., Additive manufacturing of photosensitive hydrogels for tissue engineering applications. *BioNanoMaterials*, 2014, **15**(3–4), 49.

46 Wloka, T., Gottschaldt, M., and Schubert, U.S., From light to structure: photo initiators for radical two-photon polymerization. *Chemistry-A European Journal*, 2022, **28**(32), e202104191.

47 Eibel, A., Fast, D.E., and Gescheidt, G., Choosing the ideal photoinitiator for free radical photopolymerizations: predictions based on simulations using established data. *Polymer Chemistry*, 2018, **9**(41), 5107.

48 Nikitin, A.N., Castignolles, P., Charleux, B., and Vairon, J.P., Determination of propagation rate coefficient of acrylates by pulsed-laser polymerization in the presence of intramolecular chain transfer to polymer. *Macromolecular Rapid Communications*, 2003, **24**(13), 778.

49 Ito, K., Relationship between kinetic chain-length and radical polymerization rate. *Journal of Polymer Science Part A-Polymer Chemistry*, 1976, **14**(7), 1761.

50 Williams, G. and Coe, J.S., The kinetic chain length in the polymerization of styrene by stannic chloride in carbon tetrachloride. *Chemistry & Industry (London)*, 1954, (2), 43.

51 Berger, K.C., Disproportionation and combination as termination mechanisms in radical polymerization of styrene. 2. Analysis of temperature dependencies. *Makromolekulare Chemie-Macromolecular Chemistry and Physics*, 1975, **176**(12), 3575.

52 Berger, K.C. and Meyerhoff, G., Disproportionation and combination as termination mechanisms in radical polymerization of styrene. 1. Experiments with C-14-labeled 2,2′-azoisobutyronitrile. *Makromolekulare Chemie-Macromolecular Chemistry and Physics*, 1975, **176**(7), 1983.

53 Hatada, K., Kitayama, T., and Masuda, E., Evidence for disproportionation in termination reaction of styrene polymerization by α,α′-azobisisobutyronitrile. *Polymer Journal*, 1985, **17**(8), 985.

54 Ribelli, T.G. et al., Disproportionation or combination? The termination of acrylate radicals in ATRP. *Macromolecules*, 2017, **50**(20), 7920.

55 Yan, D.Y. and Cai, G.F., Influence of termination and transfer on molecular-weight distribution of polymers. 1. Disproportionation and combination. *Journal of Macromolecular Science A-Chemistry*, 1986, **23**(1), 37.

56 Cai, G.F., Yan, D.Y., and Litt, M., A kinetic theory on cationic polymerization with termination by combination with a negative fragment. *Macromolecules*, 1988, **21**(3), 578.

57 Nakamura, Y. and Yamago, S., Termination mechanism in the radical polymerization of methyl methacrylate and styrene determined by the reaction of structurally well-defined polymer end radicals. *Macromolecules*, 2015, **48**(18), 6450.

58 Ballard, N., Hamzehlou, S., Ruiperez, F., and Asua, J.M., On the termination mechanism in the radical polymerization of acrylates. *Macromolecular Rapid Communications*, 2016, **37**(16), 1364.

59 Nakamura, Y., Lee, R., Coote, M.L., and Yamago, S., Termination mechanism of the radical polymerization of acrylates. *Macromolecular Rapid Communications*, 2016, **37**(6), 506.

60 Barth, J., Buback, M., Russell, G.T., and Smolne, S., Chain-length-dependent termination in radical polymerization of acrylates. *Macromolecular Chemistry and Physics*, 2011, **212**(13), 1366.

61 Shen, J.C., Tian, Y., Wang, G.B., and Yang, M.L., Termination process and the termination parameter-Kt of methyl-methacrylate bulk radical polymerization. *Science in China Series B-Chemistry*, 1990, **33**(9), 1040.

62 Zhu, S., Tian, Y., Hamielec, A.E., and Eaton, D.R., Radical trapping and termination in free-radical polymerization of methyl methacrylate. *Macromolecules*, 1990, **23**(4), 1144.

63 Sato, E., Zetterlund, P.B., and Yamada, B., Macromonomer synthesis using oligomers of omega-unsaturated methacrylate as addition-fragmentation chain transfer agents: increased efficiency by manipulation of steric hindrance. *Macromolecules*, 2004, **37**(7), 2363.

64 Bevington, J.C., Melville, H.W., and Taylor, R.P., The termination reaction in radical polymerizations. 2. Polymerizations of styrene at 60° and of methyl methacrylate at 0 and 60°, and the copolymerization of these monomers at 60°. *Journal of Polymer Science*, 1954, **14**(77), 463.

65 Laita, Z., The kinetics of radical polymerization of ethylene. 1. General kinetics. *Journal of Polymer Science*, 1959, **38**(133), 247.

66 Laita, Z. and Machacek, Z., The kinetics of radical polymerization of ethylene. 2. Determination of the kinetic coefficients. *Journal of Polymer Science*, 1959, **38**(134), 459.

67 Mortimer, G.A. and Arnold, L.C., Free-radical polymerization of olefins. *Journal of Polymer Science Part A-General Papers*, 1964, **2**(9), 4247.

68 Boghetich, L., Daues, G.W., and Mortimer, G.A., Free-radical copolymerization of α-olefins with ethylene. *Journal of Polymer Science*, 1962, **61**(171), 3.

69 Bukatov, G.D., Zakharov, V.A., and Barabanov, A.A., Mechanism of olefin polymerization on supported Ziegler–Natta catalysts based on data on the number of active centers and propagation rate constants. *Kinetics and Catalysis*, 2005, **46**(2), 166.

70 Yu, Y., Fu, Z.S., and Fan, Z.Q., Chain transfer reactions of propylene polymerization catalyzed by $AlEt_3$ activated $TiCl_4/MgCl_2$ catalyst under very low monomer addition rate. *Journal of Molecular Catalysis A-Chemical*, 2012, **363**, 134.

71 Corrigan, N. et al., Reversible-deactivation radical polymerization (controlled/living radical polymerization): from discovery to materials design and applications. *Progress in polymer science*, 2020, **111**, 101311.

72 Kubisa, P., Radical polymerization with reversible deactivation of active species. *Polimery*, 2000, **45**(11–12), 741.

73 Shipp, D.A., Reversible-deactivation radical polymerizations. *Polymer Reviews*, 2011, **51**(2), 99.

74 Harrisson, S., The chain length distribution of an ideal reversible deactivation radical polymerization. *Polymers*, 2018, **10**(8), 887.

75 Jenkins, A.D., Jones, R.G., and Moad, G., Terminology for reversible-deactivation radical polymerization previously called "controlled" radical or "living" radical polymerization (IUPAC recommendations 2010). *Pure and Applied Chemistry*, 2010, **82**(2), 483.

76 Stenzel, M.H. and Barner-Kowollik, C., The living dead - common misconceptions about reversible deactivation radical polymerization. *Materials Horizons*, 2016, **3**(6), 471.

77 Giammarino, G. and Villani, V., Termination mechanisms via H-transfer for polyolefin living catalysis at DFT level. *Macromolecular Theory and Simulations*, 2014, **23**(6), 365.

78 Starnes, W.H. et al., Branch structures in poly(vinyl-chloride) and the mechanism of chain transfer to monomer during vinyl-chloride polymerization. *Macromolecules*, 1983, **16**(5), 790.

79 Obrien, J.L. and Gornick, F., Chain transfer in the polymerization of methyl methacrylate. 1. Transfer with monomer and thiols - the mechanism of the termination reaction at 60°. *Journal of the American Chemical Society*, 1955, **77**(18), 4757.

80 Kukulj, D., Davis, T.P., and Gilbert, R.G., Chain transfer to monomer in the free-radical polymerizations of methyl methacrylate, styrene, and α-methylstyrene. *Macromolecules*, 1998, **31**(4), 994.

81 Yamada, B. and Kobatake, S., Radical polymerization, co-polymerization, and chain transfer of α-substituted acrylic esters. *Progress in polymer science*, 1994, **19**(6), 1089.

82 Ayrey, G. and Haynes, A.C., Chain transfer-reaction of azoisobutyronitrile during polymerization of methyl-methacrylate at 60°C. *Makromolekulare Chemie-Macromolecular Chemistry and Physics*, 1974, **175**(5), 1463.

83 Basu, S., Sen, J.N., and Palit, S.R., Degree of polymerization and chain transfer in methyl methacrylate. *Proceedings of the Royal Society of London Series A-Mathematical and Physical Sciences*, 1950, **202**(1071), 485.

84 Ghosh, P. and Mukhopadhyay, G., Chain transfer with pyridine in the polymerization of methyl-methacrylate, styrene, and vinyl-acetate using benzoyl peroxide as initiator at 60°C. *Makromolekulare Chemie-Macromolecular Chemistry and Physics*, 1979, **180**(9), 2253.

85 Fazekas, T.J. et al., Diversification of aliphatic C—H bonds in small molecules and polyolefins through radical chain transfer. *Science*, 2022, **375**(6580), 545.

86 Magee, C., Sugihara, Y., Zetterlund, P.B., and Aldabbagh, F., Chain transfer to solvent in the radical polymerization of structurally diverse acrylamide monomers using straight-chain and branched alcohols as solvents. *Polymer Chemistry*, 2014, **5**(7), 2259.

87 Balic, R., Fellows, C.M., and van Herk, A.M., Chain transfer to monomer and polymer in the radical polymerization of vinyl neo-decanoate. *Macromolecular Research*, 2004, **12**(4), 325.

88 Mayo, F.R., Chain transfer in the polymerization of styrene: the reaction of solvents with free radicals. *Journal of the American Chemical Society*, 1943, **65**, 2324.

89 Tan, S.B., Li, J.J., and Zhang, Z.C., Study of chain transfer reaction to solvents in the initiation stage of atom transfer radical polymerization. *Macromolecules*, 2011, **44**(20), 7911.

90 Sugihara, Y., O'Connor, P., Zetterlund, P.B., and Aldabbagh, F., Chain transfer to solvent in the radical polymerization of N-isopropylacrylamide. *Journal of Polymer Science Part A-Polymer Chemistry*, 2011, **49**(8), 1856.

91 Ji, Y.Y. and Klausen, R.S., Chain transfer to solvent in BN 2-vinylnaphthalene radical polymerization. *Journal of Polymer Science*, 2021, **59**(21), 2521.

92 Gregg, R.A. and Mayo, F.R., Chain transfer in the polymerization of styrene. 7. Compounds containing halogens, oxygen and nitrogen. *Journal of the American Chemical Society*, 1953, **75**(14), 3530.

93 Bovey, F.A. and Kolthoff, I.M., Studies of retarders and inhibitors in the emulsion polymerization of styrene. 2. Inhibitors. *Journal of Polymer Science*, 1950, **5**(5), 569.

94 Kolthoff, I.M. and Bovey, F.A., Studies of retarders and inhibitors in the emulsion polymerization of styrene. 1. Retarders. *Journal of the American Chemical Society*, 1948, **70**(2), 791.

95 Bovey, F.A. and Kolthoff, I.M., Inhibition and retardation of vinyl polymerization. *Chemical Reviews*, 1948, **42**(3), 491.

96 Allen, T.L., Oxygen inhibition of the polymerization of styrene. *Journal of Applied Chemistry*, 1954, **4**(6), 289.

97 Li, R.J. and Schork, F.J., Modeling of the inhibition mechanism of acrylic acid polymerization. *Industrial & Engineering Chemistry Research*, 2006, **45**(9), 3001.

98 Bartlett, P.D. and Trifan, D.S., Some experiments on sulfur as inhibitor for the polymerization of styrene. *Journal of Polymer Science*, 1956, **20**(96), 457.

99 Bartlett, P.D. and Kwart, H., Some inhibitors and retarders in the polymerization of liquid vinyl acetate. 2. 1,3,5-Trinitrobenzene and sulfur. *Journal of the American Chemical Society*, 1952, **74**(16), 3969.

100 Prescott, S.W., Chain-length dependence in living/controlled free-radical polymerizations: physical manifestation and Monte Carlo simulation of reversible transfer agents. *Macromolecules*, 2003, **36**(25), 9608.

101 Moad, G. and Solomon, D.H., Invited review - understanding and controlling radical polymerization. *Australian Journal of Chemistry*, 1990, **43**(2), 215.

102 Moad, G., A Critical survey of dithiocarbamate reversible addition-fragmentation chain transfer (RAFT) agents in radical polymerization. *Journal of Polymer Science Part A-Polymer Chemistry*, 2019, **57**(3), 216.

103 Otsu, T., Iniferter concept and living radical polymerization. *Journal of Polymer Science Part A-Polymer Chemistry*, 2000, **38**(12), 2121.

104 Otsu, T. and Yoshida, M., Role of initiator-transfer agent-terminator (iniferter) in radical polymerizations: polymer design by organic disulfides as iniferters. *Macromolecular Rapid Communications*, 2005, **26**(3), 137.

105 Charleux, B., Nicolas, J., and Guerret, O., Theoretical expression of the average activation-deactivation equilibrium constant in controlled/living free-radical copolymerization operating via reversible termination. Application to a strongly improved control in nitroxide-mediated polymerization of methyl methacrylate. *Macromolecules*, 2005, **38**(13), 5485.

106 Moad, G. et al., Advances in RAFT polymerization: the synthesis of polymers with defined end-groups. *Polymer*, 2005, **46**(19), 8458.

107 Chiefari, J. et al., Living free-radical polymerization by reversible addition-fragmentation chain transfer: the RAFT process. *Macromolecules*, 1998, **31**(16), 5559.

108 Hawthorne, D.G., Moad, G., Rizzardo, E., and Thang, S.H., Living radical polymerization with reversible addition-fragmentation chain transfer (RAFT): direct ESR observation of intermediate radicals. *Macromolecules*, 1999, **32**(16), 5457.

109 Yin, H. et al., Selection principle of RAFT chain transfer agents and universal RAFT chain transfer agents. *Progress in Chemistry*, 2022, **34**(6), 1298.

110 Ahmad, N.M. et al., Chain transfer to polymer and branching in controlled radical polymerizations of *n*-butyl acrylate. *Macromolecular Rapid Communications*, 2009, **30**(23), 2002.

111 Ahmad, N.M., Heatley, F., and Lovell, P.A., Chain transfer to polymer in free-radical solution polymerization of *n*-butyl acrylate studied by NMR spectroscopy. *Macromolecules*, 1998, **31**(9), 2822.

112 Britton, D., Heatley, F., and Lovell, P.A., Chain transfer to polymer in free-radical bulk and emulsion polymerization of vinyl acetate studied by NMR spectroscopy. *Macromolecules*, 1998, **31**(9), 2828.

113 Britton, D., Heatley, F., and Lovell, P.A., Effect of monomer feed rate on chain transfer to polymer in semibatch emulsion polymerization of vinyl acetate studied by NMR spectroscopy. *Macromolecules*, 2000, **33**(14), 5048.

114 Buback, M. and Schweer, J., Chain-length dependence of free-radical rate coefficients. 1. Termination rate coefficient in pure ethylene polymerization. *Makromolekulare Chemie-Rapid Communications*, 1988, **9**(10), 699.

115 Jensen, M.D., Yang, Q., Yu, Y.L., and McDaniel, M.P., Kinetics of long-chain branch formation in polyethylene. *ACS Catalysis*, 2018, **8**(1), 725.

116 Reinking, M.K., Orf, G., and McFaddin, D., Novel mechanism for the formation of long-chain branching in polyethylene. *Journal of Polymer Science Part A-Polymer Chemistry*, 1998, **36**(16), 2889.

117 Simon, L.C. and Soares, J.B.P., Polyethylene made with combinations of single-site-type catalysts: Monte Carlo simulation of long-chain branch formation. *Macromolecular Theory and Simulations*, 2002, **11**(2), 222.

118 Iedema, P.D., Grcev, S., and Hoefsloot, H.C.J., Molecular weight distribution modeling of radical polymerization in a CSTR with long chain branching through transfer to polymer and terminal double bond (TDB) propagation. *Macromolecules*, 2003, **36**(2), 458.

119 Huang, W.Y. et al., Studies on the branching atom transfer radical polymerization using different divinyl monomers as the branching agent. *Acta Polymerica Sinica*, 2011, (11), 1253.

120 Huang, W.Y. et al., Branching copolymerization of styrene and methyl methacrylate with divinylbenzene. *Iranian Polymer Journal*, 2010, **19**(8), 589.

121 Bannister, I. et al., Development of branching in living radical copolymerization of vinyl and divinyl monomers. *Macromolecules*, 2006, **39**(22), 7483.

122 Triacca, V.J. et al., Free-radical degradation of polypropylene - random chain scission. *Polymer Engineering and Science*, 1993, **33**(8), 445.

123 Pinheiro, L.A., Chinelatto, M.A., and Canevarolo, S.V., The role of chain scission and chain branching in high density polyethylene during thermo-mechanical degradation. *Polymer Degradation and Stability*, 2004, **86**(3), 445.

124 Kumagai, T. et al., Specific polymerization mechanism involving β-scission of mid-chain radical yielding oligomers in the free-radical polymerization of vinyl ethers. *Macromolecules*, 2008, **41**(20), 7347.

125 Lena, J.B. et al., Effects of chain transfer agent and temperature on branching and β-scission in radical polymerization of 2-ethylhexyl acrylate. *Macromolecular Chemistry and Physics*, 2018, **219**(9), 1700579.

126 Tobita, H., Markovian approach to free-radical polymerization with simultaneous long-chain branching and scission: effect of branching and scission kinetics. *Macromolecular Reaction Engineering*, 2015, **9**(3), 245.

127 Tobita, H., Free-radical polymerization with long-chain branching and scission: Markovian solution of the weight-average molecular weight. *Macromolecular Theory and Simulations*, 2014, **23**(8), 477.

128 Junkers, T. and Barner-Kowollik, C., The role of mid-chain radicals in acrylate free radical polymerization: branching and scission. *Journal of Polymer Science Part A-Polymer Chemistry*, 2008, **46**(23), 7585.

129 Szymanski, R. and Baran, J., Molecular weight distribution in living polymerization proceeding with reshuffling of polymer segments due to chain transfer to polymer with chain scission, 2(a) - Monte Carlo simulation of polymer reshuffling proceeding with disproportionation of chain functionalities. *Macromolecular Theory and Simulations*, 2002, **11**(8), 836.

130 Iedema, P.D. and Hoefsloot, H.C.J., The influence of branching architecture on scission in modeling free-radical polymerization with long chain branching and scission. *Chimia*, 2001, **55**(3), 256.

131 Bao, C.Y. et al., Synthesis of lipase-polymer conjugates by Cu(0)-mediated reversible deactivation radical polymerization: polymerization vs. degradation. *Polymer Chemistry*, 2020, **11**(7), 1386.

132 Bagheri, A., Fellows, C.M., and Boyer, C., Reversible deactivation radical polymerization: from polymer network synthesis to 3D printing. *Advanced Science*, 2021, **8**(5), 2003701.

133 Ni, Y.Y., Zhang, L.F., Cheng, Z.P., and Zhu, X.L., Iodine-mediated reversible-deactivation radical polymerization: a powerful strategy for polymer synthesis. *Polymer Chemistry - UK*, 2019, **10**(20), 2504.

134 Ouchi, M. and Sawamoto, M., Sequence-controlled polymers via reversible-deactivation radical polymerization. *Polymer Journal*, 2018, **50**(1), 83.

135 Vinciguerra, D., Tran, J., and Nicolas, J., Telechelic polymers from reversible-deactivation radical polymerization for biomedical applications. *Chemical Communications*, 2018, **54**(3), 228.

136 Yamago, S., Practical synthesis of dendritic hyperbranched polymers by reversible deactivation radical polymerization. *Polymer Journal*, 2021, **53**(8), 847.

137 Zhou, X.J. et al., Hydrogen bonding assisted reversible-deactivation radical copolymerization of 4-vinylpyridine and styrene: a facile approach for adjusting polymerization behavior, polymer composition, etc. *Polymer Chemistry UK*, 2013, **4**(12), 3575.

138 Zhou, J., Zhang, X.Y., and Su, Z.Q., Rational design of biomolecules/polymer hybrids by reversible deactivation radical polymerization (RDRP) for biomedical applications. *Chinese Journal of Polymer Science*, 2021, **39**(9), 1093.

139 Tang, H.D. et al., Highly active copper-based catalyst for atom transfer radical polymerization. *Journal of the American Chemical Society*, 2006, **128**(50), 16277.

140 Destarac, M., On the critical role of RAFT agent design in reversible addition-fragmentation chain transfer (RAFT) polymerization. *Polymer Reviews*, 2011, **51**(2), 163.

141 Zapata-González, I. and Saldívar-Guerra, E., The method of moments used in polymerization reaction engineering for 70 years: an overview, tutorial, and minilibrary. *Canadian Journal of Chemical Engineering*, 2023, **101**(9), 5324.

142 Mastan, E. and Zhu, S.P., Method of moments: a versatile tool for deterministic modeling of polymerization kinetics. *European Polymer Journal*, 2015, **68**, 139.

143 Zhou, Y.N. and Luo, Z.H., State-of-the-art and progress in method of moments for the model-based reversible-deactivation radical polymerization. *Macromolecular Reaction Engineering*, 2016, **10**(6), 516.

144 Herdan, G., Fitting of polymer distributions of molecular weight by the method of moments. *Journal of Polymer Science*, 1953, **10**(1), 1.

145 Samarin, A.F., Polymer average molecular weights, molecular weight distribution functions and moments. *Vysokomolekulyarnye Soedineniya Section A*, 1971, **13**(5), 1024.

146 Abedini, H., Rostami, M.R., and Shahsavar, S., Numerical simulation of atom transfer radical polymerization of styrene by moment and population balance models. *Polymer Bulletin*, 2022, **79**(6), 4303.

147 Lima, R.C. et al., Modeling and parameter estimation of step-growth polymerization of ethylene glycol and 2,5-furandicarboxylic acid (FDCA) using the method of moments. *Macromolecular Symposia*, 2020, **394**(1), 2000105.

148 Rosa, R.P., Ferreira, F.V., and Lona, L.M.F., Modeling of ring opening polymerization: a short review with insights on how to develop the method of moments. *Chemical Engineering Science*, 2021, **246**, 116934.

149 Zhang, J.H., Pu, J.Z., and Ren, M.F., Molecular weight distribution control for polymerization processes based on the moment-generating function. *Entropy*, 2022, **24**(4), 499.

150 Product rule. in *Wikipedia: the free encyclopedia*. 2025, Wikimedia Foundation, https://en.wikipedia.org/wiki/Product_rule.

151 Gruendling, T., Guilhaus, M., and Barner-Kowollik, C., Fast and accurate determination of absolute individual molecular weight distributions from mixtures of polymers via size exclusion chromatography-electrospray ionization mass spectrometry. *Macromolecules*, 2009, **42**(17), 6366.

152 Marxfigini, M., Tagliabue, M., and Figini, R.V., On the molecular-weight determination by vapor-pressure osmometry. 4. Experimental proof of the deviation of Mvpo from Mn. *Makromolekulare Chemie-Macromolecular Chemistry and Physics*, 1983, **184**(2), 319.

153 Liu, Y.G., Bo, S.Q., Zhu, Y.J., and Zhang, W.H., Determination of molecular weight and molecular sizes of polymers by high temperature gel permeation chromatography with a static and dynamic laser light scattering detector. *Polymer*, 2003, **44**(23), 7209.

154 Wang, J.F., Huang, H.Q., and Huang, X.Y., Molecular weight and the Mark–Houwink relation for ultra-high molecular weight charged polyacrylamide determined using automatic batch mode multi-angle light scattering. *Journal of Applied Polymer Science*, 2016, **133**(31), 43748.

155 Woo, S., Lee, J.Y., Choi, W., and Moon, M.H., Characterization of ultrahigh-molecular weight cationic polyacrylamide using frit-inlet asymmetrical flow field-flow fractionation and multi-angle light scattering. *Journal of Chromatography A*, 2016, **1429**, 304.

156 Farias, T.M., Cardozo, N.S.M., and Secchi, A.R., The use of Gauss-Hermite quadrature in the determination of the molecular weight distribution of linear polymers by rheometry. *Brazilian Journal of Chemical Engineering*, 2013, **30**(4), 909.

157 Liu, C.Y., Wei, Y.M., Yang, L.Y., and Chen, W., Determination of polymer molecular weight and gyration radius of DNA by capillary electrophoresis. *Analytical Letters*, 2013, **46**(5), 776.

158 Meister, J.J. and Richards, E.G., Determination of a polymers molecular-weight distribution by analytical ultracentrifugation. in *ACS symposium series*. 1989, American Chemical Society, pp. 58–81.

159 Mittal, V., Molecular weight distributions of polymer solutions: combination of field flow fractionation (FFF) and analytical ultracentrifugation (AUC). *Journal of Dispersion Science and Technology*, 2012, **33**(5), 631.

160 Tillman, E.S. et al., Synthesis of chromophore-labeled polymers and their molecular weight determination using UV–VIS spectroscopy. *Journal of Chemical Education*, 2006, **83**(8), 1215.

161 Melzak, K.A. et al., Acoustic determination of polymer molecular weights and rotation times. *Journal of Polymer Science Part B-Polymer Physics*, 2002, **40**(14), 1490.

162 Oh, Y.R., Lee, Y.S., Kwon, M.H., and Park, O.O., Determination of molecular weight and its distribution of rigid-rod polymers determined by phase-modulated flow birefringence technique. *Journal of Polymer Science Part B-Polymer Physics*, 2000, **38**(4), 509.

163 Todorova, D. and Oreshkov, T., The average molecular weight determination of polysulfide polymers by using IR spectroscopy. *Spectroscopy Letters*, 1999, **32**(5), 773.

164 Keating, A.R. and Wesdemiotis, C., Rapid and simple determination of average molecular weight and composition of synthetic polymers via electrospray ionization-mass spectrometry and a

Bayesian universal charge deconvolution. *Rapid Communications in Mass Spectrometry*, 2023, **37**(8), e9478.

165 Mazarin, M. et al., Use of pulsed gradient spin-echo NMR as a tool in MALDI method development for polymer molecular weight determination. *Analytical chemistry*, 2006, **78**(8), 2758.

166 Iv, A.G. and Blum, S.A., Polymer molecular weight determination via fluorescence lifetime. *Journal of the American Chemical Society*, 2022, **144**(49), 22416.

167 Voorter, P.J. et al., Solvent-independent molecular weight determination of polymers based on a truly universal calibration. *Angewandte Chemie-International Edition*, 2022, **61**(5), e202114536.

168 Gu, K.C. et al., Determination of the molecular weight of conjugated polymers with diffusion-ordered NMR spectroscopy. *Chemistry of Materials*, 2018, **30**(3), 570.

169 Monakhova, Y.B. et al., Novel method for the determination of average molecular weight of natural polymers based on 2D DOSY NMR and chemometrics: example of heparin. *Journal of Pharmaceutical and Biomedical Analysis*, 2018, **149**, 128.

170 Jeon, B.J. and Muthukumar, M., Determination of molecular weights in polyelectrolyte mixtures using polymer translocation through a protein nanopore. *ACS Macro Letters*, 2014, **3**(9), 911.

171 Tian, Y.X. et al., Molecular weight determination by counting molecules. *Journal of Physical Chemistry Letters*, 2015, **6**(6), 923.

172 Leonardi, F., Allal, A., and Marin, G., Determination of the molecular weight distribution of polymers melts. in *Progress and trends in rheology V. Proceedings of the Fifth European Rheology Conference*, 1998, Springer.

173 Shipp, D.A. and Matyjaszewski, K., Kinetic analysis of controlled/"living" radical polymerizations by simulations. 2. Apparent external orders of reactants in atom transfer radical polymerization. *Macromolecules*, 2000, **33**(5), 1553.

174 Pintauer, T., McKenzie, B., and Matyjaszewski, K., Toward structural and mechanistic understanding of transition metal-catalyzed atom transfer radical processes. *Advances in Controlled/Living Radical Polymerization*, 2003, **854**, 130.

175 Pintauer, T. et al., Determination of rate constants for the activation step in atom transfer radical polymerization using the stopped-flow technique. *Macromolecules*, 2004, **37**(8), 2679.

176 Harris, J.A., Hinojosa, O., and Arthur, J.C., Electron-spin resonance studies of propagating radicals in polymerization. 2. Acrylate propagating radicals. *Journal of Polymer Science Part A-Polymer Chemistry*, 1974, **12**(3), 679.

177 Worsfold, D.J. and Bywater, S., Anionic polymerization of styrene. *Canadian Journal of Chemistry-Revue Canadienne De Chimie*, 1960, **38**(10), 1891.

178 Criqui, A., Lalevée, J., Allonas, X., and Fouassier, J.P., Electron spin resonance spin trapping technique: application to the cleavage process of photoinitiators. *Macromolecular Chemistry and Physics*, 2008, **209**(21), 2223.

179 Kajiwara, A. et al., Time-resolved electron-spin-resonance study on radical polymerization with (2,4,6-trimethylbenzoyl)diphenylphosphine oxide - direct estimation of rate constants for addition reactions of diphenylphosphonyl radicals to vinyl monomers. *Macromolecules*, 1993, **26**(7), 1656.

180 Zetterlund, P.B. et al., High-conversion free-radical bulk polymerization of styrene: termination kinetics studied by electron spin resonance, Fourier transform near-infrared spectroscopy, and gel permeation chromatography. *Macromolecules*, 2001, **34**(22), 7686.

181 Bird, R.B., Stewart, W.E., and Lightfoot, E.N., *Transport phenomena*. 2 ed. 2001, John Wiley.

182 De la Barrera, E., On the sesquicentennial of Fick's laws of diffusion. *Nature Structural & Molecular Biology*, 2005, **12**(4), 280.

183 Fick, A., On liquid diffusion. *Poggendorff's Annalen der Physik und Chemie*, 1855, **94**, 59.

184 Tyrrell, H.J.V., The origin and present status of Fick's diffusion law. *Journal of Chemical Education*, 1964, **41**(7), 397.

185 Hamielec, A.E., Hodgins, J.W., and Tebbens, K., Polymer reactors and molecular weight distribution. 2. Free radical polymerization in a batch reactor. *AiCHE Journal*, 1967, **13**(6), 1087.

186 Ilare, J. et al., Free-radical polymerization of methacrylic acid: from batch to continuous using a stirred tank reactor. *Chemical Engineering and Processing-Process Intensification*, 2021, **165**, 108440.

187 Lecardinal, G., Germain, E., Gelus, M., and Guillon, B., Design of stirred batch polymerization reactor. *Chemical Engineering Science*, 1980, **35**(1–2), 499.

188 Campbell, J.D., Teymour, F., and Morbidelli, M., High temperature free radical polymerization. 1. Investigation of continuous styrene polymerization. *Macromolecules*, 2003, **36**(15), 5491.

189 Campbell, J.D., Teymour, F., and Morbidelli, M., High temperature free radical polymerization. 2. Modeling continuous styrene polymerization. *Macromolecules*, 2003, **36**(15), 5502.

190 Platzer, N., Design of continuous and batch polymerization processes. *Industrial and Engineering Chemistry*, 1970, **62**(1), 6.

191 Hoffman, R.F., Rosen, G., and Schreiber, S., Batch polymerization - narrowing molecular weight distribution. *Industrial and Engineering Chemistry*, 1964, **56**(5), 57.

192 Duerksen, J.H., Hamielec, A.E., and Hodgins, J.W., Polymer reactors and molecular weight distribution. I. Free radical polymerization in a continuous stirred-tank reactor. *Aiche Journal*, 1967, **13**(6), 1081.

193 Klahn, J. and Bandermann, F., Optimizing procedure of reaction of living polymers - theoretical considerations about molecular-weight distributions of living polymers in a periodically stirred tank reactor. *Angewandte Makromolekulare Chemie*, 1979, **83**(Dec), 1.

194 Zeman, R.J. and Amundson, N.R., Continuous polymerization models. I. Polymerization in continuous stirred tank reactors. *Chemical Engineering Science*, 1965, **20**(4), 331.

195 Gebert, W. and Bandermann, F., Optimal reaction conditions for synthesis of living polymers - synthesis of products with bimodal molecular-weight distribution functions in a tubular flow reactor. *Angewandte Makromolekulare Chemie*, 1977, **57**(25), 153.

196 Kim, J.S., Kweon, J.O., Lee, J.H., and Noh, S.T., Synthesis of high molecular weight poly(styrene-methyl methacrylate) using a plug flow reactor system by anionic polymerization. *Macromolecular Research*, 2015, **23**(1), 100.

197 Yoon, W.J. and Choi, K.Y., Polymerization of styrene in a continuous filled tubular reactor. *Polymer Engineering and Science*, 1996, **36**(1), 65.

198 Reis, M.H., Leibfarth, F.A., and Pitet, L.M., Polymerizations in continuous flow: recent advances in the synthesis of diverse polymeric materials. *ACS Macro Letters*, 2020, **9**(1), 123.

199 Agarwal, S.S. and Kleinstreuer, C., Analysis of styrene polymerization in a continuous-flow tubular reactor. *Chemical Engineering Science*, 1986, **41**(12).

200 Hornung, C.H., Nguyen, X., Dumsday, G., and Saubern, S., Integrated continuous processing and flow characterization of RAFT polymerization in tubular flow reactors. *Macromolecular Reaction Engineering*, 2012, **6**(11), 458.

201 Srinivas, G., Yethiraj, A., and Bagchi, B., Nonexponentiality of time-dependent survival probability and the fractional viscosity dependence of the rate in diffusion controlled reactions in a polymer chain. *Journal of Chemical Physics*, 2001, **114**(20), 9170.

202 Sharikov, Y.V., Petyukhin, K.V., and Zhitinkin, A.A., Kinetics of the formation of polymers accompanied by a substantial increase in the viscosity of the reaction medium. *Kinetics and Catalysis*, 1979, **20**(2), 401.

203 Abrams, C.F., Lee, N.K., and Johner, A., Diffusion/reaction in confined polymer chains. *Macromolecules*, 2006, **39**(10), 3655.

204 Everaers, R. and Rosa, A., Multi-scale modeling of diffusion-controlled reactions in polymers: renormalisation of reactivity parameters. *Journal of Chemical Physics*, 2012, **136**(1), 014902.

205 Srinivas, G. and Bagchi, B., Distribution of reaction times in diffusion controlled reactions in polymers. *Chemical Physics Letters*, 2000, **328**(4–6), 420.

206 Cui, X.H. et al., Probe into effect of the initiator on the thermal runaway hazards of methyl methacrylate bulk polymerization. *Journal of Loss Prevention in the Process Industries*, 2023, **83**, 105074.

207 Han, J.P. et al., Vinyl chloride suspension polymerization: kinetic modelling, thermal runaway prediction, and prevention. *Canadian Journal of Chemical Engineering*, 2023, **101**(9), 4927.

208 Yu, A.D. et al., Process hazard evaluation and thermal runaway prediction for styrene polymerization initiated by peroxide-azo composites. *Process Safety and Environmental Protection*, 2022, **166**, 257.

209 Liu, G.Y. and Wilhite, B.A., Model-based design for inhibition of thermal runaway in free radical polymerization. *Industrial & Engineering Chemistry Research*, 2019, **58**(37), 17244.

210 Fujita, M., Izato, Y., Iizuka, Y., and Miyake, A., Thermal hazard evaluation of runaway polymerization of acrylic acid. *Process Safety and Environmental Protection*, 2019, **129**, 339.

211 Kamachi, M., Influence of solvent on free radical polymerization of vinyl compounds, in *Polymerization processes*. 2005, Springer, pp. 58–87.

212 Gandhi, K.S. and Williams, M.C., Solvent effects on viscosity of moderately concentrated polymer solutions. *Journal of Polymer Science Part C-Polymer Symposium*, 1971, (35), 211.

213 VivaldoLima, E., Wood, P.E., Hamielec, A.E., and Penlidis, A., An updated review on suspension polymerization. *Industrial & Engineering Chemistry Research*, 1997, **36**(4), 939.

214 Yenalyev, V.D. and Melnichenko, V.I., Molecular-weight and molecular-weight distribution of polystyrene obtained with suspension polymerization. *ACS Symposium Series*, 1976, (24), 211.

215 Lovell, P.A. and Schork, F.J., Fundamentals of emulsion polymerization. *Biomacromolecules*, 2020, **21**(11), 4396.

216 Schneider, H.J., Particle formation in emulsion polymerization - a literature review. *Acta Polymerica*, 1981, **32**(11), 667.

217 Xiang, S.C. et al., Progress in process parameters and mechanism research of polymer emulsion preparation. *RSC Advances*, 2024, **14**(23), 16024.

218 Farías-Cepeda, L., Herrera-Ordonez, J., and Saldívar-Guerra, E., On the kinetics and particle size polydispersity of the styrene emulsion polymerization using aerosol MA80 and sodium dodecyl sulfate as surfactants. *Colloid and Polymer Science*, 2010, **288**(14–15), 1401.

219 Herrera-Ordonez, J., Controversies on the mechanism and kinetics of emulsion polymerization: an updated critical review. *Advances in Colloid and Interface Science*, 2023, **320**, 103005.

220 Moore, R. and Schork, F.J., On the average number of radicals per particle during free radical and living radical emulsion polymerization. *Industrial & Engineering Chemistry Research*, 2025, **64**(19), 9558.

221 Li, B.G. and Brooks, B.W., Prediction of the average number of radicals per particle for emulsion polymerization. *Journal of Polymer Science Part A-Polymer Chemistry*, 1993, **31**(9), 2397.

222 Asua, J.M., Emulsion polymerization: from fundamental mechanisms to process developments. *Journal of Polymer Science Part A-Polymer Chemistry*, 2004, **42**(5), 1025.

223 Urrea-Quintero, J.H., Marino, M., Hernandez, H., and Ochoa, S., Multiscale modeling of a free-radical emulsion polymerization process: numerical approximation by the finite element method. *Computers & Chemical Engineering*, 2020, **140**, 106974.

224 Fellows, C.M., Murison, R.D., and Russell, G.T., Model discrimination of radical desorption kinetics in emulsion polymerisation. *Macromolecular Theory and Simulations*, 2011, **20**(6), 425.

225 Pohn, J. et al., Stable free radical polymerization in emulsion: modeling the thermodynamics of monomer transfer between droplets and particles. *Macromolecular Theory and Simulations*, 2008, **17**(2–3), 73.

226 Dong, Y. and Sundberg, D.C., Radical entry in emulsion polymerization: estimation of the critical length of entry radicals via a simple lattice model. *Macromolecules*, 2002, **35**(21), 8185.

227 Morrison, B.R. et al., Free-radical exit in emulsion polymerization. 2. Model discrimination via experiment. *Journal of Polymer Science Part A-Polymer Chemistry*, 1994, **32**(4), 631.

228 Schneider, H.J. and Wolf, H.J., Radical transport in emulsion polymerization. 1. Model of the reaction kinetics including chain transfer to the monomer. *Acta Polymerica*, 1983, **34**(7), 413.

229 Fessenden, R.W., Measurement of short radical lifetimes by electron spin resonance methods. *Journal of Physical Chemistry*, 1964, **68**(6), 1508.

230 Attri, P. et al., Generation mechanism of hydroxyl radical species and its lifetime prediction during the plasma-initiated ultraviolet (UV) photolysis. *Scientific Reports*, 2015, **5**, 9332.

231 Reversible-deactivation radical polymerization, in *Wikipedia: the free encyclopedia*. 2025, Wikipedia Foundation, https://en.wikipedia.org/wiki/Reversible-deactivation_radical_polymerization.

232 Keating, J.J. and Plawsky, J.L., Radical lifetimes in atom transfer radical polymerization: a Monte Carlo and deterministic investigation. *Macromolecules*, 2020, **53**(17), 7224.

233 Achilias, D.S. and Kiparissides, C., On the validity of the steady-state approximations in high conversion diffusion-controlled free-radical copolymerization reactions. *Polymer*, 1994, **35**(8), 1714.

234 Spade, C.A. and Volpert, V.A., On the steady-state approximation in thermal free radical frontal polymerization. *Chemical Engineering Science*, 2000, **55**(3), 641.

235 Fischer, H., The persistent radical effect: a principle for selective radical reactions and living radical polymerizations. *Chemical Reviews*, 2001, **101**(12), 3581.

236 Achilias, D.S., A review of modeling of diffusion controlled polymerization reactions. *Macromolecular Theory and Simulations*, 2007, **16**(4), 319.

237 Achilias, D.S. and Verros, G.D., Modeling of diffusion-controlled reactions in free radical solution and bulk polymerization: model validation by DSC experiments. *Journal of Applied Polymer Science*, 2010, **116**(3), 1842.

238 Rabea, A.M. and Zhu, S.P. Controlled radical polymerization at high conversions: Diffusion controlled reactions. In *248th ACS National Meeting*, 2014, American Chemical Society.

239 Braun, D., Origins and development of initiation of free radical polymerization processes. *International Journal of Polymer Science*, 2009, **2009**, 893234.

240 Ito, K., A theory generalized to cage effects in radical formation for determining initiator efficiency. *Polymer Journal*, 1985, **17**(3), 421.

241 Stolzenberg, K. and Kirchner, K., Continuous free-radical polymerization of styrene in the stirred tank reactor - effect of inhibitors. *Angewandte Makromolekulare Chemie*, 1981, **95**(Apr), 185.

242 Tanczos, I., Foldesberezsnich, T., and Tudos, F., Kinetics of radical polymerization. 40. Investigation of the effect of molecule inhibitors in polymerization in solution. *European Polymer Journal*, 1983, **19**(3), 225.

243 Li, L. and Lee, L.J., Effects of inhibitors and retarders on low temperature free radical crosslinking polymerization between styrene and vinyl ester resin. *Polymer Engineering and Science*, 2001, **41**(1), 53.

244 Kirchner, K. and Rintelen, T., Continuous free-radical polymerization of styrene in the stirred tank reactor - effect of retarders. *Angewandte Makromolekulare Chemie*, 1986, **141**, 85.

245 Goto, A. and Fukuda, T., Kinetics of living radical polymerization. *Progress in polymer science*, 2004, **29**(4), 329.

246 Zhang, H.Q. et al., Effect of Cu(II) on the kinetics of the homogeneous atom transfer radical polymerization of methyl methacrylate. *Macromolecules*, 2001, **34**(18), 6169.

247 Krys, P. and Matyjaszewski, K., Kinetics of atom transfer radical polymerization. *European Polymer Journal*, 2017, **89**, 482.

248 Parker, J., Jones, R.G., and Holder, S.J., Direct evidence for the interaction of the mechanisms of thermally initiated and atom transfer radical polymerization. *Macromolecules*, 2000, **33**(25), 9166.

249 Matyjaszewski, K., Atom transfer radical polymerization (ATRP): current status and future perspectives. *Macromolecules*, 2012, **45**(10), 4015.

250 Matyjaszewski, K., Atom transfer radical polymerization: from mechanisms to applications. *Israel Journal of Chemistry*, 2012, **52**(3–4), 206.

251 Lamontagne, H.R. and Lessard, B.H., Nitroxide-mediated polymerization: a versatile tool for the engineering of next generation materials. *ACS Applied Polymer Materials*, 2020, **2**(12), 5327.

252 Dervan, P., Aldabbagh, F., Zetterlund, P.B., and Yamada, B., Mechanism and kinetics of the imidazolidinone nitroxide-mediated free-radical polymerization of styrene. *Journal of Polymer Science Part A-Polymer Chemistry*, 2003, **41**(2), 327.

253 Veregin, R.P.N., Odell, P.G., Michalak, L.M., and Georges, M.K., Mechanism of rate enhancement using organic acids in nitroxide-mediated living free-radical polymerizations. *Macromolecules*, 1996, **29**(11), 4161.

254 Studer, A. et al., New sterically hindered nitroxides for the living free radical polymerization: X-ray structure of an α-H-bearing nitroxide. *Macromolecules*, 2004, **37**(1), 27.

255 Tang, W., Tsarevsky, N.V., and Matyjaszewski, K., Determination of equilibrium constants for atom transfer radical polymerization. *Journal of the American Chemical Society*, 2006, **128**(5), 1598.

256 Lorandi, F. et al., New protocol to determine the equilibrium constant of atom transfer radical polymerization. *Electrochimica Acta*, 2018, **260**, 648.

257 Matyjaszewski, K., Patten, T.E., and Xia, J.H., Controlled/"living" radical polymerization. Kinetics of the homogeneous atom transfer radical polymerization of styrene. *Journal of the American Chemical Society*, 1997, **119**(4), 674.

258 Tang, W. et al., Understanding atom transfer radical polymerization: effect of ligand and initiator structures on the equilibrium constants. *Journal of the American Chemical Society*, 2008, **130**(32), 10702.

259 Zhang, M. and Ray, W.H., Modeling of "living" free-radical polymerization with RAFT chemistry. *Industrial & Engineering Chemistry Research*, 2001, **40**(20), 4336.

260 Worsfold, D.J. and Bywater, S., Anionic polymerization of styrene. *Canadian Journal of Chemistry*, 2011, **38**, 1891.

261 Hurley, S.A. and Tait, P.J.T., Anionic polymerization initiated by lithium diethylamide in organic solvents. 3. Investigation of polymerization of styrene. *Journal of Polymer Science Part A-Polymer Chemistry*, 1976, **14**(7), 1565.

262 Ceausescu, E. et al., Kinetics of anionic-polymerization of butadiene and styrene with organo-lithium catalysts in toluene. *Revue Roumaine De Chimie*, 1981, **26**(4), 509.

263 Yakimansky, A.V., Müller, A.H.E., and Van Beylen, M., Density functional theory study on the aggregation and dissociation behavior of lithium chloride in THF and its interaction with the active centers of the anionic polymerization of methyl methacrylate and styrene. *Macromolecules*, 2000, **33**(15), 5686.

264 Kojima, K., Sawamoto, M., and Higashimura, T., Living cationic polymerization of isobutyl vinyl ether by hydrogen iodide/Lewis acid initiating systems: effects of Lewis acid activators and polymerization kinetics. *Macromolecules*, 1989, **22**(4), 1552.

265 Zhang, J.H. et al., Characteristics and mechanism of styrene cationic polymerization in aqueous media initiated by cumyl alcohol/$B(C_6F_5)_{(3)}$. *Macromolecular Chemistry and Physics*, 2019, **220**(4), 1800419.

266 Kanazawa, A., Hashizume, R., Kanaoka, S., and Aoshima, S., Design of benign initiator for living cationic polymerization of vinyl ethers: facile in situ generation of vinyl ether-hydrogen halide adducts and subsequent controlled polymerization without a Lewis acid catalyst. *Macromolecules*, 2014, **47**(5), 1578.

267 Kanazawa, A. et al., Structure effects of Lewis acids on the living cationic polymerization of p-methoxystyrene: distinct difference in polymerization behavior from vinyl ethers. *Macromolecules*, 2012, **45**(19), 7749.

268 Banerjee, S., Paira, T.K., Kotal, A., and Mandal, T.K., Room temperature living cationic polymerization of styrene with HX-styrenic monomer adduct/$FeCl_3$ systems in the presence of tetrabutylammonium halide and tetraalkylphosphonium bromide salts. *Polymer*, 2010, **51**(6), 1258.

269 Lin, C.H., Xiang, J.S., and Matyjaszewski, K., Living cationic polymerization of styrene in the presence of tetrabutylammonium salts. *Macromolecules*, 1993, **26**(11), 2785.

270 Xiang, J.S., Lin, C.H., and Matyjaszewski, K. Living cationic polymerization of styrene in the presence of tetrabutyl ammonium salts. In *204th ACS National Meeting*, 1992, American Chemical Society.

271 Bermeshev, M.V. et al., Cationic polymerization of norbornene derivatives in the presence of boranes. *Macromolecules*, 2014, **47**(16), 5470.

272 Kwon, O.W., Noh, S.K., and Lyoo, W.S., Preparation of isotacticity-rich poly(tert-butyl vinyl ether) by the cationic polymerization of tert-butyl vinyl ether with dimethyl [rac-ethylenebis (indenyl)]zirconium and tri(pentafluorophenyl)borane. *Journal of Applied Polymer Science*, 2008, **107**(3), 1487.

273 Radchenko, A.V. et al., Kinetic study on the living/controlled cationic polymerization of *p*-methoxystyrene coinitiated by tris(pentafluorophenyl)borane. *Journal of Polymer Science Part A-Polymer Chemistry*, 2008, **46**(20), 6928.

274 Hammond, P. Lecture 25: "living" cationic polymerizations, examples of cationic polymerization, isobutyl rubber synthesis, polyvinyl ethers, in *Synthesis of polymers*. 2006.

275 Endres, G.F. and Overberger, C.G., Ionic polymerization. 5. Molecular termination in the cationic polymerization of styrene catalyzed by stannic chloride. *Journal of the American Chemical Society*, 1955, **77**(8), 2201.

6

Moments of Elementary Reactions in Addition Polymerization

6.1 Introduction

Polymer molecules, in the addition polymerization, are produced by the addition of monomers to the reactive propagating chains [1] followed by termination of propagating chains by termination via combination or disproportionation [2–7]. There are other elementary reactions that produce terminated chains during polymerization. These include transfer of reactive species to solvent or other small molecules in the reaction mixture [8–15], transfer of reactive species to monomer [9, 16–19], chain transfer agent [20–22], transfer to polymer chains by hydrogen abstraction [23–26], transfer to the double bond at chain-ends [27] as well as along the main chain of the terminated polymer molecules [28–30], and chain scission of the propagating reactive species [31–39]. This chapter uses the method of z-transform to relate the kth order moment of the terminated polymer molecules produced by different elementary reactions to the corresponding kth moments of the propagating chains.

6.2 Instantaneous kth Order Moments of Propagating Chains After Propagation in Terms of Propagating Chains Prior to Propagation

In the propagation step, an n-mer reactive propagating chain reacts with a monomer to form a new propagating chain with $n + 1$ repeat units, as shown below:

$$RP_n + M \xrightarrow{k_p} RP_{n+1} \tag{6.1}$$

The propagation rate or the rate of polymerization, R_P, which is the rate of change of the amount of monomer in the reaction mixture with time, is given by the following equation [40]:

$$R_p = -\frac{1}{V}\frac{d\,([M]\,V)}{d\,t} = k_p\,[M]\sum_{n=1}^{\infty}[RP_n] = k_p\,[M]\,[R] \tag{6.2}$$

where $[M]$ and $[R]$ are the concentrations of monomer and reactive species (all propagating chains), respectively, $[RP_n]$ is the concentration of reactive propagating chains with n monomer repeat units, k_p is the rate constant for propagation, V is the reaction volume, and t is time. Based on the above equation, the knowledge of the time dependence of the concentration of reactive species is required for the determination of the rate of polymerization. As the concentration of reactive species is related to chain termination reaction, we will relate the kth moment of the propagating chains to the moments of terminated chains produced by different termination reactions. The distribution and moments of the chains after propagation step are determined by the first forward difference property in the z-transform domain (see Section 4.4 of Chapter 4). According to this property, the z-transform of the difference in distribution functions $f\,(x = i + 1)$ and $f\,(x = i)$ in the x domain is equal to the z-transform of the distribution function $f\,(x)$, namely $F\,(s)$, multiplied by the factor $(s - 1)$. Therefore, the instantaneous distribution function of the propagating chains after propagation step in the z-domain is defined as follows:

$$F'_{prp}\,(s) = (s-1)\,F_{rp}\,(s) + F_{rp}\,(s) = s\,F_{rp}\,(s) \tag{6.3}$$

where $F'_{prp}\,(s)$ is the instantaneous distribution function of propagating chains after propagation step and $F_{rp}\,(s)$ is the distribution function prior to propagation. The linearity property of the z-function was used in the above equation to relate distribution functions before and after the propagation step. As defined in Chapter 4 (see Eqs. 4.8 and 4.11), the first and second order mole-fraction moments of the propagating chains before propagation, namely λ_1 and λ_2, respectively, are defined by the following equations:

$$\lambda_1 = \left.\frac{\partial F_{rp}\,(s)}{\partial s}\right|_{s=1} \tag{6.4}$$

$$\lambda_2 = \left.\frac{\partial^2 F_{rp}\,(s)}{\partial s^2}\right|_{s=1} + \left.\frac{\partial F_{rp}\,(s)}{\partial s}\right|_{s=1} \tag{6.5}$$

Based on Eq. (6.3), the instantaneous first order mole-fraction moment of the propagating chains after propagation, namely $\lambda'_{1,prp}$, is determined as follows:

$$\lambda'_{1,prp} = \left.\frac{\partial F'_{prp}\,(s)}{\partial s}\right|_{s=1} = \left.\frac{\partial}{\partial s}\left[s\,F_{rp}\,(s)\right]\right|_{s=1} = s\left.\frac{\partial F_{rp}\,(s)}{\partial s}\right|_{s=1} + F_{rp}\,(s)\big|_{s=1}\left.\frac{\partial s}{\partial s}\right|_{s=1} = \lambda_1 + 1 \tag{6.6}$$

The zero-order mole-fraction moment in the above derivation, by definition, is equal to unity, that is $F_{rp}\,(s)\big|_{s=1} = 1$. The instantaneous second order mole-fraction moment of the propagating chains after propagation, namely $\lambda'_{2,prp}$, is determined as follows:

$$\begin{aligned}
\lambda'_{2,prp} &= \left.\frac{\partial^2 F'_{prp}\,(s)}{\partial s^2}\right|_{s=1} + \left.\frac{\partial F'_{prp}\,(s)}{\partial s}\right|_{s=1} + = \left.\frac{\partial^2}{\partial s^2}\left[s\,F_{rp}\,(s)\right]\right|_{s=1} + \left.\frac{\partial}{\partial s}\left[s\,F_{rp}\,(s)\right]\right|_{s=1} \\
&= s\left\{\left.\frac{\partial^2 F_{rp}\,(s)}{\partial s^2}\right|_{s=1} + \left.\frac{\partial F_{rp}\,(s)}{\partial s}\right|_{s=1}\right\} + 2\left.\frac{\partial F_{rp}\,(s)}{\partial s}\right|_{s=1} + F_{rp}\,(s)\big|_{s=1} = \\
&= \lambda_2 + 2\lambda_1 + 1
\end{aligned} \tag{6.7}$$

Using a similar approach, the instantaneous higher-order moments of the propagating chains after propagation can be expressed in terms of the kth-order moments of the propagating chains prior to undergoing propagation.

6.3 Instantaneous *k*th Order Moments of Chains Produced by Termination in Terms of *k*th Order Moments of Propagating Chains

6.3.1 Termination by Disproportionation

In termination by disproportionation, two reactive propagating chains with n and m repeat units, namely RP_n and RP_m, react to form two terminated polymer chains TP_n and TP_m with n and m repeat units, respectively, and one unsaturated double bond at the chain-end of one of the chains by the following reaction [41, 42]:

$$RP_n + RP_m \xrightarrow{k_{td}} TP_n + TP_m + TDB \tag{6.8}$$

where k_{td} is the rate constant for termination via disproportionation, and TDB represents a chain-end double bond in a terminated polymer chain. Based on the above reaction, the distribution and moments of the terminated chains produced in the time interval between t and $t + dt$ are equal to those of the reactive propagating chains at time t during the reaction. If the kth order mole-fraction moment of the propagating chains at time t during the reaction is λ_k, the instantaneous kth order mole-fraction moment of polymer chains produced by disproportionation, namely $\mu'_{k,td}$ (the prime represents instantaneous property), in the time interval t and $t + dt$ is given by the following relation:

$$\mu'_{k,td} = \lambda_k \tag{6.9}$$

6.3.2 Termination by Combination

In termination by combination, two propagating chains RP_n and RP_m with n and m monomer repeat units, respectively, react and combine to produce a terminated polymer chain with n+m repeat units, namely TP_{n+m}, by the following reaction [43, 44]:

$$RP_n + RP_m \xrightarrow{k_{tc}} TP_{n+m} \tag{6.10}$$

where k_{tc} is the rate constant for termination via combination. Based on the above reaction, the distribution of the terminated chains produced in the time interval between t and $t + dt$ is equal to the convolution of the distributions of the two propagating chains at time t. If the distribution of the propagating chains at time t is given by $f_{rp}(y)$, the instantaneous distribution of terminated chains produced by combination, namely $f'_{tc}(x)$, in the time interval between t and $t + dt$ is given by the following relation:

$$f'_{tc}(x) = \sum_{y=1}^{x-1} f_{rp}(y)\, f_{rp}(x - y) \tag{6.11}$$

As the convolution of two functions in the z-domain is the product of the two functions (see Eq. 4.19 in Chapter 4), the z-transform of the distribution of terminated chains produced by the combination of propagating chains is given by the following equation:

$$F'_{tc}(s) = \left[F_{rp}(s)\right]^2 \tag{6.12}$$

Using the defining Eq. (4.8) in chapter 4 for the first order mole-fraction moment and Eqs. (6.4) and (6.5) for λ_1 and λ_2, respectively, the instantaneous first order mole-fraction moment of terminated chains produced by combination, namely $\mu'_{1,tc}$, is determined as follows:

$$\mu'_{1,tc} = \left.\frac{\partial F'_{tc}(s)}{\partial s}\right|_{s=1} = \frac{\partial}{\partial s}\left.\left[F_{rp}(s)\right]^2\right|_{s=1} = 2\,F_{rp}(s)\big|_{s=1}\left.\frac{\partial F_{rp}(s)}{\partial s}\right|_{s=1} = 2\lambda_1 \tag{6.13}$$

The above equation states that $\mu'_{1,tc}$ is equal to twice the first order mole-fraction moment of the propagating chains. Using the defining Eq. (4.11) in chapter 4 for the second order mole-fraction moment and Eqs. (6.4) and (6.5) for λ_1 and λ_2, respectively, the instantaneous second order mole-fraction moment of terminated chains produced by combination, namely $\mu'_{2,tc}$, is determined as follows:

$$\begin{aligned}\mu'_{2,tc} &= \left.\frac{\partial^2 F'_{tc}(s)}{\partial s^2}\right|_{s=1} + \left.\frac{\partial F'_{tc}(s)}{\partial s}\right|_{s=1} = \frac{\partial^2}{\partial s^2}\left.\left[F_{rp}(s)\right]^2\right|_{s=1} + \frac{\partial}{\partial s}\left.\left[F_{rp}(s)\right]^2\right|_{s=1} \\ &= 2\left\{\left.\frac{\partial F_{rp}(s)}{\partial s}\right|_{s=1}\right\}^2 + 2\left\{F_{rp}(s)\big|_{s=1}\right\}\left\{\left.\frac{\partial^2 F_{rp}(s)}{\partial s^2}\right|_{s=1} + \left.\frac{\partial F_{rp}(s)}{\partial s}\right|_{s=1}\right\} \\ &= 2\lambda_1^2 + 2\lambda_2\end{aligned} \tag{6.14}$$

Using a similar approach, the instantaneous higher order moments of terminated chains produced by combination can be written in terms of the kth order moments of the reactive propagating chains.

6.4 Instantaneous kth Order Moments of Chains Produced by Transfer Reactions in Terms of kth Order Moments of Propagating Chains

6.4.1 Instantaneous kth Order Moments of Terminated Chains Produced by Transfer to Small Molecules and Transfer Agent

As the concentration of small molecules in the reaction is orders of magnitude higher than the reactive species, it is more likely for a small molecule in the reaction mixture to react with a propagating chain to transfer the reactive species to the small molecule and produce a terminated polymer chain. The reactive species of a propagating chain can be transferred to monomer, solvent, polymerization initiator and catalyst, polymerization inhibitor or retarding agent, or an intentionally added chain transfer agent (CTA) [8–15]. However, it is more likely for transfer to occur to monomer, solvent, or a transfer agent because their concentration is orders of magnitude higher than those of initiator, catalyst, inhibitor, or retarder. The reactions for chain transfer to monomer [9, 16–19], solvent [12, 45], and CTA [45–49] are as follows:

$$RP_n + M \xrightarrow{k_{ctm}} TP_n + RP_1 \tag{6.15}$$

$$RP_n + S \xrightarrow{k_{cts}} TP_n + S^{\bullet} \xrightarrow{k_{pcts},\,M} RP_1 \tag{6.16}$$

$$RP_n + CTA \xrightarrow{k_{cta}} TP_n + CTA^{\bullet} \xrightarrow{k_{pcta},\,M} RP_1 \tag{6.17}$$

In the first transfer reaction, an n-mer propagating chain reacts with a monomer with rate constant k_{ctm} to form an n-mer terminated polymer chain and a new propagating chain with one monomer repeat unit. In the second reaction, an n-mer propagating chain reacts with a solvent molecule with rate constant k_{cts} to form an n-mer terminated polymer chain and a reactive solvent species $S^{\bullet}$ which subsequently reacts with a monomer with rate constant k_{pcts} to form a new propagating chain with one repeat unit. In the third reaction, an n-mer propagating chain reacts with chain transfer agent (CTA) with rate constant k_{cta} to form an n-mer terminated polymer chain and a reactive $CTA^{\bullet}$ species which subsequently reacts with a monomer with rate constant k_{pcta} to form a new propagating chain with one repeat unit. Based on the above reactions, the distribution and moments of terminated chains produced by chain transfer to small molecules are equal to those of the corresponding propagating chains. If the kth order mole-fraction moment of the propagating chains at time t during the reaction is λ_k, the instantaneous kth order mole-fraction moments of terminated chains produced by chain transfer to small molecules in the time interval between t and $t + dt$ are as follows:

$$\mu'_{k,ctm} = \mu'_{k,cts} = \mu'_{k,cta} = \lambda_k \tag{6.18}$$

where $\mu'_{k,ctm}$, $\mu'_{k,cts}$, and $\mu'_{k,cta}$ represent the instantaneous kth order mole-fraction moment of terminated chains produced by chain transfer to monomer, solvent, and chain transfer agent, respectively.

6.4.2 Instantaneous *k*th Order Moments of Dormant Chains Produced by Reaction with Transfer Agent with Reversible Deactivation

In reversible deactivation, the propagating chains are reversibly deactivated by different mechanisms to generate dormant chains [50]. In one approach (ATRP approach) [51, 52], a CTA is used in combination with an initiator $R - X$ having a labile group X (see section 2.2.1.2 in chapter two) to form a reactive $CTA - X^{\bullet}$ species. The $CTA - X^{\bullet}$ then reacts with a propagating chain to reversibly form a dormant chain by the following reaction [51]:

$$RP_n + CTA - X^{\bullet} \underset{k_{ac}}{\overset{k_{dac}}{\rightleftharpoons}} DOP_n + CTA \tag{6.19}$$

where DOP_n is an n-mer dormant chain, k_{dac} is the rate of deactivation of propagating chains to form dormant chains, and k_{ac} is the rate of activation of propagating chains by reaction with CTA to form propagating chains. If the kth order mole-fraction moment of the propagating chains at time t during the reaction is λ_k, the instantaneous kth order mole-fraction moments of dormant chains produced by reaction with the labile group-initiator complex $(CTA - X^{\bullet})$ in the time interval between t and $t + dt$ is given by the following relation:

$$\gamma'_k = \lambda_k \tag{6.20}$$

where γ'_k is the kth order mole-fraction moment of dormant chains produced by the deactivation of propagating chains. The above relationship applies if approaches other than ATRP, like NMP [53, 54] or RAFT [55–58] are used.

6.4.3 Instantaneous *k*th Order Moments of Terminated Chains Produced by Transfer to Polymer Chains

6.4.3.1 Instantaneous *k*th Order Moments of Terminated Chains Produced by Transfer to Polymer Chains via Hydrogen Abstraction

The active species in a reactive propagating chain RP_n may be transferred to an m-mer terminated polymer chain TP_m by abstracting a hydrogen atom from one of the carbons along the terminated chain, thereby producing a new polymer chain with n repeat units, namely TP_n, and a new reactive propagating chain with m repeat units as shown by the following reaction [25]:

$$
\begin{aligned}
&RP_n + TP_m \xrightarrow{k_{ctp,ha}} TP_n + SRP_m \\
&SRP_m + M \xrightarrow{k_p} RP_{m+1} + LB
\end{aligned}
\tag{6.21}
$$

where $k_{ctp,ha}$ is the rate constant for chain transfer to polymer by hydrogen abstraction, SRP_m represents an m-mer propagating chain with the reactive species attached to a secondary carbon, and LB is a long branch formed on the polymer molecule undergoing transfer by hydrogen abstraction. The overall reaction for chain transfer to polymer by hydrogen abstraction can be written as follows:

$$RP_n + TP_m + M \xrightarrow{k_{ctp,ha}} TP_n + RP_{m+1} + LB \tag{6.22}$$

According to the above reaction, the instantaneous moments of the terminated polymer chains produced by hydrogen abstraction are equal to those of the corresponding propagating chains. If the kth order mole-fraction moment of the propagating chains at time t during the reaction is λ_k, the instantaneous kth order moment of terminated chains produced by hydrogen abstraction, namely $\mu'_{k,ctpha}$, in the time interval t and $t + dt$ is given by the following relation:

$$\mu'_{k,ctpha} = \lambda_k \tag{6.23}$$

In the above chain transfer reaction to polymer, a new propagating chain RP_{m+1} having a long branch is produced. The instantaneous distribution function of RP_{m+1} chains is the first forward difference property in the z-transform domain (see Eq. 6.3 in Section 6.2 of this chapter) of the existing terminated polymer chains at time t in the reaction as follows:

$$F'_{rp,ctpha}(s) = sF_{tp}(s) \tag{6.24}$$

If polymerization is done in dilute condition such that the concentration of polymer chains is below the entanglement threshold, the instantaneous first and second order moments of the newly formed propagating chains by hydrogen abstraction are given by the following relations (see Eqs. 6.6 and 6.7):

$$\lambda'_{1,ctpha} = \mu_1 + 1 \tag{6.25}$$

$$\lambda'_{2,ctpha} == \mu_2 + 2\mu_1 + 1 \tag{6.26}$$

where $\lambda'_{1,ctpha}$ and $\lambda'_{2,ctpha}$ are the instantaneous first and second order mole-fraction moments of the new propagating chains produced by hydrogen abstraction in the time interval t and $t + dt$, respectively, and μ_1 and μ_2 are the first and second order mole-fraction moments of the terminated polymer chains at time t in the reaction. In concentrated polymer reactions, the probability of chain transfer to a polymer by hydrogen abstraction depends on chain length as longer chains have a greater number of hydrogens for abstraction [24, 59]. Therefore, in concentrated reactions, moments of polymer chains other than mole-fraction, like weight fraction, should be used to define the kth order moment of the newly formed propagating chains.

6.4.3.2 Instantaneous *k*th Order Moments of Terminated Chains Produced by Chain Transfer to Polymer by Reaction with Terminal Double Bond

In termination of propagating chains by disproportionation reaction, a propagating chain abstracts a hydrogen atom from a second propagating chain to produce two terminated polymer chains with one of the terminated chains having a double bond (TDB) at one chain end. The reactive species of a propagating chain in the reaction may react with this terminal bond to form a longer propagating chain according to the following reaction [27, 43, 44]:

$$RP_n + TDB_m \xrightarrow{k_{ctp,tdb}} SRP_{n+m} \tag{6.27}$$

$$SRP_{n+m} + M \xrightarrow{k_p} RP_{n+m+1} + LB \tag{6.28}$$

where $k_{ctp,tdb}$ is the rate constant for chain transfer to polymer by reaction with terminal double bond. The overall reaction is written as follows:

$$RP_n + TDB_m + M \xrightarrow{k_{ctp,tdb}} RP_{m+n+1} + LB \tag{6.29}$$

Based on the above reaction, the instantaneous distribution of the produced RP_{m+n+1} propagating chains in the time interval t and $t + dt$ is the first forward difference (see Eq. 6.3) of the convolution (see Eq. 6.12) of the distribution of propagating chains RP_n and polymer chains TDB_m in the z-domain at time t in the reaction as follows:

$$F'_{rp,ctptdb}(s) = s\,F_{rp}(s)\,F_{tp}(s) \tag{6.30}$$

The first and second order moments of the reactive propagating chains are given by Eqs. (6.4) and (6.5), respectively; the first and second order moments of the terminated polymer chains are defined similarly by replacing subscripts "rp" with "tp." The instantaneous first order moment of the produced RP_{m+n+1} propagating chains, namely $\lambda'_{1,ctptdb}$, is derived starting from Eq. (6.30) by using the defining equation for the first order moment (Eq. 6.4) as follows:

$$\begin{aligned}
\lambda'_{1,ctptdb} &= \left.\frac{\partial F'_{rp,ctptdb}(s)}{\partial s}\right|_{s=1} = \left.\frac{\partial}{\partial s}\left[s\,F_{rp}(s)\,F_{tp}(s)\right]\right|_{s=1} \\
&= \left.\left[F_{rp}(s)\,F_{tp}(s)\right]\right|_{s=1} + \left.\left[s\frac{\partial F_{rp}(s)}{\partial s}F_{tp}(s)\right]\right|_{s=1} + \left.\left[s\,F_{rp}(s)\frac{\partial F_{tp}(s)}{\partial s}\right]\right|_{s=1} \\
&= 1 + \lambda_1 + \mu_1
\end{aligned} \tag{6.31}$$

It can be shown that the second derivative of the function $F'_{rp,ctptdb}$ with respect to s is given by the following equation:

$$\left.\frac{\partial^2 F'_{rp,ctptdb}(s)}{\partial s^2}\right|_{s=1} = \lambda_2 + \mu_2 + 2\lambda_1\mu_1 + \lambda_1 + \mu_1 \tag{6.32}$$

Using the above relation and the defining equation for the second order moment (Eq. 6.5), the instantaneous second order moment of the RP_{m+n+1} propagating chains, namely $\lambda'_{2,ctptdb}$, is derived as follows:

$$\begin{aligned}
\lambda'_{2,ctptdb} &= \left.\frac{\partial^2 F'_{rp,ctptdb}(s)}{\partial s^2}\right|_{s=1} + \left.\frac{\partial F'_{rp,ctptdb}(s)}{\partial s}\right|_{s=1} \\
&= [\lambda_2 + \mu_2 + 2\lambda_1\mu_1 + \lambda_1 + \mu_1] + [1 + \lambda_1 + \mu_1] \\
&= \lambda_2 + \mu_2 + 2\lambda_1\mu_1 + 2\lambda_1 + 2\mu_1 + 1
\end{aligned} \tag{6.33}$$

If the number of repeat units of the propagating and terminated polymer chains are much greater than unity, that is $m + n >> 1$, then $m + n + 1 \approx m + n$, and the equations for $\lambda'_{1,ctptdb}$ and $\lambda'_{2,ctptdb}$ reduce to the following:

$$\lambda'_{1,ctptdb} = \lambda_1 + \mu_1 \qquad \in m + n >> 1 \tag{6.34}$$

$$\lambda'_{2,ctptdb} = \lambda_2 + \mu_2 + 2\lambda_1\mu_1 \qquad \in m + n >> 1 \tag{6.35}$$

According to Eq. (6.34), the instantaneous first order mole-fraction moment of propagating chains produced by chain transfer to polymer terminal double bond in the time interval t and $t + dt$ is the sum of the first order moments of the propagating and terminated chains at time t in the reaction.

6.4.3.3 Instantaneous *k*th Order Moments of Terminated Chains Produced by Chain Transfer to Polymer by Reaction with Double Bond Along Polymer Chain

Polymerization of monomers with more than one unsaturated group, like butadiene or copolymerization of mono-unsaturated monomers with monomers like butadiene produces terminated chains with pendant unsaturated double bonds along the chain. Transfer of the propagating chain to a pendant double bond along the polymer chain (PDB) can take place in the same way as to the double bond at the chain-end, with the difference that the propagation of the new chain leads to the formation of a long branch, as shown [30].

$$RP_n + PDB_m \xrightarrow{k_{ctp,pdb}} SRP_{n+m} + LB \tag{6.36}$$

$$SRP_{n+m} + M \xrightarrow{k_p} RP_{n+m+1} + LB \tag{6.37}$$

where $k_{ctp,pdb}$ is the rate constant for chain transfer to polymer by reaction with a pendant double bond along the chain. The overall reaction is written as follows:

$$RP_n + PDB_m + M \xrightarrow{k_{ctp,pdb}} RP_{m+n+1} + {}^2LB \tag{6.38}$$

The difference between transfer to terminal and pendant double bond is in the number of long branches produced; transfer to terminal double bond generates only one branch whereas transfer to pendant double bond generates two branches. Further, transfer to terminal bond is relatively independent of the polymer chain length (one terminal double bond per chain), whereas transfer to pendant bond increases with the number of pendant double bonds along the chain. Therefore, transfer to terminal bond depends on the number distribution (i.e., mole fraction) of polymer chains whereas transfer to pendant double bond depends on the weight distribution (i.e., weight fraction). In this context, the instantaneous first and second order moments of the new propagating chains produced by transfer to polymer pendant double bond are similar in form to Eqs. (6.34) and (6.35) (for $m + n >> 1$) with the difference that the mole-fraction moments of terminated chains are replaced with weight-fraction moments as shown below:

$$\lambda'_{1,ctppdb} = \lambda_1 + \nu_{1,tp} \quad \in m + n >> 1 \tag{6.39}$$

$$\lambda'_{2,ctptdb} = \lambda_2 + \nu_{2,tp} + 2\lambda_1\nu_{1,tp} \quad \in m + n >> 1 \tag{6.40}$$

where $\nu_{1,tp}$ and $\nu_{2,tp}$ are the first and second order weight-fraction moments of terminated polymer chains at time t in the reaction, respectively.

6.4.4 Instantaneous *k*th Order Moments of Terminated Chains Produced by Intramolecular Hydrogen Abstraction or Backbiting

In a backbiting reaction, a reactive propagating chain PR_n abstracts a hydrogen from one of the secondary carbon bonds along its chain, transfers the active species to the secondary carbon, and forms a short branch as shown below [60–62]:

$$RP_n \xrightarrow{k_{ctbb}} SRP_n \tag{6.41}$$

$$SRP_n + M \xrightarrow{k_p} RP_{n+1} + SB \tag{6.42}$$

where k_{ctbb} is the rate constant for chain transfer by backbiting and SB is a short branch formed in the propagating chain. The overall reaction for chain backbiting is written as follows:

$$RP_n + M \xrightarrow{k_{ctbb}} RP_{n+1} + SB \tag{6.43}$$

The instantaneous distribution of RP_{n+1} chains formed by backbiting in the time interval t and $t + dt$ in the z-domain is the first forward difference distribution of the propagating chains RP_n at time t in the reaction (see Eq. 3 in section 6.2). Therefore, the instantaneous first and second order mole-fraction moments of a chain undergoing backbiting are given by the following relations (see Eqs. 6.6 and 6.7):

$$\lambda'_{1,ctbb} = \lambda_1 + 1 \tag{6.44}$$

$$\lambda'_{2,ctbb} == \lambda_2 + 2\lambda_1 + 1 \tag{6.45}$$

where $\lambda'_{1,ctbb}$ and $\lambda'_{2,ctbb}$ are the first and second order moments of the reactive propagating chains formed by backbiting in the time interval t and $t + dt$, respectively.

6.5 Instantaneous *k*th Order Moments of Chains Produced by Chain Scission

Chain scission reactions occur at high temperatures [63], upon exposure of the polymer chains to radiation [64–67], or solvent swelling of polymer chains [68]. Polymer chain scission can also be induced by flow or deformation [69–71]. Frequently, a chain scission reaction is preceded by a transfer reaction in which the active species on a propagating chain is transferred to a secondary or tertiary carbon of a chain [34, 72, 73]. As active species on a secondary or tertiary carbon is less stable compared to reactive species on a primary carbon, the bond adjacent to a secondary/tertiary carbon undergoes scission to increase stability. There are two types of chain scission, namely β_2 and β_3. In β_2 scission, the active species is transferred to a secondary carbon, whereas in β_3 scission, the active species is transferred to a tertiary carbon.

6.5.1 Instantaneous *k*th Order Moments of Chains Produced by β_2 Scission

A β_2 scission is preceded by a reaction (chain transfer) to transfer the reactive species from a primary carbon to a secondary carbon. The transfer reaction is followed by a scission reaction to form a new shorter propagating chain and a terminated polymer chain having a terminal double bond. The two reactions leading to a β_2 chain scission are as follows:

$$RP_n + TP_x\overset{\text{H}}{\underset{\text{H}}{\text{C}}}\text{—}\overset{\text{H}}{\underset{\text{H}}{\text{C}}}TP_{y-1} \xrightarrow{k_{ctp}} TP_n + SRP_x\overset{\text{H}}{\underset{\bullet}{\text{C}}}\text{—}\overset{\text{H}}{\underset{\text{H}}{\text{C}}}SRP_{y-1} \tag{6.46}$$

$$SRP_x\overset{\text{H}}{\underset{\bullet}{\text{C}}}\text{—}\overset{\text{H}}{\underset{\text{H}}{\text{C}}}SRP_{y-1} \xrightarrow{k_{\beta 2}} RP_x + \overset{\text{H}}{\underset{\text{H}}{\text{C}}}\text{=}\overset{\text{H}}{\text{C}}TDB_{y-1} \tag{6.47}$$

In the first reaction, a propagating chain RP_n abstracts a hydrogen from a carbon of one of the monomers on a terminated polymer chain TP_{x+y} to form a new n-mer terminated polymer chain TP_n and a new $(x + y)$-mer propagating chain SRP_{x+y} having the reactive species on a secondary carbon. In the second reaction, the unstable carbon having a reactive species on SRP_{x+y} chain splits from its neighboring carbon (undergoes β_2 scission) to generate an x-mer propagating chain with the reactive species on a primary carbon plus a y-mer terminated polymer chain with a terminal double bond. Based on the first reaction, the distribution of the SRP_{x+y} chains produced in the time interval t and $t + dt$ is equal to the distribution of the terminated polymer chains at time t in the reaction. Therefore, the instantaneous distribution function of SRP_{x+y} chains in the z-domain and the first and second order moments of their distribution are given by the following expressions:

$$F'_{srp,ctp}(s) = F_{tp}(s) \tag{6.48}$$

$$\lambda'_{1,srp,ctp} = \mu_1 \quad \lambda'_{2,srp,ctp} = \mu_2 \tag{6.49}$$

where $F'_{srp,ctp}(s)$ is the instantaneous distribution function of the SRP_{x+y} chains in the z-domain produced by chain transfer to polymer and $\lambda'_{1,srp,ctp}$ and $\lambda'_{2,srp,ctp}$ are the corresponding instantaneous first and second order moments of the new chain, respectively. Based on the second reaction, the distribution of SRP_{x+y} chains (before scission) is the convolution of the distribution of the produced propagating RP_x and terminated TDB_y chains. As the hydrogen abstraction by the propagating chains occurs randomly from any carbon on the polymer chain, the distributions of RP_x and TDB_y chains after scission would have to be identical. If the instantaneous distribution function of the produced propagating chain RP_x and terminated polymer chain TDB_y following β_2 scission is $f'_{\beta 2}(x)$, the instantaneous distribution function of SRP_{x+y} chains, namely $f'_{srp,ctp}(x)$, is as follows:

$$f'_{srp,ctp}(x) = \sum_{y=1}^{x-1} f'_{\beta 2}(y)\, f'_{\beta 2}(x - y) \tag{6.50}$$

Based on Eq. (4.19) in chapter 4, the z-transform of the above equation is as follows:

$$F'_{srp,ctp}(s) = \left[F'_{\beta 2}(s)\right]^2 \tag{6.51}$$

Upon substitution for $F'_{srp,ctp}(s)$ from Eq. (6.48), the instantaneous z-transform distribution function of the produced RP_x and TDB_y chains is as follows:

$$F'_{\beta 2}(s) = \left[F'_{srp,ctp}(s)\right]^{1/2} = \left[F_{tp}(s)\right]^{1/2} \tag{6.52}$$

Using the defining Eq. (6.4) for the first moment, Eq. (6.49) for $\lambda'_{1,srp,ctp}$, and the above equation, the instantaneous first moment of the propagating chain RP_x, $\lambda'_{1,\beta2}$, and the first order moment of the terminated chain TDB_y, $\mu'_{1,\beta2}$, produced by β_2 chain scission are derived as follows:

$$\begin{aligned}\lambda'_{1,\beta2} = \mu'_{1,\beta2} = \left.\frac{\partial F'_{\beta2}(s)}{\partial s}\right|_{s=1} &= \left.\frac{\partial}{\partial s}\left[F_{tp}{}^{1/2}(s)\right]\right|_{s=1} \\ &= \left.\left[(1/2)\,F_{tp}{}^{-1/2}(s)\,\frac{\partial F_{tp}(s)}{\partial s}\right]\right|_{s=1} = \mu_1/2\end{aligned} \tag{6.53}$$

where μ_1 is the first order moment of the terminated polymer chains at time t in the reaction. Using the defining Eq. (6.5) for the second order moment, Eq. (6.49) for $\lambda'_{2,srp,ctp}$, and the above equation, the instantaneous second order moment of the propagating chain RP_x, $\lambda'_{2,\beta}$, and the second order moment of the terminated chain TDB_y, $\mu'_{2,\beta2}$, produced by β_2 scission are derived as follows:

$$\begin{aligned}\lambda'_{2,\beta2} = \mu'_{2,\beta2} &= \left.\frac{\partial^2 F'_{\beta2}(s)}{\partial s^2}\right|_{s=1} + \left.\frac{\partial F'_{\beta2}(s)}{\partial s}\right|_{s=1} \\ &= \left.\frac{\partial^2}{\partial s^2}\left[F_{tp}(s)\right]^{1/2}\right|_{s=1} + \left.\frac{\partial}{\partial s}\left[F_{tp}(s)\right]^{1/2}\right|_{s=1} \\ &= \left.\left[(1/2)\left[F_{tp}(s)\right]^{-1/2}\left(\frac{\partial^2 F_{tp}(s)}{\partial s^2} + \frac{\partial F_{tp}(s)}{\partial s}\right)\right]\right|_{s=1} + \left.\left[(-1/4)\left[F_{tp}(s)\right]^{-3/2}\left(\frac{\partial F_{tp}(s)}{\partial s}\right)^2\right]\right|_{s=1} \\ &= \frac{\mu_2}{2} - \frac{\mu_1^2}{4}\end{aligned} \tag{6.54}$$

where μ_2 is the second order moment of the terminated polymer chains at time t in the reaction. In the above derivation, it was assumed that chain scission occurs under dilute conditions such that chain transfer to polymer is independent of the length of terminated chains. In concentrated reactions, the probability of chain transfer to terminated chains depends on chain length as longer chains have greater number of bonds for hydrogen abstraction [24, 59]. Therefore, in concentrated solutions, moments of the polymer chains other than mole-fraction moments may need to be used to define the kth order moment of the SRP_{x+y} chains.

6.5.2 Instantaneous *k*th Order Moments of Chains Produced by β_3 Scission

A β_3 scission, like β_2, is preceded by a chain transfer reaction but with the difference that the reactive species is transferred from a primary carbon on a propagating chain to a tertiary carbon on a terminated polymer chain. The transfer reaction is followed by a scission reaction to form a new shorter propagating chain and a terminated polymer chain having a terminal double bond and a side chain. The two reactions leading to a β_3 chain scission are as follows:

$$RP_n + TP_z\underset{H}{\overset{H}{C}}\!-\!\underset{H}{C}\left\langle{}^{TP_x}_{TP_{y-1}}\right. \xrightarrow{k_{ctp}} TP_n + TRP_z\underset{H}{\overset{H}{C}}\!-\!\overset{\bullet}{C}\left\langle{}^{TRP_x}_{TRP_{y-1}}\right. \tag{6.55}$$

$$TRP_z\underset{H}{\overset{H}{C}}\!-\!\overset{\bullet}{C}\left\langle{}^{TRP_x}_{TRP_{y-1}}\right. \xrightarrow{k_{\beta3}} RP_z + \underset{H}{\overset{H}{C}} = C\left\langle{}^{TRP_x}_{TRP_{y-1}}\right. \tag{6.56}$$

In the first reaction, a propagating chain RP_n abstracts a hydrogen from a tertiary carbon of a branch point on a terminated polymer chain TP_{x+y+z} to form a new n-mer terminated polymer chain TP_n and a new $(x + y + z)$-mer propagating chain TRP_{x+y+z} having the reactive species on a tertiary carbon. The "TRP" represents a propagating chain with the reactive species on a tertiary carbon atom. In the second reaction, the unstable carbon having a reactive species on the TRP_{x+y+z} chain splits from its neighboring carbon (undergoes β_3 scission) to generate a z-mer propagating chain with the reactive species on a primary carbon plus a $(x + y)$-mer terminated polymer chain TDB_{x+y} with a terminal double bond. Based on the first reaction, the instantaneous distribution of TRP_{x+y+z} chains produced in the time interval t and $t + dt$ is equal to the distribution of the terminated polymer chains at time t in the reaction. Therefore, the distribution function of TRP_{x+y+z} chains in the z-domain and the first and second order moments of their distribution are given by the following expressions:

$$F'_{trp,ctp}(s) = F_{tp}(s) \tag{6.57}$$

$$\lambda'_{1,trp,ctp} = \mu_1 \qquad \lambda'_{2,trp,ctp} == \mu_2 \tag{6.58}$$

There are three cases for the moments of the produced propagating chains and terminated chains in the second reaction depending on the length of TRP_{x+y+z} polymer chains from the first reaction. In case one, the side chain of TP_{x+y+z} is short which becomes the new propagating chain RP_z. Since a short side chain ranges 2-9 repeat units in length, which is much smaller than the length of the terminated polymer chain, its length can be assumed to be equal to unity as compared to the other chains. On this basis, the instantaneous kth order moment of the produced propagating chains after β_3 scission is equal to unity and the instantaneous kth order moment of the produced terminated chains is equal to that of TP_{x+y+z} polymer chains as follows:

$$\lambda'_{1,\beta 3} = 1 \qquad \lambda'_{2,\beta 3} = 1 \qquad \in \text{ case 1} \tag{6.59}$$

$$\mu'_{1,\beta 3} = \mu_1 \qquad \mu'_{2,\beta 3} = \mu_2 \qquad \in \text{ case 1} \tag{6.60}$$

where $\lambda'_{1,\beta 3}$ and $\lambda'_{2,\beta 3}$ are the instantaneous first and second order moments of the produced RP_z chains after β_3 scission, and $\mu'_{1,\beta 3}$ and $\mu'_{2,\beta 3}$ are the instantaneous first and second order moments of the produced TDB_{x+y} chains. In case two, the side chain of the TP_{x+y+z} is short but it becomes a part of the new terminated chain TDB_{x+y}. The case two is like a β_2 scission in which the distribution function of the terminated polymer chains before scission is the convolution of the instantaneous distribution functions of the produced propagating and terminated chains after scission. In this case, according to Eqs. (6.52) and (6.53), the first and second order moments of the produced propagating RP_z and terminated TBD_{x+y} chains after scission are as follows:

$$\lambda'_{1,\beta 3} = \mu'_{1,\beta 3} = \mu_1/2 \qquad \in \quad case\ 2 \tag{6.61}$$

$$\lambda'_{2,\beta 3} = \mu'_{2,\beta 3} = \frac{\mu_2}{2} - \frac{\mu_1{}^2}{4} \qquad \in \quad case\ 2 \tag{6.62}$$

In case three, the side chain is long which means all three branches of the polymer chain before chain scission are relatively long chains. As a result, if the instantaneous distribution functions of the produced propagating chains RP_z and terminated polymer chains TDB_{x+y} are $f'_{rp,\beta 3}(x)$ and $f'_{tdb,\beta 3}(x)$ following β_3 scission, respectively, the instantaneous distribution function of SRP_{x+y+z} chains, namely $f'_{trp,ctp}(x)$ is as follows:

$$f'_{trp,ctp}(x) = \sum_{z=1}^{x-1} f'_{rp,\beta 3}(z)\, f'_{tdb,\beta 3}(x-z) \tag{6.63}$$

where $f'_{\beta 3,tdb}(x-z)$ itself is the convolution of the distribution functions of two propagating RP_z chains with identical distributions as given below:

$$f'_{tdb,\beta 3}(x-z) = \sum_{y=1}^{x-z-1} f'_{rp,\beta 3}(y)\, f'_{rp,\beta 3}(x-z-y) \tag{6.64}$$

Therefore, the distribution functions of propagating RP_z and terminated polymer chains TDB_{x+y} after chain scission in the z-domain are given by the following equations:

$$F'_{\beta 3,rp}(s) = \left[F'_{trp,ctp}(s)\right]^{1/3} = \left[F_{tp}(s)\right]^{1/3} \qquad \in \ case\ 3 \tag{6.65}$$

$$F'_{\beta 3,tdb}(s) = \left[F'_{trp,ctp}(s)\right]^{2/3} = \left[F_{tp}(s)\right]^{2/3} \qquad \in \ case\ 3 \tag{6.66}$$

Using the defining Eq. (6.4) and the above equations, the instantaneous first order moments of the propagating RP_z and terminated TDB_{x+y} chains after β_3 scission in the time interval t and $t+dt$, namely $\lambda'_{1,\beta 3}$ and $\mu'_{1,\beta 3}$, are given by the following equations:

$$\lambda'_{1,\beta 3} = \left.\frac{\partial F'_{\beta 3,rp}(s)}{\partial s}\right|_{s=1} = \left.\frac{\partial}{\partial s}\left[F_{tp}(s)\right]^{1/3}\right|_{s=1} = \mu_1/3 \qquad \in \ case\ 3 \tag{6.67}$$

$$\mu'_{1,\beta 3} = \left.\frac{\partial F'_{\beta 3,tdb}(s)}{\partial s}\right|_{s=1} = \left.\frac{\partial}{\partial s}\left[F_{tp}(s)\right]^{2/3}\right|_{s=1} = 2\mu_1/3 \qquad \in \ case\ 3 \tag{6.68}$$

Using the defining Eq. (6.5) and Eqs. (6.65) and (6.66), the instantaneous second order moments of the propagating RP_z and terminated TDB_{x+y} chains after β_3 scission in the time interval t and $t+dt$, namely $\lambda'_{2,\beta 3}$ and $\mu'_{2,\beta 3}$, are given by the following equations:

$$\begin{aligned} \lambda'_{2,\beta 3} &= \left.\frac{\partial^2 F'_{\beta 3,rp}(s)}{\partial s^2}\right|_{s=1} + \left.\frac{\partial F'_{\beta 3,rp}(s)}{\partial s}\right|_{s=1} \\ &= \left.\frac{\partial^2}{\partial s^2}\left[F_{tp}(s)\right]^{1/3}\right|_{s=1} + \left.\frac{\partial}{\partial s}\left[F_{tp}(s)\right]^{1/3}\right|_{s=1} = \frac{\mu_2}{3} - \frac{2\mu_1^2}{9} \qquad \in \ case\ 3 \end{aligned} \tag{6.69}$$

$$\begin{aligned} \mu'_{2,\beta 3} &= \left.\frac{\partial^2 F'_{\beta 3,tdb}(s)}{\partial s^2}\right|_{s=1} + \left.\frac{\partial F'_{\beta 3,tdb}(s)}{\partial s}\right|_{s=1} \\ &= \left.\frac{\partial^2}{\partial s^2}\left[F_{tp}(s)\right]^{2/3}\right|_{s=1} + \left.\frac{\partial}{\partial s}\left[F_{tp}(s)\right]^{2/3}\right|_{s=1} = \frac{2\mu_2}{3} - \frac{2\mu_1^2}{9} \qquad \in \ case\ 3 \end{aligned} \tag{6.70}$$

In the derivation of the above equations for β_3 scission, it was assumed that the scission reaction is carried out under conditions such that transfer to polymer is independent of the chain length. Under conditions that the rate of chain transfer to polymer is dependent on chain length [24, 59], moments of polymer chains other than mole-fraction distribution, like weight-based distribution, may need to be used to define the kth order moment of the TRP_{x+y+z} chains.

6.6 Summary

The polymerization reaction, at any given time t, consists of propagating and dormant chains, and terminated polymer chains. In the time interval between t and $t + dt$, new propagating and polymer chains are produced or consumed by different elementary reactions. To model the polymerization reaction, the kth order moments of produced or consumed chains by different elementary reactions during any time interval t and $t + dt$ need to be related to the kth order moments of the existing propagating and terminated chains at time t. This chapter applied the method of z-transform, described in chapter 4, to mathematically relate the kth order moments of chains produced or consumed by different elementary reactions in the time interval t and $t + dt$ to the kth order moments of existing propagating and terminated chains in the reaction at time t. The elementary reactions modeled in this chapter included propagation, termination by disproportionation and combination, transfer to small molecules and transfer agent, transfer to polymer by hydrogen abstraction, by reaction with the terminal double bond, or by reaction with double bonds along a terminated chain, transfer by chain backbiting, and chain scission. The relations developed for the kth order moments of different elementary reactions in terms of the kth moments of the existing chains will be used in the next chapter to model polymerization reactions.

6.7 List of Symbols and Abbreviations

CTA: chain transfer agent

$CTA^{\bullet}$: activated chain transfer agent that reacts with a monomer to form a new propagating chain

$CTA - X^{\bullet}$: reactive complex formed between a chain transfer agent and the labile group X of an initiator which can reversibly react with a propagating chain to form a formant chain

DOP_n: dormant polymer chain with n repeats

dt: infinitesimal change in reaction time (s)

$f(x)$: distribution function of a random discrete variable in the x domain where x is the degree of polymerization

$f_{tc}(x)$: distribution function for the degree of polymerization of chains produced by termination via combination

$f'_{tc}(x)$: instantaneous distribution function for the degree of polymerization of chains produced in differential time dt by termination via combination

$f_{rp}(x)$: distribution function for the degree of polymerization of the reactive propagating chains

$f'_{srp,ctp}(x)$: instantaneous distribution function for the degree of polymerization of propagating chains with reactive species on a secondary carbon produced by chain transfer to polymer

$f'_{trp,ctp}(x)$: instantaneous distribution function for the degree of polymerization of propagating chains with reactive species on a tertiary carbon produced by chain transfer to polymer

$f'_{\beta 2}(x)$: instantaneous distribution function for the degree of polymerization of propagating chains or terminated polymer chains produced by β_2 chain scission

$f'_{rp,\beta 3}(x)$: instantaneous distribution function for the degree of polymerization of propagating chains produced by β_3 chain scission

$f'_{tdb,\beta 3}(x)$: instantaneous distribution function for the degree of polymerization of terminated polymer chains with a terminal double bond produced by β_3 chain scission

$F(s)$: s-transform of the distribution function $f(x)$ in the s domain where $s = 1/z$

$F'_{tc}(s)$: s-transform of the instantaneous distribution function for the degree of polymerization of chains produced by termination via combination

$F_{rp}(s)$: s-transform of the distribution function for the degree of polymerization of reactive propagating chains at time t in the reaction

$F_{tp}(s)$: s-transform of the distribution function for the degree of polymerization of terminated polymer chains at time t in the reaction

$F'_{prp}(s)$: s-transform of the instantaneous distribution function for the degree of polymerization of chains produced by propagation reaction

$F'_{rp,ctpha}(s)$: s-transform of the instantaneous distribution function for the degree of polymerization of reactive propagating chains produced by transfer to terminated polymer chains by hydrogen abstraction

$F'_{rp,ctptdb}(s)$: s-transform of the instantaneous distribution function for the degree of polymerization of reactive propagating chains produced by transfer to terminal double bond of terminated polymer chains

$F'_{srp,ctp}(s)$: s-transform of the instantaneous distribution function for the degree of polymerization of propagating chains with the reactive species on a secondary carbon atom produced by transfer to terminated polymer chains

$F'_{\beta 2}(s)$: s-transform of the instantaneous distribution function for the degree of polymerization of propagating chains or terminated polymer chains produced by β_2 chain scission

$F'_{rp,\beta 3}(x)$: s-transform of the instantaneous distribution function for the degree of polymerization of propagating chains produced by β_3 chain scission

$F'_{tdb,\beta 3}(x)$: s-transform of the instantaneous distribution function for the degree of polymerization of terminated polymer chains with a terminal double bond produced by β_3 chain scission

k_p: rate constant for propagation reaction between a propagating chain and a monomer (m^3 $mole^{-1}$ s^{-1})

k_{td}: rate constant for termination of two propagating chains by disproportionation (m^3 $mole^{-1}$ s^{-1})

k_{tc}: rate constant for termination of two propagating chains by combination (m^3 $mole^{-1}$ s^{-1})

k_{cts}: rate constant for chain transfer to solvent in addition polymerization (s^{-1})

k_{pcts}: rate constant for the reaction of an activated solvent molecule ($S^{\bullet}$) with a monomer to form a propagating chain with one repeat unit (m^3 $mole^{-1}$ s^{-1})

k_{cta}: rate constant for transfer of a propagating species to a chain transfer agent in addition polymerization (m^3 $mole^{-1}$ s^{-1})

k_{pcta}: rate constant for the reaction of an activated chain transfer agent ($CTA^{\bullet}$) with a monomer to form a propagating chain with one repeat unit (m^3 $mole^{-1}$ s^{-1})

k_{ctm}: rate constant for chain transfer to monomer (m^3 $mole^{-1}$ s^{-1})

k_{dac}: rate constant for deactivation of propagating chains to form dormant chains in reversible deactivation radical polymerization (m^3 $mole^{-1}$ s^{-1})

k_{ac}: rate constant for activation of dormant chains to form an active propagating chains in reversible deactivation radical polymerization (m^3 $mole^{-1}$ s^{-1})

k_{ctp}: rate constant for chain transfer to polymer (m^3 $mole^{-1}$ s^{-1})

$k_{ctp,ha}$: rate constant for radical transfer to polymer by hydrogen abstraction (m^3 $mole^{-1}$ s^{-1})

k_{ctbb}: rate constant for chain transfer by intramolecular backbiting in addition polymerization (m^3 $mole^{-1}$ s^{-1})

$k_{ctp,tdb}$: rate constant for chain transfer to terminal double bond on a terminated polymer chain to form a long branch (m^3 $mole^{-1}$ s^{-1})

$k_{ctp,pdb}$: rate constant for chain transfer to a pendant double bond along the main chain of a terminated polymer chain to form two long branches (m^3 $mole^{-1}$ s^{-1})

$k_{\beta 2}$: rate constant for β_2 chain scission of an n-mer propagating chain by splitting the bond between α and β carbons, where α is the carbon carrying a secondary reactive species to form two shorter chains (s^{-1})

$k_{\beta 3}$: rate constant for β_2 chain scission of an n-mer propagating chain by splitting the bond between α and β carbons where α is the carbon carrying a tertiary reactive species to form two shorter chains (s^{-1})

LB: a long side chain on a polymer chain

$[M]$: monomer concentration

n: number of repeat units on a polymer chain

PDB_n: n-mer terminated polymer chain with pedant double bonds or a pendant double bond on an n-mer terminated polymer chain participating in chain transfer reaction to polymer

$[R]$: concentration of primary radicals formed by dissociation of an initiator

R_p: rate of chain propagation reaction (mole m^{-3} s^{-1})

RP_1: reactive propagating chain with one repeat unit

RP_n: reactive propagating chain with n repeat units

$[RP_n]$: concentration of reactive propagating chains with n repeat units (mole cm^{-3})

s: dependent variable in the z transfer domain where $s = 1/z$

S: solvent molecule

SB: a short side chain on a polymer chain

$S^{\bullet}$: activated solvent species that reacts with a monomer to form a new propagating chain

SRP_n: reactive propagating chain with the reactive species on a secondary carbon with n repeat units

t: reaction time (s)

TDB: terminal double bond on a polymer chain

TDB_n: n-mer terminated polymer chain with a terminal double bond

TP_n: terminated polymer chain n repeat units

TRP_n: reactive propagating chain with the reactive species on a tertiary carbon with n repeat units

V: reaction volume (cm^3)

x: dependent variable in the domain of degree of polymer (DP)

β_2: reactive propagating chain with the reactive species on a secondary carbon splitting from its neighboring carbon to generate a shorter propagating chain with the reactive species on a primary carbon plus a terminated polymer chain with a terminal double bond

β_3: reactive propagating chain with the reactive species on a tertiary carbon splitting from its neighboring carbon to generate a shorter propagating chain with the reactive species on a primary carbon plus a terminated polymer chain with a terminal double bond

γ'_k: kth order mole-fraction moment of dormant polymer chains produced by deactivation of propagating chains in reversible deactivation radical polymerization

λ_1, λ_2: first and second order mole-fraction moment of reactive propagating chains at time t in the reaction

λ_k: kth order mole-fraction moment of reactive propagating chains at time t in the reaction

λ'_k: instantaneous kth order mole-fraction moment of reactive propagating chains in time interval dt in the reaction

$\lambda'_{1,prp}, \lambda'_{2,prp}$: instantaneous first and second order mole-fraction moment of reactive propagating chains produced by propagation reaction in time interval dt in the reaction

$\lambda'_{1,ctpha}, \lambda'_{2,ctpha}$: instantaneous first and second order mole-fraction moment of reactive propagating chains produced by chain transfer to polymer chains via hydrogen abstraction in time interval dt

$\lambda'_{1,ctptdb}$, $\lambda'_{2,ctptdb}$: instantaneous first and second order mole-fraction moment of reactive propagating chains produced by chain transfer to polymer chains via terminal double bond of the polymer in time interval dt

$\lambda'_{1,ctppdb}$, $\lambda'_{2,ctppdb}$: instantaneous first and second order mole-fraction moment of reactive propagating chains produced by chain transfer to polymer chains via pendant double bonds of the polymer in time interval dt

$\lambda'_{1,ctbb}$, $\lambda'_{2,ctbb}$: instantaneous first and second order mole-fraction moment of reactive propagating chains produced by chain transfer to intramolecular backbiting in time interval dt

$\lambda'_{1,\beta 2}$, $\lambda'_{2,\beta 2}$: instantaneous first and second order mole-fraction moment of reactive propagating chains produced by β_2 chain scission in time interval dt

$\lambda'_{1,\beta 3}$, $\lambda'_{2,\beta 3}$: instantaneous first and second order mole-fraction moment of reactive propagating chains produced by β_3 chain scission in time interval dt

μ_1, μ_2: first and second order mole-fraction moment of terminated polymer chains at time t in the reaction

$\mu'_{k,prp}$: instantaneous kth order mole-fraction moment of terminated polymer chains produced by termination via disproportionation in time interval dt

$\mu'_{1,tc}$, $\mu'_{2,tc}$: instantaneous first and second order mole-fraction moment of terminated polymer chains produced by termination via combination in time interval dt

$\mu'_{k,ctm}$: instantaneous kth order mole-fraction moment of terminated polymer chains produced by chain transfer to monomer in time interval dt

$\mu'_{k,cts}$: instantaneous kth order mole-fraction moment of terminated polymer chains produced by chain transfer to solvent in time interval dt

$\mu'_{k,cta}$: instantaneous kth order mole-fraction moment of terminated polymer chains produced by transfer to chain transfer agent in time interval dt

$\mu'_{k,ctpha}$: instantaneous kth order mole-fraction moment of terminated polymer chains produced by chain transfer to polymer via hydrogen abstraction in time interval dt

$\mu'_{1,\beta 2}$, $\mu'_{2,\beta 2}$: instantaneous first and second order mole-fraction moment of terminated polymer chains produced by β_2 chain scission in time interval dt

$\mu'_{1,\beta 3}$, $\mu'_{2,\beta 3}$: instantaneous first and second order mole-fraction moment of terminated polymer chains produced by β_3 chain scission in time interval dt

$\nu_{1,tp}$, $\nu_{2,tp}$: first and second order weight-fraction moment of terminated polymer chains at time t in the reaction

6.8 Practice Problems

Problem 6.1 In chain transfer to polymer via terminal double bond, the reactive species of a propagating chain RP_n combines with the terminal double bond of a polymer chain TDB_m to form a new propagating chain SRP_{n+m} with the reactive species on a secondary carbon. Next, the SRP_{n+m} chain reacts with a monomer to form an (n+m+1)-mer propagating chain and a long branch. The overall reaction is given by Eq. (6.29). The instantaneous distribution function of RP_{n+m+1} chains in the z-domain in terms of the distribution functions of RP_n and TDB_m is described by the following equation (Eq. 6.30):

$$F'_{rp,ctptdb}(s) = s\,F_{rp}(s)\,F_{tp}(s)$$

Based on the above equation and the defining equations for first and second order mole-fraction moments of chains (see Eqs. 6.4 and 6.5), derive step-by-step expressions for the produced instantaneous first and second order mole-fraction moments of RP_{n+m+1} chains, namely $\lambda'_{1,ctptdb}$ and $\lambda'_{2,ctptdb}$, respectively, in terms of first and second order mole-fraction moments of RP_n chains, namely λ_1 and λ_2, and first and second order mole-fraction moments of TDB_m chains, namely μ_1 and μ_2 starting from the following equations (see Eqs. 6.31 and 6.33):

$$\lambda'_{1,ctptdb} = \left.\frac{\partial F'_{rp,ctptdb}(s)}{\partial s}\right|_{s=1}$$

$$\lambda'_{2,ctptdb} = \left.\frac{\partial^2 F'_{rp,ctptdb}(s)}{\partial s^2}\right|_{s=1} + \left.\frac{\partial^2 F'_{rp,ctptdb}(s)}{\partial s^2}\right|_{s=1}$$

Problem 6.2 A β_2 scission consists of two reactions. In the first reaction, the reactive species is transferred from a primary carbon to a secondary carbon by chain transfer to polymer chain TP_{x+y} via hydrogen abstraction. In the second reaction, the unstable carbon having a reactive species on a secondary carbon of the SRP_{x+y} chain produced in the first reaction splits from its neighboring carbon to generate a propagating chain RP_x with the reactive species on a more stable primary carbon plus a terminated polymer chain TDB_y. The two reactions are given by Eqs. (6.46) and (6.47). If $F_{tp}(s)$ is the distribution function of polymer chains in the z-domain, $F'_{srp,ctp}(s)$ is the instantaneous distribution function of the SRP_{x+y} chains in the z-domain produced by chain transfer to polymer, the instantaneous distribution function of the propagating RP_x or terminated TDB_y chains in the z-domain after scission, namely $F'_{\beta 2}(s)$, is given by the following equation (see Eq. 6.52):

$$F'_{\beta 2}(s) = \left[F'_{srp,ctp}(s)\right]^{1/2} = \left[F_{tp}(s)\right]^{1/2}$$

Based on the above relations and the defining Equations for the second order mole-fraction moment of chains (see Eq. 6.5), derive step-by-step an expression for the instantaneous second order mole-fraction moments of the propagating RP_x or terminated TDB_y chains, namely $\lambda'_{2,\beta 2}$ and $\mu'_{2,\beta 2}$, in terms of first and second order mole-fraction moments of RP_n chains, namely λ_1 and λ_2, and first and second order mole-fraction moments of TP_{x+y} chains, namely μ_1 and μ_2 starting from the following equation (see Eq. 6.54):

$$\lambda'_{2,\beta 2} = \mu'_{2,\beta 2} = \left.\frac{\partial^2 F'_{\beta 2}(s)}{\partial s^2}\right|_{s=1} + \left.\frac{\partial F'_{\beta 2}(s)}{\partial s}\right|_{s=1}$$

Problem 6.3 In the first reaction in β_3 scission, the reactive species is transferred from a primary carbon on a propagating chain RP_n to a tertiary carbon on a propagating chain TRP_{x+y+z} by chain transfer to polymer chain TP_m. In the second reaction, the unstable carbon having a reactive species on a tertiary carbon of the TRP_{x+y+z} chain splits from its neighboring carbon to generate a propagating chain RP_z with the reactive species on a more stable primary carbon plus a terminated polymer chain TDB_{x+y}. The two reactions are given by Eqs. (6.55) and (6.56). If the tertiary carbon is part of a long branch of a polymer chain and if $F_{tp}(s)$ is the distribution function of TP_{x+y+z} polymer chains in the reaction in the z-domain, and if $F'_{trp,ctp}(s)$ is the instantaneous distribution function of the TRP_{x+y+z} chains in the z-domain produced by chain transfer to polymer in time interval dt, then, the instantaneous distribution function of the propagating RP_z chains and the terminated TDB_{x+y} polymer chains with a terminal double bond in the z-domain after scission, namely $F'_{\beta 3,rp}(s)$ and $F'_{\beta 3,tdb}(s)$, are given by the following equations (see Eqs. 6.65 and 6.66):

$$F'_{\beta 3,rp}(s) = \left[F'_{trp,ctp}(s)\right]^{1/3} = \left[F_{tp}(s)\right]^{1/3}$$

$$F'_{\beta 3,tdb}(s) = \left[F'_{trp,ctp}(s)\right]^{2/3} = \left[F_{tp}(s)\right]^{2/3}$$

Based on the above equations and the defining equations for the second order mole-fraction moment of the chains (see Eq. 6.5), derive step-by-step expressions for the instantaneous second order mole-fraction moments of the propagating RP_z and terminated TDB_{x+y} chains, namely $\lambda'_{2,\beta 2}$ and $\mu'_{2,\beta 2}$, in terms of first and second order mole-fraction moments of RP_n chains, namely λ_1 and λ_2, and first and second order mole-fraction moments of TP_{x+y+z} chains, namely μ_1 and μ_2 starting from the following equations (see Eqs. 6.69 and 6.70):

$$\lambda'_{2,\beta 3} = \left.\frac{\partial^2 F'_{\beta 3,rp}(s)}{\partial s^2}\right|_{s=1} + \left.\frac{\partial F'_{\beta 3,rp}(s)}{\partial s}\right|_{s=1}$$

$$\mu'_{2,\beta 3} = \left.\frac{\partial^2 F'_{\beta 3,tdb}(s)}{\partial s^2}\right|_{s=1} + \left.\frac{\partial F'_{\beta 3,tdb}(s)}{\partial s}\right|_{s=1}$$

References

1 Nikitin, A.N., Castignolles, P., Charleux, B., and Vairon, J.P., Determination of propagation rate coefficient of acrylates by pulsed-laser polymerization in the presence of intramolecular chain transfer to polymer. *Macromolecular Rapid Communications*, 2003, **24**(13), 778.

2 Berger, K.C., Disproportionation and combination as termination mechanisms in radical polymerization of styrene. 2. Analysis of temperature dependencies. *Makromolekulare Chemie-Macromolecular Chemistry and Physics*, 1975, **176**(12), 3575.

3 Berger, K.C. and Meyerhoff, G., Disproportionation and combination as termination mechanisms in radical polymerization of styrene. 1. Experiments with C-14-labeled 2,2'-azoisobutyronitrile. *Makromolekulare Chemie-Macromolecular Chemistry and Physics*, 1975, **176**(7), 1983.

4 Hatada, K., Kitayama, T., and Masuda, E., Evidence for disproportionation in termination reaction of styrene polymerization by α,α'-azobisisobutyronitrile. *Polymer Journal*, 1985, **17**(8), 985.

5 Ribelli, T.G. et al., Disproportionation or combination? The termination of acrylate radicals in ATRP. *Macromolecules*, 2017, **50**(20), 7920.

6 Yan, D.Y. and Cai, G.F., Influence of termination and transfer on molecular-weight distribution of polymers. 1. Disproportionation and combination. *Journal of Macromolecular Science A-Chemistry*, 1986, **23**(1), 37.

7 Cai, G.F., Yan, D.Y., and Litt, M., A kinetic-theory on cationic polymerization with termination by combination with a negative fragment. *Macromolecules*, 1988, **21**(3), 578.

8 Fazekas, T.J. et al., Diversification of aliphatic C–H bonds in small molecules and polyolefins through radical chain transfer. *Science*, 2022, **375**(6580), 545.

9 Kukulj, D., Davis, T.P., and Gilbert, R.G., Chain transfer to monomer in the free-radical polymerizations of methyl methacrylate, styrene, and alpha-methylstyrene. *Macromolecules*, 1998, **31**(4), 994.

10 Magee, C., Sugihara, Y., Zetterlund, P.B., and Aldabbagh, F., Chain transfer to solvent in the radical polymerization of structurally diverse acrylamide monomers using straight-chain and branched alcohols as solvents. *Polymer Chemistry*, 2014, **5**(7), 2259.

11 Balic, R., Fellows, C.M., and van Herk, A.M., Chain transfer to monomer and polymer in the radical polymerization of vinyl neo-decanoate. *Macromolecular Research*, 2004, **12**(4), 325.

12 Mayo, F.R., Chain transfer in the polymerization of styrene: the reaction of solvents with free radicals. *Journal of the American Chemical Society*, 1943, **65**, 2324.

13 Tan, S.B., Li, J.J., and Zhang, Z.C., Study of chain transfer reaction to solvents in the initiation stage of atom transfer radical polymerization. *Macromolecules*, 2011, **44**(20), 7911.

14 Sugihara, Y., O'Connor, P., Zetterlund, P.B., and Aldabbagh, F., Chain transfer to solvent in the radical polymerization of N-isopropylacrylamide. *Journal of Polymer Science Part a-Polymer Chemistry*, 2011, **49**(8), 1856.

15 Ji, Y.Y. and Klausen, R.S., Chain transfer to solvent in BN 2-vinylnaphthalene radical polymerization. *Journal of Polymer Science*, 2021, **59**(21), 2521.

16 Starnes, W.H. et al., Branch structures in poly(vinyl-chloride) and the mechanism of chain transfer to monomer during vinyl-chloride polymerization. *Macromolecules*, 1983, **16**(5), 790.

17 Obrien, J.L. and Gornick, F., Chain transfer in the polymerization of methyl methacrylate. 1. Transfer with monomer and thiols - the mechanism of the termination reaction at 60-degrees. *Journal of the American Chemical Society*, 1955, **77**(18), 4757.

18 Yu, Y., Fu, Z.S., and Fan, Z.Q., Chain transfer reactions of propylene polymerization catalyzed by $AlEt_3$ activated $TiCl_4/MgCl_2$ catalyst under very low monomer addition rate. *Journal of Molecular Catalysis a-Chemical*, 2012, **363**, 134.

19 Bukatov, G.D., Zakharov, V.A., and Barabanov, A.A., Mechanism of olefin polymerization on supported Ziegler-Natta catalysts based on data on the number of active centers and propagation rate constants. *Kinetics and Catalysis*, 2005, **46**(2), 166.

20 Prescott, S.W., Chain-length dependence in living/controlled free-radical polymerizations: physical manifestation and Monte Carlo simulation of reversible transfer agents. *Macromolecules*, 2003, **36**(25), 9608.

21 Moad, G. and Solomon, D.H., Invited review - understanding and controlling radical polymerization. *Australian Journal of Chemistry*, 1990, **43**(2), 215.

22 Moad, G., A critical survey of dithiocarbamate reversible addition-fragmentation chain transfer (RAFT) agents in radical polymerization. *Journal of Polymer Science Part A-Polymer Chemistry*, 2019, **57**(3), 216.

23 Ahmad, N.M. et al., Chain transfer to polymer and branching in controlled radical polymerizations of n-butyl acrylate. *Macromolecular Rapid Communications*, 2009, **30**(23), 2002.

24 Ahmad, N.M., Heatley, F., and Lovell, P.A., Chain transfer to polymer in free-radical solution polymerization of n-butyl acrylate studied by NMR spectroscopy. *Macromolecules*, 1998, **31**(9), 2822.

25 Britton, D., Heatley, F., and Lovell, P.A., Chain transfer to polymer in free-radical bulk and emulsion polymerization of vinyl acetate studied by NMR spectroscopy. *Macromolecules*, 1998, **31**(9), 2828.

26 Britton, D., Heatley, F., and Lovell, P.A., Effect of monomer feed rate on chain transfer to polymer in semibatch emulsion polymerization of vinyl acetate studied by NMR spectroscopy. *Macromolecules*, 2000, **33**(14), 5048.

27 Iedema, P.D., Grcev, S., and Hoefsloot, H.C.J., Molecular weight distribution modeling of radical polymerization in a CSTR with long chain branching through transfer to polymer and terminal double bond (TDB) propagation. *Macromolecules*, 2003, **36**(2), 458.

28 Huang, W.Y. et al., Studies on the branching atom transfer radical polymerization using different divinyl monomers as the branching agent. *Acta Polymerica Sinica*, 2011, (11), 1253.

29 Huang, W.Y. et al., Branching copolymerization of styrene and methyl methacrylate with divinylbenzene. *Iranian Polymer Journal*, 2010, **19**(8), 589.

30 Bannister, I. et al., Development of branching in living radical copolymerization of vinyl and divinyl monomers. *Macromolecules*, 2006, **39**(22), 7483.

31 Triacca, V.J. et al., Free-radical degradation of polypropylene - random chain scission. *Polymer Engineering and Science*, 1993, **33**(8), 445.

32 Pinheiro, L.A., Chinelatto, M.A., and Canevarolo, S.V., The role of chain scission and chain branching in high density polyethylene during thermo-mechanical degradation. *Polymer Degradation and Stability*, 2004, **86**(3), 445.

33 Kumagai, T. et al., Specific polymerization mechanism involving beta-scission of mid-chain radical yielding oligomers in the free-radical polymerization of vinyl ethers. *Macromolecules*, 2008, **41**(20), 7347.

34 Lena, J.B. et al., Effects of chain transfer agent and temperature on branching and beta-scission in radical polymerization of 2-ethylhexyl acrylate. *Macromolecular Chemistry and Physics*, 2018, **219**(9), 1700579.

35 Tobita, H., Markovian approach to free-radical polymerization with simultaneous long-chain branching and scission: effect of branching and scission kinetics. *Macromolecular Reaction Engineering*, 2015, **9**(3), 245.

36 Tobita, H., Free-radical polymerization with long-chain branching and scission: Markovian solution of the weight-average molecular weight. *Macromolecular Theory and Simulations*, 2014, **23**(8), 477.

37 Junkers, T. and Barner-Kowollik, C., The role of mid-chain radicals in acrylate free radical polymerization: branching and scission. *Journal of Polymer Science Part A-Polymer Chemistry*, 2008, **46**(23), 7585.

38 Szymanski, R. and Baran, J., Molecular weight distribution in living polymerization proceeding with reshuffling of polymer segments due to chain transfer to polymer with chain scission, 2(a) - Monte Carlo simulation of polymer reshuffling proceeding with disproportionation of chain functionalities. *Macromolecular Theory and Simulations*, 2002, **11**(8), 836.

39 Iedema, P.D. and Hoefsloot, H.C.J., The influence of branching architecture on scission in modeling free-radical polymerization with long chain branching and scission. *Chimia*, 2001, **55**(3), 256.

40 Ito, K., Relationship between kinetic chain-length and radical polymerization rate. *Journal of Polymer Science Part A-Polymer Chemistry*, 1976, **14**(7), 1761.

41 Ballard, N., Hamzehlou, S., Ruiperez, F., and Asua, J.M., On the termination mechanism in the radical polymerization of acrylates. *Macromolecular Rapid Communications*, 2016, **37**(16), 1364.

42 Nakamura, Y., Lee, R., Coote, M.L., and Yamago, S., Termination mechanism of the radical polymerization of acrylates. *Macromolecular Rapid Communications*, 2016, **37**(6), 506.

43 Shen, J.C., Tian, Y., Wang, G.B., and Yang, M.L., Termination process and the termination parameter-Kt of methyl-methacrylate bulk radical polymerization. *Science in China Series B-Chemistry*, 1990, **33**(9), 1040.

44 Zhu, S., Tian, Y., Hamielec, A.E., and Eaton, D.R., Radical trapping and termination in free-radical polymerization of methyl methacrylate. *Macromolecules*, 1990, **23**(4), 1144.

45 Gregg, R.A. and Mayo, F.R., Chain transfer in the polymerization of styrene. 7. Compounds containing halogens, oxygen and nitrogen. *Journal of the American Chemical Society*, 1953, **75**(14), 3530.

46 Furuncuoglu, T., Ugur, I., Degirmenci, I., and Aviyente, V., Role of chain transfer agents in free radical polymerization kinetics. *Macromolecules*, 2010, **43**(4), 1823.

47 Gatzke, A.L., Advances in polymer science. *Journal of Polymer Science Part A-1-Polymer Chemistry*, 1969, **7**(8), 2281.

48 Peng, L., Haucourt, N., and Goethals, E.J., Acetals as chain transfer agents in the cationic polymerization of vinyl ethers. *Macromolecular Rapid Communications*, 1994, **15**(3), 197.

49 Benedicto, A.D., Claverie, J.P., and Grubbs, R.H., Molecular-weight distribution of living polymerization involving chain-transfer agents - computational results, analytical solutions, and experimental investigations using ring-opening metathesis polymerization. *Macromolecules*, 1995, **28**(2), 500.

50 Harrisson, S., The chain length distribution of an ideal reversible deactivation radical polymerization. *Polymers*, 2018, **10**(8), 887.

51 Charleux, B., Nicolas, J., and Guerret, O., Theoretical expression of the average activation-deactivation equilibrium constant in controlled/living free-radical copolymerization operating via reversible termination. Application to a strongly improved control in nitroxide-mediated polymerization of methyl methacrylate. *Macromolecules*, 2005, **38**(13), 5485.

52 Tang, H.D. et al., Highly active copper-based catalyst for atom transfer radical polymerization. *Journal of the American Chemical Society*, 2006, **128**(50), 16277.

53 Otsu, T., Iniferter concept and living radical polymerization. *Journal of Polymer Science Part A-Polymer Chemistry*, 2000, **38**(12), 2121.

54 Otsu, T. and Yoshida, M., Role of initiator-transfer agent-terminator (iniferter) in radical polymerizations: polymer design by organic disulfides as iniferters. *Macromolecular Rapid Communications*, 2005, **26**(3), 137.

55 Moad, G. et al., Advances in RAFT polymerization: the synthesis of polymers with defined end-groups. *Polymer*, 2005, **46**(19), 8458.

56 Chiefari, J. et al., Living free-radical polymerization by reversible addition-fragmentation chain transfer: the RAFT process. *Macromolecules*, 1998, **31**(16), 5559.

57 Hawthorne, D.G., Moad, G., Rizzardo, E., and Thang, S.H., Living radical polymerization with reversible addition-fragmentation chain transfer (RAFT): direct ESR observation of intermediate radicals. *Macromolecules*, 1999, **32**(16), 5457.

58 Yin, H. et al., Selection principle of RAFT chain transfer agents and universal RAFT chain transfer agents. *Progress in Chemistry*, 2022, **34**(6), 1298.

59 Heatley, F., Lovell, P.A., and Yamashita, T., Chain transfer to polymer in free-radical solution polymerization of 2-ethylhexyl acrylate studied by NMR spectroscopy. *Macromolecules*, 2001, **34**(22), 7636.

60 Ishihara, Y. et al., Back biting reactions during the catalytic decomposition of polyethylene. *Bulletin of the Chemical Society of Japan*, 1991, **64**(12), 3585.

61 Elliott, J.R., Myers, R.L., and Roedel, G.F., Suspension polymerization of chlorotrifluoroethylene. *Industrial and Engineering Chemistry*, 1953, **45**(8), 1786.

62 McCord, E.F., Shaw, W.H., and Hutchinson, R.A., Short-chain branching structures in ethylene copolymers prepared by high-pressure free-radical polymerization: an NMR analysis. *Macromolecules*, 1997, **30**(2), 246.

63 Qian, S.Y., Igarashi, T., and Nitta, K., Thermal degradation behavior of polypropylene in the melt state: molecular weight distribution changes and chain scission mechanism. *Polymer Bulletin*, 2011, **67**(8), 1661.

64 Wollersheim, O. et al., In situ ESR investigations of synchrotron radiation-induced degradations of polymers. *Radiation Physics and Chemistry*, 1999, **55**(1), 103.

65 Torikai, A., Photoinduced and radiation-induced degradation of synthetic-polymers - polymer structure and stability. *Angewandte Makromolekulare Chemie*, 1994, **216**, 225.

66 Wilski, H., The radiation-induced degradation of polymers. *Radiation Physics and Chemistry*, 1987, **29**(1), 1.

67 Torikai, A., Geetha, R., Nagaya, S., and Fueki, K., Radiation-induced degradation of polyethylene - polymer structure and stability. *Journal of Polymer Science Part A-Polymer Chemistry*, 1990, **28**(13), 3639.

68 Messmer, D. et al., Main-chain scission of individual macromolecules induced by solvent swelling. *Chemical Science*, 2019, **10**(24), 6125.

69 Vanapalli, S.A., Ceccio, S.L., and Solomon, M.J., Universal scaling for polymer chain scission in turbulence. *Proceedings of the National Academy of Sciences of the United States of America*, 2006, **103**(45), 16660.

70 Buchholz, B.A. et al., Flow-induced chain scission as a physical route to narrowly distributed, high molar mass polymers. *Polymer*, 2004, **45**(4), 1223.

71 Horn, A.F. and Merrill, E.W., Extensional flow-induced chain scission of macromolecules in dilute-solution - effect of solvent power in a mixed-solvent system. *Polymer Communications*, 1987, **28**(6), 172.

72 Baran, J. et al., Intermolecular chain transfer to polymer with chain scission: general treatment and determination of k(p)/k(tr) in L,L-lactide polymerization. *Macromolecular Rapid Communications*, 1997, **18**(4), 325.

73 Penczek, S., Kinetics of chain transfer to monomer and intramolecular termination in cationic polymerization of alpha-epichlorohydrin. *Bulletin De L Academie Polonaise Des Sciences-Serie Des Sciences Chimiques*, 1970, **18**(1), 39.

7

Conservation of Moments Applied to Polymerization Reactions

7.1 Introduction

This chapter applies the methods developed in previous chapters to mathematically model radical polymerization and predict concentrations of reactants and products, average chain length, and polydispersity of the produced polymer chains with reaction time. The equations of conservation of moments of polymer chains described in Chapter 5 will be used to relate the kth moments of propagating and terminated chains at time t in the reaction to the instantaneous kth moments of chains produced or consumed via different elementary reactions. The z-transform methodology described in Chapter 6 will be used to relate the instantaneous kth moment of chains produced or consumed via different elementary reactions to the moments of propagating and terminated chains to solve the conservation of moments equations. The equations described in Chapter 4 will be used to determine different molecular weights of the produced polymer chains from the kth moments. Most polymerization systems modeled in this chapter are based on batch reactions [1–5]. It is also assumed that polymerization is not controlled by diffusion of reactants but controlled by the intrinsic rates of polymerization reactions [6–15]. As the reaction systems in this chapter are not diffusion controlled, the assumption of quasi-steady state (QSSA) for reactive species is applied to the modeled polymerization systems [6, 16–18]. QSSA assumes reactive species do not accumulate in the system, or reactive species produced in a differential time step at any time in the reaction are consumed in the same differential time step. QSSA is applied to both small molecule primary radicals produced by initiator dissociation as well as reactive propagating chains. Modeling of diffusion-controlled polymerization reactions will be covered in the next chapter. The polymerization systems modelled include chemically initiated radical polymerization with or without chain transfer, polymerization with transfer to intentionally added chain transfer agent, polymerization with transfer to polymer chains via hydrogen abstraction, atom-transfer radical polymerization, ionic polymerization, anionic and cationic polymerization, polymerization with chain scission, and photoinitiated polymerization.

7.2 General Conservation Equations for kth Order Moment of Propagating, Dormant, and Terminated Chains

The general moment conservation equations for the chains in a polymerization reaction were derived in Section 5.5 in Chapter 5 in the Cartesian coordinate system and given by Eq. (5.14A) for propagating chains, Eq. (5.17) for terminated chains, Eq. (5.18) for dormant chains, and Eq. (5.19) for small-molecule reactants. This chapter applies the general conservation equations to modeling polymerizations that are kinetically controlled (i.e., not limited by molecular diffusion) in the absence of convective mass transfer and those polymerizations for which the quasi-steady state assumption (QSSA) for all reactive propagating species (i.e., propagating small molecules and chains) is valid. Diffusion controlled polymerizations will be covered in the next chapter. The QSSA assumption implies that the propagating species do not accumulate in the reaction, thus $\frac{d\left(\lambda_k Q_{0,rp}\right)}{dt} = 0$. Under the above assumptions, the conservation equations for kth order molar concentration moment of propagating, dormant, and terminated chains, and small molecules simplify to the following expressions:

$$\sum_i \lambda'_{k,i} R_{rp,prd,i} - \sum_l \lambda'_{k,l} R_{rp,con,l} = 0 \tag{7.1A}$$

$$\frac{d\left(\gamma_k Q_{0,dop}\right)}{dt} = \sum_i \gamma'_{k,i} R_{dop,prd,i} - \sum_l \gamma'_{k,l} R_{dop,con,l} \tag{7.1B}$$

$$\frac{d\left(\mu_k Q_{0,tp}\right)}{dt} = \sum_i \mu'_{k,i} R_{tp,prd,i} - \sum_l \mu'_{k,l} R_{tp,con,l} \tag{7.1C}$$

$$\frac{dC_{sm}}{dt} = \sum_i R_{sm,prd,i} - \sum_l R_{sm,con,l} \tag{7.1D}$$

where γ_k and μ_k are the cumulative (at time t) kth order mole-fraction moment of dormant and terminated chains, respectively; the prime in $\lambda'_{k,i}$, $\gamma'_{k,i}$, $\mu'_{k,i}$ indicates the instantaneous kth order moment of the propagating, dormant, and terminated chains produced/consumed by a reaction; $Q_{0,dop}$, $Q_{0,tp}$, and C_{sm} are the molar concentrations of dormant and polymer chains and the small molecule; and R_{rp}, R_{dop}, R_{tp}, and R_{sm} are the rate of production/consumption of the propagating, dormant, and terminated chains and the small-molecule by a reaction.

7.3 Modeling Radical Polymerization with Termination by Disproportionation in the Absence of Chain Transfer

This is the simplest case of radical polymerization. The mechanism of termination (disproportionation versus combination) depends, among others, on monomer and solvent type, size and chemistry of the monomer's side group, reaction temperature and impurities, reaction viscosity, and degree of polymerization [19–25]. As an example, the predominant mechanism of chain termination in the polymerization of acrylates, like methyl methacrylate, at ambient temperature is by disproportionation but termination by disproportionation decreases with increasing temperature [20]. Conversely, propagating chains in polymerization of styrene terminate predominantly by combination due to the favorable energetic π–π interaction between the neighboring phenyl side chains [21]. Termination by disproportionation is favored over combination at high reaction viscosities for both methyl methacrylate and styrene [25]. The elementary reactions for this polymerization system are as follows:

$$I \xrightarrow{k_d} 2R \tag{7.2A}$$

$$R + R \xrightarrow{k_c} I^* \tag{7.2B}$$

$$R + M \xrightarrow{k_{ci}} RP_1 \tag{7.2C}$$

$$RP_n + M \xrightarrow{k_p} RP_{n+1} \tag{7.2D}$$

$$RP_n + RP_m \xrightarrow{k_{td}} TP_n + TP_m + TDB \tag{7.2E}$$

In the above reactions, I, R, and I^* are the initiator, primary radical, and the inactive fragment of radical combination, respectively; M, RP_n, and TP_n are the monomer, n-mer reactive propagating chain, and n-mer terminated chain; and TDB is a terminal double bond in a chain terminated by disproportionation. Rate constants k_d, k_c, k_{ci}, k_p, and k_{td} represent rates of initiator dissociation, primary radical combination, chain initiation, propagation, and chain termination by disproportionation, respectively. Reactions 7.2A–E correspond to the initiator dissociation to form primary reactive species, primary reactive species combination which reduces initiation efficiency, chain initiation by addition of the first monomer, chain propagation to produce longer chains, and chain termination to produce an unreactive terminated chain (polymer product). The termination by disproportionation generates a TDB at one chain end of one of the terminated chains. This model assumes negligible chain transfer to monomer and polymer, no chain scission, and termination solely by disproportionation.

Based on the kinetic theory of elementary reactions [26–28], the rate of initiator dissociation is $R_d = k_d\,[I]$ where $[I]$ is the molar initiator concentration; the rate of primary radical recombination is $R_c = k_c\,[R]^2$ where $[R]$ is the molar concentration of primary radicals; the rate of chain initiation reaction is $R_{ci} = k_{ci}\,[R][M]$ where $[M]$ is the molar monomer concentration; the rate of chain propagation is $R_p = k_p\,[RP][M]$ where $[RP]$ is the molar concentration of propagating chains; and the rate of chain termination by disproportionation is $R_{td} = k_{td}\,[RP]^2$.

7.3.1 Modeling Radical Polymerization with Termination by Disproportionation in a Batch Stirred Vessel Reactor (BSVR)

In BSVR, the reactor is closed after the reactants are charged in the reactor. There are no streams entering or exiting the reactor, and the reaction is well mixed. Therefore, the concentration of reactants and products is a function of time and independent of space. Upon applying the conservation equation for small molecules in the reaction (Eq. 7.1D) to the initiator, the differential equation for time dependence of the initiator concentration is given by the following equation:

$$\frac{d[I]}{dt} = -k_d\,[I] \tag{7.3A}$$

where $[I]_0$ is the initial initiator concentration and t is time. Upon integration from $[I] = [I]_0$ at $t = 0$ to time t, the time dependence of initiator concentration is given as follows:

$$[I] = [I]_0\, e^{-k_d t} \tag{7.3B}$$

Upon applying the conservation equation for the reactive species in the reaction (Eq. 7.1A) to the primary reactive species R, assuming validity of QSSA, the R concentration is related to the monomer and initiator concentrations by the following equation:

$$[R] = \frac{2k_d f_0\,[I]}{k_{ci}\,[M]} = \left[\frac{2\,k_d\,(1 - f_0)\,[I]}{k_c}\right]^{1/2} \tag{7.4A}$$

where f_0 is the initiator efficiency (see Eq. 5.29E in Chapter 5). According to the above equation, $[R]$ at a given time t in the reaction depends on initiator to monomer ratio $\left(\frac{[I]}{[M]}\right)$. Increasing $[M]$ or reducing $[I]$ will reduce the primary radical concentration. As there are many elementary reactions in radical polymerization that deactivate the primary reactive species, like recombination, reaction with oxygen, inhibitor, retarder, and other impurities [29–39], f_0 is rarely equal to one. As discussed in Chapter 5, if $f_0 = 1$, then $k_{ci}[R]\,[M] = rk_d\,[I]$, and all primary reactive species are instantaneously converted to reactive propagating chains (RP). Upon substitution for $[I]$ from Eq. (7.3B) in the above equation, the time dependence of $[R]$ is given by the following equation:

$$[R] = \frac{2k_d f_0\,[I]}{k_{ci}\,[M]} = \left[\frac{2k_d\,(1-f_0)\,[I]_0}{k_c}\right]^{1/2} e^{\frac{-k_d t}{2}} \quad f_0 < 1 \tag{7.4B}$$

The concentration of primary radicals decreases exponentially with time in response to the exponential decrease in initiator concentration.

7.3.1.1 Time Dependence of Concentration of Propagating Chains in Radical Polymerization with Termination by Disproportionation

The reactive propagating chains RP have a distribution of degree of polymerization from $n = 1$ to $n = \infty$. Upon applying the conservation equation for the reactive species in the reaction (Eq. 7.1A) to the zero-order moment of the propagating chains RP, assuming validity of QSSA, the RP concentration, $Q_{0,rp}$, is given by the following equation for this batch radical polymerization system:

$$\sum\nolimits_i \lambda'_{0,i} R_{rp,prd,i} = \sum\nolimits_l \lambda'_{0,l} R_{rp,con,l} \tag{7.5A}$$

where λ_0 and $Q_{0,rp}$ are the zero-order mole-fraction moment and the concentration of the propagating chains at time t, respectively; $\lambda'_{0,i}$ and $\lambda'_{0,l}$ are the instantaneous zero-order mole-fraction moment of the produced and consumed propagating chains by the ith and lth elementary reactions in the time interval dt; and $R_{rp,prd,i}$ and $R_{rp,con,l}$ are the ith and lth reaction rates for the production and consumption of RP. As the zero-order mole-fraction moment, by definition, is equal to one, that is, $\lambda'_{0,i} = \lambda'_{0,l} = 1$, the above equation simplifies as follows:

$$\sum\nolimits_i R_{rp,prd,l} = \sum\nolimits_l R_{rp,con,l} \tag{7.5B}$$

In Rxns 7.2A–E, the elementary reactions to produce the propagating chains are Rxns 7.2C, D and those to consume are Rxns 7.2D, E. Therefore, the left and right sides of the above equation are given by the following equations:

$$\sum\nolimits_i R_{rp,prd,l} = k_{ci}\,[M]\,[R] + k_p\,[M]\,[RP] \tag{7.5C}$$

$$\sum\nolimits_l R_{rp,con,l} = k_{td}[RP]^2 + k_p\,[RP]\,[M] \tag{7.5D}$$

where $[RP] = \sum_{j=1}^{\infty} [RP_j]$ and $[RP_j]$ in $\sum [RP_j]$ is the concentration of j-mer propagating chains. Upon substitution of the above relations in Eq. (7.5B), $[RP]$ is related to monomer and primary radical concentrations by the following equation:

$$[RP] = \left(\frac{k_{ci}\,[M][R]}{k_{td}}\right)^{\frac{1}{2}} \tag{7.5E}$$

Substitution for k_{ci} $[M]$ $[R]$ from Eq. (7.4B) and for $[I]$ in terms of time from Eq. (7.3B) results in the following expression relating $[RP]$ to time:

$$[RP] = \left(\frac{2k_d f_0 [I]}{k_{td}}\right)^{\frac{1}{2}} = \left[\frac{2k_d f_0 [I]_0}{k_{td}}\right]^{1/2} e^{-\frac{k_d t}{2}} \tag{7.5F}$$

when $f_0 = 1$, $[RP]$ is maximum; hence, according to Eq. (7.4A), the concentration of primary radical species $[R] = 2k_d [I]/k_{ci} [M]$. This implies that the primary radicals totally convert to propagating chains when initiation efficiency is equal to one.

7.3.1.2 Time Dependence of Monomer Concentration in Radical Polymerization with Termination by Disproportionation

As the monomer is consumed during the polymerization, its concentration decreases with time in a BSVR. Upon applying the conservation equation for small molecules in the reaction (Eq. 7.1D) to the monomer, the differential equation for time dependence of the monomer concentration is given by the following equation:

$$\frac{d[M]}{dt} = \sum_i R_{m,prd,i} - \sum_l R_{m,con,l} \tag{7.6A}$$

where $R_{m,prd,i}$ and $R_{m,con,l}$ are the rates of monomer production and consumption by reactions i and l respectively. In Rxns 7.2A–E, no monomer is produced, and monomer is consumed by Rxns. 7.2C, D. Therefore, the equations for monomer production and consumption are given by the following equations:

$$\sum_i R_{m,prd,i} = 0 \tag{7.6B}$$

$$\sum_i R_{m,con,i} = (k_{ci} [M] [R] + k_p [M] [RP]) \tag{7.6C}$$

Upon substitution for the production and consumption terms in Eq. (7.6A) from the above equations followed by the separation of variables results in the following differential equation:

$$\frac{d[M]}{[M]} = -(k_p [RP] + k_{ci} [R]) \, dt \tag{7.6D}$$

The term k_p $[RP]$ on the right side of the above equation must be much larger than k_{ci} $[R]$ to form polymer chains, that is k_p $[RP] >> k_{ci}$ $[R]$. Upon neglecting k_{ci} $[R]$ to produce polymer chains followed by substitution for $[RP]$ from Eq. (7.5F) results in the following differential equation:

$$\frac{d[M]}{[M]} = -k_p \left(\frac{2k_d f_0 [I]_0}{k_{td}}\right)^{1/2} e^{-\frac{k_d t}{2}} dt \tag{7.6E}$$

Integration of the above equation from $[M] = [M]_0$ at $t = 0$ to $[M]$ at time t results in the following relation for time dependence of monomer concentration:

$$\int_{[M]_0}^{[M]} d[M] / [M] = -k_p (2k_d f_0 [I]_0/k_{td})^{1/2} \int_{t=0}^{t} e^{-k_d t/2} dt$$
$$[M] = [M]_0 \exp\left[-(8k_p^2 f_0 [I]_0/k_d k_{td})^{1/2} (1 - e^{-k_d t/2})\right] \tag{7.6F}$$

If the double exponential term on the right side is expanded for short times using the Taylor series and using the first two terms of the expansion series, that is $e^{-x} \approx 1 - x$, the above equation simplifies to the following form for short reaction times:

$$[M] = [M]_0 \left[1 - \left(\frac{2k_p^2 k_d f_0 [I]_0}{k_{td}} \right)^{1/2} t \right] \tag{7.6G}$$

If reaction conditions such as temperature, pH, solvent, co-initiator concentration, initiator type, and concentration are chosen such that the rate of initiator dissociation compared to propagation is slow, the change in $[I]$ with time is negligible, then $[R]$ and $[RP]$ can be assumed to be constant. Therefore, $[I] \approx [I]_0$, the time dependence of $[M]$ is given by the following simplified equation:

$$[I] \approx [I]_0 \Rightarrow In \left(\frac{[M]}{[M]_0} \right) = -k_p \left(\frac{2k_d f_0 [I]_0}{k_{td}} \right)^{1/2} t \tag{7.6H}$$

The above equation is commonly used to study kinetics of batch radical polymerization as the relation between $ln\left(\frac{[M]}{[M]_0}\right)$ versus time is linear and the rate constant for propagation, k_p, can be determined from the slope of the linear line [40, 41].

7.3.1.3 First-order Moment of Propagating Chains in Radical Polymerization with Termination by Disproportionation

The number-average chain length or the degree of polymerization of the propagating chains $\left(\overline{DP}_{n,rp}\right)$ is determined from the first order mole-fraction moment of the propagating chains (λ_1). Upon applying the conservation equation for the reactive species in the reaction (Eq. 7.1A) to the first order mole-fraction moment of the propagating chains, assuming the validity of QSSA, the conservation equation for λ_1 is given by the following equation:

$$\sum_i \lambda'_{1,i} R_{rp,prd,i} = \sum_l \lambda'_{1,l} R_{rp,con,l} \tag{7.7A}$$

where $\lambda'_{1,i}$ and $\lambda'_{1,l}$ are the instantaneous first order mole-fraction moment of the produced and consumed propagating chains by the ith and lth elementary reaction in the time interval dt, respectively. Propagating chains are produced by Rxns. 7.2C, D with their production rates given by Eq. (7.5C). Propagating chains are consumed by Rxns 7.2D, E with their consumption rates given by Eq. (7.5D). Substitution from Eqs. (7.5C, D) in the above equation results in the following expression:

$$\begin{aligned}\left(\lambda'_{1,ci}\right) k_{ci} [M][R] + \left(\lambda'_{1,prp,prd}\right) k_p [M][RP] = \left(\lambda'_{1,td}\right) k_{td} [RP]^2 \\ + \left(\lambda'_{1,prp,con}\right) k_p [RP][M]\end{aligned} \tag{7.7B}$$

where $\lambda'_{1,ci}$ is the first instantaneous moment of 1-mer propagating chains produced by initiation; $\lambda'_{1,prp,prd}$ and $\lambda'_{1,prp,con}$ are the first instantaneous moments of the propagating chains produced and consumed by the propagation, respectively; and $\lambda'_{1,td}$ is the first instantaneous moment of the propagating chains consumed by termination via disproportionation. The $\lambda'_{1,ci}$ is evaluated as follows:

$$\lambda'_{1,ci} = \sum_{n=1}^{1} (n)^1 f_{rp}(n) = (1)^1 f_{rp}(n=1) = 1 \tag{7.7C}$$

The $\lambda'_{1,prp,prd}$ was derived in Section 6.2 of Chapter 6 using the first forward difference property of the z-transform distribution function of the propagating chains. According to Eq. (6.6) in Chapter 6, $\lambda'_{1,prp,prd}$ is related to λ_1 as follows:

$$\lambda'_{1,prp,prd} = \lambda_1 + 1 \tag{7.7D}$$

The instantaneous first moment of the propagating chains consumed in the propagation $\left(\lambda'_{1,prp,cons}\right)$ and termination $\left(\lambda'_{1,td}\right)$ reactions are just equal to the first moment of the propagating chains at time t as follows:

$$\lambda'_{1,prp,con} = \lambda'_{1,td} = \lambda_1 \tag{7.7E}$$

Substitution of Eqs. (7.7C–E) in Eq. (7.7B) and substitution for $k_{ci}[M][R]$ and $k_{td}[RP]^2$ from Eqs. (7.4A) to (7.5E), respectively, yields the following relation for λ_1 in terms of $[M]$, $[I]$, and $[RP]$:

$$\lambda_1 = \frac{k_p[M][RP]}{2k_d f_0 [I]} + 1 = \frac{k_p[M][RP] + k_{ci}[R][M]}{2k_d f_0 [I]} \tag{7.7F}$$

The numerator in the above equation is the rate of monomer consumption, whereas the denominator is the rate of chain initiation. This ratio is known as the kinetic chain length, L, as follows [42–46]:

$$\lambda_1 = \frac{k_p[M][RP] + k_{ci}[R][M]}{2k_d f_0 [I]} = L \tag{7.7G}$$

Substitution for $[I]$, $[RP]$, and $[M]$ from Eqs. (7.3B), (7.5F), and (7.6F), respectively, and assuming $L >> 1$, yields the following equation relating λ_1 or L to reaction time:

$$\lambda_1 = L = \left[\frac{[M]_0^2 k_p^2}{2k_d f_0 [I]_0 k_{td}}\right]^{1/2} \left(e^{\frac{k_d t}{2}}\right) \exp\left[-\left(\frac{8k_p^2 f_0 [I]_0}{k_d k_{td}}\right)^{1/2} \left(1 - e^{-\frac{k_d t}{2}}\right)\right] \quad \in L >> 1 \tag{7.7H}$$

At the start of the polymerization when $[M] \approx [M]_0$ and $[I] \approx [I]_0$, the kinetic chain length is given by the following equation:

$$\lambda_1 = L = \frac{k_p[M]_0}{\left(2k_d k_{td} f_0 [I]_0\right)^{1/2}} \quad \in L >> 1 \quad t \approx 0 \tag{7.7I}$$

The initial kinetic chain length increases linearly with $[M]_0$ and $k_p$whereas it decreases with the square root of k_d, k_{td}, and $[I]_0$. The number-average degree of polymerization of the propagating chains $\left(\overline{DP}_{n,rp}\right)$, as defined by Eq. (3.42) in Chapter 3, is equal to the kinetic chain length as follows:

$$\overline{DP}_{n,rp} = \frac{\lambda_1}{\lambda_0} = \lambda_1 = L \tag{7.7J}$$

The above equation states that $\overline{DP}_{n,rp}$ is nearly independent of time at low conversions in BSVR in the absence of chain transfer or chain scission reactions or termination by combination.

7.3.1.4 Second-order Moment of Propagating Chains in Radical Polymerization with Termination by Disproportionation

The weight-average degree of polymerization of the propagating chains $\left(\overline{DP}_{w,rp}\right)$ is determined from the knowledge of first and second order mole-fraction moments of the chains. Upon applying the approach used for derivation of λ_1 in the previous section using Eq. (7.1A), the conservation equation for the second order mole-fraction moment of the propagating chains is given as follows:

$$\left(\lambda'_{2,ci}\right) k_{ci} [M] [R] + \left(\lambda'_{2,prp,prd}\right) k_p [M] [RP] = \left(\lambda'_{2,td}\right) k_{td} [RP]^2 + \left(\lambda'_{2,prp,con}\right) k_p [RP] [M] \quad (7.8A)$$

where $\lambda'_{2,ci}$ is the second order instantaneous moment of 1-mer propagating chains produced by initiation, $\lambda'_{2,prp,prd}$ and $\lambda'_{2,prp,con}$ are the second order instantaneous moments of the propagating chains produced and consumed by propagation reaction, respectively, and $\lambda'_{2,td}$ is the second order instantaneous moment of the propagating chains consumed by termination via disproportionation. The $\lambda'_{2,ci}$ is evaluated as follows:

$$\lambda'_{2,ci} = \sum_{n=1}^{1} (n)^2 f_{rp}(n) = (1)^2 f_{rp}(n=1) = 1 \quad (7.8B)$$

The $\lambda'_{2,prp,prd}$ was derived in Section 6.2 of Chapter 6 using the first forward difference property of the z-transform distribution function of the propagating chains. According to Eq. (6.7) in Chapter 6, $\lambda'_{2,prp,prd}$ is related to λ_1 and λ_2 by the following equation:

$$\lambda'_{2,prp,prd} = \lambda_2 + 2\lambda_1 + 1 \quad (7.8C)$$

The second moment of the propagating chains consumed in the propagation $\left(\lambda'_{2,prp,cons}\right)$ and termination $\left(\lambda'_{2,td}\right)$ reactions are equal to the second moment of the propagating chains at time t in the reaction as follows:

$$\lambda'_{2,prp,con} = \lambda'_{2,td} = \lambda_2 \quad (7.8D)$$

Substitution of Eqs. (7.8B–D) in Eq. (7.8A) and substitution for $k_{ci} [M] [R]$ and $k_{td} [RP]^2$ from Eqs. (7.4A) and (7.5E), respectively, yields the following relation for λ_2 in terms of $[M]$, $[I]$, and $[RP]$:

$$\lambda_2 = \frac{(2\lambda_1 + 1) k_p [M] [RP]}{2k_d f_0 [I]} + 1 \quad (7.8E)$$

Upon applying $\frac{k_p [M] [RP]}{2k_d f_0 [I]} = L - 1$ from Eq. (7.7F), $\lambda_1 = L$ from the defining Eq. (7.7G), and using $2L^2 >> L + 1$ for large L values, the equation for λ_2 simplifies to the following:

$$\lambda_2 = \frac{(2\lambda_1 + 1) k_p [M] [RP]}{2k_d f_0 [I]} + 1 = 2L^2 - L \approx 2L^2 \quad \in L >> 1 \quad (7.8F)$$

where L is given in terms of time by Eq. (7.7H). Using the above equation and Eq. (7.7G) for $\lambda_1 = L$, the weight-average degree of polymerization of the propagating chains $\overline{DP}_{w,rp}$, defined by Eq. (3.45) in Chapter 3, is related to the kinetic chain length by the following equation:

$$\overline{DP}_{w,rp} = \frac{\lambda_2}{\lambda_1} = \frac{2L^2}{L} \approx 2L \quad \in L >> 1 \quad (7.8G)$$

Based on the above equation, $\overline{DP}_{w,rp}$ for long chains at low conversions in the absence of chain transfer or scission reactions and termination by disproportionation is twice the value of $\overline{DP}_{n,rp}$.

The polydispersity index of the propagating chains, namely PDI_{rp}, as defined by Eq. (3.66) in Chapter 3, is given as follows for BSVR:

$$PDI_{rp} = \frac{\overline{DP}_{w,rp}}{\overline{DP}_{n,rp}} \approx \frac{2L}{L} \approx 2 \quad \in L >> 1 \tag{7.8H}$$

Although L, $\overline{DP}_{w,rp}$, and $\overline{DP}_{n,rp}$ are functions of time in BVSR, the polydispersity index of the propagating chains, which is the ratio of $\overline{DP}_{w,rp}$ to $\overline{DP}_{n,rp}$, is nearly constant.

7.3.1.5 Concentration of Terminated Polymer Chains in Radical Polymerization with Termination by Disproportionation

Unlike propagating chains that are in a quasi-steady state, terminated polymer chains accumulate in the reaction, and their concentration increases with reaction time. The concentration of terminated chains is equal to the zero-order concentration moment of the terminated chains. Upon applying the conservation equation for the terminated chains in the reaction (Eq. 7.1C) to the zero-order order mole-fraction moment of the terminated chains, the $Q_{0,tp} = [TP]$ conservation equation is given by the following equation:

$$\frac{d\left(\mu_0 Q_{0,tp}\right)}{dt} = \sum_i \mu'_{0,i} R_{tp,prd,i} - \sum_l \mu'_{0,l} R_{tp,con,l} \tag{7.9A}$$

where μ_0 is the zero-order mole-fraction moment of the polymer chains at time t, $\mu'_{0,i}$ and $\mu'_{0,l}$ are the instantaneous zero-order moment of the polymer chains produced and consumed by elementary reactions i and l, respectively. As defined by Eq. (3.14) in Chapter 3, the zero-order mole-fraction moments are equal to unity, that is $\mu_0 = \mu'_{0,i} = \mu'_{0,l} = 1$. The $Q_{0,tp}$ is equal to the concentration of polymer chains, $[TP]$ (see Eq. 3.16 in Chapter 3). Based on the elementary Rxns. 7.2A–E, terminated polymer chains are produced by termination via disproportionation (Rxn. 7.2E), and they are not consumed in the reaction. As a result, the above conservation equation simplifies to the following:

$$\frac{dQ_{0,tp}}{dt} = \frac{d\,[TP]}{dt} = R_{td} = k_{td}\,[RP]^2 \tag{7.9B}$$

Substitution for $[RP]$ from Eq. (7.5F) results in following differential equation relating $[TP]$ to time:

$$\frac{d\,[TP]}{dt} = 2\,k_d f_0\,[I]_0\,e^{-k_d t} \tag{7.9C}$$

Upon integrating the above equation from the initial concentration $[TP]_0 = 0$ at $t = 0$ to any time t, the $[TP]$ is given in terms of time by the following equation:

$$[TP] = Q_{0,tp} = 2k_d f_0\,[I]_0 \int_{t=0}^{t} e^{-k_d t} = 2 f_0\,[I]_0 \left(1 - e^{-k_d t}\right) \tag{7.9D}$$

The above equation states that the $[TP]$ reaches the fraction of dissociated initiator molecules that undergo chain initiation as time approaches infinity. The concentration of double bonds at one end of the terminated polymer chains, $[TDB]$, produced by termination via disproportionation, is equal to one-half of the concentration of terminated polymer chains, that is $[TDB] = [TP]/2 = f_0\,[I]_0\left[1 - e^{-k_d t}\right]$.

7.3.1.6 First-order Moment of Terminated Polymer Chains in Radical Polymerization with Termination by Disproportionation

The number-average degree of polymerization of terminated polymer chains $\left(\overline{DP}_{n,tp}\right)$ is equal to the first order mole-fraction moment of the polymer chains, μ_1 (see Eq. 3.18 in Chapter 3). Upon applying the conservation equation for the terminated chains in the reaction (Eq. 7.1C) to the first order mole-fraction moment of the terminated chains with terminated chains produced only by termination via disproportionation (Rxn. 7.2E) and no consumption, the μ_1 conservation equation is given by the following equation:

$$\frac{d\left(\mu_1 Q_{0,tp}\right)}{dt} = \frac{d\left(Q_{1,tp}\right)}{dt} = \mu'_{1,td} R_{td} \tag{7.10A}$$

where $Q_{1,tp}$ is the first molar-concentration moment of the terminated polymer chains and $\mu'_{1,td}$ is the first instantaneous mole-fraction moment of the terminated chains produced by termination via disproportionation. According to Eq. (6.9) in Section 6.3.1 of Chapter 6, the instantaneous kth order moment of the terminated chains produced by disproportionation is equal to the kth moment of the propagating chains at time t, hence, $\mu'_{1,td} = \lambda_1$. Upon substitution for $\lambda_1 = L$ (Eq. 7.7G), the rate of chain termination by $R_{td} = k_{td}\,[RP]^2$ from Eq. (7.5E), the rate of change of $Q_{1,tp}$ with time is given by the following equation:

$$\begin{aligned} \frac{d\left(Q_{1,tp}\right)}{dt} &= \frac{d\left(\mu_1 Q_{0,tp}\right)}{dt} = \mu'_{1,td} R_{td} = \lambda_1 k_{td}\,[RP]^2 \\ &= \left(\frac{k_p\,[M]\,[RP] + k_{ci}\,[M]\,[R]}{k_{ci}\,[M]\,[R]}\right)\left(k_{ci}\,[M]\,[R]\right) \\ &= \left(k_p\,[M]\,[RP] + k_{ci}\,[M]\,[R]\right) \end{aligned} \tag{7.10B}$$

The last term in the above equalities, according to Eq. (7.6D), is the rate of change of monomer concentration with time as follows:

$$\frac{d\left(Q_{1,tp}\right)}{dt} = \left(k_p\,[M]\,[RP] + k_{ci}\,[M]\,[R]\right) = -\frac{d\,[M]}{dt} \tag{7.10C}$$

The above equation states that the rate of change of $Q_{1,tp}$ with time is inversely related to the rate of change of monomer concentration with time. Integration of the above differential equation from $Q_{1,tp} = 0$ and $[M] = [M]_0$ at $t = 0$ to time t results in the following relation between $Q_{1,tp}$ and $[M]$:

$$\int_0^{Q_{1,tp}} \frac{d\left(QP_{1,tp}\right)}{dt} = -\int_{[M]_0}^{[M]_t} \frac{d\,[M]}{dt} \Rightarrow Q_{1,tp} = [M]_0 - [M] = [M]_{con} \tag{7.10D}$$

The above equation states that the first concentration moment of the terminated polymer chains is equal to the concentration of consumed monomers, $[M]_{con}$. The first order mole-fraction moment of the terminated chains μ_1, which is equal to $\overline{DP}_{n,tp}$ (see Eq. 3.20), is given as a function of time for BSVR by substituting for $[M]$ from Eq. (7.6F) and $[TP]$ from Eq. (7.9D) as follows:

$$\mu_1 = \overline{DP}_{n,tp} = \frac{Q_{1,tp}}{Q_{0,tp}} = \frac{[M]_0\left\{1 - \exp\left[-\left(8k_p^2 f_0\,[I]_0\,/k_d k_{td}\right)^{1/2}\left(1 - e^{-k_d t/2}\right)\right]\right\}}{2f_0\,[I]_0\left(1 - e^{-k_d\,t}\right)} \tag{7.10E}$$

The above equation states that μ_1 is equal to the ratio of monomer consumed to the produced terminated chains. Initially, in the reaction when $[M] \approx [M]_0$ and $[I] \approx [I]_0$, the above equation for $\overline{DP}_{n,tp} = \mu_1$ is given by the following simplified expression:

$$\overline{DP}_{n,tp} = \mu_1 = \frac{k_p [M]_0}{(2k_d k_{td} f_0 [I]_0)^{1/2}} \in L >> 1 \quad t \approx 0 \tag{7.10F}$$

The $\overline{DP}_{n,tp}$ like L increases linearly with $[M]$ and k_p, and decreases with the square root of k_d, k_{td}, and $[I]$.

7.3.1.7 Second-order Moment of Terminated Polymer Chains in Radical Polymerization with Termination by Disproportionation

The weight-average degree of polymerization of terminated chains, $\overline{DP}_{w,tp}$, is determined from the knowledge of first and second order moments of the terminated chains. Upon applying the conservation equation for the terminated chains in the reaction (Eq. 7.1C) to the second order mole-fraction moment of the terminated chains with terminated chains produced only by disproportionation (Rxn. 7.2E) and no consumption of terminated chains, the μ_2 conservation equation is given by the following equation:

$$\frac{d(\mu_2 Q_{0,tp})}{dt} = \frac{d(Q_{2,tp})}{dt} = \mu'_{2,td} R_{td} = \mu'_{2,td} k_{td} [RP]^2 \tag{7.11A}$$

Where $Q_{2,tp}$ is the second concentration moment of the terminated chains and $\mu'_{2,td}$ is the instantaneous second order mole-fraction moment of the terminated chains produced by disproportionation. According to Eq. (6.9) in Section 6.3.1 of Chapter 6, $\mu'_{k,td} = \lambda_k$; thus $\mu'_{2,td} = \lambda_2$. Upon substitution for λ_2 from Eqs. (7.7H) and (7.8F) and $k_{td} [RP]^2$ from Eq. (7.5F), the rate of change of $Q_{2,tp}$ with time given by the following equation:

$$\frac{d(Q_{2,tp})}{dt} = \left(2[M]_0^2 k_p^2 / k_{td}\right) \exp\left[-2\left(8k_p^2 f_0 [I]_0 / k_d k_{td}\right)^{1/2} \left(1 - e^{-k_d t/2}\right)\right] \tag{7.11B}$$

where it is assumed in the above equation that $L >> 1$. Integration of the above equation from $Q_{2,tp} = 0$ at $t = 0$ to time t results in the following expression for $Q_{2,tp}$

$$Q_{2,tp} = \left(2[M]_0^2 k_p^2 / k_{td}\right) \int_{t=0}^{t} \exp\left[-2\left(8k_p^2 f_0 [I]_0 / k_d k_{td}\right)^{1/2} \left(1 - e^{-k_d t/2}\right)\right] dt \tag{7.11C}$$

The second order mole-fraction moment of terminated polymer chains, μ_2, as defined by Eq. (3.24) in Chapter 3 is determined from $Q_{2,tp}$ as follows:

$$\mu_2 = \frac{Q_{2,tp}}{Q_{0,tp}} = \frac{Q_{2,tp}}{[TP]} = \frac{Q_{2,tp}}{2f_0 [I]_0 [1 - e^{-k_d t}]} \tag{7.11D}$$

where Eq. (7.9D) is used for $[TP]$ in terms of time. Using the first and second order moments of the terminated chains, $\overline{DP}_{w,tp}$ as defined by Eq. (3.45) in Chapter 3, is determined as follows:

$$\overline{DP}_{w,tp} = \frac{\mu_2}{\mu_1} = \frac{Q_{2,tp}}{Q_{1,tp}}$$

$$= \frac{\left(2[M]_0 k_p^2 / k_{td}\right) \int_{t=0}^{t} \exp\left[-2\left(8k_p^2 f_0 [I]_0 / k_d k_{td}\right)^{1/2} \left(1 - e^{-k_d t/2}\right)\right] dt}{\left\{1 - \exp\left[-\left(8k_p^2 f_0 [I]_0 / k_d k_{td}\right)^{1/2} \left(1 - e^{-k_d t/2}\right)\right]\right\}} \tag{7.11E}$$

The polydispersity (PDI) of the terminated polymer chains, as defined by Eq. (3.66) in Chapter 3, is given by the following equation:

$$PDI_{tp} = \frac{\overline{DP}_{w,tp}}{\overline{DP}_{n,tp}} \tag{7.11F}$$

where $\overline{DP}_{w,tp}$ and $\overline{DP}_{n,tp}$ are given by Eqs. (7.11E) and (7.10F), respectively.

7.3.1.8 (1+a)-order Moment of the Terminated Chains in Radical Polymerization with Termination by Disproportionation

The kth order conservation equation for mole-fraction moment of the polymer chains can be used not only for integer k values, but also for non-integer k values with a decimal like $(1 + a)$ order with $a < 1$. As derived in Section 3.6 of Chapter 3, the viscosity-average degree of polymerization of terminated chains $\left(\overline{DP}_{\eta,tp}\right)$ is determined from the (1+a)th and first order molar moments of the terminated chains using the following equation:

$$\overline{DP}_{\eta,tp} = \left[\frac{\mu_{1+a,tp}}{\mu_{1,tp}}\right]^{1/a} = \left[\frac{Q_{1+a,tp}}{Q_{1,tp}}\right]^{1/a} \tag{7.12A}$$

where the constant "a" is the exponent in Mark–Houwink–Sakurada (MHS) equation, as defined in Chapter 3 [47–49]. The $\mu_{1+a,tp}$ and $Q_{1+a,tp}$ in the above equation are $(1 + a)$th order mole-fraction and molar concentration moments of the terminated chains, respectively. Upon applying the conservation equation for the terminated chains (Eq. 7.1C) to the $(1 + a)$th order mole-fraction moment of the terminated chains produced by disproportionation (Rxn. 7.2E), the μ_{1+a} conservation equation is given as follows:

$$\frac{d\left(\mu_{1+a,t}Q_{0,tp}\right)}{dt} = \frac{d\left(QP_{1+a,tp}\right)}{dt} = \mu'_{1+a,td}R_{td} = \lambda_{1+a}k_{td}\,[RP]^2 \tag{7.12B}$$

where $\mu'_{1+a,td}$ is the instantaneous $(1 + a)$th order moment of the terminated chains produced by disproportionation and λ_{1+a} is $(1 + a)$th order mole-fraction moment of the propagating chains at time t. The equality $\mu'_{1+a,td} = \lambda_{1+a}$ is based on Eq. (6.9) in Chapter 6 for termination by disproportionation. Upon substitution for $[RP]$ from Eq. (7.5F), the rate of change of $QP_{1+a,tp}$ with time is given by the following equation:

$$\frac{d\left(Q_{1+a,tp}\right)}{dt} = \lambda_{1+a}k_{td}\,[RP]^2 = 2k_d f_0\,[I]_0\,e^{-k_d t/2}\lambda_{1+a} \tag{7.12C}$$

The relation between λ_{1+a} and the kinetic chain length, L, can be derived if the chain distribution of the propagating chains is known. If the propagating chains follow the Flory–Schultz most probable distribution [50–53], that is $f_{RP}\,(n = i) = (1 - p)\,p^i$, where p is the probability for propagation as opposed to termination by disproportionation and $1 < i < \infty$ (see Section 3.9.2 in Chapter 3), then it can be shown that λ_{1+a} is related to L by the following equation:

$$\lambda_{1+a} = \sum_{i=1}^{+\infty} i^{1+a} f_{RP}(i) = \left(\frac{1}{L+1}\right)\sum_{i=1}^{+\infty} i^{1+a}\left(\frac{L}{L+1}\right)^i \tag{7.12D}$$

Upon substitution of the above equation in Eq. (7.12C) followed by separation of variables and integration ($Q_{1+a,tp} = 0$ for $t = 0$), $QP_{1+a,tp}$ as a function of time is given by the following equation (see Eq. 7.7H for time dependence of L):

$$Q_{1+a,tp} = 2k_d f_0 [I]_0 \int_{t=0}^{t} e^{-k_d t/2} \lambda_{1+a} dt$$
$$= 2k_d f_0 [I]_0 \int_{t=0}^{t} \frac{e^{-k_d t/2}}{L+1} \left(\sum_{n=1}^{+\infty} i^{1+a} (L/L+1)^i \right) dt \tag{7.12E}$$

Based on Eq. (3.20) in Chapter 3, $\mu_{1+a} = Q_{1+a,tp}/Q_{0,tp}$. Upon substitution for $Q_{1+a,tp}$ from the above equation and $Q_{0,tp}$ from Eq. (7.9D), the time dependence of μ_{1+a} is given by the following equation:

$$\mu_{1+a} = \frac{Q_{1+a,tp}}{Q_{0,tp}} = \left[\frac{2k_d f_0 [I]_0 \int_{t=0}^{t} \frac{e^{-k_d t/2}}{L+1} \left(\sum_{n=1}^{+\infty} i^{1+a} (L/L+1)^i \right) dt}{2f_0 [I]_0 [1 - e^{-K_d t}]} \right] \tag{7.12F}$$

The viscosity-average degree of polymerization $\left(\overline{DP}_{\eta,tp}\right)$ for this polymerization system in a BSVR is given by the following equation as a function of time:

$$\overline{DP}_{\eta,tp} = \left[\frac{2k_d f_0 [I]_0 \int_{t=0}^{t} \frac{e^{-k_d t/2}}{L+1} \left(\sum_{n=1}^{+\infty} i^{1+a} (L/L+1)^i \right) dt}{[M]_0 \left\{ 1 - \exp \left[-\left(8k_p^2 f_0 [I]_0 / k_d k_{td}\right)^{1/2} \left(1 - e^{-k_d t/2}\right) \right] \right\}} \right]^{1/a} \tag{7.12G}$$

The melt viscosity of a polymer increases with $\overline{DP}_{\eta,tp}$ and decreases with temperature. High $\overline{DP}_{\eta,tp}$ values result in excessively high melt viscosities which requires higher temperatures or higher pressures to make the polymer flow during processing. High pressures, which excessively increase the power requirements or high temperatures, which cause thermal and oxidative degradation of polymers, are not desirable in a polymerization reaction [54–56].

7.3.1.9 Example Problem 1

This example illustrates the application of the equations derived in Section 7.3.1 to thermally initiated radical polymerization of styrene in toluene as the solvent with no chain transfer agent and no chain transfer to solvent. Since the reaction is done in a solvent, it can be assumed that QSSA holds for the primary radicals (R) as well as the reactive propagating chains (RP). The elementary reactions are given by Rxns (7.2A–E) in Section 7.3. Furthermore, it can be assumed that the rate of propagating chain initiation, that is k_i, is equal to the rate of chain propagation k_p. The reaction initially consists of 20% by mole styrene and 80% toluene, and the initial ratio of the monomer to initiator concentrations is $[M]_0 / [I]_0 = 1000$. The initiator efficiency $f_0 = 1$ and assumed to be independent of reaction time. The molecular weights of styrene and toluene monomers are 104 and 92 g mol^{-1}, respectively; and their respective densities are 0.909 and 0.867 g ml^{-1}. The rate constants for initiator dissociation, chain initiation, propagation, and termination by disproportionation are as follows:

$$k_d = 1.38 \times 10^{-4} \text{s}^{-1} \quad @T = 85\,°\text{C}$$

$$k_p = k_{ci} = 7.24 \times 10^2 \text{ l mol}^{-1} \text{ s}^{-1} \quad @T = 85\,°\text{C}$$

$$k_{td} = 3.95 \times 10^7 \text{ l mol}^{-1} \text{ s}^{-1} \quad @T = 85\,°\text{C}$$

Assuming the solvent and monomer volumes are additive, determine the concentrations $[I]$, $[M]$, $[R]$, $[RP]$ and $[TP]$, kinetic chain length L, degree of polymerizations $\overline{DP}_{n,rp}$, $\overline{DP}_{w,rp}$, $\overline{DP}_{n,tp}$ and $\overline{DP}_{w,tp}$, and polydispersity index PDI_{tp} as a function of time.

Solution: We choose as a reference one mole of solution (styrene and toluene). The volume of monomer in the reaction is calculated from its density and molecular weight as follows:

$$V_M = \frac{(0.2\ \text{mol})(104\ \text{g/mol})}{(0.909\ \text{g/ml})(1000\ \text{ml/l})} = 0.023\ \text{l}$$

The initial concentrations of monomer (styrene) and initiator are calculated by the following equations:

$$[M]_0 = \frac{m_M}{V_M} = \frac{0.2\ \text{mol}}{0.023\ \text{l}} = 8.7\ \text{mol l}^{-1} \quad [I]_0 = \frac{[M]_0}{1000} = 8.7 \times 10^{-3}\ \text{mol l}^{-1}$$

The initiator $[I]$, primary radical $[R]$, and reactive propagating chain $[RP]$ concentrations are given by Eqs. (7.3B), (7.4B), and (7.5F), respectively. Using the data provided, $[I]$, $[R]$ and $[RP]$ are given by the following relations as a function of time:

$$[I] = [I]_0\, e^{-k_d t} = 8.7 \times 10^{-3}\, e^{-1.38\times10^{-4} t}\ \frac{\text{mol}}{\text{l}} \tag{7.13A}$$

$$[R] = \left[\frac{2k_d\,(1 - f_0)\,[I]_0}{k_c}\right]^{1/2} e^{\frac{-k_d t}{2}} = 0 \ \in f_0 = 1 \tag{7.13B}$$

$$[RP] = \left[\frac{2k_d f_0\,[I]_0}{k_{td}}\right]^{1/2} e^{-\frac{k_d t}{2}} = 2.47 \times 10^{-7} e^{-6.9\times10^{-5} t}\ \frac{\text{mol}}{\text{l}} \tag{7.13C}$$

As the initiation is 100% efficient ($f_0 = 1$), all primary radicals formed by initiator dissociation react with a monomer and form reactive propagating chains. As a result, the concentration of primary radicals is given by $[R] = 2k_d\,[I]/k_{ci}\,[M]$. The time variation of $[I]$ and $[RP]$ are shown in Figure 7.1a, b.

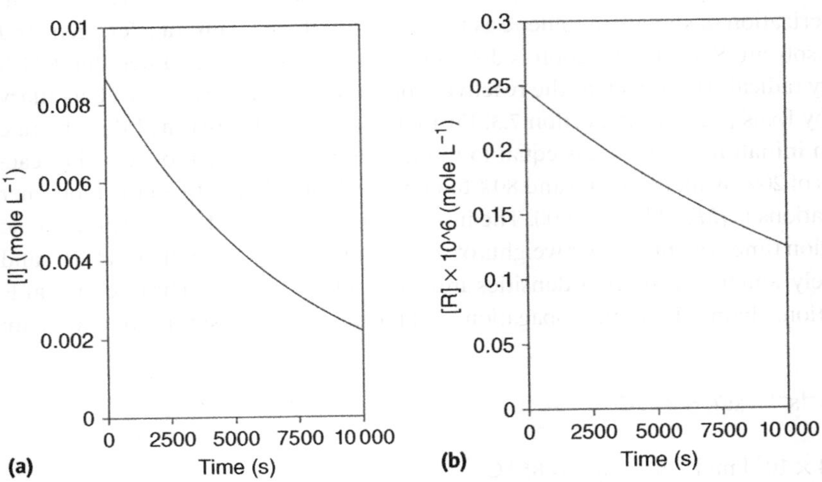

Figure 7.1 Concentrations of initiator (a) and propagating chains (b) as a function of reaction time for Example Problem 1.

The monomer concentration $[M]$ is given by Eq. (7.6F). Using the data provided, $[M]$ is given by the following relation as a function of time:

$$
\begin{aligned}
[M] &= [M]_0 \exp\left[-\left(\frac{8k_p^2 f_0 [I]_0}{k_d k_{td}}\right)^{1/2} \left(1 - e^{-\frac{k_d t}{2}}\right)\right] \\
&= 8.7 \exp\left[-0.800\left(1 - e^{-6.9\times10^{-5}t}\right)\right] \text{ mol l}^{-1}
\end{aligned} \quad (7.13D)
$$

The kinetic chain length L is given by Eq. (7.7H). Using the data provided, L is given by the following equation as a function of time:

$$
\begin{aligned}
L &= \left[\frac{[M]_0^2 k_p^2}{2k_d f_0 [I]_0 k_{td}}\right]^{1/2} \left(e^{k_d t/2}\right) \\
&\quad \exp\left[-\left(8k_p^2 f_0 [I]_0 / k_d k_{td}\right)^{1/2} \left(1 - e^{-k_d t/2}\right)\right] \\
&= 200\left(e^{6.9\times10^{-5}t}\right) \exp\left[-0.800\left(1 - e^{-6.9\times10^{-5}t}\right)\right]
\end{aligned} \quad (7.13E)
$$

The time variation of $[M]$ and L are shown in Figure 7.2a, b. As the monomer concentration decreases faster than initiator, the kinetic chain length monotonically decreases with reaction time (Figure 7.2b).

The number-average $\overline{DP}_{n,rp}$ and weight-average $\overline{DP}_{w,rp}$ of the reactive propagating chains are given by Eqs. (7.7J) and (7.8G), respectively. Using the data provided, $\overline{DP}_{n,rp}$ and $\overline{DP}_{w,rp}$ are given by the following relations as a function of time:

$$\overline{DP}_{n,rp} = \lambda_1 = L = 200\left(e^{6.9\times10^{-5}t}\right) \exp\left[-0.800\left(1 - e^{-6.9\times10^{-5}t}\right)\right] \quad (7.13F)$$

$$\overline{DP}_{w,rp} \approx 2L = 400\left(e^{6.9\times10^{-5}t}\right) \exp\left[-0.800\left(1 - e^{-6.9\times10^{-5}t}\right)\right] \quad (7.13G)$$

The time variation of $\overline{DP}_{n,rp}$ and $\overline{DP}_{w,rp}$ are shown in Figure 7.3a, b. The decline in $\overline{DP}_{n,rp}$ and $\overline{DP}_{w,rp}$ is related to the faster consumption of the monomer compared to the initiator with time.

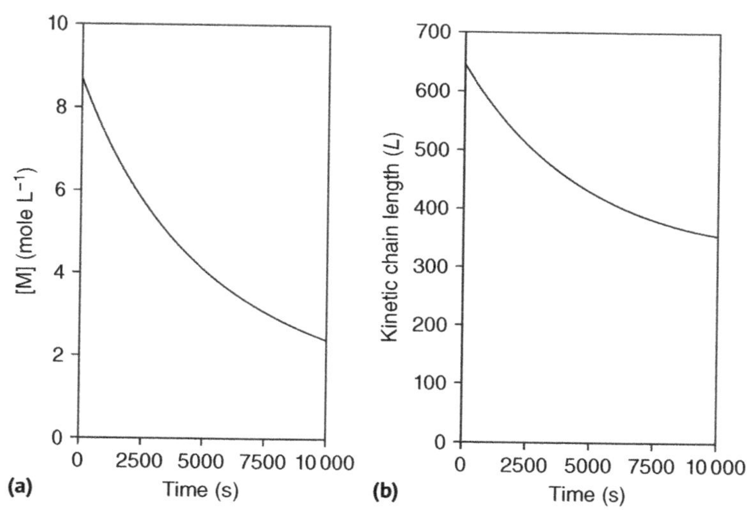

Figure 7.2 Variation of monomer concentration (a) and the kinetic chain length L (b) as a function of reaction time for Example Problem 1.

The concentration of terminated polymer chains $[TP]$ is given by Eq. (7.9D). Using the data provided, $[TP]$ is given by the following relation as a function of time, and the relation is plotted in Figure 7.4.

$$[TP] = 2f_0\,[I]_0\left(1 - e^{-k_d t}\right) = 1.74 \times 10^{-2}\left(1 - e^{-1.38\times10^{-4}t}\right)\ \text{mol l}^{-1} \tag{7.13H}$$

The $\overline{DP}_{n,tp}$ and $\overline{DP}_{w,tp}$ of the terminated polymer chains are given by Eqs. (7.10F) and (7.11E), respectively. Using the data provided, $\overline{DP}_{n,tp}$ and $\overline{DP}_{w,tp}$ are given by the following relations as a function of time and the relations are plotted in Figure 7.5.

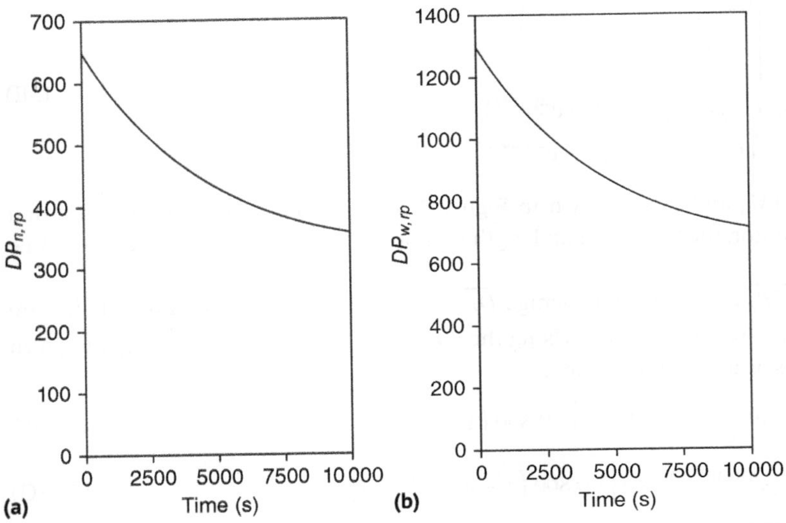

Figure 7.3 Variation of $\overline{DP}_{n,rp}$ (a) and $\overline{DP}_{w,rp}$ (b) of propagating chains as a function of reaction time for Example Problem 1.

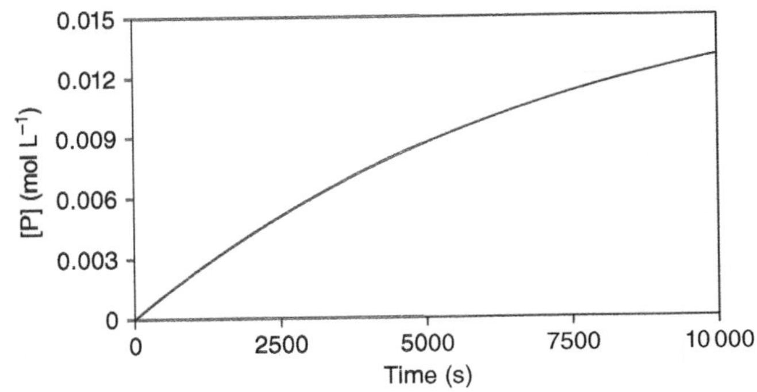

Figure 7.4 Variation of $[TP]$ with reaction time for Example Problem 1.

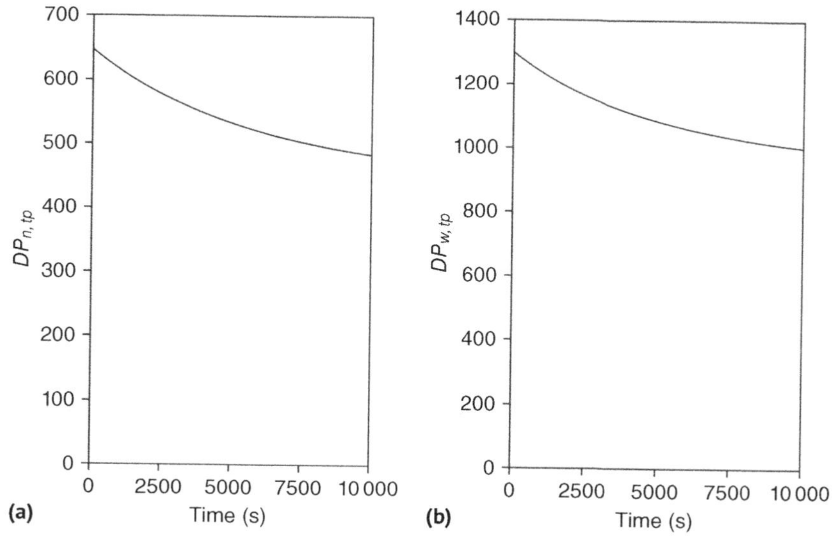

Figure 7.5 Variation of $\overline{DP}_{n,tp}$ (a) and $\overline{DP}_{w,tp}$ (b) of the terminated polymer chains with reaction time for Example Problem 1.

$$\overline{DP}_{n,tp} = \frac{[M]_0\left\{1-\exp\left[-\left(8k_p^2 f_0 [I]_0/k_d k_{td}\right)^{1/2}\left(1-e^{-k_d t/2}\right)\right]\right\}}{2f_0[I]_0\left(1-e^{-k_d t}\right)} \tag{7.13I}$$
$$= 500\left\{1-\exp\left[-0.800\left(1-e^{-6.9\times10^{-5}t}\right)\right]\right\}\left(1-e^{-1.38\times10^{-4}t}\right)^{-1}$$

$$\overline{DP}_{w,tp} = \frac{\left(2[M]_0 k_p^2/k_{td}\right)\displaystyle\int_{t=0}^{t}\exp\left[-2\left(8k_p^2 f_0[I]_0/k_d k_{td}\right)^{1/2}\left(1-e^{-k_d t/2}\right)\right]dt}{\left\{1-\exp\left[-\left(8k_p^2 f_0[I]_0/k_d k_{td}\right)^{1/2}\left(1-e^{-k_d t/2}\right)\right]\right\}}$$
$$= 0.022\left\{\int_{t=0}^{t}\exp\left[-1.6\left(1-e^{-6.9\times10^{-5}t}\right)\right]dt\right\}\left\{1-\exp\left[-0.8\left(1-e^{-1.38\times10^{-4}t}\right)\right]\right\}^{-1} \tag{7.13J}$$

Polydispersity index of terminated polymer chains PDI_{tp} which is the ratio of $\overline{DP}_{w,tp}$ to $\overline{DP}_{n,tp}$ is given as follows and plotted in Figure 7.6.

$$PDI_{tp} = \frac{4.4\times10^{-5}\left\{\displaystyle\int_{t=0}^{t}\exp\left[-1.6\left(1-e^{-6.9\times10^{-5}t}\right)\right]dt\right\}\left(1-e^{-1.38\times10^{-4}t}\right)}{\left\{1-\exp\left[-0.800\left(1-e^{-6.9\times10^{-5}t}\right)\right]\right\}^2} \tag{7.13K}$$

The PDI_{tp} increases with time because shorter terminated polymer chains are produced with time as monomers are consumed at a faster rate compared to the rate of formation of propagating chains. The above example illustrates the use of the method of moments in combination with the method of z-transform to model and predict the number and weight average degree of polymerization and the chain polydispersity in radical polymerization.

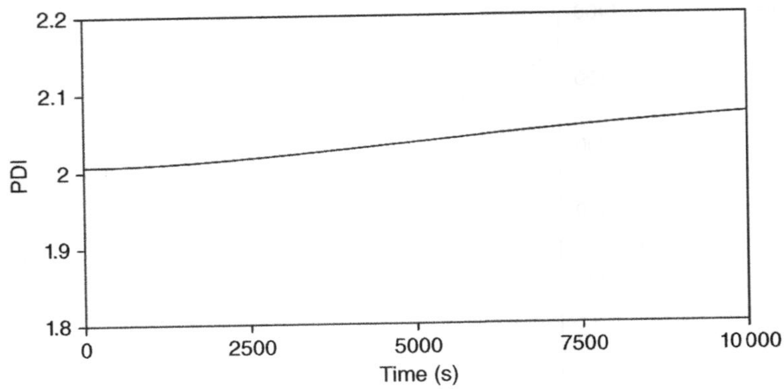

Figure 7.6 Variation of PDI_{tp} with time for Example Problem 1.

7.3.2 Modeling Radical Polymerization with Termination by Disproportionation in a Continuous Stirred Tank Reactor (CSTR)

In a CSTR, the reactor is open and at steady state, as shown in Figure 5.6 of Chapter 5. Assuming the change in density with reaction is negligible, the reactant and product streams flow in and out of the reactor at equal and constant volumetric flow rate and composition. In a CSTR, the composition of the product stream is identical to the composition inside the reactor. The general conservation equations for the kth mole-fraction moment of the propagating chains, terminated chains, and small molecule reactants for a CSTR are given by Eqs. (5.21C–E) in Chapter 5, respectively. In this section, we apply the conservation equations for a CSTR to polymerization with elementary reactions 7.2A–E. In elementary reactions 7.2A–E, initiator is not produced, and it is consumed by the first reaction. Using the conservation equation for small molecules, the equation for the initiator concentration with consumption by the first reaction is given by the following relation:

$$[I]_{\text{out}} = [I]_{\text{in}} - \tau_{CSTR} R_d = [I]_{\text{in}} - \tau_{CSTR} k_d [I]_{\text{out}} \tag{7.14A}$$

Where $[I]_{\text{in}}$ and $[I]_{\text{out}}$ are the initiator concentration in the inlet and outlet streams, respectively, $\tau_{CSTR} = \frac{V_r}{\dot{V}}$ is the mean residence time of the reactants in the reactor, V_r and $\dot{V}$ are the reactor volume and volumetric flow rate of reactant/product stream, and $k_d [I]_{\text{out}}$ is the rate of initiator consumption in the reactor. Upon rearrangement of the above equation, $[I]_{\text{out}}$ for a CSTR is given by the following equation in terms of $[I]_{\text{in}}$:

$$I_{\text{out}} = \frac{I_{\text{in}}}{1 + \tau_{CSTR} k_d} \tag{7.14B}$$

The assumption of quasi-steady state (QSSA) of the reactive species is valid in a CSTR if the reaction is not diffusion controlled. If QSSA is valid, it infers that the concentration of reactive species in the inlet reactant and outlet product streams, due to their short half-life, is equal to zero and it exists only in the reactor. Upon applying the conservation equation for the propagating species (zero-order moment), the equation for the primary radical concentration with production by the first reaction and consumption by the second and third reaction is given by the following relation:

$$\begin{aligned} 0 &= \tau_{CSTR} (R_d - R_c - R_{ci}) \\ &= \tau_{CSTR} \left(k_d [I]_{\text{out}} - k_c [R]^2 - k_{ci} [R] [M]_{\text{out}} \right) \end{aligned} \tag{7.15A}$$

where $[R]$ is the concentration of primary radical in the reactor and $[M]_{\text{out}}$ is the monomer concentration in the outlet stream. Upon applying the initiation efficiency (see Eq. 5.29E in Chapter 5), f_0, to eliminate k_c and rearrangement of the above equation, $[R]$ for a CSTR is given by the following equation:

$$[R] = \frac{2k_d f_0 [I]_{\text{out}}}{k_{ci} [M]_{\text{out}}} \tag{7.15B}$$

The reactive propagating chains, RP, are produced by the chain initiation reaction 7.2C, produced and consumed by the propagation reaction 7.2D, and consumed by the termination reaction 7.2E. Upon applying the conservation equation for the propagating chains (zero-order), with $\lambda_0' = \lambda_0 = 1$, and the production and consumption rates given by Eqs. (7.5C, D), the equation for the concentration of propagating chains is given by the following relation:

$$0 = \tau_{CSTR} \left(R_{ci} + R_p - R_p - R_{td}\right) = \tau_{CSTR} \left(k_{ci} [M]_{\text{out}} [R] - k_{td} [RP]^2\right) \tag{7.16A}$$

Upon rearrangement of the above equation and substitution for $[R]$ from Eq. (7.15B), the $[RP]$ of the reaction system in a CSTR is given by the following equation:

$$[RP] = \left(\frac{k_{ci} [M]_{\text{out}} [R]}{k_{td}}\right)^{\frac{1}{2}} = \left(\frac{2k_d f_0 [I]_{\text{out}}}{k_{td}}\right)^{\frac{1}{2}} \tag{7.16B}$$

The above Eq. (7.15B) for $[R]$ and 7.16B for $[PR]$ in a CSTR are like those derived in a BSVR (Eqs. 7.4A and 7.5F) with the difference that $[I]_{\text{out}}$ and $[M]_{\text{out}}$ are at steady state in a CSTR and independent of time. The monomer is consumed by the chain initiation and propagation reactions (Rxns. 7.2C, D) and it is not produced. Upon applying the conservation equation for the small molecules and using the Eqs. (7.6B, C) for monomer production and consumption, respectively, the equation for the monomer concentration is given by the following relation:

$$\begin{aligned} [M]_{\text{out}} &= [M]_{\text{in}} - \tau_{CSTR} \left(R_{ci} + R_p\right) \\ &= [M]_{\text{in}} - \tau_{CSTR} \left(k_{ci} [M]_{\text{out}} [R] + k_p [RP] [M]_{\text{out}}\right) \end{aligned} \tag{7.17A}$$

Upon rearrangement of the above equation, substitution for $[R]$ from Eq. (7.15B) and for $[RP]$ from Eq. (7.16B), the $[M]_{\text{out}}$ in the CSTR product stream of the reaction system is given by the following equation:

$$[M]_{\text{out}} = \frac{[M]_{\text{in}} - \tau_{CSTR} 2k_d f_0 [I]_{\text{out}}}{\left[1 + \tau_{CSTR} k_p \left(\frac{2k_d f_0 [I]_{\text{out}}}{k_{td}}\right)^{\frac{1}{2}}\right]} \tag{7.17B}$$

If the kinetic chain length $L >> 1$, then $R_p >> R_{ci}$, as a result $k_p [RP] [M]_{\text{out}} >> k_{ci} [M]_{\text{out}} [R]$, and the equation for $[M]_{\text{out}}$ simplifies to the following relation:

$$[M]_{\text{out}} = \frac{[M]_{\text{in}}}{\left[1 + \tau_{CSTR} k_p \left(\frac{2k_d f_0 [I]_{\text{out}}}{k_{td}}\right)^{\frac{1}{2}}\right]} \tag{7.17C}$$

The propagating chains are produced by chain initiation and propagation reactions and consumed by propagation and termination reactions. Applying the conservation equation for the propagating chains (first order), with production and consumption rates given by Eqs. (7.5C, D), the equation for the first order mole-fraction moment of the propagating chains is given by the following relation:

$$0 = \tau_{CSTR}\left[\lambda'_{1,ci}R_{ci} + \left(\lambda'_{1,prp,prd} - \lambda'_{1,prp,con}\right)R_p - \lambda'_{1,td}R_{td}\right]$$
$$= \tau_{CSTR}\left[\lambda'_{1,ci}k_{ci}\,[M]_{out}\,[R] - \lambda'_{1,td}k_{td}\,[RP]^2\right]$$
$$+\tau_{CSTR}\left(\lambda'_{1,prp,prd} - \lambda'_{1,prp,con}\right)k_p\,[M]_{out}\,[RP] \quad (7.18A)$$

Upon using Eq. (7.7C) for $\lambda'_{1,ci}$, Eq. (7.7D) for $\lambda'_{1,prp,prd}$, and Eq. (7.7E) for $\lambda'_{1,prp,con}$ and $\lambda'_{1,td}$, using Eq. (7.16B) for $k_{td}\,[RP]^2$, and rearrangement, the equation for λ_1 or the kinetic chain length (L) of the propagating chains is given by the following relation:

$$\lambda_1 = \frac{k_p\,[M]_{out}\,[RP] + 2k_d f_0\,[I]_{out}}{2k_d f_0\,[I]_{out}} = \frac{k_p\,[M]_{out}\,[RP]}{2k_d f_0\,[I]_{out}} + 1 = L + 1 \quad (7.18B)$$

If $L >> 1$ (i.e., long chains), then $k_p\,[M]_{out}\,[RP] >> 2k_d f_0\,[I]_{out}$, the above equation for L simplifies to the following relation:

$$\lambda_1 = L \approx \frac{k_p\,[M]_{out}}{\left(2k_d k_{td} f_0\,[I]_{out}\right)^{1/2}} \quad \in L >> 1 \quad (7.18C)$$

Similarly, the equation for the second mole-fraction moment of the propagating chains is given by the following relation:

$$0 = \tau_{CSTR}\left[\lambda'_{2,ci}k_{ci}\,[M]_{out}\,[R] - \lambda'_{2,td}k_{td}\,[RP]^2\right]$$
$$+\tau_{CSTR}\left(\lambda'_{2,prp,prd} - \lambda'_{2,prp,con}\right)k_p\,[M]_{out}\,[RP] \quad (7.19A)$$

Upon using Eq. (7.8B) for $\lambda'_{2,ci}$, Eq. (7.8C) for $\lambda'_{2,prp,prd}$, and Eq. (7.8D) for $\lambda'_{2,prp,con}$ and $\lambda'_{2,td}$, using Eq. (7.16B) for $k_{td}\,[RP]^2$ and rearrangement, the equation for λ_2 of the propagating chains is given by the following relation:

$$\lambda_2 = \frac{(2\lambda_1 + 1)\,k_p\,[M]_{out}}{\left(2k_d k_{td} f_0\,[I]_{out}\right)^{1/2}} + 1 \quad (7.19B)$$

If $L >> 1$ and using Eq. (7.18B), the equation for λ_2 of the reaction system in a CSTR simplifies as follows:

$$\lambda_2 = \frac{(2\lambda_1 + 1)\,k_p\,[M]_{out}}{\left(2k_d k_{td} f_0\,[I]_{out}\right)^{1/2}} + 1 = (2L+1)(L-1) \approx 2L^2 \quad \in L >> 1 \quad (7.19C)$$

In reactions 7.2A–E, terminated chains are produced by the last termination reaction. Applying the conservation equation for the terminated chains (zero-order), with $\mu'_0 = \mu_0 = 1$ and the production rate given by $R_{td} = k_{td}\,[RP]^2$, the concentration of terminated chains, $Q_{0,tp} = [TP]$ is given by the following equation:

$$Q_{0,tp,out} = [TP] = \tau_{CSTR}k_{td}\,[RP]^2 \quad (7.20A)$$

Upon substitution for $[RP]$ from Eq. (7.16B) and $[I]_{out}$ from Eq. (7.14B), the $Q_{0,tp} = [TP]$ is given as follows:

$$Q_{0,tp,out} = \tau_{CSTR}2k_d f_0\,[I]_{out} = \tau_{CSTR}\left(\frac{2k_d f_0 I_{in}}{1 + \tau_{CSTR}k_d}\right) \quad (7.20B)$$

According to Eq. (6.9) in Chapter 6, $\mu'_{1,td} = \lambda_1$. Applying the conservation equation for the terminated chains (first order), with $\mu'_{1,td} = \lambda_1$ and $R_{td} = k_{td}\,[RP]^2$, λ_1 from Eq. (7.18B), $[RP]$ from Eq. (7.16B), and $[M]_{\text{out}}$ from Eq. (7.17C), the first order concentration moment of the terminated chains in the outlet stream $Q_{1,tp,\text{out}} = \mu_1 Q_{0,tp,\text{out}}$ is given by the following equation:

$$Q_{1,tp,\text{out}} = \tau_{CSTR}\mu'_{1,td}R_{td} = \tau_{CSTR}\lambda_1 k_{td}\,[RP]^2 = [M]_{\text{in}} - [M]_{\text{out}} \tag{7.21A}$$

where $[M]_{\text{out}}$ is given in terms of $[M]_{\text{in}}$ by Eq. (7.17B). Upon substitution for $Q_{0,tp,\text{out}}$ from Eq. (7.20A) and λ_1 from Eq. (7.18B), the first order mole-fraction moment of the terminated chains, μ_1, for the reaction system in a CSTR is given by the following equation:

$$\mu_1 = \frac{Q_{1,tp,\text{out}}}{Q_{0,tp,\text{out}}} = \frac{\tau_{CSTR}\lambda_1 k_{td}\,[RP]^2}{\tau_{CSTR}k_{td}\,[RP]^2} = \lambda_1 = L - 1 = \frac{k_p\,[M]_{\text{out}}}{\left(2k_d k_{td} f_0\,[I]_{\text{out}}\right)^{1/2}} - 1 \tag{7.21B}$$

If $L >> 1$, the μ_1, which is equal to $\overline{DP}_{n,tp}$, for the reaction system in a CSTR simplifies as follows:

$$\mu_1 = L - 1 \approx L \approx \frac{k_p\,[M]_{\text{out}}}{\left(2k_d k_{td} f_0\,[I]_{\text{out}}\right)^{1/2}} \quad \in L >> 1 \tag{7.21C}$$

According to Eq. (6.9) in Chapter 6, $\mu'_{k,td} = \lambda_k$; thus $\mu'_{2,td} = \lambda_2$. Applying the conservation equation for the terminated chains (second order), with $\mu'_{2,td} = \lambda_2$ and $R_{td} = k_{td}\,[RP]^2$, λ_2 from Eq. (7.19C), λ_1 from Eq. (7.18B), $[RP]$ from Eq. (7.16B), and $[M]_{\text{out}}$ from Eq. (7.17C), the second order concentration moment of the terminated chains in the outlet stream $Q_{2,tp,\text{out}} = \mu_2 Q_{0,tp,\text{out}}$ is given by the following equation:

$$\begin{aligned} Q_{2,tp,\text{out}} &= \mu_2 Q_{0,tp,\text{out}} = \tau_{CSTR}\mu'_{2,td}R_{td} = \tau_{CSTR}\lambda_2 2k_d f_0\,[I]_{\text{out}} \\ &= \tau_{CSTR}\left[\frac{(2\lambda_1 + 1)\,k_p\,[M]_{\text{out}}}{\left(2k_d k_{td} f_0\,[I]_{\text{out}}\right)^{1/2}} + 1\right] 2k_d f_0\,[I]_{\text{out}} \end{aligned} \tag{7.22A}$$

If $L >> 1$ and substitution for λ_1 from Eq. (7.18C), the second order mole-fraction moment of the terminated chains, μ_2, for the reaction system in a CSTR is given by the following equation:

$$\begin{aligned} \mu_2 &= \frac{Q_{2,tp,\text{out}}}{Q_{0,tp,\text{out}}} = \frac{\tau_{CSTR}\lambda_2 k_{td}\,[RP]^2}{\tau_{CSTR}k_{td}\,[RP]^2} = \lambda_2 \\ &= \frac{(2\lambda_1 + 1)\,k_p\,[M]_{\text{out}}}{\left(2k_d k_{td} f_0\,[I]_{\text{out}}\right)^{1/2}} + 1 \approx \frac{(2\lambda_1 + 1)\,k_p\,[M]_{\text{out}}}{\left(2k_d k_{td} f_0\,[I]_{\text{out}}\right)^{1/2}} \quad \in L >> 1 \end{aligned} \tag{7.22B}$$

The weight-average molecular weight of the polymer chains, $\overline{DP}_{w,tp} = \frac{\mu_2}{\mu_1}$, for the reaction system in a CSTR for $L >> 1$ is given by the following equation:

$$\overline{DP}_{w,tp} = \frac{\mu_2}{\mu_1} = \frac{Q_{2,tp,\text{out}}}{Q_{1,tp,\text{out}}} = \frac{\lambda_2}{\lambda_1} \approx \frac{2\lambda_1 + 1}{\lambda_1} \approx 2 \tag{7.22C}$$

The equations for μ_2, $\overline{DP}_{n,tp}$, and $\overline{DP}_{w,tp}$ of the reaction system in CSTR are similar in form to those in BSVR with the difference that the polymer concentration and its different molecular weights in CSTR are at steady state, thus they are not a function of time.

7.3.3 Modeling Radical Polymerization with Termination by Disproportionation in a Plug Flow Tubular Reactor (PFTR)

In PFTR, the reactants flow inside a tubular reactor as an ideal plug flow, that is, with a uniform velocity in the direction perpendicular to the flow and no mixing in the flow direction, as shown in Figure 5.7 of Chapter 5. The reactants undergo polymerization as they move in a plug flow inside the tube. The reaction is at steady state at every position in the tubular reactor but conversion of monomer to polymer increases in the flow direction. Therefore, a PFTR resembles many small CSTR reactors in series, with the outlet flow of one CSTR as the inlet flow to the next CSTR. The volume of each CSTR is $\Delta V = A\Delta z$ with length Δz in the flow direction and tubular cross-sectional area A. The reaction is at steady state with respect to time, but composition changes in the flow direction. The general conservation equations for the kth order concentration moment of the propagating and terminated chains, and small molecule reactants for PFTR, derived in Chapter 5, are given by Eqs. (5.22D, E, F) in Chapter 5, respectively. The following two equations compare the kth order concentration moment of the propagating chains for batch stirred vessel reactor (BSVR) with that of PFTR:

$$\frac{d\left(Q_{k,rp}\right)}{dt} = \sum_i \lambda'_{k,i} R_{rp,prd,i} - \sum_l \lambda'_{k,l} R_{rp,con,l} \tag{7.23A}$$

$$\frac{1}{\tau_{PFTR}} \frac{d\left(Q_{k,rp}\right)}{dz'} = \sum_i \lambda'_{k,i} R_{rp,prd,i} - \sum_l \lambda'_{k,l} R_{rp,con,l} \tag{7.23B}$$

The difference between the above two equations is that the differential time dt in BSVR is replaced with $\tau_{PFTR} dz'$ where τ_{PFTR} is the mean residence time of the reactants/products in the reactor, $z' = \frac{z}{L_{PFTR}}$ is the normalized length of the reactor in the z-direction, and L_{PFTR} is the reactor length.

For the radical polymerization with elementary reactions 7.2A–E in a BSVR, the concentrations for initiator $[I]$, primary radical $[R]$, reactive propagating chains $[RP]$, monomer $[M]$, and terminated chains $[TP]$, derived in Section 7.3.1, are given by Eqs. (7.3B), (7.4B), (7.5F), (7.6F), and (7.9D). By replacing dt with $\tau_{PFTR} dz'$, the corresponding equations for the concentrations in a PFTR are given as follows:

$$[I] = [I]_0\, e^{-k_d \tau_{PFTR} z'} \tag{7.24A}$$

$$[R] = \left[\frac{2k_d\left(1 - f_0\right)[I]_0}{k_c}\right]^{1/2} e^{\frac{-k_d \tau_{PFTR} z'}{2}} \tag{7.24B}$$

$$[RP] = \left[\frac{2k_d f_0\, [I]_0}{k_{td}}\right]^{1/2} e^{-k_d \tau_{PFTR} z'/2} \tag{7.24C}$$

$$[M] = [M]_0 \exp\left[-\left(8k_p^2 f_0\, [I]_0 / k_d k_{td}\right)^{1/2}\left(1 - e^{-k_d \tau_{PFTR}\, z'/2}\right)\right] \tag{7.24D}$$

$$[TP] = 2 f_0\, [I]_0 \left(1 - e^{-k_d \tau_{PFTR} z'}\right) \tag{7.24E}$$

Likewise, the first, λ_1, and second, λ_2, mole-fraction moments of the propagating chains, and the first, $\mu_1 = \overline{DP}_{n,tp}$, and second, $\mu_2 = \mu_1 \overline{DP}_{w,tp}$, mole-fraction moments of the terminated chains in a BSVR are given by Eqs. (7.7H), (7.8F), (7.10E), and (7.11D). By replacing dt with $\tau_{PFTR}dz'$, the corresponding equations for the mole-fraction moments in a PFTR are given as follows:

$$\lambda_1 = L = \left[\frac{[M]_0^2 k_p^2}{2k_d f_0 [I]_0 k_{td}}\right]^{1/2} \left(e^{k_d \tau_{PFTR} z'/2}\right) \exp\left[-\left(8k_p^2 f_0 [I]_0 / k_d k_{td}\right)^{1/2} \left(1 - e^{-k_d \tau_{PFTR} z'/2}\right)\right] \tag{7.25A}$$

$$\lambda_2 = 2L^2 - L \approx 2L^2 \quad \in L >> 1 \tag{7.25B}$$

$$\mu_1 = \overline{DP}_{n,tp} = \frac{[M]_0 \left\{1 - \exp\left[-\left(8k_p^2 f_0 [I]_0 / k_d k_{td}\right)^{1/2} \left(1 - e^{-k_d \tau_{PFTR} z'/2}\right)\right]\right\}}{2 f_0 [I]_0 \left(1 - e^{-k_d \tau_{PFTR} z'}\right)} \tag{7.25C}$$

$$\mu_2 = \frac{\left([M]_0^2 k_p^2 / k_{td}\right) \int_{z'=0}^{z'} \exp\left[-2\left(8k_p^2 f_0 [I]_0 / k_d k_{td}\right)^{1/2} \left(1 - e^{-k_d \tau_{PFTR} z'/2}\right)\right] dz'}{f_0 [I]_0 \left[1 - e^{-k_d \tau_{PFTR} z'}\right]} \tag{7.25D}$$

7.4 Modeling Radical Polymerization with Chain Transfer Agent

A chain transfer agent (CTA) is a small molecule added to the polymerization, which reacts with a propagating chain to form a terminated polymer chain and a new primary radical that, in turn, can react with monomer to generate a new propagating chain [57–65]. In some polymerizations, like ethylene, the monomer or impurities like methane, ethane, or propane act as a chain transfer agent [57]. A chain transfer agent is added to the polymerization reaction to control the degree of polymerization. The elementary reactions for radical polymerization with termination by disproportionation and transfer to CTA are as follows:

$$I \xrightarrow{k_d} {}^2R_1 \tag{7.26A}$$

$$R_1 + R_1 \xrightarrow{k_c} I^{\bullet} \tag{7.26B}$$

$$R_1 + M \xrightarrow{k_{ci}} RP_1 \tag{7.26C}$$

$$RP_n + M \xrightarrow{k_p} RP_{n+1} \tag{7.26D}$$

$$RP_n + CTA \xrightarrow{k_{cta}} TP_n + R_2 \tag{7.26E}$$

$$R_2 + M \xrightarrow{k_{ci}} RP_1 \tag{7.26F}$$

$$RP_n + RP_m \xrightarrow{k_{td}} TP_n + TP_m + TDB \tag{7.26G}$$

where R_2 is the new primary reactive species produced by chain transfer reaction and k_{tra} is the rate constant for the chain transfer reaction. In Rxn. 7.26F, it is assumed that the rate constant for chain initiation reaction between a monomer and R_2 is the same as that for a monomer with R_1. If chain transfer agent is the monomer, then Rxn. 7.26C represents both chain transfer and new chain initiation reactions. The time dependence of the initiator dissociation is the same as the polymerization without transfer agent (see Eq. 7.3B). The molar mass balance for the primary reactive species is given by the following expression:

$$\left\{\begin{array}{l}\text{Rate of change}\\ \text{in the number}\\ \text{of moles of } R\\ \text{with time}\end{array}\right\} = \left\{\begin{array}{l}\text{Rate of}\\ \text{production}\\ \text{of } R1 \text{ by I}\\ \text{dissociation}\end{array}\right\} - \left\{\begin{array}{l}\text{Rate of}\\ \text{consumption}\\ \text{of } R1 \text{ by}\\ \text{recombination}\end{array}\right\} \tag{7.27A}$$

$$- \left\{\begin{array}{l}\text{Rate of}\\ \text{consumption}\\ \text{of } R1 \text{ by}\\ \text{chain initiation}\end{array}\right\} + \left\{\begin{array}{l}\text{Rate of}\\ \text{production}\\ \text{of } R2 \text{ by}\\ \text{chain transfer}\end{array}\right\} - \left\{\begin{array}{l}\text{Rate of}\\ \text{consumption}\\ \text{of } R2 \text{ by}\\ \text{chain initiation}\end{array}\right\}$$

The above mole conservation balance in mathematical form for a constant volume reaction is given by the following equation:

$$\frac{d\,([R])}{dt} = +2k_d f_0\,[I] - k_{ci}\,[R_1]\,[M] + k_{tra}\,[RP]\,[CTA] - k_{ci}\,[R_2]\,[M] \tag{7.27B}$$

where $[R] = [R_1] + [R_2]$. Using QSSA for primary reactive species and rearranging the above equation, the following equation is obtained for $[R]$ in terms of the initiator, monomer, and transfer agent concentrations:

$$[R] = \frac{2k_d f_0\,[I] + k_{cta}\,[RP]\,[CTA]}{k_{ci}\,[M]} \tag{7.27C}$$

Comparison of the above equation with Eq. (7.4A) for radical polymerization in the absence of CTA shows that $[R]$ is higher in the presence of CTA.

7.4.1 Concentration of Propagating Chains in Radical Polymerization with Chain Transfer Agent

The concentration of reactive propagating chains, $[RP]$, is determined from the conservation equation for the zero-order moment of the propagating chains (see Eq. 7.1A). Using QSSA for the propagating chains and with $\lambda'_{0,i} = \lambda'_{0,l} = 1$ by definition, the terms for the rates of production and consumption in the conservation equation for the zero-order moment of propagating chains are given by the following relations:

$$\sum_i R_{rp,prd,i} = k_{ci}\,[M]\,[R_1] + k_{ci}\,[M]\,[R_2] + k_p\,[M]\,[RP] \tag{7.28A}$$

$$\sum_l R_{rp,con,l} = k_{td}[RP]^2 + k_p\,[RP]\,[M] + k_{tra}\,[RP]\,[CTA] \tag{7.28B}$$

Using the equality $[R] = [R_1] + [R_2]$ and Eq. (7.27B), and canceling equal terms results in the following expression for $[RP]$:

$$[RP] = \left(\frac{2k_d f_0 [I]}{k_{td}}\right)^{1/2} = \left[\frac{2k_d f_0 [I]_0}{k_{td}}\right]^{1/2} e^{-k_d t/2} \tag{7.28C}$$

The above equation shows that the concentration of propagating chains is the same with or without the addition of transfer agent to the reaction because the transfer agent terminates an n-mer and initiates a new 1-mer propagating chain.

7.4.2 Concentration of Chain Transfer Agent in Radical Polymerization with Chain Transfer Agent

CTA is not produced during the reaction, but it is consumed by elementary reaction 7.26E. Based on the general conservation equation for small molecules in a BSVR system (Eq. 7.1D), the time dependence of CTA is determined from the molar conservation equation as follows:

$$\frac{d[CTA]}{dt} = -R_{cta} = -k_{cta}[CTA][RP] \tag{7.29A}$$

where R_{cta} is the rate of transfer of reactive species from the propagating chain to CTA. Upon separation of variables, substitution for $[RP]$ from Eq. (7.28C), and integration, the time dependence of $[CTA]$ is expressed by the following equation:

$$[CTA] = [CTA]_0 \exp\left[-\left(\frac{8k_{tra}^2 f_0 [I]_0}{k_d k_{td}}\right)^{1/2} \left(1 - e^{-k_d t/2}\right)\right] \tag{7.29B}$$

7.4.3 Monomer Concentration in Radical Polymerization with Chain Transfer Agent

Monomer is consumed by Rxns. 7.26C, D, F. Based on the general conservation equation for small molecules in a BSVR with constant reaction volume (Eq. 7.1D), the rate of change of the monomer concentration with time is expressed by the following equation:

$$\begin{aligned}\frac{d[M]}{dt} &= -\left(k_p[RP][M] + k_{ci}[R_1][M] + k_{ci}[R_2][M]\right)\\ &= -\left(k_p[RP][M] + 2f_0 k_d [I] + k_{cta}[RP][CTA]\right)\end{aligned} \tag{7.30A}$$

The application of QSSA to R_1 and R_2 results in $k_{ci}[R_1][M] = 2f_0 k_d [I]$ and $k_{ci}[R_2][M] = k_{cta}[RP][CTA]$, respectively. For producing a high molecular weight polymer, it can be assumed that $k_p[RP][M] >> 2f_0 k_d [I] + k_{cta}[RP][CTA]$. After substitution for $[RP]$ from Eq. (7.28C) followed by separation of variables and integration, the time dependence of $[M]$ is given by the following equation:

$$[M] = [M]_0 \exp\left[-\left(8k_p^2 f_0 [I]_0 / k_d k_{td}\right)^{1/2} \left(1 - e^{-k_d t/2}\right)\right] \quad \in L >> 1 \tag{7.30B}$$

The above equation is the same as the one derived for radical polymerization in the absence of chain transfer agent (see Eq. 7.6F). This is logical considering that CTA does not increase the concentration of propagating chains which does not change the rate of monomer consumption in the reaction.

7.4.4 *k*th Order Moments of Propagating Chains in Radical Polymerization with Chain Transfer Agent

In the presence of a transfer agent, the propagating chains are produced by Rxns. 7.26C, D, F and consumed by Rxns. 7.26D–F. Based on the general conservation equation for the moments of the propagating chains in a BSVR system under QSSA (Eq. 7.1A), the conservation equation for the first mole-fraction moment of the propagating chains is given by the following expression:

$$\begin{aligned} &k_{ci}\,[M]\,[R] + (\lambda_1 + 1)\,k_p\,[RP]\,[M] + k_{ci}\,[M]\,[R_2] \\ &= \lambda_1 k_p\,[RP]\,[M] + \lambda_1 k_{cta}\,[RP]\,[CTA] + \lambda_1 k_{td}\,[RP]^2 \end{aligned} \tag{7.31A}$$

where λ_1 is the first mole-fraction moment of the propagating chains. Upon rearrangement, λ_1 is given by the following expression:

$$\lambda_1 = \frac{k_p\,[RP]\,[M] + k_{ci}\,[R_1]\,[M] + k_{ci}\,[R_2]\,[M]}{k_{cta}\,[RP]\,[CTA] + k_{td}\,[RP]^2} \tag{7.31B}$$

The numerator in the above equation is the rate of propagation, whereas the denominator is the rate of termination. Using the equality $k_{cta}\,[RP]\,[CTA] + k_{td}\,[RP]^2 = k_{ci}\,[R_1]\,[M] + k_{ci}\,[R_2]\,[M]$, which is obtained by applying QSSA to the production and consumption of all reactive species, it is apparent that the denominator in the above equation is the rate of chain initiation. Therefore, the first moment is also the kinetic chain length, L, of the propagating chains. Assuming $L >> 1$, λ_1 can be expressed by the following equation:

$$\begin{aligned} \lambda_1 &= \frac{k_p\,[RP]\,[M]}{k_{cta}\,[RP]\,[CTA] + k_{td}\,[RP]^2} + 1 \approx \frac{k_p\,[M]}{k_{cta}\,[CTA] + k_{td}\,[RP]} = L \\ &= \frac{k_p\,[M]_0\, e^{\left[-\left(8k_p^2 f_0 [I]_0/k_d k_{td}\right)^{1/2}\left(1-e^{-k_d t/2}\right)\right]}}{k_{cta}\,[CTA]_0\, e^{\left[-\left(8k_{tra}^2 f_0 [I]_0/k_d k_{td}\right)^{1/2}\left(1-e^{-k_d t/2}\right)\right]} + \left(2k_d k_{td} f_0\,[I]_0\right)^{1/2} e^{-k_d t/2}} \end{aligned} \tag{7.31C}$$

where Eqs. (7.28C), (7.29B), and (7.30B) were used to substitute for $[RP]$, $[CTA]$, and $[M]$ in terms of time, respectively. Comparison of the above equation with the equation for L without chain transfer agent (see Eq. 7.7H) indicates that L is reduced with the addition of transfer agent as a result of addition of the term $k_{cta}\,[RP]\,[CTA]$ in the denominator. Based on the general conservation equation for the kth moment of the propagating chains (Eq. 7.1A), it can be shown that the second mole-fraction moment of the propagating chains is given by the following expression (see practice problems of the chapter):

$$\lambda_2 = \frac{(2\lambda_1 + 1)\,k_p\,[M]\,[RP]}{k_{cta}\,[RP]\,[CTA] + k_{td}\,[RP]^2} + 1 = 2L^2 - L \approx 2L^2 \quad \in L >> 1 \tag{7.31D}$$

7.4.5 Concentration of Terminated Polymer Chains in Radical Polymerization with Chain Transfer Agent

The concentration of terminated chains is given by the zero-order concentration moment of the terminated chains (Eq. 7.1C). In the reactions, terminated chains are produced by chain transfer Rxn. 7.26E and termination Rxn. 7.26G. Based on the moment conservation Eq. (7.1C) for terminated chains, the differential equation for the concentration of terminated chains is given by the following expression:

$$\frac{dQP_{0,t}}{dt} = \frac{d\,[TP]}{dt} = R_{td} + R_{cta} = k_{td}\,[RP]^2 + k_{cta}\,[RP]\,[CTA] \tag{7.32A}$$

where R_{cta} is the rate of chain transfer by transfer agent. It can be shown using an overall molar mass conservation balance of the reactive species and application of QSSA that $k_{td}\,[RP]^2 = 2f_0 k_d\,[I]$. Using Eqs. (7.3B), (7.28C), and (7.29B) for the time dependence of $[I]$, $[RP]$, and $[CTA]$, respectively, in the above equation followed by integration, $[TP]$ as a function of time for radical polymerization in the presence of transfer agent is given as follows:

$$\begin{aligned}[TP] &= 2f_0\,[I]_0\left(1 - e^{-k_d t}\right)\\ &+[TA]_0\left[\frac{2k_{tra}^2 k_d f_0\,[I]_0}{k_{td}}\right]^{1/2}\int_{t=0}^{t} e^{\left[-\left(8k_{tra}^2 f_0 [I]_0/k_d k_{td}\right)^{1/2}\left(1-e^{-k_d t/2}\right)\right]} e^{-k_d t/2}\,dt\end{aligned} \tag{7.32B}$$

Comparison of the above equation with the equation for $[TP]$ without transfer agent (see Eq. 7.9D) indicates that $[TP]$ increases with the addition of a transfer agent to the reaction due to the term $k_{cta}\,[RP]\,[CTA]$ for chain transfer to the term $k_{td}\,[RP]^2$ for termination.

7.4.6 Higher Moments of Terminated Chains in Radical Polymerization with Chain Transfer Agent

The number-average DP of the terminated chains in radical polymerization with chain transfer agent, $\overline{DP}_{n,tp}$, is determined from the conservation equation for the first order moment of the terminated chains (Eq. 7.1C). Based on the conservation Eq. (7.1C), Eq. (7.31C) for λ_1, and using Eq. (7.30A) for $\frac{d\,[M]}{dt}$, the conservation equation for the first concentration moment of the terminated chains is given as follows:

$$\frac{d\left(Q_{1,tp}\right)}{dt} = \lambda_1\left(k_{cta}\,[RP]\,[CTA] + k_{td}\,[RP]^2\right) = -\frac{d\,[M]}{dt} \tag{7.33A}$$

Upon using the equality $\overline{DP}_{n,tp} = Q_{1,tp}/Q_{0,tp} = Q_{1,tp}/\,[TP]$ and integration of the above equation, the $\overline{DP}_{n,tp}$ for the terminated chains is given by the following equation:

$$\overline{DP}_{n,tp} = \frac{Q_{1,tp}}{Q_{0,tp}} = \frac{[M]_0 - [M]}{[TP]} \tag{7.33B}$$

where $[M]$ and $[TP]$ are given as a function of time by Eqs. (7.30B) and (7.32B), respectively. Although the functional form of $[M]$ is the same in radical polymerization with or without transfer agent, $[TP]$ is higher with chain transfer agent which reduces $\overline{DP}_{n,tp}$. It can be shown that $Q_{2,tp}$ and $\overline{DP}_{w,tp}$ of the terminated chains in radical polymerization with chain transfer agent are given by the following equations:

$$\begin{aligned}Q_{2,tp} &= \int_{t=0}^{t} \lambda_2\left(k_{cta}\,[RP]\,[CTA] + k_{td}[RP]^2\right)dt\\ &= 2k_p\int_{t=0}^{t} L\,[M]\,[RP]\,dt\end{aligned} \tag{7.33C}$$

$$\overline{DP}_{w,tp} = \frac{Q_{2,tp}}{Q_{1,tp}} = \frac{2k_p \int_{t=0}^{t} L\,[M]\,[RP]\,dt}{[M]_0 - [M]} \tag{7.33D}$$

where $[RP]$, $[M]$, and L are given as a function of time by Eqs. (7.28C), (7.30B), and (7.31C) respectively. The polydispersity index, PDI, is determined from the ratio of $\frac{\overline{DP}_{w,tp}}{\overline{DP}_{n,tp}}$. In general, the addition of a chain transfer agent reduces both $\overline{DP}_{n,tp}$ and $\overline{DP}_{w,tp}$ in agreement with the function of the chain transfer agent in polymerization to control the polymer molecular weight [65–69]. The addition of chain transfer agent to the polymerization reaction can potentially reduce polydispersity as well as the molecular weight [68].

7.5 Modeling Radical Polymerization with Chain Transfer to the Terminated Polymer Chains

Chain transfer to polymer is studied extensively in polymerization of acrylates and acetates, like n-butyl acrylate and vinyl acetate [70–76]. Chain transfer to polymer is also important in polymerization of ethylene which causes the formation of short and long branches [76, 77]. The elementary reactions for radical polymerization with chain transfer to polymer are as follows:

$$I \xrightarrow{k_d} 2R \tag{7.34A}$$

$$R + R \xrightarrow{k_c} I^{\bullet} \tag{7.34B}$$

$$R + M \xrightarrow{k_{ci}} RP_1 \tag{7.34C}$$

$$RP_n + M \xrightarrow{k_p} RP_{n+1} \tag{7.34D}$$

$$RP_n + TP_m \xrightarrow{k_{ctp}} TP_n + RP'_m \tag{7.34E}$$

$$RP'_m + M \xrightarrow{k'_p \approx k_p} RP_{n+1} + LB \tag{7.34F}$$

$$RP_n + RP_m \xrightarrow{k_{td}} TP_n + TP_m + TDB \tag{7.34G}$$

where k_{ctp} is the rate constant for the chain transfer to polymer, k'_p is the rate constant for the reaction of a monomer with the reactive species on a tertiary carbon of a propagating chain, that is RP', to form a reactive species on a secondary carbon of that propagating chain and a long branch. It is assumed in this reaction system that $k'_p \approx k_p$ because for every propagation reaction with rate constant k'_p there are order of magnitude higher propagation reactions with rate constant k_p in long polymer chains. In this system, we assume that chain transfer to polymer is not dependent on chain length which is a good assumption at relatively low polymer concentrations. This also implies that a polymer chain undergoing chain transfer holds one reactive species, that is, there is no polyradical chain and no gelation in the polymerization [78, 79]. Examination of the elementary reactions indicates that the presence of chain transfer to polymer does not affect initiator dissociation, formation of primary reactive species, reactive propagating chains, and consumption of the monomer. Therefore, $[I]$, $[R]$, $[RP]$, and $[M]$ as a function of time are given by Eqs. (7.3B), (7.4B), (7.5F), and (7.6F), respectively. As a terminated polymer chain is consumed and another terminated chain is produced in the chain transfer to polymer reaction (Rxn. 7.34E), $[TP]$ is also not affected by chain transfer to polymer and is given by Eq. (7.9D). Using QSSA for the reactive species, a conservation of molar mass balance for the RP' propagating chains yields the following equation for the concentration of RP':

$$[RP'] = \frac{k_{ctp}}{k_p} \frac{[RP]\,[TP]}{[M]} \tag{7.35}$$

The above equation infers that $[RP']$ is proportional to the ratio k_{ctp}/k_p. Furthermore, the above equation indicates that chain transfer to polymer and branching increases with conversion as monomer is consumed and polymer is produced.

7.5.1 Higher Moments of Propagating Chains in Radical Polymerization with Chain Transfer to Terminated Chains

In the presence of chain transfer to polymer, the propagating chains are produced by Rxns. 7.34C–F and consumed by Rxns. 7.34D–G. Applying QSSA and using the general conservation equation for the kth order mole-fraction moment of the propagating chains (Eq. 7.1A), λ_1, is given by the following expression:

$$\begin{aligned} k_{ci}\,[M]\,[R] + (\lambda_1 + 1)\,k_p\,[RP]\,[M] + (\mu_1)\,k_{ctp}\,[RP]\,[TP] \\ + (\mu_1 + 1)\,k_p\,[RP']\,[M] = \lambda_1 k_p\,[RP]\,[M] \\ + \lambda_1 k_{ctp}\,[RP]\,[TP] + (\mu_1)\,k_p\,[RP']\,[M] + \lambda_1 k_{td}\,[RP]^2 \end{aligned} \tag{7.36A}$$

Upon canceling like terms and rearrangement, the above expression for λ_1 simplifies by the following equation:

$$\lambda_1 = \frac{k_p\,[RP]\,[M] + k_{ci}\,[R_1]\,[M] + k_p\,[RP']\,[M] + (\mu_1 - \lambda_1)\,k_{ctp}\,[RP]\,[TP]}{k_{td}\,[RP]^2} \tag{7.36B}$$

The numerator in the above equation is the rate of addition of monomers to the propagating chains, and the denominator is the rate of chain termination (or the rate of initiation since QSSA is valid) at any time t, which is equal to the kinetic chain length or L. Taking the term involving λ_1 from the numerator to the denominator, substituting for $k_p\,[RP']\,[M]$ from Eq. (7.35), and using the equality $k_{ci}\,[R_1]\,[M] = k_{td}\,[RP]^2$ (see Eq. 7.5E), $\lambda_1 = L$ is given by the following equation:

$$\lambda_1 = L = \frac{k_p\,[RP]\,[M] + \mu_1 k_{ctp}\,[RP]\,[TP]}{k_{ctp}\,[RP]\,[TP] + k_{td}\,[RP]^2} + 1 \tag{7.36C}$$

where $[RP]$, $[M]$, and $[TP]$ are given by Eqs. (7.5F), (7.6F), (7.6D), and (7.9D), respectively, and μ_1 is determined by the conservation equation for the first order mole-fraction moment of the terminated polymer chains. The above equation shows that L is affected by the degree of polymerization of the terminated polymer chains. In a similar approach, using the general conservation Eq. (7.1A), it can be shown that the second order mole-fraction moment of the propagating chains in the presence of chain transfer to polymer is given by the following equation:

$$\lambda_2 = \frac{(2\lambda_1 + 1)\,k_p\,[RP]\,[M] + (\mu_2 + 2\mu_1)\,k_{ctp}\,[RP]\,[TP]}{k_{ctp}\,[RP]\,[TP] + k_{td}\,[RP]^2} + 1 \tag{7.36D}$$

where μ_2 is determined in the next section by the conservation equation for the second order mole-fraction moment of the terminated polymer chains.

7.5.2 Higher Moments of Terminated Polymer Chains in Radical Polymerization with Chain Transfer to Terminated Chains

In Rxns. 7.34A–G, terminated chains are produced by transfer to polymer Rxn. 7.34E with first and second order mole-fraction moments λ_1 and λ_2, respectively; terminated polymer chains are consumed by transfer to polymer reaction with first and second order mole-fraction moments μ_1 and μ_2; and terminated chains are produced by termination by disproportionation with first and second order mole-fraction moments λ_1 and λ_2. Therefore, the conservation equations for the first and second order mole-fraction moments of the terminated polymer chains for this reaction system are as follows:

$$\frac{d\left(Q_{1,tp}\right)}{dt}=\frac{d\left(\mu_1\left[TP\right]\right)}{dt}=\left[RP\right]\left(\lambda_1 k_{ctp}\left[TP\right]+\lambda_1 k_{td}\left[RP\right]-\mu_1 k_{ctp}\left[TP\right]\right)\tag{7.37A}$$

$$\frac{d\left(Q_{2,tp}\right)}{dt}=\frac{d\left(\mu_2\left[TP\right]\right)}{dt}=\left[RP\right]\left(\lambda_2 k_{ctp}\left[TP\right]+\lambda_2 k_{td}\left[RP\right]-\mu_2 k_{ctp}\left[TP\right]\right)\tag{7.37B}$$

where $[RP]$, $[TP]$ and λ_1 are given by Eq. (7.5F), (7.32B), and (7.36C), respectively. The above nonlinear equations can be solved numerically with initial condition $\mu_{1,tp}=\mu_{2,tp}=0$ at $t=0$ to determine μ_1 and μ_2 at any time t. Once the first and second mole-fraction moments of the terminated chains are determined, then the $\overline{DP}_{n,tp}$, $\overline{DP}_{w,tp}$, and PDI of the terminated chains in radical polymerization with chain transfer to terminated chains are given by $\overline{DP}_{n,tp}=\mu_1$, $\overline{DP}_{w,tp}=\mu_2/\mu_1$, and $PDI=\overline{DP}_{w,tp}/\overline{DP}_{n,tp}$.

7.5.3 Long Branching Density in Radical Polymerization with Chain Transfer to Terminated Chains

Elementary reaction 7.34F results in the formation of long branches in the polymer chains, which affects properties and processing of the polymer [80–86]. Branching in polymer chains is characterized by branching density which is the number of branches per mole of polymer chains or n_{LB}. It is the ratio of the concentration of long branches, $[LB]$, to the concentration of terminated chains, $[TP]$. For a constant volume reaction, the molar conservation equation for long branches is given by the following expression:

$$\left\{\begin{matrix}\text{Rate of change}\\ \text{in the number of}\\ \text{moles of } LB\\ \text{with time}\end{matrix}\right\}=\left\{\begin{matrix}\text{Rate of molar}\\ \text{production of}\\ LB\end{matrix}\right\}-\left\{\begin{matrix}\text{Rate of molar}\\ \text{consumption}\\ \text{of } LB\end{matrix}\right\}\tag{7.38A}$$

Long branches are produced by the elementary Rxn. 7.34F. Therefore, the above conservation equation is expressed mathematically as follows:

$$\frac{d\left[LB\right]}{dt}=k_p\left[RP'\right]\left[M\right]=k_{ctp}\left[RP\right]\left[TP\right]\tag{7.38B}$$

where Eq. (7.35) is used to substitute for $[RP']$ in terms of $[RP]$. Upon substitution for $[RP]$ and $[TP]$ from Eqs. (7.5F) and (7.9D), respectively, the differential equation relating $[LB]$ to reaction time is given as follows:

$$\frac{d\left[LB\right]}{dt}=\left(8k_{ctp}^2 k_d f_0^3 [I]_0^3/k_{td}\right)^{1/2}\left(e^{-k_d t/2}-e^{-3k_d t/2}\right)\tag{7.38C}$$

After integration using the initial condition $[LB]=0$ at $t=0$, $[LB]$ is given with respect to time by the following equation:

$$\left[LB\right]=\left(8k_{ctp}^2 f_0^3 [I]_0^3/k_{td}k_d\right)^{1/2}\left[\frac{4}{3}+\frac{2}{3}e^{-3k_d t/2}-2e^{-k_d t/2}\right]\tag{7.38D}$$

Using Eq. (7.9D) for [TP], the branching density, n_{LB} which is the ratio of $[LB]$ to $[TP]$, is given by the following equation:

$$n_{LB} = \frac{[LB]}{[TP]} = \frac{\left(2k_{ctp}^{2} f_0 [I]_0 / k_{td} k_d\right)^{1/2} \left[\frac{4}{3} + \frac{2}{3} e^{-3k_d t/2} - 2e^{-k_d t/2}\right]}{(1 - e^{-k_d t})} \tag{7.38D}$$

At long reaction times or as $t \to \infty$, the above equation predicts the following relation between branching density and the reaction rate constants:

$$n_{LB}\big|_{t\to\infty} = \frac{4}{3} k_{ctp} \left(\frac{2 f_0 [I]_0}{k_{td} k_d}\right)^{1/2} \tag{7.38E}$$

The above equation asserts that, at long times, branching density increases linearly with k_{ctp} and decreases with the square root of k_{td}.

7.6 Modeling Radical Polymerization with β_2 Chain Scission

This radical polymerization system consists of a thermal initiator, chain termination by disproportionation, and chain transfer to polymer to generate reactive species on a secondary carbon as the first step toward β_2 chain scission but no other chain transfer reaction (see Section 2.2.8 in Chapter 2) [63, 82, 87–100]. The elementary reactions for this polymerization system are as follows:

$$I \xrightarrow{k_d} 2R \tag{7.39A}$$

$$R + R \xrightarrow{k_c} I^{\bullet} \tag{7.39B}$$

$$R + M \xrightarrow{k_{ci}} RP_1 \tag{7.39C}$$

$$RP_n + M \xrightarrow{k_p} RP_{n+1} \tag{7.39D}$$

$$RP_n + TP_m \xrightarrow{k_{ctp}} TP_n + RP'_m \tag{7.39E}$$

$$RP'_m \xrightarrow{k_{\beta 2}} RP_x + TP_{m-x} + TDB \tag{7.39F}$$

$$RP_n + RP_m \xrightarrow{k_{td}} TP_n + TP_m + TDB \tag{7.39G}$$

In the first step of a β_2 scission in Rxn. 7.39E, the reactive species on a primary carbon atom of a propagating chain is transferred to a terminated polymer chain with m repeat units to generate an unstable reactive species on a secondary carbon atom. In the second step in Rxn. 7.39F, the reactive species on a secondary carbon of a chain with m repeat units, namely RP'_m, splits due to its instability into two chains by scission of the chemical bond between the α and β carbons of the reactive species to form a shorter propagating chain RP_x with the reactive species on a primary carbon and a terminated polymer chain TP_{m-x} with a terminal double bond (TDB). It is assumed that the active species on a secondary carbon, due to steric hindrance, does not terminate by disproportionation or combination, and undergoes chain scission by Eq. (7.39F). It is further assumed that the chain scission reaction has no effect on the production or consumption of primary radicals during the reaction. As the chain scission reaction does not affect initiator dissociation, the formation of primary reactive species, concentration of primary propagating chains, and monomer consumption,

$[I]$, $[R]$, $[RP]$, and $[M]$ concentrations as a function of time are given by Eqs. (7.3B), (7.4B), (7.5F), and (7.6F), respectively. However, β_2 scission affects first and higher order moments of the reactive propagating chains and all moments of the terminated polymer chains. The QSSA assumption applies to the secondary RP' as well as the primary RP propagating chains. A molar mass conservation equation for the propagating RP' chains under QSSA (see Eq. 7.1A) results in the following expression for quasi-steady state RP' concentration:

$$[RP'] = \frac{k_{ctp}}{k_{\beta 2}} [RP] [TP] \tag{7.40}$$

The above equation shows that $[RP']$ is proportional to the ratio $k_{ctp}/k_{\beta 2}$. Furthermore, the above equation indicates that β_2 scission increases with conversion as more terminated polymer chains are produced.

7.6.1 Higher Moments of Propagating Chains in Radical Polymerization with β_2 Chain Scission

There are two types of propagating chains in this polymerization, namely RP with the reactive species on a primary carbon and RP' with the reactive species on a secondary carbon. As QSSA can be applied to both RP and RP' propagating chains, their instantaneous kth order mole-fraction moments are the same as their overall kth order moments. The symbols λ_n and $\lambda_{n,RP'}$ are used to denote the kth order mole-fraction moments of the RP and RP' chains, respectively. Based on Rxn. 7.39E in which RP' is formed from a terminated polymer chain and based on Eq. (6.49) in Section 6.5.1 of Chapter 6, $\lambda_{n,RP'}$ is given by the following relation:

$$\lambda_{n,RP'} = \mu_n \tag{7.41A}$$

where μ_n is the kth order mole-fraction moment of the terminated polymer chains at time t in the reaction. In the next Rxn. 7.39F, the secondary propagating chain RP' with $\lambda_{n,RP'} = \mu_n$ splits and forms a shorter primary propagating chain and a shorter terminated polymer chain. As the chain transfer to polymer is a random process and it happens equally to all secondary carbons in the polymer chain, plus based on Eqs. (6.53) and (6.54) in Section 6.5.1 of Chapter 6, the instantaneous first and second order mole-fraction moments of RP_x and TP_{m-x} chains produced by scission in Rxn. 7.39F are given by the following relations:

$$\begin{cases} \lambda'_{1,\beta 2} = \mu'_{1,\beta 2} = \dfrac{\mu_1}{2} \\ \lambda'_{2,\beta 2} = \mu'_{2,\beta 2} = \dfrac{\mu_2}{2} - \dfrac{\mu_1^2}{4} \end{cases} \tag{7.41B}$$

where $\lambda'_{1,\beta 2}$ and $\lambda'_{2,\beta 2}$ are the first and second order instantaneous mole-fraction moments of the propagating RP_x chains, respectively, and $\mu'_{1,\beta 2}$ and $\mu'_{2,\beta 2}$ are the first and second order instantaneous mole-fraction moments of the terminated TP_{m-x} chains produced by β_2 scission. In this reaction system, the propagating chains are produced by Rxns. 7.39C–F and consumed by Rxns. 7.39D–G. Applying QSSA, the conservation equation for the first order mole-fraction moment of the propagating chains (see general Eq. 7.1A) is given by the following expression:

$$\begin{aligned} k_{ci} [M] [R] &+ (\lambda_1 + 1) k_p [RP] [M] + (\mu_1) k_{ctp} [RP] [TP] \\ &+ (\mu_1/2) k_{\beta 2} [RP'] = \lambda_1 k_p [RP] [M] \\ &+ (\lambda_1) k_{ctp} [RP] [TP] + (\mu_1) k_{\beta 2} [RP'] + (\lambda_1) k_{td} [RP]^2 \end{aligned} \tag{7.41C}$$

Upon canceling like terms and rearrangement, λ_1 is given by the following equation:

$$\lambda_1 = \frac{k_p [RP][M] + k_{ci} [R][M] + (\mu_1 - \lambda_1) k_{ctp} [RP][TP] - \mu_1 k_{\beta 2} [RP']/2}{k_{td} [RP]^2} \tag{7.41D}$$

In the above equation, the numerator is the rate of monomer addition to the propagating chains, whereas the denominator is the rate of termination of propagating chains, and the ratio is the kinetic chain length. Taking the term involving λ_1 from the numerator to the denominator, substituting for $k_{\beta 2} [RP']$ from Eq. (7.40), and using the equality $k_{ci} [R_1][M] = k_{td} [RP]^2$ (see Eqs. 7.5E), results in the following equation for λ_1:

$$\lambda_1 = L = \frac{k_p [RP][M] + k_{td} [RP]^2 + \mu_1 k_{ctp} [RP][TP]/2}{k_{ctp} [RP][TP] + k_{td} [RP]^2} \tag{7.41E}$$

where $[RP]$, $[M]$, and $[TP]$ are given by Eqs. (7.5F), (7.6F), and (7.9D), respectively. In a similar approach, it can be shown that the second order mole-fraction moment of the propagating chains in the presence of chain scission is given by the following equation:

$$\lambda_2 = \frac{(2\lambda_1 + 1) k_p [RP][M] + k_{td} [RP]^2 + (\mu_2/2 - \mu_1^2/4) k_{ctp} [RP][TP]}{k_{ctp} [RP][TP] + k_{td} [RP]^2} \tag{7.41F}$$

where μ_2 and μ_1 are determined in the next section using conservation equations for the first and second order mole-fraction moments of the terminated polymer chains, respectively.

7.6.2 Concentration of Terminated Polymer Chains in Radical Polymerization with β_2 Chain Scission

Terminated polymer chains are produced by Rxns. 7.39E–G, whereas they are consumed by Rx. 7.39E. The conservation equation of the zero-order moment of the terminated polymer chains at constant volume (see Eq. 7.1C) is expressed as follows:

$$\begin{aligned} \frac{d[TP]}{dt} &= \sum_i R_{tp,prd,i} - \sum_l R_{tp,con,l} \\ &= k_{ctp} [RP][TP] + k_{\beta 2} [RP'] + k_{td} [RP]^2 - k_{ctp} [RP][TP] \end{aligned} \tag{7.42A}$$

Deleting like terms in the above equation, substituting for $[RP']$ from Eq. (7.40), and substituting for $[RP]$ from Eq. (7.5F) results in the following equation for the rate of change of $[TP]$ with time:

$$\begin{aligned} \frac{d[TP]}{dt} &= k_{td} [RP]^2 \left(1 + \frac{k_{ctp}}{k_{td}} \frac{[TP]}{[RP]}\right) \\ &= (2k_d f_0 [I]_0) e^{-k_d t} \left[1 + (2k_{td} k_d f_0 [I]_0 / k_{ctp}^2)^{-1/2} e^{k_d t/2} [TP]\right] \end{aligned} \tag{7.42B}$$

The above non-linear differential equation can be solved numerically subject to the initial condition $[TP] = 0$ at $t = 0$ (no polymer initially) to determine $[TP]$ as a function of time.

7.6.3 Higher Moments of Terminated Polymer Chains in Radical Polymerization with β_2 Chain Scission

In this reaction system, terminated chains are produced by chain transfer to polymer Rxn. 7.39E with first and second order mole-fraction moments λ_1 and λ_2, respectively; terminated polymer chains are consumed by chain transfer to polymer reaction with first and second order mole-fraction moments $\mu_1$and μ_2; terminated chains are also produced by β_2 scission with first and second order mole-fraction moments $\frac{\mu_1}{2}$ and$\left(\frac{\mu_2}{2}-\frac{\mu_1^2}{4}\right)$; and terminated chains are produced by termination by disproportionation with first and second order mole-fraction moments λ_1 and λ_2. Therefore, the conservation equation for the first mole-fraction moments of the terminated chains for Rxns. 7.39A–G is given as follows:

$$\frac{d\,(Q_{1,tp})}{dt}=\frac{d\,(\mu_1\,[TP])}{dt} \tag{7.43A}$$
$$=\lambda_1 k_{ctp}\,[TP]\,[RP]-\mu_1 k_{ctp}\,[TP]\,[RP]+\mu_1 k_{\beta 2}\,[RP']\,/2+\lambda_1 k_{td}[RP]^2$$

After substitution for $[RP']$ from Eq. (7.40) and factoring common terms, the conservation equation for the first order mole-fraction moment of the terminated chains is given as follows:

$$\frac{d\,(Q_{1,tp})}{dt}=\frac{d(\mu_1\,[TP])}{dt}=(\lambda_1-\mu_1/2)\,k_{ctp}\,[TP]\,[RP]+\lambda_1 k_{td}[RP]^2 \tag{7.43B}$$

Using a similar approach, the conservation equation for the second order mole-fraction moment of the terminated chains is given as follows:

$$\frac{d\,(Q_{2,tp})}{dt}=\frac{d\,(\mu_2\,[TP])}{dt} \tag{7.43C}$$
$$=(\lambda_2-\mu_2/2-\mu_1^2/4)\,\lambda_2 k_{ctp}\,[TP]\,[RP]+\lambda_2 k_{td}\,[RP]^2$$

where λ_1 and λ_2 are given by Eqs. (7.41E, F), respectively, and $[RP]$ and $[TP]$ are given by Eqs. (7.5F) and (7.9D). The above non-linear equations can be solved numerically to determine the time dependence of μ_1 and μ_2. Once μ_1 and μ_2 are determined, then the $\overline{DP}_{n,tp}$, $\overline{DP}_{w,tp}$, and PDI for radical polymerization with β_2 chain scission are determined by relations $\overline{DP}_{n,tp}=\mu_1$, $\overline{DP}_{w,tp}=\frac{\mu_2}{\mu_1}$, and $PDI=\frac{\overline{DP}_{w,tp}}{\overline{DP}_{n,tp}}$.

7.6.4 Density of Terminal Double Bonds in Terminated Polymer Chains in Radical Polymerization with β_2 Chain Scission

Like branching, TDB is defined as the number of terminal double bonds per mole of polymer chains or n_{TDB} which is the ratio of the concentration of the terminal double bonds, $[TDB]$, to the concentration of terminated chains, $[TP]$. The conservation equation for TDB in polymer chains for a constant volume reaction is expressed as follows (see general Eq. 7.1D):

$$\left\{\begin{matrix}\text{Rate of change}\\ \text{in the number of}\\ \text{moles of } TDB\\ \text{with time}\end{matrix}\right\}=\left\{\begin{matrix}\text{Rate of molar}\\ \text{production of}\\ TDB\end{matrix}\right\}-\left\{\begin{matrix}\text{Rate of molar}\\ \text{consumption}\\ \text{of } TDB\end{matrix}\right\} \tag{7.44A}$$

In Rxns. 7.39A–G, TDB on polymer chains is formed by the scission Rxn. 7.39F and termination by disproportionation Rxn. 7.39G. Therefore, the above conservation equation is expressed mathematically as follows:

$$\frac{d\,[TDB]}{dt}=k_{\beta 2}\,[RP']+k_{td}[RP]^2=k_{ctp}\,[RP]\,[TP]+k_{td}[RP]^2 \tag{7.44B}$$

where Eq. (7.40) is used in the above equation to substitute for $[RP']$ in terms of $[RP]$. Upon substitution for $[RP]$ and $[TP]$ from Eqs. (7.5F) and (7.9D), respectively, the differential equation relating $[TDB]$ to reaction time is expressed as follows:

$$[TDB] = 2f_0\,[I]_0 \left\{ \left(\frac{2k_{ctp}^2 f_0\,[I]_0}{k_d k_{td}}\right)^{1/2} \left(\frac{4}{3} + \frac{2}{3}e^{-3k_d t/2} - 2e^{-k_d t/2}\right) + \left(1 - e^{-k_d t}\right) \right\} \tag{7.44C}$$

The TDB density or n_{TDB} which is the ratio of $[TDB]$ to $[TP]$, is given by the following equation for radical polymerization with β_2 chain scission:

$$n_{TDB} = \frac{[TDB]}{[TP]} = \frac{\left(\frac{2k_{ctp}^2 f_0 [I]_0}{k_d k_{td}}\right)^{1/2} \left(\frac{4}{3} + \frac{2}{3}e^{-3k_d t/2} - 2e^{-k_d t/2}\right) + \left(1 - e^{-k_d t}\right)}{\left(1 - e^{-k_d t}\right)} \tag{7.44D}$$

As $t \to \infty$ (times much longer than the initiator half-life), the above equation for n_{TDB} is simplified to the following relation:

$$n_{TDB}\big|_{t\to\infty} = \frac{4}{3}k_{ctp}\left(\frac{2f_0\,[I]_0}{k_d k_{td}}\right)^{1/2} + 1 \tag{7.44E}$$

7.7 Modeling Radical Polymerization with Photoinitiation

This section models polymerization with photoinitiation, which is a different type of initiator. Unlike thermal initiation, photoinitiated polymerization is limited by the depth of penetration of light through the sample [101–104]. As a result, photoinitiation is mostly used for polymerization of thin films, membranes, coatings, and in three-dimensional layer-by-layer printing [105–109]. Figure 7.7 shows a schematic diagram for the photoinitiated polymerization of a monomer.

The light intensity is highest on the surface of the liquid monomer film, expressed by $I_{\nu,0}$, but the intensity decreases as the light travels through the film, and it is absorbed by the photo-initiator dissolved in the liquid monomer. Therefore, the extent of polymerization is both time and depth-dependent. The elementary reactions for this polymerization system are as follows:

$$I + h\nu \longrightarrow I' \xrightarrow{\phi} {}^{2}R \tag{7.45A}$$

$$R + R \xrightarrow{k_c} I^{\bullet} \tag{7.45B}$$

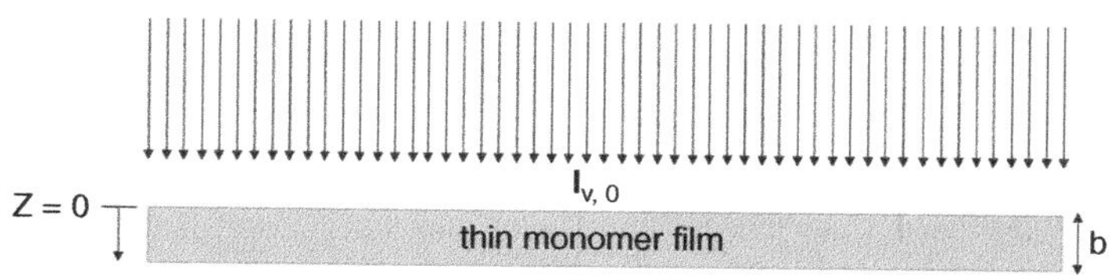

Figure 7.7 Schematic diagram for photoinitiated polymerization.

$$R + M \xrightarrow{k_{ci}} RP_1 \tag{7.45C}$$

$$RP_n + M \xrightarrow{k_p} RP_{n+1} \tag{7.45D}$$

$$RP_n + RP_m \xrightarrow{k_{td}} TP_n + TP_m + TDB \tag{7.45E}$$

In the first reaction above, the photo-initiator dissolved in the reaction mixture absorbs one or more photon of energy, $h\nu$, which elevates the initiator to the higher energy state I'. The activated, unstable initiator dissociates with efficiency ϕ to generate two primary radicals R. The generated primary radicals may recombine to generate an inactive species $I^{\cdot}$ and emit a photon of energy which is scattered in the medium.

The primary radical R undergoes chain initiation to form a reactive propagating chain with one repeat unit, which propagates in the next reaction to form an n-mer propagating chain. In the last reaction, two n-mer and m-mer propagating chains terminate by disproportionation to form two terminated polymer chains with a TDB at chain-end of one of the terminated chains. The most common photo-initiators are benzophenone, acetophenone, benzoin and its ethers, and benzyl peroxide and its ketals [110–114].

7.7.1 Absorbed Radiation in Radical Polymerization with Photoinitiation

A general molar conservation equation for the photo-initiator without convective flow (molecular diffusion only) is written as follows (see the general Eq. (5.19) in Chapter 5 for the conservation equation of molar mass of small molecules in the Cartesian coordinate system):

$$\begin{Bmatrix} \text{Rate of} \\ \text{change} \\ \text{of molar} \\ \text{mass of} \\ \text{initiator } I \\ \text{with} \\ \text{reaction} \\ \text{time} \end{Bmatrix} = \begin{Bmatrix} \text{Rate of molar} \\ \text{mass of} \\ \text{initiator } I \\ \text{entering} \\ \text{infinitesimal} \\ \text{volume at} \\ z \text{ in the } z \\ \text{direction} \end{Bmatrix} - \begin{Bmatrix} \text{Rate of molar} \\ \text{mass of} \\ \text{initiator } I \\ \text{leaving} \\ \text{infinitesimal} \\ \text{volume at} \\ z + dz \text{ in the} \\ \text{z direction} \end{Bmatrix}$$
$$+ \begin{Bmatrix} \text{Rate of molar} \\ \text{mass of} \\ \text{initiator I} \\ \text{entering} \\ \text{infinitesimal} \\ \text{volume at} \\ x \text{ in the } x \\ \text{direction} \end{Bmatrix} - \begin{Bmatrix} \text{Rate of molar} \\ \text{mass of} \\ \text{initiator } I \\ \text{leaving} \\ \text{infinitesimal} \\ \text{volume at} \\ x + dx \text{ in the} \\ x \text{ direction} \end{Bmatrix} + \begin{Bmatrix} \text{Rate of molar} \\ \text{mass of} \\ \text{initiator } I \\ \text{entering} \\ \text{infinitesimal} \\ \text{volume} \\ \text{at } y \text{ in the } y \\ \text{direction} \end{Bmatrix} \tag{7.46A}$$
$$- \begin{Bmatrix} \text{Rate of molar} \\ \text{mass of} \\ \text{initiator } I \\ \text{leaving} \\ \text{infinitesimal} \\ \text{volume at} \\ y + dy \text{ in the} \\ y \text{ direction} \end{Bmatrix} + \begin{Bmatrix} \text{Rate of} \\ \text{molar} \\ \text{production} \\ \text{of initiator } I \\ \text{in the} \\ \text{differential} \\ \text{volume} \\ \text{element} \end{Bmatrix} - \begin{Bmatrix} \text{Rate of} \\ \text{molar} \\ \text{consumption} \\ \text{of initiator } I \\ \text{in the} \\ \text{differential} \\ \text{volume} \\ \text{element} \end{Bmatrix}$$

The above conservation equation in mathematical form is written as follows:

$$\frac{\Delta(m_I)}{\Delta t} = \dot{m}_{I,z}\big|_z - \dot{m}_{I,z}\big|_{z+\Delta z} + \dot{m}_{I,x}\big|_x - \dot{m}_{I,x}\big|_{x+\Delta x} + \dot{m}_{I,y}\big|_y - \dot{m}_{I,y}\big|_{y+\Delta y} + R_{prd,I}\Delta V - R_{con,I}\Delta V \tag{7.46B}$$

where m_I is the molar mass of initiator, $\dot{m}_I$ is the rate of molar mass flowing in a certain direction, and $R_{prd,I}$ and $R_{con,I}$ are the rates of production and consumption of the initiator in the differential volume. After substitution for m_I in terms of concentration $[I]$ and $\dot{m}_I$ in terms of molar flux N_I, the above equation changes to the following form:

$$\frac{\Delta\left([I]\,\Delta V\right)}{\Delta t} = \left(N_{I,z}\Delta A_{xy}\right)\big|_z - \left(N_{I,z}\Delta A_{xy}\right)\big|_{z+\Delta z} + \left(N_{I,x}\Delta A_{yz}\right)\big|_x - \left(N_{I,x}\Delta A_{yz}\right)\big|_{x+\Delta x} + \left(N_{I,y}\Delta A_{xz}\right)\big|_y - \left(N_{I,y}\Delta A_{xz}\right)\big|_{y+\Delta y} + R_{prd,I}\Delta V - R_{con,I}\Delta V \tag{7.46C}$$

Substituting for $\Delta V = \Delta x\Delta y\Delta z$, ΔA in terms of Δx, Δy and Δz, and assuming no change in volume during the reaction, the above equation changes to the following form:

$$\Delta x\Delta y\Delta z\frac{\Delta\left([I]\right)}{\Delta t} = \Delta x\Delta y\left(N_{I,z}\right)\big|_z - \Delta x\Delta y\left(N_{I,z}\right)\big|_{z+\Delta z} + \Delta y\Delta z\left(N_{I,x}\right)\big|_x - \Delta y\Delta z\left(N_{I,x}\right)\big|_{x+\Delta x} + \Delta x\Delta y\left(N_{I,y}\right)\big|_y - \Delta x\Delta y\left(N_{I,y}\right)\big|_{y+\Delta y} + \Delta x\Delta y\Delta z\left(R_{prd,I} - R_{con,I}\right) \tag{7.46D}$$

Upon division of both sides of the above equation by $\Delta x\Delta y\Delta z$, the above equation simplifies to the following form:

$$\frac{\Delta\left([I]\right)}{\Delta t} = -\frac{N_{I,z}\big|_{z+\Delta z} - N_{I,z}\big|_z}{\Delta z} - \frac{N_{I,x}\big|_{x+\Delta x} - N_{I,x}\big|_x}{\Delta x} - \frac{N_{I,y}\big|_{y+\Delta y} - N_{I,y}\big|_y}{\Delta y} + R_{prd,I} - R_{con,I} \tag{7.46E}$$

After taking the limit as Δx, Δy, and Δz approaches zero, the above equation becomes the following differential form:

$$\lim_{\Delta t\to 0}\frac{\Delta\left([I]\right)}{\Delta t} = -\lim_{\Delta z\to 0}\frac{N_{I,z}\big|_{z+\Delta z} - N_{I,z}\big|_z}{\Delta z} - \lim_{\Delta x\to 0}\frac{N_{I,x}\big|_{x+\Delta x} - N_{I,x}\big|_x}{\Delta x} - \lim_{\Delta y\to 0}\frac{N_{I,y}\big|_{y+\Delta y} - N_{I,y}\big|_y}{\Delta y} + R_{prd,I} - R_{con,I} \tag{7.46F}$$

$$\Rightarrow \frac{\partial\,[I]}{\partial t} = -\frac{\partial N_{I,z}}{\partial z} - \frac{\partial N_{I,x}}{\partial x} - \frac{\partial N_{I,y}}{\partial y} + R_{prd,I} - R_{con,I} \tag{7.46G}$$

where $N_{I,x}$, $N_{I,y}$, and $N_{I,z}$ are the molecular flux of the initiator molecules in the x, y, and z directions, respectively, due to the gradients in the concentration of initiator in those directions. The above differential equation states that the rate of change of the molar mass of $[I]$ with time in dV of the thin film at rest (no flow) is equal to minus the molecular flux of $[I]$ out of dV in the z-direction, minus the molecular flux of $[I]$ out of dV in the x-direction, minus the molecular flux of $[I]$ out of dV in the y-direction, plus the rate of production of $[I]$ in dV, and minus the rate of consumption of $[I]$ in dV. As the film is infinite in the x and y directions as compared to the film thickness in the z-direction, the flux of the molecules in the x and y directions is negligible. There is also no production of $[I]$ in the differential volume (only consumption via initiator dissociation by the incident radiation), the above conservation equation simplifies as follows:

$$\frac{\partial [I]}{\partial t} = -\frac{\partial N_{I,z}}{\partial z} - R_{con,I} \tag{7.46H}$$

Substitution for $N_{I,z}$ in terms of Fick's law of diffusion, that is $N_{I,z} = -D_I \partial [I] / \partial z$, results in the following equation:

$$\frac{\partial [I]}{\partial t} = D_I \frac{\partial^2 [I]}{\partial z^2} - R_{con,I} \tag{7.46I}$$

where $\partial [I] / \partial z$ is the gradient of the initiator concentration in the z-direction and D_I is the diffusion coefficient of the initiator in the reaction medium. The above equation is called a reaction-diffusion model of photoinitiation because it accounts for the changes in $[I]$ with time due to molecular diffusion and disappearance by reaction. If the rate of diffusion of initiator molecules is much faster than the rate of their disappearance by photodissociation (radical formation), then any concentration gradient in the z-direction disappears instantaneously on the scale of the reaction time. Under the condition of fast diffusion compared to reaction, $D_I \partial^2 [I] / \partial z^2 << R_{con,I}$, the above equation simplifies to the following:

$$\frac{\partial [I]}{\partial t} = -R_{con,I} \tag{7.46J}$$

The above equation is called the reaction model of photoinitiation because it assumes negligible gradient of $[I]$ in any direction due to fast diffusion. The above derivation implies that $[I]$ is a function of time but not the thickness direction z. Photopolymerization is often limited by diffusion as it is performed in bulk and the reaction is relatively fast [115], but the reaction-diffusion model is beyond the scope of this chapter. Polymerization reactions that are diffusion limited and require a reaction-diffusion model will be discussed in the next chapter (see chapter 8 and reference [116]). The equations in the following subsections for determining $[R]$, $[RP]$, $[TP]$, λ_1, λ_2, μ_1, μ_2 are derived based on the reaction model of photoinitiation. The rate of consumption of $[I]$, that is $R_{con,I}$, depends on the amount of light absorbed by the photo-initiator and the efficiency of radical formation, given by the following expression [115, 117, 118]:

$$\frac{\partial [I]}{\partial t} = -R_{con,I} = -2\phi I_{a,z} \tag{7.46K}$$

where $[I]$ is photo-initiator concentration, and $I_{a,z}$ is the absorbed radiation at depth z, and ϕ is the dissociation quantum yield of photo-initiator (number of initiators dissociated per candela of absorbed light per unit volume). It is assumed in the above equation that two reactive species R are formed from the dissociation of a photo-initiator molecule. The intensity I_a is the intensity of radiation absorbed in the unit volume of the reaction mixture per unit time, that is $I_a = I_{\nu,0} - I_{\nu,z}$. In this equation, $I_{\nu,z}$ and $I_{\nu,0}$ are the intensities of light at depth z and at the surface of the thin film, respectively. It should be noted that I_a is a function of z even though $[I]$ is not. The change in the intensity of light at depth z is a function of transmitted light intensity and the initiator concentration as follows:

$$\frac{dI_{\nu,z}}{dz} = -\varepsilon [I] I_{\nu,z} \tag{7.46L}$$

where ε is the molar extinction coefficient of initiator in photoinitiation in units of m^3 $mole^{-1}$ m^{-1}. Upon substitution for $I_a = I_{\nu,0} - I_{\nu,z}$, the differential of I_a with respect to z is given by the following equation:

$$\frac{dI_a}{dz} = +\varepsilon [I] (I_{\nu,0} - I_a) \tag{7.46M}$$

The above differential equation can be integrated subject to the boundary condition $I_\nu = I_{\nu,0}$, $I_a = 0$ at $z = 0$ and the absorbed intensity I_a at $z = b$ where b is the film thickness (see Figure 7.7) to obtain the following equality:

$$I_a = I_{\nu,0} - I_{\nu,0}e^{-\varepsilon[I]b} = I_{\nu,0} \left(1 - e^{-\varepsilon[I]b}\right) \tag{7.46N}$$

The above equation indicates that there is a film thickness beyond which there is no incident radiation for absorption and initiator dissociation. The maximum thickness b_{max} is defined as the thickness with 95% of the incident radiation absorbed. Upon substitution for $I_a \approx 0.95 I_{\upsilon,0}$ at $b = b_{max}$ and rearrangement, b_{max} is given by the following equation:

$$b_{max} = \frac{3.0}{\varepsilon [I]} \tag{7.46O}$$

The values for ε and $[I]$ for a typical photo-initiator like Irgacure-907 are 35×10^3 L mol^{-1} cm^{-1} and 1×10^{-3} mol L^{-1}, respectively, leading to a typical $b_{max} \approx 100 - 1000\,\mu m$. The maximum thickness decreases with increasing extinction coefficient and the initiator concentration. If the film is thicker than b_{max}, there is effectively no polymerization beyond that thickness. Upon substitution of Eq. (7.46N) in Eq. (7.46K), the change in $[I]$ with time is given by the following equation:

$$\frac{d[I]}{dt} = -2\phi I_{\nu,0} \left(1 - e^{-\varepsilon[I]b}\right) \tag{7.46P}$$

Integrating the above equation with respect to time, subject to the initial condition $[I] = [I]_0$ at $t = 0$, results in the following expression for $[I]$ as a function of time:

$$[I] = [I]_0 + (1/\varepsilon b) \ln\left[\left(1 - e^{-\varepsilon[I]_0 b}\right) / \left(1 - e^{-\varepsilon[I]b}\right)\right] - 2\phi I_{\nu,0} t \tag{7.46Q}$$

The above equation is non-linear with respect to $[I]$ and should be solved numerically to predict $[I]$ as a function of reaction time.

7.7.2 Concentration of Primary Radicals and Propagating Chains in Radical Polymerization with Photoinitiation

Primary radicals are produced by Rxn. 7.45A and consumed by Rxns. 7.45B, C. Applying QSSA to the primary radicals, the conservation equation for $[R]$ is as follows (see general Eq. 7.1D):

$$\frac{d[R]}{dt} = 2\phi I_a - k_c [R]^2 - k_{ci} [M] [R] = 0 \tag{7.47A}$$

The initiator efficiency, f_0, is defined as the ratio of the rate of chain initiation to the rate of primary radical formation by photoinitiation. The rate of chain initiation is $k_{ci} [M] [R]$ whereas the rate of primary radical formation is $2\phi I_a$ (Eq. 7.46K). Therefore, the initiator efficiency for photoinitiation is defined by the following equation:

$$f_0 = \frac{k_{ci} [M] [R]}{2\phi I_a} \tag{7.47B}$$

Substitution of the above equation in Eq. (7.47A) to eliminate $k_c\,[R]^2$ results in the following equation for $[R]$:

$$[R] = \frac{2\phi f_0 I_a}{k_{ci}\,[M]} = \frac{2\phi f_0 I_{\nu,0}\left(1 - e^{-\varepsilon[I]b}\right)}{k_{ci}\,[M]} \tag{7.47C}$$

where $[I]$ is defined by Eq. (7.46Q) as a function of time. The concentration of reactive propagating chains, $[RP]$, is determined from the conservation equation for the zero-order molar moment of the propagating chains (see the general Eq. 7.1A). Propagating chains are produced by Rxns. 7.45C, D, whereas they are consumed by Rxns. 7.45D, E. Applying QSSA, the conservation equation for the zero-order concentration moment of the propagating chains is as follows:

$$\frac{d\,[RP]}{dt} = k_{ci}\,[M]\,[R] + k_p\,[M]\,[RP] - k_p\,[M]\,[RP] - k_{td}\,[RP]^2 = 0 \tag{7.48A}$$

After canceling like terms and substituting for $[R]$ from Eq. (7.47C), $[RP]$ is given by the following equation:

$$[RP] = \left(\frac{k_{ci}\,[M]\,[R]}{k_{td}}\right)^{1/2} = \left(\frac{2\phi f_0 I_a}{k_{td}}\right)^{1/2} \tag{7.48B}$$

In thermal initiation, $[RP]$ is a function of f_0 and $[I]$ whereas $[RP]$ in photoinitiation is a function of ϕ, f_0, and I_a. After substitution for I_a from Eq. (7.46N), $[RP]$ as a function of the depth of light penetration b is given as follows:

$$[RP] = \left[2\phi f_0 I_{\nu,0}\left(1 - e^{-\varepsilon[I]b}\right)/k_{td}\right]^{1/2} \tag{7.48C}$$

If the film is thin, i.e., $b \approx 1 - 10\ \mu\text{m} << b_{\max}$, then $e^{-\varepsilon[I]b} \approx 1 - \varepsilon\,[I]\,b$, thus $1 - e^{-\varepsilon[I]b} \approx \varepsilon\,[I]\,b$, and Eq. (7.46Q) simplifies to $[I] = [I]_0 - 2\phi I_{\nu,0} t$, which results in the following relation for $[RP]$ with reaction time:

$$[RP] = \left[2\phi f_0 I_{\nu,0}\varepsilon b\,[I]_0/k_{td}\right]^{1/2}\left(1 - \frac{2\phi I_{\nu,0}}{[I]_0}t\right)^{1/2} \quad \in b << b_{\max} \tag{7.48D}$$

According to the above equation, the plot of $[RP]$ versus $\left(1 - 2\phi I_{\nu,0}t/\,[I]_0\right)^{1/2}$ for reaction controlled photopolymerization (no diffusion limitation) should be linear for a polymerizing thin film. Furthermore, $[RP]$ decreases with time as the initiator is consumed and increases with the square root of intensity of incident radiation, $I_{\nu,0}$.

7.7.3 Monomer Concentration in Radical Polymerization with Photoinitiation

Monomer is consumed by Rxns. 7.45C, D. For constant reaction volume, the molar conservation equation for the monomer in photopolymerization is given as follows (see Eq. 7.1D):

$$\frac{d\,([M])}{dt} = -k_{ci}\,[M]\,[R] - k_p\,[M]\,[RP] \approx -k_p\,[M]\,[RP] \quad L >> 1 \tag{7.49A}$$

The first term $k_{ci}\,[M]\,[R]$ is much smaller than $k_p\,[M]\,[RP]$ for long chains. After substitution for $[RP]$ from Eq. (7.48C) and separation of variables, the time dependence of the monomer concentration is given by the following differential equation:

$$\frac{d\,([M])}{[M]} = -\left(2\phi k_p^2 f_0 I_{\nu,0}/k_{td}\right)^{1/2}\left(1 - e^{-\varepsilon[I]b}\right)^{1/2} dt \quad L >> 1 \tag{7.49B}$$

where the variation of $[I]$ with time is given by Eq. (7.46Q). Upon integration of the above equation from $[M] = [M]_0$ at $t = 0$ to time t, the time dependence of the monomer concentration is given by the following expression:

$$\ln\left([M]/[M]_0\right) = -\left(2\phi k_p^2 f_0 I_{\nu,0} \varepsilon b / k_{td}\right)^{1/2} \int_{t=0}^{t} \left(1 - e^{-\varepsilon [I] b}\right)^{1/2} dt \quad L >> 1 \tag{7.49C}$$

If $b << b_{\max}$, Eq. (7.48D) can be used to relate $[RP]$ to time. Therefore, for $b << b_{\max}$, using Eq. (7.48D) in Eq. (7.49B) followed by integration, the time dependence of the monomer concentration is given by the following equation:

$$\begin{aligned} \ln\left([M] / [M]_0\right) &= -\left(2\phi k_p^2 f_0 I_{\nu,0} \varepsilon b\, [I]_0 / k_{td}\right)^{1/2} \int_{t=0}^{t} \left(1 - \frac{2\phi I_{\nu,0}}{[I]_0} t\right)^{1/2} dt \\ &= -\left(2\phi k_p^2 f_0 I_{\nu,0} \varepsilon b\, [I]_0 / k_{td}\right)^{1/2} \left(\frac{[I]_0}{3\phi I_{\nu,0}}\right) \left\{1 - \left(1 - \frac{2\phi I_{\nu,0}}{[I]_0} t\right)^{3/2}\right\} \quad \in b << b_{\max} \end{aligned} \tag{7.49D}$$

7.7.4 Higher Moments of Propagating Chains in Radical Polymerization with Photoinitiation

Propagating chains are produced by Rxns. 7.45C, D, and they are consumed by Rxns. 7.45D, E. Applying QSSA, the conservation equation for the first order mole-fraction moment of the propagating chains is given by the following expression (see general Eq. 7.1A):

$$(1)\, k_{ci}\, [M]\, [R] + (\lambda_1 + 1) k_p\, [M]\, [RP] = (\lambda_1)\, k_p\, [M]\, [RP] + (\lambda_1)\, k_{td}\, [RP]^2 \tag{7.50A}$$

After canceling like terms and rearrangement, λ_1 which is the first order mole-fraction moment of the propagating chains is given by the following equation:

$$\lambda_1 = \frac{k_p\, [RP]\, [M] + k_{ci}\, [R]\, [M]}{k_{td}\, [RP]^2} \tag{7.50B}$$

As the kinetic chain length is defined by the ratio of the rate of addition of monomer to the propagating chains (numerator) to the rate of termination (denominator), then $\lambda_1 = L$. After substitution for $k_{ci}\, [R]\, [M]$ from Eq. (7.48B), the kinetic chain length is expressed as follows:

$$\lambda_1 = L = \frac{k_p\, [M]}{k_{td}\, [RP]} + 1 \tag{7.50C}$$

where $[M]$ and $[RP]$ are expressed in terms of t by Eqs. (7.49D) and (7.48D) for $b << b_{\max}$, respectively. It can be shown that the second order mole-fraction moment of the propagating chains is given by the following equation (see general conservation Eq. 7.1A):

$$\lambda_2 = \frac{(2L + 1)\, k_p\, [M]}{k_{td}\, [RP]} + 1 \tag{7.50D}$$

7.7.5 Concentration of Terminated Polymer Chains in Radical Polymerization with Photoinitiation

The TP concentration is determined from the zero-order concentration moment of the terminated chains. The general conservation equation for the kth order moment of the terminated chains in mathematical form in the absence of flow (no convective mass transfer) is written as follows (see general Eq. 5.17 in Chapter 5):

$$\frac{d\left(\mu_k Q_{0,tp}\right)}{dt} = -\frac{\partial\left(\mu'_{k,x} N_{tp,x}\right)}{\partial x} - \frac{\partial\left(\mu'_{k,y} N_{tp,y}\right)}{\partial y} - \frac{\partial\left(\mu'_{k,z} N_{tp,z}\right)}{\partial z} + \sum_i \mu'_{k,i} R_{tp,prd,i} - \sum_l \mu'_{k,l} R_{tp,con,l} \quad (7.51A)$$

where μ_k is the kth order mole-fraction moment of the terminated chains; $Q_{0,tp}$ is the zero-order molar concentration moment of the chains; $\mu'_{k,x}$, $\mu'_{k,y}$, and $\mu'_{k,z}$ are the kth order instantaneous mole-fraction moments of the terminated chains in directions x, y, and z, respectively; $N_{tp,x}$, $N_{tp,y}$, and $N_{tp,z}$ are the molar fluxes (by molecular diffusion) of the chains in x, y, and z directions; $\mu'_{k,i}$ and $\mu'_{k,l}$ are the kth order instantaneous mole-fraction moments of the terminated chains produced and consumed by ith and lth reactions, respectively; and $R_{tp,prd,i}$ and $R_{tp,con,l}$ are the production and consumption rates of the chains by the ith and lth reactions. For thin films that are commonly polymerized by photo-initiation, there is no gradient in the x and y directions. Hence, the above conservation equation reduces to the following expression:

$$\frac{d\left(\mu_k Q_{0,tp}\right)}{dt} = -\frac{\partial\left(\mu'_{k,z} N_{tp,z}\right)}{\partial z} + \sum_i \mu'_{k,i} R_{tp,prd,i} - \sum_l \mu'_{k,l} R_{tp,con,l} \quad (7.51B)$$

Like the derivation of the conservation equation for the photo-initiator, the above equation is the reaction-diffusion model of photopolymerization because it accounts for changes in $[TP]$ with time due to molecular diffusion. If the rate of diffusion of the terminated chains is much faster than their reaction rates, the above equation is simplified to the following expression:

$$\frac{d\left(\mu_k Q_{0,tp}\right)}{dt} = \sum_i \mu'_{k,i} R_{tp,prd,i} - \sum_l \mu'_{k,l} R_{tp,con,l} \quad (7.51C)$$

The above equation is called the reaction model of photopolymerization because it assumes no gradients of $[TP]$ in any direction. Applying the above general conservation equation to the zero-order mole-fraction moment of the terminated chains with no consumption and production only by termination via disproportionation, it results in the following equation:

$$\frac{d\left(\mu_0 Q_{0,tp}\right)}{dt} = \frac{d\,[TP]}{dt} = k_{td}[RP]^2 = 2\phi f_0 I_{v,0}\left(1 - e^{-\varepsilon[I]b}\right)$$
$$\Rightarrow [TP] = \left(2\phi f_0 I_{v,0}\right)\int_{t=0}^{t}\left(1 - e^{-\varepsilon[I]b}\right)dt \quad (7.51D)$$

where relations $\lambda_0 = 1$ and Eq. (7.48C) for $[RP]$ are used in deriving the above equation. The $[I]$ in the above equation is a function of t as given by Eq. (7.46Q). Upon integration of the above equation subject to the initial condition $[TP] = 0$ at $t = 0$, the concentration of terminated chains as a function of time is given by the following expression:

$$[TP] = (2\phi f_0 I_{v,0}) \int_{t=0}^{t} (1 - e^{-\varepsilon[I]b})\, dt \tag{7.51E}$$

For thin films, i.e., $b << b_{\max}$ (see Eq. 7.48D), $1 - e^{-\varepsilon[I]b} \approx \varepsilon[I]\, b$, hence Eq. (7.46Q) reduces to $[I] = [I]_0 - 2\phi I_{v,0} t$, which results in the following relation for $[TP]$ as a function of reaction time:

$$\begin{aligned} \frac{d\,[TP]}{dt} &= (2\phi f_0 I_{v,0}\,[I]_0\,\varepsilon b)\left(1 - \frac{2\phi I_{v,0}}{[I]_0} t\right) \qquad \in b << b_{\max} \\ &\Rightarrow [TP] = (2\phi f_0 I_{v,0}\,[I]_0\,\varepsilon b)\left[t - (\phi I_{v,0}/\,[I]_0)\, t^2\right] \end{aligned} \tag{7.51F}$$

The above equation predicts that the plot of $[TP]$ versus $\left[t - (\phi I_{v,0}/\,[I]_0)\, t^2\right]$ is linear for the reaction controlled photopolymerization (no diffusion limitation) with slope equal to $(2\phi f_0 I_{v,0}\,[I]_0\,\varepsilon b)$.

7.7.6 Higher Moments and Average Degree of Polymerization of Terminated Polymer Chains in Radical Polymerization with Photoinitiation

In photopolymerization Rxns. 7.45A–E, terminated chains are produced by Rxn. 7.45E at the rate of $k_{td}\,[RP]^2$ and with first and second order moments μ_1 and μ_2. It can be shown, using the kth order moment conservation equations (see general conservation Eq. 7.1C), that the first and second order concentration moments of the terminated polymer chains are given by the following equations:

$$Q_{1,tp} = [M]_0 - [M] \tag{7.52A}$$

$$Q_{2,tp} = k_p \int_{t=0}^{t} \{(2L + 1)\,[M]\}\, dt \tag{7.52B}$$

where $[M]$ and L are given by Eqs. (7.49D) and (7.50C), respectively. Once $Q_{1,tp}$ and $Q_{2,tp}$ are determined, then $\overline{DP}_{n,tp}$, $\overline{DP}_{w,tp}$, and PDI of the terminated chains in radical photopolymerization are given by $\overline{DP}_{n,tp} = Q_{1,tp}/\,[TP]$, $\overline{DP}_{w,tp} = Q_{2,tp}/Q_{1,tp}$, and $PDI = \overline{DP}_{w,tp}/\overline{DP}_{n,tp}$.

7.8 Modeling Reversible-deactivation Radical Polymerization by Atom transfer Radical Polymerization

7.8.1 Atom Transfer Radical Polymerization in the Absence of Chain Termination

As discussed in Section 1.2.4.1 of Chapter 1, ATRP uses a halide initiator and a redox transition metal catalyst, with two oxidation states and with affinity for a halogen, to reversibly deactivate the propagating radicals [119–123]. For living polymerization, the transition metal catalyst is chosen in a way that the propagating chains will be in the dormant state most of the time. The elementary reactions for ATRP are as follows:

$$RX + LM^{I}X \underset{k_{daci}}{\overset{k_{aci}}{\rightleftharpoons}} LM^{II}X_2 + R \tag{7.53A}$$

$$R + M \xrightarrow{k_{ci}} RP_1 \tag{7.53B}$$

$$RP_n + M \xrightarrow{k_p} RP_{n+1} \tag{7.53C}$$

$$RP_n + LM^{II}X_2 \underset{k_{ac}}{\overset{k_{dac}}{\rightleftharpoons}} DOP_nX + LM^IX \tag{7.53D}$$

where RX is the halide initiator, X is the halide atom, $LM^IX/LM^{II}X$ is the transition metal catalyst (TMC), L is the ligand associated with the TMC, and R is the primary radical formed by halide transfer from the initiator to the TMC. It is assumed that the reaction is perfectly living with only the production of dormant chains and no termination reaction. The reversible initiation reaction is strongly shifted toward deactivation, as conferred in Section 5.7.2 of Chapter 5. The equilibrium constant for initiation in ATRP, $K_{ATRP,i}$, is defined by the following equation [41, 124, 125]:

$$K_{ATRP,i} = \frac{[R]\,[LM^{II}X_2]}{[RX]\,[LM^IX]} \tag{7.54A}$$

In a typical ATRP, $K_{ATRP,i} \approx 1\times10^{-7}$, $[RX] \approx 1\times10^{-3}$ M, $[LM^IX] \approx 0.1\,[RX]$, and the ratio $[LM^{II}X_2]\,/\,[LM^IX] \approx 0.1$. Under these conditions, $[R] \approx 1\times10^{-9}$ M [41, 124, 125]. These calculations show that $[R]$ is a tiny fraction of the initiator concentration, that is $[R] <<< [RX]$. Therefore, it can be assumed that the radical concentration is relatively constant and given by Matyjaszewski's equation (see Eq. 5.35B in Section 5.7.2. in Chapter 5) as follows [126]:

$$[R] \approx [R]_0 = K_{ATRP,i}\,[RX]_0\,[LM^IX]_0\,/\,[LM^{II}X_2]_0 \tag{7.54B1}$$

Prior to the addition of monomer, based on Rxn. 7.53A, R is the active species and RX is the inactive species, and the relation between R and RX is defined by the equilibrium Eq. (7.54A). After the addition of monomer, primary radical R reacts with monomer to form RP_1, followed by RP_{n+1} as shown by Rxn. 7.53C, and the relation between RP and DOP is defined by the following equilibrium relation between RP and DOP in Rxn. 7.53D:

$$K_{ATRP} = \frac{[RP]\,[LM^{II}X_2]}{[DOP]\,[LM^IX]} \tag{7.54B2}$$

7.8.1.1 Conversion of Primary Radicals to Propagating Chains in ATRP in the Absence of Chain Termination

The primary radical R reacts with a monomer to form a 1-mer propagating chain RP_1. R is produced by the equilibrium initiation Rxn. 7.53A and consumed by the chain initiation Rxn. 7.53B. For a constant volume reaction and given that QSSA does not apply to reversible-deactivation radical polymerization, the molar conservation equation for the primary radicals R, based on Eq. (7.1A) for the general kth order moment conservation equation of the reactive/propagating species, is given by the following expression:

$$\begin{aligned}\frac{d\,[R]}{dt} &= +k_{aci}\,[RX]\,[LM^IX] - k_{daci}\,[RP]\,[LM^{II}X_2] - k_{ci}\,[R]\,[M] \\ &= -k_{ci}\,[R]\,[M] \approx -k_{ci}\,[R]\,[M]_0\end{aligned} \tag{7.54C}$$

The first two terms on the right cancel each other because Rxn. 7.54A is at equilibrium. Given that $[R] \approx 1\times10^{-9}$ M and $[M]_0 \approx 1$ M [41, 124, 125], the addition of monomer to the radicals does not significantly change the monomer concentration [41, 124, 125]. Hence, $[M] \approx [M]_0$ in the above equation. After separation of variables and integration subject to the initial condition $[R] = [R]_0$ (see Eq. 7.54B1) at $t = 0$, $[R]$ is given by the following equation as a function of time:

$$[R] = [R]_0\,e^{-k_{ci}[M]_0t} \approx [R]_0 \tag{7.54D}$$

where $[R]_0$ is defined by Eq. (7.54B1). According to the above equation, $[R]$ decreases exponentially as a function of time after the addition of monomer as it is converted to 1-mer propagating chains. Given that $k_{ci} \approx 1 \times 10^3\ M^{-1}s^{-1}$, $[M]_0 \approx 1$ M with polymerization time of approximately 1 h, the factor $k_{ci}\ [R]\ [M] \approx 1 \times 10^3 >> 1$ even after a short time of $t = 1$ s; hence the exponential term in the above equation approaches zero even for short times after the addition of monomer. This implies that $[R] \approx 0$ right after monomer addition to the reaction. It can be shown using a conservation of molar mass equation for the propagating chains that $[RP]$ is given by the following relation in terms of time:

$$[RP] = \sum_{n=1}^{\infty} [RP]_n = [R]_0 \left(1 - e^{-k_{ci}[M]_0 t}\right) \approx [R]_0 \tag{7.54E}$$

where $[R]_0$ is defined by Eq. (7.54B). Given that the factor $k_{ci}\ [R]\ [M] >> 1$ even at short times, then $[RP] \approx [R]_0$ after monomer addition to the reaction. Therefore, it can be assumed after the addition of monomer $[R] \approx 0$ and $[RP] \approx [R]_0$.

7.8.1.2 Monomer Concentration in ATRP in the Absence of Chain Termination

Monomer is consumed by Rxns. 7.53B, C. For constant reaction volume, using the conservation equation for small molecule reactants (see Eq. 7.1D), the molar conservation equation for the monomer is given by the following differential equation:

$$\begin{aligned}\frac{d\,[M]}{dt} &= -k_{ci}\ [R]\ [M] - k_p\ [M] \left(\sum_{n=2}^{\infty} [RP]_n\right) = -k_p\ [M] \left(\sum_{n=1}^{\infty} [RP]_n\right) \\ &= -k_p\ [M]\ [RP] \approx -k_p\ [M]\ [R]_0\end{aligned} \tag{7.55A}$$

where Eq. (7.54E) is used in the above equation for $[RP]$. Assuming $K_{ATRP} \approx K_{ATRP,i}$ and integrating the above equation subject to the initial condition $[M] = [M]_0$, the time dependence of $[M]$ is given by as follows:

$$[M] = [M]_0\ e^{-k_p[R]_0 t} = [M]_0\ e^{-\left(k_p K_{ATRP}[RX]_0\left[LM^I X\right]_0 t/\left[LM^{II}X_2\right]_0\right)} \tag{7.55B}$$

Based on the above equation, $[M]$ decreases exponentially with time or the plot of $\ln\,[M]$ is linear with a negative slope.

7.8.1.3 Higher Moments of the Propagating Chains in ATRP in Absence of Chain Termination

Propagating chains are produced by Rxns. 7.53B–D and consumed by Rxns. 7.53C, D. As mentioned previously, QSSA is not valid in RDRP as the polymerization is living. For constant reaction volume, using the conservation equation for the kth order moment of the reactive species (see Eq. 7.1A), the molar conservation equation for the first order mole-fraction moment of the propagating chains is given as follows:

$$\begin{aligned}\frac{d\,(\lambda_1\,[RP])}{dt} &= -\,(1)\,k_{ci}\ [R]\ [M] - (\lambda_1)\,k_p\ [M] \left(\sum_{n=2}^{\infty} [RP]_n\right) + (\lambda_1 + 1)\,k_p\ [M] \left(\sum_{n=1}^{\infty} [RP]_n\right) \\ &\quad + (\gamma_1)\,k_{ac}\ [DOP]\left[LM^I X\right] - (\lambda_1)\,k_p\ [M]\ [RP] - (\lambda_1)\,k_{dac}\ [RP]\left[LM^{II}X_2\right]\end{aligned} \tag{7.56A}$$

where γ_1 is the first order mole-fraction moment of the dormant chains. As dormant chains are formed from the propagating chains by reacting with $LM^{II}X_2$, the moments of the dormant chains are equal to the moments of the propagating chains, that is $\gamma_n = \lambda_n$. After substituting for γ_n with λ_n and canceling like terms, the above conservation equation simplifies to the following expression:

$$\frac{d\,(\lambda_1\,[RP])}{dt} = k_p\,[M]\left(\sum_{n=1}^{\infty}[RP]_n\right) = k_p\,[M]\,[RP] \tag{7.56B}$$

The right side of the above equation, according to Eq. (7.55A), is the negative of the derivative of monomer concentration with respect to time. Hence, the above expression simplifies to the following equation:

$$\frac{d\,(\lambda_1\,[RP])}{dt} = k_p\,[M]\,[RP] = -\frac{d\,[M]}{dt} \tag{7.56C}$$

Integration of the above equation subject to the initial condition $\lambda_1\,[RP] = 0$ and $[M] = [M]_0$ at $t = 0$ results in the following equation for λ_1:

$$\lambda_1 = \frac{[M]_0 - [M]}{[RP]} \tag{7.56D}$$

Upon substitution for $[M]$ from Eq. (7.55B) and $[RP]$ from Eq. (7.54E), the λ_1 which is also the kinetic chain length, L, and $\overline{DP}_{n,rp}$, is given as a function of time by the following equation for ATRP in the absence of chain termination:

$$\lambda_1 = \overline{DP}_{n,rp} = L = \frac{[M]_0}{[R]_0}\left(1 - e^{-k_p[R]_0 t}\right)$$
$$= \frac{[M]_0\,[LM^{II}X_2]_0}{K_{ATRP}\,[RX]_0\,[LM^{I}X]_0}\left(1 - e^{-\frac{k_p K_{ATRP}[RX]_0[LM^I X]_0}{[LM^{II}X_2]_0}t}\right) \tag{7.56E}$$

As expected, $L \to [M]_0\,/\,[R]_0$ as $t \to \infty$. As equilibrium is shifted toward the dormant state and chains remain dormant most of time, chains in ATRP as well as other RDRP methods grow gradually at a slow pace. For example, in a typical polymerization, $[R] \approx 1 \times 10^{-9}$ M and $k_p \approx 1 \times 10^3\ M^{-1}s^{-1}$ [41, 124, 125], hence only 1.1% of the monomer has reacted after 3 h in ATRP and $L = 0.011\,L_{t\to\infty}$. In ATRP in the absence of chain termination, it takes ≈200 h for 50% of the monomer to react and reach $L = 0.5\,L_{t\to\infty}$. At short times, like $[M]\,/\,[M]_0 < 0.1$, $e^{-k_p[R]_0 t} \approx 1 - k_p\,[R]_0\,t$ and the above expression for the kinetic chain length simplifies to the following equation:

$$\lambda_1 = L \approx [M]_0\,k_p t \quad \in [M]/[M]_0 < 0.1 \tag{7.56F}$$

For short times (in the order of minutes to an hour for ATRP), the kinetic chain length linearly increases with time with a slope equal to $[M]_0\,k_p$. The conservation equation for second-order mole-fraction moment of the propagating chains is written as follows (see conservation Eq. 7.1A):

$$\frac{d\,(\lambda_2\,[RP])}{dt} = (1)\,k_{ci}\,[M]\,[R] + (\lambda_2 + 2\lambda_1 + 1)\,k_p\,[M]\,[RP]$$
$$+ (\gamma_2)\,k_{ac}\,[DOP]\,[LM^I X] - (\lambda_2)\,k_p\,[M]\,[RP] - (\lambda_2)\,k_{dac}\,[RP]\,[LM^{II}X_2] \tag{7.56G}$$

After substitution of $\gamma_2 = \lambda_2$, canceling like terms, using $k_p\,[M]\,[RP] >> k_{ci}\,[M]\,[R]$ for $L >> 1$, and substituting for $k_p\,[M]\,[RP]$ from Eq. (7.55A), the above conservation equation simplifies to the following:

$$\frac{d\,(\lambda_2\,[RP])}{dt} = (2\lambda_1 + 1)\,k_p\,[M]\,[RP] = -(2\lambda_1 + 1)\,\frac{d\,[M]}{dt} \quad L >> 1 \tag{7.56H}$$

After substitution for $d[M]/dt$ from Eq. (7.55B), λ_1 from Eq. 7.56E, and using the approximation $2\lambda_1 + 1 \approx 2\lambda_1$, the $d(\lambda_2[RP])/dt$ is expressed in terms of time by the following equation:

$$\frac{d(\lambda_2[RP])}{dt} = 2k_p[M]_0^2\left(e^{-k_p[R]_0 t} - e^{-2k_p[R]_0 t}\right) \quad L >> 1 \tag{7.56I}$$

Upon integration of the above equation subject to the boundary condition $\lambda_2[RP] = 0$ at $t = 0$ and substitution for $[RP]$ from Eq. (7.54E), λ_2 is related to time by the following equation:

$$\lambda_2 = \frac{[M]_0^2}{[R]_0^2}\left(1 - 2e^{-k_p[R]_0 t} + e^{-2k_p[R]_0 t}\right) \quad L >> 1 \tag{7.56J}$$

Using the above equation and Eq. (7.56F) for λ_1, $\overline{DP}_{w,rp}$ as defined by Eq. (3.45) in Chapter 3, is given as follows:

$$\overline{DP}_{w,rp} = \frac{\lambda_2}{\lambda_1} = \frac{[M]_0}{[R]_0}\frac{\left(1 - 2e^{-k_p[R]_0 t} + e^{-2k_p[R]_0 t}\right)}{\left(1 - e^{-k_p[R]_0 t}\right)} \quad L >> 1 \tag{7.56K}$$

7.8.1.4 Higher Moments of Dormant Chains in ATRP in the Absence of Chain Termination

Dormant chains are produced by deactivation of the propagating chains via the equilibrium Rxn. 7.53D and equilibrium Eq. (7.54B2). Upon rearrangement of Eq. (7.54B2), the concentration of dormant chains in ATRP in absence of chain termination, $[DOP]$, is defined by the following equation:

$$[DOP] = \frac{[RP]\left[LM^{II}X_2\right]}{K_{ATRP}\left[LM^{I}X\right]} = K_{ATRP,i}[RX]_0 / K_{ATRP} \tag{7.57A}$$

Assuming $\left[LM^{II}X_2\right] / \left[LM^{I}X\right] \approx \left[LM^{II}X_2\right]_0 / \left[LM^{I}X\right]_0$ and using Eq. (7.54B1) to substitute for the ratio, and substituting for $[RP]$ from Eq. (7.54E), $[DOP]$ is related to time as follows:

$$[DOP] = K_{ATRP,i}[RX]_0\left(1 - e^{-k_{ci}[M]_0 t}\right) / K_{ATRP} \tag{7.57B}$$

Given that the factor $k_{ci}[R][M] >> 1$ even for short times, it follows that $e^{-k_{ci}[M]_0 t} \approx 0$ shortly after the addition of the monomer. Therefore, $[DOP]$ in ATRP in the absence of chain termination is practically independent of time and given by the following simple equation:

$$[DOP] \approx K_{ATRP,i}[RX]_0 / K_{ATRP} \tag{7.57C}$$

Dormant chains are produced by the coupling of propagating chains with one of the halides on $LM^{II}X_2$. Therefore, the dormant and propagating chains have identical chain length distribution. The first and second order mole-fraction moments of the dormant chains in ATRP are $\gamma_1 = \lambda_1$ and $\gamma_2 = \lambda_2$ with λ_1 and λ_2 given by Eqs. (7.56F, J), respectively. Furthermore, the number-average and weight-average degree of polymerization of the dormant chains in ATRP in absence of chain termination are $\overline{DP}_{n,dop} = \overline{DP}_{n,rp} = \lambda_1$ and $\overline{DP}_{w,dop} = \overline{DP}_{w,rp} = \frac{\lambda_2}{\lambda_1}$. The polydispersity index of the chains which is the ratio of $\overline{DP}_{w,dop}$ to $\overline{DP}_{n,dop}$ is given by the following relation:

$$PDI = \frac{\overline{DP}_{w,rp}}{\overline{DP}_{n,rp}} = \frac{\lambda_2}{\lambda_1^2} = \frac{\left(1 - 2e^{-k_p[R]_0 t} + e^{-2k_p[R]_0 t}\right)}{\left(1 - e^{-k_p[R]_0 t}\right)^2} \quad L >> 1 \tag{7.57D}$$

As $t \to \infty$, $\overline{DP}_{w,rp} = \overline{DP}_{n,rp}$ and $PDI \to 1$, hence, dormant chains become monodisperse at long times or high monomer conversions. The above analysis demonstrates that, after the establishment of equilibrium, the active and dormant states are maintained either in the form of R/RX prior to the addition of monomer or in the form of RP/DOP after the monomer addition. The polymerization in ATRP proceeds essentially by the insertion of monomer into the labile C · ·X bond between the carbon in the propagating chain and the halogen in $LM^{II}X_2$.

7.8.2 Atom Transfer Radical Polymerization in the Presence of Chain Termination

In the absence of chain termination, the propagating chains in RDRP are either active or dormant, but chains do not exit the polymerization by termination reactions. At time t in conventional radical polymerization in which QSSA applies, primary radicals that are produced in the time interval dt_i undergo propagation and termination in the same time interval dt_i. In the next time interval dt_j, new primary radicals are formed which undergo propagation and termination in the time interval dt_j. As a result, in conventional radical polymerization, polymer chains with high molecular weight are produced throughout the reaction including right after the start of the reaction. In RDRP without termination, QSSA does not apply because the primary radicals formed in a time interval remain a part of the whole polymerization by shifting between active and dormant states, but they do not exit the reaction. Hence, the propagating chains continue to add monomer and form longer chains with reaction time while they are in the active state. This implies that DP in RDRP increases gradually during the reaction and polymers with long DP are produced at high monomer conversions. If termination is present in RDRP polymerization, the elementary reactions are as follows:

$$RX + LM^{I}X \underset{k_{daci}}{\overset{k_{aci}}{\rightleftharpoons}} LM^{II}X_2 + R \tag{7.58A}$$

$$R + M \xrightarrow{k_{ci}} RP_1 \tag{7.58B}$$

$$RP_n + M \xrightarrow{k_p} RP_{n+1} \tag{7.58C}$$

$$RP_n + LM^{II}X_2 \underset{k_{ac}}{\overset{k_{dac}}{\rightleftharpoons}} DOP_nX + LM^{I}X \tag{7.58D}$$

$$RP_n + RP_m \xrightarrow{k_{td}} TP_n + TP_m + TDB \tag{7.58E}$$

The first four reactions are identical to those in ATRP without termination. The last reaction is termination of two active propagating chains to form two terminated polymer chains and a terminal double bond at one chain end. For simplicity, it is assumed that chains terminate only by disproportionation. In RDRP as discussed in the previous section, $[RP] \approx 1\times10^{-9}$ M, $[M] \approx 1$ M, $k_p \approx 1\times10^{3}$ M^{-1}s^{-1} [41, 124, 125], and $k_t \approx 1\times10^{7}$ M^{-1}s^{-1} [127, 128]. Based on these concentrations and rate constants, the rate of propagation is equal to $R_p = k_p\,[RP]\,[M] \approx 1\times10^{-6}$ M s^{-1} and the rate of termination is equal to $R_t = k_t\,[RP]^2 \approx 1\times10^{-11}$ M s^{-1}. The propagation reaction is part of the equilibrium Rxns. 7.58A–D between $\frac{R}{RX}$ and $\frac{RP}{DOP}$. As the termination Rxn. 7.58E removes propagating chains from the reaction with five-orders of magnitude slower rate than propagation, equilibrium Rxns. 7.58A–D are well-established when termination reaction is taking place. This implies that the concentration $[RP]$ in the termination Rxn. 7.58E is established at a given time by the above equilibrium reactions. In the previous section, it was shown that $[RP] \approx [R]_0$ (Eq. 7.54E) in the absence of termination because the primary radical R converts almost instantaneously to RP_1 at the start of the reaction to form 1-mer propagating chains in equilibrium with the dormant chains. In the presence of termination, $[RP]$ decreases with each termination as a fraction of the propagating chains exit the reaction.

7.8.2.1 Concentration of Propagating Chains in ATRP in the Presence of Termination Reaction

The molar conservation equation for the propagating chains, RP, is written as follows (see conservation Eq. 7.1A):

$$\left\{\begin{matrix}\text{Rate of change}\\ \text{in the number}\\ \text{of moles of } RP\\ \text{with time}\end{matrix}\right\} = \left\{\begin{matrix}\text{Rate of}\\ \text{production}\\ \text{of } RP1 \text{ by}\\ \text{Rxn. 7.58B}\end{matrix}\right\} + \left\{\begin{matrix}\text{Rate of}\\ \text{production}\\ \text{of } RP2 - RPn\\ \text{by Rxn. 7.58C}\end{matrix}\right\} - \left\{\begin{matrix}\text{Rate of}\\ \text{consumption}\\ \text{of } RP1 - RPn\\ \text{by Rxn. 7.58C}\end{matrix}\right\}$$

$$+ \left\{\begin{matrix}\text{Rate of}\\ \text{production}\\ \text{of } RP \text{ by}\\ \text{Rxn. 7.58D}\end{matrix}\right\} - \left\{\begin{matrix}\text{Rate of}\\ \text{consumption}\\ \text{of } RP \text{ by}\\ \text{Rxn. 7.58D}\end{matrix}\right\} - \left\{\begin{matrix}\text{Rate of}\\ \text{consumption of}\\ RP \text{ by termination}\\ \text{(Rxn. 7.58E)}\end{matrix}\right\} \tag{7.59A}$$

For constant volume reaction, the above conservation equation is written in mathematical form as follows:

$$\begin{aligned}\frac{d\,[RP]}{dt} &= +R_{RP_1} + \sum_{n=2}^{\infty} R_{RP_n} - \sum_{n=1}^{\infty} R_{RP_n} - k_{dac}\,[RP]\,[LM^{II}X_2] \\ &\quad + k_{ac}\,[DOP]\,[LM^{I}X] - k_{td}\,[RP]^2 \\ &= +R_{RP} - R_{RP} - k_{td}\,[RP]^2 = -k_{td}\,[RP]^2\end{aligned} \tag{7.59B}$$

where R_{RP_1} is the rate of production of RP_1 by Rxn. 7.58B, $\sum_{n=2}^{\infty} R_{RP_n}$ is the rate of production of RP with $n = 2 \to \infty$, $-\sum_{n=2}^{\infty} R_{RP_n}$ is the rate of consumption of RP with $n = 1 \to \infty$. Therefore, the first three terms on the right cancel each other. The fourth and fifth terms on the right cancel each other as they are the forward and reverse rates of equilibrium Rxn. 7.58D. Upon integration of the above equation subject to the initial condition $[RP] \approx [R]_0$ (see Eq. 7.54D) at $t \approx 0$, the concentration of propagating chains with time is given by the following equation in the presence of termination reaction:

$$\frac{1}{[RP]} = \frac{1}{[R]_0} + k_{td}t = \frac{[LM^{II}X]_0 \,/\, [LM^{I}X_2]_0}{K_{ATRP,i}\,[RX]_0} + k_{td}t \tag{7.59C}$$

where Eq. (7.54B1) is used in the above expression to relate $[R]_0$ to the initial concentrations of initiator and catalyst. It can be shown that the monomer concentration in the presence of termination reaction has the same functional form as that without termination (see Eq. 7.55B) except that $[R]_0$ is replaced with $[RP]$ from the above equation as follows:

$$[M] = [M]_0\, e^{-k_p[RP]t} = [M]_0\, e^{-\frac{k_p t}{k_{td}t + 1/[R]_0}} \tag{7.59D}$$

where $[R]_0$ is defined by Eq. (7.54B1). Based on the above equation, the rate of decrease of monomer concentration with time is reduced, as compared to no termination, because the termination reaction removes some of the propagating chains from the reaction.

7.8.2.2 Higher Moments of Propagating Chains in ATRP in the Presence of Termination Reaction

Applying the same approach used to derive Eq. (7.56A) in Section 7.8.1.3, the molar conservation equation for the first order mole-fraction moment of the propagating chains with termination reaction is given by the following expression (see conservation Eq. 7.1A):

$$\frac{d(\lambda_1[RP])}{dt} = -(1)R_{RP_1} - \sum_{n=2}^{\infty}\lambda_1 R_{RP_n} + \sum_{n=1}^{\infty}(\lambda_1+1)R_{RP_n} + (\gamma_1)k_{ac}[DOP][LM^IX] - (\lambda_1)k_{dac}[RP][LM^{II}X_2] - \lambda_1 k_{td}[RP]^2 \tag{7.60A}$$

where the last term on the right side is the rate of consumption of first order concentration moment of the propagating chains by termination Rxn. 7.58E. After substitution for $\gamma_n = \lambda_n$ and canceling like terms, as was done for the case with no termination reaction (see Eq. 7.56B), the above conservation equation simplifies to the following:

$$\begin{aligned}\frac{d(\lambda_1[RP])}{dt} &= \sum_{n=1}^{\infty}R_{RP_n} - \lambda_1 k_{td}[RP]^2 = k_p[M]\sum_{n=1}^{\infty}[RP] - \lambda_1 k_{td}[RP]^2\\ &= k_p[M][RP] - \lambda_1 k_{td}[RP]^2\end{aligned} \tag{7.60B}$$

$$\Rightarrow \lambda_1 = \frac{1}{[RP]}\int_{t=0}^{t}\left(k_p[M][RP] - \lambda_1 k_{td}[RP]^2\right)dt$$

where $R_{RP_n} = k_p[M][RP]_n$ is the rate of polymerization of n-mer propagating chains and $[RP] = \sum_{n=1}^{\infty}[RP]_n$. The above equation can be integrated numerically and solved for λ_1 subject to the initial condition that $\lambda_1[RP] = 0$ at $t = 0$. The termination rate is much slower than propagation in ATRP, that is $k_t[RP]^2 \approx 1\times10^{-11}\ \mathrm{M\,s^{-1}} << k_p[RP][M] \approx 1\times10^{-6}\ \mathrm{M\,s^{-1}}$, which implies that the last term on the right side in the above equation can be neglected as compared with the other term. Hence, the differential equation for the conservation of first order mole-fraction moment of the propagating chains with termination becomes identical to that given by Eq. (7.56B) with no termination. Therefore, according to Eq. (7.56D), λ_1 can be approximated by the following equation for ATRP with termination reaction:

$$\lambda_1 = \frac{[M]_0 - [M]}{[RP]} \tag{7.60C}$$

For ATRP with termination, $[M]$ and $[RP]$ are given by Eqs. (7.59D, C), respectively. After substitution for $[M]$ and $[RP]$, λ_1 or the kinetic chain length in ATRP in the presence of termination reaction is given by the following equation:

$$\lambda_1 = \frac{[M]_0 - [M]}{[RP]} = [M]_0\left(1 - e^{-\frac{k_p t}{k_{td}t+1/[R]_0}}\right)\left(\frac{1}{[R]_0} + k_{td}t\right) \tag{7.60D}$$

where $[R]_0$ is defined by Eq. (7.54B1). Applying the same approach used for the derivation of λ_2 in ATRP with no termination (see Eqs. 7.56G–7.56J) but using Eqs. (7.59C, D), and (7.60D) for $[RP]$, $[M]$, and λ_1, the time dependent differential equation for λ_2 in ATRP with termination is given as follows:

$$\frac{d(\lambda_2[RP])}{dt} = (2\lambda_1+1)k_p[M][RP] - \lambda_2 k_{td}[RP]^2 \quad L >> 1 \tag{7.60E}$$

The above nonlinear equation can be solved for λ_2 numerically subject to the initial condition that $\lambda_2[RP] = 0$ at $t = 0$. If $k_t[RP]^2 << k_p[RP][M]$, hence neglecting the second term on the right side of the above equation compared to the first term, and $\lambda_1 >> 1$, then λ_2 is approximately given by the following relation:

$$\lambda_2 = \frac{2}{[RP]} \int_{t=0}^{t} (2\lambda_1 + 1)\,[M]\,[RP]\,dt \quad L >> 1 \tag{7.60F}$$

where $[RP]$ and $[M]$ are given by Eqs. (7.59C, D), respectively.

7.8.2.3 Concentration and Higher Moments of Dormant Chains in ATRP in the Presence of Termination Reaction

The products from ATRP polymerization with termination are dormant and terminated polymer chains. There will also be some propagating chains, but their concentration is lower than the dormant chains by orders of magnitude. The concentration of dormant chains is defined by the equilibrium Eq. (7.57A) with or without termination. After substitution for $[RP]$ from Eq. (7.59C) and assuming $\frac{[LM^{II}X_2]}{[LM^{I}X]} \approx \frac{[LM^{II}X_2]_0}{[LM^{I}X]_0}$, $[DOP]$ is given by the following equation:

$$[DOP] = \frac{[LM^{II}X_2]_0}{[LM^{I}X]_0} \frac{1/K_{ATRP}}{1/[R]_0 + k_{td}t} = \frac{K_{ATRP,i}\,[RX]_0\,/K_{ATRP}}{1 + K_{ATRP,i}k_{td}\,[RX]_0\,t} \tag{7.61A}$$

$[DOP]$ is independent of time without termination (Eq. 7.57C) whereas, according to the above equation, $[DOP]$ decreases with time in the presence of termination reaction. The first and second order mole-fraction moments of the DOP chains for ATRP with termination $\gamma_1 = \lambda_1 = L = \overline{DP}_{n,dop}$ and $\gamma_2 = \lambda_2$ as given by Eqs. (7.60B, E), respectively. For ATRP with termination, factoring in that termination is much slower than propagation, γ_1 and γ_2 are given by Eqs. (7.60D, F), respectively, hence $\overline{DP}_{w,dop}$ and PDI of the dormant chains in ATRP in presence of termination reaction are $\overline{DP}_{w,dop} = \gamma_2/\gamma_1$ and $PDI = \overline{DP}_{w,dop}/\overline{DP}_{n,dop}$.

7.8.2.4 Concentration and Higher Moments of Terminated Chains in ATRP in the Presence of Termination Reaction

The terminated chains are produced by Rxn. 7.58E and there is no reaction consuming the terminated chains. The conservation equation for the zero-order concentration moment of the terminated chains, $[Q_{0,tp}]$, is written as follows (see conservation Eq. 7.1C):

$$\frac{dQ_{0,tp}}{dt} = \frac{d\,(\mu_0\,[TP])}{dt} = \frac{d\,[TP]}{dt} = +k_{td}\,[RP]^2 \tag{7.62A}$$

where $[TP]$ is the concentration of terminated chains and μ_0 is the zero-order mole-fraction moment of the terminated chains, which is equal to $\mu_0 = 1$ (see Eq. 3.14 in Chapter 3). After substitution for $[RP]$ from Eq. (7.59C) and rearrangement, $\frac{d\,[TP]}{dt}$ is given as a function of time by the following equation:

$$\frac{d\,[TP]}{dt} = +k_{td}\frac{1}{[1/[R]_0 + k_{td}t]^2} = \frac{1}{k_{td}} \frac{1}{[1/k_{td}\,[R]_0 + t]^2} \tag{7.62B}$$

Integration of the above equation subject to the initial condition $[TP] = 0$ at $t = 0$ results in the following expression for $[TP]$:

$$[TP] = \frac{1}{k_{td}} \int_{t=0}^{t} \frac{1}{[1/k_{td}\,[R]_0 + t]^2} dt = \frac{1}{k_{td}} \left[\frac{-1}{1/k_{td}\,[R]_0 + t} \right]_{t=0}^{t} = \frac{[R]_0\,t}{(1/k_{td}\,[R]_0) + t} \tag{7.62C}$$

where $[R]_0$ is defined in terms of $[RX]_0$ by Eq. (7.54B1). Based on the above equation, $[TP]$ increases with time at a decreasing rate. As $t \to \infty$, $[TP]_{t\to\infty} = [R]_0$. As the equilibrium Rxn. 7.58D is strongly shifted to the right, $[DOP] >> [RP] \approx [R]_0$, which implies that $[TP] << [DOP]$ throughout the reaction. The conservation equation for the first order mole-fraction moment of the terminated chains is as follows (see conservation Eq. 7.1C):

$$\frac{dQ_{1,tp}}{dt} = \frac{d\left(\mu_1 Q_{0,tp}\right)}{dt} = \frac{d\left(\mu_1 [TP]\right)}{dt} = +\mu_1' k_{td} [RP]^2 = \lambda_1 k_{td} [RP]^2 \tag{7.62D}$$

where μ_1 is the first order mole-fraction moment of the terminated chains and μ_1' is the instantaneous first order mole-fraction moment of the produced terminated chains in the time interval t and $t + dt$. As terminated chains are produced from termination of propagating chains by disproportionation, then $\mu_1' = \lambda_1$ (see Eq. 6.9 in Section 6.3.1 in Chapter 6). After substitution for λ_1 from Eq. (7.60D) and $[RP]$ from Eq. (7.59C), $\frac{d[\mu_1 TP]}{dt}$ is given by the following expression in terms of time:

$$\frac{d[\mu_1 TP]}{dt} = +\mu_1' k_{td} [RP]^2 = \frac{[M]_0 k_{td}\left(1 - e^{-[k_p t/(k_{td}t + 1/[R]_0)]}\right)}{[(1/[R]_0) + k_{td}t]} \tag{7.62E}$$

Integration of the above equation subject to the initial condition $[\mu_1 TP] = 0$ at $t = 0$ and substitution for $[TP]$ from Eq. (7.62C) yields the following relation for μ_1 as a function of time:

$$\mu_1 = \overline{DP}_{n,tp} = \frac{[M]_0 k_{td}[(1/k_{td}[R]_0) + t]}{[R]_0 t} \int_{t=0}^{t} \left[\left(1 - e^{-[k_p t/(k_{td}t + 1/[R]_0)]}\right) / \left(1/[R]_0 + k_{td}t\right)\right] dt \tag{7.62F}$$

where $\overline{DP}_{n,tp}$ is the number-average DP of the terminated chains in ATRP. In RDRP, the $\overline{DP}_{n,tp}$ increases gradually as the reaction proceeds. At short times, the DP distribution of the terminated chains is relatively narrow whereas the distribution is moderately and highly broad at intermediate and long reaction times, respectively. As the integral on the right side of the above equation is the summation of the first order moment of terminated chains, the average and standard deviation (breadth) of the terminated chain distribution increases with time. As a result, production of terminated chains should be constrained, and the main product should be the dormant chains to produce a low polydispersity polymer. Upon applying the conservation equation for the second order mole-fraction moment of the terminated chains, the $\overline{DP}_{w,tp}$ in ATRP with chain termination reaction is expressed as follows:

$$\overline{DP}_{w,tp} = \frac{\mu_2}{\mu_1} = \frac{\mu_2}{\overline{DP}_{n,tp}} = \frac{k_{td}\int_{t=0}^{t} [RP]^2 \left[\int_{t=0}^{t} (2\lambda_1 + 1)[M][RP]\,dt\right] dt}{[TP]\,\mu_1} \tag{7.62G}$$

where μ_1 is given by Eq. (7.62F) and $PDI_{tp} = \overline{DP}_{w,tp}/\overline{DP}_{n,tp}$. The reaction mixture after completion of the polymerization consists of dormant and terminated chains with orders of magnitude higher amounts of dormant chains than terminated chains (depending on the value of K_{ATRP} for the equilibrium Rxn. 7.58D). Based on the mixing theory of polymer chains covered in Section 5.3 in Chapter 5, The $\overline{DP}_{n,ATRP}$, $\overline{DP}_{w,ATRP}$ and PDI_{ATRP} for the overall product mixture are given by the following equations:

$$\overline{DP}_{n,ATRP} = \frac{\gamma_{1,DOP}[DOP] + \mu_{1,tp}[TP]}{[DOP] + [TP]} \tag{7.62H}$$

$$\overline{DP}_{w,ATRP} = \frac{\gamma_{2,DOP}\,[DOP] + \mu_{2,tp}\,[TP]}{\gamma_{1,DOP}\,[DOP] + \mu_{1,tp}\,[TP]} \tag{7.62I}$$

$$PDI_{ATR} = \frac{\overline{DP}_{w,ATRP}}{\overline{DP}_{n,ATRP}} = \frac{(\gamma_{2,DOP}\,[DOP] + \mu_{2,tp}\,[TP])\,([DOP] + [TP])}{(\gamma_{1,DOP}\,[DOP] + \mu_{1,tp}\,[TP])^2} \tag{7.62J}$$

7.9 Modeling Ionic Polymerization

7.9.1 Modeling Anionic Polymerization with Butyl Lithium Initiation

The dissociation of lithium chloride in aqueous solution into chloride anion and lithium cation is given by the following reaction:

$$\mathrm{LiCl} \xrightarrow{k_d} \mathrm{Cl}^- + \mathrm{Li}^+ \tag{7.63A}$$

The solubility product constant at saturation for lithium chloride is $K_{sp} = [\mathrm{Cl}^-][\mathrm{Li}^+] = 180\ \mathrm{M}^2$ at 20 °C. Hence, $[Cl^-] = [Li^+] = 13.42\,M$ [129]. Therefore, the solubility of LiCl in aqueous solution at 20 °C is $S = 569\ \mathrm{g\,L}^{-1} = 13.42\ \mathrm{M}$ [129]. This implies that >99% of LiCl dissociates into ions at 20 °C. Accordingly, the equilibrium dissociation Rxn. 7.63A for LiCl is entirely shifted to the right and the ions are separated and stabilized by solvation. Next, consider dissociation of butyl lithium (BuLi), which is widely used as the initiator in anionic polymerization, into ions in moderately polar aprotic solvents like ether or tetrahydrofuran (THF) given by the following reaction:

$$\mathrm{CH_3CH_2CH_2CH_2{-}Li} \xrightarrow{k_d} \mathrm{CH_3CH_2CH_2CH_2}^- + \mathrm{Li}^+ \tag{7.63B}$$

Butyl lithium forms dimeric, tetrameric, and even higher order clusters in ethyl ether or THF [130–132] with the lithium cations forming the core and the butyl anions forming the shell with the carbon carrying the negative charge at the interface. The clustering is attributed to the low stability of the individual lithium cations in the solvent (i.e., lack of resonance structure to stabilize the charge). The clustering confirms that BuCl dissociates into butyl anion and lithium cation in polar aprotic solvents due to the instability of the carbon-lithium bond. The clustering of lithium ions implies that significant amounts of the butyl anions are free and not bound to lithium cation in solution as ion pairs. This is further validated by the relatively large negative value for logarithm of the dissociation constant $pK_b = -36$ for BuLi with butyl anion as a strong base with >55% charge separation [133]. As a large fraction of the butyl anions of BuLi in solution are free and stable, the addition of monomer to solution results in the formation of living propagating chains under dry conditions and nitrogen atmosphere. During the reaction, the living chains continue to grow with the addition of more monomer to the reaction. As a result, polymers produced by anionic polymerization have a narrow chain distribution with $PDI < 1.1$ and nearly equal $\overline{DP}_{n,anp}$ and $\overline{DP}_{w,anp}$ given by the following equation:

$$\overline{DP}_{n,anp} \approx \overline{DP}_{w,anp} = \frac{\left(\sum_t [M]_0\right) - [M]_t}{[\mathrm{anion}^-]_0} \approx \frac{\left(\sum_t [M]_0\right) - [M]_t}{[I]_0} \tag{7.63C}$$

where $\overline{DP}_{n,anp}$ and $\overline{DP}_{w,anp}$ are the number- and weight-average degree of polymerization of the chains in anionic polymerization; the term $\sum_t [M]_0$ is the sum of all monomers added to the system either initially at $t = 0$ or during the reaction at different times; $[M]_t$ is the unreacted monomer at time t; and the terms $[\mathrm{anion}^-]_0$ and $[I]_0$ are the initial concentrations of anion and initiator, respectively. As the initiator in anionic polymerization is fully dissociated, the cations form clusters for greater stability, the anions are stable in free form without tight pairing with cations, hence $[\text{anion}^-]_0 \approx [I]_0$.

7.9.2 Modeling Cationic Polymerization with HI/ZnI$_2$ Initiation

The anion in anionic polymerization, which has an extra electron, has little desire to interact with another electron cloud. Consequently, an anion as a reactive species is relatively more stable as a free ion in solution as compared to a cation that lacks an electron. A cation, due to electron deficiency, has the propensity to interact with another electron cloud even with the electron cloud of another cation for resonance stabilization. In fact, the cluster formation by lithium cations in butyllithium in solution is attributed to resonance stabilization [130–132]. Moreover, more polar solvents containing oxygen or nitrogen atoms with unpaired electrons tend to stabilize cations by resonance to a greater extent than anions. In contrast to anions that tend to be stable as free ions in the solution, cations exhibit different degrees of interaction with the opposite charge in solution. These include as a freely acting cation ($C^+ + A^-$), strongly interacting cation (C^+A^-), weakly interacting cation ($C^+ \ldots A^-$), and inactive nonionic dormant species in solution ($C - A$) where C and A are the fragments of $C - A$ molecule that forms the corresponding cationic C^+ and anionic A^- species [134–138]. A suitable system to show the different degrees of interaction of cationic species with the oppositely charged anionic species is the cationic polymerization of styrene with hydrogen iodide (HI) as the initiator, a Lewis acid like zinc iodide (ZnI_2) as the initiation activator, and tetrabutylammonium halide (nBu_4NX, X=I, Cl, or Br) as the initiation co-activator in a polar solvent like methylene chloride [135]. Assuming the absence of any termination reaction, the elementary reactions for this cationic polymerization are given as follows [135]:

$$\mathrm{M + HI} \xrightarrow{k_i} \mathrm{HM - I} \tag{7.64A}$$

$$\mathrm{HM - I + ZnI_2} \underset{k_{dac,ai}}{\overset{k_{ac,ai}}{\rightleftharpoons}} \mathrm{HM^+..^-I.ZnI_2} \tag{7.64B}$$

$$\mathrm{HM^+..^-I.ZnI_2} \underset{k_{dac,fi}}{\overset{k_{ac,fi}}{\rightleftharpoons}} \mathrm{HM^+ + ^-I.ZnI_2} \tag{7.64C}$$

$$\mathrm{HM_n^+..^-I.ZnI_2 + M} \xrightarrow{k_{p,ai}} \mathrm{HM_{n+1}^+..^-I.ZnI_2} \tag{7.64D}$$

$$\mathrm{HM_n^+ + ^-I.ZnI_2 + M} \xrightarrow{k_{p,fi}} \mathrm{HM_{n+1}^+ + ^-I.ZnI_2} \tag{7.64E}$$

$$\mathrm{HM_i^+ + nBu_4NCl} \underset{k_{ac,dr}}{\overset{k_{dac,dr},\ ^-\mathrm{I.ZnI_2}}{\rightleftharpoons}} \mathrm{DOP_i + nBu_4NI} \tag{7.64F}$$

In the first reaction, the initiator HI reacts with a monomer with rate constant k_i to form the relatively inactive alkyl iodide; the second reaction reversibly forms a 1-mer propagating associated cation–anion pair with forward and reverse rate constants $k_{ac,ai}$ and $k_{dac,ai}$, respectively; the third reaction reversibly forms a 1-mer propagating dissociated cation–anion pair, as a relatively free cation, with forward and reverse rate constants $k_{ac,fi}$ and $k_{dac,fi}$; the fifth reaction adds a monomer to the n-mer propagating associated cation–anion pair; the sixth reaction adds a monomer to the n-mer propagating dissociated free cation; and the seventh reaction reversibly deactivates the i-mer propagating dissociated free cation with forward and reverse rate constants $k_{dac,dr}$ and $k_{ac,dr}$, to form an i-mer inactive nonionic dormant chain. According to the above reactions, there exists multiple different propagating species like the associated and dissociated propagating cations each having a unique chain distribution as well as the dormant species that remove the dissociated, free n-mer propagating cations from the reaction by deactivation. Therefore, cationic polymerization has traits of both anionic polymerization (i.e., stable dissociated free propagating cations) and RDRP (i.e., reversibly deactivated dormant species) as well as its own trait of associated versus dissociated propagating cations. Due to the presence of multiple types of propagating species, *PDI* of the polymer produced by cationic polymerization is generally higher than anionic or RDRP but less than conventional radical polymerization. It is possible to produce polymers with $PDI < 1.1$ by cationic polymerization by matching the monomer with a suitable solvent, reaction temperature, and initiation activator. For example, methoxy styrene (MOS) polymerized with HI/ZnI_2 in the absence of nBu_4NI in methylene chloride at −15 °C produces a bimodal p(MOS) with $PDI > 1.1$ whereas the same monomer polymerized in the presence of nBu_4NI produces a unimodal p(MPOS) with $PDI < 1.1$ [135]. This effect is attributed to deactivation of the free dissociated i-mer cations to dormant chains by Rxn. 7.64F, which makes the n-mer associated cation as the only propagating cation [135]. The i-mers in Rxn. 7.64F are relatively shorter chains because they are dormant most of the time during the reaction (i.e., $k_{dac} << k_{ac}$). The reaction of a monomer with HI to form $HM - I$ is fast, thus it occurs in the beginning of the reaction when $[M] = [M]_0$. Applying conservation of molar mass to HI and $HM - I$ (see conservation Eq. 7.1D for small molecules) subject to the initial condition that $[HI] = [HI]_0$ and $[M] \approx [M]_0$ at $t = 0$, the $[HMI]$ is given by the following equation as a function of time:

$$[\mathrm{HM-I}] = [\mathrm{HI}]_0\left(1 - e^{-k_i[M]_0 t}\right) \tag{7.65A}$$

The first reaction is fast because the addition of monomer to $[HI]$ is exothermic (i.e., thermodynamically favored). Therefore, it is a good assumption to set $[HM - I] \approx [HI]_0$ for the analysis of Rxns. 7.64A–F. Rxns. 7.64B, C, F are equilibrium reactions. Assuming these three reactions are at equilibrium, the concentrations of associated propagating cations $[\mathrm{HM^+..^-I.ZnI_2}]$, dissociated propagating cations $[HM^+]$, and dormant chains $[DOP]$ during the reaction are given by the following equations:

$$[\mathrm{HM^+..^-I.ZnI_2}]_e = \frac{k_{ac,ai}}{k_{dac,ai}}[\mathrm{HM-I}]_e[\mathrm{ZnI_2}]_e = \mathrm{K_{ai}}[\mathrm{HM-I}]_e[\mathrm{ZnI_2}]_e \tag{7.65B}$$

$$[\mathrm{HM^+}]_e = \left(\frac{k_{ac,fi}}{k_{dac,fi}}\right)^{1/2}[\mathrm{HM^+..^-I.ZnI_2}]_e^{1/2} = \left(K_{ai}K_{fi}\right)^{1/2}[\mathrm{HM-I}]_e^{1/2}[\mathrm{ZnI_2}]_e^{1/2} \tag{7.65C}$$

$$\begin{aligned}[DOP]_e &= \frac{1}{\frac{k_{ac}}{k_{dac}}}\frac{[\mathrm{HM^+}]_e[\mathrm{nBu_4NCl}]_e}{[\mathrm{nBu_4NI}]_e}\\ &= \left(\frac{K_{ai}K_{fi}}{K_{dr}^2}\right)^{1/2}\frac{[\mathrm{nBu_4NCl}]_e}{[\mathrm{nBu_4NI}]_e}[\mathrm{HM-I}]_e^{1/2}[\mathrm{ZnI_2}]_e^{1/2}\end{aligned} \tag{7.65F}$$

where $[\mathrm{HM-I}]_e$, $[\mathrm{ZnI_2}]_e$, $[\mathrm{nBu_4NCl}]_e$ and $[\mathrm{nBu_4NI}]_e$ are the equilibrium concentrations of HM − I, ZnI_2, nBu_4NCl, and nBu_4NI in the reaction; $[\mathrm{HM^+..^-I.ZnI_2}]_e$, $[\mathrm{HM^+}]_e$ and $[DOP]_e$ are the equilibrium concentrations of the associated propagating cation $\mathrm{HM^+..^-I.ZnI_2}$, dissociated free propagating cation HM^+, and dormant species *DOP*, respectively; K_{ai}, K_{fi}, and K_{dr} are the equilibrium constants for the formation of $\mathrm{HM^+..^-I.ZnI_2}$ associated propagating cation, $\mathrm{HM^+}$ dissociated propagating cation, and deactivated $\mathrm{HM^+}$ propagating chains to create dormant chains. These equations are based on the premise that the reactants HM − I, ZnI_2, nBu_4NCl, and nBu_4NI, and species $\mathrm{HM^+..^-I.ZnI_2}$, $\mathrm{HM^+}$, and *DOP* reach their equilibrium concentrations immediately following the addition of reactants (monomer, initiator, and activators) to the reaction independent of time (i.e., fast equilibration). Following formation and equilibration, the monomer is added to the propagating species in a time dependent manner until all monomers are consumed.

7.9.2.1 Modeling Cationic Polymerization with HI/ZnI$_2$ Initiation in the Absence of nBu$_4$NCl

In the absence of nBu$_4$NCl, free dissociated propagating cations are not deactivated to dormant chains by Rxn. 7.64F. Therefore, there are two living (i.e., no termination) propagating species, namely HM$^+$..$^-$I.ZnI$_2$ and HM$^+$ species, that propagate by reacting with monomer according to Rxns. 7.64D, E, respectively. For two different propagating species with concentrations given by equilibrium relations 7.65B, C, the concentration of reactive propagating species, $[RP]$, is given as follows:

$$\begin{aligned}[RP] &= [RP_{ai}] + [RP_{fi}] = [\mathrm{HM^+..^-I.ZnI_2}]_e + [\mathrm{HM^+}]_e \\ &= K_{ai}\,[\mathrm{HMI}]_e\,[\mathrm{ZnI_2}]_e + (K_{ai}K_{fi})^{1/2}[\mathrm{HMI}]_e^{1/2}[\mathrm{ZnI_2}]_e^{1/2}\end{aligned} \tag{7.66A}$$

where $[RP_{ai}]$ and $[RP_{fi}]$ are the concentrations of associated and dissociated (free) propagating cations, respectively; $[HM^+..^-\mathrm{I.ZnI_2}]_e$ and $[HM^+]_e$ are the equilibrium concentrations of associated and dissociated propagating cations; $[HM-I]_e$ and $[\mathrm{ZnI_2}]_e$ are the equilibrium concentrations of $HM-I$ and ZnI$_2$. The molar conservation equation for the monomer at constant reaction volume for this cationic polymerization in the absence of nBu$_4$NCl is as follows (see conservation Eq. 7.1D for small molecules):

$$\begin{aligned}\frac{d\,[M]}{dt} &= -k_{p,ai}\,[RP_{ai}]\,[M] - k_{p,fi}\,[RP_{fi}]\,[M] \\ &= -\left[k_{p,ai}K_{ai}\,[\mathrm{HMI}]_e\,[\mathrm{ZnI_2}]_e + k_{p,fi}\,(K_{ai}K_{fi})^{1/2}\,[\mathrm{HMI}]_e^{1/2}[\mathrm{ZnI_2}]_e^{1/2}\right][M]\end{aligned} \tag{7.66B}$$

After separation of variables and assuming fast equilibration compared to propagation, and subject to the initial condition $[M] = [M]_0$ at $t = 0$, the monomer concentration is given by the following equation as a function time:

$$[M] = [M]_0\,e^{-k_{p,eff}t} \tag{7.66C}$$

where $k_{p,eff}$ is defined as follows:

$$k_{p,eff} = k_{p,ai}K_{ai}\,[\mathrm{HMI}]_e\,[\mathrm{ZnI_2}]_e + k_{p,fi}\,(K_{ai}K_{fi})^{1/2}\,[\mathrm{HMI}]_e^{1/2}[\mathrm{ZnI_2}]_e^{1/2} \tag{7.66D}$$

where $k_{p,eff}$ is the effective rate constant for the chain propagation reaction. Based on the above equation, the rate of polymerization is first order with respect to monomer and $[M]$ decreases exponentially with time. The rate of cationic polymerization with HI/ZnI_2 initiation in absence of nBu$_4$NCl is to the order of 0.5–1 with respect to HI and ZnI$_2$ concentrations depending on which species dominates propagation. If the associated HM$^+$..$^-$I.ZnI$_2$ species dominates, the propagation rate is first order with respect to HI and ZnI$_2$. Conversely, if the dissociated HM^+ species dominates, the propagation rate is proportional to square root of HI and ZnI$_2$.

7.9.2.1.1 First-Order Moment of Propagating Chains in Cationic Polymerization with HI/ZnI$_2$ Initiation in the Absence of nBu$_4$NCl

Based on the general conservation equation for the kth order mole-fraction moment of the reactive species (see conservation Eq. 7.1A), the conservation equation for the first order mole-fraction moment of the propagating chains for cationic polymerization with HI/ZnI$_2$ initiation in the absence of nBu$_4$NCl is as follows (assuming the kinetic chain length $L >> 1$):

$$\begin{aligned}\frac{d\,(\lambda_1\,[RP])}{dt} &= \{-\lambda_{1,ai} + (\lambda_{1,ai}+1)\}\,k_{p,ai}\,[RP_{ai}]\,[M] + \{-\lambda_{1,fi} + (\lambda_{1,fi}+1)\}\,k_{p,fi}\,[RP_{fi}]\,[M] \\ &= k_{p,ai}\,[RP_{ai}]\,[M] + k_{p,fi}\,[RP_{fi}]\,[M] = k_{p,eff}\,[M] = -\frac{d\,[M]}{dt}\end{aligned} \tag{7.67A}$$

where $\lambda_{1,ai}$ and $\lambda_{1,fi}$ are the first order mole-fraction moment of the associated and dissociated (free) propagating cations, respectively. Furthermore, $\lambda_1\,[RP]$ is defined in terms of $\lambda_{1,ai}$ and $\lambda_{1,fi}$ as follows:

$$\lambda_1\,[RP] = \lambda_{1,ai}\,[RP_{ai}] + \lambda_{1,fi}\,[RP_{fi}] \tag{7.67B}$$

where $[RP_{ai}]$ and $[RP_{fi}]$ are defined by Eq. (7.66A). It can be shown, by making separate first order mole-fraction moment conservation balances on RP_{ai} and RP_{fi} propagating chains, that $\lambda_{1,ai}$ and $\lambda_{1,fi}$ are expressed in terms of $[M]$ and time by the following equations:

$$\lambda_{1,ai} = (k_{p,ai}/k_{p,eff})([M]_0 - [M]) = [M]_0\,(k_{p,ai}/k_{p,eff})(1 - e^{-k_{p,eff}\,t}) \tag{7.67C}$$

$$\lambda_{1,fi} = (k_{p,fi}/k_{p,eff})([M]_0 - [M]) = [M]_0\,(k_{p,fi}/k_{p,eff})(1 - e^{-k_{p,eff}\,t}) \tag{7.67D}$$

Upon substitution for $\lambda_{1,ai}$ and $\lambda_{1,fi}$ from the above equations in Eq. (7.67A), for $[RP]$ from Eq. (7.66A) followed by integration subject to the initial condition $\lambda_1 = 1$ and $[M] = [M]_0$ at $t = 0$, $\lambda_1 = L = \overline{DP}_{n,rp}$ is given by the following relation:

$$\lambda_1 = L = \overline{DP}_{n,rp} = \frac{[M]_0\,(1 - e^{-k_{p,eff}t})}{K_{ai}\,[\mathrm{HMI}]_e\,[\mathrm{ZnI_2}]_e + (K_{ai}K_{fi})^{1/2}\,[\mathrm{HMI}]_e^{1/2}[\mathrm{ZnI_2}]_e^{1/2}} \tag{7.67E}$$

The above equation indicates that the kinetic chain length is the average of the kinetic chain lengths of the two independent chain distributions, namely one distribution produced by the associated cations and the other produced by the dissociated free cations. As the propagation rate constant of the dissociated free cations is higher than the associated cations (i.e., $k_{p,fi} > k_{p,ai}$), chains produced by dissociated fee ions possess higher degree of polymerization in the bimodal distribution of chains.

7.9.2.1.2 Higher Moments of the Propagating Chains in Cationic Polymerization with HI/ZnI_2 Initiation in Absence of nBu_4NCl

The conservation equation for the second order moment of the propagating chains (see conservation Eq. 7.1A) for cationic polymerization with HI/ZnI_2 initiation in the absence of nBu_4NCl is given as follows (assuming the kinetic chain length $2\lambda_1 >> 1$):

$$\begin{aligned}\frac{d\,(\lambda_2\,[RP])}{dt} &= \{-\lambda_{2,ai} + (\lambda_{2,ai} + 2\lambda_{1,ai} + 1)\}\,k_{p,ai}\,[RP_{ai}]\,[M] \\ &\quad + \{-\lambda_{2,fi} + (\lambda_{2,fi} + 2\lambda_{1,fi} + 1)\}\,k_{p,fi}\,[RP_{fi}]\,[M] \\ &= 2\lambda_{1,ai}k_{p,ai}\,[RP_{ai}]\,[M] + 2\lambda_{1,fi}k_{p,fi}\,[RP_{fi}]\,[M] \\ &= (2/k_{p,eff})\,[M]\,([M]_0 - [M])\,\{k^2_{p,ai}\,[RP_{ai}] + k^2_{p,fi}\,[RP_{fi}]\} \\ &= (2[M]_0^2/k_{p,eff})\,\{k^2_{p,ai}\,[RP_{ai}] + k^2_{p,fi}\,[RP_{fi}]\}\,(e^{-k_{p,eff}t} - e^{-2k_{p,eff}t})\end{aligned} \tag{7.68A}$$

where Eqs. (7.66C, D) are used to substitute for $[M]$ in terms of time. After integration of the above differential equation subject to the initial condition $\lambda_2\,[RP] = 0$ at $t = 0$, λ_2 is given by the following expression:

$$\lambda_2 = \left(2[M]_0^{\,2}/k^2_{p,eff}\right)\left\{\frac{k^2_{p,ai}\,[RP_{ai}] + k^2_{p,fi}\,[RP_{fi}]}{[RP]}\right\}(0.5 + 0.5e^{-2k_{p,eff}t} - e^{-k_{p,eff}t}) \tag{7.68B}$$

The above equation shows that λ_2, like λ_1 in Eq. (7.67E), is the average of the second order moments of two propagating chain distributions with one corresponding to the associated cations and the other corresponding to the dissociated free cations. Therefore, concurrent propagation of the two cations with different concentrations and propagation rate constants results in a bimodal distribution of the propagating chains. The $\overline{DP}_{w,rp}$ and PDI_{rp} of the propagating chains for cationic polymerization with HI/ZnI_2 initiation in the absence of nBu_4NCl are determined from λ_1 and λ_2 using $\overline{DP}_{w,rp} = \lambda_2/\lambda_1$ and $PDI_{rp} = \overline{DP}_{w,rp}/\overline{DP}_{n,rp}$. It is possible for the propagating cations to show higher modality like trimodal distribution because the associated cation–anion pairs may be tightly or weakly associated [136]. In the tightly associated pair, the intimate association of the counterion with the cation hinders the reaction of monomer with the propagating chain. In the weakly associated pair, solvent molecules separate the cation from its counterion which assists the reaction of monomer with the propagating chain. Following monomer consumption, the living cationic polymerization can be terminated by the addition of a nucleophilic terminating agent like water [139–142]. In the absence of termination, the concentration and size distribution of the polymer chains are the same as the propagating chains. This implies that $[TP] = [RP]$, $\overline{DP}_{n,tp} = \overline{DP}_{n,rp}$, $\overline{DP}_{w,tp} = \overline{DP}_{w,rp}$, and $PDI_{tp} = PDI_{rp}$.

7.9.2.2 Modeling Cationic Polymerization with HI/ZnI$_2$ Initiation in the Presence of nBu$_4$NCl

If nBu_4NCl is added to the polymerization, the following reversible reaction with the dissociated free cation occurs:

$$\mathrm{HM_i^+} + \mathrm{nBu_4NCl} \underset{k_{ac,dr}}{\overset{k_{dac,dr},\ ^{-}\mathrm{I.ZnI_2}}{\rightleftharpoons}} DOP_i + \mathrm{nBu_4NI} \tag{7.69A}$$

In the above reaction, the Cl atom is transferred from nBu_4NCl to the free HM_i^+ cation to form $HM_i - Cl$, which deactivates the free propagating cation to a dormant chain DOP_i. Like RDRP in that the rate constant for deactivation of the propagating chains by transient metal catalyst to form dormant chains was a few orders of magnitude higher than the rate constant for activation ($k_{dac} >> k_{ac}$, Rxn. 7.53D in section 7.8.1), the rate constant for deactivation of HM_i^+ cations by nBu_4NCl in the above reaction is a few orders of magnitude higher than the activation reaction, that is $k_{dac,dr} >> k_{ac,dr}$. This implies that the dissociated free HM_i^+ cations are practically in the dormant state during the reaction whereas the associated $HM^+..^-I.ZnI_2$ cations continue to propagate. As a result, the following relations hold in the presence of nBu_4NCl (see Eq. 7.66A):

$$\begin{aligned} [HM_i^+] \approx 0 \ \Rightarrow\ & [RP] = [\mathrm{HM^+..^-I.ZnI_2}]_e \qquad \text{with nBu}_4\text{NCl} \\ \Rightarrow\ & [RP] = K_{ai}\,[\mathrm{HM-I}]_e\,[\mathrm{ZnI_2}]_e \end{aligned} \tag{7.69B}$$

The expressions for $k_{p,eff}$, $[M]$, λ_1, and λ_2 given by Eqs. (7.66D), (7.66C), (7.67E), and (7.68B) simplify to the following relations in the presence of nBu_4NCl:

$$[M] = [M]_0\, e^{-k_{p,ai}K_{ai}[\mathrm{HMI}]_e[\mathrm{ZnI_2}]_e t} \tag{7.69C}$$

$$\lambda_1 = L = \overline{DP}_{n,rp} = \frac{[M]_0\left(1 - e^{-k_{p,ai}K_{ai}[\mathrm{HMI}]_e[ZnI_2]_e t}\right)}{K_{ai}\,[\mathrm{HMI}]_e\,[\mathrm{ZnI_2}]_e} \tag{7.69D}$$

$$\lambda_2 = \frac{2[M]_0^{\,2}}{\left(K_{ai}\,[\mathrm{HMI}]_e\,[\mathrm{ZnI_2}]_e\right)^2}\left(0.5 + 0.5e^{-2k_{p,ai}K_{ai}[\mathrm{HMI}]_e[\mathrm{ZnI_2}]_e t} - e^{-k_{p,ai}K_{ai}[\mathrm{HMI}]_e[\mathrm{ZnI_2}]_e t}\right) \tag{7.69E}$$

In the presence of nBu_4NCl, polymerization is living with only one type of cationic propagating chain in the reaction. Therefore, the distribution of propagating cations is unimodal in the presence of nBu_4NCl. As discussed in the previous section (in the absence of nBu_4NCl), the concentration, size distribution of the produced polymer chains, and the averages of the chain distribution are the same as the propagating chains.

7.10 Summary

In this chapter, we developed mathematical models for polymerizations with different sets of elementary reactions to predict concentrations of the reactants and products, kinetic chain length, number-average and weight-average degree of polymerization, and polydispersity of the produced polymer chains. First, conventional radical polymerization was modeled in the absence of chain transfer using the assumption of quasi-steady state of the propagating radical species. The basic premise in conventional radical polymerization is that the produced radicals are transient, that is, the radicals undergo propagation and terminate within a differential time interval in the reaction, and they are not carried over to the next interval. The conventional radical polymerization was modeled in different types of reactors including batch stirred vessel reactor (BSVR), continuous stirred tank reactor (CSTR), and plug flow tubular reactor (PFTR). Next, conventional radical polymerization was modeled with chain transfer to transfer agent and terminated polymer chains. Modeling showed that chain transfer to the transfer agent serves to limit uncontrolled growth of the propagating chains, especially if bimolecular termination of propagating chains is slow. Further, modeling showed that branching density increases linearly with the rate constant for chain transfer to polymer and decreases with the square root of the rate constant for termination. Then, conventional radical polymerization with β_2 chain scission was modeled which occurs following the transfer of a radical from a primary to a secondary carbon via chain transfer to polymer. A β_2 chain scission results in the formation of a double bond at one chain-end of the terminated chains produced by chain scission. The modeling showed that the density of terminal double bonds at long times scales linearly with k_{ctp} and to ${k_{td}}^{-1/2}$ but is independent of the rate constant for chain scission. Next, conventional radical polymerization with photo-initiation was modeled in the absence of chain transfer reactions. A unique feature of photopolymerization is that polymerization is limited by the depth of penetration of light in the thin polymerizing film to generate primary reactive species. The modeling showed that the maximum depth of light penetration or the maximum film thickness is inversely related to the extinction coefficient of initiator and its concentration. Then, radical polymerization with reversible deactivation (RDRP) of reactive species by atom transfer polymerization was modeled. Unlike conventional radical polymerization in which radicals are at quasi-steady state, the radicals in RDRP shift between active and dormant states throughout the reaction; hence the radicals do not undergo termination. Further, the product in RDRP can be dormant chains. As a result, the kinetic chain length of the chains in RDRP increases gradually with increasing monomer conversion. The modeling showed that the concentration of radical species is controlled by the equilibrium constant between active and dormant species. Further, the modeling showed that the kinetic chain length is short early in the reaction but reaches the ratio of the concentration of monomer to active species at long times. Next, anionic polymerization was modeled and compared to other polymerization mechanisms. Contrary to conventional radical polymerization, all anionic reactive species in an anionic polymerization reaction form instantly, remain living, are stable with little side reactions, do not terminate, and all reactive species propagate concurrently until monomer is depleted. In addition, monomers can be added at any time to the reaction during anionic polymerization to further increase the kinetic chain length of the propagating anions. The modeling showed that anionic polymerization produces a nearly monodisperse polymer. Then cationic polymerization was modeled. In contrast to anionic polymerization, cationic reactive species are less stable than anionic species and tend to exist as ion pairs with the counterion. As a result, there can be multiple propagating species in cationic polymerization with different propagation rate constants due to the different degrees of association of propagating cations with their counterions. The modeling showed that cationic polymerization is somewhat like RDRP as the reactive species concentration depends on the equilibrium constant between tightly/weakly (with the counterion) associated cationic species and the bonded (with the counterion) inactive species. Like RDRP, the kinetic chain length is short early in the reaction and approaches the ratio of monomer to active cationic species concentrations at long times. The models developed in this chapter demonstrate the importance of the kinetic chain length of the propagating reactive species in quantifying the properties of the produced polymer.

7.11 List of Symbols and Abbreviations

a: exponent of the Mark–Houwink–Sakurada equation relating intrinsic viscosity of a polymer to its degree of polymerization

$[\text{anion}]_0$: initial concentration of anion in anionic polymerization

A: cross-sectional area of the resin perpendicular to the direction of penetration of photons in photopolymerization

ATRP: atom transfer radical polymerization

b: Penetration depth of incident radiation in a polymerizing resin in photopolymerization

b_{max}: penetration depth of incident radiation in a resin corresponding to absorption of 95% of incident radiation in photopolymerization

$C^+ + A^-$: free cation and anion stabilized by very strong interaction with solvent molecules in ionic polymerization

$C^+..A^-$: weakly interacting cation and anion stabilized by strong interaction with solvent molecules in ionic polymerization

C^+A^-: strongly interacting cation and anion stabilized by weak interaction with solvent molecules in ionic polymerization

$C - A$: inactive dormant cation–anion species due to no stabilization of anion/cation species by solvent molecules in ionic polymerization

CTA: chain transfer agent

$[CTA]$: concentration of chain transfer agent in reaction at time t

$[CTA]_0$: initial concentration of chain transfer agent in reaction

D_I: diffusion coefficient of initiator used if polymerization reaction is diffusion controlled

DOP_nX, DOP_iX: dormant polymer chain with n or i repeat units in ARTP

$\overline{DP}_{n,anp}$: number-average degree of polymerization of living polymer chains in anionic polymerization

$\overline{DP}_{w,anp}$: weight-average degree of polymerization of living polymer chains in anionic polymerization

$\overline{DP}_{n,ATRP}$: number-average degree of polymerization of terminated+dormant polymer chains in ATRP

$\overline{DP}_{w,ATRP}$: weight-average degree of polymerization of terminated+dormant polymer chains in ATRP

$\overline{DP}_{n,dop}$: number-average degree of polymerization of dormant polymer chains in ATRP

$\overline{DP}_{w,dop}$: weight-average degree of polymerization of dormant polymer chains in ATRP

$\overline{DP}_{n,rp}$: number-average degree of polymerization of reactive propagating chains

$\overline{DP}_{w,rp}$: weight-average degree of polymerization of reactive propagating chains
$\overline{DP}_{n,tp}$: number-average degree of polymerization of terminated polymer chains
$\overline{DP}_{w,tp}$: weight-average degree of polymerization of terminated polymer chains
$\overline{DP}_{\eta,tp}$: viscosity-average degree of polymerization of terminated polymer chains
$[DOP]$: concentration of dormant polymer chains in ARTP
$[DOP]_e$: equilibrium concentration of dormant polymer chains in ARTP
dt: infinitesimal change in reaction time
f_0: initiator efficiency to produce 1-mer reactive propagating chains
$f(t)$: function of time in photopolymerization
$f(z)$: function with depth of penetration in z direction in photopolymerization
$f_{rp}(n)$: number-fraction n-mer distribution of the degree of polymerization of reactive propagating chains
HI: hydrogen iodide initiator used in cationic polymerization
$HM-I$: monomer added between tightly bound hydrogen cation and iodide anion in cationic polymerization
HM^+: monomer added hydrogen cation in cationic polymerization
$HM^+..I^-.ZnCl_2$: monomer added between weakly interacting hydrogen cation and iodide anion stabilized by complexation of the anion with a Lewis acid, like zinc chloride, in cationic polymerization
$HM_n^+..I^-.ZnCl_2$: n-mer propagating cation weakly interacting with anion stabilized by complexation of the anion with a Lewis acid, like zinc chloride, in cationic polymerization
$HM_n^+ + I^-.ZnCl_2$: monomer added between free hydrogen cation and iodide anion stabilized by very strong complexation of the anion with a Lewis acid, like zinc chloride, in cationic polymerization
$[HI]$: concentration of hydrogen iodide initiator used in cationic polymerization
$[HI]_0$: initial concentration of hydrogen iodide initiator used in cationic polymerization
$[HM^+]_e$: equilibrium concentration of monomer added hydrogen cation in cationic polymerization
$[HMI]$: concentration of monomer added between tightly bound hydrogen cation and iodide anion in cationic polymerization
$[HMI]_e$: equilibrium concentration of monomer added between tightly bound hydrogen cation and iodide anion in cationic polymerization
$[HM^+..I^-.ZnCl_2]_e$: equilibrium concentration of monomer added between weakly interacting hydrogen cation and iodide anion stabilized by complexation of the anion with a Lewis acid, like zinc chloride, in cationic polymerization
$h\nu$: photon energy of radiation where h is frequency of photons and ν is Planck's constant
I: polymerization initiator
I': initiator molecule in the excited state after absorption of a photon of energy
$I^\bullet$: inactive species produced by recombination of reactive species following initiator dissociation
I_a: absorbed radiation energy in a polymerizing resin in photopolymerization
$I_{a,z}$: absorbed radiation energy in a polymerizing resin in the direction of resin thickness z
$I_{\nu=0}$: incident radiation energy in the direction of resin thickness z at z=0
$I_{\nu=z}$: incident radiation energy in the direction of resin thickness z at any thickness z
$[I]$: initiator concentration in polymerization at time t
$[I]_0$: initial initiator concentration in polymerization at time t
k_{ac}: rate constant for activation of dormant chains to form an active propagating chains in reversible deactivation radical polymerization

k_{dac}: rate constant for deactivation of propagating chains to form dormant chains in reversible deactivation radical polymerization

k_{aci}: forward (activation) rate constant for reversible transfer of the leaving group (halogen) from an initiator to a transition metal catalyst to form a primary radical (activation) in ATRP

k_{daci}: reverse (deactivation) rate constant for reversible transfer of the leaving group (halogen) from a transition metal catalyst to an initiator to regenerate the initiator in ATRP

$k_{ac,ai}$: forward rate constant for reversible addition of a propagating radical to a RAFT chain transfer agent to form a RAFT adduct propagating chain in the initiation step in RAFT polymerization

$k_{dac,ai}$: reverse rate constant for reversible addition of a propagating radical to a RAFT chain transfer agent to form a RAFT adduct propagating chain in the initiation step in RAFT polymerization

$k_{ac,dr}$: forward rate constant for reversible activation of n-mer propagating dissociated cations by exchange of halide with tetrabutylammonium halide in cationic polymerization

$k_{dac,dr}$: reverse rate constant for reversible deactivation of n-mer propagating dissociated cations by exchange of halide with tetrabutylammonium halide in cationic polymerization

$k_{ac,fi}$: forward rate constant for reversible fragmentation of a RAFT adduct propagating chain to form a new polymeric RAFT chain transfer agent and a new propagating radical in the initiation step in RAFT polymerization

$k_{dac,fi}$: reverse rate constant for reversible fragmentation of a RAFT adduct propagating chain to form a new polymeric RAFT chain transfer agent and a new propagating radical in the initiation step in RAFT polymerization

k_c: rate constant for recombination reaction of the dissociated fragments of the polymerization initiator

k_{ci}: rate constant for the reaction of dissociated initiator fragment with a monomer to produce a 1-mer propagating chain

k_{cta}: rate constant for transfer of a propagating species to a chain transfer agent in addition polymerization

k_{ctp}: rate constant for chain transfer to polymer

k_d: rate constant for the dissociation reaction of polymerization initiator

k_i: rate constant for initiation/activation of 1-mer propagating cation in cationic polymerization

k_p: rate constant for propagation reaction between a propagating chain and a monomer

$k_{p,ai}$: rate constant for propagation reaction of $HM_n^+..I^-.ZnCl_2$ with monomer in the weakly associated form stabilized by complexation of the anion with a Lewis acid in cationic polymerization

$k_{p,eff}$: effective rate constant for propagation of associated and free propagating cations in cationic polymerization

$k_{p,fi}$: rate constant for propagation reaction of $HM_n^+ + I^-.ZnCl_2$ with monomer in the free form stabilized by very strong complexation of the anion with a Lewis acid in cationic polymerization

k_{td}: rate constant for termination of two propagating chains by disproportionation

$k_{\beta 2}$: rate constant for β_2 chain scission of an n-mer propagating chain by splitting the bond between α and β carbons where α is the carbon carrying a secondary reactive species to form two shorter chains

K_{ai}: equilibrium constant between HM_n^+ n-mer propagating cation and its anion in the associated form of $HM_n^+..I^-.ZnCl_2$ stabilized by complexation of the anion with a Lewis acid in cationic polymerization

K_{ATRP}: equilibrium constant between the reactive propagating chains and the dormant polymer chains in ATRP

$K_{ATRP,i}$: equilibrium constant for initiation reaction in ATRP

K_{fi}: equilibrium constant between HM_n^+ n-mer propagating cation and its anion in the free form of $HM_n^+ + I^-.ZnCl_2$ stabilized by very strong complexation of the anion with a Lewis acid in cationic polymerization

K_{sp}: solubility product constant for dissociation of initiator, like lithium chloride into cation and anion in ionic polymerization

LB: a long side chain or branch on a polymer chain

$[LB]$: concentration of long branches in polymer chains at time t in the reaction

LM^IX: transition metal catalyst in ATRP with M as the metal, L as the ligand, and X as the leaving group in low oxidation state (I)

$LM^{II}X$: transition metal catalyst in ATRP with M as the metal, L as the ligand, and X as the leaving group in high oxidation state (II)

$[LM^IX]$: concentration of transition metal catalyst in low oxidation state (I) in ATRP

$[LM^{II}X]$: concentration of transition metal catalyst in high oxidation state (II) in ATRP

$[LM^IX]_0$: initial concentration of transition metal catalyst in low oxidation state (I) in ATRP

$[LM^{II}X]_0$: initial concentration of transition metal catalyst in high oxidation state (II) in ATRP

m_I: molar mass of initiator molecules in photopolymerization

$\dot{m}_{I,x}$, $\dot{m}_{I,y}$, $\dot{m}_{I,x}$: rate of molar mass flow of initiator molecules in x, y, or z direction in photopolymerization

m_M: initial moles of monomer in reaction

M: Polymerization monomer

MHS: Mark–Houwink–Sakurada equation relating intrinsic viscosity of a polymer to its degree of polymerization

MOS: methoxy styrene

$[M]$, $[M]_t$: monomer concentration in polymerization at time t

$[M]_0$: initial concentration of the monomer in polymerization at time t

$[M]_{con}$: concentration of monomer consumed in the reaction at time t

n: number of repeat units on a polymer chain

n_{LB}: molar density of long branches in a polymer

n_{TDB}: molar density of terminated double bonds in a polymer

nBu_4NX: tetrabutylammonium halide used as co-activator with X as the halide in cationic polymerization

$N_{I,x}$, $N_{I,y}$, $N_{I,z}$: molar flux of initiator molecules in x, y, or z direction in photopolymerization, respectively

$N_{tp,x}$, $N_{tp,y}$, $N_{tp,z}$: molar flux of terminated polymer chains in x, y, or z direction in photopolymerization

p: probability of propagation reaction in a polymerization reaction

$p_{DPn,tp}$ (i): mole-fraction probability distribution of terminated polymer chains in polymerization at time t

PDI_{ATRP}: polydispersity index of terminated+dormant polymer chains in ATRP

PDI_{dop}: polydispersity index of dormant polymer chains in ATRP

PDI_{rp}: polydispersity index of reactive propagating chains

PDI_{tp}: polydispersity index of terminated polymer chains

pK_b: logarithm of dissociation constant for initiator, like lithium chloride, into anion and cation in ionic polymerization

$p\ (MOS)$: poly(methoxy styrene)

$Q_{k,rp}$: kth order molar concentration moment of reactive propagating chains at time t in polymerization

$Q_{k,tp}$: kth order molar concentration moment of terminated polymer chains at time t in polymerization

$Q_{1+a,tp}$: (1+a)th order molar concentration moment of terminated polymer chains at time t in polymerization

QSSA: quasi-steady state assumption of reactive species in addition polymerization

R, R_1: reactive species generated by dissociation of a polymerization initiator; R is also hydrocarbon reactive segment of initiator RX in ATRP

R_2: second reactive species generated by reactive species transfer to a chain transfer agent

R_d: rate of dissociation of the initiator in polymerization

R_c: rate of recombination reaction of reactive species produced by initiator dissociation

R_{ci}: rate of reaction of the dissociated initiator fragment with a monomer to produce a 1-mer propagating chain

R_{cta}: rate of reaction of a reactive propagating chain with a chain transfer agent

$R_{m,con,l}$: rate of consumption of monomer by the lth consumption reaction in polymerization at time t

$R_{m,prd,i}$: rate of production of monomer by the ith production reaction in polymerization at time t

R_p: rate constant for propagation reaction between a propagating chain and a monomer

$R_{prd,I}$: rate of production of initiator in a differential volume at time t in photopolymerization

$R_{con,I}$: rate of consumption of initiator in a differential volume at time t in photopolymerization

R_{RP1}: rate of chain initiation by production of 1-mer propagating chains in ATRP

R_{RPn}: rate of production of n-mer propagating chains in ATRP

$R_{rp,con,l}$: rate of consumption of reactive propagating chains by the lth consumption reaction in polymerization at time t

$R_{rp,prd,i}$: rate of production of reactive propagating chains by the ith production reaction in polymerization at time t

R_{td}: rate of termination reaction between two propagating chains by disproportionation

$R_{tp,con,l}$: rate of consumption of terminated polymer chains by the lth consumption reaction in polymerization at time t

$R_{tp,prd,i}$: rate of production of terminated polymer chains by the ith production reaction in polymerization at time t

RDRP: reversible deactivation radical polymerization

RP': reactive propagating chain with reactive species on a tertiary carbon atom

RP'_m: reactive propagating chain with m repat units with reactive species on a tertiary carbon atom

RP_n: reactive propagating chain with n repeat units

RP_x: reactive propagating chain with x repeat units produced by chain scission of an n-mer reactive propagating chain

RP: reactive propagating chain

RP_{ai}: reactive cationic propagating chain associated with its anion in cationic polymerization

RP_{fi}: reactive cationic propagating chain free from its anion in cationic polymerization

RX: halide initiator with X as the halide and R as the hydrocarbon in ATRP

$[R]$: concentration of primary radicals formed by dissociation of an initiator

$[R]_0$: initial concentration of primary radicals formed by dissociation of an initiator; initial concentration of reactive species formed by dissociation of RX in ATRP

$[RP]$: concentration of reactive propagating chains in polymerization at time t

$[RP']$: concentration of reactive propagating chains with reactive species on tertiary carbon atom in polymerization at time t

$[RP_n]$: concentration of reactive propagating chains with n repeat unit

$[RP_{ai}]$: concentration of reactive cationic propagating chains associated with anion in cationic polymerization

$[RP_{fi}]$: concentration of reactive cationic propagating chain free from anion in cationic polymerization

t: reaction time

TDB: terminal double bond on a polymer chain

TDB_n: n-mer terminated polymer chain with a terminal double bond

$[TDB]$: concentration of terminal double bonds on polymer chains in reaction at time t

TMC: transition metal catalyst in ATRP

TP_n: terminated polymer chain with n repeat units

TPC: terminated polymer chains

$[TP]$: concentration of terminated polymer chains in reaction at time t

V_M: initial volume of monomers in reaction

V_R: reaction volume

x: dependent variable in the domain of degree of polymer (DP)

z: direction of depth of penetration of photons in the resin in

ZnX_2: Lewis acid, zinc halide, used as activator in cationic polymerization with X as the halide

$[ZnX_2]_e$: equilibrium concentration of Lewis acid, zinc halide, as activator in cationic polymerization

β_2: reactive propagating chain with the reactive species on a secondary carbon splitting from its neighboring carbon to generate a shorter propagating chain with the reactive species on a primary carbon plus a terminated polymer chain with a terminal double bond

γ_k, $\gamma_{k,dop}$: kth order mole-fraction moment of dormant polymer chains produced by deactivation of propagating chains in ATRP or RDRP

ε: molar extinction coefficient of initiator in photopolymerization

λ_k: kth order mole-fraction moment of reactive propagating chains at time t in the reaction

$\lambda_{k,ai}$: kth order mole-fraction moment of reactive propagating cations in associated form with its anion in cationic polymerization

$\lambda_{k,fi}$: kth order mole-fraction moment of reactive propagating cations in free form separated from its anion in cationic polymerization

λ_{1+a}: (1+a)th order mole-fraction moment of reactive propagating chains at time t in the reaction

$\lambda'_{k,i}$: instantaneous kth order mole-fraction moment of reactive propagating chains produced by ith production reaction in time interval dt

$\lambda'_{k,l}$: instantaneous kth order mole-fraction moment of reactive propagating chains consumed by lth consumption reaction in time interval dt

$\lambda'_{k,ci}$: instantaneous kth order mole-fraction moment of reactive propagating produced by chain initiation reaction

$\lambda'_{k,prp,con}$: instantaneous kth order mole-fraction moment of reactive propagating chains consumed by propagation reaction in time interval dt

$\lambda'_{k,prp,prd}$: instantaneous kth order mole-fraction moment of reactive propagating chains produced by propagation reaction in time interval dt

$\lambda_{k,RP'}$: kth order mole-fraction moment of reactive propagating chains with reactive species on a tertiary carbon atom

$\lambda'_{k,td}$: instantaneous kth order mole-fraction moment of reactive propagating chains consumed by termination by disproportionation reaction in time interval dt

$\lambda'_{k,\beta 2}$: instantaneous kth order mole-fraction moment of reactive propagating chains produced by β_2 chain scission in time interval dt

μ_{1+a}, $\mu_{1+a,tp}$: (1+a)th order mole-fraction moment of terminated polymer chains at time t in the reaction

μ_k, $\mu_{k,t}$: kth order mole-fraction moment of terminated polymer chains at time t in the reaction

μ_n: nth order mole-fraction moment of terminated polymer chains at time t used in photopolymerization

$\mu'_{1+a,i}$: instantaneous (1+a)th order mole-fraction moment of terminated polymer chains produced by the ith production reaction in time interval dt

$\mu'_{1+a,l}$: instantaneous (1+a)th order mole-fraction moment of terminated polymer chains consumed by the lth consumption reaction in time interval dt

$\mu'_{1+a,td}$: instantaneous (1+a)th order mole-fraction moment of terminated polymer chains produced by termination via disproportionation reaction in time interval dt

$\mu'_{k,i}$, $\mu'_{n,i}$: instantaneous kth or nth order mole-fraction moment of terminated polymer chains produced by the ith production reaction in time interval dt

$\mu'_{k,l}$, $\mu'_{n,l}$: instantaneous kth or nth order mole-fraction moment of terminated polymer chains consumed by the lth consumption reaction in time interval dt

$\mu'_{k,td}$: instantaneous kth order mole-fraction moment of terminated polymer chains produced by termination via disproportionation reaction in time interval dt

$\mu'_{k,\beta 2}$: instantaneous kth order mole-fraction moment of terminated polymer chains produced by β_2 chain scission in time interval dt

$\mu'_{n,x}$, $\mu'_{n,y}$, $\mu'_{n,z}$: instantaneous nth order mole-fraction moment of terminated polymer chains produced in time interval dt in the x, y, or z direction, respectively

ϕ: initiator efficiency in photopolymerization

7.12 Practice Problems

Problem 7.1 The decrease in monomer concentration with reaction time in radical polymerization of monomer M, chain initiation with a thermal initiator producing two primary radicals, chain termination by disproportionation and no chain transfer reaction is given by Eq. (7.6F) in Section 7.3.1.2. Starting from Eq. (7.6F) derive the following expression (Eq. 7.6G in Section 7.3.1.2) which shows that monomer concentration decreases linearly at short reaction times or low conversion:

$$[M] = [M]_0 \left[1 - \left(\frac{2k_p^2 k_d f_0 [I]_0}{k_{td}} \right)^{1/2} t \right]$$

where $[M]$ and $[M]_0$ are monomer concentrations at time t and initially, respectively; f_0 and $[I]_0$ are initiator efficiency and initial concentration of initiator; and k_d, k_p, and k_{td} are the rates of initiator dissociation, chain propagation, and chain termination by disproportionation.

Problem 7.2 The distribution of the propagating chains can be described by the Flory–Schulz distribution function or the "most probable distribution function" as defined by the following equation (see Eq. 3.87 in Section 3.9.2 of Chapter 3 or see Section 7.3.1.8) [51]:

$$f_{rp,n}(n) = (1-p)p^n$$

Where $f_{rp}(n)$ is the mole-fraction distribution of DP of the propagating chains and $0 \le p \le 1$ is the probability of chain growth or propagation, as defined by the following expression:

$$p = \frac{k_p [RP][M]}{k_p [RP][M] + k_{td} [RP]^2} = \frac{k_p [RP][M]}{k_p [RP][M] + 2f_0 k_d [I]} = \frac{L}{L+1}$$

where $[I]$, $[RP]$ and $[M]$ are the concentrations of initiator, propagating chain, and monomer, respectively; k_p and k_d are the rate constants for propagation and initiator dissociation; f_0 is the initiator efficiency; and L is the average DP of the propagating chains or the kinetic chain length. The viscosity-average DP $\left(\overline{DP}_\eta\right)$, as defined in Section 3.6 of Chapter 3, is given by the following expression for the propagating chains:

$$\overline{DP}_\eta = \left[\sum_{i=1}^{\infty} i^a f_{rp,w}(i)\right]^{1/a} = \left[\frac{\sum_{i=1}^{\infty} i^{1+a} f_{rp,n}(i)}{\lambda_1}\right]^{1/a} = \left[\frac{\lambda_{1+a}}{\lambda_1}\right]^{1/a}$$

Where $f_{rp,w(i)}$ is the weight-fraction distribution function of the propagating chains, a is the MHS constant (see Section 3.6 in Chapter 3), and λ_{1+a} is the (1+a)th mole-fraction moment of the propagating chains. Assuming QSSA is valid for all propagating chains irrespective of DP for the reaction system given by Eqs. (7.2A–E) in Section 7.3 (radical polymerization with no chain transfer and termination by disproportionation) and using the above information, derive an expression for λ_{1+a} and $\overline{DP}_\eta$ in terms of the kinetic chain length L.

Problem 7.3 For radical polymerization with termination by disproportionation and radical transfer to small-molecule chain transfer agent (see Section 7.4) as defined by elementary reactions 7.26A-G, derive an expression for the second mole-fraction moment of the propagating chains (λ_2) in terms the kinetic chain length L for $L >> 1$.

Problem 7.4 For radical polymerization with termination by disproportionation and radical transfer to small-molecule chain transfer agent (see Section 7.4) as defined by elementary reactions 7.26A–G, derive an expression for the second molar concentration moment of the terminated polymer chains $\left(Q_{2,tp}\right)$ and the weight-average degree of polymerization of the terminated chains $\left(\overline{DP}_{w,tp}\right)$ in terms of the kinetic chain length L, monomer concentration $[M]$, concentration of the propagating chains $[RP]$, and reaction rate constants:

Problem 7.5 For photo-initiated polymerization with no chain transfer and termination by disproportionation only (see Section 7.7) as defined by elementary reactions 7.45A–E, derive an expression for the second order mole-fraction moment of the propagating chains in terms of the kinetic chain length L, monomer concentration $[M]$, propagating chains concentration $[RP]$, and rate constants for propagation and termination.

Problem 7.6 For photo-initiated polymerization with no chain transfer reaction and termination by disproportionation (see Section 7.7) as defined by elementary reactions 7.45A–E, derive expressions for the first and second molar concentration moment of the terminated polymer chains, $Q_{1,tp}$ and $Q_{2,tp}$, respectively, in in terms of the kinetic chain length L, monomer concentration $[M]$, and the propagation rate constant k_p.

Problem 7.7 For RDRP by atom transfer radical polymerization (ATRP) with chain termination (see Section 7.8.2) as defined by elementary reactions 7.58A–E, show that the time dependence of monomer concentration in ATRP with termination reaction has the same functional form as the case without termination reaction (see Eq. 7.55B in Section 7.8.1) except that $[R]_0$ is replaced with $[RP]$ as follows:

$$\text{without termination} \Rightarrow [M] = [M]_0\, e^{-k_p[R]_0 t}$$

$$\text{with termination} \Rightarrow [M] = [M]_0 e^{-k_p[RP]t} = [M]_0 e^{-\frac{k_p t}{k_{td}t + 1/[R]_0}}$$

Problem 7.8 For RDRP by atom transfer radical polymerization (ATRP) with chain termination (see Section 7.8.2) as defined by elementary reactions 7.58A to 7.58E, drive the following expressions:

a. derive an expression for the weight-average DP of the terminated chains $\overline{DP}_{w,tp}$ in terms of the propagating chains concentration $[RP]$, monomer concentration $[M]$, terminated polymer chains concentration $[TP]$, first order mole-fraction moment of the propagating chains λ_1, and first order mole-fraction moment of the terminated polymer chains μ_1.

b. As the products of the polymerization reaction in ATRP with termination is a mixture of dormant (DOP) and terminated (TP) polymer chains, derive using the mixing theory (see Section 5.3 in Chapter 5) expressions for the overall number-average $\overline{DP}_{n,ATRP}$ and weight-average $\overline{DP}_{w,ATRP}$ degree of polymerization of the produced chains in ATRP comprising DOP and TP chains in terms of DOP and TP concentrations, first order mole-fraction moments of DOP and TP chains, and second order mole-fraction moments of DOP and TP chains.

Problem 7.9 There are at least two ionic propagating species in cationic polymerization of monomer M with hydrogen iodide (HI) as the initiator, a Lewis acid like zinc iodide (ZnI_2) as the activator in the absence of a tetrabutylammonium halide with elementary reactions 7.64A–F (see Section 7.9.2). These species include the associated propagating cation RP_{ai} and the dissociated free propagating cation RP_{fi}. These associated and free cations have different equilibrium concentrations $[HM^+..^-I.ZnI_2]_e$ and $[HM^+]_e$, respectively, and different propagation rate constants $k_{p,ai}$ and $k_{f,ai}$. Drive expression for $\lambda_{1,ai}$ and $\lambda_{1,fi}$ as a function of time in terms of initial monomer concentration $[M]_0$ and effective propagation rate constant $k_{p,eff}$ (see Eq. 7.66D) using separate conservation equations for the first order molar concentration moment of RP_{ai} and RP_{fi} propagating species.

Problem 7.10 A radical polymerization reaction is described by the following four elementary reactions:

$$I \xrightarrow{k_d} 2R$$

$$R + M \xrightarrow{k_p} RP_1$$

$$RP_n + M \xrightarrow{k_p} RP_{n+1}$$

$$RP_n + RP_m \xrightarrow{k_{td}} TP_n + TP_m + TDB$$

There is no recombination of primary radicals in the reaction and QSSA is valid for all radicals. The concentration of monomer throughout the reaction is constant and equal to $[M] = [M]_0 = 1\,M$.

a. derive an expression for the concentration of initiator, $[I]$, as a function of time.

b. derive an expression for the concentration of propagating chains, $[RP]$, as a function of time.

c. derive an expression for number-average DP of the propagating chains, $\overline{DP}_{n,rp}$, as a function of time.
d. derive an expression for the concentration of terminated polymer chains, $[TP]$, as a function of time.
e. derive an expression for monomer conversion, α, as a function of time.
f. derive an expression for the number-average DP of the terminated chains, $\overline{DP}_{n,tp}$, after all initiator molecules are consumed.

Problem 7.11 A radical polymerization reaction is described by the elementary reactions in the previous problem 7.10. There is no recombination of primary radicals in the reaction and QSSA is valid for all radicals and all times. The concentration of initiator and monomer are constant throughout the reaction and equal to $[M] = [M]_0 = 1$ M and $[I] = [I]_0 = 1 \times 10^{-4}$ M, respectively. Derive expressions for $[RP]$, kinetic chain length L, $[TP]$, conversion α, and $\overline{DP}_{n,tp}$.

References

1 Chen, Z.Q. et al., Parallel calculation methods for molecular weight distribution of batch free radical polymerization. *Computers & Chemical Engineering*, 2013, **48**, 175.
2 Shao, J. et al., Monte Carlo simulation on kinetics of batch and semi-batch free radical polymerization. *Macromolecular Research*, 2015, **23**(11), 1042.
3 Awad, M., Dhib, R., and Duever, T., Kinetics of MMA atom transfer radical polymerization initiated by reducing agent in stirred batch emulsion reactor. *Macromolecular Reaction Engineering*, 2021, **15**(2), 2000052.
4 Ilare, J. et al., Free-radical polymerization of methacrylic acid: from batch to continuous using a stirred tank reactor. *Chemical Engineering and Processing-Process Intensification*, 2021, **165**, 108440.
5 Pineda-Torres, G.M., Durán-Valencia, C., Barragán-Aroche, F., and López-Ramírez, S., Impact of fluid flow on free radical polymerization in a batch reactor, in *Modeling and simulation in engineering sciences*. 2016, IntechOpen, p. 267.
6 Achilias, D.S. and Kiparissides, C., On the validity of the steady-state approximations in high conversion diffusion-controlled free-radical copolymerization reactions. *Polymer*, 1994, **35**(8), 1714.
7 Achilias, D.S. and Verros, G.D., Modeling of diffusion-controlled reactions in free radical solution and bulk polymerization: model validation by DSC experiments. *Journal of Applied Polymer Science*, 2010, **116**(3), 1842.
8 Al-Harthi, M., Soares, J.B.P., and Simon, L.C., Modeling of atom transfer radical polymerization with bifunctional initiators: diffusion effects and case studies. *Macromolecular Chemistry and Physics*, 2006, **207**(5), 469.
9 Dorfschmid, M., Mullen, K., Zumbusch, A., and Woll, D., Translational and rotational diffusion during radical bulk polymerization: a comparative investigation by full correlation fluorescence correlation spectroscopy (fcFCS). *Macromolecules*, 2010, **43**(14), 6174.
10 Stempfle, B. et al., Single molecule diffusion and its heterogeneity during the bulk radical polymerization of styrene and methyl methacrylate. *Polymer Chemistry*, 2012, **3**(9), 2456.
11 Verros, G.D., Latsos, T., and Achilias, D.S., Development of a unified framework for calculating molecular weight distribution in diffusion controlled free radical bulk homo-polymerization. *Polymer*, 2005, **46**(2), 539.

12 Zetterlund, P.B., Compartmentalization in atom transfer radical polymerization to high conversion in dispersed systems: effects of diffusion-controlled reactions. *Macromolecules*, 2010, **43**(3), 1387.

13 Wang, A.R. and Zhu, S.P., ESR study on diffusion-controlled atom transfer radical polymerization of methyl methacrylate and ethylene glycol dimethacrylate. *Macromolecules*, 2002, **35**(27), 9926.

14 Wang, A.R. and Zhu, S.P., Effects of diffusion-controlled radical reactions on RAFT polymerization. *Macromolecular Theory and Simulations*, 2003, **12**(2-3), 196.

15 Kreft, T. and Reed, W.F., Direct monitoring of the cross-over from diffusion controlled to decomposition controlled initiation in free radical polymerization. *Macromolecular Chemistry and Physics*, 2008, **209**(24), 2463.

16 Fukuda, T., Goto, A., and Ohno, K., Mechanisms and kinetics of living radical polymerizations. *Macromolecular Rapid Communications*, 2000, **21**(4), 151.

17 Fukuda, T. et al., Mechanisms and kinetics of living radical polymerization: absolute comparison of theory and experiment, in *Advances in Controlled/Living Radical Polymerization*. 2003, American Chemical Society, pp. 24–39.

18 Goto, A. and Fukuda, T., Kinetics of living radical polymerization. *Progress in Polymer Science*, 2004, **29**(4), 329.

19 Ballard, N., Hamzehlou, S., Ruiperez, F., and Asua, J.M., On the Termination Mechanism in the Radical Polymerization of Acrylates. *Macromolecular Rapid Communications*, 2016, **37**(16), 1364.

20 Nakamura, Y., Lee, R., Coote, M.L., and Yamago, S., Termination mechanism of the radical polymerization of acrylates. *Macromolecular Rapid Communications*, 2016, **37**(6), 506.

21 Nakamura, Y. and Yamago, S., Termination mechanism in the radical polymerization of methyl methacrylate and styrene determined by the reaction of structurally well-defined polymer end radicals. *Macromolecules*, 2015, **48**(18), 6450.

22 Berger, K.C., Disproportionation and combination as termination mechanisms in radical polymerization of styrene. 2. Analysis of Temperature Dependencies. *Makromolekulare Chemie-Macromolecular Chemistry and Physics*, 1975, **176**(12), 3575.

23 Berger, K.C. and Meyerhoff, G., Disproportionation and combination as termination mechanisms in radical polymerization of styrene. 1. Experiments with C-14-labeled 2,2'-azoisobutyronitrile. *Makromolekulare Chemie-Macromolecular Chemistry and Physics*, 1975, **176**(7), 1983.

24 Zifferer, G. and Olaj, O.F., Simulation of pseudostationary radical polymerization. 3. Periodic modulation of the termination process for termination by disproportionation. *Macromolecular Theory and Simulations*, 1997, **6**(1), 53.

25 Nakamura, Y. et al., Control of the termination mechanism in radical polymerization by viscosity: selective disproportionation in viscous media. *Chemistry-a European Journal*, 2017, **23**(6), 1299.

26 Keszei, E., Theories of elementary reactions, in *Reaction kinetics: an introduction*. 2021, Springer International Publishing, pp. 5–37.

27 Clark, I.D. and Wayne, R.P., The theory of elementary reactions in solution, in *Comprehensive chemical kinetics*. 1969, Elsevier, pp. 302–376.

28 Schaal, R., Kinetic theory of elementary reactions, in *Chemical kinetics of homogeneous systems*. 1974, Springer Netherlands, pp. 36–50.

29 Stolzenberg, K. and Kirchner, K., Continuous free-radical polymerization of styrene in the stirred tank reactor - effect of inhibitors. *Angewandte Makromolekulare Chemie*, 1981, **95**(Apr), 185.

30 Tanczos, I., Foldesberezsnich, T., and Tudos, F., Kinetics of radical polymerization. 40. Investigation of the effect of molecule inhibitors in polymerization in solution. *European Polymer Journal*, 1983, **19**(3), 225.

31 Li, L. and Lee, L.J., Effects of inhibitors and retarders on low temperature free radical crosslinking polymerization between styrene and vinyl ester resin. *Polymer Engineering and Science*, 2001, **41**(1), 53.

32 Allen, T.L., Oxygen inhibition of the polymerization of styrene. *Journal of Applied Chemistry*, 1954, **4**(6), 289.

33 Li, R.J. and Schork, F.J., Modeling of the inhibition mechanism of acrylic acid polymerization. *Industrial & Engineering Chemistry Research*, 2006, **45**(9), 3001.

34 Bartlett, P.D. and Trifan, D.S., Some experiments on sulfur as inhibitor for the polymerization of styrene. *Journal of Polymer Science*, 1956, **20**(96), 457.

35 Bartlett, P.D. and Kwart, H., Some inhibitors and retarders in the polymerization of liquid vinyl acetate. 2. 1,3,5-trinitrobenzene and sulfur. *Journal of the American Chemical Society*, 1952, **74**(16), 3969.

36 Kirchner, K. and Rintelen, T., Continuous free-radical polymerization of styrene in the stirred tank reactor - effect of retarders. *Angewandte Makromolekulare Chemie*, 1986, **141**, 85.

37 Bovey, F.A. and Kolthoff, I.M., Studies of retarders and inhibitors in the emulsion polymerization of styrene. 2. Inhibitors. *Journal of Polymer Science*, 1950, **5**(5), 569.

38 Kolthoff, I.M. and Bovey, F.A., Studies of retarders and inhibitors in the emulsion polymerization of styrene. 1. Retarders. *Journal of the American Chemical Society*, 1948, **70**(2), 791.

39 Bovey, F.A. and Kolthoff, I.M., Inhibition and retardation of vinyl polymerization. *Chemical Reviews*, 1948, **42**(3), 491.

40 Matyjaszewski, K., Patten, T.E., and Xia, J.H., Controlled/living radical polymerization. Kinetics of the homogeneous atom transfer radical polymerization of styrene. *Journal of the American Chemical Society*, 1997, **119**(4), 674.

41 Tang, W. et al., Understanding atom transfer radical polymerization: effect of ligand and initiator structures on the equilibrium constants. *Journal of the American Chemical Society*, 2008, **130**(32), 10702.

42 Williams, G. and Coe, J.S., The kinetic chain length in the polymerization of styrene by stannic chloride in carbon tetrachloride. *Chemistry & Industry (London)*, 1954, No. 2, 43.

43 Allen, P.E.M., Burnett, G.M., Downer, J.M., and Melville, H., Initiation of emulsion polymerisation by gamma-irradiation - polymer grafting in a system of unlimited kinetic chain length. *Makromolekulare Chemie*, 1960, **38**, 72.

44 Ito, K., Relationship between kinetic chain-length and radical polymerization rate. *Journal of Polymer Science Part a-Polymer Chemistry*, 1976, **14**(7), 1761.

45 Li, X.S., Kinetic chain length in UV cationic polymerization of cycloaliphatic epoxides, in *203rd ACS National Meeting*. 2003, American Chemical Society, p. U712.

46 Lyu, J. et al., Active kinetic chain length: guide for control in atom transfer radical polymerization. *Chemical Engineering Journal*, 2023, **474**, 145548.

47 Kamide, K. and Kataoka, A., Theoretical relationships between parameters in Khun-Mark-Houwink-Sakurada equation. *Makromolekulare Chemie*, 1969, **128**(Oct), 217.

48 Dobbin, C.J.B., Rudin, A., and Tchir, M.F., Evaluation of GPC methods for estimation of Mark-Houwink-Sakurada constants. *Journal of Applied Polymer Science*, 1980, **25**(12), 2985.

49 Zhongde, X., Mingshi, S., Hadjichristidis, N., and Fetters, L.J., Method for gel-permeation chromatography calibration and the evaluation of Mark-Houwink-Sakurada constants. *Macromolecules*, 1981, **14**(5), 1591.

50 Soares, J.B.P., The use of instantaneous distributions in polymerization reaction engineering. *Macromolecular Reaction Engineering*, 2014, **8**(4), 235.

51 Flory, P.J., Molecular size distribution in linear condensation polymers. *Journal of the American Chemical Society*, 1936, **58**, 1877.

52 Flory, P.J., *Principles of polymer chemistry*. 1953, Cornell University Press, 688.

53 Flory, P.J. and Yoon, D.Y., Moments and distribution functions for polymer-chains of finite length. 1. Theory. *Journal of Chemical Physics*, 1974, **61**(12), 5358.

54 Albareeki, M.M., Sorroche, J.V., and Barry, C.F., Energy consumption of high speed multiscrew extruders. *Proceedings of the 35th International Conference of the Polymer Processing Society*, 2020, **2205**, 35.

55 Capone, C. et al., Thermal and mechanical degradation during polymer extrusion processing. *Polymer Engineering and Science*, 2007, **47**(11), 1813.

56 Munir, N. et al., Interpretable machine learning methods for monitoring polymer degradation in extrusion of polylactic acid. *Polymers (Basel)*, 2023, **15**(17), 3566.

57 Furuncuoglu, T., Ugur, I., Degirmenci, I., and Aviyente, V., Role of chain transfer agents in free radical polymerization kinetics. *Macromolecules*, 2010, **43**(4), 1823.

58 Gregg, R.A. and Mayo, F.R., Chain transfer in the polymerization of styrene. 7. Compounds containing halogens, oxygen and nitrogen. *Journal of the American Chemical Society*, 1953, **75**(14), 3530.

59 Grotewold, J. and Hirschler, M.M., Triethylaluminum as a concentration-dependent co-initiator and chain-transfer agent of free-radical polymerization of methyl-methacrylate in presence of benzoquinone. *Journal of Polymer Science Part a-Polymer Chemistry*, 1977, **15**(2), 393.

60 Meijs, G.F., Rizzardo, E., and Le, T.P.T., New chain transfer agents for free-radical polymerizations. *Polymer International*, 1991, **26**(4), 239.

61 Wendler, K., Wurtenberg, R., and Fedtke, M., Effect of dialkyl xanthogenedisulfide as a chain transfer agent on the latex stability in radical emulsion polymerization of butadiene. *Plaste Und Kautschuk*, 1984, **31**(10), 367.

62 Salomatina, E. et al., Synthesis of block copolymers by radical polymerization in presence of Si-based chain transfer agent. *Polymer Chemistry UK*, 2023, **14**(28), 3294.

63 Lena, J.B. et al., Effects of chain transfer agent and temperature on branching and beta-scission in radical polymerization of 2-ethylhexyl acrylate. *Macromolecular Chemistry and Physics*, 2018, **219**(9), 1700579.

64 Ballard, N., de la Cal, J.C., and Asua, J.M., The role of chain transfer agent in reducing branching content in radical polymerization of acrylates. *Macromolecules*, 2015, **48**(4), 987.

65 Ahmad, H. and Tauer, K., Effects of chain transfer agent on the radical polymerization of styrene in non-aqueous dispersion. *Colloid and Polymer Science*, 2003, **281**(7), 686.

66 Çetinkaya, O., Demirci, G., and Mergo, P., Effect of the different chain transfer agents on molecular weight and optical properties of poly(methyl methacrylate). *Optical Materials*, 2017, **70**, 25.

67 Yuan, C.M. and Farina, M., Effect of polyfunctional chain transfer agents on molecular-weight distribution in free-radical polymerization. 2. The case $C<=1$. *Macromolecular Theory and Simulations*, 1994, **3**(1), 203.

68 Semikolenova, N.V., Matsko, M.A., Zakharov, V.A., and Sun, W.H., Unusual effect of α-olefins as chain transfer agents in ethylene polymerization over the catalyst with nonsymmetrical bis(imino)pyridine complex of Fe(II) and modified methylalumoxane (MMAO) cocatalyst. *International Journal of Molecular Sciences*, 2022, **23**(22), 14384.

69 Whang, B.C.Y. et al., The effects of a chain transfer agent on the kinetics of the emulsion polymerization of styrene. *Journal of Polymer Science Part C-Polymer Letters*, 1980, **18**(11), 711.

70 Lovell, P.A., Shah, T.H., and Heatley, F., Chain transfer to polymer in emulsion polymerization of normal-butyl acrylate studied by C-13 NMR-spectroscopy and GPC. *Polymer Communications*, 1991, **32**(4), 98.

71 Lovell, P.A., Shah, T.H., and Heatley, F., *Correlation of the extent of chain transfer to polymer with reaction conditions for emulsion polymerization of normal-butyl acrylate*. 1992. pp. 188–202.

72 Ahmad, N.M., Heatley, F., and Lovell, P.A., Chain transfer to polymer in free-radical solution polymerization of n-butyl acrylate studied by NMR spectroscopy. *Macromolecules*, 1998, **31**(9), 2822.

73 Ahmad, N.M. et al., Chain transfer to polymer and branching in controlled radical polymerizations of n-butyl acrylate. *Macromolecular Rapid Communications*, 2009, **30**(23), 2002.

74 Britton, D., Heatley, F., and Lovell, P.A., Chain transfer to polymer in free-radical bulk and emulsion polymerization of vinyl acetate studied by NMR spectroscopy. *Macromolecules*, 1998, **31**(9), 2828.

75 Britton, D., Heatley, F., and Lovell, P.A., Effect of monomer feed rate on chain transfer to polymer in semibatch emulsion polymerization of vinyl acetate studied by NMR spectroscopy. *Macromolecules*, 2000, **33**(14), 5048.

76 Filley, J., McKinnon, J.T., Wu, D.T., and Ko, G.H., Theoretical study of ethylene-vinyl acetate free-radical copolymerization: reactivity ratios, penultimate effects, and relative rates of chain transfer to polymer. *Macromolecules*, 2002, **35**(9), 3731.

77 Small, P.A., Long-chain branching in polymers, in *Macroconformation of polymers*. 1975, Springer-Verlag Berlin, p. 1–64.

78 Iedema, P.D. and Hoefsloot, H.C.J., Role of polyradicals in predicting chain length distribution and gelation for radical polymerization with transfer to polymer. *Macromolecules*, 2004, **37**(26), 10155.

79 Iedema, P.D. and Hoefsloot, H.C.J., Computing the trivariate chain length/degree of branching/number of combination points distribution for radical polymerization with transfer to polymer and recombination termination. *Macromolecular Theory and Simulations*, 2005, **14**(8), 505.

80 Santamaria, A., Influence of long-chain branching in melt rheology and processing of low-density polyethylene. *Materials Chemistry and Physics*, 1985, **12**(1), 1.

81 Wang, W.J., Kharchenko, S., Migler, K., and Zhu, S.P., Triple-detector GPC characterization and processing behavior of long-chain-branched polyethylene prepared by solution polymerization with constrained geometry catalyst. *Polymer*, 2004, **45**(19), 6495.

82 Tobita, H., Continuous tanks-in-series process for free-radical polymerization with long-chain branching and scission: effect of the order of a large tank. *Macromolecular Reaction Engineering*, 2015, **9**(6), 556.

83 Duscher, B. and Schausberger, A., Influence of processing on the flow properties of long-chain branched polypropylene. *Polymer Engineering and Science*, 2018, **58**(9), 1596.

84 Drabek, J. and Zatloukal, M., Influence of long chain branching on fiber diameter distribution for polypropylene nonwovens produced by melt blown process. *Journal of Rheology*, 2019, **63**(4), 519.

85 Embabi, M. et al., Effect of ethylene comonomer content on the foam processing window of long-chain branched polypropylene. *Acs Applied Polymer Materials*, 2023, **5**(4), 2417.

86 Iiba, K., Takarada, W., and Kikutani, T., Effect of introducing long chain branching on fiber diameter and fiber diameter distribution in melt blowing process of polypropylene. *International Polymer Processing*, 2021, **36**(4), 403.

87 Baran, J. et al., Intermolecular chain transfer to polymer with chain scission: general treatment and determination of k(p)/k(tr) in L,L-lactide polymerization. *Macromolecular Rapid Communications*, 1997, **18**(4), 325.

88 Szymanski, R., Molecular weight distribution in living polymerization proceeding with reshuffling of polymer segments due to chain transfer to polymer with chain scission, 1. Determination of kp/ktr ratio from DPw/DPn data. Ideal reproduction of polymer chain activities. *Macromolecular Theory and Simulations*, 1998, **7**(1), 27.

89 Szymanski, R. and Baran, J., Molecular weight distribution in living polymerization proceeding with reshuffling of polymer segments due to chain transfer to polymer with chain scission, 2(a) – Monte Carlo simulation of polymer reshuffling proceeding with disproportionation of chain functionalities. *Macromolecular Theory and Simulations*, 2002, **11**(8), 836.

90 Kumagai, T. et al., Specific polymerization mechanism involving beta-scission of mid-chain radical yielding oligomers in the free-radical polymerization of vinyl ethers. *Macromolecules*, 2008, **41**(20), 7347.

91 Junkers, T. and Barner-Kowollik, C., The role of mid-chain radicals in acrylate free radical polymerization: branching and scission. *Journal of Polymer Science Part a-Polymer Chemistry*, 2008, **46**(23), 7585.

92 Tobita, H., Molecular weight development during simultaneous chain scission, long-chain branching and crosslinking, 2 – free-radical polymerization. *Macromolecular Theory and Simulations*, 2003, **12**(1), 32.

93 Tobita, H., Free-radical polymerization with long-chain branching and scission in a continuous stirred-tank reactor. *Macromolecular Reaction Engineering*, 2013, **7**(5), 181.

94 Tobita, H., Free-radical polymerization with long-chain branching and scission: Markovian solution of the weight-average molecular weight. *Macromolecular Theory and Simulations*, 2014, **23**(8), 477.

95 Tobita, H., Continuous free-radical polymerization with long-chain branching and scission in a tanks-in-series model. *Macromolecular Theory and Simulations*, 2014, **23**(3), 182.

96 Tobita, H., Markovian approach to free-radical polymerization with simultaneous long-chain branching and scission: effect of branching and scission kinetics. *Macromolecular Reaction Engineering*, 2015, **9**(3), 245.

97 Tobita, H., Model-based reactor design in free-radical polymerization with simultaneous long-chain branching and scission. *Processes*, 2015, **3**(4), 731.

98 Yoon, J.S., Prediction of the number- and weight-average degree of polymerization of polymers subjected to random chain scission degradation. *Polymer-Korea*, 2016, **40**(1), 124.

99 Brubaker, K.S. et al., Structural remodeling of polymeric material via diffusion controlled polymerization and chain scission. *Chemistry of Materials*, 2018, **30**(22), 8126.

100 Kawatani, R., Hagiwara, K., Tanaka, A., and Kohsaka, Y., Polythioethers bearing side groups for efficient degradation by E1cB reaction: reaction design for polymerization and main-chain scission. *Rsc Advances*, 2023, **13**(30), 20782.

101 Kurdikar, D.L. and Peppas, N.A., A kinetic-model for diffusion-controlled bulk cross-linking photopolymerizations. *Macromolecules*, 1994, **27**(15), 4084.

102 Goodner, M.D. and Bowman, C.N., Development of a comprehensive free radical photopolymerization model incorporating heat and mass transfer effects in thick films. *Chemical Engineering Science*, 2002, **57**(5), 887.

103 Wen, M. and McCormick, A.V., A kinetic model for radical trapping in photopolymerization of multifunctional monomers. *Macromolecules*, 2000, **33**(25), 9247.

104 Miller, G.A., Gou, L., Narayanan, V., and Scranton, A.B., Modeling of photobleaching for the photoinitiation of thick polymerization systems. *Journal of Polymer Science Part a-Polymer Chemistry*, 2002, **40**(6), 793.

105 Kaya, K., Jockusch, S., and Yagci, Y., Mussel-inspired coatings by photoinduced electron-transfer reactions: photopolymerization of dopamine under UV, visible, and daylight under oxygen-free conditions. *Macromolecules*, 2021, **54**(13), 5991.

106 Favre-Nicolin, C.D. and Lub, J., Stable anisotropic films obtained by in-situ photopolymerization of discotic liquid crystalline acrylates. *Macromolecules*, 1996, **29**(19), 6143.

107 Dobson, A.L. and Bowman, C.N., A comprehensive, multidimensional first-principles model for free-radical photopolymerizations in bulk and thin films. *Advanced Functional Materials*, 2024, **34**(22), 2312607.

108 Ravi, P. and Chen, V.C.P., A focused simulation-based optimization of print time and material usage with respect to orientation, layer height and support settings for multi-pathological anatomical models in inverted vat photopolymerization 3D printing. *3d Printing in Medicine*, 2021, **7**(1), 23.

109 Zhang, J. and Xiao, P., 3D printing of photopolymers. *Polymer Chemistry UK*, 2018, **9**(13), 1530.

110 Yoshida, H. et al., Initiation mechanism of radiation induced polymerization. *Advances in Chemistry Series*, 1968, (82), 548.

111 Gunzler, F. et al., Quantifying the efficiency of photoinitiation processes in methyl methacrylate free radical polymerization via electrospray ionization mass spectrometry. *Macromolecules*, 2009, **42**(5), 1488.

112 Voll, D., Hufendiek, A., Junkers, T., and Barner-Kowollik, C., Quantifying photoinitiation efficiencies in a multiphotoinitiated free-radical polymerization. *Macromolecular Rapid Communications*, 2012, **33**(1), 47.

113 Bao, Y.Y., Recent trends in advanced photoinitiators for vat photopolymerization 3D printing. *Macromolecular Rapid Communications*, 2022, **43**(14), 2002202.

114 Bagheri, A. and Jin, J.Y., Photopolymerization in 3D printing. *Acs Applied Polymer Materials*, 2019, **1**(4), 593.

115 Mateo, J.L., Serrano, J., and Bosch, P., Photopolymerization of di- and tetrafunctional methacrylic monomers in a polymeric medium: kinetics and evidence of reaction diffusion throughout the photopolymerization reaction. *Macromolecules*, 1997, **30**(5), 1285.

116 Bielykh, S.P., Galstian, T.V., and Reshetnyak, V.Y., Theoretical study of the reaction–diffusion model of a three-component photopolymerizable system taking into account the shrinking effect. *Results in Optics*, 2023, **11**, 100389.

117 Lang, M., Hirner, S., Wiesbrock, F., and Fuchs, P., A review on modeling cure kinetics and mechanisms of photopolymerization. *Polymers*, 2022, **14**(10), 2074.

118 Lin, J.T., Lalevee, J., and Cheng, D.C., A critical review for synergic kinetics and strategies for enhanced photopolymerizations for 3D-printing and additive manufacturing. *Polymers (Basel)*, 2021, **13**(14), 2325.

119 Tang, H.D. et al., Highly active copper-based catalyst for atom transfer radical polymerization. *Journal of the American Chemical Society*, 2006, **128**(50), 16277.

120 Krys, P. and Matyjaszewski, K., Kinetics of atom transfer radical polymerization. *European Polymer Journal*, 2017, **89**, 482.

121 Parker, J., Jones, R.G., and Holder, S.J., Direct evidence for the interaction of the mechanisms of thermally initiated and atom transfer radical polymerization. *Macromolecules*, 2000, **33**(25), 9166.

122 Matyjaszewski, K., Atom transfer radical polymerization (ATRP): current status and future perspectives. *Macromolecules*, 2012, **45**(10), 4015.

123 Matyjaszewski, K., Atom transfer radical polymerization: from mechanisms to applications. *Israel Journal of Chemistry*, 2012, **52**(3-4), 206.

124 Tang, W., Tsarevsky, N.V., and Matyjaszewski, K., Determination of equilibrium constants for atom transfer radical polymerization. *Journal of the American Chemical Society*, 2006, **128**(5), 1598.

125 Lorandi, F. et al., New protocol to determine the equilibrium constant of atom transfer radical polymerization. *Electrochimica Acta*, 2018, **260**, 648.

126 Zhang, H.Q. et al., Effect of Cu(II) on the kinetics of the homogeneous atom transfer radical polymerization of methyl methacrylate. *Macromolecules*, 2001, **34**(18), 6169.

127 Kuo, J.F. and Chen, C.Y., Determination of the absolute rate constants for free-radical polymerization. *Polymer Journal*, 1981, **13**(5), 453.

128 Mahabadi, H.K. and Odriscoll, K.F., Absolute rate constants in free-radical polymerization. 3. Determination of propagation and termination rate constants for styrene and methyl-methacrylate. *Journal of Macromolecular Science-Chemistry*, 1977, **A11**(5), 967.

129 Harned, H.S. and Copson, H.R., The dissociation of water in lithium chloride solutions. *Journal of the American Chemical Society*, 1933, **55**, 2206.

130 Darensbourg, M., Kimura, B.Y., Hartwell, G.E., and Brown, T.L., Cross-association of t-butyllithium. Kinetics of t-butyllithium dissociation. *Journal of the American Chemical Society*, 1970, **92**(5), 1236.

131 Reich, H.J., Role of organolithium aggregates and mixed aggregates in organolithium mechanisms. *Chemical Reviews*, 2013, **113**(9), 7130.

132 Bauer, W., Winchester, W.R., and Schleyer, P.V., Monomeric organolithium compounds in tetrahydrofuran - tert-butyllithium, sec-butyllithium, supermesityllithium, mesityllithium, and phenyllithium - carbon lithium coupling-constants and the nature of carbon lithium bonding. *Organometallics*, 1987, **6**(11), 2371.

133 Elschenbroich, C., *Organometllics*, 3e. 2006, Wiley-VCH, 817.

134 Kojima, K., Sawamoto, M., and Higashimura, T., Living cationic polymerization of isobutyl vinyl ether by hydrogen iodide/Lewis acid initiating systems: effects of Lewis acid activators and polymerization kinetics. *Macromolecules*, 1989, **22**(4), 1552.

135 Kojima, K., Sawamoto, M., and Higashimura, T., Living cationic polymerization of p-methoxystyrene by the Hi/Zni2 and Hi/I2 initiating systems – effects of tetrabutylammonium halides in a polar-solvent. *Macromolecules*, 1990, **23**(4), 948.

136 Lin, C.H., Xiang, J.S., and Matyjaszewski, K., Living cationic polymerization of styrene in the presence of tetrabutylammonium salts. *Macromolecules*, 1993, **26**(11), 2785.

137 Banerjee, S., Paira, T.K., Kotal, A., and Mandal, T.K., Room temperature living cationic polymerization of styrene with HX-styrenic monomer adduct/$FeCl_3$ systems in the presence of tetrabutylammonium halide and tetraalkylphosphonium bromide salts. *Polymer*, 2010, **51**(6), 1258.

138 Choi, W.O., Sawamoto, M., and Higashimura, T., Lifetime of living polymers in cationic polymerization. 1. Methodology and application to the Hi/I-2-initiated living cationic polymerization. *Macromolecules*, 1990, **23**(1), 48.

139 Endres, G.F. and Overberger, C.G., Ionic polymerization. 5. Molecular termination in the cationic polymerization of styrene catalyzed by stannic chloride. *Journal of the American Chemical Society*, 1955, **77**(8), 2201.

140 Tobisch, S., Mechanism of the chain termination of the allylnickel(II)-catalyzed polymerization of 1,3-butadiene. A density functional investigation for the cationic [Ni-II(RC3H4)(cis-C4H6)L](+) active catalyst. *Macromolecules*, 2003, **36**(16), 6235.

141 Cai, G.F., Yan, D.Y., and Litt, M., A kinetic-theory on cationic polymerization with termination by combination with a negative fragment. *Macromolecules*, 1988, **21**(3), 578.

142 Kennedy, J.P., Chung, D.Y.L., and Reibel, L.C., Cationic polymerizations by aromatic initiating systems. 3. Model reactions for initiation and termination using the Tert-Bucl-Tert-Phi-3al and Cl-Tert-R-Cl-Tert-Phi-3al systems. *Journal of Polymer Science Part a-Polymer Chemistry*, 1981, **19**(11), 2721.

8

Diffusion Controlled Polymerization Reactions

8.1 Introduction

The rate constants in elementary reactions between a propagating chain and a monomer for propagation and between two propagating chains for termination were assumed in previous chapters to depend only on intrinsic factors like the free energy of the reacting molecules. Consequently, the rates of these reactions were expressed as the product of the concentration of reacting species and the intrinsic rate constant. The intrinsic rate constant of a reaction is dependent on thermodynamic variables like temperature and pressure. For example, the rate of propagation R_p is written as $R_p = k_p\ [RP]\ [M]$ where $[RP]$ and $[M]$ are the concentrations of propagating chains and monomer, respectively, and k_p is the propagation rate constant, which is assumed to be only a function of temperature and pressure but independent of concentration of reactive species. The above rate equation in which the rate constant is assumed to be only a function of thermodynamic variables of temperature and pressure is referred to as reaction-controlled or reaction-limited polymerization. It is implicit in a reaction limited polymerization that the rate of diffusion of the reactants is much faster than the intrinsic rate of the reaction. These reaction-controlled rate constants are designated as the "intrinsic rate" of a reaction and a model based on the intrinsic rate constant is referred to as a reaction model of polymerization. As the reaction proceeds and long polymer chains are produced, the viscosity of the mixture increases sharply in bulk as well as in solution polymerization [1–5]. This in turn affects the diffusion time of the reactants in bimolecular reactions or the translation of one reactant with respect to the other reactant prior to collision and reaction. When diffusion is the limiting step, the measured rate constant is described as the "apparent reaction rate constant" because it does not reflect the intrinsic or the kinetic rate of the reaction. The diffusivity of small molecule reactants in solution is generally high on the order of 1×10^{-5} $cm^2\ s^{-1}$ [6, 7], and the apparent rate constant reflects the intrinsic rate. Conversely, the diffusivity of large reactive molecules like propagating chains varies widely from 1×10^{-6} to 1×10^{-16} $cm^2\ s^{-1}$ depending on their molecular weight and viscosity of the reaction mixture, thus the apparent rate constant may be much lower than the intrinsic rate. Polymerization models based on the apparent rate constant are referred to as the reaction-diffusion model of polymerization. The following quantitative analysis illustrates the effect of diffusion on the time constant of a polymerization reaction. The average separation distance between small molecules like monomers, $\overline{sd}_m$, in solution is in the range of $\overline{sd}_m = 0.554\, n^{-1/3} \approx 0.3 - 0.6$ nm where n is in units of molecules per unit volume [8]. The time constant for propagation reaction between a monomer (monomer diffusion coefficient $D_{m,\text{dilute}} \approx 1 \times 10^{-5}\ cm^2\ s^{-1}$) [6, 7] and a propagating chain at low conversion or in dilute solution, regardless of DP because the smaller monomer is the diffusing molecule, is $\tau_{\text{prop,dilute}} = \frac{\overline{sd}_m^{\,2}}{D_{m,\text{dilute}}} \approx 2 \times 10^{-10}\text{s} = 200$ ps. At moderate conversions ($50 < \alpha < 90\%$), the diffusion coefficient of monomer in the reaction mixture above the polymer glass transition temperature (T_g) is approximately $D_{m,\text{conc}} \approx 1 \times 10^{-7}$ cm^2 s^{-1} [9] with a corresponding $\tau_{\text{prop,conc}} \approx 2 \times 10^{-8}\text{s} = 20$ ns. At high conversions ($\alpha > 90\%$), monomer diffusion may be as much as two orders of magnitude slower than the intrinsic reaction rate of propagation. As time constant for a typical bimolecular reaction is in the range of 20 – 200 ps, propagation reactions at low or moderately high conversions can be assumed not to be limited by diffusion. For termination reaction, the two reacting molecules are propagating chains with concentrations in the range of $[RP] \approx 1 \times 10^{-8} - 1 \times 10^{-3}$ M [10]. The separation distance between propagating chains for an average concentration of 1×10^{-6} M is $\overline{sd}_{rp} = 0.554\, n^{-1/3} \approx 60$ nm. The diffusion coefficient of a propagating chain with 220 kDa molecular weight is $D_{rp,\text{dilute}} \approx 9 \times 10^{-7}$ cm^{2}s^{-1} [9], and the corresponding time constant for termination reaction is $\tau_{\text{term,dilute}} = \frac{\overline{sd}_{rp}}{D_{rp}} \approx 40$ μs which is four orders of magnitude longer than the intrinsic time constant of 200 ps. The diffusion coefficient of a propagating chain at high conversion in bulk above the polymer T_g is approximately $D_{rp,\text{conc}} \approx 1 \times 10^{-11}$ cm^2 s^{-1} [11] with corresponding $\tau_{\text{term,conc}} \approx 3$ s. Hence, termination rate constants are limited by diffusion not only in concentrated regime but also in dilute regime.

As the reaction viscosity increases exponentially with conversion and the diffusivity of the propagating chains is reduced sharply, their termination rate is decreased, which leads to their accumulation in the reaction and the breakdown of quasi-steady state assumption (QSSA). This chapter covers the role of viscosity on relaxation time and diffusivity of propagating chains and its effect on elementary reactions of initiation, propagation, and termination, and the transition from reaction-controlled to diffusion-controlled polymerization. The method of moments will be used to model the degree of polymerization, polydispersity and concentration of reactants and products in diffusion-controlled reactions.

8.2 Relating Diffusion Coefficient and Viscosity to Molecular Weight

8.2.1 Dilute Polymerization Regime Below the Onset of Chain Entanglement

The diffusion coefficient of a particle/polymer (D_p) in a liquid medium is described by Einstein's equation for diffusivity as follows [12–14]:

$$D_p = \frac{k_B T}{\xi_p} = k_B T \mu_p \tag{8.1}$$

where k_B and T are Boltzmann constant and temperature, respectively, ξ_p is the particle friction coefficient between the diffusing polymeric particle and the liquid medium, and μ_p is the particle mobility. Particle friction coefficient (F_d) is defined as the drag force exerted on a diffusing particle by the liquid medium divided by the diffusing particle velocity (u_p). The force exerted on diffusing particle, F_d, is given by Stokes' equation as follows [15, 16]:

$$F_d = 6\pi\, R_H\, \eta_s\, u_p \Rightarrow \xi_p = F_d/u_p = 6\pi\, R_H\, \eta_s \tag{8.2}$$

where R_H is the equivalent hydrodynamic radius of diffusing particle and η_s is the solution viscosity. For a polymeric particle with many monomeric segments, the friction coefficient between a chain segment and the liquid medium is given by the following relation:

$$\xi_0 = \xi_p/N = \xi_p/DP \tag{8.3}$$

where N and DP are the number of segments in a polymer chain and the degree of polymerization of the chain, respectively. In using N and DP, it is assumed for now that the polymer chains are monodisperse. Substitution of Eqs. (8.2) and (8.3) in Eq. (8.1) results in the following expression, known as the Stokes–Einstein equation for polymer diffusivity in dilute solution [13]:

$$D_p = \frac{k_B T}{\xi_p} = \frac{k_B T}{\xi_0 DP} = \frac{k_B T}{6\pi R_H \eta_s} \tag{8.4}$$

The above equation describes the diffusivity of a particle in a solvent in dilute solution in which the particle motion is not affected by other particles. Knowing particle diffusivity, the time constant for diffusion of polymer chains, $\tau_{diff,p}$, is defined by the following relation:

$$\tau_{diff,p} = \frac{\overline{sd}_p^{\,2}}{6D_p} \tag{8.5}$$

where $\overline{sd}_p$ is the average separation distance between two polymeric particles in solution and D_p is the polymer diffusion coefficient. The particle in solution could be a small molecule like solvent molecule, a reactive propagating chain, a dormant chain, or a terminated polymer chain. According to Eq. (8.4), the diffusion coefficient in a liquid medium is inversely proportional to the medium viscosity and to the radius of the diffusing particle or molecule, assuming the particle can be represented by an equivalent sphere. A propagating chain dissolved in its monomer in a reaction mixture can be assumed to behave as if the chain is dissolved in a theta solvent with Mark–Houwink–Sakurada exponent (MHS) [17–19] exponent "a = 0.5" (see Eq. 3.50 in Chapter 3 for definition of MHS equation and its exponent) [20]; For $a = 0.5$, the polymer takes on a random coil conformation. The R_H in Eqs. (8.2) and (8.4) can be replaced with the radius of gyration of polymer chain, R_g, with a random coil conformation defined by the following equation for a linear chain [21–25]:

$$R_g \approx R_H = \frac{b}{6} N^{1/2} = \frac{b}{6} DP^{1/2} \tag{8.6}$$

where b is the length of a monomer segment of the chain. Substitution of Eq. (8.6) in Eq. (8.4) results in the following relation for diffusion coefficient of polymer:

$$D_p = \frac{k_B T}{6\pi R_g \eta_s} = \frac{k_B T}{\pi b DP^{1/2} \eta_s} \tag{8.7}$$

This equation is used to predict polymer diffusion coefficient in dilute solution as a function of DP if the dependence of solution viscosity on polymer concentration and degree of polymerization is known [26]. The following Lyons–Tobolsky equation can be used to predict the viscosity of a polymer solution, η_s[27–30]:

$$\eta_s = \eta_{slv} \left\{ 1 + \eta_{\text{int}} [P] e^{\left(\frac{k_{LT}(\eta_{\text{int}}[P])}{1 - b_{LT}[P]} \right)} \right\} \tag{8.8}$$

where η_{slv} is the solvent viscosity in solution polymerization and monomer viscosity in bulk polymerization, $\eta_{\text{int}}$is the intrinsic viscosity of polymer, $[P]$ is the concentration of polymer chains in the reaction. The constants k_{LT} and b_{LT} are constants in the Lyons–Tobolsky equation which depend on the nature of polymer and solvent and temperature [27]. The above equation applies to solutions in which the polymer molecular weight is <150 kDa [30]. For higher molecular weights, the predicted viscosity diverges from the actual value. Another predictive equation for viscosity of a polymer solution is the following extended Martin equation by Fedors [31–33]:

$$\eta_s = \eta_{slv} \left[1 + \eta_{\text{int}} [P] e^{\left\{ d_{EM,0} + d_{EM,1}(\eta_{\text{int}}[P]) + d_{EM,2}(\eta_{\text{int}}[P])^2 \right\}} \right] \tag{8.9}$$

where η_{slv}, η_{int}, and $[P]$ are as defined by the above Lyons–Tobolsky equation, and $d_{EM,0}$, $d_{EM,1}$, and $d_{EM,2}$ are the three constants of the extended Martin equation. Like the Lyons–Tobolsky equation, the above equation applies to polymer solutions with molecular weight of <150 kDa. For a moderately concentrated PMMA solution, the dimensionless constants $d_{EM,0}$, $d_{EM,1}$, and $d_{EM,2}$ are 0.135, 0.177, and 1.69×10^{-3}, respectively, and the intrinsic viscosity (in $m^3\ kg^{-1}$) for PMMA solution (MW > 35 kDa) is $\eta_{int} = 5.2 \times 10^{-6} \overline{M}_w^{0.76}$ by MHS equation [33, 34]. Using this data, the solution viscosity of PMMA can be predicted as a function of polymer concentration for molecular weights <150 kDa (see Practice Problem 8.3).

The Vrentas–Duda equation, which is based on free volume theory of polymers, is used to predict the viscosity of polymer solutions as a function of polymer concentration [35–40]. Based on this theory, the diffusion coefficient of a polymer in a solution of its monomer is given by the following expression:

$$\eta_s = \eta_{int} e^{\left[\frac{\gamma_P}{\varsigma_P}\left\{\frac{(w_m \hat{V}_m^* + w_p \hat{V}_p^* \xi_p)}{\upsilon_{fv,mix}} - \frac{1}{\upsilon_{fv,m}}\right\}\right]} \tag{8.10A}$$

$$\varsigma_p = (\hat{V}_m^* M_{jump,m}) \,/\, (\hat{V}_p^* M_{jump,p}) \tag{8.10B}$$

In the above equations, γ_p is a constant for a given polymer-solvent system; $\hat{V}_m^*$ and $\hat{V}_p^*$ are monomer partial specific volume and polymer partial specific volume in solution, respectively; $M_{jump,m}$ and $M_{jump,p}$ are the molecular weight of the jumping unit for diffusion to occur for the monomer and polymer; w_m and w_p are weight fractions of the monomer and polymer; and ς_p, defined by Eq. (8.10B), is the ratio of the volume available for monomer diffusion divided by the volume available for polymer diffusion. The parameters $\upsilon_{fv,m}$ is the free volume fraction of pure monomer at a specified temperature and $\upsilon_{fv,mix}$ is the free volume fraction of the polymer solution, which is defined by the following relation [35–39]:

$$\upsilon_{fv,mix} = w_m \hat{V}_m^* \upsilon_{fv,m} + w_p \hat{V}_p^* \upsilon_{fv,p} \tag{8.11}$$

where $\upsilon_{fv,p}$ is free volume fraction for pure polymer at a specified temperature. For example, for polymethyl methacrylate-methyl methacrylate (PMMA-MMA) system, $\gamma_p = 0.763$, $\hat{V}_m^* = 0.822\ cm^3\ g^{-1}$, $\hat{V}_p^* = 0.770\ cm^3 g^{-1}$, $M_{jump,m} = 100\ g\ mole^{-1}$, $M_{jump,p} = 150\ g\ mole^{-1}$, $\varsigma_p = 0.712$, and $\upsilon_{fv,m}$ and $\upsilon_{fv,p}$ are given as a function of temperature by the following expressions [33, 41]:

$$\begin{aligned} &\upsilon_{fv,m} = 0.149 + 2.9 \times 10^{-4}\, T \quad T\,(°C) \quad \text{MMA} \\ &\upsilon_{fv,p} = 0.0149 + 1.3 \times 10^{-4}\, (T - 105) \quad T\,(°C) \quad \text{PMMA} \end{aligned} \tag{8.12}$$

The three empirical Lyons–Tobolsky, extended Martin, and Vrentas–Duda equations predict an exponential increase in solution viscosity with increasing polymer conversion or polymer concentration. As a high molecular weight polymer is produced in conventional radical polymerization at low conversions, reaction viscosity increases exponentially early in the reaction [42–45]. In the following section, the Lyons–Tobolsky equation is used to relate diffusivity to polymer concentration in dilute and concentrated regimes. In a similar approach, the extended Martin and Vrentas–Duda equations may be used to relate diffusivity to polymer concentration in concentrated regimes.

As viscosity increases exponentially with polymer concentration, the second term on the right of Eq. (8.8) becomes orders of magnitude greater than unity early in the reaction, thus Eq. (8.8) simplifies to the following:

$$\eta_s/\eta_{slv} = \eta_{\text{int}}\,[P]\,e^{\left(\frac{K_{LT}(\eta_{\text{int}}[P])}{1-b_{LT}[P]}\right)} \tag{8.13}$$

The above equation is comprised of the linear term $\eta_{\text{int}}\,[P]$ and the exponential term $\exp\,[K_{LT}\,(\eta_{\text{int}}\,[P])\,/\,(1-b_{LT}\,[P])]$. The above equation with the exponential term should be used above the critical concentration for the onset of chain entanglement. Below the critical concentration, the exponential term is negligible, and the solution viscosity is approximated by the linear term as follows:

$$\frac{\eta_s}{\eta_{slv}} \approx \eta_{\text{int}}\,[P] \text{ below chain entanglement} \tag{8.14}$$

The intrinsic viscosity, η_{int}, is related to the polymer molecular weight by the MHS equation [33, 34]. The MHS equation for polydisperse polymers was defined in Section 3.6 of Chapter 3 (see Eq. 3.50) as follows:

$$\eta_{\text{int}} = K_\eta \overline{M_\eta}^{\,a} = K_\eta M_0^a \overline{DP_\eta}^{\,a} \tag{8.15}$$

Where K_η and a are constants that depend on the specific polymer-solvent mixture and temperature; M_0 is the molecular weight of polymer repeat unit; $\overline{DP_\eta}$ and $\overline{M_\eta}$ are the viscosity-average degree of polymerization and viscosity-average molecular weight, respectively. In the above equation, $a = 0.5$ corresponds to a theta solvent in which the polymer chain has a random coil conformation, $a = 0.8$ corresponds to a good solvent in which the polymer chain expands, and $a = 2.0$ corresponds to a completely stretched rigid rod. Substitution of Eq. (8.15) in Eq. (8.14) gives the following relation for polymer solution viscosity below the onset of chain entanglement (dilute solution) in terms of $\overline{DP_\eta}$ and $[P]$:

$$\eta_{s,\text{dilute}}/\eta_{slv} = K_\eta M_0^a \overline{DP_\eta}^{\,a}\,[P] \tag{8.16}$$

The above equation can be substituted in Eq. (8.7) to relate the polymer diffusion coefficient in dilute regime (below chain entanglement) to $\overline{DP_\eta}$ and $[P]$:

$$D_{p,\text{dilute}} = \frac{k_B T}{(\pi b \eta_{slv} K_\eta {M_0}^a)\,[P]\,\overline{DP_\eta}^{\,0.5+a}} \tag{8.17}$$

Based on the above equation, diffusion coefficient of the polymer chains (propagating or terminated chains) in the concentration regime below chain entanglement (Rouse model for polymer diffusion [46–48]) scales with $\overline{DP_\eta}$ to the power of $-(0.5 + \text{a})$. If the chains are in a theta solvent or own monomer ($a = 0.5$), and D_p is inversely proportional to the first power of $\overline{DP_\eta}$. The characteristic diffusion time of polymer chain in dilute regime, $\tau_{diff,p,\text{dilute}}$, is defined by Eq. (8.5). Upon substitution of Eq. (8.17) in Eq. (8.5), $\tau_{diff,p,\text{dilute}}$ is given by the following equation in terms of $\overline{DP_\eta}$ and $[P]$:

$$\tau_{diff,p,\text{dilute}} = \frac{(\pi b\, \eta_{slv}\, K_\eta\, M_0^a)\,\overline{sd_p}^{\,2}\,[P]\,\overline{DP_\eta}^{\,(0.5+a)}}{6\,k_B T} \tag{8.18}$$

where $\overline{sd_p}$, as defined above, is the average separation distance between two chains in the reaction. Based on the above equation, the characteristic diffusion time of the chains in the dilute regime below chain entanglement scales with $\overline{DP_\eta}$ to the power of $(0.5 + a)$.

8.2.2 Semi-dilute or Concentrated Regime Below Threshold Molecular Weight for Reptation

In this regime, the polymer solution is concentrated, but the molecular weight is below the threshold for dominance of diffusion by reptation. In this regime, there is significant matrix relaxation around a given chain such that both Rouse and reptation motions contribute to chain diffusion [49–51]. In this situation, the exponential term in Eq. (8.13) is significant and higher than the linear term $\eta_{\text{int}}\,[P]$. Therefore, in semi-dilute solution or concentrated regime below the molecular weight for the onset of reptation, D_p and $\tau_{d,p}$ are given by the following equations:

$$D_{p,\text{semidilute}} = \frac{k_B T}{\left(\pi b\, \eta_{slv}\, K_\eta\, M_0^a\right) [P]\, \overline{DP}_\eta^{\,(0.5+a)}} e^{-\left(\frac{K_{LT}\left(K_\eta\, M_0^a\, \overline{DP}_\eta^{\,a}\,[P]\right)}{1 - b_{LT}\,[P]}\right)} \tag{8.19}$$

$$\tau_{diff,p,\text{semidilute}} = \frac{\left(\pi b\, \eta_{slv}\, K_\eta\, M_0^a\right) [P]\, d_{p-p}^2\, \overline{DP}_\eta^{\,(0.5+a)}}{6 k_B T} e^{\left(\frac{K_{LT}\left(K_\eta\, M_0^a \overline{DP}_\eta^{\,a}\,[P]\right)}{1 - b_{LT}[P]}\right)} \tag{8.20}$$

Based on the above equations, D_p in semi-dilute regime or concentrated regime below the molecular weight for dominance of reptation decreases exponentially with polymer concentration and molecular weight whereas $\tau_{diff,p}$ increases exponentially. Another very useful and interesting model in the semi-dilute or concentrated regime below the threshold for reptation is the blob model, which is a fusion of Rouse and reptation models of macromolecular diffusion, but it is beyond the scope of this chapter.

8.2.3 Concentrated Regime Above Threshold Molecular Weight for Reptation

For polymer concentrations and molecular weights above entanglement (MW>150 kDa), chains overlap to such an extent that the matrix appears frozen on the time scale of chain diffusion [52–55]. In the reptation regime, chains are confined to a tube formed by the neighboring polymer chains, as proposed by P.G. de Gennes [46, 56–58]. As a result, the polymer chains diffuse like a snake within the tube, and the contribution of Rouse-like motion to the overall diffusion of chains becomes negligible. In the reptation regime, the polymer solution viscosity predicted by Lyons–Tobolsky, extended Martin, or Vrentas–Duda models deviate from the actual viscosities. This deviation for poly(methyl methacrylate)-methyl methacrylate (PMMA-MMA) polymerization system occurs at $C_p \overline{M}_w^{\,0.76} \approx 1 \times 10^4$ in *CGS* units where C_p is polymer concentration and $\overline{M}_w$ is the weight-average molecular weight of the polymer [33, 59]. In general, the deviation from Rouse or mixed regime to reptation occurs at $C_p \overline{M}_w^{\,a} \approx 1 \times 10^4$ where a is the exponent in MHS equation [33, 34] (see Eq. 8.15) and the C_p in the above equation is related to reaction conversion α by the following relation [33]:

$$\alpha = \frac{C_p/\rho_m}{1 + (C_p/\rho_m)\left[1 - (\rho_m/\rho_p)\right]} \tag{8.21}$$

where ρ_m and ρ_p are densities of pure monomer and polymer, respectively. For example, according to the above equation, for PMMA with $\overline{M}_w = 1.6 \times 10^5$ g mol^{-1} in methyl methacrylate (MMA) monomer with $a = 0.76$, $C_p \approx 1 \times 10^4 / (1.6 \times 10^5)^{0.76} = 1.11\, g/cm^3$. Using the values of $\rho_{MMA} = 0.88$ g cm^{-3} and $\rho_{\text{PMMA}} = 1.2$ g cm^{-3}, Eq. (8.21) predicts a transition from Rouse to reptation regime at conversion $\alpha = 94\%$.

Based on the reptation theory [46, 56–58], the polymer chain mobility, $\mu_{p,rep}$, is related to $\overline{DP}_\eta$ by the following equation:

$$\mu_{p,rep} = \frac{\mu_{seg}\, l_{rs}^2}{3\, b^2}\frac{1}{\overline{DP}^2} = \frac{l_{rs}^2}{3\xi_{seg} b^2}\frac{1}{\overline{DP}_\eta^{\,2}} \tag{8.22}$$

where μ_{seg} is chain segment mobility which is the inverse of chain segment friction coefficient ξ_{seg}, l_{rs} is the reptation step length of a diffusing chain, and b is the monomer segment length as defined in Eq. (8.6). In the above equation, $\mu_{p,rep}$ is related to $\overline{DP}_\eta$ because mobility is related to molecular motion which in turn is described by viscosity. After substitution of the above equation in Eq. (8.1), $D_{p,rep}$ is given in terms of $\overline{DP}_\eta$ as follows:

$$D_{p,rep} = k_B T\, \mu_{p,rep} = \frac{k_B T\, l_{rs}^2}{3\, \xi_{seg}\, b^2}\frac{1}{\overline{DP}_\eta^{\,2}} \tag{8.23}$$

The unknown parameter in the above equation is the reptation step length, l_{rs}, which is related to the number of repeat units for the onset of chain entanglement, i.e. N_e, by the following expression [60]:

$$l_{rs} = \left(4\, b^2 N_e/5\right)^{1/2} \tag{8.24}$$

Substitution for l_{rs} from the above equation in Eq. (8.23) results in the following expression relating $D_{p,rep}$ to $\overline{DP}_\eta$:

$$D_{p,rep} = \frac{4\, k_B T\, N_e}{15\, \xi_{seg}}\frac{1}{\overline{DP}_\eta^{\,2}} \tag{8.25}$$

The characteristic diffusion time of polymer chains in the reaction, $\tau_{diff,p,rep}$, is obtained by substitution for D_p from the above equation in Eq. (8.5) as follows:

$$\tau_{diff,p,rep} = \frac{15\, \bar{d}_{p-p}^{\,2}\, \xi_{seg} \overline{DP}_\eta^{\,2}}{24\, k_B T\, N_e} \tag{8.26}$$

According to the above equation, the reptation theory predicts that diffusion coefficient scales inversely with the second power of $\overline{DP}_\eta$ and the characteristic diffusion time is directly proportional to the second power of $\overline{DP}_\eta$. Reptation significantly influences the rates of elementary reactions in radical polymerization because long chains are produced early in the reaction. In condensation reactions, the diffusion of chains follows the Rouse model in the early stages when DP is below the entanglement limit whereas reptation dominates at high conversions.

8.3 Role of Diffusion in Bulk Polymerization of Methyl Methacrylate

8.3.1 Diffusion Time of Monomer and Other Small Molecules

In bulk polymerization of methyl methacrylate (MMA), the reactor is charged with monomer and initiator, and the initiator is activated as discussed in the first two chapters to start the reaction. Upon dissociation of initiator and radical formation, the monomer (M) reacts with the primary radicals (R) to form reactive propagating chains (PR). Two propagating chains react by combination or disproportionation to form terminated polymer chains (TP) [61–66]. In bulk polymerization of methyl methacrylate, the rate of polymerization is affected by diffusivity and mobility of the reactants, especially at high conversions. The self-diffusion coefficient of MMA is equal to 8×10^{-6} cm^2 s^{-1} at 50 °C [67, 68]. In general, the diffusion coefficient of small molecules like solvent, monomer, initiators, and initiator fragments, inhibitor, and retarding agent in liquid reaction medium near or above ambient temperature is approximately 1×10^{-5} cm^2 s^{-1} [6]. As the concentration of PMMA in its monomer solution at 50 °C is increased from zero to 10%, 50%, and 80% by weight, the diffusion coefficient of MMA decreases from 0.8×10^{-5} to 4.0×10^{-6}, 2.0×10^{-6}, and 1.0×10^{-7} cm^2 s^{-1}, respectively [6, 67]. Therefore, the diffusion coefficient of small molecules like the monomer in polymerization changes by two orders of magnitude from ~1×10^{-5} cm^2 s^{-1} in the dilute regime (low conversion) to 1×10^{-7} cm^2 s^{-1} in the concentrated regime (high conversion). Assuming the temperature is above T_g of the polymer, the farthest distance the monomer would have to diffuse to react with a propagating chain is the radius of gyration of propagating chain, $R_{g,rp}$. As monomer is the faster diffusing molecule compared to the propagating chain, it diffuses toward the radical of the propagating chain in the propagation reaction. Based on Eq. (8.6) and assuming the propagating chain can be represented as a freely jointed chain (FJ) with $R_{g,FJ} = b\overline{DP}_{rp}^{\,1/2}/6$ [69], the diffusion time of monomer in propagation reaction is given by the following equation:

$$\tau_{diff,m} = \frac{R_{g,rp}^2}{6\,D_m} = \frac{1}{6}\frac{\left[(b/6)\,\overline{DP}_{rp}^{\,1/2}\right]^2}{6\,D_m} = \frac{b^2\overline{DP}_{rp}}{216\,D_m} \tag{8.27}$$

where $R_{g,FJ}$ is the radius of gyration of freely jointed chain (propagating chain) with average degree of polymerization $\overline{DP}_{rp}$ and monomer diffusion coefficient D_m. Using $b = 0.58$ nm for PMMA [70], $\overline{DP}_{rp} \approx 1200$ (see next Section 8.3.2), and MMA diffusion coefficient ranging from 1×10^{-5} cm^2 s^{-1} in dilute polymerization regime to 1×10^{-7} cm^2 s^{-1} in concentrated regime [6, 67], the time constant for diffusion of small molecules like monomer in the polymerization reaction is in the range of 2 ns in the dilute regime to 200 ns in the concentrated regime. The upper limit of the intrinsic rate constant for a typical bimolecular reaction between small molecules is $k_{bm} = 1 \times 10^{10}$ M^{-1}s^{-1} [71]. For methyl methacrylate with maximum (bulk) concentration of 4.8 M and solution concentration of 0.48 M (10% v/v), the time constant for a typical bimolecular reaction is in the range of $\tau_{r,bm} = 1/k_{bm}\,[C] \approx 0.02 - 0.2$ ns. Hence, the propagation reaction at low conversions can be assumed not to be limited by diffusion. At moderately high to high conversions, monomer diffusion of 50 ns can be slower by as much as two orders of magnitude compared to the intrinsic time constant of 0.2 ns.

8.3.2 Diffusion Time of Propagating and Terminated Polymer Chains

The rates of reactions for termination or chain transfer to polymer are affected by diffusion of propagating or polymer chains in dilute regime below the critical conversion for chain entanglement, concentrated regime with chain entanglement, and highly concentrated regime in which reptation is the dominant mode of chain diffusion. To determine the effect of diffusion on reactions of polymer chains, the critical conversion for the onset of chain entanglement needs to be calculated for PMMA polymerization. Initially in bulk polymerization, the reaction mixture consists of mostly monomer and a small initiator fraction. The density of MMA at a typical polymerization temperature of 50 °C is 0.893 g cm^{-3}. Therefore, the initial monomer concentration in the reaction is:

$$[M]_0 = \frac{\rho_{MMA}}{M_{MMA}} = \frac{0.893\text{ g cm}^{-3}}{100.12\text{ g mol}^{-1}} = 8.9 \times 10^{-3}\text{mol cm}^{-3} = 8.9\text{ M} \tag{8.28}$$

where ρ_{MMA} and M_{MMA} are density and molecular weight of MMA, respectively. In a typical MMA radical polymerization at 50 °C, the initial concentration of initiator, $[I]_0$, initiator efficiency, f_0, dissociation rate constant, k_d, and propagation rate constant, k_p, and termination rate constant by disproportionation, k_{td}, are given as follows [33, 70, 72–77]:

$$[I]_0 = 0.001\ \text{M} \quad f_0 = 1 \quad k_d = 5 \times 10^{-5}\ \text{s}^{-1} \tag{8.29A}$$

$$k_{td} = 5 \times 10^7\ \text{lit mol}^{-1}\ \text{s}^{-1} \quad k_p = 300\ \text{lit mol}^{-1}\ \text{s}^{-1} \tag{8.29B}$$

The equation for concentration of propagating chains, $[RP]$, under the assumption of quasi-steady state was derived in Section 7.3.1.1 of Chapter 7. Based on Eq. (7.5F), $[RP]_0$ is determined as follows:

$$[RP]_0 = \left(\frac{2k_d f_0 [I]_0}{k_{td}}\right)^{\frac{1}{2}} = \left[\frac{2(5 \times 10^{-5}/\text{s})(1)\left(1 \times 10^{-3}\ \text{mol lit}^{-1}\right)}{\left(5 \times 10^7\ \text{lit mol}^{-1}\ \text{s}^{-1}\right)}\right]^{1/2} = 4.47 \times 10^{-8}\ \text{M} = 2.69 \times 10^{13}\ \frac{\#}{\text{cm}^3} \tag{8.30}$$

The average separation distance between two particles at a specified concentration is calculated from the probability distribution $p_{nnd}(r)$ for a neighboring second particle to be at a distance between r and $r + dr$ from the first particle, which has been derived previously, and given by the following equation [8]:

$$p_{nnd}(r) = 4\pi r^2 (n/V)\, e^{-4\pi r^3 n/V} \tag{8.31}$$

where n is the number of particles, V is volume, and r is the distance between the particles. The average or most probable distance between the particles, $\overline{sd}_p$, is calculated by integrating the above equation from $r = 0$ to $r \to \infty$ as follows [8]:

$$\overline{sd}_p = \langle r \rangle = \int_0^{\infty} r^1\, p_{nnd}(r)\, dr = \int_0^{\infty} r\, \pi r^2 (n/V)\, e^{-4\pi r^3 n/V}\, dr = 0.554\,(n/V)^{-1/3} \tag{8.32}$$

The $\overline{sd}_{rp}$ for propagating chains with initial concentration given by Eq. (8.30) is determined as follows:

$$\overline{sd}_{rp} = 0.554\left(\frac{\text{n}}{\text{V}}\right)^{-1/3} = 0.554\left(\frac{2.69 \times 10^{13}\#}{1\ \text{cm}^3}\right)^{-1/3} = 1.87 \times 10^{-5}\text{cm} \approx 200\ \text{nm}$$

Based on the above analysis, the average distance that a PMMA propagating chain would have to diffuse to react with another propagating chain to form two terminated polymer chains is approximately 200 nm. Next, the average number of repeat units or the kinetic chain length (L) of the propagating chains, which was derived in Section 7.3.1.3 of Chapter 7, needs to be calculated. Based on Eq. (7.71), the initial L in MMA polymerization is determined as follows:

$$L = \frac{k_p [RP]_0 [M]_0}{2k_d f_0 [I]_0} = \frac{\left(300\,\text{lit mol}^{-1}\text{s}^{-1}\right)\left(4.47 \times 10^{-8}\ \text{mol lit}^{-1}\right)\left(8.9\,\text{mol lit}^{-1}\right)}{2\,(5 \times 10^{-5}/\text{s})\,(1)\,(1 \times 10^{-3}\text{mol lit}^{-1})} \sim 1200 \tag{8.33}$$

Assuming the propagating chain can be represented as a freely jointed chain with radius of gyration $R_{g,FJ} = b\overline{DP}_{n,rp}^{1/2}/6$ [69, 78] (b is the segment length defined in Section 8.2.3), the average diameter of the propagating chains, $\overline{d}_{rp}$, is determined as follows:

$$\overline{d}_{rp} = 2R_{g,FJ} = \frac{2b\overline{DP}_{n,rp}^{1/2}}{6} = \frac{bL^{1/2}}{3} = \frac{(0.58\ \text{nm})(1200)^{1/2}}{3} = 6.8\ \text{nm} \tag{8.34}$$

The ratio of $\overline{sd}_{rp}$ to $\overline{d}_{rp}$ is a normalized measure of the distance the propagating chains would have to diffuse to react with another propagating chain to form a terminated polymer chain. Using Eqs. (8.33) and (8.34) for MMA polymerization, this initial normalized distance is $\frac{\overline{sd}_{rp}}{\overline{d}_{rp}} \approx 30$. This value indicates that the propagating PMMA chain would have to diffuse a distance of 30-fold its diameter to react with another chain for termination. This average diffusion distance between two propagating chains persists because the concentration of propagating chains is a few orders of magnitude lower than the monomer throughout the reaction. Next, concentration $[TP]$ and $\overline{DP}_{n,tp}$ of the terminated polymer chains produced initially in MMA polymerization need to be determined. Assuming termination is dominated by disproportionation (mechanism of termination does not significantly change the following analysis) [79–82], the initial $\overline{DP}_{n,tp}$ of the terminated chains would be the same as the initial kinetic chain length. The concentration of terminated chains initially in the reaction is $[TP]_0 = 0$. Therefore, $[TP]$ will be determined using Eq. (7.9D) in Section 7.3.1.5 of Chapter 7 at t = 10 s near zero time as follows:

$$\begin{aligned}[TP]_{t=100\,\text{s}} &= 2f_0\,[I]_0\left(1 - e^{-k_d t}\right)\\ &= 2 \times 1 \times 10^{-3}\text{M}\left(1 - e^{-5\times10^{-5}\,\text{s}^{-1}\times 100\,\text{s}}\right)\\ &= 1 \times 10^{-5}\text{M} \approx 6 \times 10^{15}\#/\text{cm}^3\end{aligned} \tag{8.35A}$$

$$\overline{DP}_{n,tp} = L = \lambda_1 \approx 1200 \tag{8.35B}$$

Using Eqs. (8.33) and (8.34), the average diameter, $\overline{d}_{tp}$, and the average distance, $\overline{sd}_{tp}$, between terminated polymer chains are determined for MMA polymerization as follows:

$$\overline{d}_{tp} = \frac{2b\overline{DP}_{n,tp}^{1/2}}{6} = \frac{bL^{1/2}}{3} = \frac{(0.58\ \text{nm})\,(1200)^{1/2}}{3} = 6.8\ \text{nm} \tag{8.36A}$$

$$\overline{sd}_{tp} = 0.554\left(\frac{\text{n}}{\text{V}}\right)^{-1/3} = 0.554\left(\frac{6 \times 10^{15}\#}{1\ \text{cm}^3}\right)^{-1/3} = 3.1 \times 10^{-6}\ \text{cm} \approx 30\ \text{nm} \tag{8.36B}$$

The above analysis demonstrates that the average separation distance between terminated chains is less than an order of magnitude higher than the size of the chain. As the reaction proceeds, more chains terminate and the separation distance between the polymer chains in the reaction continues to decrease. The ratio of $\overline{d}_{tp}$ to $\overline{sd}_{tp}$, which is a function of time, is related to the extent of chain entanglement and referred to as the entanglement index or EI as follows:

$$EI = \frac{\overline{d}_{tp}}{\overline{sd}_{tp}} \tag{8.37}$$

At start of the reaction, $\overline{sd}_{tp} = 30$ nm, and $EI = 0.23$, the average separation distance between the terminated chains is large, chains are not entangled, the chains are mainly in contact with monomer, their diffusion is mostly independent of the other chains, and chain diffusion follows the Rouse model [83, 84]. As the reaction proceeds, the concentration of terminated chains continues to increase, their separation distance decreases, and EI increases. At a critical concentration $[TP]_{cr}$, the center-to-center distance of the neighboring chains becomes equal to their hydrodynamic diameter. In other words, the available volume to a chain equals its hydrodynamic volume. At this critical concentration, $\overline{sd}_{tp} = \overline{d}_{tp}$, and $EI = 1$. Using Eq. (8.36A) for $\overline{d}_{tp}$ and Eq. (8.36B) for $\overline{sd}_{tp}$, $[TP]_{cr}$ is expressed in terms of b and $\overline{DP}_{n,tp}$ as follows:

$$\begin{aligned} \overline{d}_{tp} &= \frac{2\,b\,\overline{DP}_{n,tp}^{1/2}}{6} = \overline{sd}_{tp} = 0.554\left(\frac{n}{V}\right)^{-1/3} \quad \in EI = 1 \\ \Rightarrow \frac{2\,b\,\overline{DP}_{n,tp}^{1/2}}{6} &= 0.554\left(\frac{n}{V}\right)^{-1/3} = 0.554\left(\frac{[TP]_{cr} \times N_{AV}\#}{1 \times 10^{-24}\,nm^3/L}\right)^{-1/3} \\ \Rightarrow [TP]_{cr} &= \frac{8.053}{b^3\,\overline{DP}_{n,tp}^{3/2}} \quad \in EI = 1 \end{aligned} \tag{8.38}$$

where $N_{AV} = 6.022 \times 10^{23}$ #/mole is the Avogadro's number, b is the polymer segment length, and $\overline{DP}_{n,tp}$ is the number-average degree of polymerization of the polymer chains. Using Eq. (7.10D) in Section 7.3.1.6 of Chapter 7 which relates $[TP]$ to $[M]$, the equation for determination of critical monomer concentration at $EI = 1$ is derived as follows:

$$\begin{aligned} \mu_1 = \overline{DP}_{n,tp} &= \frac{Q_{1,tp}}{Q_{0,tp}} = \frac{[M]_0 - [M]}{[TP]} \Rightarrow \overline{DP}_{n,tp} = \frac{[M]_0 - [M]_{cr}}{[TP]_{cr}} \\ \Rightarrow [M]_{cr} &= [M]_0 - \overline{DP}_{n,tp}\,[TP]_{cr} = [M]_0 - \frac{8.053}{b^3\,\overline{DP}_{n,tp}^{1/2}} \end{aligned} \tag{8.39}$$

Using the definition of reaction conversion as $\alpha = \dfrac{([M]_0 - [M])}{[M]_0}$, the critical conversion at $EI = 1$ is given by the following equation:

$$\alpha_{cr} = 1 - ([M]_{cr} \,/\, [M]_0) = \frac{8.053}{b^3\,\overline{DP}_{n,tp}^{1/2}\,[M]_0} \quad \in EI = 1 \tag{8.40}$$

Using $[M]_0 = 8.9$ M, $\overline{DP}_{n,tp} = 1200$, $b = 0.58$ nm in the above equation, α_{cr} is determined as follows:

$$\alpha_{cr} = \frac{8.053}{b^3\,\overline{DP}_{n,tp}^{1/2}\,[M]_0} = \frac{8.053}{0.3^3 \times 1200^{1/2} \times 8.9} = 0.13 \quad \in EI = 1 \tag{8.41}$$

In the above derivation, the polymer chains were treated like spheres with hydrodynamic radius R_g. In the following sub-sections, we will apply the Lyons–Tobolsky model [27–30] for viscosity of polymer solutions (Eq. 8.8) to predict the diffusion time of propagating chains in methyl methacrylate polymerization in dilute, semi-dilute, reptation, and glassy regimes. The values of the parameters in Eqs. (8.8) and (8.15) (MHS model for intrinsic viscosity [33, 34]) for bulk polymerization of methyl methacrylate are as follows [33, 70, 72–77]:

$$b = 5.8 \times 10^{-10}\text{ m} \quad \overline{DP}_n = 1200 \quad [M]_0 = 8.9$$

$$T = 50\ ^\circ\text{C} = 323.15\text{ K} \quad M_0 = 100.12\text{ kg kmole}^{-1}$$

$$k_B = 1.381 \times 10^{-23}\text{ kg m}^2\text{ s}^{-2}\text{ K}^{-1}$$

$$a_{\text{PMMA}} = 0.76 \quad K_\eta = 5.2 \times 10^{-6} \quad K_\eta\, M_0^{0.76} = 1.72 \times 10^{-4}\ \text{m}^3\ \text{kg}^{-1}$$

$$k_{LT,\text{PMMA}} = 1.4898 \times 10^{-6} \quad b_{LT,\text{PMMA}} = 3.25 \times 10^{-4}\ \text{m}^3\ \text{kg}^{-1}$$

$$\begin{aligned}\log_{10}\eta_{MMA} &= -7.7825 + 7.3478 \times 10^2/\text{T} + 1.0258 \times 10^{-2}\ \text{T} - 1.1343 \times 10^{-5}\ \text{T}^2 \\ \log_{10}\eta_{MMA} &= -3.3783 \quad \Rightarrow \eta_{MMA} = 4.18 \times 10^{-4}\ \text{kg m}^{-1}\ \text{s}^{-1} \quad \in T = 323.15\ \text{K}\end{aligned} \tag{8.42}$$

8.3.2.1 Diffusion Time of PMMA Propagating Chains in Dilute Regime

The monomer conversion of 10% ($\alpha = 0.1$) is within the dilute regime considering $\alpha_{cr} = 0.13$. In the dilute regime, the Lyons–Tobolsky model for solution viscosity reduces to Eq. (8.16) with diffusion coefficient given by Eq. (8.17). For $\alpha = 0.1$, based on Eq. (7.10D) in Chapter 7 and $\alpha = \dfrac{([M]_0 - [M])}{[M]_0}$, the polymer concentration is determined as follows:

$$[TP]_{\alpha=0.1} = \frac{[M]_0 - [M]}{\overline{DP}_n} = \frac{[M]_0}{\overline{DP}_n}\alpha = \frac{8.9\ \text{M}}{1200}0.1 = 7.42 \times 10^{-4}\ \text{M} = 74\ \text{kg m}^{-3} \tag{8.43}$$

Using Eq. (8.16), the solution viscosity corresponding to $\alpha = 0.1$ is determined as follows:

$$\begin{aligned}\eta_{s,\text{dilute}} &= \eta_{MMA} K_\eta M_0^a \overline{DP_n}^{\,a}\ [TP] \\ &= 4.18 \times 10^{-4}\ \text{kg m}^{-1}\text{s}^{-1} \times 5.2 \times 10^{-6} \times 100^{0.76} \times\ 1200^{0.76} \times 74\ \text{kg m}^{-3} \\ &= 1.2 \times 10^{-3}\ \text{kg m}^{-1}\text{s}^{-1}\end{aligned} \tag{8.44}$$

Knowing the solution viscosity, the diffusion coefficient of the propagating chains is determined using Eq. (8.7) as follows:

$$\begin{aligned}D_{rp,\text{dilute}} &= \frac{k_B T}{\pi b \overline{DP}_n{}^{1/2}\eta_s} = \frac{1.381 \times 10^{-23}\ \text{kg m}^2\ \text{s}^{-2}\text{K}^{-1} \times 323\ \text{K}}{\pi \times 5.8 \times 10^{-10}\ \text{m} \times 1200^{1/2} \times 1.2 \times 10^{-3}\ \text{kg s}^{-1}\text{s}^{-1}} \\ &= 6.7 \times 10^{-11}\ \text{m}^2\text{s}^{-1} = 6.7 \times 10^{-7}\ \text{cm}^2\text{s}^{-1}\end{aligned} \tag{8.45}$$

The diffusion coefficient of propagating chains in dilute solution is six times lower than that of MMA (4.0×10^{-6} cm^2 s^{-1}; 10 wt% PMMA [9]) but it is of the same order of magnitude. The characteristic diffusion time as defined by Eq. (8.5), with $\overline{d}_{rp}$ from Eq. (8.34), is determined as follows:

$$\tau_{diff,rp,\text{dilute}} = \frac{\overline{d}_{rp}^2}{6D_{rp,\text{dilute}}} = \frac{(200 \times 10^{-7}\ \text{cm})^2}{6 \times 6.7 \times 10^{-7}\ \text{cm}^2\text{s}^{-1}} \approx 9.8 \times 10^{-5}\text{s} \approx 100\ \mu s \tag{8.46}$$

The diffusion time of the propagating chains in the dilute regime is sixfold higher than the monomer. This difference in diffusion time in dilute methyl methacrylate polymerization leads to a slow but gradual increase in molecular weight with increasing conversion, but does not affect the validity of QSSA. In condensation polymerization, the slower diffusivity of longer reactive chains as compared to shorter chains results in faster reaction and depletion of shorter chains, and narrowing of the molecular weight distribution.

8.3.2.2 Diffusion Time of PMMA Propagating Chains in Semi-dilute Regime

In this regime, viscosity increases exponentially with polymer concentration as described by Eq. (8.8). Based on this equation, a conversion of 50% is well within the regime of exponentially increasing solution viscosity. For $\alpha = 0.5$, $[TP] = 3.71 \times 10^{-3}\ \text{M} = 370\ \text{kg m}^{-3}$ (see Eq. 8.43) and using

Eq. (8.8) $\eta_s = 1.9\ \text{kg m}^{-1}\text{s}^{-1}$. Applying Eqs. (8.7) and (8.5) results in $D_{rp,\text{conc}} = 2.5 \times 10^{-9}\ \text{cm}^2\text{s}^{-1}$ and $\tau_{diff,rp,\text{conc}} = 26$ ms. The diffusion coefficient of the propagating chains is three orders of magnitude lower than that of MMA monomer (~10 μs) at 50% conversion in the semi-dilute regime. Therefore, termination reaction, as compared to propagation, is significantly affected by slower diffusion of the propagating chains. This difference leads to the accumulation of propagating chains in the reaction in the semi-dilute regime in MMA which affects the validity of QSSA. However, QSSA may be valid in this regime by increased chain transfer to the monomer to offset the accumulation of reactive propagating species. The mechanism of diffusion in this regime is a combination of Rouse and reptation models with reptation contribution growing with increasing conversion.

8.3.2.3 Diffusion Time of PMMA Propagating Chains in Reptation Regime

In this regime, the fraction of polymerized monomers is much higher than the unreacted monomer such that the entangled chains form transient tube-like structures, and chains diffuse like a snake within the formed tubes. For $\overline{DP}_n = 1200$, a conversion of 80% should be well within the reptation regime. With $\alpha = 0.8$, the polymerization mixture is above its T_g with reptation as the dominant mechanism of diffusion and the diffusion coefficient is given by Eq. (8.25). The molecular weight between chain entanglements for PMMA is $M_e = 4800\ \text{g mole}^{-1}$ corresponding to the number of monomer segments between chain entanglements $N_e = 48$ [85]. The segment friction coefficient for PMMA in MMA is $\xi_{seg} = 4.03 \times 10^{-10}\ \text{kg s}^{-1}$ [85]. Using this data in Eqs. (8.25) and (8.26), the diffusion coefficient and diffusion time of PMMA propagating chains in the reptation regime are determined as follows:

$$D_{rp,rep} = \frac{4k_BTN_e}{15\xi_{seg}\overline{DP}_n^2} = \frac{4\left(1.381 \times 10^{-23}\frac{\text{kg m}^2}{\text{s}^2\text{K}}\right)(323\text{K})(48)}{15\,(4.03 \times 10^{-10}\ \text{kg s}^{-1})\,(1200)^2}$$
$$= 9.8 \times 10^{-17}\ \text{m}^2\text{s}^{-1} \approx 1.0 \times 10^{-12}\ \text{cm}^2\text{s}^{-1} \tag{8.47}$$

$$\tau_{diff,rp,rep} = \frac{\overline{sd}_{rp}^2}{6D_{rp,rep}} = \frac{(200 \times 10^{-9}\text{m})^2}{6\,(9.8 \times 10^{-17}\text{m}^2\text{s}^{-1})} = 68\ \text{s} \tag{8.48}$$

Based on Eq. (8.47), diffusion coefficient of the propagating chains is five-orders of magnitude lower than that of MMA at 50 °C at 80% conversion ($1.0 \times 10^{-7}\ \text{cm}^2\ \text{s}^{-1}$ for MMA at 80 wt% PMMA [9]). The diffusion time of the propagating chains is seven orders of magnitude longer than that of monomer (~10 μs). This discrepancy in diffusion coefficient of the propagating chains as compared to the monomer results in very slow chain termination rates and accumulation of propagating chains in the reaction. In the reptation regime, the segmental friction coefficient is so high that the reactive species formed by initiator dissociation fail to escape the dissociation volume and recombine to form inactive species. Due to these limitations, the initiation efficiency $f_0 \ll 1$ and QSSA is not valid in the reptation regime.

8.3.2.4 Diffusion Time of PMMA Propagating Chains in Glassy Regime

At conversions $\alpha > 0.95$, the MMA content is insufficient to maintain a viscous-rubbery state, translational diffusion of the chains ceases, and the state of polymerizing mixture shifts from rubbery to glassy. Although the diffusion of the whole chain ceases, there is translation of segments of the chains [86]. For example, the diffusion coefficient of styrene in polystyrene at 30 °C below the polymer T_g is on the order of 1×10^{-12} cm^2 s^{-1} as compared to 1×10^{-8} cm^2 s^{-1} above T_g [87]. All elementary reactions, including initiator dissociation, chain propagation, and chain termination reactions, are affected at temperatures below the T_g of the polymerizing mixture.

The effect of conversion on initiator efficiency (f), concentration of propagating chains ([RP]), rate constant for propagation (k_p), and rate constant for termination (k_t) is shown in Figure 8.1. In the dilute regime, polymerization reaction is not limited by diffusion, that is, f, [RP], k_p, and k_t are relatively independent of conversion in this regime. In the semi-dilute regime, the reactions of propagating chains is controlled by diffusion, hence k_t which is the rate constant for reaction between two propagating chains is sharply reduced in this regime with increasing conversion which leads to moderate accumulation of propagating chains in the reaction. There is a slight decrease in the efficiency of initiation but k_p which is related to monomer reactivity is not affected in the semi-dilute regime. In the reptation regime, k_p remains steady and k_t remains steady after a sharp reduction in the concentrated regime. However, f sharply declines in the reptation regime which triggers a sharp decline in [RP]. In the glassy regime, all elementary reactions are controlled by lack of mobility at the molecular scale, leading to a sharp decline in k_p, k_t, f, but a rise in [RP]. The rise in [RP] in the glassy state is attributed to the slow termination of the existing propagating chains in the polymerization.

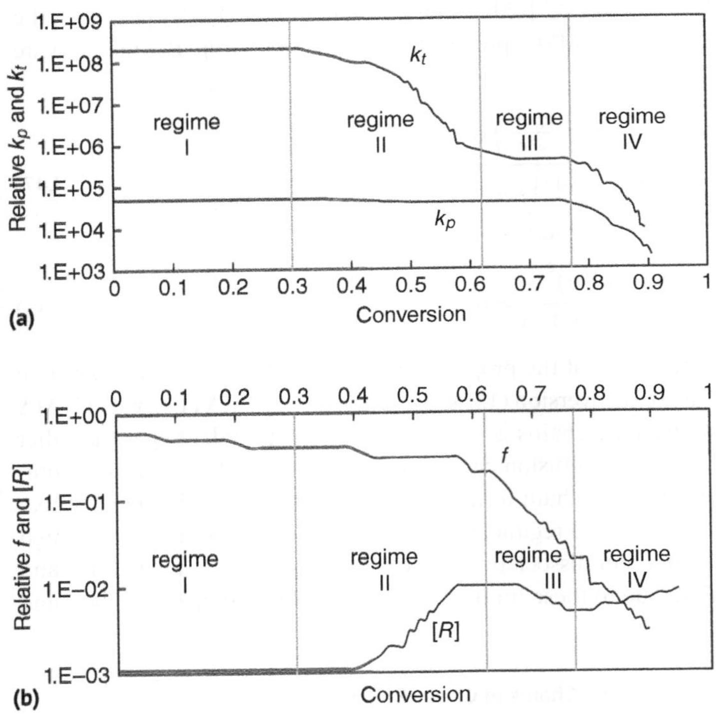

Figure 8.1 Schematic diagrams showing relative values of termination (k_t) and propagation (k_p) (plot a) and radical efficiency (f) and radical concentration ([R]) (plot b) in the four regimes in polymerization of methyl methacrylate as a function of conversion.

8.4 Termination Rate Constant in Diffusion-limited Polymerization Reactions

The termination reaction involves three steps, as shown schematically in Figure 8.2. The first step is the translation of the propagating chain 1 to the proximity of the propagating chain 2 in the diffusion layer. The second step is interdiffusion of chain segments of the propagating chain 2 in the volume occupied by the propagating chain 1, and vice versa for the reactive species on the two propagating chains to make contact. The third step is the reaction between the two reactive species to terminate the two propagating chains and form one or two terminated polymer chains depending on termination mechanism [68, 88–94]. The first step is controlled by diffusion; the second step is controlled by segmental interdiffusion, whereas the third step is kinetically controlled. The apparent rate constant for the reaction accounts for diffusivity of the reactants as well as their intrinsic reaction rate. This section will use the mass conservation equation of the propagating chains in the diffusion layer to relate the apparent termination rate constant to diffusivity and intrinsic reactivity of the propagating chains.

8.4.1 Mass Conservation Equation for Propagating Chains in Diffusion-limited Polymerization Reactions

A randomly selected propagating chain at position 2 in the bulk of the reaction in Figure 8.2 with bulk concentration $[RP]_b$ and at distance r_b from the center-of-mass of a second random propagating chain at point 1 diffuses in the diffusion layer toward the surface of the propagating chain at position 1. The center-of-mass of the propagating chain at position 1 is located at the origin of the coordinate system and its surface is at a distance R_g from its center-of-mass. After diffusing to the surface of the propagating chain at position 1, the two propagating chains react to form one or two terminated polymer chains depending on the mechanism of termination [79–82, 95, 96]. As the reaction of the reactive species of the propagating chain in position 1 is "diffusion limited," its concentration at distance R_g from its center-of-mass is denoted by $[RP]_{dl}$. The general mass conservation equation of the propagating chains for a diffusion-limited reaction, applied to a differential

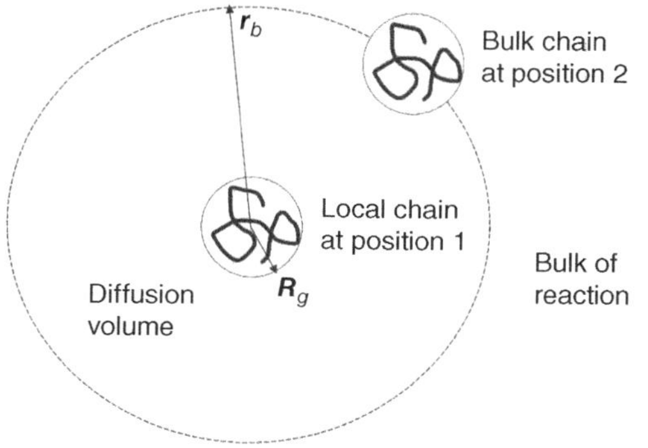

Figure 8.2 Schematic diagram of propagating chains 1 and 2 in the diffusion layer to form terminated polymer chains.

volume in the diffusion layer in the spherical coordinate system (r, θ, ϕ), containing the two propagating chains is as follows [16]:

$$
\left\{\begin{matrix}\text{Rate of change}\\ \text{of molar mass}\\ \text{of propagating}\\ \text{chains with}\\ \text{reaction time}\\ \text{in a differential}\\ \text{volume of the diffusion}\\ \text{layer}\end{matrix}\right\}
+
\left\{\begin{matrix}\text{Net rate of}\\ \text{transfer of molar}\\ \text{mass of propagating}\\ \text{chains by convection}\\ \text{in the differential}\\ \text{volume of the}\\ \text{diffusion layer}\\ \text{in the } r \text{ direction}\end{matrix}\right\}
+
\left\{\begin{matrix}\text{Net rate of}\\ \text{transfer of molar}\\ \text{mass of propagating}\\ \text{chains by convection}\\ \text{in the differential}\\ \text{volume of the}\\ \text{the diffusion layer}\\ \text{in the } \theta \text{ direction}\end{matrix}\right\}
$$

$$
+\left\{\begin{matrix}\text{Net rate of}\\ \text{transfer of molar}\\ \text{mass of propagating}\\ \text{chains by convection}\\ \text{in the differential}\\ \text{volume of the}\\ \text{diffusion layer}\\ \text{in the } \phi \text{ direction}\end{matrix}\right\}
=
\left\{\begin{matrix}\text{Net rate of}\\ \text{transfer of molar}\\ \text{mass of propagating}\\ \text{chains by molecular}\\ \text{diffusion in the}\\ \text{differential volume}\\ \text{of the diffusion layer}\\ \text{in the } r \text{ direction}\end{matrix}\right\}
+
\left\{\begin{matrix}\text{Net rate of}\\ \text{transfer of molar}\\ \text{mass of propagating}\\ \text{chains by molecular}\\ \text{diffusion in the}\\ \text{differential volume}\\ \text{of the diffusion layer}\\ \text{in the } \theta \text{ direction}\end{matrix}\right\}
\tag{8.49}
$$

$$
+\left\{\begin{matrix}\text{Net rate of}\\ \text{transfer of molar}\\ \text{mass of propagating}\\ \text{chains by molecular}\\ \text{diffusion in the}\\ \text{differential volume}\\ \text{of the diffusion layer}\\ \text{in the } \phi \text{ direction}\end{matrix}\right\}
+
\left\{\begin{matrix}\text{Net rate of}\\ \text{production of molar}\\ \text{mass of propagating}\\ \text{chains by chemical}\\ \text{reaction in the}\\ \text{differential}\\ \text{volume of the}\\ \text{diffusion layer}\end{matrix}\right\}
-
\left\{\begin{matrix}\text{Net rate of}\\ \text{consumption of molar}\\ \text{mass of propagating}\\ \text{chains by chemical}\\ \text{reaction in the}\\ \text{differential}\\ \text{volume of the}\\ \text{diffusion layer}\end{matrix}\right\}
$$

The above mass conservation equation can be written in mathematical form as follows:

$$
\begin{aligned}
&\frac{\partial [RP]}{\partial t} + \left(u_r \frac{\partial [RP]}{\partial r} + u_\theta \frac{1}{r}\frac{\partial [RP]}{\partial \theta} + u_\phi \frac{1}{r \sin\theta}\frac{\partial [RP]}{\partial \phi}\right) \\
&\quad = D_{rp}\left(\frac{1}{r^2}\frac{\partial}{\partial r}\left(r^2\frac{\partial [RP]}{\partial r}\right)\right) + \frac{1}{r^2 \sin\theta}\frac{\partial}{\partial \theta}\left(\sin\theta\frac{\partial [RP]}{\partial \theta}\right) \\
&\quad + \frac{1}{r^2 \sin^2\theta}\frac{\partial^2 [RP]}{\partial \phi^2} + R_{rp,prd} - R_{rp,con}
\end{aligned}
\tag{8.50}
$$

In the above conservation equation, $[RP]$ is the concentration of reactive propagating chains, r, θ, and ϕ are the three directions in the spherical coordinate, t is time, D_{rp} is diffusion coefficient of the propagating chains, $R_{rp,prd}$ and $R_{rp,con}$ are the rates of production and consumption of the propagating chains, respectively, in the differential volume of the diffusion layer. The u_r, u_θ, and u_ϕ are the velocity of the reaction mixture in the r, θ, and ϕ directions, respectively. The first term on the left side is the rate of change of $[RP]$ with time, the next three terms on the left side are the net rates of transfer of $[RP]$ by convection in different directions, the first three terms on the right side are the net rates of transfer of $[RP]$ by molecular diffusion in different directions, and the last two terms on the right side are the rates of production and consumption of $[RP]$ in the differential volume of the diffusion layer. If there is no fluid flow in any direction within the reactor (i.e., no convection), the second to fourth terms on the left side go to zero and the above equation is reduced to the following:

$$\frac{\partial [RP]}{\partial t} = D_{rp}\left(\frac{1}{r^2}\frac{\partial}{\partial r}\left(r^2\frac{\partial [RP]}{\partial r}\right)\right) + \frac{1}{r^2 \sin\theta}\frac{\partial}{\partial \theta}\left(\sin\theta\frac{\partial [RP]}{\partial \theta}\right) + \frac{1}{r^2 \sin^2\theta}\frac{\partial^2 [RP]}{\partial \phi^2} + R_{rp,prd} - R_{rp,con} \tag{8.51}$$

If the reaction is well mixed and there is no macroscopic gradient of $[RP]$ in any direction and considering that $[RP]$ is symmetric with respect to θ and ϕ directions, the second and third terms on the right side go to zero and the above equation is reduced to the following:

$$\frac{\partial [RP]}{\partial t} = D_{rp}\left(\frac{1}{r^2}\frac{\partial}{\partial r}\left(r^2\frac{\partial [RP]}{\partial r}\right)\right) + R_{rp,prd} - R_{rp,con} \tag{8.52}$$

Any production of RP happens in the bulk of the reaction, not in the diffusion layer, and the consumption occurs at the interface of the two propagating chains, not in the diffusion layer. Therefore, the terms for production and consumption of RP on the right side go to zero and the above equation is reduced to the following:

$$\frac{\partial [RP]}{\partial t} = D_{rp}\left(\frac{1}{r^2}\frac{\partial}{\partial r}\left(r^2\frac{\partial [RP]}{\partial r}\right)\right) \tag{8.53}$$

The $[RP]$ may change with time in the bulk of the reaction but if the reaction is homogenous (well mixed), there can be no localized accumulation or reduction of $[RP]$ in the diffusion layer; hence, the time-dependent term on the left side goes to zero and the above equation simplifies to the following:

$$D_{rp}\left(\frac{1}{r^2}\frac{\partial}{\partial r}\left(r^2\frac{\partial [RP]}{\partial r}\right)\right) = 0 \tag{8.54}$$

Neither r nor D_{rp} can be zero in the above equation. Therefore, the above equation is reduced to the following:

$$\frac{\partial}{\partial r}\left(r^2\frac{\partial [RP]}{\partial r}\right) = 0 \tag{8.55}$$

Integration of the above equation twice with respect to the radial direction r results in the following equation:

$$[RP] = -\frac{C_1}{r} + C_2 \tag{8.56}$$

where C_1 and C_2 are constants of integration subject to the boundary conditions that $[RP] = [RP]_b$ at $r = r_b$ and $[RP] = [RP]_{loc}$at $r = R_g$ in the diffusion layer. Substitution for C_1 and C_2 using the boundary conditions, results in the following equation for $[RP]$ as a function of radial direction in the diffusion layer:

$$[RP] = -R_g r_b \left(\frac{[RP]_b - [RP]_{loc}}{r_b - R_g} \right) \frac{1}{r} + \frac{[RP]_b\, r_b - [RP]_{loc}\, R_g}{r_b - R_g} \tag{8.57}$$

The first derivative of $[RP]$ with respect to r is given by the following equation:

$$\frac{d\,[RP]}{dr} = R_g r_b \left(\frac{[RP]_b - [RP]_{loc}}{r_b - R_g} \right) \frac{1}{r^2} \tag{8.58}$$

8.4.2 Diffusive Flux of Propagating Chains in Diffusion-limited Polymerization Reactions

The diffusive flux of the propagating chains is defined as the product of the diffusion coefficient times the gradient in concentration as follows:

$$J_{rp}\Big|_{r=R_g} = D_{rp} \left.\frac{d\,[RP]}{dr}\right|_{r=R_g} \tag{8.59A}$$

where $J_{rp}\big|_{r=R_g}$ is the diffusive flux of the propagating chains evaluated at the surface of the propagating chain at position 1. After substitution for $\frac{d\,[RP]}{dr}$ from Eq. (8.58) and its evaluation at $r = R_g$, the flux is given by the following expression:

$$J_{rp}\Big|_{r=R_g} = D_{rp} \left(\frac{[RP]_b - [RP]_{loc}}{r_b - R_g} \right) \frac{r_b}{R_g} \tag{8.59B}$$

The transfer rate of the propagating chains, N_{rg}, due to a gradient between the bulk of the reaction at position 2 and the diffusion limited (local) reaction site at position 1 is the diffusive flux multiplied by the reaction surface area at site 1 as follows:

$$N_{rg} = S\big|_{r=rg}\, J_{rp}\big|_{r=R_g} = \left(4\pi {R_g}^2\right) \frac{D_{rp}}{R_g} \left([RP]_b - [RP]_{loc}\right)\left(\frac{r_b}{r_b - R_g}\right) \tag{8.60}$$

where the surface area for diffusion at site 1 is the area corresponding to the radius of gyration of the propagating chain or $S\big|_{r=rg} = 4\pi {R_g}^2$. The r_b which is the distance between the centers of mass of the two propagating chains is equal to $\overline{sd}_{rp}$ (see Eq. 8.32). For a typical polymerization with initiator concentration of 0.001 M, $r_b = \overline{sd}_{rp} \approx 200$ nm. The radius of gyration of propagating chains $R_g = \frac{b\overline{DP}_{rp}^{\;1/2}}{6}$. For PMMA with a typical $\overline{DP}_{rp} = 1200$ and $b = 0.58$ nm, and $R_g = 3.3$ nm [69]. Hence, $r_b \gg R_g$ and $\frac{r_b}{(r_b - R_g)} \approx 1$, and the above equation simplifies to the following:

$$N_{rg} = S\big|_{r=rg}\, J_{rp}\big|_{r=R_g} = \left(4\pi {R_g}^2\right) \frac{D_{rp}}{R_g} \left([RP]_b - [RP]_{loc}\right) \tag{8.61}$$

8.4.3 Transfer Rate of Reactive Species on Propagating Chains in Diffusion-limited Polymerization Reactions

The second step is the interdiffusion of chain segments of one of the propagating chains in the other propagating chain's volume for the reactive species of one of the propagating chains to collide with the reactive species of the other chain. The collision probability of reactive species on chain-ends of two propagating chains, represented by $p_{rp,seg}$, depends on the ratio of the size of a segment to that of the whole propagating chain [68, 88–94, 97]. The size of the propagating chains, R_g, is defined as the radius of a sphere containing the propagating chain as a freely rotating chain. In previous sections (see Eq. 8.6 in Section 8.2.1), b was defined as the length of a monomer segment on a polymer chain. However, a monomer on a chain is not a freely rotating segment of a polymer chain due to bond angle constraints and steric hinderance of the side chain. Hence, the Kuhn length of a polymer chain b_{kuhn} is defined to represent the minimum segment length for the polymer chain to behave as an equivalent freely rotating chain of a polymer chain [98]. For example, the monomer segment lengths of poly(methyl methacrylate) and atactic polystyrene are $b_{\text{PMMA}} = 0.57$ nm and $b_{PS} = 0.58$ nm, respectively, whereas their Kuhn lengths are $b_{\text{kuhn,PMMA}} = 1.7$ nm and $b_{\text{kuhn},PS} = 1.8$ nm [98]. The probability $p_{rp,seg}$ is defined in terms of the ratio b_{kuhn}/R_g as follows [68, 89]:

$$p_{rp,seg} = \frac{b_{kuhn}^3}{16R_g^3}\left(\frac{b_{kuhn}}{R_g} + \frac{6\sqrt{2}}{\pi}a_{init}\right) \tag{8.62}$$

where a_{init} is a constant related to the chemical structure of the initiator. For example, the constant a_{init} for azobisisobutyronitrile (AIBN) initiator is 0.28 [68, 89]. For PMMA with molecular weight $\overline{DP}_{rp} = 1200$, $R_g = 3.3$ nm [69], $b_{\text{kuhn,PMMA}} = 1.7$ nm, and $a_{init} = 0.28$, $p_{rp,seg}$ is determined as follows:

$$\begin{aligned} p_{rp,seg,\text{PMMA}} &= \frac{1}{16}\left(\frac{1.7\text{ nm}}{3.3\text{ nm}}\right)^3\left(\frac{1.7\text{ nm}}{3.3\text{ nm}} + \frac{6\sqrt{2}}{\pi} \times 0.28\right) \\ &= 8.54 \times 10^{-3}\,(0.52 + 0.76) = 0.011 \end{aligned} \tag{8.63}$$

The transfer rate of the reactive species is the transfer rate of the propagating chains at position 1 (Eq. 8.61) multiplied by the probability of collision between the reactive species on the propagating chains (Eq. 8.62) as follows:

$$\begin{aligned} N_{rs-rp} &= p_{rp,seg}N_{rp} \\ &= \frac{\pi R_g D_{rp}}{4}\left(\frac{b_{kuhn}}{R_g}\right)^3\left(\frac{b_{kuhn}}{R_g} + \frac{6\sqrt{2}}{\pi}a_{init}\right)([RP]_b - [RP]_{loc}) \end{aligned} \tag{8.64}$$

8.4.4 Local Termination Rate of the Reactive Species in Diffusion-limited Polymerization Reactions

The local termination rate of the reactive species at position 1, $R_{t,loc}$, is written as follows:

$$R_{t,loc} = V_{loc,R_g}R_{t,\text{int}} = V_{loc,R_g}k_{t,\text{int}}\,[RP]_b\,[RP]_{loc} \tag{8.65}$$

where V_{loc,R_g} is the local reaction volume corresponding to the volume of the propagating chain at position 1 with radius R_g; $R_{t,int}$ is the intrinsic rate of the termination reaction if the reaction is not limited by chain diffusion and/or segmental diffusion; $k_{t,int}$ is the intrinsic rate constant for the reaction; and $[RP]_b$ and $[RP]_{loc}$ are the bulk (position 2) and local (position 1) concentrations of the reactive propagating chains. Substitution for $V_{loc,R_g} = 4\pi R_g^3/3$ results in the following expression for $R_{t,loc}$:

$$R_{t,loc} = V_{loc,R_g}R_{t,\text{int}} = (4\pi R_g^3/3)\,k_{t,\text{int}}\,[RP]_b\,[RP]_{loc} \tag{8.66}$$

8.4.5 Local Concentration of Propagating Chains in Diffusion-limited Polymerization Reactions

The local concentration of propagating chains $[RP]_{loc}$ in Eq. (8.64) for transfer rate of the reactive species is not known. At position 1, the reactive species on the propagating chains are conserved, that is their concentration change with time is equal to their diffusion minus their consumption by termination as follows:

$$\left\{\begin{array}{c}\text{Rate of change}\\ \text{of molar mass}\\ \text{of the reactive}\\ \text{species on the}\\ \text{propagating}\\ \text{chains with time}\\ \text{in the local}\\ \text{reaction volume}\end{array}\right\} = \left\{\begin{array}{c}\text{Rate of}\\ \text{transfer of}\\ \text{molar mass}\\ \text{of the reactive}\\ \text{species on the}\\ \text{propgating chains}\\ \text{to the local}\\ \text{reaction volume}\end{array}\right\} - \left\{\begin{array}{c}\text{Rate of}\\ \text{disappearance of}\\ \text{the reactive species}\\ \text{on the propagating}\\ \text{chains by}\\ \text{termination}\\ \text{in the local}\\ \text{reaction volume}\end{array}\right\} \tag{8.67}$$

The above mass conservation equation for the reactive species in the local reaction volume corresponding to R_g of the propagating chain 1 can be written in mathematical form as follows:

$$\frac{\partial\left([RP]_{loc}\, V_{loc,R_g}\right)}{\partial t} = N_{rs-rp} - R_{t,loc} \tag{8.68}$$

As the reaction is diffusion limited, the reactive species react instantaneously following segmental diffusion which implies that there is no accumulation of the reactive species in the local reaction volume or $\partial\left([RP]_{loc}\, V_{loc,R_g}\right)/\partial t = 0$. After substitution for N_{rs-rp} and $R_{t,loc}$ from Eqs. (8.64) and (8.66), respectively, the above conservation equation is written as follows:

$$\begin{aligned} &N_{rs-rp} = R_{t,loc} \;\Rightarrow p_{rp,seg} N_{rp} = V_{loc,R_g} R_{t,\text{int}} \\ &\Rightarrow 4\pi\, R_g\, D_{rp}\, p_{rp,seg}\left([RP]_b - [RP]_{loc}\right) = \left(4\pi\, R_g^3/3\right) k_{t,\text{int}}\, [RP]_b\, [RP]_{loc} \end{aligned} \tag{8.69}$$

After rearrangement of the above equation, the unknown concentration $[RP]_{loc}$ is given in terms of bulk concentration $[RP]_b$, intrinsic rate constant $k_{t,int}$, diffusion coefficient D_{rp}, and size of the propagating chains R_g as follows:

$$[RP]_{loc} = \frac{[RP]_b}{1 + \left(\dfrac{k_{t,\text{int}} R_g^2 [RP]_b}{3 p_{rp,seg} D_{rp}}\right)} = \frac{[RP]_b}{1 + \left[\dfrac{16 k_{t,\text{int}} R_g^5 [RP]_b}{3 b_{kuhn}^3 D_{rp}} \Big/ \left(\dfrac{b_{kuhn}}{R_g} + \dfrac{6\sqrt{2}}{\pi} a_{init}\right)\right]} \tag{8.70}$$

where Eq. (8.62) is used to substitute for $p_{rg,seg}$. Based on the above equation, $[RP]_{loc}$ is less than the bulk concentration $[RP]_b$ due to diffusion limitation in the reaction.

8.4.6 Apparent Termination Rate Constant in Diffusion-limited Polymerization Reactions

The rate of termination reaction, R_t, can be expressed in terms of intrinsic and apparent rate constants as follows [99, 100]:

$$R_t = R_{t,loc} = k_{t,\text{int}}\left([RP]_{loc}\, [RP]_b\right) = k_{t,app} [RP]_b^2 \tag{8.71}$$

where R_t is the observed or apparent rate of termination which is expressed in terms of the apparent rate constant, $k_{t,app}$, and the bulk concentration of the propagating chains, $[RP]_b$. The local termination rate, $R_{t,loc}$, is equal to R_t but with the difference that $R_{t,loc}$ is expressed in terms of $k_{t,int}$, $[RP]_{loc}$ and $[RP]_b$. After substitution for $[RP]_{loc}$ from Eq. (8.70) in the above equation and rearrangement, the relation between $k_{t,app}$ and $k_{t,\text{int}}$ is as follows:

$$\frac{1}{k_{t,app}} = \frac{1}{k_{t,\text{int}}} + \frac{R_g^2\,[RP]_b}{3p_{rp,seg}D_{rp}} = \frac{1}{k_{t,\text{int}}} + \left[\frac{16R_g^5\,[RP]_b}{3b_{kuhn}^3 D_{rp}} \Bigg/ \left(\frac{b_{kuhn}}{R_g} + \frac{6\sqrt{2}}{\pi}a_{init}\right)\right] \tag{8.72}$$

If diffusion is fast, then diffusion coefficient is large, $\frac{R_g^2\,[RP]_b}{3p_{rp,seg}D_{rp}} \ll \frac{1}{k_{t,int}}$, hence $k_{t,app} \approx k_{t,\text{int}}$ and termination is reaction controlled. If diffusion is slow, then diffusion coefficient is small, $R_g^2\,[RP]_b\,/3p_{rp,seg}D_{rp} \gg 1/k_{t,\text{int}}$, hence $k_{t,app} \approx 3p_{rp,seg}\,D_{rp}/R_g^2\,[RP]_b$ and termination is diffusion limited or diffusion controlled.

8.4.7 Time Constant for Termination Reaction in Diffusion-limited Polymerization Reactions

It can be shown that Eq. (8.70) can be written in terms of $\frac{k_{t,app}}{k_{t,int}}$ as follows:

$$\frac{k_{t,app}}{k_{t,\text{int}}} = \frac{1}{1 + \frac{R_g^2/3p_{rp,seg}D_{rp}}{1/k_{t,\text{int}}[RP]_b}} \tag{8.73}$$

The terms $R_g^2/3p_{rp,seg}D_{rp}$ and $1/k_{t,\text{int}}\,[RP]_b$ have units of time and represent time constants for diffusion of propagating chains and intrinsic termination reaction of propagating chains as follows:

$$\tau_{t,diff} = \frac{R_g^2}{3p_{rp,seg}D_{rp}} \qquad \tau_{t,rxn} = \frac{1}{k_{t,\text{int}}\,[RP]_b} \tag{8.74}$$

where $\tau_{t,diff}$ and $\tau_{t,rxn}$ are the diffusion and reaction time constants for termination, respectively. After substitution for $R_g^2/3p_{rp,seg}D_{rp}$ and $1/k_{t,\text{int}}\,[RP]_b$ from the above equations, $k_{t,app}/k_{t,\text{int}}$ is expressed in terms of time constants as follows:

$$\frac{k_{t,app}}{k_{t,\text{int}}} = \frac{1}{1 + \frac{\tau_{t,diff}}{\tau_{t,rxn}}} \tag{8.75}$$

Figure 8.3 shows the double logarithmic plot of $k_{t,app}/k_{t,\text{int}}$ versus $\tau_{t,diff}/\tau_{t,rxn}$. If diffusion of propagating chains is fast, $\tau_{t,diff} \ll \tau_{t,rxn}$ and $1 + \tau_{t,diff}/\tau_{t,rxn} \approx 1$, hence $k_{t,app} \approx k_{t,\text{int}}$ and $\log\left(k_{t,app}/k_{t,\text{int}}\right)$ approaches zero. If diffusion is slow, $\tau_{t,diff} \gg \tau_{t,rxn}$ and $1 + \tau_{t,diff}/\tau_{t,rxn} \approx \tau_{t,diff}/\tau_{t,rxn}$, hence $k_{t,app} \approx k_{t,\text{int}}\tau_{t,rxn}/\tau_{t,diff}$ and $\log\left(k_{t,app}/k_{t,\text{int}}\right)$ has increasingly negative values.

In a similar approach, apparent initiation efficiency, f_{app}, and apparent rate constant for propagation, $k_{p,app}$, can be defined for initiation and propagation reactions as follows:

$$\frac{f_{app}}{f_0} = \frac{1}{1 + \frac{\tau_{ci,diff}}{\tau_{ci,rxn}}} \tag{8.76}$$

$$\frac{k_{p,app}}{k_{p,\text{int}}} = \frac{1}{1 + \frac{\tau_{p,diff}}{\tau_{p,rxn}}} \tag{8.77}$$

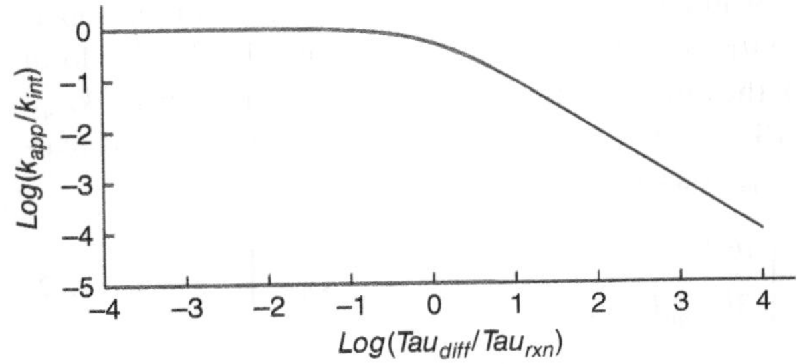

Figure 8.3 Double logarithmic plot of the ratio of apparent to intrinsic rate constants for termination versus the ratio of diffusion to reaction time constants for termination.

where f_0 is the intrinsic reaction-controlled efficiency of chain initiation and $k_{p,\text{int}}$ is the intrinsic rate constant for propagation reaction; $\tau_{ci,diff}$ and $\tau_{ci,rxn}$ are the diffusion and reaction time constants for chain initiation reaction, respectively; $\tau_{p,diff}$ and $\tau_{p,rxn}$ are the diffusion and reaction time constants for chain propagation reaction. Like the time constants for termination reaction, $\tau_{ci,diff}$ and $\tau_{ci,rxn}$ are defined by the following equations:

$$\tau_{ci,diff} = \frac{r_r^2}{3D_r} \quad \tau_{I,rxn} = \frac{1}{k_{ci,\text{int}}\,[R]} \tag{8.78}$$

where r_r and D_r are the radius and diffusion coefficient of the primary radical in the initiation reaction, $k_{ci,int}$ is the intrinsic rate constant for initiation, and $[R]$ is concentration of the primary radical. In the above equations, it is assumed that the radical in the chain initiation reaction between monomer and radical is the limiting reactant. As both reactants (monomer and primary radical) are small molecules, the probability of collision between the reactants by segmental diffusion is equal to unity, that is $p_{r,seg} = 1$ (see Eq. 8.74). Likewise, $\tau_{p,diff}$ and $\tau_{p,rxn}$ are defined as follows:

$$\tau_{p,diff} = \frac{R_g^2}{3D_m} \quad \tau_{p,rxn} = \frac{1}{k_{p,\text{int}}\,[RP]} \tag{8.79}$$

where R_g is the radius of gyration of the propagating chain reacting with monomer, D_m is the diffusion coefficient of the faster diffusing reactant (monomer), $k_{p,int}$ is the intrinsic rate constant for propagation, and $[RP]$ is concentration of the limiting reactant which is the propagating chain. Like $p_{r,seg}$ in $\tau_{ci,diff}$, the probability of collision by segmental diffusion of monomer is $p_{m,seg} = 1$. In most polymerizations, the termination reaction between two propagating chains is most likely limited by diffusion as compared to chain initiation or propagation reactions.

8.5 Quasi Steady State in Diffusion-limited Polymerization Reactions

The quasi-steady state assumption (QSSA), as described in Section 5.6 of Chapter 5, states that the rate of production of reactive species is equal to their rate of consumption in an infinitesimally small interval of time as compared to the reaction time. In other words, QSSA states that the reactive species do not accumulate in the reaction. This quasi-steady state is attributed to the high reactivity and instability of reactive species formed by dissociation of initiator [101] which may not be valid when the reaction is limited by diffusion [102, 103]. The production of reactive species is by unimolecular dissociation of initiator which is less affected by diffusion. Conversely, the disappearance of reactive species after propagation is a bimolecular reaction between two propagating species to terminate the chains by combination or disproportionation. Considering that the diffusivity of propagating chains decreases with conversion, the disappearance of propagating chains is significantly affected by diffusion as compared to their unimolecular production. As a result, the propagating chains start to accumulate in the reaction with increasing conversion, which invalidates QSSA. Here, we reexamine the role of QSSA in diffusion limited polymerization by distinguishing between small-molecule primary reactive species (R) produced by initiator dissociation and chain-like reactive propagating species (RP).

8.5.1 Mass Conservation Equation for Primary Reactive Species in Diffusion-limited Polymerization

The elementary reactions for the production and consumption of primary reactive radical species are as follows (see Section 5.7.1 in Chapter 5):

$$I \xrightarrow{k_d} {}^2R \tag{8.80}$$

$$R + R \xrightarrow{k_c} I^{\bullet} \tag{8.81}$$

$$R + M \xrightarrow{k_{ci}} RP_1 \tag{8.82}$$

where I and R are the initiator and radical reactive species formed by initiator dissociation, respectively, $I^{\bullet}$ is the inactive species generated by recombination of two reactive fragments; M is monomer, and RP_1 is a 1-mer reactive propagating chain. The rate constants k_d, k_c, and k_{ci} are for initiator dissociation, radical recombination, and chain initiation, respectively. It is assumed that initiator dissociation produces two radicals, but numbers other than two are possible. It is also assumed that recombination generates an inactive species [104–111]. The combination of a primary radical with a reactive propagating chain is not included because RP concentration is many orders of magnitude lower than that of monomer in the reaction. The general molar conservation equation for primary radicals in a well-mixed reaction (no gradients) is as follows:

$$\frac{d\left([R]\,V\right)}{dt} = 2k_d\,[I]\,V - k_c\,[R]^2\,V - k_{ci}\,[R]\,[M]\,V \tag{8.83}$$

where $[PR]$, $[I]$, and $[M]$ are concentrations of primary radicals, initiator, and monomer, respectively, t is time, and V is the reaction volume. Expansion of the term on the left side of the above equation and rearrangement results in the following equation:

$$\frac{d\,[R]}{dt} = 2k_d\,[I] - k_c\,[R]^2 - k_{ci}\,[R]\,[M] - \frac{[R]}{V}\frac{dV}{dt} \tag{8.84}$$

The last term on the right side of the above equation is the effect of volume change due to the conversion of monomer to polymer. This term may be significant at high conversions as the volume decreases with increasing conversion (i.e., polymer specific volume is lower than the monomer). For example, in bulk polymerization of MMA, the specific volume decreases from 1.07 mL g^{-1} for pure MMA to 0.85 mL g^{-1} for pure PMMA. The first term on the right is not affected by diffusion for temperatures above the T_g of the polymer because initiator dissociation to reactive species is unimolecular in conventional radical polymerization (i.e., mainly a function of reaction temperature and pressure). The second term on the right (recombination Rxn. 8.81) is generally not limited by diffusion, as the two fragments, following initiator dissociation, have not diffused away. The third term on the right (Rxn. 8.82) which is the reaction between two small molecules (reactive species and monomer) is not limited by diffusion unless the reaction temperature is below T_g of the polymerization mixture. This is attributed to the monomer diffusion coefficient changing only by one to two orders of magnitude with increasing conversion above T_g as compared to the polymer diffusion coefficient which changes by many orders of magnitude [6, 11, 67]. Therefore, QSSA is valid for primary radicals even in the concentrated and reptation regimes for reaction temperatures above the polymer T_g. After applying $d[R]/dt = 0$, the above equation is written as follows:

$$2k_d[I] - k_c[R]^2 - k_{ci}[R][M] - \frac{[R]}{V}\frac{dV}{dt} = 0 \quad \in d[R]/dt = 0 \tag{8.85}$$

Defining apparent initiation efficiency, f_{app}, in the same manner as f_0 was defined in Chapter 5 (see Eq. 5.29E in Section 5.7.1 of Chapter 5) yields the following relation:

$$f_{app} = \frac{R_{ci}}{R_d} = \frac{k_{ci}[R][M]}{2k_d[I]} \Rightarrow (1 - f_{app}) = \frac{k_c[R]^2}{2k_d[I]} \tag{8.86}$$

After substitution of the above relations in Eq. (8.85), $[R]$ is expressed in terms of monomer and initiator concentrations and the rate of volume change by the following equation:

$$[R] = \frac{2f_{app}k_d[I]}{k_{ci}[M] + (1/V)\,dV/dt} \tag{8.87}$$

where f_{app} is related to f_0 in the dilute regime in the absence of diffusion limitation by Eq. (8.76) subject to the condition that $f_{app} \le f_0$. The above relation implies that initiation efficiency decreases with decreasing diffusivity of the reactants at high conversions, which is compensated by increasing the rate of radical recombination to maintain QSSA. This decrease in efficiency in turn reduces the quasi-steady state concentration of primary radicals.

8.6 Modeling Diffusion-limited Radical Polymerization with Termination by Disproportionation and Combination and Chain Transfer to Monomer

In this diffusion-limited reaction, it is reasonable to assume that QSSA applies to small-molecule primary radicals but not to the longer n-mer reactive propagating chains. The elementary reactions for this polymerization system are as follows:

$$I \xrightarrow{k_d} {}^2R \tag{8.88}$$

$$R + R \xrightarrow{k_c} I^{\bullet} \tag{8.89}$$

$$R + M \xrightarrow{k_{ci}} RP_1 \tag{8.90}$$

$$RP_n + M \xrightarrow{k_p} RP_{n+1} \tag{8.91}$$

$$RP_n + M \xrightarrow{k_{ctm}} TP_n + RP_1 \tag{8.92}$$

$$RP_n + RP_m \xrightarrow{k_{td}} TP_n + TP_m + TDB \tag{8.93}$$

$$RP_n + RP_m \xrightarrow{k_{tc}} TP_{n+m} \tag{8.94}$$

The first three reactions are initiator dissociation, recombination, and chain initiation; the next two reactions are chain propagation and transfer to monomer; and the last two are termination by disproportionation and combination. In the above reactions, TP_n is an n-mer terminated polymer chain, TDB is a terminal double bond in a polymer chain, and k_{ctm} and k_{tc} stand for rate constants for chain transfer to monomer and termination by combination, respectively.

8.6.1 Time dependence of Reaction Volume in Diffusion-limited Polymerization

The rate of change of the reaction volume with time, dV/dt, can be expressed as the product of the rate of change of moles of reacted monomer with time multiplied by the difference in specific molar volumes of the reacted (monomer residue in polymer) and unreacted monomer as follows:

$$\frac{dV}{dt} = \left[-\frac{d\left([M]\,V\right)}{dt}\right]\left(\hat{\upsilon}_{m,r} - \hat{\upsilon}_{m,ur}\right) \tag{8.95}$$

where $-d\left([M]\,V\right)/dt$ is the rate of change of moles of the reacted monomer, $\hat{\upsilon}_{m,r}$ is the partial specific molar volume of the reacted monomer (molar volume of the monomer residue in the polymer) and $\hat{\upsilon}_{m,ur}$ is the monomer partial molar volume. $\hat{\upsilon}_{M,R}$ and $\hat{\upsilon}_{M,R}$ are inversely equal to the partial molar density of the reacted monomer, $\hat{\rho}_{m,r}$, and partial molar density of unreacted monomer, $\hat{\rho}_{m,ur}$, respectively. Assuming the volumes are additive, i.e., no change in volume upon mixing, $\hat{\rho}_{m,r}$ and $\hat{\rho}_{m,ur}$ can be replaced by molar density of the reacted, $\rho_{m,r}$, and unreacted, $\rho_{m,ur}$, monomer, respectively. The assumption of additivity of volumes is a good assumption if polymerization is carried out in bulk or in a theta solvent [112–114]. As $\rho_{m,ur}$ is just the monomer molar density (ρ_m) and $\rho_{m,r}$ is the molar density of the monomer repeat unit in the polymer chains $(\rho_{ru,p})$, the term $\hat{\upsilon}_{M,R} - \hat{\upsilon}_{M,UR}$ in the above equation can be written as follows:

$$\hat{\upsilon}_{m,r} - \hat{\upsilon}_{m,ur} = \frac{1}{\hat{\rho}_{m,r}} - \frac{1}{\hat{\rho}_{m,ur}} \simeq \frac{1}{\rho_{ru,p}} - \frac{1}{\rho_m} = -\left(\frac{\rho_{ru,p} - \rho_m}{\rho_{ru,p}}\right)\frac{1}{\rho_m} = -\frac{\varepsilon}{\rho_m} \tag{8.96}$$

where ε is the coefficient of shrinkage in polymerization, defined as the difference in molar densities of the polymer repeat unit and monomer divided by the molar density of polymer repeat unit [115, 116]. Substitution of the above relation in Eq. (8.95) results in the following equation:

$$\frac{dV}{dt} = \frac{d\left([M]\,V\right)}{dt}\frac{\varepsilon}{\rho_m} \tag{8.97}$$

Expansion of the differential term on the right using the product rule and rearrangement results in the following equation:

$$\frac{1}{V}\frac{dV}{dt} = \frac{1}{(\rho_m/\varepsilon) - [M]}\frac{d[M]}{dt} \tag{8.98}$$

The reaction conversion is defined as $\alpha = \left([M]_0\, V_0 - [M]\, V\right) / [M]_0\, V_0$ with $[M]_0$ and V_0 as monomer concentration and reaction volume at $t = 0$, respectively. Upon substitution for $[M]$ in terms of α and integration of the above equation from $[M] = [M]_0$ at $\alpha = 0$ and $V = V_0$ to any conversion α, the reaction volume, monomer concentration, and their product as a function of α are given by the following equations:

$$V = V_0 \left(1 - \frac{\varepsilon\,[M]_0}{\rho_m}\alpha\right) \tag{8.99}$$

$$[M] = (1-\alpha) \Big/ \left(\frac{1}{[M]_0} - \frac{\varepsilon}{\rho_m}\alpha\right) \tag{8.100}$$

$$[M]\,V = [M]_0\,V_0\,(1-\alpha) \tag{8.101}$$

8.6.2 Time Dependence of Initiator Concentration in Diffusion-limited Polymerization

The initiator is consumed by the elementary Rxn. 8.88 and there is no initiator production. Therefore, the molar mass conservation equation for initiator is as follows:

$$\frac{d\,([I]\,V)}{dt} = -k_d\,[I]\,V \tag{8.102}$$

As discussed in the previous section, the unimolecular dissociation reaction is not diffusion limited for temperatures above the polymer T_g. However, the diffusion of the produced reactive species away from the dissociation cage and their reaction with monomers to initiate propagating chains may be diffusion limited which is included in the apparent initiation efficiency f_{app} (see Eq. 8.76). Expansion of the differential term on the left using the product rule and substitution for V and $\frac{dV}{dt}$ from Eq. (8.99) followed by rearrangement results in the following expression:

$$\frac{1}{[I]}\frac{d\,[I]}{dt} = -k_d + \frac{1}{\left[\left(\rho_m/\varepsilon\,[M]_0\right) - \alpha\right]}\frac{d\alpha}{dt} \tag{8.103}$$

This equation may be solved numerically for initiator concentration as a function of time and conversion subject to the initial condition $[I] = [I]_0$ and $\alpha = 0$ at $t = 0$.

8.6.3 Time Dependence of Monomer Concentration in Diffusion-limited Polymerization

Monomer is consumed in Rxns. 8.90–8.92 and there is no monomer production. Therefore, the molar mass conservation equation for monomer is as follows:

$$\frac{d\,([M]\,V)}{dt} = -k_{ci,app}\,[M]\,V\,[R] - k_{p,app}\,[M]\,V \sum_{j=1}^{\infty}[RP_n] - k_{ctm,app}\,[M]\,V \sum_{j=1}^{\infty}[RP_n] \tag{8.104}$$

The term on the left side is the rate of change of moles of monomer with reaction time. The first, second, and third terms on the right side are the consumption of monomer by chain initiation, chain propagation, and chain transfer to monomer, respectively. The parameters $k_{ci,app}$, $k_{p,app}$, and $k_{ctm,app}$ are the apparent rate constants for initiation, propagation, and chain transfer to monomer, respectively, which are affected by diffusion. As a result, the rate constants may be a strong function of polymer concentration and medium viscosity. Based on Eq. (3.16) in Chapter 3, the zero-order molar concentration moment of the propagating chains is equal to $Q_{0,rp} = \lambda_0\,[RP]$. As the zero-order mole-fraction moment, by definition, is $\lambda_0 = 1$, $Q_{0,rp} = \sum_{j=1}^{\infty}[RP_n] = [RP]$. After substitution for $[RP] = Q_{0,rp}$, for $[M]\,V$ in terms of α from Eq. 8.101, and rearrangement, the above equation simplifies to the following:

$$\frac{1}{(1-\alpha)}\frac{d\alpha}{dt} = -k_{ci,app}\,[R] - \left(k_{p,app} + k_{ctm,app}\right) Q_{0,rp} \tag{8.105}$$

The terms $(1-\alpha)$ and $d\alpha/dt$ can be expressed in terms of $[M]$ using Eq. (8.100). After substitution for α in terms of $[M]$, the conservation equation for monomer is written as follows:

$$\frac{1}{(1-\varepsilon[M]/\rho_m)[M]}\frac{d[M]}{dt} = -k_{ci,app}[R] - \left(k_{p,app} + k_{ctm,app}\right) Q_{0,rp} \tag{8.106}$$

In the above equation, ε/ρ_m is related to molar densities of monomer repeat unit in the polymer and pure monomer (see Eq. 8.96) which are known for a given polymerization. The equation for the determination of $Q_{0,rp}$ will be derived in the next section. Based on defining equations for apparent rate constants (see Eqs. 8.75–8.77), $k_{p,app}$ and $k_{ctm,app}$ are given by the following equations:

$$\frac{1}{k_{p,app}} = \frac{1}{k_{p,\text{int}}}\left(1+\frac{\tau_{p,diff}}{\tau_{p,rxn}}\right) = \frac{1}{k_{p,0}}\left(1+\frac{R_g^2\, k_{p,0}\, Q_{0,rp}}{3D_m}\right) \tag{8.107}$$

$$\frac{1}{k_{ctm,app}} = \frac{1}{k_{ctm,\text{int}}}\left(1+\frac{\tau_{ctm,diff}}{\tau_{ctm,rxn}}\right) = \frac{1}{k_{ctm,0}}\left(1+\frac{R_g^2\, k_{ctm,0}\, Q_{0,rp}}{3D_m}\right) \tag{8.108}$$

where $k_{p,\text{int}}$ is the intrinsic rate constant for propagation which is equal to the propagation rate constant as polymer concentration approaches zero or $k_{p,0}$. Likewise, $k_{ctm,\text{int}}$ is the intrinsic rate constant for chain transfer to monomer which is equal to the transfer rate constant as polymer concentration approaches zero or $k_{ctm,0}$. The time constants $\tau_{p,diff}$ and $\tau_{ctm,diff}$ are the diffusion limited time constants for propagation and chain transfer to monomer, respectively; $\tau_{p,rxn}$ and $\tau_{ctm,rxn}$ are the reaction limited time constants for propagation and chain transfer to monomer; R_g is the radius of gyration of the propagating chains and D_m is the diffusion coefficient of the faster diffusing reactant which is the monomer in propagation and chain transfer to monomer. Based on the defining Eq. (8.7), the monomer diffusion coefficient is given by the following equation:

$$D_m = \frac{k_B T}{\pi\, b_m\, \eta_s} \tag{8.109}$$

where b_m is the monomer size and η_s is the viscosity of the polymer solution. The η_s is estimated using a predictive model like Lyons–Tobolsky (Eq. 8.8), extended Martin (Eq. 8.9), or Vrentas–Duda (Eq. 8.10) as discussed in Section 8.2. After solving Eq. 8.106 numerically subject to the boundary condition $[M]=[M]_0$ at $t=0$ and finding monomer concentration as a function of time, the conversion α, volume V, and moles of unreacted monomer $[M]\,V$ are determined as a function of time using Eqs. (8.99–8.101), respectively.

8.6.4 Conservation Equation for *k*th Order Molar Concentration Moments of Propagating Chains in Diffusion-limited Polymerization

The conservation equation for the kth order molar concentration moment of the propagating chains, in general form, is written as follows:

$$\frac{d\left(Q_{k,rp}V\right)}{dt} = \sum_i \lambda'_{k,i} R_{rp,prd,i} V - \sum_j \lambda'_{k,j} R_{rp,con,j} V \tag{8.110}$$

The term on the left in the above equation is the rate of change in the number of moles of the kth order moment of the propagating chains with time; first and second terms on the right are the rates of production and consumption of the kth order moment of the propagating chains by elementary reactions i and j, respectively, summed overall all production and consumption reactions. In the above equation, $Q_{k,rp}$ is the kth order molar concentration moment of the propagating chains, $\lambda'_{k,i}$

is the instantaneous kth order mole-fraction moment of the propagating chains produced by the elementary reaction i, $R_{rp,prd,i}$ is the rate of production of propagating chains by the elementary reaction i, $\lambda'_{k,j}$ is the instantaneous kth order mole-fraction moment of the propagating chains consumed by the elementary reaction j, and $R_{rp,con,j}$ is the rate of consumption of propagating chains by the elementary reaction j. For the polymerization system described by Rxns. 8.88–8.94, the chain initiation Rxn. 8.90 produces a 1-mer propagating chain; the propagating Rxn. 8.91 consumes and produces an n-mer and an (n+1)-mer propagating chains, respectively; chain transfer Rxn. 8.92 consumes and produces an n-mer and 1-mer propagating chains; and termination Rxns. 8.93 and 8.94 each consume two n-mer and m-mer propagating chains. Therefore, the rate of production of propagating chains (first term on the right in the above equation) by chain initiation, propagation, and chain transfer to monomer is expressed as follows:

$$\sum_i \lambda'_{k,i} R_{rp,prd,i} V = k_{ci,app}\,[M]\,V\,[R]\,\lambda'_{k,ci} + k_{p,app}\,[M]\,V\,[RP]\,\lambda'_{k,p,prd} + k_{ctm,app}\,[M]\,V\,[RP]\,\lambda'_{k,ctm,prd} \tag{8.111}$$

where $\lambda'_{k,ci} = \lambda'_{k,ctm,prd} = 1$ because the produced RP_1 in chain initiation and transfer to monomer reactions have only one monomer. The $\lambda'_{k,p,prd}$ and $\lambda'_{k,ctm,prd}$ are the instantaneous kth order mole-fraction moments of the chains produced by propagation and chain transfer to monomer, respectively. The rates of consumption of propagating chains (second term on the right in Eq. 8.110) by propagation, chain transfer to monomer, and termination reactions are expressed as follows:

$$\sum_j \lambda'_{k,j} R_{rp,con,j} V = +\,k_{p,app}\,[M]\,V\,[RP]\,\lambda'_{k,p,con} + k_{ctm,app}\,[M]\,V\,[RP]\,\lambda'_{k,ctm,con} + k_{td,app}[RP]^2 V\,\lambda'_{k,td} + k_{tc,app}[RP]^2 V\,\lambda'_{k,tc} \tag{8.112}$$

Substitution of Eqs. (8.111) and (8.112) in Eq. (8.110), applying the equality $Q_{0,rp} = \lambda_0\,[RP] = [RP]$, using the defining equation for apparent initiation efficiency f_{app} (see Eq. 8.86) by assuming QSSA applies to the primary radicals and expanding the differential term on the left side using the product rule and rearrangement of the terms, results in the following expression for conservation of the kth order molar concentration moment of the propagating chains:

$$\frac{dQ_{k,rp}}{dt} = 2f_{app}k_d\,[I] + k_{p,app}\,[M]\,Q_{0,rp}\left(\lambda'_{k,p,prd} - \lambda'_{k,p,con}\right) + k_{ctm,app}\,[M]\,Q_{0,rp}\left(\lambda'_{k,ctm,prd} - \lambda'_{k,ctm,con}\right) - Q_{0,rp}^2\left(k_{td,app}\lambda'_{k,td} + k_{tc,app}\lambda'_{k,tc}\right) - \frac{Q_{k,rp}}{V}\frac{dV}{dt} \tag{8.113}$$

In the above equation, f_{app} is the apparent efficiency of chain initiation, and $k_{p,app}$, $k_{ctm,app}$, $k_{td,app}$, and $k_{tc,app}$ represent apparent rate constants for chain propagation, transfer to monomer, termination by disproportionation, and termination by combination, respectively. The apparent rate constants are used in the conservation equation to account for diffusion limitation in the rates of elementary reactions. It is worth noting that the rate constant for unimolecular dissociation of initiator, k_d, is not diffusion limited whereas the rate constant for chain initiation, k_{ci}, which is a bimolecular reaction may be diffusion limited. The diffusion limitation in chain initiation reaction is accounted for through the apparent efficiency of chain initiation f_{app}. The apparent rate constants $k_{p,app}$ and $k_{ctm,app}$ were defined in Eqs. (8.107 and 8.108), respectively, in Section 8.6.3.

The apparent efficiency of chain initiation f_{app} and apparent rate constants $k_{td,app}$ and $k_{tc,app}$, based on the defining Eqs. (8.75–8.77), are given by the following equations:

$$\frac{1}{f_{app}} = \frac{1}{f_0}\left(1 + \frac{\tau_{ci,diff}}{\tau_{ci,rxn}}\right) = \frac{1}{f_0}\left(1 + \frac{b_r{}^2 k_{ci,0}\,[R]}{12D_r}\right) \tag{8.114}$$

$$\frac{1}{k_{td,app}} = \frac{1}{k_{td,\text{int}}}\left(1 + \frac{\tau_{td,diff}}{\tau_{td,rxn}}\right) = \frac{1}{k_{td,0}}\left(1 + \frac{R_g^2 k_{td,0} QR_0}{3p_{rp,seg}D_{rp}}\right) \tag{8.115}$$

$$\frac{1}{k_{tc,app}} = \frac{1}{k_{tc,\text{int}}}\left(1 + \frac{\tau_{tc,diff}}{\tau_{tc,rxn}}\right) = \frac{1}{k_{tc,0}}\left(1 + \frac{R_g^2 k_{tc,0} QR_0}{3p_{rp,seg}D_{rp}}\right) \tag{8.116}$$

In the above equations, b_r and D_r are the average size and diffusion coefficient of the primary radicals produced by initiator dissociation, respectively, and D_{rp} is the diffusion coefficient of the propagating chains. The diffusion coefficients D_r and D_{rp} are estimated using the Stokes–Einstein equation [13] (see Eq. 8.109) and a model for predicting viscosity of a polymer solution as a function of molecular weight and concentration (see Eqs. 8.8–8.10 and Eq. 8.25 in Section 8.2).

8.6.4.1 Concentration of Propagating Chains in Diffusion-limited Polymerization

The concentration of propagating chains, $Q_{0,rp}$, is determined from the conservation equation for zero-order concentration moment of the propagating chains. Based on the general kth order conservation equation (see Eq. 8.113), the zero-order moment for the propagating chains is written as follows:

$$\begin{aligned}\frac{dQ_{0,rp}}{dt} &= 2f_{app}k_d\,[I] + k_{p,app}\,[M]\,Q_{0,rp}\left(\lambda'_{0,p,prd} - \lambda'_{0,p,con}\right)\\ &\quad + k_{ctm,app}\,[M]\,Q_{0,rp}\left(\lambda'_{0,ctm,prd} - \lambda'_{0,ctm,con}\right)\\ &\quad - Q_{0,rp}^2\left(k_{td,app}\lambda'_{0,td} + k_{tc,app}\lambda'_{0,tc}\right) - \frac{Q_{0,rp}}{V}\frac{dV}{dt}\end{aligned} \tag{8.117}$$

The zero-order mole-fraction moment of chains, by definition, is zero (see Eq. 3.14 in Chapter 3), that is $\lambda'_0 = 1$. Therefore, the above differential equation simplifies to the following expression:

$$\frac{dQ_{0,rp}}{dt} = 2f_{app}k_d\,[I] - Q_{0,rp}^2\left(k_{td,app} + k_{tc,app}\right) - \frac{Q_{0,rp}}{V}\frac{dV}{dt} \tag{8.118}$$

Substitution for $(1/V)\,dV/dt$ from Eq. (8.99) and dividing both sides by $1/Q_{0,rp}$ yields the following differential equation relating the rate of change in QR_0 to the rate of change in conversion:

$$\frac{1}{Q_{0,rp}}\frac{dQ_{0,rp}}{dt} = \frac{2f_{app}k_d\,[I]}{Q_{0,rp}} - Q_{0,rp}\left(k_{td,app} + k_{tc,app}\right) + \frac{1}{\left[\left(\rho_m/\varepsilon\,[M]_0\right) - \alpha\right]}\frac{d\alpha}{dt} \tag{8.119}$$

The above equation and Eq. (8.103) for $[I]$, Eq. (8.100) for α, Eq. (8.106) for $[M]$, Eqs. (8.114–8.116) for f_{app}, $k_{td,app}$, and $k_{tc,app}$, respectively, are solved simultaneously to find $Q_{0,rp} = [RP]$ as a function of reaction time or monomer conversion.

8.6.4.2 Kinetic Chain Length of Propagating Chains in Diffusion-limited Polymerization

The kinetic chain length is determined from the first order concentration moment of the propagating chains. Based on the general conservation Eq. (8.113), the first order moment of the propagating chains is written as follows:

$$\begin{aligned}\frac{dQ_{1,rp}}{dt} &= 2f_{app}k_d\,[I] + k_{p,app}\,[M]\,Q_{0,rp}\left(\lambda'_{1,p,prd} - \lambda'_{1,p,con}\right) \\ &+ k_{ctm,app}\,[M]\,Q_{0,rp}\left(\lambda'_{1,ctm,prd} - \lambda'_{1,ctm,con}\right) \\ &- Q_{0,rp}^2\left(k_{td,app}\lambda'_{1,td} + k_{tc,app}\lambda'_{1,tc}\right) - \frac{Q_{1,rp}}{V}\frac{dV}{dt}\end{aligned} \tag{8.120}$$

The instantaneous moments of the propagating chains $\lambda'_{1,p,con}$ and $\lambda'_{1,p,prd}$ before and after propagation, respectively, are equal to λ_1 and $\lambda_1 + 1$ (see Section 6.2 in Chapter 6). As the reactive species in chain transfer to monomer is transferred from an n-mer propagating chain to a 1-mer monomer, $\lambda'_{1,ctm,con} = \lambda_1$ and $\lambda'_{1,ctm,prd} = 1$ (see Section 6.4.1 in Chapter 6). The $\lambda'_{1,td}$ and $\lambda'_{1,tc}$ which are the instantaneous first order mole-fraction moments of the propagating chains that undergo termination by disproportionation and combination, respectively, are both equal to λ_1. Upon substitution for the instantaneous moments, for $\left(\frac{1}{V}\right)\frac{dV}{dt}$ from Eq. (8.99), use of the defining equation for $\lambda_1 = \frac{Q_{1,rp}}{Q_{0,rp}}$ (see Section 3.2 of Chapter 3), and dividing both sides by $1/Q_{1,rp}$, the above differential equation simplifies to the following:

$$\begin{aligned}\frac{1}{Q_{1,rp}}\frac{dQ_{1,rp}}{dt} &= +\frac{2f_{app}k_d\,[I]}{Q_{1,rp}} + k_{p,app}\,[M]\frac{Q_{0,rp}}{Q_{1,rp}} + k_{ctm,app}\,[M]\left(\frac{Q_{0,rp}}{Q_{1,rp}} - 1\right) \\ &- Q_{0,rp}\left(k_{td,app} + k_{tc,app}\right) - \frac{1}{\left[\left(\rho_m/\varepsilon\,[M]_0\right) - \alpha\right]}\frac{d\alpha}{dt}\end{aligned} \tag{8.121}$$

where λ_1 is the first order mole-fraction moment of the propagating chains at time t corresponding to conversion α. After solving the differential Eqs. (8.119) and (8.121) for $Q_{0,rp}$ and $Q_{1,rp}$, the kinetic chain length L of the propagating chains is determined by $L = \overline{DP}_{n,rp} = Q_{1,rp}/Q_{0,rp}$.

8.6.5 Conservation Equation for *k*th Order Molar Concentration Moments of Terminated Chains in Diffusion-limited Polymerization

The conservation equation for kth order molar concentration moment of the terminated polymer chains, $Q_{k,tp}$, in general form is written as follows:

$$\frac{d\left(Q_{k,tp}V\right)}{dt} = \sum_i \mu'_{k,i}R_{tp,prd,i}V - \sum_j \mu'_{k,j}R_{tp,con,j}V \tag{8.122}$$

The term on the left in the above equation is the rate of change in kth order molar moment of the terminated polymer chains with time. The first and second terms on the right are the rates of production and consumption of the kth order molar moments of the terminated polymer chains by elementary reactions i and j summed overall all production and consumption reactions, respectively. In the above conservation equation, $\mu'_{k,i}$ and $\mu'_{k,j}$ are the instantaneous kth order mole-fraction moments of the terminated polymer chains produced and consumed by the elementary reaction i and j, respectively; and $R_{tp,prd,i}$ and $R_{tp,con,i}$ are the rate of production and consumption of terminated polymer chains by the elementary reaction i and j. As there is no consumption of terminated polymer chains in Rxns. 8.88–8.94, the second term on the right, that is $\sum_j \mu'_{k,j}R_{tp,con,j}V = 0$. In chain transfer to monomer reaction, a terminated polymer chain is produced with rate given by $R_{tp,prd,ctm} = k_{ctm,app}\,[M]\,[RP]$ and $\mu'_{k,ctm}$ equal to the kth order mole-fraction moment of the propagating chains at time t which is given by $\lambda_k = Q_{k,rp}/Q_{0,rp}$(see Section 6.4.1. in Chapter 6). In termination by disproportionation, two terminated polymer chains are produced by disproportionation of two propagating chains with rate given by $R_{tp,prd,td} = 2k_{td,app}\,[RP]^2$ (the factor two is because two terminated polymer chains are produced by disproportionation) and $\mu'_{k,td}$ equal to the kth order mole fraction moment of the propagating chains at time t given by $\mu'_{k,td} = \lambda_k = Q_{k,rp}/Q_{0,rp}$ (see Section 6.3.1 in Chapter 6). In termination by combination, a terminated polymer chain is produced by combination of two propagating chains with rate given by $R_{tp,prd,tc} = k_{tc,app}\,[RP]^2$ and with instantaneous kth order mole-fraction moment $\mu'_{k,prd,tc}$. Substitution of the above relations in the conservation Eq. (8.122), applying the equality $Q_{0,rp} = \lambda_0\,[RP] = [RP]$, expansion of the differential term on the left using the product rule, and substitution for $(1/V)\,dV/dt$ from Eq. (8.99) followed by rearrangement of the terms, yields the following differential equation:

$$\begin{aligned}\frac{1}{Q_{k,tp}}\frac{dQ_{k,tp}}{dt} &= k_{ctm,app}\,[M]\,\frac{Q_{0,rp}}{Q_{k,tp}}\lambda_k + 2k_{td,app}\frac{Q_{0,rp}^2}{Q_{k,tp}}\lambda_k \\ &+ k_{tc,app}\frac{Q_{0,rp}^2}{Q_{k,tp}}\mu'_{k,prd,tc} - \frac{1}{\left[\left(\rho_m/\varepsilon\,[M]_0\right)-\alpha\right]}\frac{d\alpha}{dt}\end{aligned} \tag{8.123}$$

where $k_{ctm,app}$, $k_{td,app}$ and $k_{tc,app}$ in the above equation are related to the intrinsic rate constants $k_{ctm,0}$, $k_{td,0}$, and $k_{tc,0}$ by Eqs. (8.108), (8.115), and (8.116), respectively.

8.6.5.1 Concentration of Terminated Chains in Diffusion-limited Polymerization

Concentration of terminated chains is equal to $Q_{0,tp}$ or the zero-order concentration moment of the polymer chains. All zero-order mole-fraction moments are equal to one, that is $\lambda_0 = \mu'_{0,prd,tc} = 1$ (see Eq. 3.14 in Chapter 3). Based on the above kth order conservation equation, the differential equation for $[TP] = Q_{0,tp}$ is given as follows:

$$\begin{aligned}\frac{1}{Q_{0,tp}}\frac{dQ_{0,tp}}{dt} &= k_{ctm,app}\,[M]\,\frac{Q_{0,rp}}{Q_{0,tp}} + \left(2k_{td,app} + k_{tc,app}\right)\frac{Q_{0,rp}^2}{Q_{0,tp}} \\ &- \frac{1}{\left[\left(\rho_m/\varepsilon\,[M]_0\right)-\alpha\right]}\frac{d\alpha}{dt}\end{aligned} \tag{8.124}$$

The above differential equation is solved simultaneously with those for $[I]$ (Eq. 8.103), α (Eq. 8.100), $[M]$ (Eq. 8.106), $Q_{0,rp}$ (Eq. 8.119) and Eqs. (8.114–8.116) for f_{app}, $k_{td,app}$, and $k_{tc,app}$ to determine $Q_{0,tp} = [TP]$ as a function of conversion.

8.6.5.2 Number-average Degree of Polymerization of Terminated Chains in Diffusion-limited Polymerization

Number-average degree of polymerization is equal to the ratio of the first order $(Q_{1,tp})$ to zero-order $(Q_{0,tp})$ concentration moments of the terminated chains. In the general conservation Eq. (8.123), the first order mole-fraction moment of the propagating chains is $\lambda_1 = Q_{1,rp}/Q_{0,rp}$, and the instantaneous first order mole-fraction moment of the produced polymer chains by combination is $\mu'_{k,prd,tc} = 2\lambda_1 = 2\frac{Q_{1,rp}}{Q_{0,rp}}$ (see Section 6.3.2 of Chapter 6). After substitution for λ_1 and $\mu'_{2,prd,tc}$

in the kth order conservation Eq. (8.123), the differential equation for determination of $Q_{1,tp}$ is as follows:

$$\frac{1}{Q_{1,tp}}\frac{dQ_{1,tp}}{dt} = k_{ctm,app}\,[M]\,\frac{Q_{1,rp}}{Q_{1,tp}} + 2\left(k_{td,app} + k_{tc,app}\right)\frac{Q_{0,rp}\,Q_{1,rp}}{Q_{1,tp}} - \frac{1}{\left[\left(\rho_m/\varepsilon\,[M]_0\right) - \alpha\right]}\frac{d\alpha}{dt} \quad (8.125)$$

After solving the above differential equation simultaneously with that of $Q_{0,tp}$ (Eq. 8.124), the number-average degree of polymerization of the polymer chains is given by $\overline{DP}_{n,tp} = Q_{1,tp}/Q_{0,tp}$.

8.6.5.3 Weight-average Degree of Polymerization of Terminated Chains in Diffusion-limited Polymerization

Weight-average degree of polymerization is equal to the ratio of the second order $(Q_{2,tp})$ to the first order $(Q_{1,tp})$ concentration moments of the terminated chains. In the general conservation Eq. (8.123), the second order mole-fraction moment of the propagating chains is $\lambda_2 = Q_{2,rp}/Q_{0,rp}$, and the instantaneous first order mole-fraction moment of the produced polymer chains by combination is $\mu'_{k,prd,tc} = 2\lambda_1^2 + 2\lambda_2 = 2\left(Q_{1,rp}/Q_{0,rp}\right)^2 + 2Q_{2,rp}/Q_{0,rp}$ (see Section 6.3.2 of Chapter 6). After substitution for λ_2 and $\mu'_{2,prd,tc}$ in the kth order conservation Eq. (8.123), the differential equation for determination of $Q_{2,tp}$ is as follows:

$$\frac{1}{Q_{2,tp}}\frac{dQ_{2,tp}}{dt} = k_{ctm,app}\,[M]\,\frac{Q_{2,rp}}{Q_{2,tp}} + 2\left(k_{td,app} + k_{tc,app}\right)\frac{Q_{0,rp}\,Q_{2,rp}}{Q_{2,tp}} + 2k_{tc,app}\frac{Q_{1,rp}^2}{Q_{2,tp}} - \frac{1}{\left[\left(\rho_m/\varepsilon\,[M]_0\right) - \alpha\right]}\frac{d\alpha}{dt} \quad (8.126)$$

After solving the above differential equation simultaneously with that of $Q_{1,tp}$ (Eq. 8.125) and $Q_{0,tp}$ (Eq. 8.124), the weight-average degree of polymerization of terminated chains is given by $\overline{DP}_{w,tp} = \dfrac{Q_{2,tp}}{Q_{1,tp}}$ and polydispersity is given by $PDI = \dfrac{\overline{DP}_{w,tp}}{\overline{DP}_{n,tp}}$.

8.7 Modeling Diffusion-limited Bulk Radical Polymerization of Methyl Methacrylate

This example illustrates the application of the equations derived in Section 8.6 of this chapter to the determination and analysis of molecular weight and polydispersity of methyl methacrylate (MMA) polymerized in bulk using a thermal initiator as a function of conversion. At low conversions, MMA polymerization is reaction controlled, whereas the reaction is diffusion limited at high conversions. The elementary reactions for MMA polymerization are as follows:

$$I \xrightarrow{k_d} 2R \quad (8.127)$$

$$R + M \xrightarrow{k_{ci}} RP_1 \quad (8.128)$$

$$RP_n + M \xrightarrow{k_p} RP_{n+1} \quad (8.129)$$

$$RP_n + RP_m \xrightarrow{k_{td}} TP_n + TP_m + TDB \quad (8.130)$$

$$RP_n + RP_m \xrightarrow{k_{tc}} TP_{n+m} \tag{8.131}$$

Although the recombination reaction of primary radicals R is not included in the elementary reactions, it is accounted for by the apparent efficiency of chain initiation, f_{app} (see Section 8.5.1 of this chapter). The reaction is maintained at constant temperature of 75 °C. The reaction is initiated thermally using azobisisobutyronitrile (AIBN) with intrinsic (zero polymer concentration) initiation efficiency of $f_0 = 0.58$ [65, 117–119] and initiator concentrations of $[I]_0 = 5,\ 15,\ 25$ mM. Although the dominant mechanism of termination in MMA polymerization is disproportionation, termination by combination as well as disproportionation are included in the modeling with intrinsic ratio $\frac{k_{tc,0}}{k_{td,0}} = 0.15$ at $T = 75$ °C [79]. It is assumed that the effect of volume changes with conversion on solution viscosity and diffusivity of the reactants is negligible, hence $dV/dt = 0$. The reaction is modeled up to conversion $\alpha = 95\%$ to maintain the reaction mixture above the T_g of PMMA. Therefore, it is assumed that QSSA is valid for the primary radicals (R) produced by initiator dissociation throughout the reaction (see Section 8.5 of this chapter), but not for the reactive propagating chains (RP). The intrinsic rate constants for chain initiation and propagation are equal, that is $k_{p,0} = k_{ci,0}$. The intrinsic rate constants for initiator dissociation $k_{d,0}$, propagation $k_{p,0}$, and termination $k_{t,0} = k_{td,0} + k_{tc,0}$ at $T = 75$ °C are 3.57×10^{-3} min^{-1}, 5.49×10^{14} L mol^{-1} min^{-1}, and 2.13×10^{9} L mol^{-1} min^{-1}, respectively [79, 95, 120, 121]. The reaction rate constants in MMA polymerization with AIBN initiator as a function of temperature are given by the following relations [122]:

$$\begin{aligned}
k_{d,0}\left[\text{min}^{-1}\right] &= 6.32 \times 10^{16} e^{\frac{-15430}{T}} \quad T\,[\text{K}] \\
k_{p,0}\left[\text{Lmol}^{-1}\ \text{min}^{-1}\right] &= k_{ci,0} = 2.97 \times 10^{17} e^{\frac{-2191}{T}} \quad T\,[\text{K}] \\
k_{t,0}\left[\text{Lmol}^{-1}\ \text{min}^{-1}\right] &= k_{tc,0} + k_{td,0} = 5.88 \times 10^{9} e^{\frac{-353}{T}} \quad T\,[\text{K}] \\
\frac{k_{tc,0}}{k_{td,0}} &= 3.956 \times 10^{-4} e^{\frac{2058}{T}} \quad T\,[\text{K}]
\end{aligned} \tag{8.132}$$

It is assumed that the concentration of the initiator, primary radical, and propagating species in the polymerization mixture are negligible as compared to those of monomer and polymer (initiator is three orders of magnitude lower than monomer) for the calculation of viscosity of the reaction medium. The model of Vrentas–Duda is used to predict viscosity of the reaction medium, η_s, consisting of MMA monomer and polymer as follows [35–40]:

$$\eta_s = \eta_{\text{int}} \exp\left\{\frac{\gamma_p}{\varsigma_p}\left[\frac{\left(w_m \widehat{V}_m^* + w_p \widehat{V}_p^* \varsigma_p\right)}{\widehat{V}_{fv,mix}} - \frac{1}{v_{fv,m}}\right]\right\} \tag{8.133A}$$

where η_{int} is intrinsic viscosity, and the dimensionless constant $\gamma_p = 0.76$. The parameter ς_p and diffusion coefficients of the primary radical and monomer are given by the following expressions:

$$\varsigma_p = \left(\widehat{V}_m^* M_{jm}\right) / \left(\widehat{V}_p^* M_{jp}\right) \tag{8.133B}$$

$$\hat{V}_{fv,mix}\ [\text{cm}^3\ \text{g}^{-1}] = (1 - w_p)\ \hat{V}^*_m \upsilon_{fv,m} + w_p \hat{V}^*_p \upsilon_{fv,p} \tag{8.133C}$$

$$D_r = \frac{k_B T}{6\pi b_r \eta_s} = \frac{D_{r,0}}{\eta_s} \tag{8.133D}$$

$$D_m = \frac{k_B T}{6\pi b_m \eta_s} = \frac{D_{m,0}}{\eta_s} \tag{8.133E}$$

where D_r and D_m are the diffusion coefficients of the primary radical and monomer, respectively, and $D_{r,0}$ and $D_{m,0}$ are their diffusion coefficients at infinite polymer dilution with values of $D_{r,0} = 3.25 \times 10^{-5}\text{cm}^2\ \text{s}^{-1}$ and $D_{m,0} = 3.45 \times 10^{-5}\text{cm}^2\text{s}^{-1}$ at $T = 75\ °\text{C}$ [68, 123]; $\hat{V}^*_m$, and $\hat{V}^*_p$ are partial specific volumes of monomer and polymer, respectively, with values of $\hat{V}^*_m = 0.82\ \text{cm}^3\text{g}^{-1}$ and $\hat{V}^*_m = 0.77\ \text{cm}^3\ \text{g}^{-1}$; M_{jm} and M_{jp} are molecular weights of the jumping units for diffusion of monomer and polymer with $M_{jm} = 100$ Da and $M_{jp} = 150$ Da; $\upsilon_{fv,m}$ and $\upsilon_{fv,p}$ are the free volume fractions of pure monomer and polymer with value of $\upsilon_{fv,m} = 0.17$ and $\upsilon_{fv,p} = 0.01$ at 75 °C; and the ratio $\varsigma_p = \dfrac{\hat{V}^*_m M_{jm}}{\hat{V}^*_p M_{jp}} = 0.71$ [68, 124–126]. The viscosity of pure monomer at 75 °C is $\eta_m = 3.52 \times 10^{-3}\ \text{g cm}^{-1}\text{s}^{-1}$ [33, 70, 127]. The diffusivity of chains (propagating or terminated chains) depends on the factor $[TP]\,\overline{M}_w^{0.76}$ where $[TP]$ is the polymer concentration and $\overline{M}_w$ is the weight-average polymer molecular weight. The diffusivity of propagating chains follows the Vrentas–Duda model for $[TP]\,\overline{M}_{w,rp}^{0.76} < 1 \times 10^4$ and the reptation model for $[TP]\,\overline{M}_{w,rp}^{0.76} \geq 1 \times 10^4$ as follows [26, 60, 128–130]:

$$\begin{aligned} D_{rp,\text{conc}} &= D_{rp,0} \exp\left\{\frac{-\gamma_p}{\varsigma_p}\left(\frac{[w_m \hat{V}^*_m + w_p \hat{V}^*_p \varsigma_p]}{\hat{V}_{fv,mix}} - \frac{1}{\upsilon_{fv,m}}\right)\right\} && \in [TP]\,\overline{M}_{w,rp}^{0.76} < 1 \times 10^4 \\ D_{rp,rep} &= \frac{4k_B T N_e}{15\xi_{seg}} \frac{1}{\overline{DP}_{w,rp}^{\,2}} && \in [TP]\,\overline{M}_{w,rp}^{0.76} \geq 1 \times 10^4 \end{aligned} \tag{8.134A}$$

where $D_{rp,\text{conc}}$ and $D_{rp,rep}$ are the diffusion coefficients of the propagating chains in the concentrated and reptation regimes, respectively, $\overline{M}_{w,rp}$ is the weight-average molecular weight of the propagating chains, the Boltzmann constant $k_B = 1.381 \times 10^{-23}\ \text{kg m}^2\ \text{s}^{-2}\text{K}^{-1}$, $T = 348.15$ K, the average number of monomer units between entanglements for PMMA $N_e = 48$, and the segmental friction coefficient $\xi_{seg} = 4.03 \times 10^{-10}\ \text{kg s}^{-1}$ [60]. Based on Stokes–Einstein equation, the diffusion coefficient of PMMA at zero dilution, $D_{p,0}$, is given by the following equation [13]:

$$D_{rp,0} = \frac{k_B T}{6\pi R_H \eta_{rp,int}} = \frac{k_B T}{\pi b \overline{DP}_{n,rp}^{1/2} \eta_s} = \frac{k_B T}{\pi b \left(\frac{Q_{1,rp}}{Q_{0,rp}}\right)^{1/2} \eta_{rp,int}} \tag{8.134B}$$

where $Q_{1,rp}$ and $Q_{0,rp}$ are the first and zero order molar concentration moments of the propagating chains, $\overline{DP}_{n,rp}$ is the number-average DP of the propagating chains, and the size of a monomer segment on polymer chains $b = b_m = 0.58$ nm for PMMA [70]. The size of a primary radical is assumed to be equal to the size of a monomer segment, that is $b_r = b_m = 0.58$ nm. The intrinsic viscosity of the propagating chains in MMA monomer is given by the following equation [33]:

$$\eta_{rp,\text{int}}\ [\text{cm}^3/\text{g}] = 5.2 \times 10^{-3} \left(\overline{DP}_{w,rp} M_0\right)^{0.76} = 0.172\ (Q_{2,rp}/Q_{1,rp})^{0.76} \tag{8.134C}$$

As disproportionation is the dominant mechanism of termination in MMA polymerization [79], it is reasonable to assume $\overline{DP}_{rp} = \overline{DP}_{tp}$ for calculation of diffusion coefficients where $\overline{DP}_{tp}$ is the average DP of the terminated chains. Based on the above data, determine A) conversion α as a function of reaction time; B) concentrations of primary radicals $[R]$, propagating chains $[RP]$, and kinetic chain length Las a function of conversion; and C) concentration of terminated chains $[TP]$, number-average $\overline{DP}_{n,tp}$ and weight-average $\overline{DP}_{w,tp}$ degree of polymerization of terminated polymer chains as a function of conversion.

8.7.1 Dependence of Polymer Diffusion Coefficient on Conversion in Bulk MMA Polymerization

Assuming the concentrations of initiator, primary radical, and propagating chains are negligible as compared to those of monomer and polymer chains, the weight fraction of polymer, w_p, under constant volume is related to conversion α as follows:

$$w_p = \frac{m_p}{m_{m,0}} = \frac{([M]_0\, M_0 V_0) - ([M]\, M_0 V_0)}{[M]_0\, M_0 V_0} = 1 - \frac{[M]}{[M]_0} = \alpha \tag{8.135}$$

where m_p, $m_{m,0}$, and M_0 are polymer mass, initial monomer mass, and molecular weight of the monomer, respectively. The specific free volume of the polymerization mixture, after substitution for w_p in terms of α using the above equation, is related to conversion as follows:

$$\begin{aligned}\hat{V}_{f,mix}\,[\mathrm{cm^3 g^{-1}}] &= (1 - w_p)\,\hat{V}^*_m \upsilon_{fv,m} + w_p \hat{V}^*_p \upsilon_{fv,p} \\ &= (1-\alpha)(0.82\ \mathrm{cm^3 g^{-1}})(0.17) + \alpha\,(0.77\ \mathrm{cm^3 g^{-1}})(0.01) \\ &= 0.14\,(1-\alpha) + 8.47\times 10^{-3}\alpha\end{aligned} \tag{8.136}$$

Based on Eqs. (8.133A–E) and substitution for w_p in terms of α from the above equation, the diffusion coefficients of primary radical, D_r, and monomer, D_m, in terms of conversion are as follows:

$$D_r\,[\mathrm{cm^2\,s^{-1}}] = 3.25\times 10^{-5}\exp\left\{-57.5\left(\frac{0.01\,(1-\alpha) + 6.67\times 10^{-3}\alpha}{0.14\,(1-\alpha) + 8.47\times 10^{-3}\alpha}\right)\right\} \tag{8.137A}$$

$$D_m\,[\mathrm{cm^2 s^{-1}}] = 3.45\times 10^{-5}\exp\left\{-62.7\left(\frac{0.01\,(1-\alpha) + 6.67\times 10^{-3}\alpha}{0.14\,(1-\alpha) + 8.47\times 10^{-3}\alpha}\right)\right\} \tag{8.137B}$$

After substitution for $\eta_{\mathrm{int},rp}$ from Eq. (8.134C) in Eq. (8.134B), $D_{rp,0}$ is given in terms of conversion and concentration moments of propagating chains by the following equation:

$$D_{rp,0}\,[\mathrm{cm^2 s^{-1}}] = 2.97\times 10^{-6}\left(\frac{Q_{1,rp}}{Q_{0,rp}}\right)^{-0.5}\left(\frac{Q_{2,rp}}{Q_{1,rp}}\right)^{-0.76} \tag{8.138}$$

After substitution for $D_{rp,0}$ from the above equation and for w_p in terms of conversion, $D_{rp,\mathrm{conc}}$ in $[TP]\,\overline{M}_{w,rp}^{0.76} < 1\times 10^4$ regime is expressed as follows:

$$D_{rp,\mathrm{conc}}\,[\mathrm{cm^2 s^{-1}}] = 2.97\times 10^{-6}\left(\frac{Q_{1,rp}}{Q_{0,rp}}\right)^{-0.5}\times\left(\frac{Q_{2,rp}}{Q_{1,rp}}\right)^{-0.76} e^{-1.07F_{rp}(\alpha)} \quad \in [TP]\,\overline{M}_{w,rp}^{0.76} < 1\times 10^4 \tag{8.139A}$$

where $F_{rp}(\alpha)$ is given by the following equation:

$$F_{rp}(\alpha) = \left(\frac{0.82\,(1-\alpha) + 0.55\alpha}{0.14\,(1-\alpha) + 8.47\times 10^{-3}\alpha} - 5.85\right) \tag{8.139B}$$

After substitution for $N_e = 48$, $T = 348.15$ K, $\xi_{seg} = 4.03 \times 10^{-10}$ kg s^{-1}, and $\overline{DP}_{w,rp} = \frac{Q_{2,rp}}{Q_{1,rp}}$ in Eq. (8.134A), $D_{rp,rep}$ for $[TP]\,\overline{M}_{w,rp}^{0.76} \geq 1 \times 10^4$ is given by the following equation:

$$D_{rp,rep}\,[\mathrm{cm^2 s^{-1}}] = \frac{4k_B T N_e}{15\xi_{seg}}\left(\frac{Q_{2,rp}}{Q_{1,rp}}\right)^{-2} \quad \in [TP]\,\overline{M}_{w,rp}^{0.76} \geq 1 \times 10^4$$
$$= \frac{4\,(1.38 \times 10^{-19}\ \mathrm{kg\,cm^2 s^{-2} K^{-1}})\,(348.15\mathrm{K})\,(48)}{15\,(4.03 \times 10^{-10}\ \mathrm{kg\,s^{-1}})}\left(\frac{Q_{2,rp}}{Q_{1,rp}}\right)^{-2} = 1.53 \times 10^{-6}\left(\frac{Q_{2,rp}}{Q_{1,rp}}\right)^{-2} \tag{8.139C}$$

In the above Eqs. (8.139A) and (8.139B), $Q_{0,rp}$, $Q_{1,rp}$, and $Q_{2,rp}$ are zero, first, and second order molar concentration moments of the propagating chains, respectively, which will be determined in the following sections.

8.7.2 Dependence of Apparent Initiation Efficiency and Rate Constants on Conversion in Bulk MMA Polymerization

Apparent initiation efficiency, f_{app}, is given by Eq. (8.114). After substitution for $f_0 = 0.58$, $b_r = 0.58$ nm, $k_{ci,0} = 2.97 \times 10^{17} e^{\frac{-2191}{T}}$, $T = 348.15\,K$, and D_r as a function of α from Eq. (8.137A), f_{app} is given as a function of α and $[PR]$ by the following expression:

$$f_{app} = 0.58\left\{1 + 21.2\,[PR]\exp\left[57.5\left(\frac{0.01\,(1-\alpha) + 6.67 \times 10^{-3}\alpha}{0.14\,(1-\alpha) + 8.47 \times 10^{-3}\alpha}\right)\right]\right\}^{-1} \tag{8.140A}$$

where $[PR]$ has unit of mole L^{-1} and f_{app} is dimensionless. The apparent rate constant for propagation, $k_{p,app}$, is given by Eq. (8.107). After substitution for $k_{p,0} = 2.97 \times 10^{17} e^{\frac{-2191}{T}}$, $T = 348.15$ K, D_m as a function of α from Eq. (8.137B), $R_g = b_m D_{n,rp}^{1/2}/6 = (b_m/6)(Q_{1,rp}/Q_{0,rp})^{1/2}$, and $b_m = 0.58$ nm, the apparent rate of propagation $k_{p,app}$ is given as a function of α and $Q_{1,rp}$ by the following expression:

$$k_{p,app} = 9.15 \times 10^{12}\left\{1 + 13.2\,Q_{1,rp}\exp\left[62.7\left(\frac{0.01\,(1-\alpha) + 6.67 \times 10^{-3}\alpha}{0.14\,(1-\alpha) + 8.47 \times 10^{-3}\alpha}\right)\right]\right\}^{-1} \tag{8.140B}$$

where $Q_{1,rp}$ and $k_{p,app}$ have units of mole L^{-1} and L mole^{-1} s^{-1}, respectively. The apparent rate constants for termination by disproportionation $k_{td,app}$ and combination $k_{tc,app}$ are given by Eqs. (8.115) and (8.116), respectively. The parameter $p_{rp,seg}$ in these equations is given by Eq. (8.62) which is equal to $p_{rp,seg} = 0.011$ for PMMA (see Eq. 8.63). For $[TP]\,\overline{M}_{w,rp}^{0.76} < 1 \times 10^4$ in semi-dilute to concentrated regime, after substitution for $k_{td,0} = 3.10 \times 10^7$ L mol^{-1} s^{-1} and $k_{tc,0} = 4.53 \times 10^6$ L mol^{-1}s^{-1} (see Eq. 8.132), $R_g = (b_m/6)(Q_{1,rp}/Q_{0,rp})^{1/2}$ with $b_m = 0.58$ nm, and $D_{rp,\mathrm{conc}}$ from Eq. (8.139A) in Eqs. (8.115) and (8.116), the apparent termination rate constants $k_{td,app}$ and $k_{tc,app}$ in the regime below the onset of reptation for MMA polymerization are as follows:

$$k_{td,app,conc} = 3.10 \times 10^7\left\{1 + 2.41 \times 10^{-6} Q_{1,rp}\left(\frac{Q_{1,rp}}{Q_{0,rp}}\right)^{0.5}\left(\frac{Q_{2,rp}}{Q_{1,rp}}\right)^{0.76} e^{-1.07F_{rp}(\alpha)}\right\}^{-1}$$

$$k_{tc,app,conc} = 4.53 \times 10^6\left\{1 + 3.53 \times 10^{-7} Q_{1,rp}\left(\frac{Q_{1,rp}}{Q_{0,rp}}\right)^{0.5}\left(\frac{Q_{2,rp}}{Q_{1,rp}}\right)^{0.76} e^{-1.07F_{rp}(\alpha)}\right\}^{-1} \tag{8.140C}$$

Where $F_{rp}(\alpha)$ is given by Eq. (8.139B). For $[TP]\overline{M}_{w,rp}^{0.76} \geq 1 \times 10^4$ in the reptation regime, after substitution for $N_e = 48$, $T = 348.15$ K, $\xi_{seg} = 4.03 \times 10^{-10}$ kg s^{-1}, and $D_{rp,rep}$ from Eq. (8.139C) in Eqs. (8.115) and (8.116), the apparent termination rate constants $k_{td,app}$ and $k_{tc,app}$ in the reptation regime for MMA polymerization are expressed as follows:

$$k_{td,app,rep} = 3.10 \times 10^7 \left[1 + 0.22Q_{1,rp}\left(Q_{2,rp}/Q_{1,rp}\right)^2\right]^{-1}$$
$$k_{tc,app,rep} = 4.53 \times 10^6 \left[1 + 0.032Q_{1,rp}\left(Q_{2,rp}/Q_{1,rp}\right)^2\right]^{-1} \quad (8.140D)$$

In the above equations, the units of $Q_{1,rp}$ *and* $Q_{2,rp}$ are mole L^{-1} and the units of $k_{td,app}$ or $k_{tc,app}$ are L mole^{-1} s^{-1}.

8.7.3 Time Dependence of Conversion in Bulk MMA Polymerization

The conservation equation for initiator is given by Eq. (8.102). For constant volume reaction, after integration and substitution for $k_d = 5.95 \times 10^{-5}$ s^{-1}, the time dependence of $[I]$ is as follows:

$$[I] = [I]_0 \, e^{-5.95\times10^{-5}\,t} \quad (8.141A)$$

where t has unit of second and $[I]_0$ takes the values of $[I]_0 = 5 \times 10^{-3}$, 15×10^{-3}, and 25×10^{-3} M. The conservation equation for $[M]$ is given by Eq. (8.104). For constant volume, no chain transfer reaction to monomer $(k_{ctm,app} = 0)$, $k_{ci}\,[M]\,[R] = 2f_{app}k_d\,[I]$ at constant volume (see Eq. 8.87), $k_d = 5.95 \times 10^{-5}$ s^{-1}, f_{app} as given by Eq. (8.86), $[I]$ as given by the above equation, $\sum_{j=1}^{\infty}[RP_n] = [RP] = Q_{0,rp}$, $k_{p,app}$ as given by Eq. (8.140B), the differential equation for rate of change of monomer concentration with time is given as follows:

$$\frac{d\,[M]}{dt} = -1.20 \times 10^{-4} f_{app}\,[I]_0\,e^{-5.95\times10^{-5}\,t} - k_{p,app}\,[M]\,Q_{0,rp} \quad (8.141B)$$

After substitution for $[M]$ in the above equation in terms of α using Eq. (8.101), the conversion as a function of time at constant reaction volume is expressed as follows:

$$\frac{d\alpha}{dt} = 1.20 \times 10^{-4} f_{app}\frac{[I]_0}{[M]_0}e^{-5.95\times10^{-5}\,t} + k_{p,app}\,(1-\alpha)\,Q_{0,rp} \quad (8.141C)$$

Where the equation for $Q_{0,rp}$ as a function of conversion is derived in the next section.

8.7.4 Time Dependence of *k*th Order Concentration Moments of Propagating Chains in Bulk MMA Polymerization

The conservation equations for $Q_{0,rp}$ and $Q_{1,rp}$ are given by Eqs. (8.119) and (8.121), respectively. For a constant volume reaction ($\varepsilon = 0$), no chain transfer reaction to monomer $(k_{ctm,app} = 0)$, $k_d = 5.95 \times 10^{-5}$ s^{-1}, f_{app} as given by Eq. (8.86), $[I]$ as given by Ew. 8.141A, $[M]$ as given by Eq. (141B), α as given by Eq. (8.141C), $k_{p,app}$ as given by Eq. (8.140B), $k_{td,app}$ and $k_{tc,app}$ as given by Eq. (8.140C-D), the differential equations for the rate of change of $Q_{0,rp}$ and $Q_{1,rp}$ with time are expressed by the following expressions:

$$\frac{1}{Q_{0,rp}}\frac{dQ_{0,rp}}{dt} = \frac{1.20 \times 10^{-4} f_{app}\,[I]_0\,e^{-5.95\times10^{-5}\,t}}{Q_{0,rp}} - Q_{0,rp}\left(k_{td,app} + k_{tc,app}\right) \quad (8.142A)$$

$$\frac{1}{Q_{1,rp}}\frac{dQ_{1,rp}}{dt} = \frac{1.20\times 10^{-4} f_{app}\,[I]_0\, e^{-5.95\times 10^{-5}\,t}}{Q_{1,rp}} + k_{p,app}\,[M]\,\frac{Q_{0,rp}}{Q_{1,rp}} - Q_{0,rp}\left(k_{td,app} + k_{tc,app}\right) \tag{8.142B}$$

In a similar approach to derivation of $Q_{0,rp}$ and $Q_{0,rp}$, it can be shown that the differential equation for the rate of change of the second order concentration moment of the propagating chains, $Q_{2,rp}$, with time is expressed as follows:

$$\frac{1}{Q_{2,rp}}\frac{d\,(QR_2)}{dt} = \frac{1.20\times 10^{-4} f_{app}\,[I]_0\, e^{-5.95\times 10^{-5}\,t}}{Q_{2,rp}} + k_{p,app}\,[M]\,\frac{\left(2Q_{1,rp} + Q_{0,rp}\right)}{Q_{2,rp}} - Q_{0,rp}\left(k_{td,app} + k_{tc,app}\right) \tag{8.142C}$$

Solving the above three differential equations in combination with those for $[I]$ and $[M]$ subject to the initial condition $\alpha = Q_{0,rp} = Q_{1,rp} = Q_{2,rp} = 0$, $[I] = [I]_0$, and $[M] = [M]_0$ at $t = 0$ yields numerical values for $Q_{0,rp}$, $Q_{1,rp}$, $Q_{2,rp}$, α, and $[M]$ as a function of time for the polymerization of MMA based on elementary Rxns. 8.127–8.131. After solving the above differentiation equations, the kinetic chain length is equal to the number-average DP of the propagating chains, that is, $L = \overline{DP}_{n,rp} = Q_{1,rp}/Q_{0,rp}$. Figure 8.4a shows the time dependence of monomer conversion for MMA polymerization at 75 °C for three initial initiator concentrations of 0.005, 0.0155, and 0.025 M. For a given reaction time, the concentration of propagating chains increased with increasing initiator concentration, which in turn increased conversion. For a given initiator concentration, the rate of change of conversion decreased with time, which is attributed to the decrease in monomer concentration. At a critical conversion of approximately $\alpha = 80\%$ corresponding to reaction times of 40, 80, and 150 min for initial initiator concentrations of 0.025, 0.0155, and 0.005 M, respectively, there was a sharp increase in conversion. This effect is attributed to the diffusion-limited increase in the concentration of propagating chains and the decrease in termination of propagating chains.

Figure 8.4b shows concentration of propagating chains $[RP] = Q_{0,rp}$ versus conversion for MMA polymerization at 75 °C for three initial initiator concentrations of 0.005, 0.0155, and 0.025 M. For all three initiator concentrations, the concentration of propagating chains remained relatively steady up to a critical conversion of 80%. This conversion coincided with the critical conversion observed in the conversion versus reaction time in Figure 8.4a. After reaching this critical conversion, the concentration of propagating chains sharply increased for all three initiator concentrations which was attributed to the diffusion-limited slow termination of propagating chains. Furthermore, for conversions >80%, the rate of change of $[RP]$ with conversion increased with increasing initial initiator concentration. Figures 8.4b shows a sharp increase in the concentration of propagating chains over a relatively narrow conversion range resulting in a much higher rate of propagation compared to termination, which is characterized by a rapid rise in the exothermic rate of polymerization and temperature of the reaction. This effect can lead to thermal runaways and reactor meltdown unless the reactor is cooled rapidly [131–135]. This can be prevented by addition of polymerization inhibitors to the reaction at high conversions [136] or utilizing more efficient cooling systems [132].

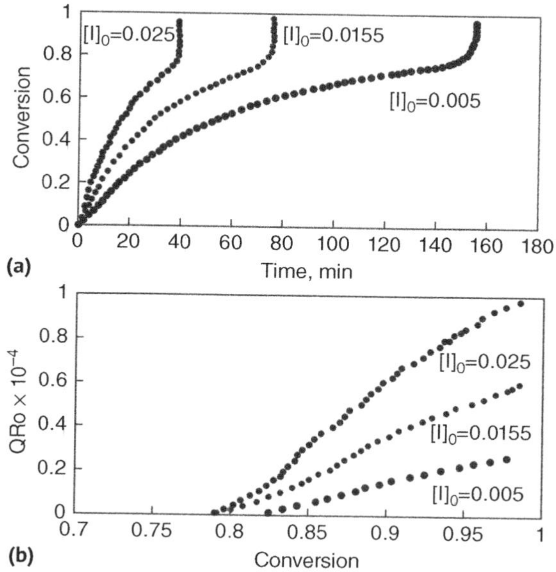

Figure 8.4 (a) Monomer conversion versus reaction time and (b) concentration of propagating chains versus conversion for MMA polymerization at 75 °C for three initial initiator concentrations of 0.005, 0.0155, and 0.025 M based on modeling of elementary Rxns. 8.127–8.131.

8.7.5 Dependence of *k*th Order Concentration Moments of Terminated Chains on Conversion in Bulk MMA Polymerization

The zero, first, and second order conservation equations for molar concentration moments of the polymer chains is given by Eqs. (8.124–8.126), respectively. For a constant volume reaction ($\varepsilon = 0$) and no chain transfer to monomer, the differential equations relating the change in $Q_{0,tp}$, $Q_{1,tp}$, and $Q_{2,tp}$ with time are given by the following equations:

$$\frac{1}{Q_{0,tp}}\frac{dQ_{0,tp}}{dt} = \left(2k_{td,app} + k_{tc,app}\right)\frac{Q_{0,rp}^2}{Q_{0,tp}} \tag{8.143A}$$

$$\frac{1}{Q_{1,tp}}\frac{dQ_{1,tp}}{dt} = 2\left(k_{td,app} + k_{tc,app}\right)\frac{Q_{0,rp}\,Q_{1,rp}}{Q_{1,tp}} \tag{8.143B}$$

$$\frac{1}{Q_{2,tp}}\frac{dQ_{2,tp}}{dt} = 2\left(k_{td,app} + k_{tc,app}\right)\frac{Q_{0,rp}\,Q_{2,rp}}{Q_{2,tp}} + 2k_{tc,app}\frac{Q_{1,rp}^2}{Q_{2,tp}} \tag{8.143C}$$

where $k_{td,app}$ and $k_{tc,app}$ are given by Eq. (8.140C-D), and $Q_{0,rp}$, $Q_{1,rp}$, and $Q_{2,rp}$ are given by Eqs. (8.142A–C), respectively. Upon solving the above differential equations, the concentration of polymer chains, number-average DP, weight-average DP, and polydispersity index are determined by $[TP] = Q_{0,tp}$, $\overline{DP}_{n,tp} = \frac{Q_{1,tp}}{Q_{0,tp}}$, $\overline{DP}_{w,tp} = \frac{Q_{2,tp}}{Q_{1,tp}}$, and $PDI = \frac{\overline{DP}_{w,tp}}{\overline{DP}_{n,tp}}$. Figure 8.5 shows the concentration of terminated polymer chains $[TP] = Q_{0,tp}$ versus conversion for MMA polymerization at 75 °C for three initial initiator concentrations of 0.005, 0.0155, and 0.025 M. The rate of increase in polymerization with conversion increased with increasing initial initiator concentration, which was attributed to higher concentration of propagating chains. For all three initiator concentrations, $[TP]$ increased linearly with conversion up to $\alpha = 80\%$. For $\alpha > 80\%$, the polymer concentration

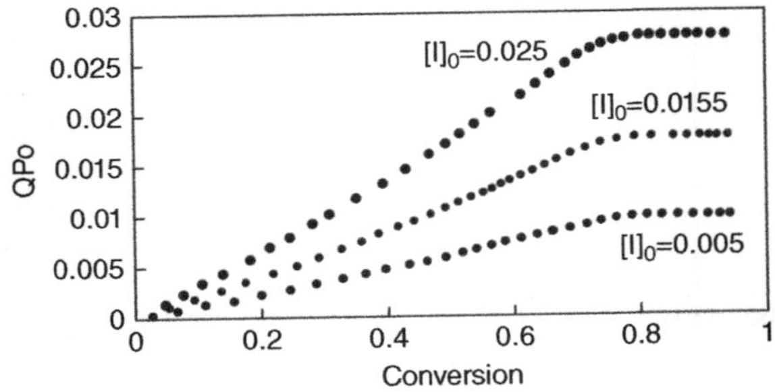

Figure 8.5 Concentration of terminated polymer chains versus monomer conversion for MMA polymerization at 75 °C for three initial initiator concentrations of 0.005, 0.0155, and 0.025 M based on modeling of elementary Rxns. 8.127–8.131.

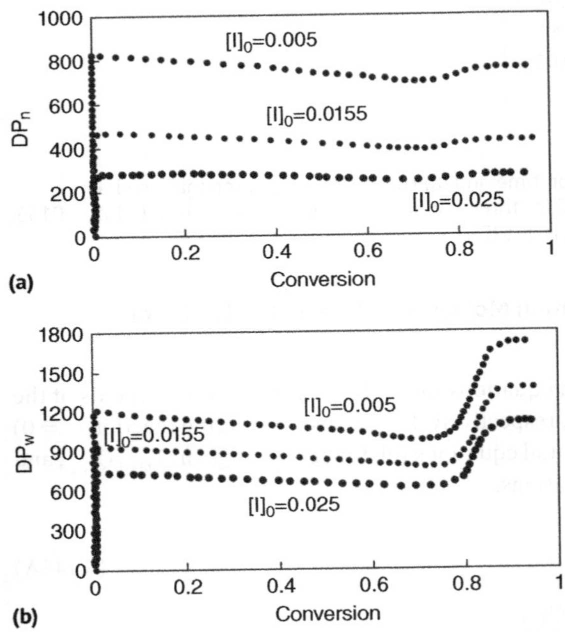

Figure 8.6 Number- (a) and weight-average (b) degree of polymerization of polymer chains versus conversion for MMA polymerization at 75 °C for three initial initiator concentrations of 0.005, 0.0155, and 0.025 M based on modeling of elementary Rxns. 8.127–8.131.

reached a plateau which was attributed to the slow, diffusion-limited termination of the propagating chains. This slow termination led to the formation of longer but fewer terminated polymer chains.

Figure 8.6a, b shows $\overline{DP}_{n,tp}$ and $\overline{DP}_{w,tp}$ of polymer chains, respectively, versus conversion for MMA polymerization at 75 °C for three initial initiator concentrations of 0.005, 0.0155, and 0.025 M based on mathematical modeling of elementary Rxns. 8.127–8.131.

The $\overline{DP}_{n,tp}$ and $\overline{DP}_{w,tp}$ reached large finite values instantly after initiation consistent with the mechanism of radical polymerization. The $\overline{DP}_{n,tp}$ and $\overline{DP}_{w,tp}$ values decreased with increasing conversion up to $\alpha = 80\%$, which was attributed to the faster rate of consumption of monomer as compared to the rate of initiator dissociation. For $\alpha = 80\%$, a sharp increase in $\overline{DP}_{n,tp}$ and $\overline{DP}_{w,tp}$ was observed, which was attributed to the slow diffusion-limited termination of the propagating chains. This slow termination led to the formation of longer but fewer polymer chains, consistent with plateauing of $[TP]$ with conversion at $\alpha = 80\%$ (see Figure 8.5). This was further confirmed by the greater increase in $\overline{DP}_{w,tp}$ as compared with $\overline{DP}_{n,tp}$ at $\alpha = 80\%$. For $\alpha > 80\%$, $\overline{DP}_{n,tp}$ and $\overline{DP}_{w,tp}$ reached a plateau which was attributed to the slow, diffusion-limited chain termination.

8.8 Summary

In this chapter, we developed mathematical models for the addition polymerization reactions that are limited by diffusion for predicting concentrations of reactants and products, kinetic chain length, number- and weight-average degree of polymerization, and polydispersity of the produced polymer. First, we defined the apparent rate constant for a reaction, k_{app}, as compared to the intrinsic rate constant, k_{int}. The intrinsic rate constant is affected by the concentration of reactants, pressure, temperature, solvent, and reaction catalyst. The apparent rate constant accounts for diffusivity of reactants in the reaction medium in addition to those factors in the intrinsic rate. If $k_{app} \simeq k_{int}$, the reaction is controlled by kinetics of the reaction, and it is not limited by diffusion of the reactants. If $k_{app} \ll k_{int}$, the reaction is not controlled by kinetics of the reaction because it is limited by diffusion of the reactants to the local reaction sites. Based on intrinsic and apparent rate constants, the time constant for diffusion, τ_{diff}, as compared to intrinsic time constant, τ_{int}, was defined for a given elementary reaction. Polymerization reactions carried out in bulk are often limited by diffusion of reactants because the viscosity of the reaction medium increases by many orders of magnitude with conversion of monomer to polymer chains. The Einstein–Stokes equation was used to relate the diffusion coefficients of the reactants to viscosity of the medium. Next, models were presented for predicting the viscosity of polymerization medium in different regimes. In the semi-dilute to moderately concentrated regime, the chains are not fully confined within a tube formed by the neighboring chains, hence the empirical equations of Lyons–Tobolsky, extended Martin, and the free volume model of Vrentas–Duda were presented for predicting the change in viscosity of polymerization reaction with conversion. In these models, viscosity increases exponentially with increasing polymer concentration. In the highly concentrated regime, the chains are fully confined within a tube formed by the neighboring chains and diffusion is by chain reptation, so the reptation model was presented for predicting the change in viscosity of polymerization reaction with conversion. Then, the conservation equation in differential form for molar mass of the diffusing species in chain termination reaction between two propagating chains was derived to relate the local, diffusion-limited concentration of the reactive propagating species to their bulk concentration. The solution of the differential equation was then combined with the equations for kinetics of the reaction to relate apparent rate constants to intrinsic rate constants for termination reaction. The derived relations between apparent and intrinsic rate constants were applied to modeling bulk, radical polymerization of methyl methacrylate. The modeling showed four distinct regimes including reaction-controlled polymerization in the dilute regime, termination limited polymerization in the semi-concentrated regime, propagation and termination limited polymerization in the concentrated regime, and initiation limited polymerization in the highly concentrated regime. The sharp increase in the concentration of propagating chains over a relatively narrow conversion range, due to diffusion-limited termination reactions, resulted in much higher rates of propagation compared to termination, which was characterized by a rapid rise in the rate of polymerization and temperature of the reaction. This rapid rise in temperature by diffusion-limited termination reaction can lead to thermal runaways unless the reaction is rapidly cooled to control temperature.

8.9 List of Symbols and Abbreviations

a: exponent of the Mark–Houwink–Sakurada equation relating intrinsic viscosity of a polymer to its degree of polymerization

a_{init}: constant related to the chemical nature of initiator in polymerization reaction

AIBN: azobisisobutyronitrile initiator

b: length of a segment or repeat unit on a polymer chain

b_{kuhn}: length of a repeat unit or monomer residue of a polymer chain if the polymer chain is represented by the Kuhn model of a freely jointed chain

b_{LT}: constant in Lyons–Tobolsky equation for determination of viscosity of a polymer solution

b_m: size of a monomer molecule

b_r: size of a reactive species produced by dissociation of initiator

C_1, C_2: integration constants in the equation relating concentration of reactive propagating chains to radial direction obtained by conservation of mass

C_p: molar concentration of polymer in a polymerization mixture

$d_{EM,0}, d_{EM,1}, d_{EM,3}$: constants in the extended Martin equation by Fedors for determination of viscosity of a polymer solution

$\overline{d}_{p-p}$: average separation distance between two polymer chains in a polymerization medium for determination of polymer diffusion time constant

$\overline{d}_{rp}$: average diameter of reactive propagating chains in a polymerization reaction

dt: infinitesimal change in reaction time (s)

$\overline{d}_{tp}$: average diameter of terminated polymer chains in a polymerization reaction

D_m: diffusion coefficient of monomer in a polymerization mixture

$D_{m,0}$: diffusion coefficient of monomer as the polymer concentration in polymerization mixture approaches zero

$D_{m,\mathrm{dilute}}$: diffusion coefficient of monomer in dilute polymerization mixture where the chains are below entanglement concentration

D_p: diffusion coefficient of a particle or a polymer chain in a liquid polymerization medium for determination of time constant for diffusion using Einstein equation

$D_{p,\mathrm{dilute}}$: diffusion coefficient of polymer chains in a polymerization mixture where the chains are below the entanglement concentration

$D_{p,rep}$: diffusion coefficient of polymer chains in a concentrated polymerization mixture where the propagating chains diffuse by reptation

$D_{p,\mathrm{semidilute}}$: diffusion coefficient of polymer chains in a polymerization mixture where the chains are above the entanglement concentration but are not fully entangled to behave like a gel

D_r: diffusion coefficient of the primary reactive species generated by initiator dissociation in a polymerization mixture

$D_{r,0}$: diffusion coefficient of the primary reactive species generated by initiator dissociation as the polymer concentration in polymerization mixture approaches zero

D_{rp}: diffusion coefficient of reactive propagating chains in a polymerization mixture

$D_{rp,0}$: diffusion coefficient of reactive propagating chains as the polymer concentration in polymerization mixture approaches zero

$D_{rp,\text{conc}}$: diffusion coefficient of reactive propagating chains in concentrated polymerization mixture where the chains are fully entangled

$D_{rp,\text{dilute}}$: diffusion coefficient of reactive propagating chains in dilute polymerization mixture where the chains are below the entanglement concentration

$D_{rp,rep}$: diffusion coefficient of reactive propagating chains in a concentrated polymerization mixture where the propagating chains diffuse by reptation

DP: number average degree of polymerization of a polymer chain

$\overline{DP}$, $\overline{DP}_n$: number-average degree of polymerization of polymer chains

$\overline{DP}_{rp}$, $\overline{DP}_{n,rp}$: number-average degree of polymerization of reactive propagating chains

$\overline{DP}_{tp}$, $\overline{DP}_{n,tp}$: number-average degree of polymerization of terminated polymer chains

$\overline{DP}_{w,rp}$: weight-average degree of polymerization of reactive propagating chains

$\overline{DP}_{w,tp}$: weight-average degree of polymerization of terminated polymer chains

$\overline{DP}_{\eta}$: viscosity-average degree of polymerization of polymer chains

EI: entanglement index which is the ratio of average diameter of terminated polymer chains to average distance between two polymer chains in a polymerization mixture

f_0: efficiency of initiator molecules to dissociate and generate reactive species

f: efficiency of initiator molecules to generate 1-mer reactive propagating chains following dissociation

f_{app}: apparent efficiency of initiator molecules to generate 1-mer reactive propagating chains following dissociation in a diffusion-limited polymerization reaction

F_d: friction coefficient which is defined as the drag force exerted on a diffusing particle or polymer chain by the liquid medium divided by the particle or polymer chain velocity

I: polymerization initiator

$I^{\bullet}$: inactive species produced by recombination of reactive species following initiator dissociation

$[I]$: initiator concentration in polymerization at time t

$[I]_0$: initial initiator concentration in polymerization at time t

J_{rp}: mass flux of reactive propagating chains in radial direction in polymerization mixture

k_{bm}: rate constant for a bimolecular reaction between two small molecules in a polymerization mixture

k_B: Boltzmann constant

k_c: rate constant for recombination reaction of the dissociated fragments of the polymerization initiator

k_{ci}: rate constant for the reaction of dissociated initiator fragment with a monomer to produce a 1-mer propagating chain

$k_{ci,app}$: apparent rate constant for chain initiation reaction of primary reactive species with monomer in a diffusion-limited polymerization reaction

$k_{ci,int}$: intrinsic (reaction controlled) rate constant for chain initiation between primary reactive species and monomer to form a 1-mer propagating chain with intrinsic (reaction controlled) rate constant in the absence of diffusion limitation

k_{ctm}: rate constant for transfer of reactive species on a propagating chain to a monomer in a polymerization mixture

$k_{ctm,0}$: rate constant for transfer of reactive species on a propagating chain to a monomer as the concentration of polymer chains in polymerization reaction approaches zero

$k_{ctm,app}$: apparent rate constant for transfer of reactive species on a propagating chain to a monomer in a polymerization mixture

$k_{ctm,int}$: intrinsic (reaction controlled) rate constant for transfer of reactive species on a propagating chain to a monomer in a polymerization mixture in the absence of diffusion limitation
k_d: rate constant for the dissociation reaction of polymerization initiator
k_{LT}: constant in Lyons–Tobolsky equation for determination of viscosity of a polymer solution
k_p: rate constant for propagation reaction between a propagating chain and a monomer
$k_{p,0}$: rate constant for propagation reaction between a propagating chain and a monomer as the concentration of polymer chains in polymerization mixture approaches zero
$k_{p,app}$: apparent rate constant for propagation reaction of propagating chains with monomer in a diffusion-limited polymerization reaction
k_t: rate constant for termination reaction of reactive propagating chains to produce terminated polymer chains
$k_{t,int}$: intrinsic (reaction controlled) rate constant for termination reaction of reactive propagating chains in the absence of diffusion limitation
k_{tc}: rate constant for termination of two propagating chains by combination
$k_{tc,0}$: rate constant for termination of two propagating chains by combination as the concentration of polymer chains in polymerization reaction approaches zero
$k_{tc,app}$: apparent rate constant for termination of reactive propagating chains by combination reaction in a polymerization mixture
$k_{tc,app,rep}$: apparent rate constant for termination of reactive propagating chains by combination reaction in a polymerization mixture where the propagating chains diffuse by reptation
$k_{tc,int}$: intrinsic (reaction controlled) rate constant for termination reaction by combination of reactive propagating chains in the absence of diffusion limitation
k_{td}: rate constant for termination of two propagating chains by disproportionation
$k_{td,0}$: rate constant for termination of two propagating chains by disproportionation as the concentration of polymer chains in polymerization reaction approaches zero
$k_{td,app}$: apparent rate constant for termination of reactive propagating chains by disproportionation reaction in a polymerization mixture
$k_{td,app,rep}$: apparent rate constant for termination of reactive propagating chains by disproportionation reaction in a polymerization mixture where the propagating chains diffuse by reptation
$k_{td,int}$: intrinsic (reaction controlled) rate constant for termination reaction by disproportionation of reactive propagating chains in the absence of diffusion limitation
K_η: constant in Mark–Houwink–Sakurada equation for determination of intrinsic viscosity of a polymer solution from viscosity -average molecular weight
l_{rs}: reptation step length of a polymer chain diffusing by reptation
L: kinetic chain length or number average degree of polymerization of reactive propagating chains in polymerization
m_M: initial moles of monomer in reaction
$m_{m,0}$: initial mass of monomer in reaction
M: Polymerization monomer
M_0: molecular weight of the monomer repeat unit or monomer residue in polymer chains
M_e: molecular weight between entanglement for a polymer in bulk or in solution
$M_{jump,m}$: molecular weight of jumping unit of monomer for diffusion in Vrentas–Duda equation for determination of viscosity of a polymerization mixture
$M_{jump,p}$: molecular weight of jumping unit of monomer for diffusion in Vrentas–Duda equation for determination of viscosity of a polymerization mixture
M_{MMA}: molecular weight of methyl methacrylate monomer
$\overline{M}_w$: weight average molecular weight of polymer chains

$\overline{M}_\eta$: viscosity average molecular weight of polymer chains

$[M]$: monomer concentration in polymerization at time t

$[M]_0$: initial concentration of monomer in polymerization at time t

$[M]_{cr}$: critical concentration of monomers in polymerization mixture for the onset of chain entanglement where the center-to-center distance of neighboring polymer chains becomes equal to their hydrodynamic diameter

MHS: Mark–Houwink–Sakurada equation relating intrinsic viscosity of a polymer to its degree of polymerization

MMA: methyl methacrylate monomer

n: number of molecules per unit volume

N: number of repeat units or monomer residues in a polymer chain for determination of hydrodynamic radius or radius of gyration of a polymer chain

N_{AV}: Avogadro's number

N_e: number of repeat units or monomer residues between two neighboring entanglement points of polymer chains in a polymerization mixture

N_{rp}: mass transport rate of reactive propagating chains in radial direction in polymerization mixture

N_{rs-rp}: mass transport rate of reactive propagating species at the chain end of the propagating chains in polymerization mixture

$p_{m,seg}$: collision probability of a monomer with the reactive species on chain-end of a propagating chain for propagation reaction in a polymerization reaction

$p_{nnd}\,(r)$: probability distribution for separation distance between two neighboring particles or polymer chains with a specified concentration with the first particle at the center of a sphere and the second particle at a distance between r and $r + dr$ with r ranging from zero to infinity

$p_{r,seg}$: collision probability of the reactive species on a primary reactive species produced by dissociation of initiator with the reactive species on chain-end of a propagating chain in a polymerization mixture

$p_{rp,seg}$: collision probability of the reactive species on chain-ends of two propagating chains by interdiffusion

$[P]$: concentration of polymer chains in a polymer solution

PDI, PDI_{tp}: polydispersity index of terminated polymer chains

PDI_{rp}: polydispersity index of reactive propagating chains

PMMA: poly(methyl methacrylate)

$Q_{k,rp}$: kth order molar concentration moment of reactive propagating chains at time t in polymerization

$Q_{k,tp}$: kth order molar concentration moment of terminated polymer chains at time t in polymerization

$QSSA$: quasi-steady state assumption of reactive species in addition polymerization

r: radius of a sphere for calculation of average separation distance between two polymer chains in polymerization mixture

$\langle r \rangle$: average separation distance between two polymer chains in polymerization mixture

r_b: radius of a sphere in which a second reactive propagating chain is at this radius representing the bulk concentration of propagating chains while a first reactive propagating chain is at the center of the sphere representing the local concentration of propagating chains in a diffusion limited polymerization mixture

r_r: radius of the primary reactive species generated by initiator dissociation in a polymerization mixture

R: primary radicals formed by dissociation of initiator in polymerization

R_{ci}: rate of reaction of the dissociated initiator fragment with a monomer to produce a 1-mer propagating chain

R_g: radius of gyration of a particle or a polymer chain as the distance from the center of mass of the particle or polymer chain which the entire mass of the particle or polymer chain can be concentrated without changing its moment of inertia

$R_{g,FJ}$: radius of gyration of a polymer chain represented by the Kuhn model of a freely jointed chain

$R_{g,rp}$: radius of gyration of reactive propagating chains in polymerization

R_H: equivalent hydrodynamic radius of a diffusing particle or a polymer chain in a liquid medium represented by a hypothetical sphere with identical volume as the particle or polymer chain

R_p: reaction rate for propagation between a propagating chain and a monomer

$R_{rp,con}$: rate of consumption of reactive propagating chains in polymerization at time t

$R_{rp,con,j}$: rate of consumption of reactive propagating chains by the jth consumption reaction in polymerization at time t

$R_{rp,prd}$: rate of production of reactive propagating chains in polymerization at time t

$R_{rp,prd,i}$: rate of production of reactive propagating chains by the ith production reaction in polymerization at time t

R_t: rate of termination reaction between two reactive propagating chains

$R_{t,loc}$: rate of termination reaction between a local reactive propagating chain at the center of spherical coordinate and a second propagating chain in the bulk of a diffusion limited polymerization mixture

R_{td}: rate of termination reaction between two propagating chains by disproportionation (m^3 $mole^{-1}$ s^{-1})

$R_{tp,con,l}$: rate of consumption of terminated polymer chains by the lth consumption reaction in polymerization at time t

$R_{tp,prd,i}$: rate of production of terminated polymer chains by the ith production reaction in polymerization at time t

$R_{tp,prd,ctm}$: rate of production of terminated polymer chains by transfer of reactive species from a propagating chain to a monomer in polymerization at time t

$R_{tp,prd,tc}$: rate of production of terminated polymer chains by termination reaction via combination in polymerization at time t

$R_{tp,prd,td}$: rate of production of terminated polymer chains by termination reaction via disproportionation in polymerization at time t

$[R]$: concentration of primary radicals formed by dissociation of an initiator

RP, RP_n: reactive propagating chain in polymerization with n repeat units

$[RP]$: concentration of reactive propagating chains in polymerization at time t

$[RP]_0$: initial concentration of reactive propagating chains in polymerization

$[RP]_b$: bulk concentration of reactive propagating chains in the polymerization mixture at time t

$[RP]_{dl}$: concentration of reactive species of the propagating chains in the diffusion-limited regime at distance R_g from the center-of-mass of the propagating chain

$[RP]_{loc}$: local concentration of reactive propagating chains at the reaction site in polymerization at time t

S: cross-sectional area for diffusion of reactive propagating chains in the conservation of mass equation

$\overline{sd}$: average separation distance between monomer molecules and reactive propagating chains in dilute polymerization mixture where the chains are below the entanglement concentration

$\overline{sd_m}$: average separation distance between small molecules like monomer in polymerization solution
$\overline{sd_p}$: average separation distance between particles or polymer chains in a liquid polymerization medium for determination of time constant for diffusion using Einstein equation
$\overline{sd_{rp}}$: average separation distance between reactive propagating chains in polymerization mixture for determination of time constant for termination reaction
$\overline{sd_{tp}}$: average separation distance between terminated polymer chains in polymerization mixture
t: reaction time
T: Temperature of polymerization reaction
T_g: polymer glass transition temperature
TDB: terminal double bond on a polymer chain
TDB_n: n-mer terminated polymer chain with a terminal double bond
TP, TP_n: terminated polymer chain with n repeat units
$[TDB]$: concentration of terminal double bonds on polymer chains in reaction at time t
$[TP]$: concentration of terminated polymer chains in reaction at time t
$[TP]_0$: initial concentration of terminated polymer chains in polymerization at time t
$[TP]_{cr}$: critical concentration of terminated polymer chains in polymerization mixture for the onset of chain entanglement where the center-to-center distance of neighboring polymer chains becomes equal to their hydrodynamic diameter
u_p: velocity of a particle or polymer chain in a liquid medium
u_r: radical component of velocity of reactive propagating chains in polymerization mixture due to convection in the mass conservation equation
u_ϕ: azimuthal angle component of velocity of reactive propagating chains in polymerization mixture due to convection in the mass conservation equation
u_θ: polar angle component of velocity of reactive propagating chains in polymerization mixture due to convection in the mass conservation equation
V: volume of a sphere for calculation of average separation distance between two polymer chains in polymerization mixture; also used as volume of the polymerization reaction
V_0: initial volume of polymerization reaction
V_{loc,R_g}: volume corresponding to the local reactive propagating chain at the center of spherical coordinate in a diffusion limited polymerization mixture
$\hat{V}_m^*$: partial specific volume of monomer in Vrentas–Duda equation for determination of viscosity of a polymerization mixture
$\hat{V}_p^*$: partial specific volume of polymer in Vrentas–Duda equation for determination of viscosity of a polymerization mixture
V_R, V_{rxn}: volume of polymerization reaction
w_m: weight fraction of monomer in Vrentas–Duda equation for determination of viscosity of a polymerization mixture
w_p: weight fraction of polymer in Vrentas–Duda equation for determination of viscosity of a polymerization mixture
α: reaction conversion or fraction of monomers in polymerization reaction converted to polymer chains
α_{cr}: critical reaction conversion in polymerization reaction for the onset of chain entanglement where the center-to-center distance of neighboring polymer chains becomes equal to their hydrodynamic diameter
γ_p: constant for a given polymer-solvent system in Vrentas–Duda equation for determination of viscosity of a polymerization mixture

ε: coefficient of shrinkage in polymerization reaction with conversion of monomer to polymer

η_s: viscosity of a polymer solution

$\eta_{s,\text{dilute}}$: viscosity of a polymer solution in dilute regime where the chains are below the entanglement concentration

η_{slv}: intrinsic viscosity of a polymer or Staudinger index or limiting viscosity number is the capacity of a polymer to increase viscosity of a solvent in the absence of polymer–polymer intermolecular interaction which is the ratio of relative viscosity to polymer concentration at infinite dilution

λ_k: kth order mole-fraction moment of reactive propagating chains at time t in the reaction

$\lambda'_{k,ci}$: instantaneous kth order mole-fraction moment of reactive propagating chains produced by chain initiation reaction in time interval dt

$\lambda'_{k,ctm}$, $\lambda'_{k,ctm,prd}$: instantaneous kth order mole-fraction moment of reactive propagating chains produced by transfer of a reactive species from a propagating chain to a monomer in polymerization mixture in time interval dt

$\lambda'_{k,ctm,con}$: instantaneous kth order mole-fraction moment of reactive propagating chains consumed by transfer of a reactive species from a propagating chain to a monomer in polymerization mixture in time interval dt

$\lambda'_{k,i}$: instantaneous kth order mole-fraction moment of reactive propagating chains produced by ith production reaction in time interval dt

$\lambda'_{k,l}$: instantaneous kth order mole-fraction moment of reactive propagating chains consumed by lth consumption reaction in time interval dt

$\lambda'_{k,prp,con}$, $\lambda'_{k,p,con}$: instantaneous kth order mole-fraction moment of reactive propagating chains consumed by propagation reaction in time interval dt

$\lambda'_{k,prp,prd}$, $\lambda'_{k,p,prd}$: instantaneous kth order mole-fraction moment of reactive propagating chains produced by propagation reaction in time interval dt

$\lambda'_{k,tc}$: instantaneous kth order mole-fraction moment of reactive propagating chains consumed by termination by combination reaction in time interval dt

$\lambda'_{k,td}$: instantaneous kth order mole-fraction moment of reactive propagating chains consumed by termination by disproportionation reaction in time interval dt

μ_k: kth order mole-fraction moment of terminated polymer chains at time t in the reaction; also used particle or polymer chain mobility in Einstein equation for determination of diffusion coefficient

$\mu'_{k,i}$: instantaneous kth order mole-fraction moment of terminated polymer chains produced by the ith production reaction in time interval dt

$\mu'_{k,l}$: instantaneous kth order mole-fraction moment of terminated polymer chains consumed by the lth consumption reaction in time interval dt

$\mu'_{k,prd,tc}$: instantaneous kth order mole-fraction moment of terminated polymer chains produced by termination reaction via combination in time interval dt

$\mu_{p,rep}$: mobility of whole polymer chain diffusing by reptation

μ_{seg}: segmental mobility of a polymer chain

$\hat{\upsilon}_{fv,m}$: free volume fraction of the pure monomer in Vrentas–Duda equation for determination of viscosity of a polymerization mixture

$\hat{\upsilon}_{fv,mix}$: free volume fraction of the mixture in Vrentas–Duda equation for determination of viscosity of a polymerization mixture

$\hat{\upsilon}_{fv,p}$: free volume fraction of the pure polymer in Vrentas–Duda equation for determination of viscosity of a polymerization mixture

$\hat{\upsilon}_{m,r}$: partial specific molar volume of the monomer residue (monomer repeat unit) in polymerization mixture

$\hat{v}_{m,ur}$: partial specific molar volume of unreacted monomer in polymerization mixture

ξ_0, ξ_{seg}: monomeric friction coefficient between a repeat units or monomer residue on a polymer chain and the liquid medium for determination of polymer diffusion coefficient by Einstein equation

ξ_p: friction coefficient between a diffusing particle or a polymer chain and the liquid medium for determination of polymer diffusion coefficient by Einstein equation

ρ_m: density of pure monomer

$\hat{\rho}_{m,r}$: partial molar density of the monomer residue (monomer repeat unit) in polymerization mixture

$\hat{\rho}_{m,ur}$: partial molar density of unreacted monomer in polymerization mixture

ρ_{MMA}: density of pure methyl methacrylate monomer

$\rho_{ru,p}$: molar density of the monomer repeat unit in the polymer chains in polymerization mixture

ρ_p: density of pure polymer

ρ_{PMMA}: density of pure methyl methacrylate polymer

ς_p: ratio of available volume for monomer diffusion by the available volume for polymer diffusion in Vrentas–Duda equation for determination of viscosity of a polymerization mixture

$\tau_{ci,diff}$: time constant for diffusion of primary reactive species, generated by initiator dissociation, for chain initiation between primary reactive species and monomer to form a 1-mer propagating chain

$\tau_{ci,rxn}$: time constant for reaction of primary reactive species, generated by initiator dissociation, for chain initiation between primary reactive species and monomer to form a 1-mer propagating chain in the absence of diffusion

$\tau_{ctm,diff}$: time constant for transfer of reactive species from a propagating chain to a monomer in a polymerization mixture

$\tau_{ctm,rxn}$: time constant for transfer of reactive species from a propagating chain to a monomer in a polymerization mixture in the absence of diffusion limitation (reaction controlled)

$\tau_{diff,m}$: time constant for diffusion of monomer in the propagation reaction in polymerization mixture

$\tau_{diff,p}$, $\tau_{d,p}$, $\tau_{p,diff}$: time constant for diffusion of a particle or a polymer chain in liquid medium calculated from diffusion coefficient determined by Einstein equation

$\tau_{diff,p,\mathrm{dilute}}$: time constant for diffusion of polymer chains in a dilute polymerization mixture where the chains are below the entanglement concentration

$\tau_{diff,p,rep}$: time constant for diffusion of polymer chains in a concentrated polymerization mixture where the propagating chains diffuse by reptation

$\tau_{diff,p,\mathrm{semidilute}}$: time constant for diffusion of polymer chains in a semidilute polymerization mixture where the chains are above the entanglement concentration but are not fully entangled to behave like a gel

$\tau_{diff,rp,\mathrm{conc}}$: time constant for diffusion of reactive propagating chains in a concentrated polymerization mixture where the chains are fully entangled

$\tau_{diff,rp,\mathrm{dilute}}$: time constant for diffusion of reactive propagating chains in a dilute polymerization mixture where the chains are below the entanglement concentration

$\tau_{diff,rp,rep}$: time constant for diffusion of reactive propagating chains in a concentrated polymerization mixture where the propagating chains diffuse by reptation

$\tau_{p,rxn}$: time constant for propagation reaction of reactive propagating chains with monomer in the absence of diffusion (reaction controlled)

$\tau_{\mathrm{prop,conc}}$: time constant for propagation reaction between reactive propagating chains and monomer in a concentrated polymerization mixture where the chains are fully entangled

$\tau_{prop,dilute}$: time constant for propagation reaction between reactive propagating chains and monomer in a dilute polymerization mixture where the chains are below the entanglement concentration

$\tau_{r,bm}$: time constant for a typical bimolecular reaction between two small molecules in a polymerization mixture

$\tau_{t,diff}$: time constant for termination reaction of reactive propagating chains if chain termination is limited by diffusion

$\tau_{t,rxn}$: time constant for termination reaction of reactive propagating chains in the absence of diffusion limitation (reaction controlled)

$\tau_{tc,diff}$: time constant for termination reaction of reactive propagating chains by combination reaction if chain termination is limited by diffusion

$\tau_{tc,rxn}$: time constant for termination reaction of reactive propagating chains by combination the absence of diffusion limitation (reaction controlled)

$\tau_{td,diff}$: time constant for termination reaction of reactive propagating chains by disproportionation reaction if chain termination is limited by diffusion

$\tau_{td,rxn}$: time constant for termination reaction of reactive propagating chains by disproportionation the absence of diffusion limitation (reaction controlled)

$\tau_{term,conc}$: time constant for termination reaction between two propagating chains diffusion in a concentrated polymerization mixture where the chains are fully entangled

$\tau_{term,dilute}$: time constant for termination reaction between two reactive propagating chains in dilute polymerization mixture the chains are below the entanglement concentration

8.10 Practice Problems

Problem 8.1 Reaction conversion α is defined by the following equation:

$$\alpha = \frac{[M]_0 V_0 - [M] V}{[M]_0 V_0}$$

where $[M]_0$ and V_0 are the initial monomer concentration and volume, respectively, and $[M]$ and V are the monomer concentration and volume at conversion α.

a. Starting from Eq. (8.98) in Section 8.6.1, derive the following equation relating the rate of change of reaction volume with time to the rate of change in conversion with time:

$$\frac{dV}{dt} = -\frac{\varepsilon [M]_0 V_0}{\rho_m} \frac{d\alpha}{dt}$$

b. Integrate the above equation subject to the initial condition and show that volume is related to conversion as follows:

$$V = V_0 \left(1 - \frac{\varepsilon [M]_0}{\rho_m} \alpha\right)$$

c. Use the above equation and the defining equation for conversion to derive the following relation between the monomer concentration and conversion:

$$[M] = (1 - \alpha) \Big/ \left(\frac{1}{[M]_0} - \frac{\varepsilon}{\rho_m} \alpha\right)$$

Problem 8.2 The molar mass conservation equation of monomer for the radical polymerization reaction described by Rxns. 8.88–8.94 with varying volume is given by the following equation (see Eq. 8.104):

$$\frac{d\left([M]\,V\right)}{dt} \cong -\,[M]\,V\,[RP]\left(k_{p,app} + k_{trm,app}\right)$$

Using the above equation as the starting point and the relation between volume and conversion (Eq. 8.99) and monomer concentration and conversion (Eq. 8.100), derive step-by-step the following equations:

$$\frac{1}{(1-\alpha)}\frac{d\alpha}{dt} \cong -Q_{0,rp}\left(k_{p,app} + k_{trm,app}\right)$$

$$\frac{1}{\left(1-\varepsilon\,[M]\,/\rho_m\right)[M]}\frac{d\,[M]}{dt} \cong -\left(k_{p,app} + k_{trm,app}\right)Q_{0,rp}$$

Problem 8.3 The Vrentas–Duda equation for the prediction of viscosity of polymer solutions as a function of polymer concentration is given by the following equations [35–40]:

$$\eta_s = \eta_{\text{int}}\exp\left\{\frac{\gamma}{\varsigma_p}\left[\frac{\left(w_m\widehat{V}^*_m + w_p\widehat{V}^*_p\xi_p\right)}{\widehat{\upsilon}_{fv,mix}} - \frac{1}{\upsilon_{fv,m}}\right]\right\}$$

$$\varsigma_p = \left(\widehat{V}^*_m M_{jump,m}\right)/\left(\widehat{V}^*_p M_{jump,p}\right)$$

The parameters of the Vrentas–Duda equation were defined in Section 8.2. The parameters of Vrentas–Duda equation for PMMA-MMA bulk polymerization system are as follows: $\gamma_p = 0.763$, $\widehat{V}^*_m = 0.822\ \text{cm}^3\text{g}^{-1}$ $\widehat{V}^*_p = 0.770\ \text{cm}^3\text{g}^{-1}$, $M_{\text{jump},m} = 100\ \text{g mole}^{-1}$, $M_{\text{jump},p} = 150\ \text{g mole}^{-1}$, and $\varsigma_p = 0.712$. Furthermore, $\upsilon_{fv,m}$ and $\upsilon_{fv,p}$ are given as a function of temperature by the following expressions [33, 41]:

$$MMA \Rightarrow \upsilon_{fv,m} = 0.149 + 2.9\times10^{-4}T \qquad T\ (°\text{C})$$

$$\text{PMMA} \Rightarrow \upsilon_{fv,p} = 0.0149 + 1.3\times10^{-4}\,(T-105) \qquad T\ (°\text{C})$$

In a typical bulk MMA radical polymerization at 50 °C, monomer density, ρ_{MMA}, polymer density, ρ_{PMMA}, initiator concentration, $[I]_0$, initiator efficiency, f_0, dissociation rate constant, k_d, propagation rate constant, k_p, and termination rate constant by disproportionation, k_{td}, are as follows [33, 70, 72–77]:

$$\rho_{MMA} = 0.88\ \text{g cm}^{-3} \quad \rho_{\text{PMMA}} = 1.2\ \text{g cm}^{-3} \quad [I]_0 = 0.001\ \text{M} \quad f_0 = 1$$

$$k_d = 5\times10^{-5}\ \text{s}^{-1} \quad k_{p,\text{int}} = 300\frac{\text{lit}}{\text{mol s}} \quad k_{td,\text{int}} = 5\times10^{7}\frac{\text{lit}}{\text{mol s}}$$

The intrinsic viscosity (in $\text{m}^3\ \text{kg}^{-1}$) for PMMA solutions (MW >35 kDa), based on MHS equation, is $\eta_{\text{int}} = 5.2\times10^{-6}\overline{M}_w^{\,0.76}$ [33, 34] and the length of a monomer segment on PMMA chains is $b = 0.58$ nm [70]. Assuming termination is by disproportionation only, determine the following:

a. determine the concentrations of monomer and polymer corresponding to 15%, 45%, and 75% monomer conversion.

b. determine number-average DP of the propagating chains, $\overline{DP}_{n,rp}$, and terminated polymer chains, $\overline{DP}_{n,tp}$, at $t \approx 0$.

c. based on the DP values, determine the average diameter of the propagating chains, $\overline{d}_{rp}$, and terminated polymer chains, $\overline{d}_{tp}$.

d. based on the concentrations in part A, determine the average separation distance between the propagating chains, $\overline{sd}_{rp}$, and terminated polymer chains, $\overline{sd}_{tp}$, in the reaction mixture corresponding to 15%, 45%, and 75% conversion.

e. assuming $\overline{M}_w \approx \overline{M}_n$ for this PMMA polymerization and $\overline{M}_n \approx \overline{M}_{n,t=0}$, determine the intrinsic viscosity of the PMMA in its monomer.

f. using the given parameters for Vrentas–Duda equation for PMMA-MMA bulk polymerization, determine the solution viscosity of PMMA-MMA polymerization at conversions of 15%, 45%, and 75%.

g. using the Stokes–Einstein relation [13] (Eq. 8.4 or 8.7), determine the diffusion coefficient of the monomer in the polymerization solution at 50 °C for conversions of 15%, 45%, and 75%.

h. using the average separation distances from part d, the diffusion coefficients in part g, and the defining equation for time constant (Eq. 8.5), determine the diffusion time constants for propagation and termination by disproportionation reactions, $\tau_{diff,p}$ and $\tau_{diff,td}$ for conversions of 15%, 45%, and 75%.

i. determine the apparent rate constants for propagation and termination by disproportionation reactions, $k_{p,app}$ and $k_{td,app}$ for conversions of 15%, 45%, and 75%.

Problem 8.4 The general conservation equation for kth order molar concentration moment of the propagating chains for the radical polymerization reaction described by Rxns. 8.88–8.94 with varying volume is given by the following equation (see Eq. 8.113):

$$\frac{dQ_{k,rp}}{dt} = 2f_{app}k_d [I] + k_{p,app} [M] Q_{0,rp} \left(\lambda'_{k,p,prd} - \lambda'_{k,p,con}\right) + k_{ctm,app} [M] Q_{0,rp} \left(\lambda'_{k,ctm,prd} - \lambda'_{k,ctm,con}\right) - Q^2_{0,rp} \left(k_{td,app}\lambda'_{k,td} + k_{tc,app}\lambda'_{k,tc}\right) - \frac{Q_{k,rp}}{V}\frac{dV}{dt}$$

The equations for zero and first order concentration moments of the propagating chains, $Q_{0,rp}$ and $Q_{1,rp}$, respectively, were derived in Sections 8.6.4.1 and 8.6.4.2 (see Eqs. 8.119 and 8.121) and used to determine the kinetic chain length or $\overline{DP}_{n,rp} = L$. The weight-average DP of the propagating chains which is defined by $\overline{DP}_{w,rp} = \frac{Q_{2,rp}}{Q_{1,rp}}$ where $Q_{2,rp}$ is the second order concentration moment of the propagating chains for the polymerization system described by Rxns. 8.88–8.94. Derive the differential equation for $Q_{2,rp}$ and show that it is given by the following expression:

$$\frac{1}{Q_{2,rp}}\frac{dQ_{2,rp}}{dt} = +\frac{2f_{app}k_d [I]}{Q_{2,rp}} + k_{p,app} [M] \left(\frac{2Q_{1,rp} + Q_{0,rp}}{Q_{2,rp}}\right) + k_{ctm,app} [M] \left(\frac{Q_{0,rp}}{Q_{2,rp}} - 1\right) - Q_{0,rp} \left(k_{td,app} + k_{tc,app}\right) - \frac{1}{[(\rho_m/\varepsilon [M]_0) - \alpha]}\frac{d\alpha}{dt}$$

References

1 Srinivas, G., Yethiraj, A., and Bagchi, B., Nonexponentiality of time dependent survival probability and the fractional viscosity dependence of the rate in diffusion controlled reactions in a polymer chain. *Journal of Chemical Physics*, 2001, **114**(20), 9170.

2 Sharikov, Y.V., Petyukhin, K.V., and Zhitinkin, A.A., Kinetics of the formation of polymers accompanied by a substantial increase in the viscosity of the reaction medium. *Kinetics and Catalysis*, 1979, **20**(2), 401.

3 Abrams, C.F., Lee, N.K., and Johner, A., Diffusion/reaction in confined polymer chains. *Macromolecules*, 2006, **39**(10), 3655.

4 Everaers, R. and Rosa, A., Multi-scale modeling of diffusion-controlled reactions in polymers: renormalisation of reactivity parameters. *Journal of Chemical Physics*, 2012, **136**(1), 014902.

5 Srinivas, G. and Bagchi, B., Distribution of reaction times in diffusion controlled reactions in polymers. *Chemical Physics Letters*, 2000, **328**(4–6), 420.

6 Karlsson, O.J., Stubbs, J.M., Karlsson, L.E., and Sundberg, D.C., Estimating diffusion coefficients for small molecules in polymers and polymer solutions. *Polymer*, 2001, **42**(11), 4915.

7 Waggoner, R.A., Blum, F.D., and Macelroy, J.M.D., Dependence of the solvent diffusion-coefficient on concentration in polymer-solutions. *Macromolecules*, 1993, **26**(25), 6841.

8 Chandrasekhar, S., Stochastic problems in physics and astronomy. *Reviews Modern Physics*, 1943, **15**(1), 1.

9 Dehara, K., Yoshizaki, T., and Yamakawa, H., Translational diffusion-coefficient of oligomethyl and poly(methyl methacrylate)s in dilute-solutions. *Macromolecules*, 1993, **26**(19), 5137.

10 Zhu, S., Tian, Y., Hamielec, A.E., and Eaton, D.R., Radical concentrations in free-radical copolymerization of MMA/EGDMA. *Polymer*, 1990, **31**(1), 154.

11 Bartels, C.R., Crist, B., and Graessley, W.W., Self-diffusion coefficient in melts of linear-polymers - chain-length and temperature-dependence for hydrogenated polybutadiene. *Macromolecules*, 1984, **17**(12), 2702.

12 Dill, K.A. and Bromberg, S., *Molecular driving forces: statistical thermodynamics in chemistry and biology*. 2003, Garland Science, 327.

13 Miller, C.C., The stokes einstein law for diffusion in solution. *Proceedings of the Royal Society of London Series a-Containing Papers of a Mathematical and Physical Character*, 1924, **106**(740), 724–749.

14 Kubo, R., The fluctuation-dissipation theorem. *Reports Progress Physics*, 1966, **29**(1), 255.

15 Stokes, G.G., On the effect of internal friction of fluids on the motion of pendulums. *Transactions of the Cambridge Philosophical Society*, 1856, **9**, 8.

16 Bird, R.B., Stewart, W.E., and Lightfoot, E.N., *Transport phenomena*. 2 ed. 2001, John Wiley, p. 545.

17 Kamide, K. and Kataoka, A., Theoretical relationships between parameters in Khun-Mark-Houwink-Sakurada equation. *Makromolekulare Chemie*, 1969, **128**(Oct), 217.

18 Dobbin, C.J.B., Rudin, A., and Tchir, M.F., Evaluation of GPC methods for estimation of Mark-Houwink-Sakurada constants. *Journal of Applied Polymer Science*, 1980, **25**(12), 2985.

19 Zhongde, X., Mingshi, S., Hadjichristidis, N., and Fetters, L.J., Method for gel-permeation chromatography calibration and the evaluation of Mark–Houwink–Sakurada constants. *Macromolecules*, 1981, **14**(5), 1591.

20 Hiemenz, P. and Lodge, T., *Polymer chemistry*. 2007, CRC Press, 608.

21 Flory, P.J., *Principles of polymer chemistry*. 1953, Cornell University Press, 688.

22 Wall, F.T., Hiller, L.A., and Wheeler, D.J., Statistical computation of mean dimensions of macromolecules. I. *Journal of Chemical Physics*, 1954, **22**(6), 1036.

23 Wall, F.T., Hiller, L.A., and Atchison, W.F., Statistical computation of mean dimensions of macromolecules. II. *Journal of Chemical Physics*, 1955, **23**(5), 913.

24 Wall, F.T., Hiller, L.A., and Atchison, W.F., Statistical computation of mean dimensions of macromolecules. III. *Journal of Chemical Physics*, 1955, **23**(12), 2314.

25 Wall, F.T., Hiller, L.A., and Atchison, W.F., Statistical computation of mean dimensions of polymer molecules. IV. *Journal of Chemical Physics*, 1957, **26**(6), 1742.

26 Masaro, L. and Zhu, X.X., Physical models of diffusion for polymer solutions, gels and solids. *Progress in Polymer Science*, 1999, **24**(5), 731.

27 Lyons, P.F. and Tobolsky, A.V., Viscosity of polypropylene oxide solutions over entire concentration range. *Polymer Engineering and Science*, 1970, **10**(1), 1.

28 Rodriguez, F., Suitability of equations for viscosity correlation over entire concentration range. *Journal of Polymer Science Part B-Polymer Letters*, 1972, **10**(6), 455.

29 Quadrat, O., Dependence of viscosity on concentration of polymer-solutions – use of Lyons–Tobolsky equation. *Collection of Czechoslovak Chemical Communications*, 1977, **42**(5), 1520.

30 Rink, M., Pavan, A., and Roccasalvo, S., Concentration-dependence of zero-shear viscosity for polymer-solutions - test of Lyons–Tobolsky equation. *Polymer Engineering and Science*, 1978, **18**(10), 755.

31 Martin, A.F., *American chemical society.*, Memphis, TN, 1942, **19**, pp. 252.

32 Fedors, R.F., Equation suitable for describing the viscosity of dilute to moderately concentrated polymer-solutions. *Polymer*, 1979, **20**(2), 225.

33 Singh, G.J. and Gupta, S.K., Viscosity of moderately concentrated solutions of polymethyl-methacrylate in methyl-methacrylate. *Journal of Applied Polymer Science*, 2008, **109**(4), 2139.

34 Wagner, H.L., The Mark-Houwink-Sakurada relation for poly(methyl methacrylate). *Journal of Physical and Chemical Reference Data*, 1987, **16**(2), 165.

35 Vrentas, J.S. and Duda, J.L., Diffusion of large penetrant molecules in amorphous polymers. *Journal of Polymer Science Part B-Polymer Physics*, 1979, **17**(6), 1085.

36 Vrentas, J.S. and Duda, J.L., Molecular-diffusion in polymer-solutions. *Aiche Journal*, 1979, **25**(1), 1.

37 Duda, J.L. and Vrentas, J.S., Free-volume analysis of molecular-diffusion in polymer melts. *Abstracts of Papers of the American Chemical Society*, 1981, **181**(3), 104.

38 Duda, J.L. and Vrentas, J.S., Diffusion in polymer solvent systems. *Journal of Polymer Science-Polymer Symposia*, 1985, **72**(1), 299.

39 Vrentas, J.S., Liu, H.T., and Duda, J.L., Estimation of diffusion-coefficients for trace amounts of solvents in glassy and molten polymers. *Journal of Applied Polymer Science*, 1980, **25**(7), 1297.

40 Zhong, C.L., Yang, C.S., and Qu, Y.X., A modified Vrentas–Duda model for the correlation of the solvent self-diffusion coefficients in polymer solutions. *Polymer Journal*, 2001, **33**(11), 842.

41 Liu, H.T., Duda, J.L., and Vrentas, J.S., Influence of solvent size on the concentration-dependence of polymer-solvent diffusion-coefficients. *Macromolecules*, 1980, **13**(6), 1587.

42 Veeravalli, M.S. and Rosen, S.L., Free-radical addition polymerization – effects of chain-length, viscosity, and solvent power on diffusion-controlled termination at low conversion. *Journal of Polymer Science Part B-Polymer Physics*, 1990, **28**(5), 775.

43 Dvornic, P.R. and Jacovic, M.S., The viscosity effect on auto-acceleration of the rate of free-radical polymerization. *Polymer Engineering and Science*, 1981, **21**(12), 792.

44 Ogo, Y. and Kyotani, T., Effect of pressure on termination rate constant in free-radical polymerization - correlation between rate constant and monomer viscosity. *Makromolekulare Chemie-Macromolecular Chemistry and Physics*, 1978, **179**(10), 2407.

45 Brooks, B.W., Viscosity effects in free-radical polymerization of methyl-methacrylate. *Proceedings of the Royal Society of London Series A-Mathematical and Physical Sciences*, 1977, **357**(1689), 183.

46 Degennes, P.G., Dynamics of entangled polymer-solutions. 1. Rouse Model. *Macromolecules*, 1976, **9**(4), 587.

47 Noolandi, J., Slater, G.W., and Allegra, G., Generalized rouse model for polymer melt dynamics. *Makromolekulare Chemie-Rapid Communications*, 1987, **8**(1), 51.

48 Bandyopadhyay, T. and Ghosh, S.K., Diffusion assisted end-to-end relaxation of a flexible Rouse polymer chain: fluorescence quenching through a model energy transfer. *Journal of Chemical Physics*, 2003, **119**(1), 572.

49 Hervet, H., Leger, L., and Rondelez, F., Self-diffusion in polymer-solutions - test for scaling and reptation. *Physical Review Letters*, 1979, **42**(25), 1681.

50 Hess, W., Self-diffusion and reptation in semidilute polymer-solutions. *Macromolecules*, 1986, **19**(5), 1395.

51 Brown, W. and Pu, Z., Self-diffusion - an experimental demonstration of the non-applicability of the reptation model to semidilute polymer solutions. *Polymer*, 1990, **31**(5), 772.

52 Doi, M. and Edwards, S.F., Dynamics of concentrated polymer systems. 4. Rheological properties. *Journal of the Chemical Society-Faraday Transactions 2: Molecular and Chemical Physics*, 1979, **75**, 38.

53 Doi, M. and Edwards, S.F., Dynamics of concentrated polymer systems. 3. Constitutive equation. *Journal of the Chemical Society-Faraday Transactions 2: Molecular and Chemical Physics*, 1978, **74**, 1818.

54 Doi, M. and Edwards, S.F., Dynamics of concentrated polymer systems. 2. Molecular-motion under flow. *Journal of the Chemical Society-Faraday Transactions 2: Molecular and Chemical Physics*, 1978, **74**, 1802.

55 Doi, M. and Edwards, S.F., Dynamics of concentrated polymer systems. 1. Brownian-motion in equilibrium state. *Journal of the Chemical Society-Faraday Transactions 2: Molecular and Chemical Physics*, 1978, **74**, 1789.

56 Degennes, P.G., Reptation of a polymer chain in presence of fixed obstacles. *Journal of Chemical Physics*, 1971, **55**(2), 572.

57 Degennes, P.G., Reptation of stars. *Journal De Physique*, 1975, **36**(12), 1199.

58 Degennes, P.G., Reptation of a heterogeneous chain. *Journal De Physique Lettres*, 1983, **44**(6), L225.

59 Stickler, M., Panke, D., and Wunderlich, W., Solution properties of poly(methyl methacrylate) in methyl-methacrylate. 1. Viscosities from the dilute to the concentrated-solution regime. *Makromolekulare Chemie-Macromolecular Chemistry and Physics*, 1987, **188**(11), 2651.

60 van Meerveld, J., A method to extract the monomer friction coefficient from the linear viscoelastic behavior of linear, entangled polymer melts. *Rheologica Acta*, 2004, **43**(6), 615.

61 Victoria-Valenzuela, D. et al., Bulk free radical polymerization of methyl methacrylate and vinyl acetate: a comparative study. *Macromolecular Reaction Engineering*, 2016, **10**(6), 577.

62 Jaso, V.M., Stoiljkovic, D.M., and Radicevic, R.Z., Effect of temperature on initial stage of methyl methacrylate free radical bulk polymerization. *Hemijska Industrija*, 2010, **64**(6), 503.

63 Stempfle, B. et al., Single molecule diffusion and its heterogeneity during the bulk radical polymerization of styrene and methyl methacrylate. *Polymer Chemistry*, 2012, **3**(9), 2456.

64 Wunderlich, W., Benfaremo, N., Klapper, M., and Mullen, K., Effect of a pi-donor on the radical bulk polymerization of methyl methacrylate. *Macromolecular Rapid Communications*, 1996, **17**(6), 433.

65 Shen, J.C., Yuan, T., Wang, G.B., and Yang, M.L., Modeling and kinetic-study on radical polymerization of methyl-methacrylate in bulk. 1. Propagation and termination rate coefficients and initiation efficiency. *Makromolekulare Chemie-Macromolecular Chemistry and Physics*, 1991, **192**(11), 2669.

66 Garrett, R.W. et al., Application of ESR spectroscopy to the kinetics of free-radical polymerization of methyl-methacrylate in bulk to high conversion. *Polymer Bulletin*, 1989, **22**(5–6), 611.

67 Faldi, A., Tirrell, M., Lodge, T.P., and Vonmeerwall, E.D., Monomer diffusion and the kinetics of propagation in free-radical polymerization. *Abstracts of Papers of the American Chemical Society*, 1991, **202**, 165.

68 Faldi, A., Tirrell, M., Lodge, T.P., and Vonmeerwall, E., Monomer diffusion and the kinetics of methyl-methacrylate radical polymerization at intermediate to high conversion. *Macromolecules*, 1994, **27**(15), 418.

69 Lei, J.Z., Probability distribution of the radius of gyration of freely jointed chains. *Journal of Chemical Physics*, 2010, **133**(10), 104903.

70 Moyses, S., Solution properties of poly(methyl methacrylate) in dimethylsulfoxide. *International Journal of Polymer Analysis and Characterization*, 2008, **13**(6), 413.

71 Reaction rate constant, in *Wikipedia: the free encyclopedia*. 2025, Wikipedia Foundation, https://en.wikipedia.org/wiki/Reaction_rate_constant.

72 Beuermann, S. and Buback, M., Rate coefficients of free-radical polymerization deduced from pulsed laser experiments. *Progress in Polymer Science*, 2002, **27**(2), 191.

73 Zoller, A., Gigmes, D., and Guillaneuf, Y., Simulation of radical polymerization of methyl methacrylate at room temperature using a tertiary amine/BPO initiating system. *Polymer Chemistry UK*, 2015, **6**(31), 5719.

74 Beuermann, S. et al., Critically evaluated rate coefficients for free-radical polymerization .2. Propagation rate coefficients for methyl methacrylate. *Macromolecular Chemistry and Physics*, 1997, **198**(5), 1545.

75 Gilbert, R.G., Critically-evaluated propagation rate coefficients in free radical polymerizations. 1. Styrene and methyl methacrylate. *Pure and Applied Chemistry*, 1996, **68**(7), 1491.

76 Zammit, M.D., Davis, T.P., and Haddleton, D.M., Determination of the propagation rate coefficient (k(p)) and termination mode in the free-radical polymerization of methyl methacrylate, employing matrix-assisted laser desorption ionization time-of-flight mass spectrometry for molecular weight distribution analysis. *Macromolecules*, 1996, **29**(1), 492.

77 Sackkouloumbris, R. and Meyerhoff, G., Radical polymerization of methyl-methacrylate over the full range of conversion - stationary and non-stationary experiments for the determination of the rates of propagation and termination. *Makromolekulare Chemie-Macromolecular Chemistry and Physics*, 1989, **190**(5), 1133.

78 Osa, M., Yoshizaki, T., and Yamakawa, H., Mean-square radius of gyration of oligo- and poly(α-methylstyrene)s in dilute solution. *Macromolecules*, 2000, **33**(13), 4828.

79 Nakamura, Y. and Yamago, S., Termination mechanism in the radical polymerization of methyl methacrylate and styrene determined by the reaction of structurally well-defined polymer end radicals. *Macromolecules*, 2015, **48**(18), 6450.

80 Ballard, N., Hamzehlou, S., Ruiperez, F., and Asua, J.M., On the termination mechanism in the radical polymerization of acrylates. *Macromolecular Rapid Communications*, 2016, **37**(16), 1364.

81 Nakamura, Y., Lee, R., Coote, M.L., and Yamago, S., Termination mechanism of the radical polymerization of acrylates. *Macromolecular Rapid Communications*, 2016, **37**(6), 506.

82 Barth, J., Buback, M., Russell, G.T., and Smolne, S., Chain-length-dependent termination in radical polymerization of acrylates. *Macromolecular Chemistry and Physics*, 2011, **212**(13), 1366.

83 Bixon, M., Dynamics of polymer molecules in dilute-solutions. *Journal of Chemical Physics*, 1973, **58**(4), 1459.

84 Bixon, M., Polymer dynamics in solution. *Annual Review of Physical Chemistry*, 1976, **27**, 65.

85 Shearmur, T.E. et al., Interdiffusion of deuterated and protonated poly(methyl methacrylate). *Polymer*, 1998, **39**(11), 2155.

86 Nam, J.E., Lee, J.K., and Mauldin, T.C., Isothermal physical aging of thin PMMA films near the glass transition temperature. *Polymer Bulletin*, 2010, **65**(8), 825.

87 Ramesh, N. and Duda, J.L., Diffusion in polymers below the glass transition temperature: comparison of two approaches based on free volume concepts. *Korean Journal of Chemical Engineering*, 2000, **17**(3), 310.

88 Friedman, B. and Oshaughnessy, B., Kinetics of intermolecular reactions in dilute polymer-solutions and unentangled melts. *Macromolecules*, 1993, **26**(21), 5726.

89 Russell, G.T., The kinetics of free-radical polymerizing systems at low conversion .1. On the rate-determining step of the bimolecular termination reaction. *Macromolecular Theory and Simulations*, 1995, **4**(3), 497.

90 Russell, G.T., Napper, D.H., and Gilbert, R.G., Termination in free-radical polymerizing systems at high conversion. *Macromolecules*, 1988, **21**(7), 2133.

91 Zhu, S. and Hamielec, A.E., Chain-length-dependent termination for free-radical polymerization. *Macromolecules*, 1989, **22**(7), 3093.

92 Buback, M. and Russell, G.T., Detailed analysis of termination kinetics in radical polymerization. *Polymer International*, 2023, **72**(10), 869.

93 Berg, O.G. and Vonhippel, P.H., Diffusion-controlled macromolecular interactions. *Annual Review of Biophysics and Biophysical Chemistry*, 1985, **14**, 131.

94 Vijaykumar, A., Bolhuis, P.G., and ten Wolde, P.R., The intrinsic rate constants in diffusion-influenced reactions. *Faraday Discussions*, 2016, **195**, 421.

95 Shen, J.C., Tian, Y., Wang, G.B., and Yang, M.L., Termination process and the termination parameter-kt of methyl-methacrylate bulk radical polymerization. *Science in China Series B-Chemistry*, 1990, **33**(9), 1040.

96 Zhu, S., Tian, Y., Hamielec, A.E., and Eaton, D.R., Radical trapping and termination in free-radical polymerization of methyl methacrylate. *Macromolecules*, 1990, **23**(4), 1144.

97 Horie, K., Mita, I., and Kambe, H., Fast reaction and micro-Brownian motion of flexible polymer molecules in solution. *Polymer Journal*, 1973, **4**(3), 341.

98 Fetters, L.J., Lohse, D.J., and Colby, R.H., Chain dimensions and entanglement spacings, in *Physical properties of polymers handbook*. 2007, Springer, pp. 447–454.

99 Mahabadi, H.K., Effects of chain-length dependence of termination rate-constant on the kinetics of free-radical polymerization. 1. Evaluation of an analytical expression relating the apparent rate-constant of termination to the number-average degree of polymerization. *Macromolecules*, 1985, **18**(6), 1319.

100 Leuschner, R., Krohn, H., and Dohrmann, J.K., Termination rates by time-resolved electron-spin-resonance - the N-hydropyridinyl radical in solution and the influence of non-uniform radical concentration on the apparent rate-constant. *Berichte Der Bunsen-Gesellschaft-Physical Chemistry Chemical Physics*, 1984, **88**(5), 462.

101 Penlidis, A., Ponnuswamy, S.R., Kiparissides, C., and Odriscoll, K.F., Polymer reaction-engineering - modeling considerations for control studies. *Chemical Engineering Journal and the Biochemical Engineering Journal*, 1992, **50**(2), 95.

102 Achilias, D.S. and Kiparissides, C., On the validity of the steady-state approximations in high conversion diffusion-controlled free-radical copolymerization reactions. *Polymer*, 1994, **35**(8), 1714.

103 Spade, C.A. and Volpert, V.A., On the steady-state approximation in thermal free radical frontal polymerization. *Chemical Engineering Science*, 2000, **55**(3), 641.

104 Bamford, C.H. and Mullik, S.U., New thermal initiators of free-radical polymerization and block copolymerization. *Polymer*, 1976, **17**(1), 94.

105 Kim, B. et al., Thermal radical initiator derivatives based on o-imino-isourea: synthesis, polymerization, and characterization. *Journal of Polymer Science Part a-Polymer Chemistry*, 2016, **54**(22), 3593.

106 Kang, S. et al., New optimized triazene radical initiators for thermal polymerization. *Journal of Industrial and Engineering Chemistry*, 2019, **79**, 274.

107 Noguchi, N. et al., 2,2'-azobis (4-methoxy-2,4-dimethylvaleronitrile), a new lipid-soluble azo initiator: application to oxidations of lipids and low-density lipoprotein in solution and in aqueous dispersions. *Free Radical Biology and Medicine*, 1998, **24**(2), 259.

108 Huang, N.J. and Sundberg, D.C., Fundamental-studies of grafting reactions in free-radical copolymerization. 2. Grafting of styrene, acrylate, and methacrylate monomers onto cis-polybutadiene using AIBN initiator in solution polymerization. *Journal of Polymer Science Part a-Polymer Chemistry*, 1995, **33**(15), 2551.

109 Huang, N.J. and Sundberg, D.C., Fundamental-studies of grafting reactions in free-radical copolymerization.4. Grafting of styrene, acrylate, and methacrylate monomers onto vinyl-polybutadiene using benzoyl peroxide and AIBN initiators in solution polymerization. *Journal of Polymer Science Part a-Polymer Chemistry*, 1995, **33**(15), 2587.

110 Yamago, S., Iida, K., Nakajima, M., and Yoshida, J., Practical protocols for organotellurium-mediated living radical polymerization by in situ generated initiators from AIBN and ditellurides. *Macromolecules*, 2003, **36**(11), 3793.

111 Berger, K.C., Deb, P.C., and Meyerhoff, G., Radical vinyl polymerization - reactions of benzoyl peroxide during bulk-polymerization of styrene with labeled initiator. *Macromolecules*, 1977, **10**(5), 1075.

112 Flory, P.J. and Krigbaum, W.R., Thermodynamics of high polymer solutions. *Annual Review of Physical Chemistry*, 1951, **2**, 383.

113 Sanchez, I.C. and Lacombe, R.H., Statistical thermodynamics of polymer-solutions. *Macromolecules*, 1978, **11**(6), 1145.

114 Hayward, R.C. and Graessley, W.W., Excluded volume effects in polymer solutions. 1. Dilute solution properties of linear chains in good and Θ solvents. *Macromolecules*, 1999, **32**(10), 3502.

115 Alnazzawi, A. and Watts, D.C., Simultaneous determination of polymerization shrinkage, exotherm and thermal expansion coefficient for dental resin-composites. *Dental Materials*, 2012, **28**(12), 1240.

116 Sanda, F., Takata, T., and Endo, T., Radical ring-opening polymerization of novel vinylcyclopropanes designed as low shrinkage monomers - structure of the polymer, mechanism of the polymerization, and volume change on the polymerization. *Macromolecules*, 1995, **28**(5), 1346.

117 Kaloforov, N.Y., AIBN-initiated polymerization of methyl-methacrylate in pyridine. *Journal of Macromolecular Science-Chemistry*, 1977, **A11**(4), 733.

118 Moad, G., A critical survey of dithiocarbamate reversible addition-fragmentation chain transfer (RAFT) agents in radical polymerization. *Journal of Polymer Science Part a-Polymer Chemistry*, 2019, **57**(3), 216.

119 Senapati, M. et al., Kinetics of polymerization of methyl-methacrylate initiated by AIBN in the presence of acetophenone. *Journal of Macromolecular Science A-Chemistry*, 1983, **20**(5–6), 567.

120 Matheson, M.S., Auer, E.E., Bevilacqua, E.B., and Hart, E.J., Rate constants in free radical polymerizations. 1. Methyl methacrylate. *Journal of the American Chemical Society*, 1949, **71**(2), 497.

121 Shen, J.C., Tian, Y., Zeng, Y.G., and Qiu, Z.W., Electron-spin-resonance study of the propagating rate-constant and radical concentration of bulk-polymerization of methyl-methacrylate. *Makromolekulare Chemie-Rapid Communications*, 1987, **8**(12), 615.

122 Kim, J.Y. and Laurence, R.L., Mathematical model and analysis of PMMA solution processes. *Korean Journal of Chemical Engineering*, 1998, **15**(3), 287.

123 Chen, W.C., Lee, S.J., and Ho, B.C., Diffusion coefficients of acrylic monomers in poly(methyl methacrylate). *Journal of Polymer Research-Taiwan*, 1998, **5**(3), 187.

124 Zielinski, J.M., Benesi, A.J., and Duda, J.L., Use of solvent C-13 relaxation to predict the temperature and concentration behavior of polymer solvent diffusion-coefficients. *Industrial & Engineering Chemistry Research*, 1992, **31**(9), 2146.

125 Zielinski, J.M. and Duda, J.L., Predicting polymer solvent diffusion-coefficients using free-volume theory. *Aiche Journal*, 1992, **38**(3), 404.

126 Sturm, D.R., Danner, R.P., Moser, J.D., and Chiu, S.W., Application of the Vrentas–Duda free-volume theory of diffusion below the glass-transition temperature: application to hypromellose acetate succinate-solvent systems. *Journal of Applied Polymer Science*, 2019, **136**(16), 47351.

127 Fan, W., Zhou, Q., Sun, J., and Zhang, S.J., Density, excess molar volume, and viscosity for the methyl methacrylate+1-butyl-3-methylimidazolium hexafluorophosphate ionic liquid binary system at atmospheric pressure. *Journal of Chemical and Engineering Data*, 2009, **54**(8), 2307.

128 Z-transform, in *Wikipedia: the free encyclopedia*. 2025, Wikimedia Foundation, https://en.wikipedia.org/wiki/Z-transform.

129 Vrentas, J.S., Duda, J.L., and Ni, L.W., Concentration-dependence of polymer self-diffusion coefficients. *Macromolecules*, 1983, **16**(2), 261.

130 Vrentas, J.S. and Duda, J.L., Diffusion in polymer-solvent systems. 2. Predictive theory for dependence of diffusion-coefficients on temperature, concentration, and molecular-weight. *Journal of Polymer Science Part B-Polymer Physics*, 1977, **15**(3), 417.

131 Cui, X.H. et al., Probe into effect of the initiator on the thermal runaway hazards of methyl methacrylate bulk polymerization. *Journal of Loss Prevention in the Process Industries*, 2023, **83**, 105074.

132 Han, J.P. et al., Vinyl chloride suspension polymerization: kinetic modelling, thermal runaway prediction, and prevention. *Canadian Journal of Chemical Engineering*, 2023, **101**(9), 4927.

133 Yu, A.D. et al., Process hazard evaluation and thermal runaway prediction for styrene polymerization initiated by peroxide-azo composites. *Process Safety and Environmental Protection*, 2022, **166**, 257.

134 Liu, G.Y. and Wilhite, B.A., Model-based design for inhibition of thermal runaway in free radical polymerization. *Industrial & Engineering Chemistry Research*, 2019, **58**(37), 17244.

135 Fujita, M., Izato, Y., Iizuka, Y., and Miyake, A., Thermal hazard evaluation of runaway polymerization of acrylic acid. *Process Safety and Environmental Protection*, 2019, **129**, 339.

136 Chen, Y.J. et al., Thermal runaway inhibition of styrene polymerization reaction by 4,6-dinitro-sec-butylphenol (DNBP). *Journal of Thermal Analysis and Calorimetry*, 2024, **149**, 4491.

9

Markov Chain Modeling of Polymerization Reactions

9.1 Introduction

In polymerization reactions, monomers are added one at a time to the propagating chains. The propagation event in the reaction follows a random process that depends on the likelihood or probability of chain propagation versus the likelihood of chain termination. Due to its random nature, chains in a polymerization reaction are polydisperse and have a distribution of degree of polymerization (DP) with a mean $\overline{DP}$. Likewise, the chain initiation event in a polymerization reaction is a random process that depends on the probability of the combination of two primary radicals versus the likelihood of addition of a first monomer to a primary radical. As a Markov chain is a stochastic model involving a random variable like DP, polymerization reactions can be modeled using a Markov process. In this chapter, we analyze the process of formation, propagation, and termination of chains in a polymer reaction using a Markov chain process.

9.2 Definition of a Markov Chain

A Markov chain is a probabilistic model in which the current state of the system is dependent only on its previous states [1, 2]. A chain consists of a finite number of successive steps, observations, or time intervals. If the system has gone through $(k-1)$ steps, the state of the system in the kth step depends only on the states in the previous $(k-1)$ steps. A Markov state in a polymerization reaction can be, among others, initiation, propagation, chain transfer, or termination. If a system described by a Markov chain has n states, namely $(E_1, E_2, \ldots E_i, \ldots, E_{k-1}, E_k)$, the probability p that the system be in state E_{i_n} in the nth step/observation/event is written by the following mathematical expression [3–6]:

$$p\left[x_n = E_{i_n} \middle| x_1 = E_{j_1}, x_2 = E_{j_2} \ldots, x_{n-1} = E_{j_{n-1}}\right] \tag{9.1}$$

In the above expression, x is an observation or a step, x_n is the state of the nth observation/step, E_{i_n} is the state of the system in the nth observation/step, x_1 is the first observation/step, E_{j1} is the state of the system in the first observation/step, x_{n-1} is the $(n-1)$ th observation/step, and $E_{j_{n-1}}$ is the state of the system in the $(n-1)$ th observation/step. A Markov chain is defined by its state space, initial probability vector, and transition probability matrix as described in the following sections.

9.3 State Space of a Markov Chain

The state space S of a Markov chain consists of all mutually exclusive states of the system and is given by the following row vector [1–4]:

$$S = [E_1, E_2, \ldots, E_i, \ldots, E_{k-1}, E_k] \tag{9.2}$$

where E_i is the ith state of the system and k is the number of states. Mutually exclusive means that two or more states of the system cannot happen concurrently in a step/observation. In practical applications, like in polymerization reactions, the set of states of the system is finite or countably infinite. As an example, in the copolymerization of monomers A and B, the states of the terminal group of a living propagating chain (i.e., no termination reaction) are A and B, which are countable. As another example, in radical polymerization and in the absence of chain transfer reactions, the states for a chain are initiation, propagation, and termination, which are countable.

9.4 Initial Probability Vector

The initial probability vector, which defines the selection probabilities for all states of the system at the beginning of the Markov chain process, is given by the following row vector [1–4]:

$$\begin{aligned}\overline{P}^{(0)} &= \left[p_{E_1}^{(0)}, p_{E_2}^{(0)}, \ldots, p_{E_i}^{(0)}, \ldots, p_{E_{k-1}}^{(0)}, p_{E_k}^{(0)}\right] \\ &= \left[p_1^{(0)}, p_2^{(0)}, \ldots, p_i^{(0)}, \ldots p_{k-1}^{(0)}, p_k^{(0)}\right] \qquad \forall E_i \in S\end{aligned} \tag{9.3}$$

In the above expression, $\overline{P}^{(0)}$ is the initial probability vector, $p_{E_i}^{(0)}$ is the initial probability for selection of the ith state of the system, and k is the number of states of the system. In the second expression, $p_{E_i}^{(0)}$ is simplified to $p_i^{(0)}$.

9.5 One-step Uniform Transition Probability Matrix

It is the matrix of probabilities, $\overline{P}^{(1)}$ or $\overline{P}$, for transitioning from state E_i of the system to state E_j in one step, one observation, one event, or an interval of time. If the one-step transition probability from any state E_i in step $n-1$ to any state E_j in step n for a system with k states is given by the following expression [1–4]:

$$p_{E_iE_j} = p\left[x_n = E_j | x_{n-1} = E_i\right] \qquad \forall E_i, E_j \in S \tag{9.4}$$

then the one-step transition probability matrix $\overline{P}$ is given by the following matrix:

$$\overline{P} = \begin{bmatrix} p_{E_1E_1} & p_{E_1E_2} & p_{E_1E_i} & p_{E_1E_j} & p_{E_1E_{k-1}} & p_{E_1E_k} \\ p_{E_2E_1} & p_{E_2E_2} & p_{E_2E_i} & p_{E_2E_j} & p_{E_2E_{k-1}} & p_{E_2E_k} \\ p_{E_iE_1} & p_{E_iE_2} & p_{E_iE_i} & p_{E_iE_j} & p_{E_iE_{k-1}} & p_{E_iE_k} \\ p_{E_jE_1} & p_{E_jE_2} & p_{E_jE_i} & p_{E_jE_j} & p_{E_jE_{k-1}} & p_{E_jE_k} \\ p_{E_{k-1}E_1} & p_{E_{k-1}E_2} & p_{E_{k-1}E_i} & p_{E_{k-1}E_j} & p_{E_{k-1}E_{k-1}} & p_{E_{k-1}E_k} \\ p_{E_kE_1} & p_{E_kE_2} & p_{E_kE_i} & p_{E_kE_j} & p_{E_kE_{k-1}} & p_{E_kE_k} \end{bmatrix} \qquad \forall E_i \in S \tag{9.5}$$

where $p_{E_iE_j}$ is the probability of one-step transition from any state E_i to any state E_j and $p_{E_iE_k}$ is the transition probability from any state E_i to the last state E_k as defined by Eq. (9.4). The subscript i stands for any state, whereas the subscript k is the number of states in the system. If $p_{E_iE_j}$ is abbreviated as p_{ij} where i and j are shorthand symbols for E_i and E_j, the above transition probability matrix simplifies to the following:

$$\overline{P} = \begin{bmatrix} p_{11} & p_{12} & p_{1i} & p_{1j} & p_{1(k-1)} & p_{1k} \\ p_{21} & p_{22} & p_{2i} & p_{2j} & p_{2(k-1)} & p_{2k} \\ p_{i1} & p_{i2} & p_{ii} & p_{ij} & p_{i(k-1)} & p_{ik} \\ p_{j1} & p_{j2} & p_{ji} & p_{jj} & p_{j[k-1]} & p_{jk} \\ p_{(k-1)1} & p_{(k-1)2} & p_{(k-1)i} & p_{(k-1)j} & p_{(k-1)(k-1)} & p_{(k-1)k} \\ p_{k1} & p_{k2} & p_{ki} & p_{kj} & p_{k(k-1)} & p_{kk} \end{bmatrix} \quad \forall E_i \in S \tag{9.6}$$

where it is understood that $p_{i(k-1)}$ is the transition probability from any state E_i to the E_{k-1} before the last in states of the system.

9.6 *n*-step Uniform Transition Probability Matrix

It is the matrix of probabilities, $\overline{P}^{(n)}$, for transitioning from any state E_i of the system to any state E_j in n-steps, n-observations, n-events, or n-intervals of time. If the n-step probability for transitioning from any state E_i in step m to any state E_j in step $m+n$ for a system with k states is given by the following expression [7–12]:

$$p_{E_iE_j}^{(n)} = p_{ij}^{(n)} = p\left[x_{m+n} = E_j | x_m = E_i\right] \quad \forall E_i, E_j \in S \tag{9.7}$$

then the n-step transition probability matrix $\overline{P}^{(n)}$ is given by the following matrix:

$$\overline{P}^{(n)} = \begin{bmatrix} p_{11}^{(n)} & p_{12}^{(n)} & p_{1i}^{(n)} & p_{1j}^{(n)} & p_{1(k-1)}^{(n)} & p_{1k}^{(n)} \\ p_{21}^{(n)} & p_{22}^{(n)} & p_{2i}^{(n)} & p_{2j}^{(n)} & p_{2(k-1)}^{(n)} & p_{2k}^{(n)} \\ p_{i1}^{(n)} & p_{i2}^{(n)} & p_{ii}^{(n)} & p_{ij}^{(n)} & p_{i(k-1)}^{(n)} & p_{ik}^{(n)} \\ p_{j1}^{(n)} & p_{j2}^{(n)} & p_{ji}^{(n)} & p_{jj}^{(n)} & p_{j(k-1)}^{(n)} & p_{jk}^{(n)} \\ p_{(k-1)1}^{(n)} & p_{(k-1)2}^{(n)} & p_{(k-1)i}^{(n)} & p_{(k-1)j}^{(n)} & p_{(k-1)(k-1)}^{(n)} & p_{(k-1)k}^{(n)} \\ p_{k1}^{(n)} & p_{k2}^{(n)} & p_{ki}^{(n)} & p_{kj}^{(n)} & p_{k(k-1)}^{(n)} & p_{kk}^{(n)} \end{bmatrix} \quad \forall E_i, E_j \in S \tag{9.8}$$

If the steps, observations, events, or intervals of time are independent of each other, then the n-step probability $pi_{ij}^{(n)}$, where "i" in pi stands for independent steps, for transitioning from any state E_i in step m to any state E_j in step $m+n$ for a system with k states is given by the following expression:

$$pi_{ij}^{(n)} = \sum_{l=1}^{k} p_{il}\, p_{lj}^{(n-1)} \quad \forall E_i, E_l, E_j \in S \quad \text{independent steps} \tag{9.9}$$

and the n-step transition probability $\overline{Pi}^{(n)}$ is given by the following matrix:

$$\overline{Pi}^{(n)} = \overline{P}\,\overline{P}^{(n-1)} = \begin{bmatrix} p_{11} & p_{12} & p_{1i} & p_{1j} & p_{1(k-1)} & p_{1k} \\ p_{21} & p_{22} & p_{2i} & p_{2j} & p_{2(k-1)} & p_{2k} \\ p_{i1} & p_{i2} & p_{ii} & p_{ij} & p_{i(k-1)} & p_{ik} \\ p_{j1} & p_{j2} & p_{ji} & p_{jj} & p_{j[k-1]} & p_{jk} \\ p_{(k-1)1} & p_{(k-1)2} & p_{(k-1)i} & p_{(k-1)j} & p_{(k-1)(k-1)} & p_{(k-1)k} \\ p_{k1} & p_{k2} & p_{ki} & p_{kj} & p_{k(k-1)} & p_{kk} \end{bmatrix} \tag{9.10}$$

$$\begin{bmatrix} p_{11}^{(n-1)} & p_{12}^{(n-1)} & p_{1i}^{(n-1)} & p_{1j}^{(n-1)} & p_{1(k-1)}^{(n-1)} & p_{1k}^{(n-1)} \\ p_{21}^{(n-1)} & p_{22}^{(n-1)} & p_{2i}^{(n-1)} & p_{2j}^{(n-1)} & p_{2(k-1)}^{(n-1)} & p_{2k}^{(n-1)} \\ p_{i1}^{(n-1)} & p_{i2}^{(n-1)} & p_{ii}^{(n-1)} & p_{ij}^{(n-1)} & p_{i(k-1)}^{(n-1)} & p_{ik}^{(n-1)} \\ p_{j1}^{(n-1)} & p_{j2}^{(n-1)} & p_{ji}^{(n-1)} & p_{jj}^{(n-1)} & p_{j(k-1)}^{(n-1)} & p_{jk}^{(n-1)} \\ p_{(k-1)1}^{(n-1)} & p_{(k-1)2}^{(n-1)} & p_{(k-1)i}^{(n-1)} & p_{(k-1)j}^{(n-1)} & p_{(k-1)(k-1)}^{(n-1)} & p_{(k-1)k}^{(n-1)} \\ p_{k1}^{(n-1)} & p_{k2}^{(n-1)} & p_{ki}^{(n-1)} & p_{kj}^{(n-1)} & p_{k(k-1)}^{(n-1)} & p_{kk}^{(n-1)} \end{bmatrix}$$

9.7 n-step Absolute Probability Vector

It is a (1 × k) matrix of probabilities, $\overline{P}_{abs}^{(n)}$, for being in any state E_j after $(m+n)$ steps, observations, or intervals of time, regardless of the initial state of the system in step m. Defining the n-step transition probability from step m to any state E_j in step $m+n$ for a system with k states is given by the following expression:

$$p_{abs,j}^{(n)} = p\left[x_n = E_j\right] = \sum_{l=1}^{k} p_{1l}^{(n)} p_{lj}^{(n)} \qquad \forall E_i, E_l \in S \tag{9.11}$$

then the n-step absolute probability vector, $\overline{P}_{abs}^{(n)}$, is given by the following (1 × k) row vector:

$$\overline{P}_{abs}^{(n)} = \left[p_{abs,1}^{(n)} \quad p_{abs,2}^{(n)} \quad p_{abs,i}^{(n)} \quad p_{abs,k-1}^{(n)} \quad p_{abs,k}^{(n)}\right] \qquad \forall E_i \in S \tag{9.12}$$

where k is the number of states of the system.

9.8 Types of States in State Space of a Markov Chain

Two types of states in a Markov process can be distinguished in the state space. A state E_i is an accessible state or transitional state if the n-step transition probability from that state to any state E_j is greater than zero. Conversely, a state E_i is an absorbing state or terminating state if the n-step transition probability from that state to any state E_j is equal to zero. Therefore, the state space S of a Markov chain, as defined by Eq. (9.2), can be written in the following vector form [13–19]:

$$S = \left[\{S_a\} \quad \{S_t\}\right] = \left[\{E_{a,1}, \ldots, E_{a,i}, \ldots, E_{a,r}\} \quad \{E_{t,1}, \ldots, E_{t,i}, \ldots, E_{t,k-r}\}\right] \tag{9.13}$$

where S_a and S_t are the absorbing and transitional spaces in the state space of the system, respectively; k is the total number of states in the system, r is the number of absorbing states, and $k-r$ is the number of transitional states; $E_{a,i}$ is the absorbing state i and $E_{t,i}$ is the transitional state i. For example, there are two general types of states for a propagating chain in free-radical polymerization, namely the propagation state and termination state. There is a finite, non-zero probability of transitioning from a propagation state to any state of polymerization, including termination state. Conversely, the probability for a terminating state (a chain in a termination state) to transition to other states is zero. Therefore, the propagation reaction in radical polymerization is a transitional state, whereas the termination reaction is an absorbing state. Similarly, the initial probability vector, as defined by Eq. (9.3), can be divided into the initial absorbing probability vector and the initial transitional probability vector as follows:

$$\overline{P}^{(0)} = \left[\overline{P}_a^{(0)} \;\; \overline{P}_t^{(0)}\right] = \left[\left\{p_{a,1}^{(0)}, \ldots, p_{a,i}^{(0)}, \ldots, p_{a,r}^{(0)}\right\} \; \left\{p_{t,1}^{(0)}, \ldots, p_{t,i}^{(0)}, \ldots, p_{t,k-r}^{(0)}\right\}\right] \tag{9.14}$$

where $\overline{P}_a^{(0)}$ and $\overline{P}_t^{(0)}$ are the initial absorbing and transitional probability vectors, respectively, $p_{a,i}^{(0)}$ is the initial probability for the absorbing state $E_{a,i}$, and $p_{t,i}^{(0)}$ is the initial probability for the transitional state $E_{t,i}$. Similarly, the one-step transition probability matrix, as defined by Eq. (9.5), can be divided into the absorbing probability matrix and the transitional probability matrix as follows:

$$\overline{P} = \begin{bmatrix} \overline{P}_{aa} & \overline{P}_{at} \\ \overline{P}_{ta} & \overline{P}_{tt} \end{bmatrix} = \begin{Bmatrix} \overline{I} & \overline{0} \\ \overline{R} & \overline{Q} \end{Bmatrix} = \begin{Bmatrix} \begin{bmatrix} 1 & 0 & 0 \\ 0 & 1 & 0 \\ 0 & 0 & 1 \end{bmatrix} & \begin{bmatrix} 0 & 0 & 0 \\ 0 & 0 & 0 \\ 0 & 0 & 0 \end{bmatrix} \\ \begin{bmatrix} p_{ta,11} & p_{ta,1i} & p_{ta,1r} \\ p_{ta,i1} & p_{ta,ii} & p_{ta,ir} \\ p_{ta,r1} & p_{ta,ri} & p_{ta,rr} \end{bmatrix} & \begin{bmatrix} p_{tt,11} & p_{tt,1i} & p_{tt,1(k-r)} \\ p_{tt,i1} & p_{tt,ii} & p_{tt,1(k-r)} \\ p_{tt,(k-r)1} & p_{tt,(k-r)i} & p_{tt,(k-r)\,(k-r)} \end{bmatrix} \end{Bmatrix} \tag{9.15}$$

where $\overline{P}_{aa}$ and $\overline{P}_{at}$ are the one-step probability matrices for transitioning from absorbing to absorbing states and absorbing to transitional states, respectively; $\overline{P}_{ta}$ and $\overline{P}_{tt}$ are the one-step probability matrices for transitioning from transitional to absorbing states and transitional to transitional states. By definition, absorbing states remain absorbing with 100% probability in subsequent steps, $\overline{P}_{aa}$ is equal to an $(r \times r)$ unity matrix $(\overline{1})$ where r is the number of absorbing states. By definition, absorbing states have a zero percent chance of changing to a transitional state in subsequent steps, $\overline{P}_{at}$ which is equal to a $(r \times k - r)$ zero matrix $(\overline{0})$ where k is the total number of states in the system. The matrix $\overline{P}_{ta}$ is hereafter referred to as the probability matrix for absorbing states and denoted by $\overline{R}$ which is a $(k - r \times r)$ matrix. The matrix $\overline{P}_{tt}$ is hereafter referred to as the probability matrix for transitional states and denoted by $\overline{Q}$ which is a $(k - r \times k - r)$ matrix.

9.9 n-step Probability of Reaching an Absorbing State

The probability matrix for reaching an absorbing state in n-steps for a Markovian chain process with k states and n-step transition probability matrix $\overline{P}^{(n)}$ is defined by the following expression:

$$p_a^{(n)} = \overline{P}_t^{(0)} \overline{Q}^{(n-1)} \left(\overline{R} \, x \, \overline{1}\right) \tag{9.16}$$

where $p_a^{(n)}$ is the probability of reaching an absorbing state in n steps; $\overline{P}_t^{(0)}$ is the initial probability vector for the transitional states; $\overline{Q}^{(n-1)}$ is the $(n-1)$-steps probability matrix for passing between the transitional states; $\overline{R}$ is the probability matrix for passage from transitional to absorbing states; and $\overline{1}$ is the unity matrix with number of rows equal to the number of columns of $\overline{R}$. As the number of steps n increases from one to infinity, $p_a^{(n)}$ becomes the probability distribution function for reaching an absorbing state in n steps. For example, the termination reaction in radical polymerization is an absorbing state that stops the growth of the propagating chain. Therefore, the function $p_a^{(n)}$ provides the distribution of degree of polymerization when applied to radical polymerization with termination reaction. Once the probability distribution of reaching an absorbing state as a function of the number of steps n is known, the kth moment of the distribution can be determined by the defining Eq. (3.7) in Chapter 3. The first $(M_{n,a,1})$ and second $(M_{n,a,2})$ mole-fraction moments of the distribution for reaching an absorbing state in n-steps, according to Eqs. (3.18) and (3.22), are given as follows:

$$M_{n,a,1} = \sum_{n=1}^{\infty} n^1 p_a^{(n)} = \overline{P}_t^{(0)} \left[\overline{I} - \overline{Q}\right]^{-1} x\, \overline{1} \tag{9.17}$$

$$M_{n,a,2} = \sum_{n=1}^{\infty} n^2 p_a^{(n)} = \overline{P}_t^{(0)} \left[2\overline{N} - \overline{I}\right] \overline{N}\, x\, \overline{1} \tag{9.18}$$

where the matrix $\overline{N}$ is defined as follows:

$$\overline{N} = \left[\overline{I} - \overline{Q}\right]^{-1} \tag{9.19}$$

In the above equations, $\overline{I}$ is the identity matrix with the same size as the $\overline{Q}$ matrix, with ones for the main diagonal elements and zeros for the other elements. Once the first and second moments $M_{n,a,1}$ and $M_{n,a,2}$ are known, the average number of steps, n, to reach an absorbing state $(\overline{n}_{n,a})$, standard deviation of the distribution of n values $(\sigma_{n,a})$, and the normalized standard deviation $(\Delta_{n,a})$, according to Eqs. (3.47), (3.64), and (3.65) of Chapter 3, respectively, are expressed as follows:

$$\overline{n}_{n,a} = M_{n,a,1} = \overline{P}_t^{(0)} \left[\overline{I} - \overline{Q}\right]^{-1} x\, \overline{1} \tag{9.20}$$

$$\sigma_{n,a} = \overline{n}_{n,a} \sqrt{\frac{M_{n,a,2}}{M_{n,a,1}^2} - 1} \tag{9.21}$$

$$\Delta_{n,a} = \sqrt{\frac{M_{n,a,2}}{M_{n,a,1}^2} - 1} \tag{9.22}$$

9.10 Matrix Operations

9.10.1 Multiplication of a Matrix by a Scalar

If $\overline{A}$ is a 2×2 matrix and c is a scalar, $c\overline{A}$ is obtained by multiplying each element of $\overline{A}$ by c as follows [20–23]:

$$c\overline{A} = c \begin{bmatrix} a_{11} & a_{12} \\ a_{21} & a_{22} \end{bmatrix} = \begin{bmatrix} c\,a_{11} & c\,a_{12} \\ c\,a_{21} & c\,a_{22} \end{bmatrix} \tag{9.23}$$

9.10.2 Addition and Subtraction of Matrices

These operations can take place only between matrices of equal sizes. If $\overline{A}$ and $\overline{B}$ are two 2 × 2 matrices, $\overline{A} + \overline{B}$ is determined by adding each element of $\overline{A}$ to the same element of $\overline{B}$ as follows:

$$\overline{A} + \overline{B} = \begin{bmatrix} a_{11} & a_{12} \\ a_{21} & a_{22} \end{bmatrix} + \begin{bmatrix} b_{11} & b_{12} \\ b_{21} & b_{22} \end{bmatrix} = \begin{bmatrix} a_{11} + b_{11} & a_{12} + b_{12} \\ a_{21} + b_{21} & a_{22} + b_{22} \end{bmatrix} \quad (9.24)$$

In a similar approach, $\overline{A} - \overline{B}$ is determined by subtracting each element of $\overline{B}$ from the same element of $\overline{A}$ as follows:

$$\overline{A} - \overline{B} = \begin{bmatrix} a_{11} & a_{12} \\ a_{21} & a_{22} \end{bmatrix} - \begin{bmatrix} b_{11} & b_{12} \\ b_{21} & b_{22} \end{bmatrix} = \begin{bmatrix} a_{11} - b_{11} & a_{12} - b_{12} \\ a_{21} - b_{21} & a_{22} - b_{22} \end{bmatrix} \quad (9.25)$$

9.10.3 Multiplication of Matrices

This operation can take place only between those matrices in which the number of columns of the first matrix is equal to the number of rows of the second matrix. Furthermore, the resulting matrix has the number of rows of the first matrix and the number of columns of the second matrix. In general, if $\overline{X}$ is an $(m \times n)$ matrix with elements x_{ij} and $\overline{Y}$ is an $(n \times p)$ matrix with elements x_{jk}, the product $\overline{Z} = \overline{X} \times \overline{Y} = \overline{XY}$ is an $(m \times p)$ matrix with elements z_{ik} given by the following expression [20–23]:

$$z_{ik} = \sum_{j=1}^{m} x_{ij} y_{jk} \quad (9.26)$$

If $\overline{A}$ and $\overline{B}$ are two 2×2 matrices with elements a_{ij} and b_{jk}, respectively, $\overline{AB}$ is determined as follows:

$$\overline{AB} = \begin{bmatrix} a_{11} & a_{12} \\ a_{21} & a_{22} \end{bmatrix} \times \begin{bmatrix} b_{11} & b_{12} \\ b_{21} & b_{22} \end{bmatrix} = \begin{bmatrix} a_{11}b_{11} + a_{12}b_{21} & a_{11}b_{12} + a_{12}b_{22} \\ a_{21}b_{11} + a_{22}b_{21} & a_{21}b_{12} + a_{22}b_{22} \end{bmatrix} \quad (9.27)$$

9.10.4 Determinant of a Matrix

The determinant is a scalar not a matrix. Only square matrices have a determinant. The determinant of an $(n \times n)$ matrix $\overline{A}$ with elements a_{ij}, represented by $\left|\overline{A}\right|$, is given by the following expression [20–23]:

$$\left|\overline{A}\right| = \sum_{j=1}^{n} (-1)^{i+j} a_{ij} \left|M_{ij}\right| \quad (9.28)$$

where $\left|M_{ij}\right|$ is the determinant of the matrix $(n-1) \times (n-1)$ resulting from removing the ith row and the jth column from the matrix $\overline{A}$. Any row "i" may be chosen to determine $\left|\overline{A}\right|$. If $\overline{A}$ is the following (3×3) matrix with elements a_{ij}:

$$\overline{A} = \begin{bmatrix} a_{11} & a_{12} & a_{13} \\ a_{21} & a_{22} & a_{23} \\ a_{31} & a_{32} & a_{33} \end{bmatrix} \quad (9.29)$$

The determinant $|\overline{A}|$, choosing $i = 1$, is determined as follows:

$$\begin{aligned}|\overline{A}| &= \sum_{j=1}^{n} (-1)^{1+j} a_{1j} |M_{1j}| \\ &= (-1)^{1+1} a_{11} \begin{vmatrix} a_{22} & a_{23} \\ a_{32} & a_{33} \end{vmatrix} + (-1)^{1+2} a_{12} \begin{vmatrix} a_{21} & a_{23} \\ a_{31} & a_{33} \end{vmatrix} + (-1)^{1+3} a_{13} \begin{vmatrix} a_{21} & a_{22} \\ a_{31} & a_{32} \end{vmatrix} \\ &= a_{11}(a_{22}a_{33} - a_{23}a_{32}) - a_{12}(a_{21}a_{33} - a_{23}a_{31}) + a_{13}(a_{21}a_{32} - a_{22}a_{31})\end{aligned} \quad (9.30)$$

9.10.5 Matrix of Cofactors

Only square matrices have a matrix of cofactors. The elements of the matrix of cofactors $(\overline{A}_{cf})$ of an (n × n) matrix $\overline{A}$ with elements a_{ij} is given by $(-1)^{i+j} |M_{ij}|$ for all $1 \le i \le n$ and $1 \le j \le n$ where $|M_{ij}|$ was defined in Section 9.10.4. If $\overline{A}$ is a (3 × 3) matrix given by Eq. (9.29) with elements a_{ij}, the matrix $\overline{A}_{cf}$ is determined as follows [20–23]:

$$\overline{A}_{cf} = \begin{bmatrix} (-1)^{1+1}|M_{11}| & (-1)^{1+2}|M_{12}| & (-1)^{1+3}|M_{13}| \\ (-1)^{2+1}|M_{21}| & (-1)^{2+2}|M_{22}| & (-1)^{2+3}|M_{23}| \\ (-1)^{3+1}|M_{31}| & (-1)^{3+2}|M_{32}| & (-1)^{3+3}|M_{33}| \end{bmatrix}$$

$$= \begin{bmatrix} \begin{vmatrix} a_{22} & a_{23} \\ a_{32} & a_{33} \end{vmatrix} & -\begin{vmatrix} a_{21} & a_{23} \\ a_{31} & a_{33} \end{vmatrix} & \begin{vmatrix} a_{21} & a_{22} \\ a_{31} & a_{32} \end{vmatrix} \\ -\begin{vmatrix} a_{12} & a_{13} \\ a_{32} & a_{33} \end{vmatrix} & \begin{vmatrix} a_{11} & a_{13} \\ a_{31} & a_{33} \end{vmatrix} & -\begin{vmatrix} a_{11} & a_{22} \\ a_{31} & a_{32} \end{vmatrix} \\ \begin{vmatrix} a_{12} & a_{13} \\ a_{22} & a_{23} \end{vmatrix} & -\begin{vmatrix} a_{11} & a_{13} \\ a_{21} & a_{23} \end{vmatrix} & \begin{vmatrix} a_{11} & a_{12} \\ a_{21} & a_{22} \end{vmatrix} \end{bmatrix} \quad (9.31)$$

$$= \begin{bmatrix} a_{22}a_{33} - a_{23}a_{32} & -(a_{21}a_{33} - a_{23}a_{31}) & a_{21}a_{32} - a_{22}a_{31} \\ -(a_{12}a_{33} - a_{13}a_{32}) & a_{11}a_{33} - a_{13}a_{31} & -(a_{11}a_{32} - a_{22}a_{31}) \\ a_{12}a_{23} - a_{13}a_{22} & -(a_{11}a_{23} - a_{13}a_{21}) & a_{11}a_{22} - a_{12}a_{21} \end{bmatrix}$$

9.10.6 Transpose of a Matrix

Any matrix regardless of size has a transpose. The transpose of an $(n \times m)$ matrix $\overline{X}$ with elements x_{ij} is an $(m \times n)$ matrix $\overline{X}^T$ with elements x_{ji}. In other words, the transpose flips the matrix over its diagonal elements. If $\overline{A}$ is a (3 × 3) matrix given by Eq. (9.29) with elements a_{ij}, the transpose matrix $\overline{A}^T$ is given by the following matrix [20–23]:

$$\overline{A}^T = \begin{bmatrix} a_{11} & a_{21} & a_{31} \\ a_{12} & a_{22} & a_{32} \\ a_{13} & a_{23} & a_{33} \end{bmatrix} \quad (9.32)$$

9.10.7 Adjugate Matrix

Only square matrices have adjugate matrices. The adjugate matrix of $\overline{A}$ is represented by $\overline{A}_{adj}$. The elements of the adjugate matrix $\overline{A}_{adj}$ of an (n × n) matrix $\overline{A}$ with elements a_{ij} is given by $(-1)^{i+j}\left|M_{ji}\right|$ for all $1 \le i \le n$ and $1 \le j \le n$ or it is the transpose of the matrix of cofactors $\overline{A}_{cf}$ with elements $(-1)^{i+j}\left|M_{ij}\right|$ as defined in Section 9.10.5. If $\overline{A}$ is a (3 × 3) matrix given by Eq. (9.29) with elements a_{ij} and with $\overline{A}_{cf}$ given by Eq. (9.31), the matrix $\overline{A}_{adj}$ is given as follows [20–23]:

$$\overline{A}_{adj} = \begin{bmatrix} a_{22}a_{33} - a_{23}a_{32} & -(a_{12}a_{33} - a_{13}a_{32}) & a_{12}a_{23} - a_{13}a_{22} \\ -(a_{21}a_{33} - a_{23}a_{31}) & a_{11}a_{33} - a_{13}a_{31} & -(a_{11}a_{23} - a_{13}a_{21}) \\ a_{21}a_{32} - a_{22}a_{31} & -(a_{11}a_{32} - a_{22}a_{31}) & a_{11}a_{22} - a_{12}a_{21} \end{bmatrix} \tag{9.33}$$

9.10.8 Inverse of a Matrix

It is defined only for square matrices. The inverse matrix of $\overline{A}$ is represented by $\overline{A}^{-1}$. The elements of the inverse matrix $\overline{A}^{-1}$ of an (n × n) matrix $\overline{A}$ with elements a_{ij} is given by the elements of $\overline{A}_{adj}$ divided by $\left|\overline{A}\right|$ or $(-1)^{i+j}\left|M_{ji}\right| / \left|\overline{A}\right|$ for all $1 \le i \le n$ and $1 \le j \le n$. If $\overline{A}$ is the (3 × 3) matrix given by Eq. (9.29) with elements a_{ij} and with $\overline{A}_{adj}$ given by Eq. (9.33), the matrix $\overline{A}^{-1}$ is given as follows [20–23]:

$$\overline{A}^{-1} = \frac{1}{\left|\overline{A}\right|}\begin{bmatrix} a_{22}a_{33} - a_{23}a_{32} & -(a_{12}a_{33} - a_{13}a_{32}) & a_{12}a_{23} - a_{13}a_{22} \\ -(a_{21}a_{33} - a_{23}a_{31}) & a_{11}a_{33} - a_{13}a_{31} & -(a_{11}a_{23} - a_{13}a_{21}) \\ a_{21}a_{32} - a_{22}a_{31} & -(a_{11}a_{32} - a_{22}a_{31}) & a_{11}a_{22} - a_{12}a_{21} \end{bmatrix} \tag{9.34}$$

where $\left|\overline{A}\right|$ is given by Eq. (9.30).

9.11 Application of Markov Chains to Radical Polymerization

This section applies Markov chain stochastic modeling to predict the distribution of degree of polymerization (DP) in radical polymerization, the average DP, and polydispersity [24–26]. Free radical polymerization consists of, among others, initiation, propagation, and termination. Assuming the ratio of monomer to primary radicals remains constant during polymerization, the DP distribution after the formation of primary radicals depends largely on the rates of propagation and termination reactions. This section focuses on propagation and termination events as the system states to predict DP distribution by Markov chain modeling. A propagation event can lead to either propagation or termination; thus, propagation is a transitional state. Conversely, a termination event stops the chain from propagating, thus, termination is an absorbing state.

9.11.1 Radical Polymerization with Termination by Disproportionation

This is the simplest case of free radical polymerization with the following elementary reactions [27–29]:

$$RP_n + M \xrightarrow{k_p} RP_{n+1} \tag{9.35}$$

$$RP_n + RP_m \xrightarrow{k_{td}} TP_n + TP_m \tag{9.36}$$

where RP_n is an n-mer reactive propagating chain, TP_n is an n-mer non-propagating or terminated polymer chain, M is monomer, and k_p and k_{td} are the rates of propagation and termination by disproportionation, respectively. The state space of the polymerization system consists of two states, namely propagation and termination by disproportionation. The propagation reaction is a transitional state, whereas termination is an absorbing state. Therefore, the state space can be expressed by the following vector:

$$S = [E_1, E_2] \tag{9.37}$$

where E_1 is the absorbing termination state and E_2 is the transitional propagation state. The one-step transition probability matrix, $\overline{P}$, is a (2 × 2) matrix with four elements, namely p_{11}, p_{12}, p_{21}, and p_{22}. The element p_{11} is the probability for a terminated chain (absorbing state) to remain terminated, which is equal to one, and p_{12} is the probability for a terminated chain to undergo propagation, which is equal to zero, as follows:

$$p_{11} = 1 \qquad p_{12} = 0 \tag{9.38}$$

The element p_{21} of the matrix $\overline{P}$ is the probability of a propagating chain to undergo termination, which is equal to the rate of termination reaction divided by the rates of termination and propagation as follows:

$$p_{21} = \frac{R_{td}}{R_p + R_{td}} = \frac{k_{td}\,[RP]^2}{k_p\,[RP]\,[M] + k_{td}\,[RP]^2} \tag{9.39}$$

where R_{td} and R_p are the rates of termination and propagation reactions, respectively, $[M]$ and $[RP]$ are the concentrations of monomer and propagating chains, and k_{td} and k_p are the rate constants for termination and propagation reactions. The rate equations for propagation and termination were derived in Section 7.3.1 of Chapter 7. Assuming quasi-steady state (QSSA) for the propagating chains and substituting for $[RP]$ from Eq. (7.5E) in Chapter 7, p_{21} can be expressed in terms of monomer and initiator concentrations as follows:

$$p_{21} = \frac{R_{td}}{R_p + R_{td}} = \frac{1}{\left\{k_p\,[M]/(2k_d k_{td} f_0[I])^{1/2}\right\} + 1} \tag{9.40}$$

where f_0 is efficiency of initiation reaction, $[I]$ is initiator concentration, and k_d is dissociation rate of the initiator. The element p_{22} of the matrix $\overline{P}$ is the probability of a propagating chain to undergo propagation, which is given by the following equation:

$$p_{22} = \frac{R_p}{R_p + R_{td}} = \frac{1}{1 + \left\{(2k_d k_{td} f_0[I])^{1/2}/k_p\,[M]\right\}} = p \tag{9.41}$$

where $p = p_{22}$ is defined as the probability of propagation reaction and $1 - p = p_{21}$. According to Eq. (9.15), the one-step transition probability matrix for this polymerization system is given by the following (2×2) matrix:

$$\overline{P} = \begin{bmatrix} \overline{I} & \overline{0} \\ \overline{R} & \overline{Q} \end{bmatrix} = \begin{bmatrix} 1 & 0 \\ 1-p & p \end{bmatrix} \tag{9.42}$$

Comparison of the above two matrices leads to the following matrices for $\overline{I}$, $\overline{0}$, $\overline{R}$, and $\overline{Q}$:

$$\begin{aligned} \overline{I} &= [1] = 1 & \overline{0} &= [0] = 0 \\ \overline{R} &= [1-p] = 1-p & \overline{Q} &= [p] = p \end{aligned} \tag{9.43}$$

Initially, all primary radicals in the reaction add a monomer to form 1-mer propagating chains, and none of the primary radicals undergoes termination. Therefore, according to Eq. (9.14), the initial probability vectors, $\overline{P}^{(0)}$, $\overline{P}_a^{(0)}$, and $\overline{P}_t^{(0)}$ are given as follows:

$$\overline{P}^{(0)} = \left[\overline{P}_a^{(0)} \;\; \overline{P}_t^{(0)}\right] = \left[0 \;\; 1\right] \quad \overline{P}_a^{(0)} = [0] = 0 \quad \overline{P}_t^{(0)} = [1] = 1 \tag{9.44}$$

Upon using $\overline{P}_t^{(0)} = 1$ from the above equation, and $\overline{Q}$ and $\overline{R}$ from Eq. (9.43), the n-step probability distribution of reaching an absorbing state, $p_a^{(n)} = p_{DP}(n)$, is determined from Eq. (9.16) as follows:

$$p_{DP}(n) = p_a^{(n)} = \overline{P}_t^{(0)} \overline{Q}^{(n-1)} \left(\overline{R} \, x \, \overline{1}\right) = (1-p)\, p^{n-1} \tag{9.45}$$

where $p_{DP}(n)$ is the DP probability distribution. According to Eq. (9.17), the first mole-fraction moment of the distribution which is equal to the average number of steps to reach an absorbing state, $M_{n,a,1} = \overline{n}_{n,a} = \overline{DP}_n$, is determined as follows:

$$\overline{DP}_n = M_{n,a,1} = \sum_{n=1}^{\infty} n^1 p_a^{(n)} = \overline{P}_t^{(0)} \left[\overline{I} - \overline{Q}\right]^{-1} x \, \overline{1} = \frac{1}{1-p} \tag{9.46}$$

According to Eq. (9.19), the matrix $\overline{N}$ is determined from matrices $\overline{I}$ and $\overline{Q}$ as follows:

$$\overline{N} = \left[\overline{I} - \overline{Q}\right]^{-1} = [1-p]^{-1} = \frac{1}{1-p} \tag{9.47}$$

The second mole-fraction moment of the distribution for reaching an absorbing state, $M_{n,a,2}$, according to Eq. (9.18), is determined as follows:

$$M_{n,a,2} = \sum_{n=1}^{\infty} n^2 p_a^{(n)} = \overline{P}_t^{(0)} \left[2\overline{N} - \overline{I}\right] \overline{N} \, x \, \overline{1} = [1] \left[\frac{2}{1-p} - 1\right] \left[\frac{1}{1-p}\right] [1] = \frac{1+p}{(1-p)^2} \tag{9.48}$$

Once the first and second moments of the distribution are determined, weight-average DP $\left(\overline{DP}_w\right)$, normalized standard deviation of DP (Δ), and polydispersity index (PDI), according to Eqs. (3.34), (3.65), and (3.66), respectively, are determined as follows:

$$\overline{DP}_w = \frac{M_{n,a,2}}{M_{n,a,1}} = \frac{(1+p)/(1-p)^2}{1/(1-p)} = \frac{1+p}{1-p} \tag{9.49}$$

$$\Delta = \sqrt{\frac{M_{n,a,2}}{M_{n,a,1}^2} - 1} = \sqrt{\frac{\overline{DP}_w}{\overline{DP}_n} - 1} = \sqrt{(1-p) - 1} = \sqrt{p} \tag{9.50}$$

$$PDI = \frac{\overline{DP}_w}{\overline{DP}_n} = \frac{(1+p)/(1-p)}{1-p} = 1 + p \tag{9.51}$$

where p in the above equations is the probability of propagation as defined by Eq. (9.41).

9.11.2 Radical Polymerization with Exclusion of Monomers from Chain Distribution

After completion of the reaction in radical polymerization, the produced polymer is separated from the unreacted monomer and purified by precipitation in a non-solvent or by physisorption [30–32]. As a result of monomer separation, most, if not all 1-mer chains are removed from the polymer distribution. This section shows how Markov chain modeling is used to determine the distribution of DP of radical polymerization and its averages when the monomer is excluded from the distribution. One approach to exclude the monomer from the distribution is to envision that the 1-mer chains do not terminate during the reaction to become a part of the terminated polymer distribution. Thus, the chains undergo only propagation as described by the following elementary reactions:

$$RP_1 + M \xrightarrow{k_{p1}>>k_{td}} RP_2 \qquad n = 1 \tag{9.52}$$

$$RP_n + M \xrightarrow{k_p} RP_{n+1} \qquad n > 1 \tag{9.53}$$

$$RP_n + RP_m \xrightarrow{k_{td}} TP_n + TP_m \qquad n, m > 1 \tag{9.54}$$

In the first reaction, the 1-mer propagating chain reacts with a monomer with rate constant k_{p1} much larger than the rate of termination, that is $k_{p1} \gg k_{td}$, such that the 1-mers propagate all the time. In the second and third reactions, the n-mers with $n > 1$ propagate with rate constant k_p or terminate with rate constant k_{td}. The state space of the reaction system is defined by the following vector:

$$S = [E_1, E_2, E_3] \tag{9.55}$$

where E_1 is the absorbing or termination state, E_2 is the state for 1-mer chains, and E_3 is the transitional propagation state for n-mer chains with $n > 1$. The one-step transition probability matrix, $\overline{P}$, is a (3×3) matrix with nine elements as follows:

$$\overline{P} = \begin{bmatrix} p_{11} & p_{12} & p_{13} \\ p_{21} & p_{22} & p_{23} \\ p_{31} & p_{32} & p_{33} \end{bmatrix} \tag{9.56}$$

The element p_{11} is the probability for a terminated chain to remain terminated, which is equal to one whereas p_{12} and p_{13} are the probabilities for a terminated chain to return to propagation which is equal to zero. The element p_{21} is the probability of a 1-mer propagating chain to undergo termination, which is zero, to be excluded from the polymer distribution; p_{22} is the probability of a 1-mer to remain in the 1-mer state which is equal to zero; and p_{23} is the probability of a propagating 1-mer to add another monomer which is equal to one. The element p_{31} is the probability of a propagating n-mer to undergo termination which is equal to $p_{31} = 1 - p$ (see Eq. 9.40); p_{32} is the probability of an n-mer to revert to a 1-mer which is equal to zero; and p_{33} is the probability of an n-mer to add another monomer which is equal to $p_{33} = p$ (see Eq. 9.41). Therefore, $\overline{P}$ is given by the following matrix:

$$\overline{P} = \begin{bmatrix} 1 & 0 & 0 \\ 0 & 0 & 1 \\ 1-p & 0 & p \end{bmatrix} = \begin{bmatrix} \overline{I} & \overline{0} \\ \overline{R} & \overline{Q} \end{bmatrix} \tag{9.57}$$

where p is the probability of propagation, and $\overline{I}$, $\overline{0}$, $\overline{R}$, and $\overline{Q}$ are given by the following matrices:

$$\overline{I} = [1] \quad \overline{0} = \begin{bmatrix} 0 & 0 \end{bmatrix} \quad \overline{R} = \begin{bmatrix} 0 \\ 1-p \end{bmatrix} \quad \overline{Q} = \begin{bmatrix} 0 & 1 \\ 0 & p \end{bmatrix} \tag{9.58}$$

Initially, all primary radicals in the reaction add a monomer to form 1-mer propagating chains. Therefore, the initial probability vectors, $\overline{P}^{(0)}$, $\overline{P}_a^{(0)}$, and $\overline{P}_t^{(0)}$ are given as follows (see Eq. 9.14):

$$\overline{P}^{(0)} = \begin{bmatrix} \overline{P}_a^{(0)} & \overline{P}_t^{(0)} \end{bmatrix} = \begin{bmatrix} 0 & 1 & 0 \end{bmatrix} \quad \overline{P}_a^{(0)} = [0] \quad \overline{P}_t^{(0)} = \begin{bmatrix} 1 & 0 \end{bmatrix} \tag{9.59}$$

Using the above $\overline{P}_t^{(0)}$ vector, $\overline{Q}$ and $\overline{R}$ from Eq. (9.58), the n-step probability distribution of reaching an absorbing state or the DP distribution, $p_a^{(n)} = p_{DP}(n)$, is determined from Eq. (9.16) as follows:

$$p_{DP}(n) = p_a^{(n)} = \overline{P}_t^{(0)} \overline{Q}^{(n-1)} \left(\overline{R} \, x \, \overline{1}\right) = \begin{bmatrix} 1 & 0 \end{bmatrix} \begin{bmatrix} 0 & 1 \\ 0 & p \end{bmatrix}^{n-1} \begin{bmatrix} 0 \\ 1-p \end{bmatrix} [1] \tag{9.60}$$

Based on Cayley–Hamilton theorem, it can be shown that the (2 × 2) $\overline{Q}^{(n-1)}$ matrix for this polymerization is equal to the following [33–40]:

$$\overline{Q}^{(n-1)} = \begin{bmatrix} 0 & 1 \\ 0 & p \end{bmatrix}^{n-1} = \begin{bmatrix} 0 & p^{n-2} \\ 0 & p^{n-1} \end{bmatrix} \tag{9.61}$$

Substitution of the above equality in Eq. (9.60) results in the following expression for $p_{DP}(n)$:

$$p_{DP}(n) = \begin{bmatrix} 1 & 0 \end{bmatrix} \begin{bmatrix} 0 & p^{n-2} \\ 0 & p^{n-1} \end{bmatrix} \left(\begin{bmatrix} \begin{bmatrix} 0 \\ 1-p \end{bmatrix} x \, [1] \end{bmatrix} \right) = (1-p)\, p^{n-2} \qquad n > 1 \tag{9.62}$$

where $p_{DP}(n)$ is the DP probability distribution. Based on the relations in Eq. (9.58), the matrix $\overline{N} = \left[\overline{I} - \overline{Q}\right]^{-1}$ is determined as follows:

$$\left[\overline{I} - \overline{Q}\right]^{-1} = \left[\begin{bmatrix} 1 & 0 \\ 0 & 1 \end{bmatrix} - \begin{bmatrix} 0 & 1 \\ 0 & p \end{bmatrix} \right]^{-1} = \begin{bmatrix} 1 & \frac{1}{1-p} \\ 0 & \frac{1}{1-p} \end{bmatrix} \tag{9.63}$$

Using the above matrix for $\left[\overline{I} - \overline{Q}\right]^{-1}$, according to Eq. (9.17), $M_{n,a,1} = \overline{n}_{n,a} = \overline{DP}_n$ is determined as follows:

$$\overline{DP}_n = M_{n,a,1} = \overline{P}_t^{(0)} \left[\overline{I} - \overline{Q}\right]^{-1} x \, \overline{1} = \begin{bmatrix} 1 & 0 \end{bmatrix} \begin{bmatrix} 1 & \frac{1}{1-p} \\ 0 & \frac{1}{1-p} \end{bmatrix} \begin{bmatrix} 1 \\ 1 \end{bmatrix} = \frac{2-p}{1-p} \tag{9.64}$$

Using Eq. (9.63) for $\overline{N} = \left[\overline{I} - \overline{Q}\right]^{-1}$, the matrix $\left[2\overline{N} - \overline{I}\right]$ is determined as follows:

$$\left[2\overline{N} - \overline{I}\right] = \left[\begin{bmatrix} 2 & \frac{2}{1-p} \\ 0 & \frac{2}{1-p} \end{bmatrix} - \begin{bmatrix} 1 & 0 \\ 0 & 1 \end{bmatrix} \right] = \begin{bmatrix} 1 & \frac{2}{1-p} \\ 0 & \frac{1+p}{1-p} \end{bmatrix} \tag{9.65}$$

Using the above matrix for $\left[2\overline{N}-\overline{I}\right]$, according to Eq. (9.18), $M_{n,a,2}=\overline{DP}_w$ is determined as follows:

$$M_{n,a,2}=\overline{P}_t^{(0)}\left[2\overline{N}-\overline{I}\right]\overline{N}\times\overline{1}=\begin{bmatrix}1 & 0\end{bmatrix}\begin{bmatrix}1 & \dfrac{2}{1-p}\\ 0 & \dfrac{1+p}{1-p}\end{bmatrix}\begin{bmatrix}1 & \dfrac{1}{1-p}\\ 0 & \dfrac{1}{1-p}\end{bmatrix}\begin{bmatrix}1\\1\end{bmatrix}=\frac{p^2-3p+4}{(1-p)^2} \tag{9.66}$$

Knowing $M_{n,a,1}$ and $M_{n,a,2}$, the $\overline{DP}_w$, Δ, and PDI of the chain distribution, according to Eqs. (3.34), (3.65), and (3.66), respectively, are determined as follows:

$$\overline{DP}_w=\frac{M_{n,a,2}}{M_{n,a,1}}=\frac{(p^2-3p+4)/(1-p)^2}{(2-p)/(1-p)}=\frac{p^2-3p+4}{(1-p)(2-p)} \tag{9.67}$$

$$\Delta=\sqrt{\frac{M_{n,a,2}}{M_{n,a,1}^2}-1}=\sqrt{\frac{\overline{DP}_w}{\overline{DP}_n}-1}=\sqrt{\frac{p^2-3p+4}{(2-p)^2}-1}=\frac{\sqrt{p}}{2-p} \tag{9.68}$$

$$PDI=\frac{\overline{DP}_w}{\overline{DP}_n}=\frac{p^2-3p+4}{(2-p)^2} \tag{9.69}$$

where p in the above equations is the probability of propagation as defined by Eq. (9.41).

9.11.3 Radical Polymerization with Termination by Disproportionation and Combination and Exclusion of Monomers from Chain Distribution

The propagating chains in radical polymerization can terminate by disproportionation or combination [29, 41–43]. For example, the propagating chains in polystyrene terminate predominantly by combination due to the attractive π–π interaction between the phenyl rings, whereas the chains in polymethyl methacrylate terminate predominantly by disproportionation due to the steric hindrance between the bulky methoxy groups [42, 44]. To distinguish between the two termination mechanisms, we can envision disproportionation as a typical propagation state in which the chain undergoes propagation or termination, whereas we can imagine combination as a special state in which the other m-mer terminating chain is generated by propagation prior to transitioning to a typical propagation state, as described by the following elementary reactions:

$$RP_1+M\xrightarrow{k_{p1}>>k_{td}}RP_2\qquad n=1 \tag{9.70}$$

$$RP_n+M\xrightarrow{k_p}RP_{n+1}\qquad n>1 \tag{9.71}$$

$$RP_n+RP_m\xrightarrow{k_{td}}TP_n+TP_m\qquad n,m>1 \tag{9.72}$$

$$RP_n+RP_m\xrightarrow{k_{tc}}TP_{n+m}\qquad n,m>1 \tag{9.73}$$

where k_{tc} is the rate constant for termination by combination. The state space of the reaction system is defined by the following vector:

$$S=[E_1,\ E_2,\ E_3,\ E_4] \tag{9.74}$$

where E_1 is the absorbing or termination state, E_2 is the state for 1-mer chains, E_3 is the combinational state to form the second m-mer chain for termination of the first n-mer chain or transition to a typical transitional propagation state, and E_4 is a typical transitional propagation state for n-mer chains with $n > 1$. The one-step transition probability matrix, $\overline{P}$, is the following (4 × 4) matrix with 16 elements:

$$\overline{P} = \begin{bmatrix} p_{11} & p_{12} & p_{13} & p_{14} \\ p_{12} & p_{22} & p_{23} & p_{24} \\ p_{31} & p_{32} & p_{33} & p_{34} \\ p_{41} & p_{42} & p_{43} & p_{44} \end{bmatrix} = \begin{bmatrix} \overline{I} & \overline{0} \\ \overline{R} & \overline{Q} \end{bmatrix} \tag{9.75}$$

As E_1 is an absorbing state, $p_{11} = 1$ and $p_{12} = p_{13} = p_{14} = 0$. The state E_2 is a 1-mer non-terminating state which only propagates, thus $p_{21} = p_{22} = p_{23} = 0$, and $p_{24} = 1$. As E_3 is a combinational state that can either remain in E_3 to propagate to form the second propagating chain for termination by combination with probability p or transition to a typical propagation state with probability $1 - p$, then $p_{31} = p_{32} = 0$, $p_{33} = p$, and $p_{34} = 1 - p$. As E_4 is a transitional state that can either propagate with probability p or terminate with probability $1 - p$, hence $p_{41} = 1 - p$, $p_{42} = p_{43} = 0$, and $p_{44} = p$. Therefore, $\overline{P}$ is given by the following matrix:

$$\overline{P} = \begin{bmatrix} 1 & 0 & 0 & 0 \\ 0 & 0 & 0 & 1 \\ 0 & 0 & p & 1-p \\ 1-p & 0 & 0 & p \end{bmatrix} = \begin{bmatrix} \overline{I} & \overline{0} \\ \overline{R} & \overline{Q} \end{bmatrix} \tag{9.76}$$

where $\overline{I}$, $\overline{0}$, $\overline{R}$, and $\overline{Q}$ are given by the following matrices:

$$\overline{I} = [1] \qquad \overline{0} = \begin{bmatrix} 0 & 0 & 0 \end{bmatrix} \qquad \overline{R} = \begin{bmatrix} 0 \\ 0 \\ 1-p \end{bmatrix} \qquad \overline{Q} = \begin{bmatrix} 0 & 0 & 1 \\ 0 & p & 1-p \\ 0 & 0 & p \end{bmatrix} \tag{9.77}$$

where the probability of propagation, p, is given by the following equation:

$$p = \frac{R_p}{R_p + R_{td} + R_{tc}} = \frac{1}{1 + \left\{ \left(2k_d \left(k_{td} + k_{tc} \right) f_0 \, [I] \right)^{1/2} / k_p \, [M] \right\}} \tag{9.78}$$

where R_p, R_{td}, and R_{tc} are the rates of propagation, termination by disproportionation, and termination by combination, respectively. As mentioned previously, chains terminating by disproportionation propagate in state E_4 to grow with probability p or transition to an absorbing state with probability $1 - p$. The chains terminating by combination will remain in combinational state E_3 with probability p or transition to a typical propagation state E_4. The chains in the system are divided into disproportionation and combination chains. The disproportionation chains start in the 1-me E_2 state and transition to the propagating E_4 to grow or terminate. The combination chains start in the E_3 state to generate the second m-mer chain prior to transition to a typical propagation state E_4 to grow or terminate. Therefore, at the start of the reaction, all radicals in the system are either in state E_2 (1-mer state for chains terminating by disproportionation) or state E_3 (chains terminating by combination). If the mole fractions of polymer chains terminated by disproportionation and termination are $f_{p,d}$ and $f_{p,c}$, respectively, the initial probability vectors, $\overline{P}^{(0)}$, $\overline{P}_a^{(0)}$, and $\overline{P}_t^{(0)}$ are given as follows (see Eq. 9.14):

$$\begin{aligned}
\overline{P}^{(0)} &= \left[\overline{P}_a^{(0)} \quad \overline{P}_t^{(0)}\right] = \left[0 \quad f_{p,d} \quad f_{p,c} \quad 0\right] \\
\overline{P}_a^{(0)} &= [0] \qquad \overline{P}_t^{(0)} = \left[f_{p,d} \quad f_{p,c} \quad 0\right]
\end{aligned} \tag{9.79}$$

If $f_{r,d}$ is the mole fraction of propagating chains terminated by disproportionation, then $f_{p,d}$ and $f_{p,c}$ are expressed in terms of $f_{r,d}$ by the following relations:

$$f_{p,d} = \frac{f_{r,d}}{f_{r,d} + (1 - f_{r,d})/2} = \frac{2f_{r,d}}{1 + f_{r,d}} \tag{9.80}$$

$$f_{p,c} = \frac{(1 - f_{r,d})/2}{f_{r,d} + (1 - f_{r,d})/2} = \frac{1 - f_{r,d}}{1 + f_{r,d}} \tag{9.81}$$

Using the above relations, the vectors $\overline{P}_a^{(0)}$ and $\overline{P}_t^{(0)}$ are expressed in terms of $f_{r,d}$ as follows:

$$\overline{P}_a^{(0)} = [0] \qquad \overline{P}_t^{(0)} = \left[\frac{2f_{r,d}}{1 + f_{r,d}} \quad \frac{1 - f_{r,d}}{1 + f_{r,d}} \quad 0\right] \tag{9.82}$$

Using Cayley–Hamilton theorem, it can be shown that the (3×3) $\overline{Q}^{(n-1)}$ matrix for this polymerization process, as defined by Eq. (9.77), is equal to the following [33–40]:

$$\overline{Q}^{(n-1)} = \begin{bmatrix} 0 & 0 & 1 \\ 0 & p & 1-p \\ 0 & 0 & p \end{bmatrix}^{n-1} = \begin{bmatrix} 0 & 0 & p^{n-2} \\ 0 & p^{n-1} & (n-1)(1-p)\,p^{n-2} \\ 0 & 0 & p^{n-1} \end{bmatrix} \tag{9.83}$$

Using the above $\overline{P}_t^{(0)}$ vector and $\overline{Q}^{(n-1)}$ matrix, and $\overline{R}$ from Eq. (9.77), the $p_{DP}(n)$ according to Eq. (9.16) is expressed as follows:

$$\begin{aligned}
p_{DP}(n) &= \left[\frac{2f_{r,d}}{1 + f_{r,d}} \quad \frac{1 - f_{r,d}}{1 + f_{r,d}} \quad 0\right] \begin{bmatrix} 0 & 0 & p^{n-2} \\ 0 & p^{n-1} & (n-1)(1-p)\,p^{n-2} \\ 0 & 0 & p^{n-1} \end{bmatrix} \begin{bmatrix} 0 \\ 0 \\ 1-p \end{bmatrix} [1] \\
&= \frac{1-p}{1 + f_{r,d}} \left[(n-1)(1-p)(1-f_{r,d}) + 2f_{r,d}\right] p^{n-2}
\end{aligned} \tag{9.84}$$

Using matrices in Eq. (9.77) and $\overline{N} = \left(\overline{I} - \overline{Q}\right)^{-1}$ (see practice problem 9.2), according to Eqs. (9.17) to (9.22), $\overline{DP}_n = M_{n,a,1}$, $\overline{DP}_w$, Δ, and PDI are given by the following expressions for this polymerization system:

$$\begin{aligned}
\overline{DP}_n = M_{n,a,1} = \overline{P}_t^{(0)} \left[\overline{I} - \overline{Q}\right]^{-1} x\overline{1} &= \left[\frac{2f_{r,d}}{1 + f_{r,d}} \quad \frac{1 - f_{r,d}}{1 + f_{r,d}} \quad 0\right] \begin{bmatrix} 1 & 0 & \dfrac{1}{1-p} \\ 0 & \dfrac{1}{1-p} & \dfrac{1}{1-p} \\ 0 & 0 & \dfrac{1}{1-p} \end{bmatrix} \begin{bmatrix} 1 \\ 1 \\ 1 \end{bmatrix} \\
&= \frac{2\left[f_{r,d}(1-p) + 1\right]}{(1 + f_{r,d})(1-p)}
\end{aligned} \tag{9.85}$$

$$\overline{DP}_w = \frac{M_{n,a,2}}{M_{n,a,1}} = \frac{2 + p + (p^2 - 4p + 2)\, f_{r,d}}{(1-p)\left[f_{r,d}\,(1-p) + 1\right]} \tag{9.86}$$

$$\Delta = \sqrt{\frac{\overline{DP}_w}{\overline{DP}_n} - 1} = \sqrt{\frac{\left[2 + p + (p^2 - 4p + 2)\, f_{r,d}\right](1 + f_{r,d})}{2\left[f_{r,d}\,(1-p) + 1\right]^2} - 1} \tag{9.87}$$

$$PDI = \frac{\overline{DP}_w}{\overline{DP}_n} = \frac{\left[2 + p + (p^2 - 4p + 2)\, f_{r,d}\right](1 + f_{r,d})}{2\left[f_{r,d}\,(1-p) + 1\right]^2} \tag{9.88}$$

where p in the above expressions is defined by Eq. (9.78). The $f_{r,d}$ is the mole fraction of propagating chains terminated by disproportionation which is related to the rates of termination by disproportionation (R_{td}) and combination (R_{tc}) (see Section 7.3.1 of Chapter 7).

9.12 Application of Markov Chain to Condensation Polymerization

In radical polymerization, only the reactive species on propagating chains, a small fraction of the molecules in the reaction mixture, react with monomers to form longer chains. Conversely, in condensation polymerization, any multifunctional molecule in the reaction mixture (monomer or polymer) can react with any other multifunctional monomer with complementary reactive groups to form longer chains [45–47]. Any molecule with complementary reactive groups, regardless of size, can participate in the reaction in condensation polymerization. The simplest condensation polymerization system consists of monomer ARB in which A- and B- are complementary reactive groups, i.e., A- and B- react with each other but A cannot react with A and B cannot react with B [48, 49]. The R in ARB is the monomer residue. This section applies Markov chain modeling to predict DP distribution and averages in condensation polymerization reactions.

9.12.1 Condensation Polymerization of ARB Monomers

The general reaction for condensation of ARB monomers is as follows [48–54]:

$$A\,(RB - A)_n\,RB + A\,(RB - A)_m\,RB \xrightarrow{k_p} A\,(RB - A)_{m+n+1}\,RB \tag{9.89}$$

According to the above reaction, a chain with (n+1) monomers, n repeat units, (n+1) R residues, or n B-A linkages react with another chain in the reaction mixture with (m+1) monomers, m repeat units, (m+1) residues, or m B-A linkages to form a longer chain with (m+n+2) monomers, (m+n+1) repeat units, (m+n+2) R residues, or (m+n+1) B-A linkages. The dash line in the chains indicates that the reaction occurs between the complementary reactive A and B groups. The A-B linkage is the functional unit, and RB-A is the repeat unit of the chain. If we move from the first R residue on the left end of a chain stepwise toward the right end, we encounter two types of groups, namely B-A groups along the chain and a -B group at the right end. Conversely, if we move from the first R residue at the right end of a chain stepwise toward the left end, we encounter two types of groups, namely B-A groups along the chain and an -A at the right end. The B-A linkage is considered a propagated functional group, whereas the unreacted -B or the unreacted -A is considered a terminal or absorbing functional group. In a Markov chain process, the B-A linkage acts as a transitional propagating state whereas the unreacted -A or -B group is considered an absorbing state. Therefore, the state space for this system is expressed by the following vector:

$$S = [E_1, E_2] \tag{9.90}$$

where E_1 is the absorbing or chain-end state and E_2 is the transitional state. The one-step transition probability matrix, $\overline{p}$, is a (2×2) matrix with elements p_{11}, p_{12}, p_{21}, and p_{22}. The element p_{11} is the probability of remaining in a terminal -A or -B group (absorbing state) after the reaction has ended which is equal to one; p_{12} is the probability for a terminal -A or -B to propagate and become a B-A or A-B which is equal to zero; p_{21} is the probability of transitioning from a B-A group to an -A or -B as we move along the chain from the chain-end on the left side to the right side, which is equal to the fraction of unreacted -A or -B groups or $1 - \alpha$; and p_{22} is the probability of transitioning from a B-A linkage to another B-A as we move along the chain, which is equal to the fraction of reacted -A or -B groups or the reaction conversion α, defined as follows for a constant volume reaction:

$$\alpha = \frac{[A]_0 - [A]}{[A]_0} = 1 - \frac{[P]}{[A]_0} \tag{9.91}$$

where $[A]_0$ is the initial concentration of unreacted -A functional groups, $[A]$ is the concentration of unreacted -A in the polymer. As every chain ends in one -A or one -B group, $[A]$ is equal to the concentration of polymer $[P]$. Therefore, the matrix $\overline{p}$ and its partitions $\overline{I}$, $\overline{0}$, $\overline{R}$, and $\overline{Q}$ are given as follows:

$$\overline{P} = \begin{bmatrix} 1 & 0 \\ 1-\alpha & \alpha \end{bmatrix} = \begin{bmatrix} \overline{I} & \overline{0} \\ \overline{R} & \overline{Q} \end{bmatrix} \quad \overline{I} = [1] \quad \overline{0} = [0] \quad \overline{R} = [1-\alpha] \quad \overline{Q} = [\alpha] \tag{9.92}$$

As we move along a chain, it is expected initially to observe a reacted B-A or A-B linkage if $\alpha \neq 0$. Therefore, the initial probability vector for this chain process, according to Eq. (9.14), is given as follows:

$$\overline{P}^{(0)} = \left[\overline{P}_a^{(0)} \;\; \overline{P}_t^{(0)}\right] = \begin{bmatrix} 0 & 1 \end{bmatrix} \quad \overline{P}_a^{(0)} = [0] \quad \overline{P}_t^{(0)} = [1] \tag{9.93}$$

Using the same approach as in Section 9.11.1, $p_{DP}(n)$, $M_{n,a,1} = \overline{DP}_n$, $M_{n,a,2}$, $\overline{DP}_w$, Δ, and PDI for condensation polymerization of ARB monomer are given as follows:

$$p_{DP}(n) = \overline{P}_t^{(0)} \overline{Q}^{(n-1)} \left(\overline{R}\, x\, \overline{1}\right) = (1-\alpha)\,\alpha^{n-1} \tag{9.94}$$

$$\overline{DP}_n = M_{n,a,1} = \overline{P}_t^{(0)} \left[\overline{I} - \overline{Q}\right]^{-1} x\, \overline{1} = \frac{1}{1-\alpha} \tag{9.95}$$

$$M_{n,a,2} = \overline{P}_t^{(0)} \left[2\overline{N} - \overline{I}\right] \overline{N}\, x\, \overline{1} = \frac{1+\alpha}{(1-\alpha)^2} \tag{9.96}$$

$$\overline{DP}_w = \frac{M_{n,a,2}}{M_{n,a,1}} = \frac{1+\alpha}{1-\alpha} \tag{9.97}$$

$$\Delta = \sqrt{\frac{M_{n,a,2}}{M_{n,a,1}^2} - 1} = \sqrt{\frac{\overline{DP}_w}{\overline{DP}_n} - 1} = \sqrt{\alpha} \tag{9.98}$$

$$PDI = \frac{\overline{DP}_w}{\overline{DP}_n} = 1 + \alpha \tag{9.99}$$

where α is conversion. As conversion increases toward one, $\overline{DP}_n$ and $\overline{DP}_w$ increase but so does Δ and PDI.

9.12.2 Condensation Polymerization of AR_1A, AR_2, BR_3B, and BR_4 Monomers

A more common condensation or step-growth polymerization system consists of two bifunctional monomers AR_1A and BR_3B with a small fraction of monofunctional AR_2 and BR_4 [47, 55–59]. The residues R_1 and R_2 are residues of the bifunctional A and B monomers, respectively, and R_3 and R_4 are the residues for monofunctional A and B monomers. The complementary functional groups A and B on AR_1A and BR_3B monomers react to form polymer chains whereas the functional groups A and B on AR_2 and BR_4 react to control the chain length. Polymer chains with the following end groups are produced in the polymerization of AR_1A, AR_2, BR_3B, and BR_4 monomers with an unequal number of A and B functional groups:

$$\begin{aligned}
&A(R_1A - BR_3B - A)_n R_1A\\
&A(R_1A - BR_3B - A)_n R_1A - BR_3B\\
&A(R_1A - BR_3B - A)_n R_1A - BR_4\\
&B(R_3B - AR_1A - B)_n R_3B\\
&B(R_3B - AR_1A - B)_n R_3B - AR_2\\
&R_2A - B(R_3B - AR_1A - B)_n R_3B - AR_2\\
&R_4B - A(R_1A - BR_3B - A)_n R_1A - BR_4
\end{aligned} \tag{9.100}$$

The first chain starts and ends with AR_1A whereas the second chain starts with AR_1A and ends with BR_3B; the third chain starts with AR_1A and ends with BR_4 whereas the fourth chain starts and ends with BR_3B; the fifth chain starts with BR_3B and ends with AR_2 whereas the sixth chain starts and ends with AR_2; and the last chain starts and ends with BR_4. The repeat unit of all chains is $R_1A - BR_3B - A$ or $R_3B - AR_1A - B$ which are identical. The terminal groups of the chains are R_1A, AR_2, R_3B, or BR_4, which can be represented by one absorbing state E_1. Unlike the ARB polymerization system, the number of A-B linkages may not be equal to the number of B-A linkages because the number of reacted Bs may not be equal to the number of reacted As. Therefore, A-B and B-A linkages should be treated as two independent transitional states E_2 and E_3. The A reacting with B to form A-B is the transitional state E_2 and the B reacting with A to form B-A is the transitional state E_3. Therefore, the state space for this system is given by the following vector:

$$S = [E_1, E_2 E_3] \tag{9.101}$$

If the reacting group A is the limiting reactant and the mole fraction of AR_1A, AR_2, BR_3B, and BR_4 monomers are x_{A2}, x_{A1}, x_{B2}, and x_{B1}, respectively, then the following inequality holds:

$$x_{A1} + 2x_{A2} \leq x_{B1} + 2x_{B2} \tag{9.102}$$

As the A group is the limiting reactant, the conversion α is defined in terms of A. If N_0 is the total number of AR_1A, AR_2, BR_3B, and BR_4 monomers, then the initial number of A $(N_{A,0})$ and B $(N_{B,0})$ groups, the number of reacted A $(N_{A,r})$ and remaining A $(N_{A,\alpha})$ are defined as follows:

$$N_{A,0} = (x_{A1} + 2x_{A2})\, N_0 \tag{9.103}$$

$$N_{B,0} = (x_{B1} + 2x_{B2})\, N_0 \tag{9.104}$$

$$N_{A,r} = (x_{A1} + 2x_{A2})\,\alpha\,N_0 \tag{9.105}$$

$$N_{A,\alpha} = (1-\alpha)(x_{A1} + 2x_{A2})\,N_0 \tag{9.106}$$

As for every reacted A, a B is consumed, the number of reacted B ($N_{B,r}$), remaining B ($N_{B,\alpha}$) groups, and the conversion of B groups (α_B) in terms of α are defined as follows:

$$N_{B,r} = (x_{A1} + 2x_{A2})\,\alpha\,N_0 \tag{9.107}$$

$$N_{B,\alpha} = N_{B,0} - N_{B,r} = [(x_{B1} + 2x_{B2}) - (x_{A1} + 2x_{A2})\,\alpha]\,N_0 \tag{9.108}$$

$$\alpha_B = \frac{N_{B,r}}{N_{B,0}} = \frac{(x_{A1} + 2x_{A2})\,\alpha N_0}{(x_{B1} + 2x_{B2})\,N_0} = \frac{(x_{A1} + 2x_{A2})\,\alpha}{(x_{B1} + 2x_{B2})} \tag{9.109}$$

The one-step transition probability matrix, $\overline{p}$, is a (3 × 3) matrix with elements p_{ij}. As the state E_1 is an absorbing state, $p_{11} = 1$ and $p_{12} = p_{13} = 0$. As A-B and B-A alternate on a polymer chain, the probability of transitioning from state A-B to itself is zero, therefore $p_{22} = 0$. The probability of transitioning from state A-B to B-A, p_{23}, depends on the reaction conversion and fraction of AR_1A as AR_2 terminates the chain. The probability of transitioning from A-B to an absorbing state, p_{21}, depends on conversion and fraction of terminating AR_2 monomer. Therefore, p_{23} and p_{21} are given by the following equations:

$$p_{23} = \frac{\alpha\,x_{A2}}{x_{A1} + x_{A2}} \tag{9.110}$$

$$p_{21} = 1 - p_{23} = 1 - \frac{\alpha\,x_{A2}}{x_{A1} + x_{A2}} \tag{9.111}$$

As B-A and A-B alternate on a polymer chain, the probability of transitioning from state B-A to itself is zero, therefore $p_{33} = 0$. The probability of transitioning from state B-A to A-B, p_{32}, depends on the conversion of B groups and fraction of BR_3B in B monomers (BR_4 terminates the chain). The probability of transitioning from state B-A to an absorbing state, p_{31}, depends on the conversion of Bs and fraction of terminating BR_4 monomer. Therefore, upon using Eq. (9.109) for α_B, p_{32} and p_{31} are given as follows:

$$p_{32} = \frac{\alpha_B\,x_{B2}}{x_{B1} + x_{B2}} = \frac{(x_{A1} + 2x_{A2})\,\alpha}{(x_{B1} + 2x_{B2})}\,\frac{x_{B2}}{x_{B1} + x_{B2}} \tag{9.112}$$

$$p_{31} = 1 - p_{32} = 1 - \frac{\alpha_B\,x_{B2}}{x_{B1} + x_{B2}} = 1 - \frac{(x_{A1} + 2x_{A2})\,\alpha}{(x_{B1} + 2x_{B2})}\,\frac{x_{B2}}{x_{B1} + x_{B2}} \tag{9.113}$$

Therefore, $\overline{p}$ is given by the following matrix:

$$\overline{P} = \begin{bmatrix} 1 & 0 & 0 \\ 1 - \dfrac{\alpha\,x_{A2}}{x_{A1} + x_{A2}} & 0 & \dfrac{\alpha\,x_{A2}}{x_{A1} + x_{A2}} \\ 1 - \dfrac{\alpha_B\,x_{B2}}{x_{B1} + x_{B2}} & \dfrac{\alpha_B\,x_{B2}}{x_{B1} + x_{B2}} & 0 \end{bmatrix} = \begin{bmatrix} \overline{I} & \overline{0} \\ \overline{R} & \overline{Q} \end{bmatrix} \tag{9.114}$$

where the partitions $\overline{I}$, $\overline{0}$, $\overline{R}$, and $\overline{Q}$ are given by the following matrices:

$$\overline{I} = [1] \quad \overline{0} = \begin{bmatrix} 0 & 0 \end{bmatrix} \quad \overline{R} = \begin{bmatrix} 1 - \dfrac{\alpha\,x_{A2}}{x_{A1} + x_{A2}} \\ 1 - \dfrac{\alpha_B\,x_{B2}}{x_{B1} + x_{B2}} \end{bmatrix} \quad \overline{Q} = \begin{bmatrix} 0 & \dfrac{\alpha\,x_{A2}}{x_{A1} + x_{A2}} \\ \dfrac{\alpha_B\,x_{B2}}{x_{B1} + x_{B2}} & 0 \end{bmatrix} \tag{9.115}$$

The initial probability matrix, $\overline{P}^{(0)}$, is a (1×3) vector with one absorbing state $\overline{P}_a^{(0)}$ and two transitional states $\overline{P}_{t1}^{(0)}$ and $\overline{P}_{t2}^{(0)}$. Initially, if monomers are in the absorbing state, no polymerization takes place, therefore $\overline{P}_a^{(0)} = 0$. The probability of having a reactive A to form an A-B linkage (state E_2) is the same as the probability of finding a polymer chain with a terminal A group. Similarly, the probability of having a reactive B to form a B-A linkage (state E_3) is the same as the probability of finding a polymer chain with a terminal B group. If N_0 is the total number of AR_1A, AR_2, BR_3B, and BR_4 monomers, then the number of terminal A $(N_{A,ter})$ and terminal B $(N_{B,ter})$ groups, and total number of terminal A and B groups $(N_{A+B,ter})$ in polymer chains in terms of α and α_B are given as follows:

$$N_{A,ter} = x_{A1}N_0 + (1-\alpha)\, x_{A2}N_0 \tag{9.116}$$

$$N_{B,ter} = x_{B1}N_0 + (1-\alpha_B)\, x_{B2}N_0 \tag{9.117}$$

$$N_{A+B,ter} = (x_{A1} + x_{B1})\, N_0 + [(1-\alpha)\, x_{A2} + (1-\alpha_B)\, x_{B2}]\, N_0 \tag{9.118}$$

Based on the above equations, $p_{t1}^{(0)}$ and $p_{t2}^{(0)}$ are given as follows:

$$p_{t1}^{(0)} = \frac{N_{A,ter}}{N_{A,ter} + N_{B,ter}} = \frac{x_{A1} + (1-\alpha)\, x_{A2}}{(x_{A1} + x_{B1}) + [(1-\alpha)\, x_{A2} + (1-\alpha_B)\, x_{B2}]} \tag{9.119}$$

$$p_{t2}^{(0)} = \frac{N_{B,ter}}{N_{A,ter} + N_{B,ter}} = \frac{x_{B1} + (1-\alpha_B)\, x_{B2}}{(x_{A1} + x_{B1}) + [(1-\alpha)\, x_{A2} + (1-\alpha_B)\, x_{B2}]} \tag{9.120}$$

The $\overline{P}^{(0)}$ matrix and its partitions are given as follows:

$$\overline{P}^{(0)} = \left[\overline{P}_a^{(0)} \;\; \overline{P}_t^{(0)}\right] \qquad \overline{P}_a^{(0)} = [0]$$

$$\overline{P}_t^{(0)} = \left[\frac{x_{A1} + (1-\alpha)\, x_{A2}}{(x_{A1} + x_{B1}) + [(1-\alpha)\, x_{A2} + (1-\alpha_B)\, x_{B2}]} \quad \frac{x_{B1} + (1-\alpha_B)\, x_{B2}}{(x_{A1} + x_{B1}) + [(1-\alpha)\, x_{A2} + (1-\alpha_B)\, x_{B2}]}\right] \tag{9.121}$$

Having defined the state space S, the transitional probability matrix $\overline{P}$, and the initial probability vector $\overline{P}^{(0)}$, the Eqs. (9.16) through (9.22) can be used to determine chain distribution $p_{DP}(n)$, averages $\overline{DP}_n$ and $\overline{DP}_w$, and breadth of the distribution Δ and PDI for condensation polymerization of AR_1A, AR_2, BR_3B, and BR_4 monomers. In the following subsection, $p_{DP}(n)$, $\overline{DP}_n$, $\overline{DP}_w$, Δ, and PDI are determined for the special case of condensation polymerization of difunctional AR_1A and BR_3B monomers in the absence of monofunctional A and B monomers (i.e., $x_{A1} = x_{B1} = 0$).

9.12.3 Condensation Polymerization of AR_1A and BR_3B Monomers with Unequal Number of A and B Groups

Polymer chains with the following end groups are produced in the polymerization of AR_1A and BR_3B monomers with an unequal number of A and B functional groups:

$$\begin{aligned} &A(R_1A - BR_3B - A)_n\, R_1A \\ &A(R_1A - BR_3B - A)_n\, R_1A - BR_3B \\ &B(R_3B - AR_1A - B)_n\, R_3B \end{aligned} \tag{9.122}$$

The first chain starts and ends with AR_1A whereas the second chain starts with AR_1A and ends with BR_3B; and the third chain starts with BR_3B and ends with BR_3B. The repeat unit of all chains is $R_1A - BR_3B - A$ or $R_3B - AR_1A - B$ which are identical. Assuming that A is the limiting functional group, α is the conversion of A groups, $x_{A2} = x$ is the mole fraction of AR_1A monomer, and $x_{B2} = 1 - x_{A2} = 1 - x$ is the mole fraction of BR_3B, then $x < 0.5$. The imbalance ratio, r, between functional groups A and B is defined as:

$$r = \frac{N_{A0}}{N_{B0}} = \frac{2x_{A2}N_0}{2x_{B2}N_0} = \frac{xN_0}{(1-x)\,N_0} = \frac{x}{1-x} \tag{9.123}$$

Using Eq. (9.109) with $x_{A1} = x_{B1} = 0$, the conversion of B functional groups, α_B, is given as follows:

$$\alpha_B = \frac{(x_{A1} + 2x_{A2})\,\alpha}{(x_{B1} + 2x_{B2})} = r\,\alpha \tag{9.124}$$

Using Eq. (9.114) with $x_{A1} = x_{B1} = 0$, $\overline{p}$ and its partitions $\overline{R}$ and $\overline{Q}$ are given by the following matrices:

$$\overline{P} = \begin{bmatrix} 1 & 0 & 0 \\ 1-\alpha & 0 & \alpha \\ 1-r\,\alpha & r\,\alpha & 0 \end{bmatrix} \quad \overline{R} = \begin{bmatrix} 1-\alpha \\ 1-r\,\alpha \end{bmatrix} \quad \overline{Q} = \begin{bmatrix} 0 & \alpha \\ r\,\alpha & 0 \end{bmatrix} \tag{9.125}$$

Using Eq. (9.121) with $x_{A1} = x_{B1} = 0$, the initial probability vector $\overline{P}^{(0)}$ and its partitions are as follows:

$$\overline{P}^{(0)} = \left[\overline{P}_a^{(0)} \;\; \overline{P}_t^{(0)}\right] \quad \overline{P}_a^{(0)} = [0] \quad \overline{P}_t^{(0)} = \left[\frac{r\,(1-\alpha)}{(1-\alpha)\,r + (1-r\alpha)} \;\; \frac{1-r\alpha}{(1-\alpha)\,r + (1-r\alpha)}\right] \tag{9.126}$$

Using Cayley–Hamilton theorem and writing out the terms, it can be shown that the $(2\times 2)\,\overline{Q}^{(n-1)}$ matrix, as $\overline{Q}$ defined by Eq. (9.125), is given as follows [33–40]:

$$\begin{cases} \overline{Q}^{(n-1)} = \begin{bmatrix} r^{\frac{n-1}{2}}\alpha^{n-1} & 0 \\ 0 & r^{\frac{n-1}{2}}\alpha^{n-1} \end{bmatrix} & n = 2k-1 \quad k = 1, 2, 3, \ldots \\ \overline{Q}^{(n-1)} = \begin{bmatrix} 0 & r^{\frac{n-2}{2}}\alpha^{n-1} \\ r^{\frac{n}{2}}\alpha^{n-1} & 0 \end{bmatrix} & n = 2k \quad k = 1, 2, 3, \ldots \end{cases} \tag{9.127}$$

Applying the above equation for $\overline{Q}^{(n-1)}$, Eq. (9.126) for $\overline{P}_t^{(0)}$, and using the defining Eq. (9.16), $p_{DP}(n)$ is given by the following expressions for odd and even DP of chains:

$$\begin{cases} p_{DP}(n) = \left[p_{t1}^{(0)}\,(1-\alpha) + p_{t2}^{(0)}\,(1-r\,\alpha)\right] r^{\frac{n-1}{2}}\alpha^{n-1} & n = 2k-1 \quad k = 1, 2, \ldots \\ p_{DP}(n) = \left[p_{t1}^{(0)}\,\{(1/r)-\alpha\} + p_{t1}^{(0)}\,(1-\alpha)\right] r^{\frac{n}{2}}\alpha^{n-1} & n = 2k \quad k = 1, 2, \ldots \end{cases} \tag{9.128}$$

where α is conversion of the limiting reactant A, r is the imbalance ratio defined by Eq. (9.123), and $p_{t1}^{(0)}$ and $p_{t2}^{(0)}$ are defined by Eq. (9.126) in terms of α and r. Using the matrices in Eq. (9.125) and $\overline{N} = \left(\overline{I} - \overline{Q}\right)^{-1}$ (see Practice Problem 9.2), $M_{n,a,1} = \overline{DP}_n$, $\overline{DP}_w = M_{n,a,2}/M_{n,a,1}$, Δ, and PDI for condensation polymerization of AR_1A and BR_3B monomers are given as follows:

$$\overline{DP}_n = \overline{P}_t^{(0)} \overline{N}^{-1} x \overline{1} = \frac{1+r}{(1-\alpha)\,r + (1-r\alpha)} = \frac{1}{1-2\alpha x} \tag{9.129}$$

$$\overline{DP}_w = \frac{M_{n,a,2}}{M_{n,a,1}} = \frac{(1-x)(4x\alpha+1)+x\alpha^2}{(1-x-x\alpha^2)} \tag{9.130}$$

$$\Delta = \sqrt{\frac{\overline{DP}_w}{\overline{DP}_n} - 1} = \sqrt{\frac{\left[(1-x)(4x\alpha+1)+x\alpha^2\right](1-2x\alpha)}{(1-x-x\alpha^2)} - 1} \tag{9.131}$$

$$PDI = \frac{\overline{DP}_w}{\overline{DP}_n} = \frac{\left[(1-x)(4x\alpha+1)+x\alpha^2\right](1-2x\alpha)}{(1-x-x\alpha^2)} \tag{9.132}$$

where $x = r/\,(1+r)$. The ranges of α, r, and x are $0 \le \alpha \le 1$, $0 < r \le 1$, and $0 < x \le 0.5$. The above equations show that $\overline{DP}_n$ and $\overline{DP}_w$ increase with increasing α, r, and x. Furthermore, Δ and PDI also increase with increasing α, r, and x.

9.13 Summary

This chapter described the Markov chain process for modeling polymers produced by addition and step growth polymerization. As a Markov chain is a process describing a sequence of stochastic events, this method is used to describe the random and discrete initiation, propagation, termination, and chain transfer steps, among others, in polymerization reactions. The Markov chain modeling approach involves defining the state space, one-step transition probability matrix, and the initial probability vector of the states. The state space contains the different events/reactions the chain could experience in each step like adding a monomer to propagate or undergoing disproportionation to terminate. The states of the system are divided into absorbing states which terminate the chain growth, and transitional states which the chain continues to add monomer and propagate. The transition matrix provides the probability of transitioning from any state of the system to another state in one step. The initial probability vector provides the initial probabilities for the states of the system. Once the state space, transition probability matrix, and initial probability vector are defined for a stochastic Markov process, this knowledge can be used to find the probability distribution of reaching an absorbing state, which is the same as the distribution of degree of polymerization of the chains in the reaction. Then, averages of the chain distribution and polydispersity index can be calculated from the DP distribution. Next, the Markov process was used in this chapter to model addition (radical) and step growth (condensation) polymerization to predict the distribution of degree of polymerization of polymer chains. Polymerization reactions modeled in this chapter demonstrate that the method of Markov chains can be used to model complex polymerization reactions like condensation polymerization of monofunctional and bifunctional AR_1A, AR_2, BR_3B, and BR_4 monomers.

9.14 List of Symbols and Abbreviations

$[A]$: molar concentration of reactive functional group A in step-growth polymerization at any conversion or time during the reaction

$[A]_0$: initial molar concentration of reactive functional group A in step-growth polymerization

$\overline{A}, \overline{B}, \overline{X}, \overline{Y}$: matrices for demonstration of matrix operation

$a_{ij}, b_{ij}, x_{ij}, y_{ij}$: an element in ith row and jth column of the matrices $\overline{A}, \overline{B}, \overline{X},$ *and* $\overline{Y}$

$\overline{A}^{-1}$ $(n \times n)$: inverse of matrix $\overline{A}$ $(n \times n)$

$\overline{A}_{adj}$ $(n \times n)$: adjugate matrix $\overline{A}$ $(n \times n)$

$\overline{A}_{cf}$ $(n \times n)$: matrix of cofactors of $\overline{A}$ $(n \times n)$

a_{ij}^{cf}: an element in ith row and jth column of the cofactor of element a_{ij} of the matrix $\overline{A} = \overline{P} - \overline{I}$

AR_1A: bifunctional monomer used in step-growth polymerization consisting of two A reactive non-complementary (do react with each other) functional groups linked via an R_1 repeat unit

AR_2: monofunctional monomer used in step-growth polymerization consisting of one A reactive functional group linked to an R_2 repeat unit

ARB: bifunctional monomer used in step-growth polymerization consisting of A and B reactive complementary (react with each other) functional groups linked via an R repeat unit

BR_3B: bifunctional monomer used in step-growth polymerization consisting of two reactive non-complementary (do not react with each other) functional groups linked via an R_3 repeat unit

BR_4: monofunctional monomer used in step-growth polymerization consisting of one B reactive functional group linked to an R_4 repeat unit

c: constant

DP: degree of polymerization of a polymer chain

$\overline{DP}_n$: number-average degree of polymerization of polymer chains

$\overline{DP}_w$: weight-average degree of polymerization of polymer chains

E_i: ith state of the system among k states in a step, observation, or time interval in a Markov chain process, as an abbreviated representation for E_{i_n}

E_{j_n}: jth state of the system in the nth step, observation, or time interval in a Markov chain process

f_0: efficiency of dissociation of initiator to generate 1-mer reactive propagating chains

$f_{p,c}$: mole fraction of terminated polymer chains that are terminated by combination

$f_{p,d}$: mole fraction of terminated polymer chains that are terminated by disproportionation

$f_{r,d}$: mole fraction of reactive propagating chains terminated by disproportionation

$[I]$: initiator concentration at any conversion or time

$\overline{I}$ $(k - r \times k - r)$: unit matrix with $(k - r)$ rows and columns where diagonal elements are equal to one and off-diagonal elements are zero

k: number of states of the system in a Markov chain process

$(k - r)$: number of transitional states of the system in a Markov chain process

k_d: rate constant for dissociation of the initiator

k_p: rate constant for propagation reaction

k_{p1}: rate constant for propagation between the 1-mer reactive propagating chain and monomer to form a 2-mer propagating chain

k_{tc}: rate constant for termination of two propagating chains by combination to form a terminated polymer chain

k_{td}: rate constant for termination of two propagating chains by disproportionation to form two terminated polymer chains

m, n: number of monomer residues or repeat units on a polymer chain or the number of steps to reach an absorbing state of a Markov chain process

M: monomer

$[M]$: monomer concentration at any conversion or time

$[M]_0$: initial monomer concentration in polymerization

$|M_{ij}|$: determinant of the $(n-1)\times(n-1)$ matrix resulting from removing the ith row and jth column from the matrix $\overline{A}\,(n \times n)$

$M_{n,a,1}$: first order, mole fraction moment of the distribution of steps to reach an absorbing state of a Markov chain process

$M_{n,a,2}$: second order, mole fraction moment of the distribution of steps to reach an absorbing state of a Markov chain process

n: successive step, observation, or time interval number in a Markov chain process

$\overline{N}\,(k-r \times k-r)$: a matrix with $(k-r)$ rows and columns equal to $\left[\overline{I}-\overline{Q}\right]^{-1}$

N_0: initial total number of moles of monomers in step-growth polymerization

$N_{A,0}$: initial number of moles of monomer of A functional group in step-growth polymerization

$N_{A,r}$: number of moles of reacted A functional group at any conversion α or time t in step-growth polymerization

$N_{A,\alpha}$: number of moles of remaining A functional group at any conversion α or time t in step-growth polymerization

$N_{A,ter}$: number of moles of terminal (at chain ends) A functional group at any conversion α or time t in step-growth polymerization

$N_{A+B,ter}$: number of moles of terminal (at chain ends) $A+B$ functional groups at any conversion α or time t

$N_{B,0}$: initial number of moles of monomer of B functional group in step-growth polymerization

$N_{B,r}$: number of moles of reacted B functional group at any conversion α or time t in step-growth polymerization

$N_{B,\alpha}$: number of moles of remaining B functional group at any conversion α or time t in step-growth polymerization

$N_{B,ter}$: number of moles of terminal (at chain ends) B functional group at any conversion α or time t in step-growth polymerization

$\overline{n}_{n,a}$: average number of steps to reach an absorbing state of a Markov chain process

p: probability of chain propagation by the reaction between an n-mer reactive propagating chain and monomer to form an $(n+1)$-mer propagating chain

$[P]$: molar concentration of polymer chains at any conversion or time during the reaction

$\overline{P}\,(k \times k)$: one-step transition probability matrix with k rows and columns for transitioning from ith state to jth state

$\overline{P}^{(0)}\,(1 \times k)$: initial probability vector with one row and k columns for states of the system in a Markov process

$\overline{P}^{(n)}\,(k \times k)$: n-steps transition probability matrix with k rows and columns for transitioning from ith state to jth state

$\overline{P}_a^{(0)}\,(1 \times k)$: initial probability vector with one row and r columns for absorbing states of the system in a Markov process

$\overline{P}_{aa}\,(r \times r)$: one-step transition probability matrix with r rows and columns for transitioning from ith absorbing state to jth absorbing state

$\overline{P}^{(n)}_{abs}\,(1 \times k) = \overline{P}^{(n)}_{a}$: n-steps absolute probability vector with one row and k columns for transitioning from any state in the system to an absorbing state in n-steps

$p^{(n)}_{abs,ij}$: n-step probability of transitioning from ith state to jth absorbing state of the system

$p^{(0)}_{a,i}$: initial probability of finding the system in ith absorbing state of the system in a Markov chain process

$\overline{P}_{at}\,(r \times k - r)$: one-step transition probability matrix with r rows and $(k - r)$ columns for transitioning from ith absorbing state to jth transitional state

$\overline{P}_{DP}\,(n) = \overline{P}^{(n)}_{a}$: n-steps probability distribution of reaching an absorbing state or distribution of degree of polymerization in a Markov chain process

$p^{(0)}_{E_i} = p^{(0)}_{i}$: initial probability of finding the system in state E_i or the ith state of the system in a Markov chain process

$p_{E_iE_j} = p_{ij}$: probability of transitioning from state E_i or ith state to state E_j or jth state of the system in one step, observation, or time interval in a Markov chain process

$p^{(n)}_{E_iE_j} = p^{(n)}_{ij}$: probability of transitioning from state E_i or ith state to state E_j or jth state of the system in n steps, observations, or time intervals in a Markov chain process

$\overline{P}^{(0)}_{t}\,(1 \times k)$: initial probability vector with one row and (k-r) columns for transitional states of the system in a Markov process

$\overline{P}_{ta}\,(k - r \times r)$: one-step transition probability matrix with $(k - r)$ rows and r columns for transitioning from ith transitional state to jth absorbing state

$p^{(0)}_{t,i}$: initial probability of finding the system in ith transitional state of the system in a Markov chain process

$\overline{P}_{tt}\,(k - r \times k - r)$: one-step transition probability matrix with $(k - r)$ rows and columns for transitioning from ith transitional state to jth transitional state

$pi^{(n)}_{E_iE_j} = pi^{(n)}_{ij}$: probability of transitioning from state E_i or ith state to state E_j or jth state of the system in n steps when steps, observations, or time intervals are independent of each other

PDI: ratio of number- to weight-average of degree of polymerization of polymer chains

$\overline{Q}$: one-step transition probability matrix with $(k - r)$ rows and columns for transitioning from ith transitional state to jth transitional state which is equal to $\overline{P}_{tt}\,(k - r \times k - r)$

$\overline{Q}^{(n-1)}$: transition probability matrix with $(k - r)$ rows and columns for transitioning from ith transitional state to jth transitional state in (n-1) steps

$QSSA$: quasi-steady state assumption of reactive species in radical polymerization

r: number of absorbing states of the system in a Markov chain process or ratio of initial number of moles of A functional group to initial B functional group in step growth polymerization or imbalance ratio

$\overline{R}$: one-step transition probability matrix with $(k - r)$ rows and r columns for transitioning from ith transitional state to jth absorbing state which is equal to $\overline{P}_{ta}\,(k - r \times r)$

R_p: rate of propagation reaction

$[RP]$: total concentration of reactive propagating chains

RP_n: reactive propagating chain with n repeat units

R_{tc}: rate of termination reaction by combination

R_{td}: rate of termination reaction by disproportionation

S: state space of a Markov chain process

S_a: state space for absorbing states of a Markov chain process

S_t: state space for transitional states of a Markov chain process

TP_n: terminated polymer chain with n repeat units

x: mole fraction of reactive A functional groups in step-growth polymerization of AR_1A and BR_3B
$\overline{X}^T$ $(m \times n)$: transpose of matrix $\overline{X}$ $(n \times m)$
x_{A1}: mole fraction of monofunctional monomer with A functional group $(A - R_2)$ in feed in step-growth polymerization
x_{A2}: mole fraction of difunctional monomer with two A functional groups $(A - R_1 - A)$ in feed in step-growth polymerization
x_{B1}: mole fraction of monofunctional monomer with on B functional group $(B - R_4)$ in feed in step-growth polymerization
x_{B2}: mole fraction of difunctional monomer with two B functional groups $(B - R_3 - B)$ in feed in step-growth polymerization
x_n: state of the system in the nth step, observation, or time interval in a Markov chain process
$\overline{Z}$: a matrix with m rows and p columns as the product of matrices $\overline{X}$ $(m \times n)$ and $\overline{Y}$ $(n \times p)$
z_{ik}: an element in ith row and kth column of the matrix $\overline{Z} = \overline{X} \times \overline{Y}$
α: conversion of reactive functional group (A or B) in step-growth polymerization defined based on the limiting reactive functional group (if $N_{A,0} < N_{B,0}$ then $\alpha = \alpha_A$)
α_B: conversion of non-limiting reactive functional group (if $N_{A,0} < N_{B,0}$ then $\alpha = \alpha_A$) in step-growth polymerization defined based on the limiting reactive functional group (if $N_{A,0} < N_{B,0}$ then B is non-limiting)
Δ: normalized standard deviation of the distribution of polymer chains
$\Delta_{n,a}$: normalized standard deviation of the distribution of steps to reach an absorbing state of a Markov chain process
$\sigma_{n,a}$: standard deviation of the distribution of steps to reach an absorbing state of a Markov chain process

9.15 Practice Problems

Problem 9.1 The mole fractions of the propagating chains terminated by disproportionation and combination in radical polymerization were denoted by $f_{r,d}$ and $f_{r,c}$, respectively, in Section 9.11.3. Show that the mole fractions of polymer chains produced by termination by disproportionation $(f_{p,d})$ and combination $(f_{p,c})$ are expressed in terms of $f_{r,d}$ by the following equations:

$$f_{p,d} = \frac{2f_{r,d}}{1 + f_{r,d}} \qquad f_{p,c} = \frac{1 - f_{r,d}}{1 + f_{r,d}}$$

Problem 9.2 The partitions of the transition probability matrix in Markov chain modeling of radical polymerization with termination by disproportionation and combination, no chain transfer reaction, and exclusion of monomers from the distribution were given by Eq. (9.77). Using the matrices $\overline{I}$, $\overline{0}$, $\overline{R}$, and $\overline{Q}$ in Eq. (9.77), derive the matrices for $\overline{N} = \left(\overline{I} - \overline{Q}\right)^{-1}$, $\left(2\overline{N} - \overline{I}\right)$, and $\left(\overline{N}x\overline{1}\right)$ for this radical polymerization.

Problem 9.3 The defining equation for the second order mole-fraction moment of reaching an absorbing state $(M_{n,a,2})$ was given by Eq. (9.18). Using $\overline{N} = \left(\overline{I} - \overline{Q}\right)^{-1}$, $\left(2\overline{N} - \overline{I}\right)$, and $\left(\overline{N}x\overline{1}\right)$ derived in Problem 9.2, show that $M_{n,a,2}$ in radical polymerization with termination by disproportionation and combination, no chain transfer reaction, and exclusion of monomers from the distribution is given by the following expression:

$$M_{n,a,2} = \frac{2\left[2 + p + (p^2 - 4p + 2)\, f_{r,d}\right]}{(1 + f_{r,d})(1 - p)^2}$$

where p is the probability of propagation, and $f_{r,d}$ is the mole fraction of the propagating chains terminated by disproportionation in radical polymerization.

Problem 9.4 The probability matrix $\overline{Q}$ of transitional states in modeling of condensation polymerization of AR_1A and BR_3B monomers with unequal number of A and B groups was given by Eq. (9.125) in Section 9.12.3. Using the Cayley–Hamilton theorem and by writing out the terms, derive the matrix $\overline{Q}^{(n-1)}$ for odd and even DP values as given by the matrices in Eq. (9.127).

Problem 9.5 In Markov chain modeling, the n-step probability distribution of reaching an absorbing state or the chain probability distribution, $p_a^{(n)} = p_{DP}(n)$ was defined by Eq. (9.16). Applying the defining equation, using Eq. (9.126) for $\overline{P_t}^{(0)}$ and the derived $\overline{Q}^{(n-1)}$ matrix from Problem 9.4, derive the following expression for $p_{DP}(n)$ for condensation polymerization of AR_1A and BR_3B monomers with unequal number of A and B groups as a function of odd and even DP values.

$$\begin{cases} p_{DP}(n) = \left[p_{t1}^{(0)}(1-\alpha) + p_{t2}^{(0)}(1 - r\alpha)\right] r^{\frac{n-1}{2}} \alpha^{n-1} & n = 2k-1 \quad k = 1, 2, \ldots \\ p_{DP}(n) = \left[p_{t1}^{(0)}\{(1/r) - \alpha\} + p_{t1}^{(0)}(1-\alpha)\right] r^{\frac{n}{2}} \alpha^{n-1} & n = 2k \qquad\quad k = 1, 2, \ldots \end{cases}$$

Problem 9.6 The partitions of the transition probability matrix in Markov chain modeling of condensation polymerization of AR_1A and BR_3B monomers with unequal number of A and B groups were given by Eq. (9.125) in Section 9.12.3. Using the matrices $\overline{I}, \overline{0}, \overline{R}$, and $\overline{Q}$ in Eq. (9.125), derive the matrices for $\overline{N} = \left(\overline{I} - \overline{Q}\right)^{-1}$, $\left(2\overline{N} - \overline{I}\right)$, and $\left(\overline{N}x\overline{1}\right)$ for this condensation polymerization.

Problem 9.7 The defining equation for the second mole-fraction moment of reaching an absorbing state $(M_{n,a,2})$ was given by Eq. (9.18). Using $\overline{N} = \left(\overline{I} - \overline{Q}\right)^{-1}$, $\left(2\overline{N} - \overline{I}\right)$, and $\left(\overline{N}x\overline{1}\right)$ derived in Problem 9.6, show that $M_{n,a,2}$ in condensation polymerization of AR_1A and BR_3B monomers with unequal numbers of A and B groups is given by the following expression:

$$M_{n,a,2} = \frac{(1-x)(4x\alpha + 1) + x\alpha^2}{(1 - 2x\alpha)(1 - x - x\alpha^2)}$$

where α is conversion of the limiting reactant A, $x = r/(1+r)$, and r is the imbalance ratio defined by. Eq. (9.123)

Problem 9.8 The Markov chain process was used in Section 9.11.2 to model radical polymerization with termination by disproportionation, no chain transfer reaction, and exclusion of monomers from the distribution. In the separation process after completion of a polymerization reaction, the monomers and dimers are removed from the produced polymer.

a. using the approach described in Section 9.11.2, define the state space (S), one-step transition probability matrix $\left(\overline{P}\right)$, and initial probability vector $\left(\overline{P}^{(0)}\right)$ for radical polymerization with termination by disproportionation, no chain transfer reaction, and exclusion of monomers and dimers from the polymer distribution.

b. use the defined S, $\overline{P}$, and $\overline{P}^{(0)}$ to derive expressions for distribution of polymer DP $\left(p_a^{(n)} = p_{DP}(n)\right)$, number-average $\left(\overline{DP}_n\right)$and weight-average $\left(\overline{DP}_w\right)$, and polydispersity index (PDI) of the distribution for this polymerization system.

Problem 9.9 The Markov chain process was used in Section 9.11.3 to model radical polymerization with termination by disproportionation and combination, no chain transfer reaction, and exclusion of monomers from the distribution. In a radical polymerization reaction with termination by disproportionation and combination, the produced polymer is used to make plastic products without monomer removal and purification.

a. using the approach described in Section 9.11.3, define the state space (S), one-step transition probability matrix $\left(\overline{P}\right)$, and initial probability vector $\left(\overline{P}^{(0)}\right)$ for radical polymerization with termination by disproportionation and combination, no chain transfer reaction, and with inclusion of monomer in the polymer product.

b. use the defined S, $\overline{P}$, and $\overline{P}^{(0)}$ to derive expressions for distribution of polymer DP $\left(p_a^{(n)} = p_{DP}(n)\right)$, number-average $\left(\overline{DP}_n\right)$, weight-average $\left(\overline{DP}_w\right)$, and polydispersity index (PDI) for this polymerization system.

Problem 9.10 The Markov chain process was used in Section 9.11.2 to model radical polymerization with termination by disproportionation, no chain transfer reaction, and exclusion of monomers from the distribution. Furthermore, it was used in Section 9.12.3. to model condensation polymerization of AR_1A and BR_3B monomers with unequal number of A and B groups. Use these two approaches to model the following polymerization system. In a condensation polymerization process, monomers AR_1A and BR_3B with unequal numbers of A and B groups are reacted to form a polymer. After polymerization, the monomer is removed from the reaction to improve the mechanical properties of the produced polymer.

a. model this condensation polymerization after monomer removal by Markov chain process by defining the state space $S, \overline{P}, \overline{P}^{(0)}$ matrices.

b. use the defined S, $\overline{P}$, and $\overline{P}^{(0)}$ matrices to derive expressions for distribution of polymer DP $\left(p_a^{(n)} = p_{DP}(n)\right)$, number-average $\left(\overline{DP}_n\right)$weight-average $\left(\overline{DP}_w\right)$, and polydispersity index (PDI) for the produced polymer.

References

1 Bellman, R.A., Markovian decision process. *Journal of Mathematics and Mechanics*, 1957, **6**(5), 679.

2 Burke, C.J. and Rosenblatt, M.A., Markovian function of a Markov-Chain. *Annals of Mathematical Statistics*, 1958, **29**(4), 1112.

3 Bernardo, M., Theory and application of extended Markovian process algebra. *Bulletin-European Association for Theoretical Computer Science*, 1999, **68**, 239.

4 Buchholz, P., *On a Markovian process algebra*. Vol. 500. 1994, Universität Dortmund, 37.

5 Salehi, S.A., Riedel, M.D., and Parhi, K.K., Markov chain computations using molecular reactions. in *2015 IEEE International Conference on Digital Signal Processing (Dsp)*, 2015, IEEE Xplore, 689–693.

6 Velikanov, M.V. and Kapral, R., Polymerase chain reaction: a Markov process approach. *Journal of Theoretical Biology*, 1999, **201**(4), 239.

7 Sharif, B., Zhang, Y.H., Gaind, N., and Kurt, M., Extrapolating discrete-time disease evolution data in limited series of transition probability matrices: a continuous- time Markov chain approach. *Medical Decision Making*, 2024, **44**(2), NP46–NP47.

8 Bezhaeva, Z.I., Weak-convergence of matrices of transition-probabilities for conditional Markov-chains. *Theory of Probability and Its Applications*, 1983, **27**(1), 59.

9 El-Shehawey, M.A. and Al-Henawy, A.S., Computer aided solving the higher-order transition probability matrix of the finite Markov chain. *Applied Mathematics and Computation*, 2010, **217**(8), 4311.

10 Kaijser, T., On Markov chains induced by partitioned transition probability matrices. *Acta Mathematica Sinica-English Series*, 2011, **27**(3), 441.

11 Nopiah, Z.M. et al., Transition probabilities matrix of Markov chain in the fatigue crack growth model. in *4th International Conference on Quantitative Sciences and Its Applications (ICOQSIA 2016)*, 2016, AIP Publishing.

12 Peng, N.F., Spectral representations of the transition probability matrices for continuous time finite Markov chains. *Journal of Applied Probability*, 1996, **33**(1), 28.

13 Rescher, N., Discrete state systems, Markov-chains, and problems in the theory of scientific explanation and prediction. *Philosophy of Science*, 1963, **30**(4), 325.

14 Achilias, D.S., A review of modeling of diffusion controlled polymerization reactions. *Macromolecular Theory and Simulations*, 2007, **16**(4), 319.

15 Crossman, R.J., Coolen-Schrijner, P., Skulj, D., and Coolen, F.P.A., Imprecise Markov chains with an absorbing state. *ISIPTA '09: Proceedings of the Sixth International Symposium on Imprecise Probability: Theories and Applications*, 2009, 119.

16 Teslenko, V.I. and Kapitanchuk, O.L., A closed description of the non-autonomous dynamics for an absorbing Markov chain with three states and random transition probabilities. *Journal of Physical Studies*, 2019, **23**(3), 3002.

17 Wang, P. et al., Estimating random effects in a finite Markov chain with absorbing states: application to cognitive data. *Statistica Neerlandica*, 2023, **77**(3), 304.

18 Watterson, G., Markov-chains with absorbing states - a genetic example. *Annals of Mathematical Statistics*, 1961, **32**, 716.

19 Zeifman, A., Chegodaev, A., and Shilova, G. Almost absorbing continuous-time Markov's chains on finite state space. in *International Conference Modelling of Business, Industrial and Transport Systems*, 2008, Taylor & Francis.

20 Graham, A., *Matrix Theory and applications for scientists and engineers*. 2018, Dover Publications, 304.

21 Jeffrey, A., *Matrix operations for engineers and scientists: an essential guide in linear algebra*. 2010, Springer Nature, 314.

22 Rothman, H. and Reilly, P., Computer systems for matrix operations. *Journal of the Structural Division*, 1964, **90**(3), 125.

23 Chasnov, J.R., *Matrix Algebra for Engineers*. 2018, The Hong Kong University of Science and Technology, p. 4., https://www.math.hkust.edu.hk/~machas/matrix-algebra-for-engineers.pdf.

24 Tobita, H., Markovian approach to nonlinear polymer formation: free-radical polymerization with chain transfer to polymer. *Journal of Polymer Science Part B-Polymer Physics*, 1998, **36**(2), 357.

25 Tobita, H., Free-radical polymerization with long-chain branching and scission: Markovian solution of the weight-average molecular weight. *Macromolecular Theory and Simulations*, 2014, **23**(8), 477.

26 Tobita, H., Markovian approach to free-radical polymerization with simultaneous long-chain branching and scission: effect of branching and scission kinetics. *Macromolecular Reaction Engineering*, 2015, **9**(3), 245.

27 Yamada, B., Westmoreland, D.G., Kobatake, S., and Konosu, O., ESR spectroscopic studies of radical polymerization. *Progress in Polymer Science*, 1999, **24**(4), 565.

28 Gromov, V.F. et al., Effect of solvent on chain propagation and termination reaction-rates in radical polymerization. *European Polymer Journal*, 1980, **16**(6), 529.

29 Nakamura, Y., Lee, R., Coote, M.L., and Yamago, S., Termination mechanism of the radical polymerization of acrylates. *Macromolecular Rapid Communications*, 2016, **37**(6), 506.

30 Lechuga-Islas, V.D. et al., Separation of volatile compounds from polymers by physisorption. *European Polymer Journal*, 2021, **159**, 110748.

31 Kodama, S. et al., Effect of purification solvent on polymer impurities and device performance. *ACS Applied Polymer Materials*, 2019, **1**(8), 2083.

32 Faucher, S., Okrutny, P., and Zhu, S.P., Facile and effective purification of polymers produced by atom transfer radical polymerization via simple catalyst precipitation and microfiltration. *Macromolecules*, 2006, **39**(1), 3.

33 Vejju, P., George, A., and Rahulkar, A.D., Comparative study on the computation of state transition matrix using La-grange's interpolation technique and Cayley–Hamilton theorem. in *10th International Conference on Intelligent Systems and Control (ISCO'16)*, 2016, IEEE Xplore.

34 Chen, L.Z., A generalization of the Cayley–Hamilton theorem. *American Mathematical Monthly*, 2012, **119**(4), 340.

35 Bernhardt, C., A proof of the Cayley–Hamilton theorem. *American Mathematical Monthly*, 2009, **116**(5), 456.

36 *Cayley–Hamilton theorem*, in *Wikipedia: The free encyclopedia*. 2025, Wikimedia Foundation, https://en.wikipedia.org/wiki/Cayley%E2%80%93Hamilton_theorem.

37 Feng, L.G., Tan, H.J., and Zhao, K.M., A generalized Cayley–Hamilton theorem. *Linear Algebra and Its Applications*, 2012, **436**(7), 2440.

38 Hwang, S.G., A generalized Cayley–Hamilton theorem for polynomials with matrix coefficients. *Linear Algebra and Its Applications*, 2011, **434**(2), 475.

39 Leonard, I.E., The matrix exponential. *Siam Review*, 1996, **38**(3), 507.

40 Máté, A. *The Cayley–Hamilton theorem*. 2016.

41 Sato, E., Zetterlund, P.B., and Yamada, B., Macromonomer synthesis using oligomers of omega-unsaturated methacrylate as addition-fragmentation chain transfer agents: increased efficiency by manipulation of steric hindrance. *Macromolecules*, 2004, **37**(7), 2363.

42 Nakamura, Y. and Yamago, S., Termination mechanism in the radical polymerization of methyl methacrylate and styrene determined by the reaction of structurally well-defined polymer end radicals. *Macromolecules*, 2015, **48**(18), 6450.

43 Bevington, J.C., Melville, H.W., and Taylor, R.P., The termination reaction in radical polymerizations. 2. polymerizations of styrene at 60-degrees and of methyl methacrylate at 0 and 60-degrees, and the copolymerization of these monomers at 60-degrees. *Journal of Polymer Science*, 1954, **14**(77), 463.

44 Catalgil-Giz, H., Giz, A., and Oncul-Koc, A., Termination mechanism of polymethyl methacrylate and polystyrene studied by ultrasonic degradation technique. *Polymer Bulletin*, 1999, **43**(2–3), 215.

45 Rogers, M.E. and Long, T.E., *Synthetic methods in step-growth polymers*. 2003, Hoboken, New Jersey: Wiley, 605.

46 Carraher, C.E. and Swift, G.G., eds. *Functional condensation polymers*. 2002, Kluwer Academic/Plenum Publishers, 315.

47 Flory, P.J., Fundamental principles of condensation polymerization. *Chemical Reviews*, 1946, **39**(1), 137.

48 Inkinen, S., Hakkarainen, M., Albertsson, A.C., and Sodergard, A., From lactic acid to poly(lactic acid) (PLA): characterization and analysis of PLA and its precursors. *Biomacromolecules*, 2011, **12**(3), 523.

49 Gan, L.M. and Koh, L.L., A discussion of the derivation of molecular-weight distribution-functions of linear step-growth polymerization. *Polymer Communications*, 1985, **26**(3), 76.

50 Pros, G. and Pintauer, T., ARB-type step-growth polymerization using CuAAC (copper catalyzed azidealkyne cycloaddition) with sequence-controlled monomers synthesized by atom transfer radical addition (ATRA) and polymerization (ATRP). in *254th ACS National Meeting*. 2017, American Chemical Society.

51 Kumar, A., Molecular-weight distribution of reversible ARB polymerization in homogeneous continuous-flow reactors with monomers exhibiting unequal reactivity. *Macromolecules*, 1987, **20**(1), 220.

52 Kumar, A., Molecular-weight distribution of reversible ARB polymerization in homogeneous continuous-flow reactors for monomer with reactivity different from higher homologs. *Polymer*, 1986, **27**(8), 1261.

53 Kumar, A., Gupta, S.K., and Saraf, R., Condensation polymerization of ARB type monomers in CSTRS wherein the monomer is *R* times more reactive than other homologs. *Polymer*, 1980, **21**(11), 1323.

54 Gupta, S.K., Mohan, D., and Kumar, A., Simulation of ARB type reversible step growth polymerization in semibatch reactors. *Journal of Applied Polymer Science*, 1985, **30**(2), 445.

55 Digiacomo, A., Size distribution and gelation in condensation polymers with one unsymmetric reactant. *Journal of Polymer Science*, 1960, **47**(149), 435.

56 Beers, K.J. and Ray, W.H., A linkage moment approach to modeling condensation polymerization with multiple monomers. I. Linear polymers. *Journal of Applied Polymer Science*, 2001, **79**(2), 246.

57 Ramakrishnan, S., Condensation polymerization. *Resonance*, 2017, **22**, 355.

58 Gupta, S.K. and Kumar, A., Invited review simulation of step growth polymerizations. *Chemical Engineering Communications*, 1983, **20**(1–2), 1.

59 Winter, J.H., Methods for the synthesis of uniform polymers. *Angewandte Chemie International*, 1966, **5**(10), 862.

10

Markov Chain Modeling of Copolymerization Reactions

10.1 Introduction

Copolymers are produced by adding two or more different monomers to the reaction. The benefit of copolymerization is that the different monomers confer complementary physical, mechanical, chemical, or biochemical properties to the produced copolymer [1–4]. During copolymerization, the addition of different monomers to the propagating chain is random or stochastic and based on the relative rates of reaction of the ith monomer type reacting with a propagating chain ending in the jth monomer type. Due to the random nature of the reaction, the copolymer chains possess a diverse distribution of different monomer types. Not including any structural or stereoisomers, there is only one sequence distribution of monomer residues on a homopolymer chain with a degree of polymerization $DP = 100$. Conversely, for a copolymer composed of 50 monomers of type A ($N_A = 50$) and 50 monomers of type B ($N_B = 50$) with $DP = 100$, there are $(DP - 1)!/(N_A! \times N_B!) = 99!/(50! \times 50!) \approx 1 \times 10^{27}$ circularly equivalent sequence distributions of A and B monomer residues on the copolymer chains. These different sequences of A and B monomer residues can potentially generate copolymers with diverse properties. As a Markov chain is a stochastic model involving a random variable like sequence distribution, copolymerization reactions can be modeled using a Markov process. In this chapter, we apply the method of Markov chain to model composition and sequence distribution of different monomer types in copolymerization reactions.

10.2 Number Representation of Copolymer Sequences

Numbers are used as a shorthand notation to label a monomer sequence on a copolymer [5]. Consider the copolymer sequence -BAAAAAB- represented by letters A and B corresponding to monomers A and B, respectively. This sequence can be written as "1000001" where "0" and "1" are assigned to A and B, respectively. Since there are two monomers and the sequence consists of numbers "0" and "1," the number "1000001" in base 2, written as $(1000001)_2$, is converted to base ten as follows:

$$(1000001)_2 = 1 \times 2^0 + 0 \times 2^1 + 0 \times 2^2 + 0 \times 2^3 + 0 \times 2^4 + 0 \times 2^5 + 1 \times 2^6 = (65)_{10} \qquad (10.1)$$

Based on the above conversion from base 2 to 10, the copolymer sequence -BAAAAAB- consisting of monomers A and B is represented by the unique number $(1000001)_2$ in base 2 or the number

$(65)_{10}$ in base 10. The probability of finding the sequence -BAAAAAB- in the copolymer is expressed as $p\,(65)_{10}$. The same approach is used to represent terpolymers by numbers instead of letters. Consider the sequence -CAAABBAC- of a terpolymer consisting of monomers A, B, and C. This sequence can be written as "20001102" where "0," "1," and "2" are assigned to monomers A, B, and C, respectively. As there are three monomers and the sequence consists of numbers "0," "1," and "2," the number "20001102" in base 3, written as $(20001102)_3$, is converted to base 10 as follows:

$$\begin{aligned}(20001102)_3 &= 2\times 3^0 + 0\times 3^1 + 1\times 3^2 + 1\times 3^3 + 0\times 3^4 + 0\times 3^5 + 0\times 3^6 + 2\times 3^7 \\ &= (4412)_{10}\end{aligned} \tag{10.2}$$

Based on the above conversion from base 3 to 10, the terpolymer sequence -CAAABBAC- consisting of monomers A, B, and C is represented by the unique number $(20001102)_3$ in base 3 or the number $(4412)_{10}$ in base 10. The probability of finding the sequence -CAAABBAC- in the terpolymers is expressed as $p\,(4412)_{10}$. As the addition of a monomer to a propagating chain in copolymerization depends on the type of previously reacted monomers in the propagating chain, the next section covers the order of copolymerization.

10.3 Order of Copolymerization

The type of monomer in a copolymerization that reacts with a propagating chain is subject to the sequence of monomer residues on the chain [5, 6]. The order of copolymerization, represented by ν, is defined as the number of monomers on the reactive end of the propagating chain that affects the type of monomer that is added to the chain. If the propagation reaction is independent of the monomer residues on the reactive end of the propagating chain, the order of copolymerization is zero ($\nu = 0$). If the propagation reaction is affected only by the last monomer residue on the reactive end of the propagating chain, the order of the polymerization is one ($\nu = 1$, terminal copolymerization). Likewise, if the propagation reaction is affected by the last two (penultimate copolymerization), three (ante-penultimate copolymerization), or four (pre-ante-penultimate copolymerization) monomer residues on the reactive end of the propagating chain, the order of the polymerization is $\nu = 2$, $\nu = 3$, or $\nu = 4$, respectively. A copolymerization reaction with $\nu = 0$, $\nu = 1$, $\nu = 2$, $\nu = 3$, $\nu = 4$, or $\nu = n$ corresponds to a zero, first, second, third, fourth, or nth order Markov chain process, respectively.

Consider the copolymerization of monomers M_0 and M_1, represented numerically by "0" and "1." If $\nu = 0$, there are two propagation reactions as follows:

$$\sim\sim\sim\sim\sim\bullet + M_0 \xrightarrow{k_{p,0}} \sim\sim\sim\sim\sim\bullet \tag{10.3}$$

$$\sim\sim\sim\sim\sim\bullet + M_1 \xrightarrow{k_{p,1}} \sim\sim\sim\sim\sim\bullet \tag{10.4}$$

where $k_{p,0}$ and $k_{p,1}$ are the rates of reaction of monomer type M_0 and M_1 with the propagating chain, respectively. As shown above, the type of terminal monomer residue need not be specified if $\nu = 0$ and the copolymerization is a zero order Markov chain process. The propagation reactions for copolymerization of monomers M_0, M_1, and M_2 represented numerically by "0," "1," and "2," respectively, with $\nu = 0$ are as follows:

$$\sim\sim\sim\sim\sim\bullet + M_0 \xrightarrow{k_{p,0}} \sim\sim\sim\sim\sim\bullet \tag{10.5}$$

$$\sim\sim\sim\sim\sim\bullet + M_1 \xrightarrow{k_{p,1}} \sim\sim\sim\sim\sim\bullet \tag{10.6}$$

$$\sim\sim\sim\sim\sim \bullet + M_2 \xrightarrow{k_{p,2}} \sim\sim\sim\sim\sim \bullet \qquad (10.7)$$

where $k_{p,0}$, $k_{p,1}$, and $k_{p,2}$ are the rates of reaction of monomer type M_0, M_1, and M_2 with the propagating chain, respectively. Copolymerization reactions in which the rate of reaction of any monomer type is independent of the type of monomer at the reactive end of the propagating chain are known as random copolymers [7–13]. The probability of finding any copolymer sequence in a random copolymer is equal to the product of the probabilities of finding the individual monomer residues. For example, the probability of finding the sequence $-M_0M_1M_1M_1M_0-$ in a random copolymer of monomers M_0 and M_1, expressed as $p\,(01110)_2$ or $p\,(14)_{10}$, is given as follows [5, 14]:

$$p\,(14)_{10}\,(\nu = 0) = p\,(01110)_2\,(\nu = 0) = p\,(0)\,p\,(1)\,p\,(1)\,p\,(1)\,p\,(0) \qquad (10.8)$$

where $p\,(0)$ and $p\,(1)$ are the probabilities of finding monomers M_0 and M_1 on the copolymer chain, respectively. In copolymerization of monomers M_0 and M_1, if the reaction of the two monomers with a propagating chain depends on the last monomer residue at the reactive end of the chain, that is $\nu = 1$, the propagation reactions are as follows [5]:

$$\sim\sim\sim\sim M_0 \bullet + M_0 \xrightarrow{k_{p,00}} \sim\sim\sim\sim M_0 \bullet \qquad (10.9)$$

$$\sim\sim\sim\sim M_0 \bullet + M_1 \xrightarrow{k_{p,01}} \sim\sim\sim\sim M_1 \bullet \qquad (10.10)$$

$$\sim\sim\sim\sim M_1 \bullet + M_0 \xrightarrow{k_{p,10}} \sim\sim\sim\sim M_0 \bullet \qquad (10.11)$$

$$\sim\sim\sim\sim M_1 \bullet + M_1 \xrightarrow{k_{p,11}} \sim\sim\sim\sim M_1 \bullet \qquad (10.12)$$

where $k_{p,ij}$ is the propagation rate constant for the reaction of monomer M_j with a propagating chain having monomer residue M_i at the reactive end. Copolymerization reactions described by propagation Rxns. 10.9–10.12 are known as terminal copolymerization, and they are modeled using a first order Markov chain [15–20]. The probability of finding the sequence $-M_0M_1M_1M_1M_0-$ in terminal copolymerization is expressed as follows:

$$\begin{aligned} p\,(14)_{10}\,(\nu = 1) &= p\,(01110)_2\,(\nu = 1) = p\,(0)\,p\,(1/0)\,p\,(1/1)\,p\,(1/1)\,p\,(0/1) \\ &= p\,(0)\,p_{01}p_{11}p_{11}p_{10} \end{aligned} \qquad (10.13)$$

where p_{ij} is the conditional probability of finding monomer M_j right after monomer M_i in a sequence. As the rate of propagation in copolymerization depends on the type of monomer residue at the reactive end, the probability of finding any sequence of monomers on a copolymer depends on conditional probabilities p_{ij}.

In the copolymerization of monomers M_0 and M_1, if the reaction of the two monomers with a propagating chain depends on the last two monomer residues at the reactive end, that is $\nu = 2$, the propagation reactions are as follows:

$$\sim\sim\sim\sim M_0M_0 \bullet + M_0 \xrightarrow{k_{p,000} \text{ or } k^{10}_{p,00}} \sim\sim\sim\sim M_0M_0 \bullet \qquad (10.14)$$

$$\sim\sim\sim\sim M_0M_0 \bullet + M_1 \xrightarrow{k_{p,001} \text{ or } k^{10}_{p,01}} \sim\sim\sim\sim M_0M_1 \bullet \qquad (10.15)$$

$$\sim\sim\sim\sim M_0M_1 \bullet + M_0 \xrightarrow{k_{p,010} \text{ or } k^{10}_{p,12}} \sim\sim\sim\sim M_1M_0\bullet \tag{10.16}$$

$$\sim\sim\sim\sim M_0M_1 \bullet + M_1 \xrightarrow{k_{p,011} \text{ or } k^{10}_{p,11}} \sim\sim\sim\sim M_1M_1\bullet \tag{10.17}$$

$$\sim\sim\sim\sim M_1M_0 \bullet + M_0 \xrightarrow{k_{p,100} \text{ or } k^{10}_{p,20}} \sim\sim\sim\sim M_0M_0\bullet \tag{10.18}$$

$$\sim\sim\sim\sim M_1M_0 \bullet + M_1 \xrightarrow{k_{p,101} \text{ or } k^{10}_{p,21}} \sim\sim\sim\sim M_0M_1\bullet \tag{10.19}$$

$$\sim\sim\sim\sim M_1M_1 \bullet + M_0 \xrightarrow{k_{p,110} \text{ or } k^{10}_{p,30}} \sim\sim\sim\sim M_1M_0\bullet \tag{10.20}$$

$$\sim\sim\sim\sim M_1M_1 \bullet + M_1 \xrightarrow{k_{p,111} \text{ or } k^{10}_{p,31}} \sim\sim\sim\sim M_1M_1\bullet \tag{10.21}$$

where $k_{p,ijk} = k^{10}_{p,lk}$ is the propagation rate constant for the reaction of monomer M_k with a propagating chain with terminal monomer sequence M_iM_j; the ij in $k^{10}_{p,lk}$ is replaced by l which is ij in base 10. Copolymerization reactions described by propagation Rxns. 10.14–10.21 are known as penultimate copolymerization, and they are modeled using a second order Markov chain [16, 21–26]. The probability of finding the sequence $-M_0M_1M_1M_1M_0-$ in penultimate copolymerization is expressed as follows:

$$\begin{aligned} p\,(14)_{10}\,(\nu = 2) &= p\,(01110)_2\,(\nu = 2) = p\,(01)\,p\,(11/01)\,p\,(11/11)\,p\,(10/11) \\ &= p\,(1)_{10}\, p^{10}_{13} p^{10}_{33} p^{10}_{32} \end{aligned} \tag{10.22}$$

where $p\,(1)_{10}$ is the probability of finding the dimer represented by 1 in base 10 ($0 \times 2^1 + 1 \times 2^0$ or M_0M_1) and p^{10}_{ij} is the conditional probability of finding the dimer j in base 10 right after the dimer i in base 10.

In general, the number of reactions in copolymerization of m different monomers with order of polymerization ν is equal to $m^{(1+\nu)}$. The number of sequences of monomer residues in the reactive end of the propagating chains that affect copolymerization is m^ν. The conditional probability matrix p^{10}_{ij} is a square matrix with m^ν rows and columns. The number of elements in the conditional probability matrix is $(m^\nu \times m^\nu)$. For example, for copolymerization of three monomers ($m = 3$) with $\nu = 2$, the number of reactions is $m^{(1+\nu)} = 27$, the number of sequences of monomer residues in the reactive end of propagating chains that affect copolymerization is $m^\nu = 9$, and the conditional probability matrix is a (9×9) matrix with $m^\nu \times m^\nu = 81$ elements.

10.4 State Space of Copolymerization Reaction

In copolymerization, we are interested in predicting composition and sequence distribution of monomers. In this regard, the states in the state space S of a copolymerization reaction consist of sequences of monomer residues in the reactive end of a propagating chain that affect the addition of the next monomer [5, 14]. If the number of monomers is m and the copolymerization order is ν, the number of these sequences or states is m^ν. Therefore, the state space of a copolymerization with m different monomers and order ν is expressed as a $(1 \times m^\nu)$ row vector as follows:

$$S = [E_1, E_2, \dots, E_i, \dots, E_{m^\nu - 1}, E_{m^\nu}] = \left[E^{10}_0, E^{10}_1, \dots, E^{10}_j, \dots, E^{10}_{m^\nu - 2}, E^{10}_{m^\nu - 1}\right] \tag{10.23}$$

These states are mutually exclusive, that is, only one of those states can happen by propagation per chain per event in the reaction. The subscript j in E^{10}_j in the above expression is a number in base 10 with different monomer types numbered starting from zero to (m−1) like M_0, M_1, M_2 to M_{m-1} where m is the number of different monomers. For example, if there are three monomers M_0, M_1, and M_2 in the copolymerization and $\nu = 2$, the $m^\nu = 9$ states in the space of the system are M_0M_0, M_0M_1, M_0M_2, M_1M_0, M_1M_1, M_1M_2, M_2M_0, M_2M_1, and M_2M_2 corresponding to E_{00}, E_{01}, E_{02}, E_{10}, E_{11}, E_{12}, E_{20}, E_{21}, and E_{22} in base three, respectively, or E_0^{10}, E_1^{10}, E_2^{10}, E_3^{10}, E_4^{10}, E_5^{10}, E_6^{10}, E_7^{10}, and E_8^{10} in base 10.

10.5 Conditional Probability Matrix for Copolymerization Reaction

The conditional probability matrix $\overline{P}$ is a square matrix with m^ν rows and columns [14, 27, 28]. The general relation between an element of the conditional probability matrix and reaction rate constants is given by the following expression:

$$p_{ij}^{10} = C\,k_{is}^{10}\,[M_i]\,[M_s] \quad 0 \le i, j \le \sum_{n=0}^{\nu-1} (m-1)\,m^n \quad \text{for any } 0 \le s \le (m-1) \tag{10.24}$$

where p_{ij}^{10} is the one-step conditional probability of finding the sequence denoted by j in base 10 right after the sequence denoted by i in base 10; C is a constant that relates the reaction rate to the conditional probability; k_{is}^{10} is the rate constant for a propagating chain ending in sequence i in base 10 reacting with the monomer s. It should be emphasized that i in k_{is}^{10} is a number in base 10, whereas s is the number designation of the monomer reacting with the chain ending in sequence i (for example, $s = 2$ for M_2 or, $s = 0$ for M_0); m is the number of different monomers in the copolymerization system, ν is the order of the copolymerization, and $[M_s]$ is the concentration of monomer designated by the number s. The above equation is subject to the following condition:

$$\sum_{s=0}^{m-1} p_{ij}^{10} = \sum_{s=0}^{m-1} \left[\frac{k_{is}^{10}\,[M_s]}{\sum_{t=0}^{m-1} k_{it}^{10}\,[M_t]} \right] = \frac{\sum_{s=0}^{m-1} k_{is}^{10}\,[M_s]}{\sum_{t=0}^{m-1} k_{it}^{10}\,[M_t]} = 1 \tag{10.25}$$

Upon subjecting Eq. (10.24) to the above condition to eliminate the constant C, the following expression is derived for p_{ij}^{10}:

$$p_{ij}^{10} = \frac{k_{p,is}^{10}\,[M_s]}{\sum_{t=0}^{m-1} k_{it}^{10}\,[M_t]} \quad 0 \le i, j \le \sum_{n=0}^{\nu-1} (m-1)\,m^n \quad \text{for any } 0 \le s \le (m-1) \tag{10.26}$$

The p_{ij}^{10} values are the elements of the one-step conditional probability matrix $\overline{P}(\nu)$ which is a $(m^\nu \times m^\nu)$ square matrix with $m^\nu \times m^\nu$ elements. The vector of initial probability of the states is not required for the determination of copolymer composition and sequence distribution, as the monomer residues of the chain ends are a small fraction of the total monomers on the propagating chains.

10.6 Composition of Copolymer

The copolymer molar composition vector, $\overline{X}\,(m^\nu)$, is a row vector with m^ν columns, like the row vector S for the state space of the system, defined as follows:

$$\overline{X}\,(m^\nu) = [X_0^{10}, X_1^{10}, \ldots, X_i^{10}, \ldots, X_{m^\nu-2}^{10}, X_{m^\nu-1}^{10}] \tag{10.27}$$

Subject to the condition:

$$\sum_{i=1}^{m^{v}} X_{i-1}^{10} = 1 \tag{10.28}$$

where elements X_0, X_1, and X_{m-1} are the mole fractions of the sequences corresponding to the reactive end sequences that affect the rate of copolymerization. For example, for a copolymerization with three monomers M_0, M_1, and M_2 and $\nu = 2$, $\overline{X}\,(3^2)$ is a (1×9) row vector with elements X_{00}, X_{01}, X_{02}, X_{10}, X_{11}, X_{12}, X_{20}, X_{21}, and X_{22} in base three corresponding to X_0^{10}, X_1^{10}, X_2^{10}, X_3^{10}, X_4^{10}, X_5^{10}, X_6^{10}, X_7^{10}, and X_8^{10} in base ten, respectively. For a given copolymer composition vector $\overline{X}\,(m^\nu)$ with elements X_i^{10} and conditional probability matrix $\overline{P}\,(m^\nu \times m^\nu)$ with elements p_{ij}^{10}, there exists an eigenvalue λ corresponding to the eigenvector $\overline{X}\,(m^\nu)$ that satisfies the following equality [5, 14, 29, 30]:

$$\overline{P}\,(m^\nu \times m^\nu)\,\overline{X}\,(1 \times m^\nu) = \lambda \overline{X}\,(1 \times m^\nu) \tag{10.29}$$

The λ is an eigenvalue of the $\overline{P}\,(m^\nu \times m^\nu)$ matrix if and only if the following condition is met:

$$\left|\overline{P}\,(m^\nu \times m^\nu) - \lambda \overline{I}\,(m^\nu \times m^\nu)\right| = 0 \tag{10.30}$$

where the matrix $\overline{I}$ is a $m^\nu \times m^\nu$ unit matrix. Based on the above expression, there exists m^ν values of λ that meet the above condition. One of the eigenvalues of $\overline{P}\,(m^\nu \times m^\nu)$ is $\lambda = 1$, which corresponds to the steady state copolymer composition of infinitely long chains, such that monomers at the chain ends have a negligible effect on the overall composition of the polymer chain. Therefore, the above equation reduces to the following under steady state conditions:

$$\left|\overline{P}\,(m^\nu \times m^\nu) - \overline{I}\,(m^\nu \times m^\nu)\right| = \left|\overline{A}\right| = 0 \quad \text{steady state composition} \tag{10.31}$$

Based on the above condition, the elements of the composition vector $\overline{X}\,(m^\nu)$ are determined using the following equation:

$$X_{i-1} = \frac{a_{ij}^{cf}}{\sum_{t=1}^{m^\nu} a_{tj}^{cf}} \qquad 1 \le i \le m^\nu \tag{10.32}$$

where a_{ij}^{cf} is the cofactor of the element a_{ij} of the matrix $\overline{A} = \overline{P} - \overline{I}$ (see Section 9.10.5 of Chapter 9 for matrix of cofactors). Note that any column j of the $\overline{A}$ matrix may be used to determine the elements of the composition matrix, but only one of the columns should be used.

10.7 Copolymer Sequence Distribution

The sequence distribution of monomers affects microstructure of the copolymer. If the monomers are randomly or alternatively distributed along the copolymer chain, then the monomers are molecularly dissolved, and it is more likely for the copolymer to exhibit uniform properties like one glass transition temperature (T_g) intermediate between the T_g of the pure homopolymers [31]. Conversely, if there are long sequences of one of the monomers in the copolymer, like $\sim (M_0)_n \sim$ or $\sim (M_1)_n \sim$, then it is more likely for the monomers to form separate domains, and the copolymer exhibits non-uniform properties like multiple T_g values close to the T_g of the pure homopolymers. Therefore, the sequence distribution of the monomers will affect the physiochemical and mechanical as well as biochemical properties of the copolymer.

10.7.1 Probability of Finding n-mer or Longer Sequences of a Monomer in Copolymer

Consider the monomer M_0 in copolymerization of m different monomers M_0, M_1, M_2 to M_{m-1} with order of copolymerization ν. The probability of finding a contiguous sequence of at least "n" M_0 monomers in the copolymer, that is, $\sim (M_0)_n \sim$, is given by the following probability function:

$$p\,(M_0)_n = p_n\,(0^n) = p_\nu\,(0^\nu)\left[p\left(0/0^\upsilon\right)\right]^{n-\nu} \tag{10.33}$$

where $p_n\,(0^n)$ is the probability of finding an n-mer sequence of M_0 monomer residues, $p_\nu\,(0^\nu)$ is the probability of finding a ν-mer sequence of M_0 monomer residues, and $p\left(0/0^\upsilon\right)$ is the probability of finding an M_0 monomer residue after a ν-mer sequence of M_0 monomer residues on the copolymer. Note that the above equation gives the probability of finding a contiguous M_0 sequence equal or larger than n. For ν values of 0, 1, 2, and 3, the probability function $p_n\,(0^n)$ is written as follows:

$$\nu = 0 \;\Rightarrow\; p_n\,(0^n) = [p\,(0)]^n \tag{10.34}$$

$$\nu = 1 \;\Rightarrow\; p_n\,(0^n) = p_1\,(0)\,[p\,(0/0)]^{n-1} \tag{10.35}$$

$$\nu = 2 \;\Rightarrow\; p_n\,(0^n) = p_2\,(00)\,[p\,(0/00)]^{n-2} \tag{10.36}$$

$$\nu = 3 \;\Rightarrow\; p_n\,(0^n) = p_3\,(000)\,[p\,(0/000)]^{n-3} \tag{10.37}$$

10.7.2 Probability of Finding n-mer Sequences of a Monomer in Copolymer

Consider the copolymerization of m monomers M_0, M_1, M_2 to M_{m-1} with order of copolymerization ν. The probability of finding a sequence of exactly "n" M_0 monomers in the copolymer, that is, $\sim M_i\,(M_0)_n\,M_j \sim\; i, j \neq 0$ or $p_{n+2}\,(i0^n j)$, for $m = 2,\ \nu = 0$, $m = 2,\ \nu = 1$ and $m \geq 2,\ \nu \geq 1$ are written as follows:

$$p\left[M_i\,(M_0)_n\,M_j\right] = p_{n+2}\,(10^n1) = p\,(1)^2\,[p\,(0)]^n \quad m = 2,\ \nu = 0 \tag{10.38}$$

$$p\left[M_i\,(M_0)_n\,M_j\right] = p_{n+2}\,(10^n1) = p\,(1)\,p_{10}\,[p\,(0)]^{n-1}\,p_{01} \quad m = 2,\ \nu = 1 \tag{10.39}$$

$$\begin{aligned} p\left[M_i\,(M_0)_n\,M_j\right] &= \sum_{i=1}^{m-1} p_{n+2}\,(i\,0^n i) + \frac{1}{2}\sum_{i\neq j=1}^{m-1} p_{n+2}\,(i\,0^n j) \qquad m \geq 2,\ \nu \geq 1 \\ &= \sum_{i=1}^{m-1} p_\nu\left(i0^{\nu-1}\right) p\left(0/i\,0^{\nu-1}\right)[p\,(0/0^\upsilon)]^{n-\nu}\,p\,(i/0^\upsilon) \\ &\quad + \frac{1}{2}\sum_{i\neq j=1}^{m-1} p_\nu\left(i0^{\nu-1}\right) p\left(0/i\,0^{\nu-1}\right)[p\,(0/0^\upsilon)]^{n-\nu}\,p\,(j/0^\iota) \end{aligned} \tag{10.40}$$

where $p_\nu\left(i0^{\nu-1}\right)$ is the probability of finding a ν-mer sequence $\sim M_i\,(M_0)_{\nu-1}$ where $M_i \neq M_0$; $p\left(0/i\,0^{\upsilon-1}\right)$ is the probability of an M_0 monomer residue after a ν-mer sequence $\sim M_i\,(M_0)_{\nu-1}$ where $M_i \neq M_0$; and $p\,(j/0^\upsilon)$ is the probability of an M_j monomer residue where $j \neq i \neq 0$ after the M_0 ν-mer sequence. Using the general Eq. (10.40), it can be shown that $p_{n+2}\,(i\,0^n\,j)$ for $m = 2$ is given by the following expressions for ν values of 1, 2, and 3:

$$\nu = 1 \;\Rightarrow\; p_{n+2}\,(1\,0^n1) = p_1\,(1)\,[p\,(0/1)]\,[p\,(0/0)]^{n-1}\,[p\,(1/0)] \tag{10.41}$$

$$\nu = 2 \;\Rightarrow\; p_{n+2}\left(1\,0^n 1\right) = p_2\left(10\right)\left[p\left(0/10\right)\right]\left[p\left(0/00\right)\right]^{n-2}\left[p\left(1/00\right)\right] \tag{10.42}$$

$$\nu = 3 \;\Rightarrow\; p_{n+2}\left(1\,0^n 1\right) = p_3\left(100\right)\left[p\left(0/100\right)\right]\left[p\left(0/000\right)\right]^{n-3}\left[p\left(1/000\right)\right] \tag{10.43}$$

Using the general Eq. (10.40), it can be shown that $p_{n+2}\left(i\,0^n\,j\right)$ for $m = 3$ is given by the following expressions for $\nu = 1$ and $\nu = 2$:

$$\begin{aligned} p_{n+2}\left(i\,0^n\,j\right) &= \sum_{i=1}^{m-1} p_{n+2}\left(i\,0^n i\right) + \frac{1}{2}\sum_{i\neq j=1}^{m-1} p_{n+2}\left(i\,0^n\,j\right) \quad \text{for } \nu = 1 \\ &= p\left(1\right)p\left(0/1\right)\left[p\left(0/0\right)\right]^{n-1} p\left(1/0\right) + p\left(2\right)p\left(0/2\right)\left[p(0/0)\right]^{n-1} p\left(2/0\right) \\ &\quad + \frac{1}{2}\left\{p\left(1\right)p\left(0/1\right)\left[p\left(0/0\right)\right]^{n-1} p\left(2/0\right) + p\left(2\right)p\left(0/2\right)\left[p\left(0/0\right)\right]^{n-1} p\left(1/0\right)\right\} \end{aligned} \tag{10.44}$$

$$\begin{aligned} \nu = 2 \;\Rightarrow\; p_{n+2}\left(i\,0^n\,j\right) &= \sum_{i=1}^{m-1} p_{n+2}\left(i\,0^n i\right) + \frac{1}{2}\sum_{i\neq j=1}^{m-1} p_{n+2}\left(i\,0^n\,j\right) \quad \text{for } \nu = 2 \\ &= p\left(10\right)p\left(0/10\right)\left[p\left(0/00\right)\right]^{n-2} p\left(1/00\right) + p\left(20\right)p\left(0/20\right)\left[p\left(0/00\right)\right]^{n-2} p\left(2/00\right) \\ &\quad + \frac{1}{2}\left\{p\left(10\right)p\left(0/10\right)\left[p\left(0/00\right)\right]^{n-2} p\left(2/00\right) + p\left(20\right)p\left(0/20\right)\left[p\left(0/00\right)\right]^{n-2} p\left(1/00\right)\right\} \end{aligned} \tag{10.45}$$

10.7.3 Fraction of n-mer Sequences of a Monomer in Copolymer

Consider the copolymerization of m monomers M_0, M_1, M_2 to M_{m-1} with order of copolymerization ν. The mole fraction of n-mer sequences among all M_0 sequences, $x_0\left(n\right)$, is written as follows:

$$x_0\left(n\right) = \frac{p_{n+2}\left(i\,0^n\,j\right)}{\sum_{n=1}^{\infty} p_{n+2}\left(i\,0^n\,j\right)} = \frac{\sum_{i=1}^{m-1} p_{n+2}\left(i\,0^n i\right) + \frac{1}{2}\sum_{i\neq j=1}^{m-1} p_{n+2}\left(i\,0^n\,j\right)}{\sum_{n=1}^{\infty}\sum_{i=1}^{m-1} p_{n+2}\left(i\,0^n i\right) + \frac{1}{2}\sum_{n=1}^{\infty}\sum_{i\neq j=1}^{m-1} p_{n+2}\left(i\,0^n\,j\right)} \quad i, j \neq 0 \tag{10.46}$$

where the denominator is the sum of the probabilities of all M_0 n-mers. Likewise, the weight fraction of n-mer sequences among all M_0 sequences, $w_0\left(n\right)$, is written as follows:

$$w_0(n) = \frac{nN_0\left(n\right)}{\sum_{n=1}^{\infty} nN_0\left(n\right)} = \frac{\sum_{i=1}^{m-1} np_{n+2}\left(i\,0^n i\right) + \frac{1}{2}\sum_{i\neq j=1}^{m-1} np_{n+2}\left(i\,0^n\,j\right)}{\sum_{n=1}^{\infty}\sum_{i=1}^{m-1} np_{n+2}\left(i\,0^n i\right) + \frac{1}{2}\sum_{n=1}^{\infty}\sum_{i\neq j=1}^{m-1} np_{n+2}\left(i\,0^n\,j\right)} \quad i, j \neq 0 \tag{10.47}$$

where the numerator is the weight probability of M_0 n-mers, and the denominator is the weight probability of all M_0 sequences.

10.7.4 Number-average of n-mer Sequences of a Monomer in Copolymer

Consider the copolymerization of m monomers M_0, M_1, M_2 to M_{m-1} with order of copolymerization ν. The number-average of M_0 n-mer sequences, that is, $\overline{X_n}\,(0^n)$, is defined as follows:

$$\overline{X_n}\,(0^n) = \frac{\sum_{n=1}^{\infty} n\, x_0\,(n)}{\sum_{n=1}^{\infty} x_0\,(n)} = \sum_{n=1}^{\infty} n\, x_0\,(n) \tag{10.48}$$

After substituting for $x_0\,(n)$ from Eq. (10.46), $\overline{X_n}\,(0^n)$ is given as follows:

$$\overline{X_n}\,(0^n) = \frac{\sum_{n=1}^{\infty}\sum_{i=1}^{m-1} n p_{n+2}\,(i\,0^n i) + \frac{1}{2}\sum_{n=1}^{\infty}\sum_{i\neq j=1}^{m-1} n p_{n+2}\,(i\,0^n j)}{\sum_{n=1}^{\infty}\sum_{i=1}^{m-1} p_{n+2}\,(i\,0^n i) + \frac{1}{2}\sum_{n=1}^{\infty}\sum_{i\neq j=1}^{m-1} p_{n+2}\,(i\,0^n j)} \tag{10.49}$$

where the numerator and denominator in the above equation are the weight and number probabilities of M_0 sequences, respectively. The $\overline{X_n}\,(0^n)$ for $m = 2$ is simplified to the following equation for any order of copolymerization ν:

$$m = 2 \;\Rightarrow\; \overline{X_n}\,(0^n) = \frac{\sum_{n=1}^{\infty} n\, p_{n+2}\,(1\,0^n 1)}{\sum_{n=1}^{\infty} p_{n+2}\,(1\,0^n 1)} = \frac{p_1\,(0)}{p_2\,(10)} \tag{10.50}$$

The numerator and denominator in the above equation are the probabilities of finding an M_0 monomer and M_1M_0 dimer, respectively. In a similar approach, the weight-average of M_0 n-mer sequences, that is, $\overline{X_w}\,(0^n)$, is defined as follows:

$$\overline{X_w}\,(0^n) = \frac{\sum_{i=1}^{m-1} n^2 p_{n+2}\,(i\,0^n i) + \frac{1}{2}\sum_{i\neq j=1}^{m-1} n^2 p_{n+2}\,(i\,0^n j)}{\sum_{n=1}^{\infty}\sum_{i=1}^{m-1} n p_{n+2}\,(i\,0^n i) + \frac{1}{2}\sum_{n=1}^{\infty}\sum_{i\neq j=1}^{m-1} n p_{n+2}\,(i\,0^n j)} \tag{10.51}$$

where the numerator and denominator in the above equation are the "z" (see Section 3.2 in Chapter 3 for z distribution) and weight probabilities of M_0 sequences, respectively. The $\overline{X_w}\,(0^n)$ for $m = 2$ is simplified to the following equation for any order of copolymerization ν:

$$m = 2 \;\Rightarrow\; \overline{X_w}\,(0^n) = \frac{\sum_{n=1}^{\infty} n^2 p_{n+2}\,(1\,0^n 1)}{p_1\,(0)} \tag{10.52}$$

10.7.5 Deviation Parameter from Random Copolymer

Consider the copolymerization of m monomers M_0, M_1, M_2 to M_{m-1} with order of copolymerization ν. A random copolymer is a copolymerization such that the reaction rate of different monomers with the propagating chain is entirely independent of the chain's terminal monomers or $\nu = 0$. The deviation parameter from a random copolymer, represented by χ, is a measure of the extent of deviation of the sequence distribution from a copolymer with $\nu = 0$ [32, 33]. For a copolymer with m monomers M_0, M_1, M_2 to M_{m-1}, the parameter χ is defined as follows:

$$\chi = \frac{p_m\,(M_0 M_1 \ldots M_{m-1})}{\prod_m \left[p\,(M_0)\, p\,(M_1) \ldots p\,(M_{m-1})\right]} \quad \text{for } m \text{ monomers} \tag{10.53}$$

where $p_m\,(M_0M_1 \ldots M_{m-1})$ is the probability of finding a contiguous m-mer sequence $M_0M_1 \ldots M_{m-1}$ of the different monomers in the copolymer, $p\,(M_k)$ is the probability of finding the kth monomer, and the symbol $\prod_m$ represents the product of $p\,(M_k)$ for m monomers. The parameter χ for a copolymer with $m = 2$ and a terpolymer with $m = 3$ is expressed as follows:

$$m = 2 \;\Rightarrow\; \chi = \frac{p_2\,(01)}{p\,(0)\,p\,(1)} \tag{10.54}$$

$$m = 3 \;\Rightarrow\; \chi = \frac{p_3\,(012)}{p\,(0)\,p\,(1)\,p\,(2)} \tag{10.55}$$

where $p_2\,(01)$ and $p_2\,(012)$ are the probabilities of finding 2-mer M_0M_1 in the copolymer and 3-mer $M_0M_1M_2$ in the terpolymer, respectively; and $p\,(0)$, $p\,(1)$ and $p\,(2)$ are the probabilities of finding monomers M_0, M_1, and M_2. As in random or ideal copolymerization $\nu = 0$, $p_m\,(M_0M_1 \ldots M_{m-1}) = \prod_m [p\,(M_0)\,p\,(M_1) \ldots p\,(M_{m-1})]$, thus $\chi = 1$ as follows:

$$\chi = \frac{\prod\limits_m [p\,(M_0)\,p\,(M_1) \ldots p\,(M_{m-1})]}{\prod\limits_m [p(M_0)p(M_1)\ldots p\,(M_{m-1})]} = 1 \quad \text{random or ideal copolymer} \tag{10.56}$$

For an alternating copolymer, like $\sim M_0M_1M_0M_1 \sim$, such that the monomer at the reactive end of the propagating chain reacts with the alternate monomer/s in the reaction, the polymerization would have to be at least a terminal copolymerization. In the case of terminal copolymerization, the probability $p_m\,(M_0M_1 \ldots M_{m-1})$ is expanded in terms of 1-mer and conditional probabilities as follows:

$$p_m\,(M_0M_1 \ldots M_{m-1}) = \prod_m p\,(M_0)\,p\,(M_1/M_0) \ldots p\,(M_{m-1}/M_{m-2}) = p\,(M_0) \tag{10.57}$$

If the monomers on the propagating chain are perfectly alternating, then $p\,(M_k) = 1/m$ and χ is simplified to the following expression:

$$\chi = \frac{\prod\limits_m [p\,(M_0)\,p\,(M_1/M_0) \ldots p\,(M_{m-1}/M_{m-2})]}{\prod\limits_m [p\,(M_0)\,p\,(M_1) \ldots p\,(M_{m-1})]} \quad \text{alternate copolymer}$$
$$= \frac{p\,(M_k)}{p(M_k)^m} = \frac{1/m}{(1/m)^m} = m^{m-1} > 1 \tag{10.58}$$

For example, $\chi = 2$ for an alternating copolymer with two different monomers, whereas $\chi = 9$ for an alternating copolymer with three monomers. In general, values of $\chi > 1$ indicate a copolymer with alternating monomers. For a block copolymer, like $\sim M_1M_0M_0M_0M_0M_0M_0M_1 \sim$, such that the monomer at the reactive end of the propagating chain reacts with a monomer of the same type, then$p_m\,(M_0M_1 \ldots M_{m-1}) \to 0$ but $p\,(M_k)$ is finite. Therefore, χ for a block copolymer is given as follows:

$$\chi = \frac{0}{\prod\limits_m [p(M_0)p(M_1)\ldots p(M_{m-1})]} = 0 < 1 \text{ block copolymer} \tag{10.59}$$

In general, values of $\chi < 1$ indicate a copolymer with long sequences of one monomer type followed by a long sequence of the other type, as each monomer favors reacting with its own type.

10.8 Modeling Terminal Copolymerization of Two Monomers

Consider terminal copolymerization of monomers M_0 and M_1 with $m = 2$ and $\nu = 1$. The elementary reactions are shown in Rxns. 10.9– 10.12 and the number of conditional probabilities is $m^{(1+\nu)} = 4$. Based on Eq. (9.5) in Section 9.5 of Chapter 9, the one-step uniform conditional probability matrix for terminal copolymerization with two states, which is the same as the transitional probability matrix for homo-polymerization, is given by the following (2×2) matrix:

$$\overline{P}\,(\upsilon = 1) = \begin{bmatrix} p_{00}^{10} & p_{01}^{10} \\ p_{10}^{10} & p_{11}^{10} \end{bmatrix} \tag{10.60}$$

The states of the propagating chain end in terminal copolymerization of monomers M_0 and M_1 before and after the addition of a monomer to the chain in bases 2 and 10 are shown in Table 10.1.

The first and second subscripts i and j in probability p_{ij}^{10} are the states of the chain end in base 10 before and after propagation, respectively; and the superscript 10 implies that subscripts i and j are in base 10. In terminal copolymerization with $m = 2$ and $\nu = 1$, the numbers in base 2 and 10 are identical; nevertheless, the base 10 numbers should be used for the subscripts of p_{ij}^{10}. The conditional probabilities p_{ij}^{10} are related to the reaction rate constants $k_{p,is}^{10}$, where subscript p is for propagation reaction, i is the state of the chain end in base 10 before propagation, s is the added monomer, and the superscript 10 implies that subscript i and s are in base 10, by the following equations:

$$\begin{cases} p_{00}^{10} = \dfrac{k_{p,00}^{10}\,[M_0]}{k_{p,00}^{10}\,[M_0] + k_{p,01}^{10}\,[M_1]} & p_{01}^{10} = \dfrac{k_{p,01}^{10}\,[M_1]}{k_{p,00}^{10}\,[M_0] + k_{p,01}^{10}\,[M_1]} \\[2ex] p_{10}^{10} = \dfrac{k_{p,10}^{10}\,[M_0]}{k_{p,10}^{10}\,[M_0] + k_{p,11}^{10}\,[M_1]} & p_{11}^{10} = \dfrac{K_{p,11}^{10}\,[M_1]}{k_{p,10}^{10}\,[M_0] + k_{p,11}^{10}\,[M_1]} \end{cases} \tag{10.61}$$

where $[M]_0$ and $[M]_1$ are the concentrations of monomer 0 and 1, respectively; and the reaction rate constants corresponding to p_{00}^{10}, p_{01}^{10}, p_{10}^{10} and p_{11}^{10} are $k_{p,00}^{10}$, $k_{p,01}^{10}$, $k_{p,10}^{10}$, and $k_{p,11}^{10}$, respectively. The above conditional probabilities are simplified by defining the following reactivity ratios:

$$x = \frac{[M_0]}{[M_1]} \quad r_1 = \frac{k_{p,00}^{10}}{k_{p,01}^{10}} \quad r_2 = \frac{k_{p,11}^{10}}{k_{p,10}^{10}} \tag{10.62}$$

Table 10.1 States of the propagating chain end in terminal copolymerization of monomers M_0 and M_1 before and after a propagation step.

State of chain end before propagation		Added monomer	State of chain end after propagation		Conditional probability in base 10
Base 2	**Base 10**		**Base 2**	**Base 10**	
$0\,(\sim M_0)$	0	$0\,(M_0)$	$0\,(M_0)$	0	p_{00}^{10}
$0\,(\sim M_0)$	0	$1\,(M_1)$	$1\,(M_1)$	1	p_{01}^{10}
$1\,(\sim M_1)$	1	$0\,(M_0)$	$0\,(M_0)$	0	p_{10}^{10}
$1\,(\sim M_1)$	1	$1\,(M_1)$	$1\,(M_1)$	1	p_{11}^{10}

where x is the ratio of the two monomer concentrations, r_1 is the reactivity ratio of the two monomers reacting with a propagating chain ending in M_0 monomer residue, and r_2 is the reactivity ratio of the two monomers reacting with a propagating chain ending in M_1. Upon substitution of the above ratios, the conditional probabilities in Eqs. (10.61) simplify as follows:

$$\begin{cases} p_{00}^{10} = \dfrac{r_1 x}{r_1 x + 1} & p_{01}^{10} = \dfrac{1}{r_1 x + 1} \\ p_{10}^{10} = \dfrac{x}{x + r_2} & p_{11}^{10} = \dfrac{r_2}{x + r_2} \end{cases} \tag{10.63}$$

10.8.1 Copolymer Composition in Terminal Copolymerization

For terminal copolymerization of monomers M_0 and M_1 with $m = 2$ and $\nu = 1$, according to Eq. (10.27), the copolymer molar composition vector, $\overline{X}\,(m^\nu = 2)$, is written as follows:

$$\overline{X}\,(m^\nu = 2) = [X_0, X_1] \tag{10.64}$$

where the superscript "10" referring to "base 10" for X_0^{10} and X_1^{10} are dropped for brevity. Based on Eq. (10.32), the elements X_0 and X_1 are related to the elements of the cofactor matrix $\overline{A} = \overline{P} - \overline{I}$, that is a_{ij}, as follows:

$$X_0 = \frac{a_{12}^{cf}}{a_{12}^{cf} + a_{22}^{cf}} \quad X_1 = \frac{a_{22}^{cf}}{a_{12}^{cf} + a_{22}^{cf}} \tag{10.65}$$

After substitution for $\overline{P}$ from Eq. (10.60), the cofactor matrix $\overline{A}$ is given by the following expression:

$$\overline{A} = \overline{P} - \overline{I} = \begin{bmatrix} p_{00} & p_{01} \\ p_{10} & p_{11} \end{bmatrix} - \begin{bmatrix} 1 & 0 \\ 0 & 1 \end{bmatrix} = \begin{bmatrix} p_{00} - 1 & p_{01} \\ p_{10} & p_{11} - 1 \end{bmatrix} \tag{10.66}$$

where the superscript "10" referring to "base 10" for p_{ij}^{10} probabilities is dropped for brevity. According to Eq. (9.30) in Section 9.10.4 of Chapter 9, the elements of the cofactor matrix $\overline{A}$ are given by $a_{ij} = (-1)^{i+j}\left|M_{ij}\right|$ where $\left|M_{ij}\right|$ is the determinant of the matrix $[(n-1)\times(n-1)]$ resulting from removing ith row and jth column from the matrix $\overline{A}$. Using Eq. (10.65), the elements of the copolymer composition vector, X_0 and X_1, are given as follows (see Problem 10.7 in Practice Problems):

$$X_0 = \frac{p_{10}}{p_{10} + p_{01}} \quad X_1 = \frac{p_{01}}{p_{10} + p_{01}} \tag{10.67}$$

where the equalities $p_{00} + p_{01} = 1$ and $p_{10} + p_{11} = 1$ were used to derive the above expressions. Upon substitution for p_{01} and p_{10} from Eq. (10.63), it can be shown that the copolymer composition is expressed in terms of reactivity ratios r_1 and r_2, and monomer ratio x as follows:

$$X_0 = \frac{x(r_1 x + 1)}{r_1 x^2 + 2x + r_2} \quad X_1 = \frac{x + r_2}{r_1 x^2 + 2x + r_2} \tag{10.68}$$

If the feed mole fractions f_0 and f_1 are written in terms of feed molar ratio of the monomers x by the following equations:

$$f_0 = \frac{x}{1+x} \quad f_1 = 1 - f_0 = \frac{1}{1+x} \quad x = \frac{f_0}{1 - f_0} = \frac{f_0}{f_1} \tag{10.69}$$

It can be shown that the copolymer composition is expressed in terms of reactivity ratios r_1 and r_2, and the feed mole fractions as follows:

$$X_0 = \frac{r_1 f_0^2 + f_0 f_1}{r_1 f_0^2 + 2 f_0 f_1 + r_2 f_1^2} \quad X_1 = \frac{r_2 f_1^2 + f_0 f_1}{r_1 f_0^2 + 2 f_0 f_1 + r_2 f_1^2} \tag{10.70}$$

10.8.1.1 Random Terminal Copolymer

For random copolymerization, the rate constant for propagation is independent of the type of reacting monomer and the monomer residue at the reactive end of the propagating chain. Random copolymerization is only affected by the feed composition of the monomers. Thus $k_{p,00}^{10} = k_{p,01}^{10} = k_{p,10}^{10} = k_{p,11}^{10}$ which results in the following relations:

$$\text{random, } \nu = 1 \;\Rightarrow\; p_{00} = p_{10} = p_0, \quad p_{01} = p_{11} = p_1 = 1 - p_0 \;\Rightarrow\; r_1 = r_2 = 1$$

$$X_{0,rnd} = \frac{f_0^2 + f_0 f_1}{f_0^2 + 2 f_0 f_1 + f_1^2} = f_0 \quad X_{1,rnd} = \frac{f_1^2 + f_0 f_1}{f_0^2 + 2 f_0 f_1 + f_1^2} = f_1 \tag{10.71}$$

10.8.1.2 Ideal Terminal Copolymer

For ideal copolymerization, the rate for propagation is independent of the monomer residue at the reactive end of the propagating chain but it depends on the type of reacting monomer. Therefore, $k_{p,00}^{10} = k_{p,10}^{10}$ and $k_{p,10}^{10} = k_{p,11}^{10}$ which results in the following relations (see Problem 10.8 in Practice Problems):

$$\text{ideal, } \nu = 1 \Rightarrow \frac{p_{00}}{p_{01}} = \frac{p_{10}}{p_{11}} \Rightarrow r_1 = \frac{1}{r_2} = r_{idl}$$
$$X_{0,idl} = \frac{r_{idl} x}{r_{idl} x + 1} = \frac{r_{idl} f_0}{r_{idl} f_0 + f_1} \quad X_{1,idl} = \frac{1}{r_{idl} x + 1} = \frac{f_1}{r_{idl} f_0 + f_1} \tag{10.72}$$

10.8.1.3 Alternating Terminal Copolymer

For alternating copolymerization, the rate of propagation of a chain ending in monomer residue M_i with its own type is much less than the other monomer M_j. This implies that $k_{p,00}^{10} << k_{p,01}^{10}$ and $k_{p,11}^{10} << k_{p,10}^{10}$, thus $r_1 = r_2 \to 0$ which results in the following relations:

$$\text{alternating, } \nu = 1 \Rightarrow p_{00} = p_{11} \to 0,\, p_{01} = p_{10} \to 1 \Rightarrow r_1 = r_2 \to 0$$

$$X_{0,alt} = \frac{r_1 f_0^2 + f_0 f_1}{r_1 f_0^2 + 2 f_0 f_1 + r_2 f_1^2} = 0.5 \tag{10.73A}$$

$$X_{1,alt} = \frac{r_2 f_1^2 + f_0 f_1}{r_1 f_0^2 + 2 f_0 f_1 + r_2 f_1^2} = 0.5 \tag{10.73B}$$

10.8.1.4 Block Terminal Copolymer

For block copolymerization, the rate of propagation of a chain ending in monomer residue M_i with its own type is much greater than the other monomer M_j. This implies that $k_{p,00}^{10} >> k_{p,01}^{10}$ and $k_{p,11}^{10} >> k_{p,10}^{10}$, thus $r_1 = r_2 \to \infty$ which results in the following relations:

$$\text{block, } \nu = 1 \Rightarrow p_{00} = p_{11} \to 1,\, p_{01} = p_{10} \to 0 \Rightarrow r_1 = r_2 \to \infty$$

$$X_0|_{r_1,r_2\to\infty} = X_{0,blk} = \frac{f_0^2}{f_0^2 + f_1^2} \tag{10.74A}$$

$$X_1|_{r_1,r_2\to\infty} = X_{1,blk} = \frac{f_1^2}{f_0^2 + f_1^2} \tag{10.74B}$$

Graphs a through (d) in Figure 10.1 show the mole fraction of monomer M_0 in the copolymer (X_0) versus its mole fraction in the feed (f_0) for random, ideal, alternating, and block terminal copolymerization of two monomers M_0 and M_1, respectively. The composition of the copolymer in random copolymerization is only a function of feed composition (Figure 10.1a). In ideal copolymerization, the copolymer composition is a function of monomer reactivity and feed composition; if the reactivity of monomer M_0 is higher than M_1, then M_0 fraction in the copolymer is higher than the feed ($r_{idl} = 5 > 1$ in Figure 10.1b); if the reactivity of M_0 is lower than M_1, then M_0 fraction in the copolymer is higher than the feed ($r_{idl} = 0.2 < 1$); and if M_0 and M_1 are equally reactive, then M_0 fraction in the copolymer is equal to that in the feed ($r_{idl} = 1$). Therefore, random copolymerization is a special case of ideal copolymerization such that the two monomers possess equal reactivity toward the propagating chain. The composition in alternating copolymerization approaches 0.5 regardless of feed composition (Figure 10.1c). The composition in block copolymerization favors the monomer that is abundant in the feed, as shown in Figure 10.1d. If the fraction of M_0 monomer in the feed is higher than M_1, then M_0 fraction will be higher in the copolymer and vice versa.

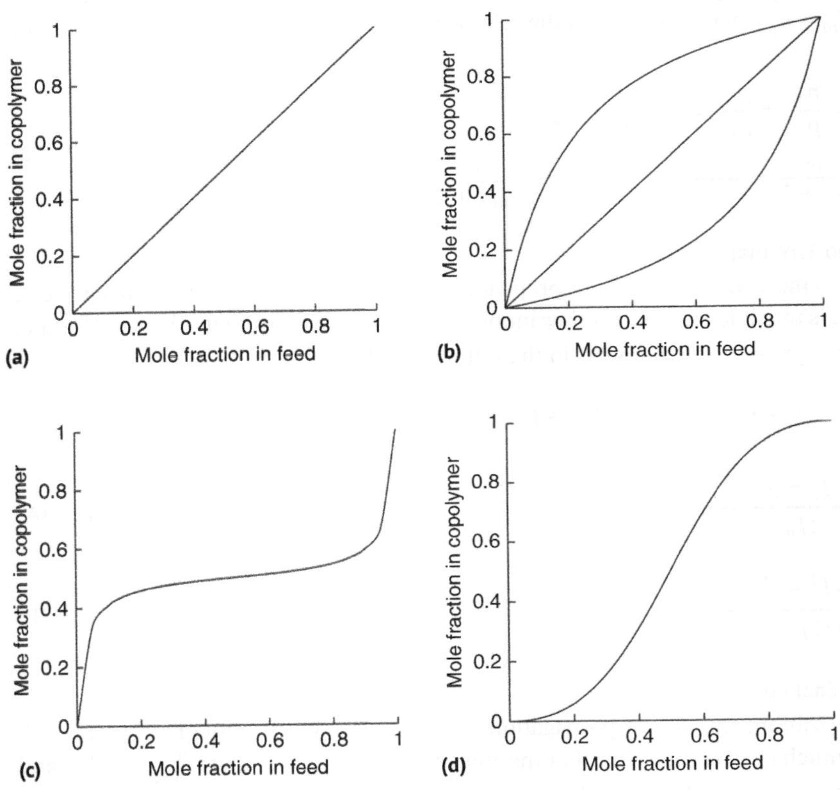

Figure 10.1 Plot of mole fraction of M_0 monomer in copolymer (X_0) versus M_0 mole fraction in feed (f_0) for random (a), ideal (b), alternate (c), block (d) copolymers.

10.8.2 Sequence Distributions in Terminal Copolymerization and Sequence Averages

For terminal copolymerization of monomers M_0 and M_1 with $m = 2$ and $\nu = 1$, the sequence distribution and the average sequence DP are determined using probabilities of $p\,(0)$ and $p\,(1)$, and conditional probabilities p_{00}, p_{01}, p_{10} and p_{11}. The expressions for p_{00}, p_{01}, p_{10}, and p_{11} in terms of molar ratio of the monomers in the feed (x) are given by Eqs. in (10.63). The expressions for $p\,(0)$ and $p\,(1)$, which are the same as X_0 and X_1, in terms of x are given by Eqs. in (10.68) and in terms of mole fraction of monomers in the feed, f_0 and f_1 by Eqs. in (10.70).

10.8.2.1 Probability of Finding n-mer Sequences of a Monomer

The probability of finding a sequence of at least "n" M_0 monomers, that is, $\sim (M_0)_n \sim$, in a terminal copolymerization is given by Eq. (10.33). After substitution for $p(0)$ and p_{00} from Eqs. (10.68) and (10.63), respectively, $p\,(M_0)_n$ is given by the following expression:

$$p_n\,(0^n) = p_1\,(0)\,[p(0/0)]^{n-1} = X_0\,[p_{00}]^{n-1} = \frac{x\,(r_1x+1)}{r_1x^2+2x+r_2}\left[\frac{r_1x}{r_1x+1}\right]^{n-1} \tag{10.75}$$

The probability of finding a sequence of exactly "n" M_0 monomers, that is, $\sim M_1(M_0)_nM_1 \sim$, in a terminal copolymer is given by Eq. (10.41). Upon substitution for $p\,(1)$ and p_{00} from Eqs. (10.68) and (10.63), $p\,[M_1(M_0)_nM_1]$ is given as follows:

$$p_{n+2}\,(10^n1) = p(1)p_{10}[p_{00}]^{n-1}p_{01} = \left[\frac{x/(r_1x+1)}{r_1x^2+2x+r_2}\right]\left(\frac{r_1x}{r_1x+1}\right)^{n-1} \tag{10.76}$$

10.8.2.2 Mole Fraction of n-mer Sequences and Number-average of Sequences of M_0 n-mers

The mole fraction of n-mer sequences among all M_0 sequences, $x_0\,(n)$, is given by Eq. (10.46). Upon substitution for $p_{n+2}\,(10^n1)$ and $p_2\,(01)$ in Eq. 10.46 from Eqs. (10.76) and (10.68), respectively, $x_0\,(n)$ is given by the following expression:

$$x_0\,(n) = \frac{p_{n+2}\,(10^n1)}{p_2\,(01)} = \left[\frac{x}{r_1x^2+2x+r_2}\right]\left(\frac{r_1x}{r_1x+1}\right)^{n-1} \tag{10.77}$$

The number-average sequence of M_0 n-mers, $\overline{X}_n(0)$, for $m = 2$ is given by Eq. (10.50). For terminal copolymerization of monomers M_0 and M_1, it simplifies to the following equation:

$$\overline{X}_n\,(0) = \frac{p_1\,(0)}{p_2\,(10)} = \frac{p_1\,(0)}{p_2\,(01)} = \frac{p_1\,(0)}{p_1\,(0)\,p_{01}} = \frac{1}{p_{01}} = r_1x+1 \tag{10.78}$$

10.8.2.3 Weight Fraction of n-mer Sequences and Weight-average of Sequences of M_0 n-mers

The weight fraction of n-mers among all M_0 sequences, $w_0\,(n)$, is given by Eq. (10.47). After substitution for $p_{n+2}\,(10^n1)$ and $p_1\,(0)$ from Eqs. (10.76) and (10.63), respectively, $w_0\,(n)$ is given by the following expression:

$$w_0\,(n) = \frac{np_{n+2}\,(10^n1)}{p_1\,(0)} = n\left(\frac{r_1x}{r_1x+1}\right)^{n-1} \tag{10.79}$$

The weight-average sequence of M_0 n-mers, $\overline{X}_w\,(0)$, for $m = 2$ is given by Eq. (10.52). Upon substitution for $p_{n+2}\,(10^n1)$ from Eq. (10.76), $p_1\,(0)$ and $p_1(1)$ from Eqs. in (10.68), respectively, and p_{ij} from Eqs. in (10.63), $\overline{X}_w\,(0)$ for $m = 2$ is given as follows:

$$\overline{X}_w\,(0) = \frac{\sum_{n=1}^{\infty} n^2p_{n+2}\,(10^n1)}{p_1\,(0)} = \left(\frac{1}{r_1x+1}\right)^2\sum_{n=1}^{\infty} n^2\left(\frac{r_1x}{r_1x+1}\right)^{n-1} \tag{10.80}$$

10.8.2.4 Deviation Parameter from Random Copolymer

The χ parameter is defined by Eq. (10.54) for $m = 2$. Using $p_1(1)$ from Eq. (10.68) and p_{01} from Eq. (10.63), the χ parameter for terminal copolymerization of monomers M_0 and M_1 is given by the following equation:

$$\chi|_{m=2,\nu=2} = \frac{p_2(01)}{p_1(0)\,p_1(1)} = \frac{p_1(0)\,p_{01}}{p_1(0)\,p_1(1)} = \frac{p_{01}}{p_1(1)} = \frac{(r_1x^2 + 2x + r_2)}{(x + r_2)(r_1x + 1)} \tag{10.81}$$

For random, ideal, alternate, and block copolymers in terminal copolymerization of monomers M_0 and M_1, the χ parameter is equal to the following:

$$\begin{cases} \text{Random} \Rightarrow r_1 = r_2 = 1 \Rightarrow \chi|_{m=2,\nu=2} = 1 \\ \text{Ideal} \Rightarrow r_1 = 1/r_2 = 1 \Rightarrow \chi|_{m=2,\nu=2} = 1 \\ \text{Alternate} \Rightarrow r_1 = r_2 = 0 \Rightarrow \chi|_{m=2,\nu=2} = 2 \\ \text{Block} \Rightarrow p_{01} = 0 \Rightarrow \chi|_{m=2,\nu=2} = 0 \end{cases} \tag{10.82}$$

Parameters for terminal copolymerization of two monomers and their corresponding equations in terms of reactivity ratios r_1 and r_2, and monomer ratio x are listed in Table 10.2.

10.8.2.5 Determination of Ratio of Reaction Rate Constants

In a copolymerization reaction, the feed composition f_0 and f_1, and their ratio x, are known but the monomer reactivity ratios r_1 and r_2 are generally not known because the reactivity ratios strongly depend on reaction conditions, solvent type, catalyst, and the polymerization process [34–43]. Accurate reactivity ratios close to the reaction conditions are often needed for reliable control of the copolymerization reaction even as the ratios may be known for some reaction conditions. The expressions derived for copolymer composition and sequence distribution are used to determine the reactivity ratios in the following sections.

Table 10.2 Parameters for terminal copolymerization of two monomers and their equations in terms of reactivity ratios r_1 and r_2, and monomer ratio x.

Parameter	Equation	Parameter	Equation
x	f_0/f_1	$p_n(0^n)$	$\frac{x(r_1x+1)}{r_1x^2+2x+r_2}\left[\frac{r_1x}{r_1x+1}\right]^{n-1}$
p_{00}	$\frac{r_1x}{r_1x+1}$	$p_n(10^n1)$	$\left[\frac{x/(r_1x+1)}{r_1x^2+2x+r_2}\right]\left(\frac{r_1x}{r_1x+1}\right)^{n-1}$
p_{01}	$\frac{1}{r_1x+1}$	$x_n(0)$	$\left[\frac{x}{r_1x^2+2x+r_2}\right]\left(\frac{r_1x}{r_1x+1}\right)^{n-1}$
p_{10}	$\frac{x}{x+r_2}$	$w_n(0)$	$n\left(\frac{r_1x}{r_1x+1}\right)^{n-1}$
p_{11}	$\frac{r_2}{x+r_2}$	$\overline{X}_n(0)$	r_1x+1
$X_0 = p_1(0)$	$\frac{x(r_1x+1)}{r_1x^2+2x+r_2}$	$\overline{X}_w(0)$	$\left(\frac{1}{r_1x+1}\right)^2\sum_{n=1}^{\infty}n^2\left(\frac{r_1x}{r_1x+1}\right)^{n-1}$
$X_1 = p_1(1)$	$\frac{x+r_2}{r_1x^2+2x+r_2}$	$\chi\|_{m=2,\nu=1}$	$\frac{(r_1x^2+2x+r_2)}{(x+r_2)(r_1x+1)}$

10.8.2.5.1 Copolymer Composition Approach

The expressions for copolymer composition, X_0 and X_1, for terminal copolymerization of monomers M_0 and M_1, given by Eqs. in (10.68), can be used to determine reactivity ratios. This is known as the Fineman–Ross method [44]. Division of X_0 by X_1 results in the following expression:

$$y = \frac{X_0}{X_1} = \left[\frac{x\,(r_1 x + 1)}{r_1 x^2 + 2x + r_2}\right] \Big/ \left[\frac{x + r_2}{r_1 x^2 + 2x + r_2}\right] = \frac{r_1 x^2 + x}{x + r_2} \tag{10.83}$$

where y is the ratio of X_0 to X_1. The above equation can be linearized with respect to r_1 and r_2 by rearrangement as follows:

$$\left(x - \frac{x}{y}\right) = r_1 \left(\frac{x^2}{y}\right) - r_2 \tag{10.84}$$

Upon defining the following parameters $G\,(x, y)$ and $F\,(x, y)$, the above equation simplifies to the following linear form:

$$\begin{cases} G\,(x, y) = x - (x/y) \\ F\,(x, y) = x^2/y \end{cases} \quad G\,(x, y) = r_1 F\,(x, y) - r_2 \tag{10.85}$$

According to the above equation, the plot of $G\,(x, y)$ versus $F\,(x, y)$ is linear with slope and intercept equal to r_1 and r_2, respectively. The following Table 10.3 provides the data for mole fraction of monomer M_1 in the feed (f_1, first column) and in the copolymer (X_1, second column) for copolymerization of vinyl chloride as monomer M_0 and methyl acrylate as monomer M_1 (see Ref. [45]).

The feed monomer ratio, $x = f_0/f_1$, is calculated from the data and is listed in the third column. The copolymer monomer ratio, $y = X_0/X_1$, is calculated and listed in the fourth column of the table. The parameters $G\,(x, y)$ and $F\,(x, y)$ are calculated from the values of x and y using Eqs (10.85) and listed in the fifth and sixth columns of the table, respectively. Figure 10.2 shows the graph of $G\,(x, y)$ versus $F\,(x, y)$ for copolymerization data in Table 10.2. The slope and intercept of the linear least squares line through the datapoints is $r_1 = 0.082$ and $-r_2 = -9.3$, respectively, with a correlation coefficient of $R^2 = 0.79$. Based on Eq. (10.85), the reactivity ratios r_1 and r_2 for this copolymerization are 0.082 and 9.3, respectively. The low correlation coefficient of the least squares line may be due to the data point "a" in Figure 10.2, which is an outlier. If the data point

Table 10.3 Data for M_1 mole fraction in the feed, monomer ratio in the feed, M_1 mole fraction in the copolymer, monomer ratio in the copolymer, and functions $G\,(x, y)$ and $F\,(x, y)$ for copolymerization of vinyl chloride with methyl acrylate.

f_1	x	X_1	y	$F\,(x,y)$	$G\,(x,y)$
0.075	12	0.44	1.3	120	2.6
0.15	5.5	0.70	0.43	70	−7.3
0.24	3.2	0.75	0.33	31	−6.5
0.33	2.1	0.83	0.21	20	−7.8
0.42	1.4	0.86	0.16	12	−7.3
0.52	0.92	0.90	0.11	7.7	−7.4
0.74	0.34	0.97	0.033	3.5	−10
0.87	0.15	0.98	0.017	1.3	−8.7

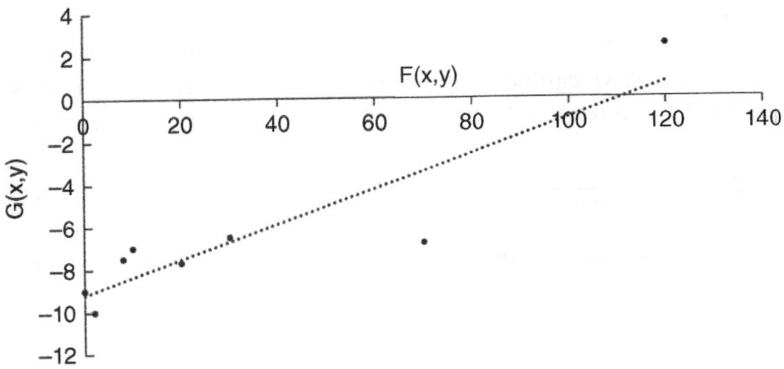

Figure 10.2 Plot of the function $G\ (x, y)$ versus $F\ (x, y)$ from the data in Table 10.3 for copolymerization of vinyl chloride with methyl acrylate.

"a" is removed, the coefficient improves to a strong correlation with $R^2 = 0.97$ with new reactivity ratios of $r_1 = 0.097$ and $r_2 = 9.2$.

There are drawbacks to using the Fineman–Ross method to determine the reactivity ratios. First, the Fineman–Ross method is not symmetric with respect to the two monomers. In other words, switching the monomers by labeling vinyl chloride as M_1 and methyl acrylate as M_0 in the data provided in Table 10.3 will change the fitted reactivity ratios. Second, if the error associated with the values of x and y is proportional to different datapoints, the error associated with values of G and F may not be proportional. This is attributed to the fact that G and F are non-linear functions of x and y. This implies that the datapoints do not equally contribute to the calculated values of reactivity ratios. Therefore, an improvement to Fineman–Ross approach to obtain reliable values of the reactivity ratios would be to use a non-linear least square method to fit the composition data to the terminal copolymerization model.

10.8.2.5.2 Copolymer Dyad Composition Approach

If the dyad or triad composition is known, it can be used with the monomer composition to determine reactivity ratios using the following procedure [46–55]. Chemical analysis methods like ^{1}H and ^{13}C Nuclear Magnetic Resonance (NMR) are used to measure sequence distribution of different monomers in copolymers [47, 49, 50, 53, 55, 56]. Using $p_1\ (0) = X_0$ and $p_1\ (1) = X_1$ and based on Eqs. (10.63), the dyad probabilities for terminal copolymerization of two monomers are given by the following equations:

$$\begin{cases} p_2\ (00) = p_1\ (0)\ p\ (0/0) = X_0 p_{00} = X_0 r_1 x / \left(r_1 x + 1\right) \\ p_2\ (01) = p_1\ (0)\ p\ (1/0) = X_0 p_{01} = X_0 / \left(r_1 x + 1\right) \\ p_2\ (10) = p_1\ (1)\ p\ (0/1) = X_1 p_{10} = X_1 x / \left(x + r_2\right) \\ p_2\ (11) = p_1\ (1)\ p(1/1) = X_1 p_{11} = X_1 r_2 / \left(x + r_2\right) \end{cases} \tag{10.86}$$

Based on the above relations, the ratios of $p_2(00)/p_2(01)$ and $p_2(10)/p_2(11)$ are determined as follows:

$$\begin{cases} p_2(00)/p_2(01) = p_{00}/p_{01} = r_1\, x \\ p_2(10)/p_2(11) = p_{10}/p_{11} = \left(1/r_2\right) x \end{cases} \tag{10.87}$$

Based on the above relations, for terminal copolymerization of M_0 and M_1, the plot of $p_2(00)/p_2(01)$ versus feed composition ratio x is linear with slope equal to r_1; likewise, the plot of $p_2(10)/p_2(11)$ versus x is linear with slope equal to $1/r_2$. Unlike the non-linear Fineman–Ross equation, the relations in Eq. (10.87) are linear functions of feed composition ratio x.

10.8.2.6 Testing Terminal Copolymerization Model

In the previous section, copolymer composition data and dyad sequence probabilities were used to determine the reactivity ratios in the terminal model. However, the determination of reactivity ratios is not a test for the validity of the terminal model for a copolymerization reaction. Consider a random copolymerization in which the propagation reaction is only a function of feed composition and independent of the monomer residues at the reactive end; that is $r_1 = r_2 = 1$. Based on Eq. (10.68), the copolymer composition is only a function of x as follows:

$$X_0 = f_0 = \frac{x}{1+x} \tag{10.88}$$

There is no reactivity ratio to be determined for random copolymerization, however, the model of random copolymerization could be tested by measuring copolymer composition as a function of monomer feed composition. According to the above equation, the copolymer is random if its composition is equal to the feed composition. Next, consider an ideal copolymerization in which the propagation reaction is a function of monomer reactivity and feed composition but independent of the monomer residues on the propagating chain end; that is $r_1 = 1/r_2 = r_{idl}$ (see Eq. 10.72). According to Eq. 10.68, r_{idl} is given as follows:

$$X_0 = \frac{r_{idl}x}{r_{idl}x+1} \Rightarrow r_{idl} = \frac{1}{x}\frac{X_0}{1-X_0} = \frac{1}{x}\frac{p_1(0)}{1-p_1(0)} \tag{10.89}$$

The above equation implies that r_{idl} can be determined by measuring copolymer composition for one monomer feed ratio. The ideal model can be tested by measuring the copolymer composition as a function of monomer feed composition. Based on the above equation, the copolymer is considered ideal if the plot of $X_0/(1-X_0)$ versus x is linear with slope equal to reactivity ratio r_{idl} and intercept equal to zero. Next, consider a terminal copolymerization in which the propagation reaction is a function of monomer reactivity, feed composition, and the terminal monomer residue at the reactive end. According to Eq. (10.85), data on copolymer composition as a function of monomer feed ratio is sufficient to determine reactivity ratios r_1 and r_2, but it is not sufficient to test the terminal model. Furthermore, according to Eq. (10.87), the dyad composition of the copolymer for one monomer feed ratio x is sufficient to determine r_1 and r_2 values, but it is not sufficient to test the terminal model. The terminal model can be tested by measuring the copolymer dyad composition as a function of monomer feed composition. The copolymer is considered terminal if the plots of $p_2(00)/p_2(01)$ and $p_2(10)/p_2(11)$ versus x are linear with slopes equal to copolymer reactivity ratios r_1 and $1/r_2$, respectively, and intercepts equal to zero. To summarize for copolymerization of two monomers ($m = 2$), the copolymer monomer composition as a function of feed ratio is necessary and sufficient to test the random or ideal copolymerization model with $\nu = 0$; the copolymer dyad composition as a function of monomer feed ratio is necessary to test the terminal model with $\nu = 1$; and the copolymer triad composition as a function of monomer feed ratio is necessary to test the penultimate model with $\nu = 2$. In general, for terminal copolymerization with order ν, the $\nu + 1$ copolymer sequence composition is necessary and sufficient to test the model.

10.8.2.7 Limitations of Terminal Copolymerization Model

As discussed in previous sections, the terminal model predicts random, ideal, alternate, and block copolymers. Although the model predicts uni-monomer alternate copolymers, like $\sim M_0M_1M_0M_1M_0M_1 \sim$, it fails to predict the following di-, tri- and higher order alternate copolymers:

$$\begin{aligned} &\text{diad alternate} \Rightarrow \sim M_0M_0M_1M_1M_0M_0M_1M_1M_0M_0M_1M_1 \sim \\ &\text{triad alternate} \Rightarrow \sim M_0M_0M_0M_1M_1M_1M_0M_0M_0M_1M_1M_1 \sim \end{aligned} \tag{10.90}$$

The modeling of copolymer sequences like the ones shown above requires higher order copolymerization models like penultimate or antepenultimate models [16, 21, 57–64].

10.9 Modeling Terminal Terpolymerization of Three Monomers

Consider terminal copolymerization of monomers M_0, M_1, and M_2 with $m = 3$ and $\nu = 1$. The number of conditional probabilities for propagation reactions is $m^{(1+\nu)} = 9$. The nine propagation reactions are as follows:

$$\sim\sim\sim\sim M_0 \bullet + M_0 \xrightarrow{k_{p,00}} \sim\sim\sim\sim M_0 \bullet \tag{10.91}$$

$$\sim\sim\sim\sim M_0 \bullet + M_1 \xrightarrow{k_{p,01}} \sim\sim\sim\sim M_1 \bullet \tag{10.92}$$

$$\sim\sim\sim\sim M_0 \bullet + M_2 \xrightarrow{k_{p,01}} \sim\sim\sim\sim M_2 \bullet \tag{10.93}$$

$$\sim\sim\sim\sim M_1 \bullet + M_0 \xrightarrow{k_{p,10}} \sim\sim\sim\sim M_0 \bullet \tag{10.94}$$

$$\sim\sim\sim\sim M_1 \bullet + M_1 \xrightarrow{k_{p,11}} \sim\sim\sim\sim M_1 \bullet \tag{10.95}$$

$$\sim\sim\sim\sim M_1 \bullet + M_2 \xrightarrow{k_{p,11}} \sim\sim\sim\sim M_1 \bullet \tag{10.96}$$

$$\sim\sim\sim\sim M_2 \bullet + M_0 \xrightarrow{k_{p,10}} \sim\sim\sim\sim M_0 \bullet \tag{10.97}$$

$$\sim\sim\sim\sim M_2 \bullet + M_1 \xrightarrow{k_{p,11}} \sim\sim\sim\sim M_1 \bullet \tag{10.98}$$

$$\sim\sim\sim\sim M_2 \bullet + M_2 \xrightarrow{k_{p,11}} \sim\sim\sim\sim M_2 \bullet \tag{10.99}$$

where $k_{p,ij}$ is the rate constant for the reaction of propagating chain ending in monomer residue M_i reacting with monomer M_j. The number of states in terminal copolymerization of three monomers is $m^\nu = 3^1 = 3$ corresponding to the states of the terminal monomer residue of a propagating chain. These states are M_0, M_1, and M_2 corresponding to the following (1×3) row matrix:

$$S = [E_0^{10}, E_1^{10}, E_2^{10}] \tag{10.100}$$

where the 0, 1, and 2 in E_0^{10}, E_1^{10} and E_2^{10} represent monomers M_0, M_1 and M_2 in base 10, respectively.

The one-step conditional probability matrix for transitioning from ith state to jth state in base 10, p_{ij}^{10}, is defined by the following (3 × 3) matrix:

$$\overline{P}\,(v=1, m=3) = \begin{bmatrix} p_{00}^{10} & p_{01}^{10} & p_{02}^{10} \\ p_{10}^{10} & p_{11}^{10} & p_{12}^{10} \\ p_{20}^{10} & p_{21}^{10} & p_{22}^{10} \end{bmatrix} \tag{10.101}$$

The states of the terminal monomer residue in the terminal copolymerization of monomers M_0, M_1, and M_2 before and after a propagation step in base 3 and 10 are shown in Table 10.4.

10.9.1 Conditional Probabilities in Terminal Terpolymerization

The conditional probabilities p_{ij}^{10} in the matrix 10.101 are related to rate constants $k_{p,is}^{10}$, where s is the reacting monomer in base 10, by the general Eq. (10.26). For terminal terpolymerization of monomers M_0, M_1, and M_2 with $m = 3$ and $v = 1$, the general equation for p_{ij}^{10} in terms of $k_{p,is}^{10}$ and monomer concentrations are as follows:

$$p_{ij}^{10} = \frac{k_{p,ij}^{10}\,[M_j]}{\sum_{l=0}^{2} k_{p,il}^{10}\,[M_l]} \qquad i, j = 0, 1, 2 \tag{10.102}$$

To further simplify the above expression for p_{ij}^{10}, the following feed composition ratios and rate constant ratios are defined:

$$x_1 = \frac{[M_0]}{[M_1]} \qquad x_2 = \frac{[M_0]}{[M_2]} \tag{10.103}$$

$$r_{01} = \frac{k_{p,00}^{10}}{k_{p,01}^{10}} \quad r_{02} = \frac{k_{p,00}^{10}}{k_{p,02}^{10}} \quad r_{11} = \frac{k_{p,10}^{10}}{k_{p,11}^{10}} \quad r_{12} = \frac{k_{p,10}^{10}}{k_{p,12}^{10}} \quad r_{21} = \frac{k_{p,20}^{10}}{k_{p,21}^{10}} \quad r_{22} = \frac{k_{p,20}^{10}}{k_{p,22}^{10}} \tag{10.104}$$

Table 10.4. States of the terminal monomer residue in terminal terpolymerization of monomers M_0, M_1, and M_2 before and after a propagation step.

State of chain end before propagation		Added monomer	State of chain end after propagation		Conditional probability in base 10
base 3	base 10		base 3	base 10	
$0\,(\sim M_0)$	0	$0\,(M_0)$	$0\,(M_0)$	0	p_{00}^{10}
$0\,(\sim M_0)$	0	$1\,(M_1)$	$1\,(M_1)$	1	p_{01}^{10}
$0\,(\sim M_0)$	0	$2\,(M_2)$	$2\,(M_2)$	2	p_{02}^{10}
$1\,(\sim M_1)$	1	$0\,(M_0)$	$0\,(M_0)$	0	p_{10}^{10}
$1\,(\sim M_1)$	1	$1\,(M_1)$	$1\,(M_1)$	1	p_{11}^{10}
$1\,(\sim M_1)$	1	$2\,(M_2)$	$2\,(M_2)$	2	p_{12}^{10}
$2\,(\sim M_2)$	2	$0\,(M_0)$	$0\,(M_0)$	0	p_{20}^{10}
$2\,(\sim M_2)$	2	$1\,(M_1)$	$1\,(M_1)$	1	p_{21}^{10}
$2\,(\sim M_2)$	2	$2\,(M_2)$	$2\,(M_2)$	2	p_{22}^{10}

After substitution from the above equations, the elements p_{ij}^{10} of the conditional probability matrix are given in terms of feed composition ratios and reactivity ratios as follows:

$$\begin{cases} p_{00}^{10} = \dfrac{1}{1+(r_{01}x_1)^{-1}+(r_{02}x_2)^{-1}} & p_{01}^{10} = \dfrac{(r_{01}x_1)^{-1}}{1+(r_{01}x_1)^{-1}+(r_{02}x_2)^{-1}} & p_{02}^{10} = \dfrac{(r_{02}x_2)^{-1}}{1+(r_{01}x_1)^{-1}+(r_{02}x_2)^{-1}} \\ p_{10}^{10} = \dfrac{1}{1+(r_{11}x_1)^{-1}+(r_{12}x_2)^{-1}} & p_{11}^{10} = \dfrac{(r_{11}x_1)^{-1}}{1+(r_{11}x_1)^{-1}+(r_{12}x_2)^{-1}} & p_{12}^{10} = \dfrac{(r_{12}x_2)^{-1}}{1+(r_{11}x_1)^{-1}+(r_{12}x_2)^{-1}} \\ p_{20}^{10} = \dfrac{1}{1+(r_{21}x_1)^{-1}+(r_{22}x_2)^{-1}} & p_{21}^{10} = \dfrac{(r_{21}x_1)^{-1}}{1+(r_{21}x_1)^{-1}+(r_{22}x_2)^{-1}} & p_{22}^{10} = \dfrac{(r_{22}x_2)^{-1}}{1+(r_{21}x_1)^{-1}+(r_{22}x_2)^{-1}} \end{cases} \tag{10.105}$$

The general expressions for p_{ij}^{10} in terms of feed composition ratios and reactivity ratios are as follows:

$$\begin{cases} p_{i0}^{10} = \dfrac{1}{1+(r_{i1}x_1)^{-1}+(r_{i2}x_2)^{-1}} & i = 0, 1, 2 \\ p_{ij}^{10} = \dfrac{(r_{ij}\,x_j)^{-1}}{1+(r_{i1}x_1)^{-1}+(r_{i2}x_2)^{-1}} & i = 0, 1, 2\; j = 1, 2 \end{cases} \tag{10.106}$$

10.9.2 Copolymer Composition in Terminal Terpolymerization

For terminal copolymerization of monomers M_0, M_1, and M_2 with $m = 3$ and $\nu = 1$, the terpolymer composition vector, $\overline{X}\,(m^\nu = 3)$, according to Eq. (10.27), is given by the following (1×3) row vector:

$$\overline{X}\,(m^\nu = 3) = [X_0, X_1, X_2] \tag{10.107}$$

Based on Eq. (10.32), the elements X_0, X_1, and X_2 are related to the elements of the cofactor matrix $\overline{A} = \overline{P} - \overline{I}$, that is a_{ij}, as follows:

$$X_0 = \frac{a_{13}^{cf}}{a_{13}^{cf}+a_{23}^{cf}+a_{33}^{cf}} \quad X_1 = \frac{a_{23}^{cf}}{a_{13}^{cf}+a_{23}^{cf}+a_{33}^{cf}} \quad X_2 = \frac{a_{33}^{cf}}{a_{13}^{cf}+a_{23}^{cf}+a_{33}^{cf}} \tag{10.108}$$

The a_{ij}^{cf} values in the above expressions are the cofactors of the elements a_{ij} of the matrix $\overline{A} = \overline{P} - \overline{I}$. After substitution for $\overline{P}$ from Eq. (10.101), the cofactor matrix $\overline{A}$ is defined as follows:

$$\overline{A} = \overline{P} - \overline{I} = \begin{bmatrix} p_{00}^{10} & p_{01}^{10} & p_{02}^{10} \\ p_{10}^{10} & p_{11}^{10} & p_{12}^{10} \\ p_{20}^{10} & p_{21}^{10} & p_{22}^{10} \end{bmatrix} - \begin{bmatrix} 1 & 0 & 0 \\ 0 & 1 & 0 \\ 0 & 0 & 1 \end{bmatrix} = \begin{bmatrix} p_{00}^{10}-1 & p_{01}^{10} & p_{02}^{10} \\ p_{10}^{10} & p_{11}^{10}-1 & p_{12}^{10} \\ p_{20}^{10} & p_{21}^{10} & p_{22}^{10}-1 \end{bmatrix} \tag{10.109}$$

According to Eqs. (9.30) and (9.31) in Section 9.10.4 of Chapter 9, the elements of the cofactor matrix $\overline{A}$ are given by $a_{ij} = (-1)^{i+j}\left|M_{ij}\right|$ where $\left|M_{ij}\right|$ is the determinant of the matrix $[(n-1)\times(n-1)]$ resulting from removing the ith row and the jth column from the matrix $\overline{A}$. Using Eq. (10.108), the elements of the copolymer composition vector, X_0, X_1, and X_2 are given as follows (see Problem 10.9 in Practice Problems):

$$X_0 = \frac{[p_{10}p_{21} - p_{20}(p_{11} - 1)]}{[p_{10}p_{21} - p_{20}(p_{11} - 1)] - [p_{21}(p_{00} - 1) - p_{01}p_{20}] + [(p_{00} - 1)(p_{11} - 1) - p_{01}p_{10}]}$$
$$X_1 = \frac{-[p_{21}(p_{00} - 1) - p_{01}p_{20}]}{[p_{10}p_{21} - p_{20}(p_{11} - 1)] - [p_{21}(p_{00} - 1) - p_{01}p_{20}] + [(p_{00} - 1)(p_{11} - 1) - p_{01}p_{10}]}$$
$$X_2 = \frac{[(p_{00} - 1)(p_{11} - 1) - p_{01}p_{10}]}{[p_{10}p_{21} - p_{20}(p_{11} - 1)] - [p_{21}(p_{00} - 1) - p_{01}p_{20}] + [(p_{00} - 1)(p_{11} - 1) - p_{01}p_{10}]} \quad (10.110)$$

10.9.2.1 Random Terminal Terpolymer

For random terpolymerization, the propagation rate constant is independent of the type of reacting monomer and the monomer residue at the reactive end of the propagating chain. Random copolymerization is only affected by the feed composition of the monomers. This results in the following relations:

$$r_{01} = r_{02} = r_{11} = r_{12} = r_{21} = r_{22} = 1 \quad \text{random}, \ m = 3, \ \nu = 1 \qquad (10.111)$$

After substitution of the above relations in Eq. (10.106), it can be shown that $p_{0j} = f_0$, $p_{1j} = f_1$, and $p_{2j} = f_2$ for $j = 0, 1, 2$, where f_0, f_1, and f_2 are the feed mole fractions of monomers M_0, M_1, and M_2, respectively. After substitution for p_{ij} from the above relations in Eqs. in (10.110), the elements of the copolymer composition vector, X_0, X_1, and X_2 are given as follows:

$$X_0 = f_0 \quad X_1 = f_1 \quad X_2 = f_2 \quad \text{random}, \ m = 3, \ \nu = 1 \qquad (10.112)$$

Based on the above equalities, the composition of a random terpolymer is the same as the feed composition.

10.9.2.2 Ideal Terminal Terpolymer

For ideal copolymerization, the propagation rate constant is independent of the monomer residue at the reactive end of the propagating chain but dependent on the type of reacting monomer and the feed composition, which results in the following:

$$\begin{cases} r_{01} = r_{11} = r_{21} = r_1 \\ r_{02} = r_{12} = r_{22} = r_2 \end{cases} \quad \text{ideal}, \ m = 3, \ \nu = 1 \qquad (10.113)$$

Substitution of the above relations in Eq. (10.106), the p_{ij} expressions and the terpolymer composition Eq. (10.110) are given as follows.

$$\begin{cases} X_0 = p_{00} = p_{10} = p_{20} = \dfrac{1}{1 + (r_1x_1)^{-1} + (r_2x_2)^{-1}} \\ X_1 = p_{01} = p_{11} = p_{21} = \dfrac{(r_1x_1)^{-1}}{1 + (r_1x_1)^{-1} + (r_2x_2)^{-1}} \\ X_2 = p_{02} = p_{12} = p_{22} = \dfrac{(r_2x_2)^{-1}}{1 + (r_1x_1)^{-1} + (r_2x_2)^{-1}} \end{cases} \quad \text{ideal } m = 3, \ \nu = 1 \qquad (10.114)$$

Based on the above equations, the chain composition of an ideal terpolymer is dependent on reactivity ratios of the reacting monomers and monomers' feed ratio, but it is independent of the terminal monomer on the propagating chain.

10.9.2.3 Alternating Terminal Terpolymer

The following is the sequence of monomers on an alternating terminal terpolymer chain:

$$\sim M_0M_1M_2M_0M_1M_2M_0M_1M_2 \sim \quad \text{alternating } m = 3,\ \nu = 1 \tag{10.115}$$

The above sequence implies that the conditional probability of a propagating chain ending in monomer M_0 to react with M_1, M_1 to react with M_2, or M_2 to react with M_0 is equal to unity whereas the other conditional probabilities are zero. Therefore, the elements of the conditional probability matrix p_{ij}, defined by Eq. (10.106), are as follows:

$$\begin{cases} p_{01} = p_{12} = p_{20} = 1 \\ p_{00} = p_{02} = p_{10} = p_{11} = p_{21} = p_{22} = 0 \end{cases} \tag{10.116}$$

Upon substitution for probabilities p_{ij} from the above relations, the terpolymer composition, defined by Eq. (10.110), is given by the following relations:

$$X_0 = \frac{1}{3} \quad X_1 = \frac{1}{3} \quad X_2 = \frac{1}{3} \quad \text{alternate } m = 3,\ \nu = 1 \tag{10.117}$$

Based on the above relations, the composition of an alternate terpolymer is independent of the reactivity ratios and feed composition.

10.9.2.4 Block Terminal Terpolymer

For a block terpolymer, the rate constant for the reaction of a propagating chain ending in monomer residue M_i with a monomer of its own type is much greater than the monomer M_j if $j \neq i$. This implies that $k^{10}_{p,00} >> k^{10}_{p,01}, k^{10}_{p,02}, k^{10}_{p,11} >> k^{10}_{p,10}, k^{10}_{p,12}$, and $k^{10}_{p,22} >> k^{10}_{p,20}, k^{10}_{p,21}$. Therefore, the following relations exist between the reactivity ratios defined in Eq. (10.104):

$$\begin{cases} r_{01} = \dfrac{k^{10}_{p,00}}{k^{10}_{p,01}} \to \infty \quad r_{11} = \dfrac{k^{10}_{p,10}}{k^{10}_{p,11}} \to 0 \quad r_{21} = \dfrac{k^{10}_{p,20}}{k^{10}_{p,21}} \to 0 \\ r_{02} = \dfrac{k^{10}_{p,00}}{k^{10}_{p,02}} \to \infty \quad r_{12} = \dfrac{k^{10}_{p,10}}{k^{10}_{p,12}} \to 0 \quad r_{22} = \dfrac{k^{10}_{p,20}}{k^{10}_{p,22}} \to 0 \end{cases} \quad \text{block } m = 3,\ \nu = 1 \tag{10.118}$$

After substitution for r_{ij} from the above relations, the elements of the conditional probability matrix p_{ij}, defined by Eq. (10.106), and the elements of the terpolymer composition vector, X_0, X_1, and X_2, are given as follows:

$$\begin{cases} p_{00} = p_{11} = p_{22} = 1 \\ p_{01} = p_{02} = p_{10} = p_{12} = p_{20} = p_{21} = 0 \end{cases} \quad \text{block } m = 3,\ \nu = 1$$

$$X_0 = \frac{f_0^2}{f_0^2 + f_1^2 + f_2^2} \quad X_1 = \frac{f_1^2}{f_0^2 + f_1^2 + f_2^2} \quad X_2 = \frac{f_2^2}{f_0^2 + f_1^2 + f_2^2} \tag{10.119}$$

10.9.3 Terpolymer Sequence Distributions and Sequence Averages

For terminal polymerization of monomers M_0, M_1, and M_2 with $m = 3$ and $\nu = 1$, the expressions for p_{ij} are given in terms of monomer ratios and reactivity ratios by Eqs. (10.106). The expressions for $p(0)$, $p(1)$ and $p(2)$ which are equal to X_0, X_1 and X_2, respectively, are given by Eqs. (10.110) in terms of p_{ij}.

10.9.3.1 Probability of Finding an n-mer Sequence of a Monomer

The probability of finding a sequence of at least "n" M_0 monomer, that is, $\sim (M_0)_n \sim$, in a terminal terpolymerization is given by Eq. (10.33). After substitution for $p\,(0)$ and p_{00} from Eqs. (10.110) and (10.105), respectively, $p\,(M_0)_n$ is given by the following expression:

$$p_n\,(0^n) = p_1\,(0)\,[p\,(0/0)]^{n-1} = X_0\,[p_{00}]^{n-1} \quad m = 3,\ \nu = 1$$
$$= \frac{[p_{10}p_{21} - p_{20}\,(p_{11} - 1)]\,[p_{00}]^{n-1}}{[p_{10}p_{21} - p_{20}\,(p_{11} - 1)] - [p_{21}\,(p_{00} - 1) - p_{01}p_{20}] + [(p_{00} - 1)\,(p_{11} - 1) - p_{01}p_{10}]} \tag{10.120}$$

The probability of finding a sequence of exactly "n" M_0 monomers, that is $\sim M_1\,(M_0)_n\,M_1 \sim$, $\sim M_2\,(M_0)_n\,M_2 \sim$, or $\sim M_1\,(M_0)_n\,M_2 \sim$ in a terminal terpolymer is given, according to Eq. (10.40), as follows:

$$m = 3,\ \nu = 1 \ \Rightarrow\ p_{n+2}\,(i\,0^n\,j) = p_{n+2}\,(10^n1) + p_{n+2}\,(20^n2) + p_{n+2}\,(10^n2) \tag{10.121}$$

It should be noted that $p_{n+2}\,(20^n1) = p_{n+2}\,(10^n2)$ for an infinite chain; Thus, only one of the two probabilities is included in the above equation to avoid counting it twice. Expansion of the probabilities in terms of conditional probabilities p_{ij} results in the following expression:

$$\begin{aligned} p_{n+2}\,(i\,0^n\,j) &= p\,(1)\,p_{10}p_{00}^{n-1}p_{01} + p\,(2)\,p_{20}p_{00}^{n-1}p_{02} + p\,(1)\,p_{10}p_{00}^{n-1}p_{02} \\ &= [X_1p_{10}p_{01} + X_2p_{20}p_{02} + X_1p_{10}p_{02}]\,p_{00}^{n-1} \end{aligned} \tag{10.122}$$

where X_i values in the above equation are related to p_{ij} by Eqs. (10.110) and p_{ij} values are related to monomer ratios and reactivity ratios by Eqs. (10.106).

10.9.3.2 Mole Fraction of n-mer Sequences and Number-average of Sequences of M_0 n-mers

The mole fraction of n-mer sequences among all M_0 sequences, $x_0\,(n)$, is given by Eq. (10.46). Upon substitution for $p_{n+2}\,(i\,0^n\,j)$ in that equation from Eqs. (10.122), $x_0\,(n)$ is given by the following expression:

$$x_0(n) = \frac{p_{n+2}\,(10^n1) + p_{n+2}\,(20^n2) + \frac{1}{2}p_{n+2}\,(10^n2) + \frac{1}{2}p_{n+2}\,(20^n1)}{\sum_{n=1}^{\infty} p_{n+2}\,(10^n1) + \sum_{n=1}^{\infty} p_{n+2}\,(20^n2) + \frac{1}{2}\sum_{n=1}^{\infty} p_{n+2}\,(10^n2) + \frac{1}{2}\sum_{n=1}^{\infty} p_{n+2}\,(20^n1)} \tag{10.123}$$

Upon substitution for $p_{n+2}\,(i\,0^n\,j)$ in the above equation from Eq. (10.40), $x_0\,(n)$ is simplified as follows:

$$x_0\,(n) = \frac{p_{00}^{n-1}}{\sum_{m=0}^{\infty} p_{00}^{m}} = \frac{p_{00}^{n-1}}{1/(1 - p_{00})} = (1 - p_{00})\,p_{00}^{n-1} \tag{10.124}$$

where the expression for geometric series $\sum_{n=0}^{\infty} x^n = 1/(1 - x)$ is used in the above equation. After substitution for p_{00} from Eq. (10.106), $x_0\,(n)$ is expressed in terms of feed composition and reactivity ratios by the following equation:

$$x_0\,(n) = \left(\frac{(r_{01}x_1)^{-1} + (r_{02}x_2)^{-1}}{1 + (r_{01}x_1)^{-1} + (r_{02}x_2)^{-1}}\right)\left(\frac{1}{1 + (r_{01}x_1)^{-1} + (r_{02}x_2)^{-1}}\right)^{n-1} \tag{10.125}$$

For a random terpolymer, it can be shown that $x_0(n)$ simplifies to the following in terms of f_0, the mole fraction of M_0 monomer in the feed:

$$x_0(n) = (1 - f_0)\, f_0^{n-1} \quad \text{random}, \ m = 3, \ \nu = 1 \tag{10.126}$$

The number-average sequence of M_0 n-mers, $\overline{X}_n(0)$, for $m = 3$ is given by Eq. (10.48). After substitution for $x_0(n)$ from 10.124, $\overline{X}_n(0)$ is given by the following equation:

$$\overline{X}_n(0) = \sum_{n=1}^{\infty} n\, x_0(n) = (1 - p_{00}) \sum_{n=1}^{\infty} n\, p_{00}^{n-1} = \frac{1}{(1 - p_{00})} \tag{10.127}$$

where the expression for geometric series $\sum_{n=1}^{\infty} nx^n = x/(1-x)^2$ is used in the above equation. After substitution for p_{00} from Eq. (10.106), $\overline{X}_n(0)$ is expressed in terms of feed composition and reactivity ratios as follows:

$$m = 3, \ \nu = 1 \ \Rightarrow \ \overline{X}_n(0) = \frac{1}{(1 - p_{00})} = \frac{1 + (r_{01}x_1)^{-1} + (r_{02}x_2)^{-1}}{(r_{01}x_1)^{-1} + (r_{02}x_2)^{-1}} \tag{10.128}$$

For a random terpolymer, it can be shown that $\overline{X}_n(0)$ simplifies to the following equation in terms of M_0 mole fraction:

$$\overline{X}_n(0) = \frac{1}{1 - f_0} \quad \text{random}, \ m = 3, \ \nu = 1 \tag{10.129}$$

10.9.3.3 Weight Fraction of n-mer Sequences and Weight-average of Sequences of M_0 n-mers

The weight fraction of n-mers among all M_0 sequences, $w_0(n)$, is given by Eq. (10.47). For terminal terpolymerization of M_0, M_1 and M_2 monomers, Eq. (10.47) simplifies as follows:

$$w_0(n) = \frac{np_{n+2}(10^n1) + np_{n+2}(20^n2) + \frac{1}{2}np_{n+2}(10^n2) + \frac{1}{2}np_{n+2}(20^n1)}{\sum_{n=1}^{\infty} np_{n+2}(10^n1) + \sum_{n=1}^{\infty} np_{n+2}(20^n2) + \frac{1}{2}\sum_{n=1}^{\infty} np_{n+2}(10^n2) + \frac{1}{2}\sum_{n=1}^{\infty} np_{n+2}(20^n1)} \tag{10.130}$$

after substitution for $p_{n+2}(i\,0^n\,j)$ from Eq. (10.40), it can be shown that the above equation for $w_0(n)$ simplifies to the following:

$$w_0(n) = \frac{\left[X_1p_{10}\left(p_{01} + \frac{p_{02}}{2}\right) + X_2p_{20}\left(p_{02} + \frac{p_{01}}{2}\right)\right] np_{00}^{n-1}}{\left[X_1p_{10}\left(p_{01} + \frac{p_{02}}{2}\right) + X_2p_{20}\left(p_{02} + \frac{p_{01}}{2}\right)\right] \sum_{n=1}^{\infty} np_{00}^{n-1}} = (1 - p_{00})^2\, np_{00}^{n-1} \tag{10.131}$$

where the expression for series $\sum_{n=1}^{\infty} nx^n = x/(1-x)^2$ is used in the above equation. After substitution for p_{00} from Eq. (10.106), $w_0(n)$ is expressed in terms of feed composition and reactivity ratios by the following relation:

$$w_0(n) = n\left(\frac{(r_{01}x_1)^{-1} + (r_{02}x_2)^{-1}}{1 + (r_{01}x_1)^{-1} + (r_{02}x_2)^{-1}}\right)^2 \left(\frac{1}{1 + (r_{01}x_1)^{-1} + (r_{02}x_2)^{-1}}\right)^{n-1} \tag{10.132}$$

For a random terpolymer, it can be shown that $w_0(n)$ further simplifies to the following equation:

$$x_0(n) = n\,(1 - f_0)^2\, f_0^{n-1} \quad \text{random}, \ m = 3, \ \nu = 1 \tag{10.133}$$

The weight-average sequence of M_0 n-mers, $\overline{X}_w(0)$, is given by Eq. (10.51). After substitution for $w_0(n)$ from Eq. (10.131), it can be shown that $\overline{X}_w(0)$ for terminal terpolymerization is given as follows:

$$\overline{X}_w(0) = \sum_{n=1}^{\infty} n\, w_0(n) = (1 - p_{00})^2 \sum_{n=1}^{\infty} n^2 p_{00}^{n-1} = \frac{(1 + p_{00})}{(1 - p_{00})} \tag{10.134}$$

where the expression for series $\sum_{n=1}^{\infty} n^2 x^n = x(1+x)/(1-x)^3$ is used in the above equation. Upon substitution for p_{00} from Eq. (10.106), $\overline{X}_w(0)$ is expressed in terms of feed composition and reactivity ratios by the following equation:

$$\overline{X}_w(0) = \frac{(1 + p_{00})}{(1 - p_{00})} = \frac{2 + (r_{01}x_1)^{-1} + (r_{02}x_2)^{-1}}{(r_{01}x_1)^{-1} + (r_{02}x_2)^{-1}} \tag{10.135}$$

For a random terpolymer, it can be shown that $\overline{X}_w(0)$ simplifies to the following equation in terms of M_0 mole fraction:

$$\overline{X}_w(0) = \frac{1 + f_0}{1 - f_0} \quad \text{random, terpolymer} \tag{10.136}$$

10.9.3.4 Polydispersity Index for Sequences of M_0 n-mers

According to Eq. (3.65) in Chapter 3, the normalized standard deviation or statistical measure of polydispersity index, $\Delta_{X_n(0)}$, after substitution for $\overline{X}_w(0)$ and $\overline{X}_n(0)$ from Eqs. (10.134) and (10.127), respectively, is given by the following equation for M_0 n-mers:

$$\Delta_{X_n(0)} = \frac{\sigma_{X_n(0)}}{\overline{X}_n(0)} = \sqrt{\frac{\overline{X}_w(0)}{\overline{X}_n(0)} - 1} = p_{00}^{1/2} = \left[1 + (r_{01}x_1)^{-1} + (r_{02}x_2)^{-1}\right]^{-1/2} \tag{10.137}$$

The commonly used polydispersity index for sequence of M_0 n-mers, $PDI_{X_n(0)}$, defined by Eq. (3.67) in Chapter 3, is given as follows

$$PDI_{X_n(0)} = \frac{\overline{X}_w(0)}{\overline{X}_n(0)} = \frac{2 + (r_{01}x_1)^{-1} + (r_{02}x_2)^{-1}}{1 + (r_{01}x_1)^{-1} + (r_{02}x_2)^{-1}} \tag{10.138}$$

For a random terpolymer with $\nu = 1$, it can be shown that $\Delta_{X_n(0)}$ and $PDI_{X_n(0)}$ simplify to the following relations in terms of the mole fraction of monomer M_0:

$$\begin{cases} \Delta_{X_n(0)} = f_0^{1/2} \\ PDI_{X_n(0)} = f_0 \end{cases} \quad \text{random, } m = 3,\ \nu = 2 \tag{10.139}$$

10.9.4 Deviation Parameter from Random Terpolymer

The χ parameter, defined by Eq. (10.55) for $m = 3$, for terminal terpolymerization of M_0, M_1 and M_2 monomers is given as follows:

$$\chi\Big|_{\substack{m=3\\ \nu=1}} = \frac{p_2(012)}{p_1(0)\,p_1(1)\,P_1(2)} = \frac{p_1(0)\,p_{01}\,p_{12}}{p_1(0)\,p_1(1)\,p_1(2)} = \frac{p_{01}\,p_{12}}{X_1 X_2} \tag{10.140}$$

where p_{01} and p_{12} are given in terms of composition and reactivity ratios by Eqs. (10.106), and X_1 and X_2 by Eqs. (10.110). It can be shown that χ for random, ideal, alternate, and block terpolymerization simplifies to the following relations:

$$\begin{cases} \chi|_{\substack{m=3\\ \nu=1}} = 1 \quad \text{random} \quad \Rightarrow \quad r_{ij} = 1 \ i,\ j = 1, 2, 3 \\ \chi|_{\substack{m=3\\ \nu=1}} = 1 \quad \text{ideal} \quad \Rightarrow \quad r_{i1} = r_1,\ r_{i2} = r_2 \ i = 1, 2, 3 \\ \chi|_{\substack{m=3\\ \nu=1}} \to 9 \quad \text{alternate} \quad \Rightarrow \quad p_{01} = p_{12} = p_{20} \to 1 \\ \chi|_{\substack{m=3\\ \nu=1}} \to 0 \quad \text{block} \quad \Rightarrow \quad p_{ij} \to 0 \ for\ i \neq j \end{cases} \tag{10.141}$$

10.9.5 Determination of Ratio of Reaction Rate Constants

In terpolymerization of M_0, M_1, and M_2 monomers, the feed molar ratios of the monomers x_1 and x_2 are known, but the six reactivity ratios r_{i1} and r_{i2} for $i = 0, 1, 2$ need to be evaluated for the reaction conditions. In terminal terpolymerization, the terpolymer composition equations X_0, X_1, and X_2 (Eqs. in 10.110) are non-linear functions of the two composition ratios (x_1 and x_2) and the six reactivity ratios (r_{01}, r_{02} r_{11}, r_{12}, r_{21} and r_{22}). Therefore, the copolymer composition approach is not useful for determination of the six reactivity ratios. However, experimental data on terpolymer dyad composition of the three monomers in the terpolymer in combination with feed composition can be used for determination of the reactivity ratios using the following procedure. According to the equations for conditional probabilities p_{ij} (Eqs. 10.106), the dyad ratios are given as follows:

$$\begin{cases} \dfrac{p_2(00)}{p_2(01)} = \dfrac{p_1(0)\,p_{00}}{p_1(0)\,p_{01}} = \dfrac{p_{00}}{p_{01}} = r_{01}\,x_1 = r_{01} f_0/f_1 \\ \dfrac{p_2(00)}{p_2(02)} = \dfrac{p_1(0)\,p_{00}}{p_1(0)\,p_{02}} = \dfrac{p_{00}}{p_{02}} = r_{02}\,x_2 = r_{01} f_0/f_2 \\ \dfrac{p_2(10)}{p_2(11)} = \dfrac{p_1(1)\,p_{10}}{p_1(1)\,p_{11}} = \dfrac{p_{10}}{p_{11}} = r_{11}\,x_1 = r_{11} f_0/f_1 \\ \dfrac{p_2(10)}{p_2(12)} = \dfrac{p_1(1)\,p_{10}}{p_1(1)\,p_{12}} = \dfrac{p_{10}}{p_{12}} = r_{12}\,x_2 = r_{12} f_0/f_2 \\ \dfrac{p_2(20)}{p_2(21)} = \dfrac{p_1(2)\,p_{20}}{p_1(2)\,p_{21}} = \dfrac{p_{20}}{p_{21}} = r_{21}\,x_1 = r_{21} f_0/f_1 \\ \dfrac{p_2(20)}{p_2(22)} = \dfrac{p_1(2)\,p_{20}}{p_1(2)\,p_{22}} = \dfrac{p_{20}}{p_{22}} = r_{22}\,x_2 = r_{22} f_0/f_2 \end{cases} \tag{10.142}$$

Based on the above equations, r_{i1} can be determined from the measurement of copolymer dyad ratios $p_2(i0)/p_2(i1)$ at one composition ratio of f_0/f_1 for $i = 0, 1, 2$; similarly r_{i2} can be determined from the measurement of copolymer dyad ratios $p_2(i0)/p_2(i2)$ at one composition ratio of f_0/f_2 for $i = 0, 1, 2$.

10.9.6 Testing Terminal Terpolymerization Model

The measurement of terpolymer dyad ratios $p_2(i0)/p_2(i1)$ and $p_2(i0)/p_2(i2)$ for $i = 0, 1, 2$ at one feed composition is sufficient to determine the six reactivity ratios in terminal terpolymerization model. The measurement of terpolymer dyad ratios at different feed compositions is necessary to test the terminal polymerization of M_0, M_1 and M_2 monomers. Based on Eqs. (10.142), the terpolymerization is considered terminal if the plots of $p_2(i0)/p_2(i1)$ and $p_2(i0)/p_2(i2)$ for $i = 0, 1, 2$ versus f_0/f_1 and f_0/f_2, respectively, are linear with slope equal to r_{i1} and r_{i2}, respectively, and intercept equal to zero. Other approaches are used in special cases like random or ideal terpolymers. For a random terpolymer, the reactivity ratios are independent of the chain's terminal monomer or the monomer reacting with the chain. Therefore, the reactivity ratios $r_{i1} = r_{i2} = 1$ for, $i = 0, 1, 2$ and the copolymer composition is equal to feed composition. A random terpolymer does not require the determination of reactivity ratios, but measurement of terpolymer composition at different feed compositions can be used to test the random model. If the terpolymer composition is the same as the feed for all tested feed compositions, the terpolymer is considered random. For an ideal terpolymer, the reactivity ratios are independent of the chain's terminal monomer but dependent on the monomer reactivity. As a result, the reactivity ratios $r_{i1} = r_1$ and $r_{i2} = r_2$ for $i = 0, 1, 2$ and the copolymer composition are given by Eqs. (10.114). Division of X_0 by X_1 or X_2 results in the following relations:

$$\frac{X_0}{X_1} = r_1 x_1 \quad \frac{X_0}{X_2} = r_2 x_2 \quad \text{ideal } m = 3, \nu = 1 \tag{10.143}$$

The above relations imply that measuring terpolymer composition corresponding to one monomer feed ratio is sufficient to determine r_1 and r_2 for an ideal terpolymer. The ideal terpolymer model can be tested by measuring the copolymer composition as a function of monomer feed composition. Based on the relations in Eq. (10.143), the terpolymer is considered ideal if plots of X_0/X_1 and X_0/X_2 versus x_1 and x_2, respectively, are linear and the intercept is zero. Under these conditions, the slopes are equal to the reactivity ratios r_1 and r_2. In general, for terpolymerization with order ν, the $\nu + 1$ sequence composition is required to test the model.

10.10 Summary

This chapter applied the method of Markov chain to model and predict the composition and sequence distribution of different monomer types in copolymerization reactions. To make the model useful for data analysis, the copolymer Markov chain in this chapter is assumed to be an infinitely long chain as though the monomer residues at chain ends do not affect the composition and sequence distribution of the copolymer. As the chains are assumed to be infinitely long, the state space of the Markovian process contains only propagating states, that is, no absorbing state. In modeling copolymerization, the size of the state space depends on the number of different monomers in the reaction and the order of copolymerization. The order of copolymerization is defined as the number of monomer residues on the reactive end of the propagating chain that affects the propagation reaction. If number of monomers is m and copolymerization order is ν, the number of states is equal to m^ν and the one-step transition probability matrix p_{ij} is a $m^\nu \times m^\nu$ square matrix. The one-step transition probability matrix relates conditional probabilities to the feed composition ratios and monomer reactivity ratios. Then, the copolymer composition is predicted from the knowledge of state space and transition probability matrix if the feed composition and monomer reactivity ratios are known. Next, the steady state copolymer composition is determined from the eigenvector of the one-step transition probability matrix corresponding to eigenvalue $\lambda = 1$. The copolymer sequence distribution of different monomers and the average of n-mer sequences are determined from the knowledge of copolymer composition, conditional probability matrix, and feed composition. Furthermore, if the reactivity ratios are not known, the experimental data on copolymer composition or sequence distribution can be fitted to a linearized form of the model to determine the reactivity ratios. If the copolymer composition and reactivity ratios are known, the model can be fitted to the experimental sequence distribution data to test the validity of the model. The experimental sequence distribution data can also be used to determine if the copolymer is random, ideal, alternate, or block. In general, for copolymerization with order ν, the copolymer $\nu + 1$ sequence composition is required to test the model.

10.11 List of Symbols and Abbreviations

a_{ij}: an element in ith row and jth column of the matrix $\overline{A} = \overline{P} - \overline{I}$

a_{ij}^{cf}: an element in ith row and jth column of the of the cofactor of element a_{ij} of the matrix $\overline{A} = \overline{P} - \overline{I}$

$\overline{A}$ $(m^\nu \times m^\nu)$: square matrix with m^ν rows and columns where $\overline{A} = \overline{P} - \overline{I}$

C: a constant that relates an element of conditional probability matrix to its reaction rate

DP: degree of polymerization of a polymer chain

E_i^j: a state represented by integer number i in base j in the state space of a copolymerization reaction

f_0, f_1: feed mole fractions of monomers M_0 and M_1 in copolymerization reaction of the two monomers

f_0, f_1, f_2: feed mole fractions of monomers M_0, M_1, and M_2 in terpolymerization reaction of the three monomers

$F\,(x, y)$: a function, defined for copolymerization of monomers M_0 and M_1 with $\nu = 1$, as the abscissas to linearize the equation describing the relation between copolymer monomer ratio to feed monomer ratio

$G\,(x, y)$: a function, defined for copolymerization of monomers M_0 and M_1 with $\nu = 1$, as the ordinate to linearize the equation describing the relation between copolymer monomer ratio to feed monomer ratio

$\overline{I}$ $(m^\nu \times m^\nu)$: unit matrix with m^ν rows and columns where diagonal elements are equal to one and off-diagonal elements are zero

$k_{p,j}$: rate constant for propagation reaction of a copolymer chain with monomer type j in copolymerization with $\nu = 0$ (i.e., $k_{p,0}$, $k_{p,1}$, $k_{p,2}$)

$k_{p,ij}$: rate constant for propagation reaction of a copolymer chain ending in monomer residue type i with a monomer type j in a copolymerization with $\nu = 1$

$k_{p,ijk}$: rate constant for propagation reaction of a copolymer chain ending in last two monomer residues i and j reacting with monomer type k in copolymerization with $\nu = 2$

$k_{p,lk}^j$: rate constant of propagation reaction for a copolymer chain ending with $\nu - mer$ sequence represented by integer number l in base j before the reaction to ending with $\nu - mer$ sequence k in base j after the reaction

m: number of different monomer types in a copolymerization reaction

M_j: monomer j in polymerization of more than one monomer type with $j = 0,\ 1$ for copolymerization and $j = 0,\ 1,\ 2$ for terpolymerization

$[M_i]$: concentration of monomer type i in reaction or concentration of polymer chains ends with sequence i in base 10

n: number of monomer residues in a sequence in a copolymer

$N_i\,(n)$: moles of n-mer sequences of ith monomer type among all sequences of ith monomer type

$p\,(i/j^{\upsilon})$: conditional probability of an ith monomer type appearing after a $\nu-mer$ sequence of jth monomer type in a copolymer chain or conditional probability of an ith monomer type reacting with a propagating chain ending with a $\nu-mer$sequence of jth monomer type in copolymerization

$p_{n+2}\,(ij^nk)$: probability of finding an $(n+2)$-mer sequence beginning with an ith monomer type followed by an n-mer sequence of jth monomer type and ending in kth monomer type in a copolymer chain

$p\,(j)$: probability of finding monomer j in a copolymer chain

$p\,(j/i) = p_{ij}$: probability of monomer j reacting with a propagating chain ending in monomer residue i in a copolymerization or probability of finding monomer residue j after residue i in a copolymer chain

p^j_{lk}: probability of propagation reaction for a copolymer chain ending with $\nu-mer$ sequence represented by integer number l in base j before the reaction to ending with $\nu-mer$ sequence k in base j after the reaction

$p\,(M_i)_n = p_n\,(i^n)$: probability of finding an n-mer sequence of monomer M_i or ith monomer type

$p\,(Z)_j$: probability of finding a monomer sequence represented by integer number Z in base j

$PDI_{X_n(0)}$: ratio of number- to weight-average of n-mer sequences of ith monomer type in a copolymer as a practical measure of polydispersity

$\overline{P}\,(m^\nu \times m^\nu)$: conditional probability matrix with m^ν rows and columns

r_1, r_2: ratios of rate constants $k^{10}_{p,00}$ to $k^{10}_{p,01}$ (r_1) and $k^{10}_{p,11}$ to $k^{10}_{p,10}$ (r_2) in copolymerization of two monomers M_0 and M_1 with $\nu = 1$

r_{i1}, r_{i2}: ratios of rate constants $k^{10}_{p,i0}$ to $k^{10}_{p,i1}$ (r_{i1}) and $k^{10}_{p,i0}$ to $k^{10}_{p,i2}$ (r_{i2}) in terpolymerization of three monomers M_0, M_1, and M_2 with $\nu = 1$

r_{idl}: rate constant ratio for an ideal copolymerization of monomers M_0 and M_1 with $\nu = 1$ where $r_{idl} = r_1 = 1/r_2$

S: state space of a copolymerization reaction

T_g: glass transition temperature of a copolymer

$w_i\,(n)$: weight fraction of n-mer sequences of ith monomer type among all sequences of ith monomer type

x: ratio of feed monomer concentrations M_0 to M_1 in copolymerization of the two monomers with $\nu = 1$

x_1, x_2: ratio of feed monomer concentrations M_0 to M_1 (x_1) and M_0 to M_2 (x_2) in copolymerization of the three monomers with $\nu = 1$

$x_i\,(n)$: mole fraction of n-mer sequences of ith monomer type among all sequences of ith monomer type

X_0, X_1: elements of the molar composition matrix or mole fractions of monomers M_0 and M_1 in copolymer chains with $\nu = 1$

X_0, X_1, X_2: elements of the molar composition matrix or mole fractions of monomers M_0, M_1, and M_2 in terpolymer with $\nu = 1$

$X_{0,blk}$, $X_{1,blk}$: elements of the molar composition matrix or mole fractions of monomers M_0 and M_1 in block copolymerization of the two monomers with $\nu = 1$

$X_{0,idl}$, $X_{1,idl}$: elements of the molar composition matrix or mole fractions of monomers M_0 and M_1 in ideal copolymerization of the two monomers with $\nu = 1$

X^{10}_i: i^{th} element of $\overline{X}\,(1 \times m^\nu)$ vector represented by an integer in base 10

$\overline{X}\,(1 \times m^\nu)$: a vector with one row and m^ν columns with elements corresponding to molar composition of the propagating chain end $\nu-mer$sequences in copolymerization

$\overline{X}_n\,(i^n)$: number-average of n-mer sequences of ith monomer type in a copolymer

$\overline{X}_w$ (i^n): weight-average of n-mer sequences of ith monomer type in a copolymer

y: ratio of mole fraction of monomers M_0 to M_1 in the copolymer of the two monomers with $\nu = 1$

$\Delta_{X_n(0)}$: normalized standard deviation of the distribution of n-mer sequences of ith monomer type in a copolymer as a statistical measure of polydispersity as a normalized statistical measure of polydispersity

λ: an eigenvalue of the matrix $\overline{P}$ $(m^\nu \times m^\nu)$ such that $\lambda = 1$ corresponds to steady state copolymer composition of infinitely long copolymer chains

ν: order of copolymerization with $\nu = 0$, 1, 2 for copolymerization independent of monomer residue, dependent on last residue, and dependent on last two residues on reactive chain end

$\sigma_{X_n(0)}$: standard deviation of the distribution of n-mer sequences of ith monomer type in a copolymer as a statistical measure of polydispersity

$\chi|_{m,\nu}$: deviation parameter from a random copolymer of m different monomers with copolymerization order ν

10.12 Practice Problems

Problem 10.1 Acrylonitrile (A), butadiene (B), and styrene (S) monomers are copolymerized to form an impact-resistant thermoplastic terpolymer (ABS) for making molded rigid products. Acrylonitrile provides chemical and thermal stability, butadiene confers toughness, and styrene affords a glossy finish to the final product. The properties of the terpolymer depend strongly on the sequence distribution of A, B, and S monomers in the terpolymer. Define the following ABS sequences by numbers in base 3 and in base 10 by using "0" for A, "1" for B, and "2" for S:

- **a.** -(ABSABSABSABS)-
- **b.** -(AAAABBBBSSSS)-
- **c.** -(BBBASSABBASB)-
- **d.** -(AABBSSAABBSS)-

Problem 10.2 Lactide (L) and glycolide (G) monomers are copolymerized to form hydrolytically degradable polymers with varying rate of degradation. The degradation rate depends on the randomness of the copolymer. A random LG copolymer hinders crystallization, resulting in more water uptake and faster degradation. Conversely, a block LG copolymer increases crystallization resulting in less water uptake and slower degradation. If the sequence –(0011001100)- with "0" and "1" representing L and G, respectively, is used to measure randomness, answer the following questions regarding $p(0011001100)_2$:

- **a.** write $p(0011001100)_2$ in base 10.
- **b.** if copolymerization order is $\nu = 0$ (random model), write the expression for $p(0011001100)_2$ in terms of monomer probabilities in base 2 and 10.
- **c.** if copolymerization order is $\nu = 1$ (terminal model), write the expression for $p(0011001100)_2$ in terms of monomer and conditional probabilities in base 2 and 10.
- **d.** if copolymerization order is $\nu = 2$ (penultimate model), write the expression for $p(0011001100)_2$ in terms of dimer and conditional probabilities in base 2 and 10.
- **e.** if copolymerization order is $\nu = 3$ (antepenultimate model), write the expression for $p(0011001100)_2$ in terms of timer and conditional probabilities in base 2 and 10.

Problem 10.3 For copolymerization of Lactide (L) and glycolide (G) monomers with number "0" for L and "1" for G, define the states in the state space S for the copolymerization reaction for the following cases:

a. random copolymerization ($\nu = 0$)
b. terminal copolymerization ($\nu = 1$)
c. penultimate copolymerization ($\nu = 2$)
d. antepenultimate copolymerization ($\nu = 3$)

Problem 10.4 Using the general Eq. (10.40), show that $p_{n+2}\left(10^n 1\right)$ for $m = 2$ and $\nu = 3$ is given by the following expression:

$$p_{n+2}\left(10^n 1\right)\Big|_{\substack{m=2\\ v=3}} = p_3(100)\,[p(0/100)]\,[p(0/000)]^{n-3}\,[p(0/100)]$$

Problem 10.5 Using the general Eq. (10.40), show that $p_{n+2}\left(i\,0^n j\right)$ for $m = 3$ and $\nu = 2$ is given by the following expression:

$$\begin{aligned}p_{n+2}\left(i0^n j\right)\Big|_{\substack{m=3\\ v=2}} &= p(10)\,p(0/10)\,[p(0/00)]^{n-2}\,p(1/00)\\ &+ p(20)\,p(0/20)\,[p(0/00)]^{n-2}\,p(2/00)\\ &+\frac{1}{2}\left\{p(10)\,p(0/10)\,[p(0/00)]^{n-2}\,p(2/00) + p(20)\,p(0/20)\,[p(0/00)]^{n-2}\,p(1/00)\right\}\end{aligned}$$

Problem 10.6 Using the defining Eq. (10.40) for the probability of finding a sequence of exactly "n" M_0 monomers in a copolymer, $p_{n+2}\left(i\,0^n j\right)$, and the defining Eqs. (10.48) and (10.51) for number- and weight-average of M_0 n-mer sequences, respectively, derive the most simplified expressions for $\overline{X_n}\left(0^n\right)$ and $\overline{X_w}\left(0^n\right)$ for $m = 3$ and copolymerization order $\nu = 1$.

Problem 10.7 For terminal copolymerization of monomers M_0 and M_1 with $m = 2$ and $\nu = 1$, use Eqs. (10.65) and (10.66), and the elements of the cofactor matrix $\overline{A}$, $a_{ij} = (-1)^{i+j}\left|M_{ij}\right|$, to derive the following expressions:

a. show that the elements of the copolymer molar composition vector, X_0 and X_1, are related to the conditional probabilities p_{ij} as follows:

$$X_0 = \frac{p_{10}}{p_{10} + p_{01}} \quad X_1 = \frac{p_{01}}{p_{10} + p_{01}}$$

b. use Eqs. (10.63) to show that X_0 and X_1 are related to the reactivity ratios r_1 and r_2, and monomer ratio x as follows:

$$X_0 = \frac{x\left(r_1 x + 1\right)}{r_1 x^2 + 2x + r_2} \quad X_1 = \frac{x + r_2}{r_1 x^2 + 2x + r_2}$$

c. use Eqs. (10.69) to show that X_0 and X_1 are related to the reactivity ratios r_1 and r_2, and feed mole fractions f_0 and f_1 as follows:

$$X_0 = \frac{r_1 f_0^2 + f_0 f_1}{r_1 f_0^2 + 2 f_0 f_1 + r_2 f_1^2} \quad X_1 = \frac{r_2 f_1^2 + f_0 f_1}{r_1 f_0^2 + 2 f_0 f_1 + r_2 f_1^2}$$

Problem 10.8 In ideal terminal copolymerization with $m = 2$, the rate for propagation is independent of the monomer residue at the end of propagating chain but it depends on the type of reacting monomer. Therefore, $k_{p,00}^{10} = k_{p,10}^{10}$ and $k_{p,01}^{10} = k_{p,11}^{10}$, thus $r_1 = 1/r_2 = r_{idl}$. Show that for this copolymerization, the conditional probabilities are expressed in terms of r_{idl} and x as follows:

$$p_{00} = p_{10} = \frac{r_{idl}\, x}{r_{idl}\, x + 1} = \quad p_{01} = p_{11} = \frac{1}{r_{idl}\, x + 1}$$

Problem 10.9 For terminal copolymerization of monomers M_0, M_1, and M_2 with $m = 2$ and $\nu = 1$, use Eqs. (10.108) and (10.109), and elements of the cofactor matrix, $a_{ij} = (-1)^{i+j}\left|M_{ij}\right|$, to show the following:

a. show that the elements of the copolymer molar composition vector, X_0, X_1, and X_2 are related to the conditional probabilities p_{ij} by the following equations:

$$X_0 = \frac{[p_{10}p_{21} - p_{20}(p_{11} - 1)]}{[p_{10}p_{21} - p_{20}(p_{11} - 1)] - [p_{21}(p_{00} - 1) - p_{01}p_{20}] + [(p_{00} - 1)(p_{11} - 1) - p_{01}p_{10}]}$$

$$X_1 = \frac{-[p_{21}(p_{00} - 1) - p_{01}p_{20}]}{[p_{10}p_{21} - p_{20}(p_{11} - 1)] - [p_{21}(p_{00} - 1) - p_{01}p_{20}] + [(p_{00} - 1)(p_{11} - 1) - p_{01}p_{10}]}$$

$$X_2 = \frac{[(p_{00} - 1)(p_{11} - 1) - p_{01}p_{10}]}{[p_{10}p_{21} - p_{20}(p_{11} - 1)] - [p_{21}(p_{00} - 1) - p_{01}p_{20}] + [(p_{00} - 1)(p_{11} - 1) - p_{01}p_{10}]}$$

b. using Eqs. (10.105), derive expressions for X_0, X_1, and X_2 in terms of feed composition ratios x_1 and x_2, and reactivity ratios r_{i1} and r_{i2} for $i = 0, 1, 2$.

c. Derive expressions for X_0, X_1, and X_2 in terms of the feed mole fractionsf_0, f_1, and f_2, and reactivity ratios r_{i1} and r_{i2} for $i = 0, 1, 2$.

Problem 10.10 Starting from Eq. (10.102) for conditional probabilities, Eqs. (10.103) for feed composition ratios, and Eqs. (10.104) for propagation rate constant ratios in terminal copolymerization of monomers M_0, M_1, and M_2 with $m = 3$ and $\nu = 1$, derive step-by-step the following relations for the terpolymer molar composition if the polymerization reaction is ideal.

$$X_0 = \frac{1}{1 + (r_1x_1)^{-1} + (r_2x_2)^{-1}} \quad X_1 = \frac{(r_1x_1)^{-1}}{1 + (r_1x_1)^{-1} + (r_2x_2)^{-1}} \quad X_2 = \frac{(r_2x_2)^{-1}}{1 + (r_1x_1)^{-1} + (r_2x_2)^{-1}}$$

Problem 10.11 The feed and copolymer composition mole fraction data for copolymerization of vinyl chloride as monomer M_0 and methyl acrylate as monomer M_1 (see Table 10.3 and Ref. [45]) was fitted to the linearized copolymer composition equation (see Eqs. 10.83–10.85 in Section 10.8.2.5.1), which resulted in the values of reactivity ratios $r_1 = 0.097$ and $r_2 = 9.2$ with $R^2 = 0.97$ (after removing the outlier point, see Figure 10.2 in Section 10.8.2.5.1). Based on the fitted values of reactivity ratios, determine the following for the copolymerization of vinyl chloride and methyl acrylate with M_1 (methyl acrylate) mole fraction of $f_1 = 0.10$, $f_1 = 0.25$, $f_1 = 0.50$, $f_1 = 0.75$, and $f_1 = 0.90$.

a. determine p_{00}, p_{01}, p_{10}, p_{11}.

b. determine X_0 and X_1.

c. determine $p_3\,(000)$ and $p_3\,(111)$.

d. determine $p_3\,(10001)$ and $p_3\,(01110)$.

e. determine x_n (000), x_n (111), x_w (000), and x_w (111).
f. determine $\overline{X}_n$ (0) and $\overline{X}_w$ (0).
g. determine $\chi|_{m=2,\nu=1}$ and explain the mode of copolymerization (random, ideal, alternate, block).

Problem 10.12 Starting from Eq. (10.123) for the mole fraction of n-mer sequences of monomer M_0 and Eq. (10.40) for $p_{n+2}\,(i\,0^n\,j)$, derive the following relations:

a. show that $x_0\,(n)$ is given as follows for terminal terpolymerization of monomers M_0, M_1, and M_2:

$$x_0(n)\big|_{\substack{m=3\\ \nu=1}} = (1-p_{00})\,p_{00}^{n-1} = \left(\frac{(r_{01}x_1)^{-1}+(r_{02}x_2)^{-1}}{1+(r_{01}x_1)^{-1}+(r_{02}x_2)^{-1}}\right)\left(\frac{1}{1+(r_{01}x_1)^{-1}+(r_{02}x_2)^{-1}}\right)^{n-1}$$

b. show that $x_0\,(n)$ is given as follows for a random terpolymer:

$$x_0\,(n)\big|_{\substack{m=3\\ \text{random}}} = (1-f_0)\,f_0^{n-1}$$

Problem 10.13 Starting from Eq. (10.130) for the weight fraction of n-mer sequences of monomer M_0 and Eq. (10.40) for $p_{n+2}\,(i\,0^n\,j)$, derive the following relations:

a. show that $w_0\,(n)$ is given as follows for terminal terpolymerization of monomers M_0, M_1, and M_2.

$$w_0(n)\big|_{\substack{m=3\\ \nu=1}} = (1-p_{00})^2\,np_{00}^{n-1} = n\left(\frac{(r_{01}x_1)^{-1}+(r_{02}x_2)^{-1}}{1+(r_{01}x_1)^{-1}+(r_{02}x_2)^{-1}}\right)^2\left(\frac{1}{1+(r_{01}x_1)^{-1}+(r_{02}x_2)^{-1}}\right)^{n-1}$$

b. show that $w_0\,(n)$ is given as follows for a random terpolymer:

$$w_0\,(n)\big|_{\substack{m=3\\ \text{random}}} = n\,(1-f_0)^2\,f_0^{n-1}$$

Problem 10.14 The three monomers triisopropylsilyl acrylate (SIA), methyl methacrylate (MMA), and butyl acrylate (BA) are polymerized by bulk radical polymerization at 70 °C with different monomer feed ratios. The composition of the resulting polymer at low conversion was measured by ^{1}H-NMR, and the resulting data were fitted to the terminal model to estimate the six reactivity ratios (see Ref. [65]). If SIA, MMA, and BA are monomers M_0, M_1, and M_2, respectively, with $m = 3$ and $\nu = 1$, the six reactivity ratios are given as follows:

$$r_{01} = \frac{k_{p,00}^{10}}{k_{p,01}^{10}} = 0.39 \quad r_{02} = \frac{k_{p,00}^{10}}{k_{p,02}^{10}} = 1.00$$

$$r_{11} = \frac{k_{p,10}^{10}}{k_{p,11}^{10}} = 1.33 \quad r_{12} = \frac{k_{p,10}^{10}}{k_{p,12}^{10}} = 1.72$$

$$r_{21} = \frac{k_{p,20}^{10}}{k_{p,21}^{10}} = 0.74 \quad r_{22} = \frac{k_{p,20}^{10}}{k_{p,22}^{10}} = 0.41$$

determine the following for this terpolymerization for three molar (SIA:MMA:BA) feed compositions of (80:10:10), (10:80:10), and (10:10:80).

a. determine $p_{00}, p_{01}, p_{02}, p_{10}, p_{11}, p_{12}, p_{20}, p_{21}$, and p_{22}.
b. determine X_0, X_1, and X_2.
c. determine p_3 (000), p_3 (111), and p_3 (222).
d. determine p_3 $(i000j)$ with $i, j \neq 0$, p_3 $(i111j)$ with $i, j \neq 1$, and p_3 $(i222j)$ with $i, j \neq 2$.
e. determine x_n (000), x_n (111), and x_n (222).
f. determine x_w (000), x_w (111), and x_w (222).
g. determine $\overline{X}_n$ (0), $\overline{X}_n$ (1), and $\overline{X}_n$ (2).
h. determine $\overline{X}_w$ (0), $\overline{X}_w$ (1), and $\overline{X}_w$ (2).
i. determine $\chi|_{m=3,\nu=1}$ and explain the mode of copolymerization (random, ideal, alternate, block).

References

1 Dau, H. et al., Linear block copolymer synthesis. *Chemical Reviews*, 2022, **122**(18), 14471.
2 Feng, H. et al., Block copolymers: synthesis, self-assembly, and applications. *Polymers (Basel)*, 2017, **9**, 494.
3 Nishimori, K. and Ouchi, M., AB-alternating copolymers chain-growth polymerization: synthesis, characterization, self-assembly, and functions. *Chemical Communications*, 2020, **56**(24), 3473.
4 Mayo, F.R. and Walling, C., Copolymerization. *Chemical Reviews*, 1950, **46**(2), 191.
5 Price, F.P., Copolymerization mathematics and description of stereoregular polymers. *Journal of Chemical Physics*, 1962, **36**(1), 209.
6 Dubé, M.A., Saldívar-Guerra, E., and Zapata-González, I., *Copolymerization*, in *Handbook of polymer synthesis, characterization, and processing*. 2013, Wiley Online Library, pp. 105–125.
7 Soteros, C.E. and Whittington, S.G., The statistical mechanics of random copolymers. *Journal of Physics a-Mathematical and General*, 2004, **37**(41), R279.
8 Dayantis, J., Chain statistics and heat of reaction in random copolymerization. *Computational and Theoretical Polymer Science*, 2000, **10**(6), 507.
9 Hu, G.J. et al., Random copolymerization of trimethylene carbonate with l-lactide initiated by amine-bridged bis(phenolate) neodymium alkoxides. *Chemical Communications*, 2023, **59**(45), 6921.
10 Chae, C.G. et al., Bis(β-ketoimino)nickel(II) complexes for random copolymerization of norbornene and methyl 5-norbornene-2-carboxylate with controlled ester group incorporation. *Macromolecules*, 2022, **55**(16), 7049.
11 Shen, Y. et al., Ultrafast ring-opening copolymerization of lactide with glycolide toward random poly(lactic-glycolic acid) copolymers by an organophosphazene base and urea binary catalysts. *Polymer Chemistry UK*, 2022, **13**(13), 1861.
12 Wang, W.Q. et al., Degradable vinyl random copolymers via photocontrolled radical ring-opening cascade copolymerization. *Angewandte Chemie-International Edition*, 2022, **61**(8), e202113302.
13 Eom, K.H., Baek, S., and Kim, I., Heterocyclic carbene-catalyzed random copolymerization of carboxyanhydrides of α-amino acids. *Polymers*, 2021, **13**(21), 3674.
14 Pyun, C.W., Some probability relations in multicomponent copolymers. *Journal of Polymer Science Part A-2: Polymer Physics*, 1971, **9**(4), 577.
15 Tenorio-López, J.A. et al., Mathematical description of the RAFT copolymerization of styrene and glycidyl methacrylate using the terminal model. *Polymers (Basel)*, 2022, **14**(7), 1448.

16 Hamzehlou, S., Reyes, Y., Hutchinson, R., and Leiza, J.R., Copolymerization of butyl acrylate and styrene: terminal vs penultimate model. *Macromolecular Chemistry and Physics*, 2014, **215**(17), 1668.

17 Karssenberg, F.G., Wang, B., Friederichs, N., and Mathot, V.B.F., Terminal and penultimate reactivity ratios in single-site ethene/propene copolymerizations: comparison of kakugo and direct peak methods. *Macromolecular Chemistry and Physics*, 2005, **206**(16), 1675.

18 Kaim, A. and Oracz, P., Solvent effects on true terminal reactivity ratios for styrene-methyl methacrylate copolymerization system. *Polymer*, 1999, **40**(25), 6952.

19 Kaim, A., Terminal and penultimate reactivity ratios in the styrene-acrylonitrile free-radical copolymerization system in bulk. *Journal of Macromolecular Science-Pure and Applied Chemistry*, 1998, **A35**(4), 577.

20 Kaim, A., Styrene-methyl methacrylate copolymer microstructure according to terminal and penultimate model for free-radical copolymerization. *Macromolecular Theory and Simulations*, 1997, **6**(5), 907.

21 Machado, F., Modeling of the penultimate unit effect in chain-growth copolymerizations. *International Journal of Polymer Science*, 2019, **2019**, 2912417.

22 ten Brummelhuis, N. and Stutz, M.B., Penultimate effects in the copolymerization of maleic anhydride with electron-rich styrene derivatives. *Journal of Polymer Science Part A-Polymer Chemistry*, 2016, **54**(18), 2932.

23 Kuchanov, S.I., Bogdanov, A.N., and Tarasevich, K.V., Theoretical consideration of phase behavior of the products of free-radical copolymerization described by the penultimate model. *Macromolecular Theory and Simulations*, 2015, **24**(5), 442.

24 Ito, H., Trinque, B.C., Kasai, P., and Willson, C.G., Penultimate effect in radical copolymerization of 2-trifluoromethylacrylates. *Journal of Polymer Science Part A-Polymer Chemistry*, 2008, **46**(5), 1559.

25 Losio, S. et al., Penultimate-unit effect in ethene/4-methyl-1-pentene copolymerization for a "Sequential" distribution of comonomers. *Macromolecules*, 2008, **41**(4), 1104.

26 Li, D.H., Li, N., and Hutchinson, R.A., High-temperature free radical copolymerization of styrene and butyl methacrylate with depropagation and penultimate kinetic effects. *Macromolecules*, 2006, **39**(13), 4366.

27 Dayantis, J., The conditional probabilities approach to the copolymerization process of two or more monomer species. *Polymer International*, 2004, **53**(11), 1644.

28 Lopezserrano, F., Castro, J.M., Macosko, C.W., and Tirrell, M., Recursive approach to copolymerization statistics. *Polymer*, 1980, **21**(3), 263.

29 Peller, L., Matrix formulation of copolymerization statistics. *Journal of Chemical Physics*, 1962, **36**(11), 2976.

30 Peller, L., Matrix formulation of copolymerization statistics. II. extensions to multimonomeric systems. *Journal of Chemical Physics*, 1965, **43**(7), 2355.

31 Daimon, H., Okitsu, H., and Kumanotani, J., Glass-transition behaviors of random and block copolymers and polymer blends of styrene and cyclododecyl acrylate .1. glass-transition temperatures. *Polymer Journal*, 1975, **7**(4), 460.

32 Stouten, J. et al., Investigation of monomer reactivity, polymer microstructure and solubility in the copolymerization of 1,5-dioxepan-2-one with alkyl substituted lactones. *European Polymer Journal*, 2023, **200**, 112523.

33 Engler, M.S., Scheubert, K., Schubert, U.S., and Böcker, S., New statistical models for copolymerization. *Polymers (Basel)*, 2016, **8**(6), 240.

34 Chen, Y. and Sen, A., Effect of Lewis acids on reactivity ratios for (Meth)acrylate/nonpolar alkene copolymerizations. *Macromolecules*, 2009, **42**(12), 3951.

35 Filley, J., McKinnon, J.T., Wu, D.T., and Ko, G.H., Theoretical study of ethylene-vinyl acetate free-radical copolymerization: reactivity ratios, penultimate effects, and relative rates of chain transfer to polymer. *Macromolecules*, 2002, **35**(9), 3731.

36 Yamada, K., Nakano, T., and Okamoto, Y., Free-radical copolymerization of vinyl esters using fluoroalcohols as solvents: the solvent effect on the monomer reactivity ratio. *Journal of Polymer Science Part A-Polymer Chemistry*, 2000, **38**(1), 220.

37 Fernández-García, M. et al., Solvent effects on the free-radical copolymerization of styrene with butyl acrylate. I. monomer reactivity ratios. *Journal of Polymer Science Part A-Polymer Chemistry*, 2000, **38**(1), 60.

38 Semchikov, Y.D., Smirnova, L.A., Kopylova, N.A., and Izvolenskii, V.V., The effect of monomer preferential solvation in radical copolymerization: reactivity ratios and compositional distribution. *European Polymer Journal*, 1996, **32**(10), 1213.

39 Klumperman, B. and Vonk, G., Solvent effects on the copolymerization of styrene with maleic-anhydride - determination of apparent reactivity ratios from the penultimate unit model. *European Polymer Journal*, 1994, **30**(8), 955.

40 Barson, C.A. and Fenn, D.R., A method for determining reactivity ratios in copolymerizations in the presence of a ceiling temperature effect. *European Polymer Journal*, 1987, **23**(11), 835.

41 Schrijver, J., Ammerdorffer, J.L., and German, A.L., Effect of pressure on free-radical copolymerization kinetics. 3. an improved method of measuring monomer reactivity ratios under high-pressure conditions. *Journal of Polymer Science Part A-Polymer Chemistry*, 1982, **20**(9), 2693.

42 Ponratnam, S. and Kapur, S.L., Effect of pH on reactivity ratios in aqueous-solution copolymerization of acrylic-acid and N-vinylpyrrolidone. *Journal of Polymer Science Part A-Polymer Chemistry*, 1976, **14**(8), 1987.

43 Overberger, C.G., Ehrig, R.J., and Tanner, D., Ionic polymerization .3. ion pair chain mechanisms - the effect of different Lewis acid catalysts on the monomer reactivity ratios in copolymerization. *Journal of the American Chemical Society*, 1954, **76**(3), 772.

44 Fineman, M. and Ross, S.D., Linear method for determining monomer reactivity ratios in copolymerization. *Journal of Polymer Science*, 1950, **5**(2), 259.

45 Wu, Y.Y. et al., Synthesis of a vinyl chloride-methyl acrylate random copolymer with controllable copolymer composition and its processing features in rigid PVC. *Industrial & Engineering Chemistry Research*, 2024, **63**(17), 7939.

46 Corno, C., Roggero, A., Salvatori, T., and Mazzei, A., Evaluation of reactivity ratios and sequence distribution in episulfide copolymers by C-13 NMR. *European Polymer Journal*, 1977, **13**(2), 77.

47 Corno, C., Priola, A., Spallanzani, G., and Cesca, S., Cationic copolymers of isobutylene. 7. reactivity ratio determination by sequence distributions obtained through NMR analysis. *Macromolecules*, 1984, **17**(1), 37.

48 Odriscoll, K.F., Computation of reactivity ratios from sequence distribution information. *Journal of Polymer Science Part A-Polymer Chemistry*, 1980, **18**(8), 2747.

49 Borchardt, J.K., Calculation of reactivity ratios and sequence distributions in copolymers from monomers C-13-Nmr data. *Journal of Macromolecular Science-Chemistry*, 1985, **A22**(12), 1711.

50 Aerdts, A.M., German, A.L., and Vandervelden, G.P.M., Determination of the reactivity ratios, sequence distribution and stereoregularity of butyl acrylate-methyl methacrylate copolymers by means of proton and C-13 NMR. *Magnetic Resonance in Chemistry*, 1994, **32**, S80.

51 Hatada, K. et al., NMR measurement of identical polymer samples by round robin method. 4. analysis of composition and monomer sequence distribution in poly(methyl methacrylate-co-acrylonitrile) leading to determinations of monomer reactivity ratios. *Polymer Journal*, 1995, **27**(11), 1104.

52 Pazhanisamy, P., Ariff, M., and Anwaruddin, Q., Copolymers of alpha-methylstyrene with N-cyclohexylacrylamide: synthesis, monomer reactivity ratios, and mean sequence length. *Journal of Macromolecular Science-Pure and Applied Chemistry*, 1997, **A34**(6), 1045.

53 Brar, A.S. and Dutta, K., Reactivity ratios and sequence determination of acrylonitrile hexyl methacrylate copolymers by one and two dimensional NMR spectroscopy. *Macromolecular Chemistry and Physics*, 1998, **199**(9), 2005.

54 Çatalgil-Giz, H. et al., Online monitoring of composition, sequence length, and molecular weight distributions during free radical copolymerization, and subsequent determination of reactivity ratios. *Macromolecules*, 2002, **35**(17), 6557.

55 Alam, M.M. et al., Reactivity ratios and sequence distribution characterization by quantitative C NMR for RAFT synthesis of styrene-acrylonitrile copolymers. *Journal of Polymer Science Part A-Polymer Chemistry*, 2017, **55**(5), 919.

56 Momose, H. et al., Statistical determination of chemical composition and monomer sequence distribution of poly(methyl methacrylate-butyl methacrylate)s by multivariate analysis of ^{13}C NMR spectra. *Polymer Journal*, 2012, **44**(8), 808.

57 Jones, S.A., Prementine, G.S., and Tirrell, D.A., Model copolymerization reactions - direct observation of a penultimate effect in a model styrene acrylonitrile copolymerization. *Journal of the American Chemical Society*, 1985, **107**(18), 5275.

58 Brown, A.S. and Fujimori, K., Application of the penultimate model in the copolymerization of styrene with maleic-anhydride in ethyl methyl ketone at 50-degrees-C. *Makromolekulare Chemie-Macromolecular Chemistry and Physics*, 1987, **188**(9), 2177.

59 Moad, G., Solomon, D.H., Spurling, T.H., and Stone, R.A., How powerful are composition data in discriminating between the terminal and penultimate models for binary copolymerization. *Macromolecules*, 1989, **22**(3), 1145.

60 Piton, M.C., Winnik, M.A., Davis, T.P., and Odriscoll, K.F., Copolymerization kinetics of 4-methoxystyrene with methyl-methacrylate and 4-methoxystyrene with styrene - a test of the penultimate model. *Journal of Polymer Science Part A-Polymer Chemistry*, 1990, **28**(8), 2097.

61 Schweer, J., Penultimate model description of the propagation kinetics for the free-radical copolymerization of styrene and methyl-methacrylate. *Makromolekulare Chemie-Theory and Simulations*, 1993, **2**(3), 485.

62 Burke, A.L., Duever, T.A., and Penlidis, A., An experimental verification of statistical discrimination between the terminal and penultimate copolymerization models. *Journal of Polymer Science Part A-Polymer Chemistry*, 1996, **34**(13), 2665.

63 Coote, M.L. and Davis, T.P., Copolymerization propagation kinetics of para-substituted styrenes: a critical test of the implicit penultimate model. *Macromolecules*, 1999, **32**(11), 3626.

64 Liu, C.X., Kong, X.Z., Zhang, S.X., and Wang, H.Z., Derivation of copolymer composition equation through kinetic-theory in binary copolymerization systems with antepenultimate effect. *Journal of Macromolecular Science A-Chemistry*, 1991, **28**(10), 1063.

65 Yousefi, F.K. et al., Terpolymerization of triisopropylsilyl acrylate, methyl methacrylate, and butyl acrylate: reactivity ratio estimation. *Macromolecular Reaction Engineering*, 2019, **13**(4), 1900014.

11

Monte Carlo Simulation of Polymerization Reactions

11.1 Introduction

The conservation of moments method for modeling polymerization reactions, covered in previous chapters, has limitations. First, the equations describing the kinetic chain length, DP distribution of the polymer chains, and averages of the distribution tend to become complex even for relatively simple reactions [1–4]. In some polymerizations involving chain transfer to polymer or chain scission, the kth order conservation equations for the propagating chains become nonlinear [4–7]. Aside from nonlinearity and complexity, a separate set of conservation equations would have to be written for chains with a different number of reactive species, but the range of the number of reactive species on the propagating chains is unknown a priori for writing the conservation equations. As a result, the method of moments is applied most effectively to those polymerizations in which the propagating chains carry only one reactive species [4]. Another limitation is that it does not have the capability to adequately model reactions like backbiting because propagating chains are not distinguishable individually in the method of moments [8–13]. Similarly, the process of network formation and gelation cannot effectively be modeled by the method of moments because the conservation equations diverge as the molecular weight approaches infinity. The Monte Carlo simulation method [11, 14–25] overcomes these limitations. This chapter describes the method of Monte Carlo simulation for predicting the molecular weight distribution of polymers.

11.2 Theory of Monte Carlo Simulation

Any polymerization reaction, regardless of complexity, can be simulated by Monte Carlo because polymer chains are individually initiated, propagated, and terminated to calculate the average properties [16, 26, 27]. Thus, individual polymer chains are distinguishable in the Monte Carlo method. In this method, the polymerization consisting of initiator, monomer and solvent molecules, primary radicals, propagating and terminated polymer chains is simulated in a representative microscopic reaction volume by allowing individual initiator molecules to dissociate and form primary radicals, followed by primary radicals adding monomer to form propagating chains, propagating chains terminating to form inactive polymer chains, polymer chains undergoing transfer of reactive species to form propagating chains, or polymer chains undergoing scission to form shorter chains [19, 28]. Within this microscopic volume, the individual chains are simulated at the molecular scale in a random stochastic process by assigning probability to different events/steps from initiation to propagation, chain transfer, scission, termination, or crosslinking. Therefore, the Monte Carlo method allows the simulation of any elementary reaction in the polymerization from intramolecular chain transfer to chain transfer to terminated polymer chains, n-radical propagating chains, and n-chain networks with “n” approaching countable infinity [16, 26–28].

In Monte Carlo simulation, the reaction is divided into short intervals, Δt, corresponding to time for an event to occur. The event is a propagating chain undergoing one of the elementary reactions like propagation, termination by combination, or chain transfer to monomer. Next, a probability value is assigned to each event or elementary reaction based on the relative rates of the reactions. For example, if the permissible elementary reactions are propagation, chain transfer to monomer, and termination by combination, the probability values p_p, p_{ctm}, and p_{tc} are assigned to propagation, chain transfer, and termination reactions such that the sum of the three probabilities would equal unity, that is, $p_p + p_{ctm} + p_{tc} = 1$. Then, a uniformly distributed random number, n_{random}, between zero and one is generated using a random number generator [29–38]. If the generated random number falls between zero and p_p $(0 < n_{\text{random}} \leq p_p)$, the propagation event occurs, and a monomer is added randomly to one of the propagating chains in the simulation. If the random number falls between p_p and $p_p + p_{ctm}$ $(p_p < n_{\text{random}} \leq p_p + p_{ctm})$, the chain transfer to monomer event occurs, which terminates one of the propagating chains in the simulation and starts a new propagating chain with one monomer residue. If the random number falls between $p_p + p_{ctm}$ and $p_p + p_{ctm} + p_{tc}$ $(p_p + p_{ctm} < n_{\text{random}} \leq p_p + p_{ctm} + p_{tc})$, the chain termination by combination occurs, two propagating chains with n and m monomer residues are selected randomly by the simulation to simultaneously terminate and form a polymer chain with $(n + m)$ residues. Clearly, the generation of random numbers that are uniform and independent of each other is immensely important in simulating a polymerization reaction by the Monte Carlo method. The next section demonstrates the use of Monte Carlo simulation with a simple problem to estimate the value of a definite integral of a function.

11.3 Estimation of an Integral Function with Monte Carlo Method

Consider the following integral of a function:

$$I = \int_a^b f(x)\,dx = \int_0^1 (x^2 + 2x + 1)\,dx \tag{11.1}$$

where I is the actual value of the integral of the function $f(x)$, which is a function of independent variable x evaluated from $x = a = 0$ to $x = b = 1$. The actual value of the integral is determined by integrating and evaluating the function between the lower and upper limits of $a = 0$ and $b = 1$ as follows:

$$I = \int_0^1 (9x^2 + 4x + 1)\,dx = \left[3x^3 + 2x^2 + x\right]_0^1 = 6 \tag{11.2}$$

The Monte Carlo estimation of an integral function like the one above, $\hat{I}_{m.c.}$, which is an estimate of the actual value, is given by the following equation:

$$\hat{I}_{m.c.} = \frac{b-a}{N}\sum_{i=1}^{N} f(\varepsilon_i) \tag{11.3}$$

where ε_i is the ith randomly selected number between a and b, and N is the number of random samplings of the function $f(x)$. The Eq. (11.3) states that an estimate of the integral of $f(x)$ may be obtained by taking the average of the function f randomly sampled N times subject to the condition that $\widehat{I}_{m.c.}$ approaches I as N approaches infinity, which is written mathematically as follows:

$$I = \lim \widehat{I}_{N\to\infty} \tag{11.4}$$

If the function is sampled randomly one time ($N = 1$), with random x-value $\varepsilon_1 = 0.233$, obtained using a uniformly distributed random number generator, the estimated value of the integral according to Eq. (11.3) is equal to the following:

$$\widehat{I}_{N=1} = \frac{(1-0)}{1}\sum_{i=1}^{N=1} f(0.233) = f(0.233) = 2.421 \tag{11.5}$$

If the function is sampled randomly five time ($N = 5$) with random x-values $\varepsilon_1 = 0.811$, $\varepsilon_2 = 0.202$, $\varepsilon_3 = 0.095$, $\varepsilon_4 = 0.178$, and $\varepsilon_5 = 0.258$, the estimated value of the integral is as follows:

$$\begin{aligned}\widehat{I}_{N=5} &= \frac{(1-0)}{5}\sum_{i=1}^{N=5} f(\varepsilon_i)\\ &= \frac{1}{5}\left[f(0.811) + f(0.202) + f(0.095) + f(0.178) + f(0.258)\right] = 3.690\end{aligned} \tag{11.6}$$

The estimated value of the integral sampled five times is closer to the actual value than the value with only one sampling. If the function is sampled randomly 25 times ($N = 25$) with random x-values $\varepsilon_1 = 0.319$, $\varepsilon_2 = 0.706$, $\varepsilon_3 = 0.625$, $\varepsilon_4 = 0.235$, $\varepsilon_5 = 0.522$, $\varepsilon_6 = 0.630$, $\varepsilon_7 = 0.545$, $\varepsilon_8 = 0.086$, $\varepsilon_9 = 0.706$, $\varepsilon_{10} = 0.356$, $\varepsilon_{11} = 0.408$, $\varepsilon_{12} = 0.763$, $\varepsilon_{13} = 0.209$, $\varepsilon_{14} = 0.240$, $\varepsilon_{15} = 0.025$, $\varepsilon_{16} = 0.671$, $\varepsilon_{17} = 0.150$, $\varepsilon_{18} = 0.349$, $\varepsilon_{19} = 0.898$, $\varepsilon_{20} = 0.342$, $\varepsilon_{21} = 0.134$, $\varepsilon_{22} = 0.631$, $\varepsilon_{23} = 0.354$, $\varepsilon_{24} = 0.690$, and $\varepsilon_{25} = 0.149$, the estimated value of the integral is as follows:

$$\begin{aligned}\widehat{I}_{N=25} &= \frac{(1-0)}{25}\sum_{i=1}^{N=25} f(\varepsilon_i) = \frac{1}{25}\left[3.19 + 8.31 + 7.02 + 2.44 + 5.54\right]\\ &= \frac{1}{25}\left[3.19 + 8.31 + 7.02 + 2.44 + 5.54 + 7.09 + 5.85 + 1.41 + 8.31\right]\\ &+ \frac{1}{25}\left[3.56 + 4.13 + 9.29 + 2.23 + 2.48 + 1.11 + 7.74 + 1.80 + 3.49\right]\\ &+ \frac{1}{25}\left[11.8 + 3.42 + 1.70 + 7.11 + 3.54 + 8.04 + 1.80\right] = 4.90\end{aligned} \tag{11.7}$$

As the number of random x-values is increased, the Monte Carlo estimate of the integral approaches the actual value. When the function was sampled randomly 1000 times ($N = 1000$), the estimated value of the integral was $\widehat{I}_{1000} = 5.96$ which is very close to the actual value $A = 6.00$. Clearly, as $N \to \infty$, the Monte Carlo estimate of the integral approaches the actual value.

11.4 Probability Distributions of Random Numbers in Monte Carlo Simulation

The estimated Monte Carlo value of a function, as demonstrated in the previous section, strongly depends on the number of randomly sampled independent variable x and the randomness as well as the uniformity of the sampled values. Random number generators like any other stochastic process, generate random numbers with a probability distribution. In this section, various probability distributions for analyzing the randomness of the generated random numbers are described.

11.4.1 Rectangular Probability Distribution

The rectangular probability distribution of the function $p(x)$ in the range of $x = a$ to $x = b$ is given by the following equation [39–41]:

$$pdf = p(x) = \begin{cases} \dfrac{1}{b-a} & a \leq x \leq b \\ 0 & x < a \text{ and } x > b \end{cases} \tag{11.8}$$

Random x-values in Monte Carlo simulations should have a rectangular distribution, that is, all values of x should be equally represented in the range of $a \leq x \leq b$ and x-values outside that range should have zero selection probability.

11.4.2 Normal Probability Distribution

The normal distribution of the function $p(x)$ is given by the following equation [42–46]:

$$pdf = p(x) = \frac{1}{\sqrt{2\pi\sigma^2}} e^{\frac{-(x-\mu)^2}{2\sigma^2}} \quad -\infty < x < +\infty \tag{11.9}$$

where μ is the mean value of x, and σ is the standard deviation from the mean value. Standard deviation is a measure of the spread of the distribution of x values. There is inherent deviation from the rectangular distribution in randomly generated x-values, and these deviations follow a normal distribution with a mean and standard deviation of probability distribution from the mean.

11.4.3 Poisson Probability Distribution

The Poisson probability distribution of function $p(x)$ with discrete nonnegative integer random variable x is defined by the following equation [47–50]:

$$pdf = p(x = k) = \frac{(\lambda t)^k \, e^{-\lambda t}}{k!} \quad k = 0, 1, 2, \ldots \quad 0 < t < \infty \tag{11.10}$$

where x is the time rate of the number of events and λ is the expected value of the probability distribution as well as the variance of the distribution. If x is defined as the number of events, the Poisson distribution of random numbers reduces to the following relation:

$$pdf = p(x = k) = \frac{\lambda^k e^{-\lambda}}{k!} \quad k = 0, 1, 2, \ldots \tag{11.11}$$

11.4.4 χ^2 Probability Distribution

The chi-square probability distribution is used extensively in testing a hypothesis to be true or false [51–53]. If there are Z_1 through Z_k independent normally distributed random variables with theoretical expectation α_1 through α_k and experimental expectation μ_1 through μ_k, and the function F is defined by the following equality [54, 55]:

$$F = \sum_{i=1}^{k} \frac{(\mu_i - \alpha_i)^2}{\alpha_i} \tag{11.12}$$

Then, the function F has the following χ_k^2 distribution with $\upsilon = k - 1$ degrees of freedom of dataset as follows:

$$pdf = p\left(\chi_k^2\right) = \frac{x^{\left(\frac{\upsilon}{2}-1\right)} e^{-\frac{x}{2}}}{2^{\frac{\upsilon}{2}} \Gamma\left(\frac{\upsilon}{2}\right)} \qquad \upsilon = 1, 2, \ldots, k-1 \qquad 0 < x < \infty \tag{11.13}$$

where Γ is the gamma function. The same approach can be used for k independent identically distributed random variables. If there are Z_1 through Z_k independent identically distributed random variables with theoretical expectation $\overline{z}$ and experimental values z_1 through z_k, and the function F is defined by the following equation:

$$F = \sum_{i=1}^{k} \frac{\left(z_i - \overline{z}\right)^2}{\overline{z}} \tag{11.14}$$

Then, the function F has a χ_k^2 distribution with $\upsilon = k - 1$ degrees of freedom as given by Eq. (11.13).

11.5 Error Estimation in Monte Carlo Simulation

The expected value of a function $h(x)$, $\overline{h}$, is defined as $h(x)$ for a given x-value multiplied by its probability distribution $p(x)$ integrated over all x-values from $x = a$ to $x = b$ as follows [38]:

$$E\,|h(x)| = \overline{h} = \int_{x=a}^{x=b} h(x)\, p(x)\, dx \tag{11.15}$$

The function $p(x)$ is the weighing factor for each x value in $h(x)$ such that $0 \le p(x) \le 1$. Consider the integral of function $f(x)$, defined by Eq. (11.1). In estimating the integral by Monte Carlo simulation, x-values were selected by a random number generator represented by some stochastic function $f_{r.n.}(x)$ in the range $x = a$ to $x = b$ with the expected value I given by the following equation:

$$I = \int_{a}^{b} f(x)\, f_{r.n.}(x)\, dx \tag{11.16}$$

where the function $f_{r.n.}(x)$ is used as a weighing factor for each x value.

Comparison of Eqs. (11.15) and (11.16) indicates that the expected value of $h(x)$ becomes equal to the integral of $f(x)$ by finding a suitable $h(x)$ function or $f_{r.n.}(x)$ stochastic function such that it satisfies the following equality:

$$E\,|h(x)| = \overline{h} \simeq I = \int_{a}^{b} f(x)\, dx \tag{11.17}$$

According to Eq. (11.3), the Monte Carlo simulated expected value of $h(x)$, represented by $\widehat{h}$, is defined as follows:

$$\widehat{h} = \frac{b-a}{N} \sum_{i=1}^{N} h(\varepsilon_i)\, f(\varepsilon_i) \tag{11.18}$$

subject to the following condition as N approaches infinity:

$$\overline{h} = \lim_{N\to\infty} \hat{h} \tag{11.19}$$

The above equations simply state that the integral of a function $f(x)$ for x-values in the range of $x = a$ to $x = b$ can be approximated by the expectation of a suitable function $h(x)$ randomly sampled using a stochastic variable X with a uniformly distributed function $f_{r.n.}(x) = \frac{1}{(b-a)}$ between $x = a$ and $x = b$.

Consider now the theoretical and approximate Monte Carlo expected values of the function $h(x)$, namely $\overline{h} = E\,|h(x)|$ and $\hat{h}$, respectively. The central limit theorem [56] states that the $\hat{h}$ values calculated by randomly sampling $h(x)$ function N times are normally distributed with the following mean (μ) and variance (V):

$$\mu = E\,|h(x)| \simeq \frac{(b-a)}{N}\sum_{i=1}^{N} h(\varepsilon_i)\, f(\varepsilon_i) = \overline{h} \tag{11.20}$$

$$\begin{aligned} V &= E\left|h(x)^2\right| - \{E\,|h(x)|\}^2 \\ &\simeq \frac{(b-a)}{N}\left[\left\{\sum_{i=1}^{N} h^2(\varepsilon_i)\, f(\varepsilon_i)\right\} - \left\{\sum_{i=1}^{N} h^2(\varepsilon_i)\, f(\varepsilon_i)\right\}^2\right] = \sigma^2 \end{aligned} \tag{11.21}$$

where σ is the standard deviation of $\overline{h}$ from the mean value μ. Applying the above equations to the Monte Carlo estimate of the integral of function $f(x)$ defined by Eq. (11.2) (see Eq. 11.16), the Monte Carlo estimate $\hat{I}$ of the function $f(x)$ randomly sampled N times is normally distributed with the following mean μ and variance V:

$$\mu = E\,|f(x)| \simeq \frac{(b-a)}{N}\sum_{i=1}^{N} f(\varepsilon_i) = \hat{I} \tag{11.22}$$

$$V = E\left|f(x)^2\right| - \{E\,|f(x)|\}^2 \simeq \frac{b-a}{N}\left[\left\{\sum_{i=1}^{N} f^2(\varepsilon_i)\right\} - \left\{\sum_{i=1}^{N} f(\varepsilon_i)\right\}^2\right] = \sigma^2 \tag{11.23}$$

where σ is the standard deviation of $\hat{I}$ from the mean value. According to the central limit theorem, the extent of deviation of the Monte Carlo estimate $\hat{I}$ from the theoretical value I is given by the normalized function $\frac{(\hat{I}-I)}{\sqrt{\frac{\sigma^2}{N}}}$ which has a normal distribution with $\mu = 0$ and $V = 1$ [56]. Based on the definition of a normal distribution (Eq. 11.9), the probability of finding $-\beta \le \left\{\frac{(\hat{I}-I)}{\sqrt{\frac{\sigma^2}{N}}}\right\} \le +\beta$ is given by the following equality:

$$p\left\{\left|\frac{(\hat{I}-I)}{\sqrt{\frac{\sigma^2}{N}}}\right| \le \beta\right\} = \frac{1}{\sqrt{2\pi}}\int_{-\beta}^{+\beta} e^{-\frac{x^2}{2}}\,dx \tag{11.24}$$

Table 11.1 Commonly used confidence levels and $\beta_{CL\%}$ values.

CL%	α%	$\beta_{CL\%}$
90	10	1.65
95	5	1.96
98	2	2.33
99	1	2.58
99.9	0.1	3.29

with a percent confidence level ($CL\% = 100 - \alpha\%$), the integral on the right side of the above equation is used to find β corresponding to $CL\%$ of the area under the normal distribution ($\beta_{CL\%}$). Table 11.1 provides the values of β corresponding to a given $CL\%$ or $\alpha\%$.

According to Eq. (11.24), for a specified $CL\%$, the condition for the Monte Carlo estimate $\hat{I}$ to fall within $\beta_{CL\%}$ is given as follows:

$$\left| \frac{(\hat{I} - I)}{\sqrt{\sigma^2/N}} \right| \leq \beta_{CL\%} \tag{11.25}$$

If percent error of the estimate value $\hat{I}$ with respect to the theoretical value I is defined by the following equation:

$$E = \left| \frac{100\,(\hat{I} - I)}{\hat{I}} \right| \tag{11.26}$$

The percent error limit of the estimate, EL, according to Eq. (11.25), is given by the following algebraic expression:

$$EL \leq \frac{100\beta_{CL\%}\sigma}{\hat{I}\sqrt{N}} \tag{11.27}$$

The above expression sets a limit on the percent error above which the specified percent confidence level is not satisfied. The above equation can be used to find the Monte Carlo sample size of random numbers for a specified percent error limit and a given confidence level.

11.6 Testing Uniformity of Random Numbers

11.6.1 Test of Uniformity

Random numbers between zero and one generated by random number generation algorithms are used in Monte Carlo simulations as well as other applications [57–60]. Before using a random number generator for simulation, it is important to sample the random numbers and evaluate the numbers for uniformity and correlation. To test for uniformity, the kth order moment of distribution of random numbers, $\mu_{rn,k}$, is evaluated using the following equation [60–63]:

$$\mu_{rn,k} = \frac{\sum_{i=1}^{N} x_i^k}{N} \tag{11.28}$$

where x_i is the random number i in a sample of N numbers. The 1st, 2nd, and 3rd order moments of the generated numbers and their corresponding theoretical values for random numbers between zero and one are given by the following equations:

$$\mu_{rn,1} = \frac{\sum_{i=1}^{N} x_i^1}{N} \qquad \mu_{rn,1}\,|_{N\to\infty} = 1/2 \tag{11.29}$$

$$\mu_{rn,2} = \frac{\sum_{i=1}^{N} x_i^2}{N} \qquad \mu_{rn,2}\,|_{N\to\infty} = 1/3 \tag{11.30}$$

$$\mu_{rn,3} = \frac{\sum_{i=1}^{N} x_i^3}{N} \qquad \mu_{rn,3}\,|_{N\to\infty} = 1/4 \tag{11.31}$$

The difference between the actual and theoretical values of $\mu_{rn,1}$ is a measure of the overall uniformity and symmetry of the random numbers about the mean value; the difference for $\mu_{rn,2}$ is a measure of the skewness of the numbers about the mean value to the right if $\mu_{rn,1} > 0.5$ or left if $\mu_{rn,1} < 0.5$; and the difference for $\mu_{rn,3}$ is a measure of number cluster formation within the sample very close to zero or one. The error limit, EL, defined in the previous section, is used to quantify the overall uniformity for a given sample of random numbers using the percent confidence level as follows (see Eq. 11.27 in Section 11.5):

$$EL \le \frac{100\beta_{CL\%}\sigma}{\sqrt{N}\mu_{rn,1}} = \frac{100\beta_{CL\%}V^{1/2}}{N^{1/2}\mu_{rn,1}} \tag{11.32}$$

where σ and V are standard deviation and variance of the distribution of the sample of random numbers about the mean, respectively. Upon substitution for V from Eq. (11.23) in the above equation, the following expression is obtained for EL in terms of $\mu_{rn,1}$ and $\mu_{rn,2}$:

$$EL \le \frac{100\beta_{CL\%}V^{1/2}}{N^{1/2}\mu_{rn,1}} = \frac{100\beta_{CL\%}\left\{\mu_{rn,2} - \mu_{rn,1}^2\right\}^{1/2}}{N^{1/2}\mu_{rn,1}} \tag{11.33}$$

Based on the above equation, $\mu_{rn,1}$ and $\mu_{rn,2}$ of the generated sample of random numbers and the sample size are sufficient to evaluate the percent error limit of a sample of random numbers for a specified confidence level.

11.6.2 Example Problem 1

Determine $\mu_{rn,1}$, $\mu_{rn,2}$, $\mu_{rn,3}$ and the percent error limit EL for the following sequence of random numbers ($N = 16$) between zero and one and $CL = 99\%$.

0.12	0.28	0.31	0.42	0.93	0.82	0.71	0.56
0.45	0.16	0.91	0.25	0.32	0.30	0.62	0.71

Solution:

$$\mu_{rn,1} = \frac{\sum_{i=1}^{N} x_i^1}{N} = \frac{7.87}{16} = 0.49 < 0.5$$

$$\mu_{rn,2} = \frac{\sum_{i=1}^{N} x_i^2}{N} = \frac{4.2}{16} = 0.31 < 1/3 \tag{11.34}$$

$$\mu_{rn,3} = \frac{\sum_{i=1}^{N} x_i^3}{N} = \frac{3.54}{16} = 0.22 < 0.25$$

The random numbers are slightly biased toward values < 0.5. According to Table 11.1, if $CL =$ 99% then $\beta_{99\%} = 2.58$, and $EL\%$ can be calculated using Eq. (11.33) as follows:

$$EL \leq \frac{100\beta_{CL\%}\left\{\mu_{rn,2} - \mu_{rn,1}^2\right\}^{1/2}}{N^{1/2}\mu_{rn,1}} = \frac{(100)(2.58)\left\{0.31 - 0.49^2\right\}^{1/2}}{(16)^{1/2}(0.49)} = 33.4\% \tag{11.35}$$

As the theoretical mean of the random numbers is 0.5, the percent error E is calculated using the following equation (analogous to Eq. 11.28):

$$E = \frac{100\left(\mu_{rn,1} - \mu_{rn,1}\mid_{N\to\infty}\right)}{\mu_{rn,1}} = \frac{(100)(0.49 - 0.50)}{0.49} = 1.62\% \tag{11.36}$$

As $E < EL$, the condition $CL = 99\%$ is satisfied. This implies that the actual mean of the generated random numbers is not significantly different from the theoretical mean with confidence level of 99% (i.e., 99% of the area under distribution of random numbers is selected).

11.6.3 Test of Interval Correlation

For testing interval correlation, the interval of random numbers is divided into M equal smaller subintervals, and the number of random numbers in each subinterval j, $n_{si,j}$ where subscript "si" stands for subinterval, is computed [61, 64–66]. If the subintervals are equal in size, the theoretical number of random numbers in the subintervals should be the same and designated by $\bar{n}_{si}$. Next, the values of $n_{si,j}$ for all subintervals and $\bar{n}_{si}$ are used to find the interval correlation function, F_{ic}, using the following equation:

$$F_{ic} = \sum_{j=1}^{M} \frac{\left(n_{si,j} - \bar{n}_{si}\right)^2}{\bar{n}_{si}} \tag{11.37}$$

The above function has a χ^2 distribution (see Eq. 11.14 in Section 11.4.4) with degrees of freedom $k = M - 1$. For a specified percent confidence level $CL\%$, the critical values of ${\chi^2}_{CL\%,k=M-1}$ and ${\chi^2}_{(100-CL\%),k=M-1}$ are found from a χ^2 distribution table [67]. Table 11.2 provides the critical values for χ^2 test for k values from one to five plus values of 9 and 81 and $CL\%$ of 99%, 95% and 90%.

If the calculated F_{ic} satisfies the following condition:

$${\chi^2}_{CL\%,k=M-1} \leq F_{ic} \leq {\chi^2}_{(100-CL\%),k=M-1} \tag{11.38}$$

the generated random numbers pass the test of interval correlation for the specified k and $CL\%$.

Table 11.2 Critical values for χ^2 test as a function of degrees of freedom k and percent confidence level $CL\%$.

	CL%			100-CL%		
K = M−1	99%	95%	90%	10%	5%	1%
1	0.000157	0.00393	0.0157	2.706	3.841	6.635
2	0.0201	0.103	0.211	4.605	5.991	9.210
3	0.115	0.352	0.584	6.251	7.815	11.345
4	0.297	0.711	1.064	7.779	9.488	13.277
5	0.554	1.145	1.610	9.236	11.070	15.086
9	2.088	3.325	4.168	14.684	16.919	21.666
81	54.357	61.261	65.176	97.680	103.010	113.512

Table 11.3 Values of $n_{si,j}$ and $\overline{n}_{si}$ for example Problem 11.3.

Subinterval, j	$n_{si,j}$	$\overline{n}_{si}$
0–0.25	3	4
0.25–5.0	6	4
0.5–0.75	4	4
0.75–1.0	3	4

11.6.4 Example Problem 2

For the random numbers in Example Problem 1, determine whether the random numbers pass the test of interval correlation for four equal subintervals of 0–0.24999, 0.25–0.4999, 0.5–0.74999, and 0.75–1 with 99% confidence level.

Solution: For the given random numbers, $N = 16$, $M = 4$, $k = M - 1 = 3$, $\bar{n}_{si} = 16/4 = 4$, and the number of random numbers in each subinterval is provided in Table 11.3.

$$F_{ic} = \sum_{j=1}^{M} \frac{(n_{si,j} - \overline{n}_{si})^2}{\overline{n}_{si}} = \frac{(-1)^2 + (2)^2 + 0 + (-1)^2}{4} = 1.5$$

$$\chi^2{}_{CL\%,\, k=M-1} = \chi^2{}_{99\%,\, 3} = 0.12 \tag{11.39}$$

$$\chi^2{}_{(100-CL\%),\, k=M-1} = \chi^2{}_{1\%,\, 3} = 11.34$$

As $0.12 \leq F_{ic} = 1.5 \leq 11.34$, the random numbers pass the test of interval correlation, meaning that the numbers in the four intervals are not correlated (i.e., randomly selected) with confidence level of 99%.

11.6.5 Test of Serial Correlation

For testing serial correlation, the interval of the random numbers is divided into M equal subintervals. Next, the number of times that a random number x_i in subinterval j is followed in h steps by a random number x_{i+h} which is in subinterval k are counted to generate a table of $n_{sc,h,jk}$ values. In the $n_{sc,h,jk}$ abbreviation, "sc" stands for serial correlation, "h" for h-step correlation, and "jk" for the first random number in subinterval j and the h-step number in subinterval k. If $h = 1$, then serial correlation is tested between consecutively selected random numbers. If the subintervals are equal, the theoretical values of $n_{sc,h,jk}$ for all j and k are equal and designated by $\overline{n}_{sc,h}$. Next, the serial correlation function, F_{sc}, is defined based on the value of $n_{sc,h,jk}$ and $\overline{n}_{sc,h}$:

$$F_{sc} = \sum_{k=1}^{M} \sum_{j=1}^{M} \frac{\left(n_{sc,h,jk} - \bar{n}_{sc,h}\right)^2}{\bar{n}_{sc,h}} \tag{11.40}$$

The function F_{sc} like the interval correlation function has a χ^2 distribution (see Eq. 11.14 in Section 11.4.4) with degrees of freedom $k = (M - 1)^2$. For a specified percent confidence level $CL\%$, the critical values of $\chi^2{}_{CL\%,\, k=(M-1)^2}$ and $\chi^2{}_{(100-CL\%),\, k=(M-1)^2}$ are found from a χ^2 distribution table [67]. If the calculated F_{sc} satisfies the following condition:

$$\chi^2{}_{CL\%,\, k=(M-1)^2} \leq F_{sc} \leq \chi^2{}_{(100-CL\%),\, k=(M-1)^2} \tag{11.41}$$

Then, the generated random numbers pass the test of serial correlation for the specified k and $CL\%$.

11.6.6 Example Problem 3

For the random numbers in Example Problem 1, determine whether the random numbers pass the test of 1-step interval correlation for four equal subintervals of 0–0.2499, 0.25–0.4999, 0.5–0.7499, and 0.75–1 with 99% confidence level.

$x_1 = 0.12$	0.28	0.31	0.42	0.93	0.82	0.71	$x_8 = 0.56$
$x_9 = 0.45$	0.16	0.91	0.25	0.32	0.30	0.62	$x_{16} = 0.71$

Solution: For the given random numbers, $N = 16$, $M = 4$, $h = 1$, $k = (M-1)^2 = 9$, $\bar{n}_{sc,1} = \frac{16}{16} = 1$, and $n_{sc,1,jk}$ values are listed in Table 11.4.

$$F_{sc} = \sum_{k=1}^{M}\sum_{j=1}^{M} \frac{(n_{sc,1,jk} - \bar{n}_{sc,1})^2}{\bar{n}_{sc,1}} = 15$$
$$\chi^2_{(100-CL\%),k=(M-1)^2} = \chi^2_{1\%,9} = 21.66 \tag{11.42}$$
$$\chi^2_{CL\%,k=(M-1)^2} = \chi^2_{99\%,9} = 2.09$$

As $2.088 \le F_{sc} = 15 \le 21.666$, the random numbers pass the test of serial correlation, meaning that the numbers are not 1-step serially correlated with a confidence level of 99% (i.e., considering 99% of the area of the distribution of the random numbers).

11.6.7 Example Problem 4

An algorithm generates 1×10^4 random numbers between zero and one. The first, second, and third order moments of the random numbers are as follows:

$$\mu_{RN,1} = 0.4991 < 0.5 \qquad \mu_{RN,2} = 0.3331 \simeq \frac{1}{3} \qquad \mu_{RN,3} = 0.2499 \simeq 0.25$$

The actual and theoretical number of random numbers in ten subintervals are given in Table 11.5. The values of $n_{sc,1,jk}$ for the subintervals are given in the following Table 11.6.

Table 11.4 Values of $n_{sc,1,jk}$ for example Problem 11.3.

	k			
j	**0–0.25**	**0.25–0.50**	**0.50–0.75**	**0.75–1.0**
0–0.2499	0	1	0	1
0.25–0.4999	1	4	1	1
0.5–0.7499	0	1	2	0
0.75–1.0	0	1	1	1

Table 11.5 Values of $n_{ic,j}$ and $\bar{n}_{ic}$ for example Problem 11.4.

Subinterval j	$n_{ic,j}$	$\bar{n}_{ic}$	Subinterval j	$n_{ic,j}$	$\bar{n}_{ic}$
0–0.1	1050	1000	0.5–0.6	1038	1000
0.1–0.2	996	1000	0.6–0.7	1035	1000
0.2–0.3	941	1000	0.7–0.8	975	1000
0.3–0.4	996	1000	0.8–0.9	1011	1000
0.4–0.5	977	1000	0.9–1.0	981	1000

Table 11.6 Values of $n_{sc,1,jk}$ for example Problem 11.4.

	k									
j	0–0.1	0.1–0.2	0.2–0.3	0.3–0.4	0.4–0.5	0.5–0.6	0.6–0.7	0.7–0.8	0.8–0.9	0.9–1.0
0–0.1	103	96	98	112	113	98	92	100	112	107
0.1–0.2	120	117	96	92	100	105	100	102	96	110
0.2–0.3	96	98	93	106	97	86	97	94	113	92
0.3–0.4	106	94	100	87	95	103	90	94	100	82
0.4–0.5	94	118	95	109	109	104	103	104	98	97
0.5–0.6	113	96	85	85	114	104	93	104	107	96
0.6–0.7	95	100	114	83	102	105	94	103	77	100
0.7–0.8	110	92	92	91	88	94	104	89	130	97
0.8–0.9	104	114	108	99	113	96	95	108	108	88
0.9–1.0	90	113	91	87	99	102	106	90	91	118

Determine whether the random numbers pass the test of overall uniformity (EL), interval correlation (F_{ic}), and serial correlation (F_{sc}).

Solution: For the random numbers generated by the algorithm, $N = 1 \times 10^4$, $M = 10$, $h = 1$, k for interval correlation $k_{\text{int}} = (M - 1) = 9$, k for serial correlation $k_{ser} = (M - 1)^2 = 81$, and $CL = 99\%$. Based on $\mu_{rn,1} = 0.4991$, $\mu_{rn,2} = 0.3331$, and $\mu_{rn,3} = 0.2499$, the generated random numbers are uniformly distributed with no skewness with respect to the mean value. According to Table 11.1, if $CL = 99\%$ then $\beta_{99\%} = 2.58$, and EL is determined using Eq. (11.33) as follows:

$$EL = \frac{100\,\beta_{CL\%}\left\{\mu_{rn,2} - \mu^2_{Rrn,1}\right\}^{1/2}}{N^{1/2}\mu_{rn,1}} = \frac{(100)(2.58)\left\{0.3331 - 0.4991^2\right\}^{1/2}}{\sqrt{1 \times 10^4}\,(0.3331)} = 2.21 \tag{11.43}$$

$$E = \left|\frac{\mu_{rn,1} - 0.5}{\mu_{rn,1}}\right| = \left|\frac{0.4991 - 0.5}{0.4911}\right| = 0.0018$$

As $E < EL$, the actual mean of the random numbers is not different from the theoretical mean with 99% confidence level (considering 99% of the area under the distribution of random numbers). The interval correlation function F_{ic} is evaluated using Eq. (11.37) with the data in Table 11.5 and $\bar{n}_{si} = 10000/10 = 1000$:

$$F_{ic} = \sum_{j=1}^{M} \frac{\left(n_{si,j} - \bar{n}_{si}\right)^2}{\bar{n}_{si}} = 10.38 \tag{11.44}$$

$$\chi^2{}_{1\%,9} = 21.66 \qquad \chi^2{}_{99\%,9} = 2.09$$

As $2.09 \le F_{ic} = 10.38 \le 21.66$, the generated random numbers do pass the test of interval correlation with 99% confidence level. The serial 1-step correlation function F_{sc} is evaluated using Eq. (11.40) with the data in Table 11.6 and $\bar{n}_{sc,1} = 10000/100 = 100$ as follows:

$$F_{sc} = \sum_{k=1}^{M}\sum_{j=1}^{M} \frac{(n_{sc,1,jk} - \bar{n}_{sc,1})^2}{\bar{n}_{sc,1}} = 88.90 \tag{11.45}$$

$$\chi^2_{1\%,81} = 113.51 \qquad \chi^2_{99\%,9} = 54.36$$

As $54.36 \le F_{sc} = 88.90 \le 113.51$, the random numbers do pass the test of serial correlation with 99% confidence level.

11.7 Monte Carlo Simulation of Initiator Dissociation

In this section, we apply the Monte Carlo simulation approach to a relatively simple decomposition of azobisisobutyronitrile (AIBN) thermal initiator at 70 °C as follows [68, 69]:

$$(H_3C)_2\overset{CN}{C}{-}N{=}N{-}\overset{CN}{C}(CH_3)_2 \xrightarrow{k_d \Delta T} (H_3C)_2\overset{CN}{C^{\bullet}} + (H_3C)_2\overset{CN}{C^{\bullet}} + N_2\uparrow \tag{11.46}$$

The dissociation of AIBN generates two 2-cyanoprop-2-yl radicals and a nitrogen molecule. The reaction is written symbolically as follows:

$$I \xrightarrow{k_d} {}^2R + N_2\uparrow \tag{11.47}$$

where I is initiator and R is the reactive radical. At the given temperature, the initiator half-life, $t_{1/2}$, is 4.8 h [70]. The following differential equation describes the dissociation of initiator (see Eq. 5.28B in Section 5.7.1 in Chapter 5):

$$\frac{d[I]}{dt} = -k_d[I] \tag{11.48}$$

where $[I]$ is the initiator concentration, $k_d = \frac{\ln(2)}{t_{1/2}}$ is the dissociation rate constant, and t is time. The above equation is subject to the initial condition $[I] = [I]_0$ at $t = 0$. The above equation can be integrated to yield an analytical expression for initiator decomposition with time. However, we demonstrate how the decomposition is simulated by the Monte Carlo method with the knowledge of the dissociation rate constant k_d. For a small interval of time Δt, as compared to the initiator half-life, the above differential equation may be written in the following difference form:

$$\Delta[I] = -k_d[I]\Delta t = -\left[\frac{\ln(2)}{t_{1/2}}\right][I]\Delta t \tag{11.49}$$

An event in the simulation consists of dissociation or non-dissociation. At any Δt, one of the two events occurs randomly with probability p_{dis} for dissociation and p_{nodis} for non-dissociation. The probability of dissociation p_{dis} is related to the rate constant for dissociation or the initiator half-life as follows:

$$p_{dis} = \left|\frac{\Delta[I]}{[I]}\right| = \left[\frac{\ln(2)}{t_{1/2}}\right]\Delta t = k_d\Delta t \tag{11.50}$$

As the sum of the two probabilities should be equal to one, the probability of non-dissociation $p_{nodis} = 1 - p_{dis}$. For a small $\Delta t = 10$ s and $t_{1/2} = 4.8\text{ h} = 1.73 \times 10^4$ s, $p_{dis} = 4.01 \times 10^{-4}$ and $p_{nodis} = 0.9996$. As the process of dissociation at the molecular level is a random process, one of the two events is randomly selected for initiator molecule i in time interval j using a random number x_{ij} from a sample of uniformly distributed random numbers based on the following criteria:

$$\begin{aligned} &0 \le x_{ij} \le 4.01 \times 10^{-5}\Delta t \;\Rightarrow\; \text{dissociation} \\ &4.01 \times 10^{-5}\Delta t < x_{ij} \le 1 \;\Rightarrow\; \text{non-dissociation} \end{aligned} \tag{11.51}$$

The Monte Carlo approach requires the size of the simulation box and the initial number of initiator molecules within the box, which depend on the initial initiator molar concentration $[I]_0$. The initial number of initiator molecules should represent the random nature of the reaction with countably infinite number of initiator molecules. If the initial molar concentration is $[I]_0$ and the initial number of initiator molecules is chosen to be $N_{I,0}$, then the simulation volume is fixed by the following relation:

$$V_{sim} = \frac{(N_{I,0} \times 10^{24})}{(N_{AV}\,[I]_0)} \tag{11.52}$$

Where N_{AV} is the Avogadro number. A typical thermal initiator concentration is $[I]_0 = 0.001$ M. If $N_{I,0} = 10$ and $[I]_0 = 0.001$ M, then the simulated volume is 1.66×10^4 nm^3 ($25 \times 25 \times 25$ nm^3); if $N_{I,0} = 100$, the simulated volume increases to 1.66×10^5 nm^3 ($55 \times 55 \times 55$ nm^3); and if $N_{I,0} = 1000$, the volume increases to 1.66×10^6 ($120 \times 120 \times 120$ nm^3). Knowing $[I]_0$ and k_d, and choosing a Δt, p_{dis} is calculated using Eq. (11.50). Next, an $N_{I,0}$ is chosen followed by calculation of V_{sim} using Eq. (11.52). All $N_{I,0}$ molecules are initially added to the list of non-dissociated initiator molecules. Then, a random number with a uniform distribution from zero to one is generated using a random number generator and one initiator molecule in the simulation volume is tested for dissociation based on the criterion given in Eq. (11.51). If the initiator is selected to dissociate, then the initiator molecule is removed from the non-dissociated list and added to the dissociated list. This process is repeated in the ensuing time intervals with the non-dissociated molecules until all molecules are dissociated. The simulation of initiator dissociation may be done in the MATLAB software program environment. Figure 11.1a, b shows two Monte Carlo simulations of the initiator dissociation both with $N_{I,0} = 10$ and $\Delta t = 10$s. The two profiles are discontinuous and different because the simulation volume of $25 \times 25 \times 25$ nm^3 corresponding to $N_{I,0} = 10$ is too small and not representative of the random nature of the reaction with large number of initiator molecules.

Figure 11.2a, b shows two Monte Carlo simulations of the initiator dissociation ($\Delta t = 10$ s) with $N_{I,0} = 100$ and $N_{I,0} = 1000$, respectively. With increasing $N_{I,0}$, the dissociation profiles become continuous and exponentially decreasing with time. The simulation with $N_{I,0} = 1000$ is indistinguishable from the theoretical, exponentially decreasing, dissociation curve of AIBN with $t_{1/2,I} = 4.8$ h, indicating that a simulation box of $120 \times 120 \times 120$ nm^3 with $N_{I,0} = 1000$ is representative of the real system.

11.8 Monte Carlo Simulation of Methyl Methacrylate Polymerization

In this section, we apply the Monte Carlo simulation approach to bulk polymerization of methyl methacrylate (MMA) to predict the concentration and molecular weight of the produced polymer [71–74]. The assumptions for the reaction system are a) the quasi-steady state assumption (QSSA) is valid for primary radicals as well as the propagating chains, b) polymerization reaction is not diffusion controlled, c) the rate constants for propagation and termination reactions are independent of the degree of polymerization [75], and d) propagating chains terminate by disproportionation and combination. Termination in MMA polymerization is dominated by disproportionation due to steric hinderance of bulky methacrylate group and to a lesser extent by combination [71, 76–78]. The elementary reactions for the polymerization are as follows [71–74]:

$$I \xrightarrow{k_d} 2R \tag{11.52}$$

$$R + M \xrightarrow{k_p} RP_1 \tag{11.53}$$

$$RP_n + M \xrightarrow{k_p} RP_{n+1} \tag{11.54}$$

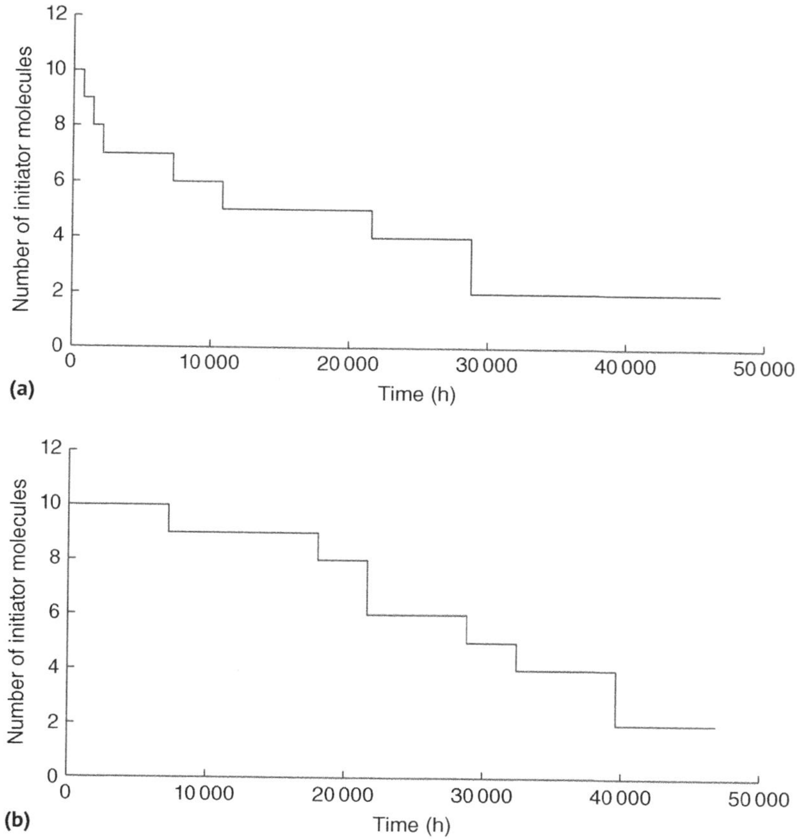

Figure 11.1 Two Monte Carlo simulations (A and B) of the initiator dissociation both with $N_{I,0} = 10$ and $\Delta t = 10$ s for AIBN with $t_{1/2} = 4.8$ h.

$$RP_n + RP_m \xrightarrow{k_{tc}} TP_{n+m} \tag{11.55}$$

$$RP_n + RP_m \xrightarrow{k_{td}} TP_n + TP_m \tag{11.56}$$

In the first reaction above, initiator dissociates, as described in the previous section, with a rate constant k_d to generate two equally reactive primary radicals; in the second reaction, a primary radical reacts with a monomer with a rate constant k_p to generate a 1-mer reactive propagating chain; in the third reaction, an n-mer propagating chain reacts with a monomer with a rate constant k_p to generate a longer $(n + 1)$-mer chain; and in the last two reactions, two n-mer and m-mer propagating chains terminate by combination or disproportionation with rate constants k_{tc} or k_{td} to form one or two terminated polymer chains, respectively. Based on MMA molecular weight and density (see Eq. 8.28 in Chapter 8), the initial molar concentration of MMA for bulk polymerization is $[M]_0 = 8.9$ M. A typical molar ratio of initiator to monomer in MMA polymerization is 1:1000, resulting in initiator concentration of $[I]_0 = 8.9 \times 10^{-3}$ M. The rate constants used for simulating the reaction are given in Table 11.7 from reference [79]:

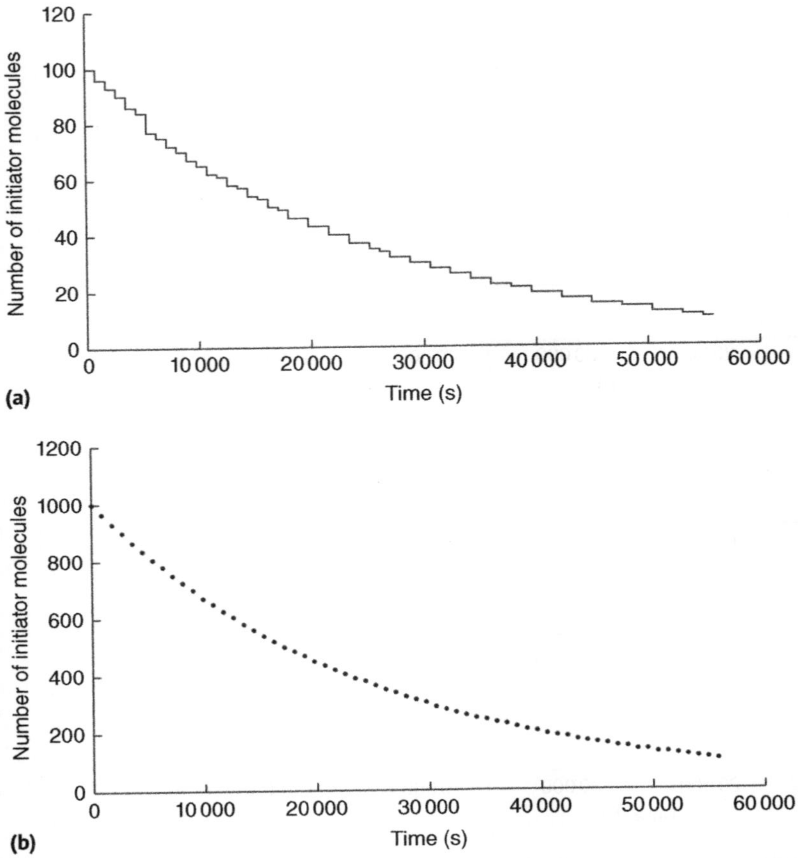

Figure 11.2 The Monte Carlo simulations of AIBN dissociation with (a) $N_{I,0} = 100$ and (b) $N_{I,0} = 1000$.

Table 11.7 Rate constants for radical polymerization of MMA.

Rate constant	Value
k_d	$1 \times 10^{-7}\ \text{s}^{-1}$
k_p	$200\ \text{L mol}^{-1}\ \text{s}^{-1}$
k_{tc}	$1 \times 10^{6}\ \text{L mol}^{-1}\ \text{s}^{-1}$
k_{td}	$9 \times 10^{6}\ \text{L mol}^{-1}\ \text{s}^{-1}$

The dissociation of the initiator is simulated as described in the previous section. Following initiation, in a time interval Δt, the generated radical species undergo one of the following events: propagation, termination by disproportionation, or termination by combination with rates R_p, R_{tc}, and R_{td}, respectively, defined by the following equations:

$$\begin{aligned} R_p &= k_p\,[RP]\,[M] \\ R_{tc} &= k_{tc}\,[RP]^2 \\ R_{td} &= k_{td}\,[RP]^2 \end{aligned} \tag{11.57}$$

where $[RP] = \sum_{n=1}^{\infty} [RP]_n$ is the total concentration of n-mer propagating chains with $1 \leq n < \infty$. As the sum of probabilities should be equal to one, the probability of propagation, the probability of termination by disproportionation, and the probability of termination by combination as the events in the simulation are given by the following relations:

$$P_p = \frac{R_p}{R_p + R_{tc} + R_{td}} = \frac{k_p [M]}{k_p [M] + k_{tc} [RP] + k_{td} [RP]} \quad (11.58)$$

$$P_{tc} = \frac{R_{tc}}{R_p + R_{tc} + R_{td}} = \frac{k_{tc} [RP]}{k_p [M] + k_{tc} [RP] + k_{td} [RP]} \quad (11.59)$$

$$P_{td} = \frac{R_{td}}{R_p + R_{tc} + R_{td}} = \frac{k_{td} [RP]}{k_p [M] + k_{tc} [RP] + k_{td} [RP]} \quad (11.60)$$

The initial initiator concentration is used to determine the simulation volume, V_{sim}, by the ratio of initial moles of initiator to its initial molar concentration as defined by Eq. (11.52) in the previous section (Section 11.7). Based on the simulation results in Section 11.7, the simulated dissociation profile with $N_{I,0} = 1000$ represents the random, stochastic nature of the initiator dissociation at the molecular scale. For $N_{I,0} = 1000$ and $[I]_0 = 8.9 \times 10^{-3}$M, Eq. (11.52) predicts a $V_{sim} = 6.77 \times 10^6$ nm^3 corresponding to a $190 \times 190 \times 190$ nm simulation box; for $N_{I,0} = 1 \times 10^4$ and the same initial initiator concentration, the simulation volume is $V_{sim} = 6.77 \times 10^7$nm^{3}corresponding to a $407 \times 407 \times 407$ nm box. As the polymerization reaction is simulated with larger simulation volumes with higher number of propagating chains, the statistical validity of the simulation results improves. The initial number of monomers, $N_{M,0}$, is determined using the monomer molar concentration, $[M]_0$, as follows:

$$N_{M,0} = \left([M]_0 \times 10^{-24}\right) N_{AV} V_{sim} \quad (11.61)$$

where the unit of V_{sim} in the above equation is nm^3. For $N_{I,0} = 1000$, $[M]_0 = 8.9$ M, then $V_{sim} = 6.77 \times 10^6$ nm^3 using Eq. (11.52), and $N_{M,0} = 3.63 \times 10^7$ based on Eq. (11.61); for $N_{I,0} = 1 \times 10^4$, $[M]_0 = 8.9\ M$, $V_{sim} = 6.77 \times 10^7$nm^3, and $N_{M,0} = 3.63 \times 10^8$. After determination of V_{sim} and $N_{M,0}$ with the knowledge of $N_{I,0}$, the first time interval Δt_0 is selected to determine the initial number of generated reactive species, $N_{R,0}$, by Monte Carlo simulation as described in the previous Section 11.7. As QSSA is valid in this polymerization, the radicals do not accumulate in the reaction volume and all generated radicals will terminate in the same time interval, thus the number of propagating radical chains, $N_{PR,0} = N_{R,0}$. Upon determination of $N_{R,0}$ by simulation, the concentration of propagating chains in the time interval Δt_0, $[RP]_0$, is calculated using the following equation:

$$[RP]_0 = \frac{N_{RP,0} \times 10^{24}}{V_{sim} N_{AV}} = \frac{N_{R,0} \times 10^{24}}{V_{sim} N_{AV}} \quad (11.62)$$

Next, using the specified $[I]_0$ and $[M]_0$, $[RP]_0$ calculated from the above equation, and rate constants given in Table 11.7, the probability for propagation, the probability for termination by combination, and the probability for termination by disproportionation as events in the simulation in the time interval Δt_0, that is P_p, P_{tc}, and P_{td}, are calculated using Eqs. (11.58)–(11.60), respectively. Then, for a given time interval $j = \Delta t_0$ and a selected random number $x_{rn,ij}$ uniformly distributed between zero and one, a propagating chain i with k repeat units in the simulation volume is tested for undergoing one of the three possible events (propagation, combination, disproportionation) based on the following criteria:

$$\begin{aligned} 0 \le x_{rn,ij} \le P_p &\Rightarrow \text{Propagation} \\ P_p < x_{rn,ij} \le P_p + P_{tc} &\Rightarrow \text{Combination} \\ P_p + P_{tc} < x_{rn,ij} \le P_p + P_{tc} + P_{td} &\Rightarrow \text{Disproportionation} \end{aligned} \tag{11.63}$$

Based on the above criteria, if $0 \le x_{rn,ij} \le P_p$, chain propagation is selected, the number of repeat units on the propagating chain i is increased by one from k to $k + 1$, and the chain is retained in the list of active chains; if $P_p < x_{rn,ij} \le P_p + P_{tc}$, termination by combination is selected, a second propagating chain with l repeat units is selected randomly from the list of active chains in the simulation box, both propagating chains with k and l repeat units are removed from the list of active chains, their number of repeat units are added to form a chain with $k + l$ repeat units, and the new chain is added to the list of terminated polymer chains; if $P_p + P_{tc} < x_{rn,ij} \le P_p + P_{tc} + P_{td}$, termination by disproportionation is selected, a second propagating chain with l repeat units is selected randomly, both propagating chains are removed from the list of active chains, and the two selected chains with k and l repeat units are added to the list of terminated polymer chains. This process is continued for all chains in the list of propagating chains as well as all the time intervals. Figure 11.3a, b shows the simulated concentrations of monomer and polymer, respectively, as a function of reaction time using a simulation volume $V_{sim} = 6.77 \times 10^9$ nm^3 with $N_{I,0} = 1 \times 10^6$.

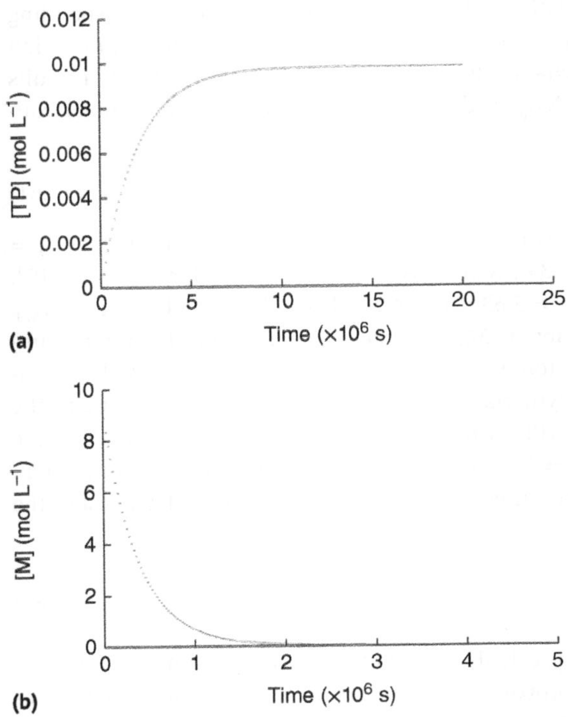

Figure 11.3 Monte Carlo simulation of (a) terminated polymer and (b) monomer concentrations for MMA polymerization with a simulation volume $V_{sim} = 6.77 \times 10^9$ nm^3 with $N_{I,0} = 1 \times 10^6$.

The differential equations describing monomer and polymer concentrations with time for the above MMA polymerization may be integrated to obtain a closed-form analytical solution using conservation of moments equations, as described in Chapters 7 and 8. The analytic solutions to monomer and polymer concentrations with time for MMA polymerization are given by the following expressions:

$$\ln\left(\frac{[M]}{[M]_0}\right) = \left[\frac{8\,k_p^2\,[I]_0}{k_d\,(k_{tc} + k_{td})}\right]^{1/2} \left(e^{-\frac{k_d t}{2}} - 1\right) \tag{11.64}$$

$$[TP] = 2\,[I]_0 \left[\frac{k_{td} + \left(\frac{k_{tc}}{2}\right)}{k_{td} + k_{tc}}\right] \left(1 - e^{-\frac{k_d t}{2}}\right) \tag{11.65}$$

The simulated concentrations closely match the theoretical concentrations. Based on simulation and analytic results, the monomer is consumed early in the reaction to form long propagating chains, thus producing long terminated polymer chains. The concentration profiles in Figure 11.3 show that polymerization should be stopped after 2×10^6 s when most of the monomer is consumed. As monomer concentration sharply decreases, polymerization rate also decreases beyond the time of 2×10^6 s which lowers the rate of polymer production. The change in DP of polymer chains with time depends on the change in the ratio of monomer to initiator concentrations with reaction time. The simulation of MMA polymerization in this section demonstrates that the Monte Carlo method is especially useful for polymerization reactions with coupled, nonlinear differential equations where there is no closed-form analytic solution to the concentration and DP of the produced polymer chains with reaction time.

11.9 Monte Carlo Simulation of Branching in Butadiene Emulsion Polymerization

Emulsion radical polymerization is used commercially to produce high molecular weight polymers [80, 81]. In emulsion polymerization, the reaction occurs within submicron micellar phase dispersed in the continuous aqueous phase with on average only one propagating chain in each micelle. As the butadiene monomer is sparingly soluble in the continuous aqueous phase, it is stored in the monomer-filled microdroplet phase during the reaction. The monomer is continuously transferred from the microdroplet phase to the micellar phase as the reaction proceeds. As termination in radical polymerization requires two propagating chains for disproportionation or combination, termination rates in emulsion polymerization are slow as compared to propagation, resulting in the production of a polymer with relatively high DP. For example, in tire industry, polybutadiene (PBD) rubber with high DP is desirable to minimize the coefficient of friction [82–86]. As DP of the rubber is increased, the density of PBD chain-ends decreases, which in turn reduces the coefficient of friction and wear rate of the rubber.

Many attempts have been made in the past few decades to model and simulate molecular weight and distribution, the extent of branching and crosslinking in emulsion polymerization. The method of moments and the finite difference method have been used to model radical emulsion polymerization to study the effects of radical entry, radical desorption, and transfer from micelles, and chain transfer to monomer on DP distribution of the produced polymer assuming no more than one reactive species per propagating chain [87–92]. Branching is also possible in emulsion polymerization, but branched chains are generated from propagating chains with multiple reactive species. The Monte Carlo method is better suited to handle polymerization reactions involving polyradical propagating chains [14, 16, 28, 93–99]. In this section, the Monte Carlo method is used to simulate branching in emulsion polymerization of butadiene (BD) by including propagating chains with multiple radicals in the reaction.

There are two types of branching in BD polymerization, namely tri-functional branching and tetra-functional crosslinking [28, 100–106]. The transfer of a radical from a propagating chain to a C—H bond in PBD results in a trifunctional branch, whereas the transfer of a radical from a propagating chain to a pendant C═C bond along the PDB chain results in a tetrafunctional crosslink. Theoretical models do not adequately predict branching in emulsion polymerization of dienes like BD with high degree of branching and interchain crosslinking. This section uses Monte Carlo simulation to predict the effect of trifunctional branching and tetrafunctional crosslinking on DP distribution and microgel formation in emulsion polymerization of BD.

11.9.1 Elementary Reactions

The elementary reactions for radical polymerization of butadiene are as follows [28]:

$$I \xrightarrow{f=1,k_d} 2R \tag{11.66}$$

$$PR + M \xrightarrow{k_{ci}} RP_1 \tag{11.67}$$

$$RP_n + M \xrightarrow{k_p} RP_{n+1} \tag{11.68}$$

$$RP_n + RP_m \xrightarrow{k_{td}} TP_n + TP_m \tag{11.69}$$

$$RP_n + M \xrightarrow{k_{ctm}} TP_n + RP_1 \tag{11.70}$$

$$RP_n + CTA \xrightarrow{K_{cta}} TP_n + CTA* \tag{11.71}$$

$$TP_n + RP_m \xrightarrow{k_{ctpha}} RP_n + TP_m + BR \tag{11.72}$$

$$TP_n + RP_m \xrightarrow{K_{ctpdb}} RP_{n+m} + CL \tag{11.73}$$

In the first reaction above, an initiator molecule I dissociates with efficiency $f = 1$ and rate constant k_d to form a reactive species R. In the second reaction, the reactive species R reacts with a BD monomer M with rate constant k_{ci} to initiate a 1-mer propagating chain RP_1. In the third reaction, an n-mer propagating chain RP_n propagates by reacting with a monomer with rate constant k_p to form an (n+1)-mer propagating chain RP_{n+1}. In the fourth reaction, two propagating chains RP_n and RP_m with n and m repeat units, respectively, undergo termination by disproportionation with rate constant k_{td} to form two terminated polymer chains TP_n and TP_m. Termination in BD polymerization is predominantly by disproportionation [102, 107, 108]. In the fifth reaction, an n-mer propagating chain RP_n transfers its radical to a monomer with rate constant k_{ctm} to form an n-mer terminated polymer chain TP_n and a 1-mer propagating chain RP_1. In the sixth reaction, an n-mer propagating chain RP_n reacts with the chain transfer agent CTA with rate constant k_{cta} to form an n-mer terminated polymer chain TP_n and an inactive agent $TA*$. Due to slow termination in emulsion polymerization, transfer agents are used to control the degree of polymerization [109]. In the seventh reaction, an m-mer propagating chain RP_m abstracts a hydrogen atom from one of the carbons on an n-mer terminated polymer chain TP_n with rate constant K_{ctpha} to form an m-mer terminated polymer chain TP_m and an n-mer propagating chain RP_n with a trifunctional branch BR. In the final reaction above, an m-mer propagating chain RP_m reacts with a pendant C=C or a main-chain C=C of an n-mer terminated polymer chain TP_n to form a new longer propagating chain RP_{n+m} with $(n + m)$ repeat units and a tetrafunctional crosslink CL. The Rxns. 11.66 and 11.67 take place predominantly in the aqueous continuous phase of the emulsion whereas Rxns. 11.68–11.73 occur primarily in the growing micellar phase.

11.9.2 Reaction Conditions and Rate Constants

The conditions and rate constants for emulsion polymerization of BD are given in Table 11.8 [28].

In the above table, T is reaction temperature, ρ_{BD} and ρ_{PBD} are the densities of BD and PBD, respectively, $\frac{[M]}{[I]}$ is the molar ratio of monomer to initiator, GM is the number of simulated growing micelles, $\frac{I}{GM}$ is the number ratio of initiator molecules to growing micelles, V_{sim} is the Monte Carlo simulation volume, and $t_{1/2,I}$ is the initiator half-life. The k_p value for BD polymerization at 25 °C was calculated using Arrhenius equation using $k_p = 100$ L mol^{-1} s^{-1}@60 °C and activation energy of 9.3 kcal mol^{-1}, and activation constant of 1.2×10^8 L mol^{-1} s^{-1} from reference [110]. The calculated k_p value was consistent with that reported at 60 °C by Odian [111] and the rate constant $k_p = 8.4$ L mol^{-1} s^{-1}@10 °C reported by Polymer Handbook [112]. The radical-bearing carbon in BD may be in two active states. In one state, the carbon is attached to a hydrogen (—CH) and a vinyl group (—HC=CH_2), which bears a resemblance to polystyrene. In the other state, the carbon is attached to two hydrogens (—CH_2) which resemble polyethylene. The BD radical is less hindered sterically compared to polystyrene but more hindered compared to polyethylene. Furthermore, the BD radical is relatively more stable than polyethylene radical but less stable than the polystyrene. Therefore, based on steric hindrance, k_{td} of the BD radical is approximated using the geometric mean of k_{td} values of polystyrene and polyethylene radicals. The reported values of k_{td} for polystyrene and polyethylene are 6.0×10^7 and 5.4×10^8 L mol^{-1} s^{-1} at 60 °C, respectively, with activation energies of 8.0 and 1.3 kJ mol^{-1} [111]. Using the Arrhenius equation,

Table 11.8 Reaction conditions and rate constants for radical emulsion polymerization of butadiene.

Conditions	Value	Rate constant	Value
T	25 °C	$t_{1/2,I}$	1800 s
ρ_{BD}	0.62 g cm^{-3} @25 °C	k_p	18 L mol^{-1} s^{-1}
ρ_{PBD}	0.96 g cm^{-3} @25 °C	k_{td}	1.5×10^8 L mol^{-1} s^{-1}
$\frac{[M]}{[I]}$	500	k_{ctm}	3.1×10^{-3} L mol^{-1} s^{-1}
GM #	400	k_{cta}	0
$\frac{I}{GM}$	2.5	k_{ctpha}	2.0×10^{-2} L mol^{-1} s^{-1}
V_{sim}	10^5 nm^3	k_{ctpdb}	1.8×10^1 L mol^{-1} s^{-1}

k_{td} values of 4.3×10^7 and 5.1×10^8 L mol^{-1} s^{-1} are calculated at 25 °C for polystyrene and polyethylene, respectively, with a geometric mean of 1.5×10^8 L mol^{-1} s^{-1}. Hence, k_{td} value of 1.5×10^8 L mol^{-1} s^{-1} is used for PBD in the simulation.

As the chemical structure of butadiene is close to that of 1-butene, k_{tm} of BD was approximated by that of 1-butene. The $C_m = k_{tm}/k_p$ values of 3.1×10^{-4}, 5.1×10^{-4}, and 7.3×10^{-4} at temperatures of 40, 50, and 60 °C, respectively, are reported for 1-butene in the Polymer Handbook [112]. Using Arrhenius equation, k_p values of 40, 60, and 100 L mol^{-1} s^{-1} are reported by Flory for BD polymerization at 40, 50, and 60 °C [110]. Using the relation $C_m = \frac{k_{tm}}{k_p}$, k_{tm} values of 1.24×10^{-2}, 3.06×10^{-2}, and 7.30×10^{-2} L mol^{-1} s^{-1} are calculated for BD polymerization at 40, 50, and 60 °C, respectively. Hence, using Arrhenius equation, the estimated k_{tm} for BD polymerization is 3.1×10^{-3} L mol^{-1} s^{-1} and this value is used in the simulation. The simulation of BD emulsion polymerization is done without a chain transfer agent, that is $k_{cta} = 0$. The $C_{ctp} = \frac{k_{ctp}}{k_p}$ value of 1.1×10^{-3} is reported for chain transfer constant to polymer in BD polymerization at 50 °C [110, 112]. Assuming the above C_{ctp} accounts only for chain transfer to C-H groups of PBD, then $C_{ctpha} = \frac{k_{ctpha}}{k_p} = 1.1 \times 10^{-3}$. In the absence of values at other temperatures, the C_{ctpha} at 50 °C is used for PBD simulation at 25 °C. Using k_p of 18 L mol^{-1} s^{-1} for BD polymerization at 25 °C, the $k_{ctpha} = 2.0 \times 10^{-2}$L mol^{-1} s^{-1} is calculated for BD polymerization at 25 °C and this value is used in the simulation. For estimation of k_{ctpdb}, the reactivity of PBD double bonds is assumed to be controlled kinetically, not by diffusion. Based on this assumption, $k_{ctpdb} = k_p$. Hence, $k_{ctpdb} = 18$ L mol^{-1}s^{-1} is used in PDB simulation at 25 °C. The densities of BD and PBD at 25 °C were obtained from the Encyclopedia of Polymer Science and Technology. The number of growing micelles in the simulation was set at 400, and the ratio of the number of initiator molecules to growing micelles was 2.5, which corresponds to a simulation volume of 1×10^5 nm^3. As the number of growing micelles in the simulation was increased to values >400, no significant change in DP distribution of simulated polymer chains was observed. Therefore, the number of growing micelles in the simulation was set constant at 400.

11.9.3 Simulation Assumptions

The simulation assumptions for emulsion polymerization of BD are (a) as BD monomer is sparingly soluble in water, BD and water form an emulsion in the presence of a surfactant with water as the continuous phase and BD-filled microdroplets and micelles as the dispersed phases; (b) the reactive species formed in the aqueous phase by dissociation of initiator react with a BD monomer in the aqueous phase to form the 1-mer propagating radical RP_1. Since the solubility of BD in the aqueous phase is low (180 ppm at 25 °C), the produced RP_1 radicals become insoluble in the aqueous phase and they are taken up by the micelles; hence chain propagation takes place exclusively in the growing micelles; (c) as the micelles with a nanometer size are orders of magnitude smaller than the microdroplets, the surface area of the micelles is orders of magnitude larger than that of microdroplets which means that RP_1 radicals are predominantly taken up by the micelles and propagate within micelles; hence, polymerization within microdroplets may be neglected; (d) the micelles are formed initially after mixing BD monomer in the aqueous phase and the growth of propagating chains only affects the size distribution of micelles not the number of micelles; hence, it is assumed that the number of micelles are constant during simulation; (e) as solubility of the propagating chains in the aqueous phase is extremely low, their desorption from the growing micelles to the aqueous phase is assumed to be negligible; (f) the growing micelles coalesce but their fusion only affects the size distribution of micelles, not the microstructure of the propagating chains in the micelles; (g) there is no phase separation in micelles with chain propagation or the micelles are homogeneous; hence, Flory–Huggins equation is assumed to be valid for determination of equilibrium BD concentration in micelles; (h) the effect of free energy of the micelle surface on equilibrium concentration of molecules in the growing micelles is assumed to be negligible which is a good assumption for micelles > 50 nm in size; (i) the reactions are not diffusion controlled and the rate constants for propagation, chain transfer, termination, branching and crosslinking are independent of time or conversion; this assumption is reasonable considering that the glass transition temperature (T_g) of PBD at <−90 °C is well below the reaction temperature of 25 °C [113]; (j) butadiene microdroplets, which serve as reservoirs of BD monomer for the growing micelles, are present throughout the reaction; hence the growing micelles are saturated with monomer throughout the reaction; (k) the reaction is performed in a batch reactor but the reaction volume may change with conversion; (l) as $T_{g,PBD} < -90$ °C, all C—H and C═C bonds in PBD are equally reactive for trifunctional branching and tetrafunctional crosslinking, respectively; hence, steric hinderance and excluded volume effects on reactivity of C—H and C═C bonds of PBD are assumed to be negligible; and finally m) cyclization in C—H trifunctional branching and C═C tetrafunctional crosslinking and their dependence on PBD chain length are not considered in this simulation. The above assumptions allow us to focus on the role of trifunctional branching and tetrafunctional crosslinking on DP distribution of PBD without undue interference of other reaction conditions.

11.9.4 Simulation Method

The flowchart for Monte Carlo simulation of emulsion polymerization of BD at 25 °C is shown in Figure 11.4. The inputs to the simulation are the number of micelles and the number of initiator and monomer molecules, initiator half-life, and the rate constants for the elementary reactions (see Table 11.8). The other inputs are random numbers, generated by a random number generator algorithm, that pass the tests of uniformity and interval, and serial correlation.

11.9.4.1 Aqueous Phase Reactions

To start the simulation, initiator molecules in the aqueous phase are tested stochastically for dissociation using random numbers. A time interval Δt_i much shorter than the initiator half-life is selected, and the probability of initiator dissociation, $p_{dis,i}$, is calculated using the following equation (see also Eq. 11.50 in Section 11.7):

$$p_{dis,i} = \left|\frac{\Delta [I]}{[I]}\right| = \left(\frac{ln\ 2}{t_{I,1/2}}\right)\Delta t_i = k_d \Delta t_i \tag{11.74}$$

A number $x_{rn,j}$ between zero and one is randomly selected by a random number generator. If $0 \le x_{rn,j} \le p_{dis,i}$, the jth initiator molecule is dissociated, the initiator is removed from the list of initiators, and two radicals are added to the list of reactive species. If $p_{dis,i} < x_{rn,j} \le 1 - p_{dis,i}$, the jth initiator remains in the list of initiator molecules in the time interval Δt_i. This process continues until all initiator molecules are tested for dissociation. After initiation, all generated radical species react spontaneously with a monomer in the aqueous phase to generate 1-mer propagating chains, RP_1. The chain initiation reaction is assumed to be spontaneous, as opposed to random, as the initiation efficiency $f = 1$ (see Rxn. 11.66). As BD monomer is hydrophobic, RP_1 is less soluble than the primary radical in the aqueous phase, thus RP_1 reactive species are randomly taken up by the micelles to continue propagating within the micelles.

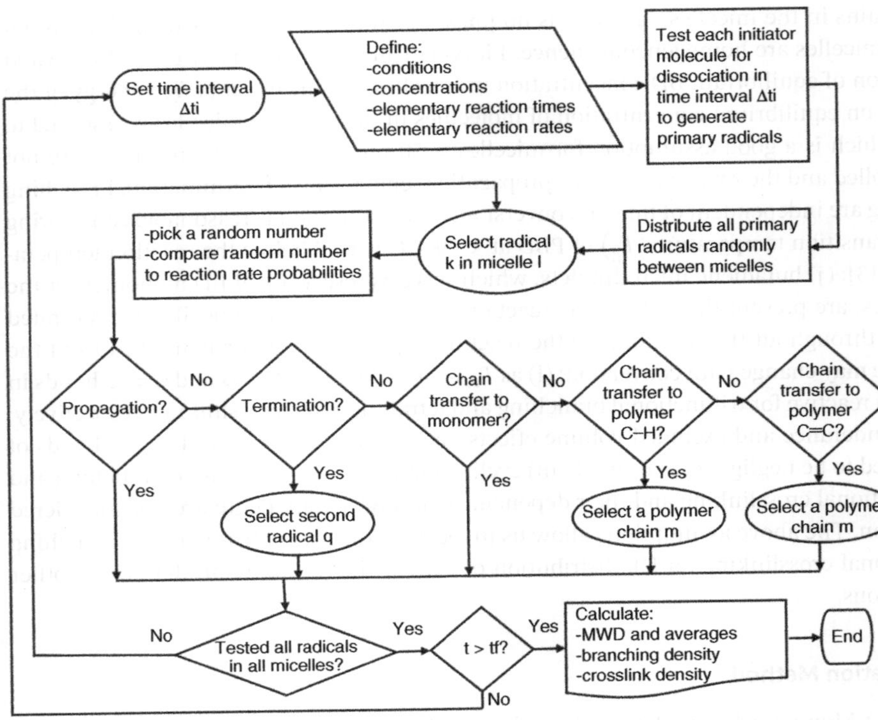

Figure 11.4 The algorithm for simulation of emulsion radical polymerization of BD by the Monte Carlo method (redrawn from Ref. [28]).

The RP_1 initiating chains generated in the time interval Δt_i are randomly assigned to one of the growing micelles using the following approach. The interval between zero and one is divided into 400 subintervals to represent the 400 growing micelles, and each subinterval k is assigned to the lth micelle. A number x_{rn} between zero and one is randomly selected by a random number generator for assignment of the kth RP_1 to one of the micelles. If x_{rn} falls within the values in the subinterval k, the kth RP_1 is assigned to the lth micelle. This process is repeated until all RP_1 initiating chains generated in time interval Δt_i are assigned to one of the 400 micelles. As micelles are equilibrated with BD, propagation of RP_1 chains continue in their assigned growing micelle.

11.9.4.2 Micelle Phase Reactions

After assigning the kth initiating chain RP_1 to the lth micelle, polymerization of the kth RP_1 proceeds randomly through one of the following elementary reactions: propagation, transfer to monomer, transfer to CTA, transfer to a PBD C—H bond, transfer to a PBD C=C bond, or termination by disproportionation. The rates of the above elementary reactions are as follows:

$$\begin{aligned}
R_{p,gm,lk} &= k_p\left[RS_{gm,lk}\right]\left[M_{gm,l}\right]\\
R_{ctm,gm,lk} &= k_{ctm}\left[RS_{gm,lk}\right]\left[M_{gm,l}\right]\\
R_{cta,gm,lk} &= k_{cta}\left[RS_{gm,lk}\right]\left[CTA_{gm,l}\right]\\
R_{ctpha,gm,lk} &= k_{ctpha}\left[RS_{gm,lk}\right]\left[MTP_{gm,lq}\right]\\
R_{ctpdb,gm,lk} &= k_{ctpdb}\left[RS_{gm,lk}\right]\left[MTP_{gm,lq}\right]\\
R_{td,gm,lk} &= k_{td}\left[RS_{gm,lk}\right]^2
\end{aligned} \tag{11.75}$$

where $R_{p,gm,lk}$, $R_{ctm,gm,lk}$, $R_{cta,gm,lk}$, $R_{ctpha,gm,lk}$, $R_{ctpdb,gm,lk}$, and $R_{td,gm,lk}$ are the rates of propagation, transfer to monomer, transfer to CTA, transfer to PBD C—H bond (branching), transfer to PBD C=C bond (crosslinking), and termination by disproportionation for the kth propagating chain in the lth micelle, respectively; $[RS_{gm,lk}]$ is the concentration of reactive species on the kth propagating chain in the lth micelle; $[M_{gm,l}]$ and $[CTA_{gm,l}]$ are the monomer and CTA concentrations in the lth micelle, respectively; and $[MTP_{gm,lq}]$ is the concentration of polymerized monomer in the qth chain selected randomly for transfer to polymer in the lth micelle. It should be noted that the monomer concentration in the micelles is dictated by equilibrium with propagating chains and terminated polymer chains, which is different from those in microdroplets and the aqueous solution. Furthermore, $[RS_{gm,lk}]$ may not be the same as $[RP_{gm,lk}]$ if the kth propagating chain in the lth micelle is a polyradical. If the kth propagating chain in the l^{th}micelle is a polyradical, then $[RS_{gm,lk}] > [RP_{gm,lk}]$. The $[RS_{gm,lk}]$ and $[M_{gm,l}]$ are defined by the following equations:

$$
\begin{aligned}
[RS_{gm,lk}] &= \frac{n_{RS,gm,lk}}{N_{AV}} V_{gm,l} \\
[M_{gm,l}] &= \frac{n_{M,gm,l}}{N_{AV}} V_{gm,l} \\
[CTA_{gm,l}] &= \frac{n_{CTA,gm,l}}{N_{AV}} V_{gm,l} \\
[MTP_{gm,lq}] &= \frac{n_{MTP,gm,lq}}{N_{AV}} V_{gm,l}
\end{aligned}
\tag{11.76}
$$

where $n_{RS,gm,lk}$ is the number of reactive species of the kth propagating chain in the lth micelle; $n_{M,gm,l}$ and $n_{CTA,gm,l}$ are the number of monomers and transfer agents in the lth micelle, respectively; $n_{MTP,gm,lq}$ is the number of monomers on the qth terminated polymer chain; N_{AV} is the Avogadro's number, and $V_{gm,l}$ is the volume of the lth micelle. An event is defined as one of the following elementary reactions occurring to one of the reactive species on the kth propagating chain in the lth micelle: propagation, transfer to monomer, transfer to CTA, transfer to PBD C—H bond, transfer to PBD C=C bond, and termination by disproportionation with probabilities $P_{p,gm,lk}$, $P_{ctm,gm,lk}$, $P_{cta,gm,lk}$, $P_{ctpha,gm,lk}$, $P_{ctpdb,gm,lk}$, and $P_{td,gm,lk}$, respectively, where the sum of all probabilities is equal to one, that is $\sum_{all\ i} p_{i,gm,lk} = 1$. These probabilities are related to the rates of elementary reactions, given by Eqs. (11.75), as follows:

$$
\begin{aligned}
P_{p,gm,lk} &= \frac{R_{p,gm,lk}}{R_{T,gm,lk}} \\
P_{ctm,gm,lk} &= \frac{R_{tm,gm,lk}}{R_{T,gm,lk}} \\
P_{cta,gm,lk} &= \frac{R_{cta,gm,lk}}{R_{T,gm,lk}} \\
P_{ctpha,gm,lk} &= \frac{R_{ctpha,gm,lk}}{R_{T,gm,lk}} \\
P_{ctpdb,gm,lk} &= \frac{R_{ctpdb,gm,lk}}{R_{T,gm,lk}} \\
P_{td,gm,lk} &= \frac{R_{td,gm,lk}}{R_{T,gm,lk}}
\end{aligned}
\tag{11.77}
$$

The $R_{T,gm,lk}$ which is the sum of all reaction rates is given by the following equation:

$$
R_{T,gm,lk} = R_{p,gm,lk} + R_{ctm,gm,lk} + R_{cta,gm,lk} + R_{ctpha,gm,lk} + R_{ctpdb,gm,lk} + R_{td,gm,lk} \tag{11.78}
$$

As concentrations change with conversion, the above probabilities in the time interval Δt_i, change with reaction time. One of the above reactions is selected randomly as follows. If the number x_{rn} between zero and one selected by a random number generator satisfies the following criteria:

$$0 \leq x_{rn} < P_{p,gm,lk} \tag{11.79}$$

Then the event propagation is selected for the kth propagating chain in the lth growing micelle, a monomer is added to the kth propagating chain, and the concentration of monomer in the lth micelle is recalculated using the Flory–Huggins equation [114]. If x_{rn} satisfies the following criteria:

$$P_{p,gm,lk} \leq x_{rn} < P_{p,gm,lk} + P_{ctm,gm,lk} \tag{11.80}$$

Then, the event chain transfer to monomer is selected, the propagating reactive species on the kth propagating chain is terminated. The kth propagating chain remains in the list of propagating chains if it is a polyradical. Otherwise, the chain is transferred to the list of terminated polymer chains without a change in its number of repeat units, and a new 1-mer propagating chain is added to the set of propagating chains in the lth micelle. If x_{rn} satisfies the following criteria:

$$P_{p,gm,lk} + P_{ctm,gm,lk} \leq x_{rn} < P_{p,gm,lk} + P_{ctm,gm,lk} + P_{cta,gm,lk} \tag{11.81}$$

The event transfer to CTA is selected, the reactive species on the kth propagating chain is terminated by removing it from the set of reactive species. The chain remains in the list of propagating chains if it is a polyradical. Otherwise, the chain is transferred to the list of terminated polymer chains without a change in its number of repeat units. As the CTA radical is assumed to be unreactive, no new radical is added to the list of reactive species in the lth micelle. If x_{rn} satisfies the following criteria:

$$P_{p,gm,lk} + P_{ctm,gm,lk} + P_{cta,gm,lk} \leq x_{rn} < P_{p,gm,lk} + P_{ctm,gm,lk} + P_{cta,gm,lk} + P_{ctpha,gm,lk} \tag{11.82}$$

Then, the event transfer to PBD C—H bonds is selected, an m-mer terminated polymer chain q is selected randomly in the lth micelle followed by random selection of one of the PBD C—H bonds, the radical on the propagating chain is transferred to the selected PBD C—H bond, a new m-mer propagating chain is added to the set of propagating chains with the addition of a branch. The n-mer chain remains in the list of propagating chains if it is a polyradical. Otherwise, the chain is transferred to the list of terminated polymer chains without a change in its number of repeat units. If x_{rn} satisfies the following criteria:

$$\begin{aligned} &P_{p,gm,lk} + P_{ctm,gm,lk} + P_{cta,gm,lk} + P_{ctpha,gm,lk} \leq x_{rn} \\ &\quad < P_{p,gm,lk} + P_{ctm,gm,lk} + P_{cta,gm,lk} + P_{ctpha,gm,lk} + P_{ctpdb,gm,lk} \end{aligned} \tag{11.83}$$

Then, the event transfer to the PBD C═C bonds of a terminated polymer chain is selected, an m-mer terminated polymer chain q is selected randomly in the lth micelle followed by random selection of one of the PBD C═C bonds, the radical on the propagating chain reacts with the C═C bond and combines to form a longer $(m+n)$-mer propagating chain and a crosslink. The n-mer chain remains in the list of propagating chains but its number of repeat units is increased to $(n+m)$ with the addition of a crosslink. Furthermore, the m-mer polymer chain is removed from the list of terminated polymer chains. If x_{rn} satisfies the following criteria:

$$P_{p,gm,lk} + P_{ctm,gm,lk} + P_{cta,gm,lk} + P_{ctpha,gm,lk} + P_{ctpdb,gm,lk} + P_{td,gm,lk} \leq x_{rn} \leq 1 \tag{11.84}$$

Then, another propagating chain in the lth micelle is randomly selected, and both chains are terminated by disproportionation by removing them from the list of propagating chains and adding them to the list of terminated polymer chains. The above process is repeated for all propagating chains in all micelles and for all Δt_i time intervals until the final reaction time is reached.

11.9.4.3 Determination of Reaction Times of Elementary Reactions

The time interval Δt is orders of magnitude smaller than the reaction time. It can be related to the elapsed time of individual elementary reactions using the following relations:

$$\begin{aligned}
\delta t_{dis,i} &= \frac{1}{(R_{dis,i} V_{aq} N_{AV})} \\
\delta t_{p,gm,lk} &= \frac{1}{(R_{p,gm,lk} V_{gm,l} N_{AV})} \\
\delta t_{ctm,gm,lk} &= \frac{1}{(R_{ctm,gm,lk} V_{gm,l} N_{AV})} \\
\delta t_{cta,gm,lk} &= \frac{1}{(R_{cta,gm,lk} V_{gm,l} N_{AV})} \\
\delta t_{ctpha,gm,lk} &= \frac{1}{(R_{ctpha,gm,lk} V_{gm,l} N_{AV})} \\
\delta t_{ctpdb,gm,lk} &= \frac{1}{(R_{ctpdb,gm,lk} V_{gm,l} N_{AV})} \\
\delta t_{td,gm,lk} &= \frac{1}{(R_{td,gm,lk} V_{gm,l} N_{AV})}
\end{aligned} \tag{11.85}$$

where $\delta t_{dis,i}$ is the time for dissociation of ith initiator in the time interval Δt_i; $\delta t_{p,gm,lk}$, $\delta t_{ctm,gm,lk}$, $\delta t_{cta,gm,lk}$, $\delta t_{ctpha,gm,lk}$ and $\delta t_{ctpdb,gm,lk}$ are the times for propagation, transfer to monomer, CTA, PBD C—H bond, and PBD C═C bond of the kth propagating chain in the lth micelle; and $\delta t_{td,gm,lk}$ is the time for termination by disproportionation; V_{aq} and $V_{gm,l}$ are the volumes of the aqueous phase and the lth micelle, respectively. The overall reaction time, t_{rxn}, is calculated from the reaction times of the elementary reactions as follows:

$$\begin{aligned}
t_{rxn} = &\sum_{\Delta t_i} \sum_{init\ j} \delta t_{dis,ij} + \sum_{\Delta t_i} \sum_{gm\ l} \sum_{chain\ k} (\delta t_{p,gm,ilk} + \delta t_{ctm,gm,ilk} + \delta t_{cta,gm,ilk}) \\
&+ \sum_{\Delta t_i} \sum_{gm\ l} \sum_{chain\ k} (\delta t_{ctpha,gm,ilk} + \delta t_{ctpdb,gm,ilk} + \delta t_{td,gm,ilk})
\end{aligned} \tag{11.86}$$

The time is also given by $t = \sum_i \Delta t_i$. If $t_{rxn} < t$ in a given time interval Δt_i, the polymerization reaction will continue for the kth propagating chain in the lth micelle until $t_{rxn} = t$. If $t_{rxn} = t$ in a given time interval Δt_i, a new Δt_{i+1} is selected and the polymerization reaction will continue with initiator dissociation and chain propagation until $t_{rxn} = t$ in the new time interval. After reaching the desired final reaction time t_f (see Figure 11.4), the reaction conversion is calculated using the equation $\alpha = 1 - \left(\frac{[M]}{[M]_0}\right)$.

11.9.4.4 Simulation Results

Chain transfer to polymer reactions in emulsion polymerization can lead to a bimodal DP distribution of terminated polymer chains. The best mode to visually illustrate the bimodality of a distribution is to plot the weight fraction distribution of chains, $P_w(n)$, versus the logarithm of DP of the chains $\log_{10}(n)$ [115, 116]. Figure 11.5 a–c shows the Monte Carlo simulated plots of $P_w(n)$ versus $\log_{10}(n)$ for emulsion polymerization of BD at 25 °C for conversions of 20%, 55%, and 75%, respectively. At low conversions, propagation and chain transfer to monomer are the dominant reactions

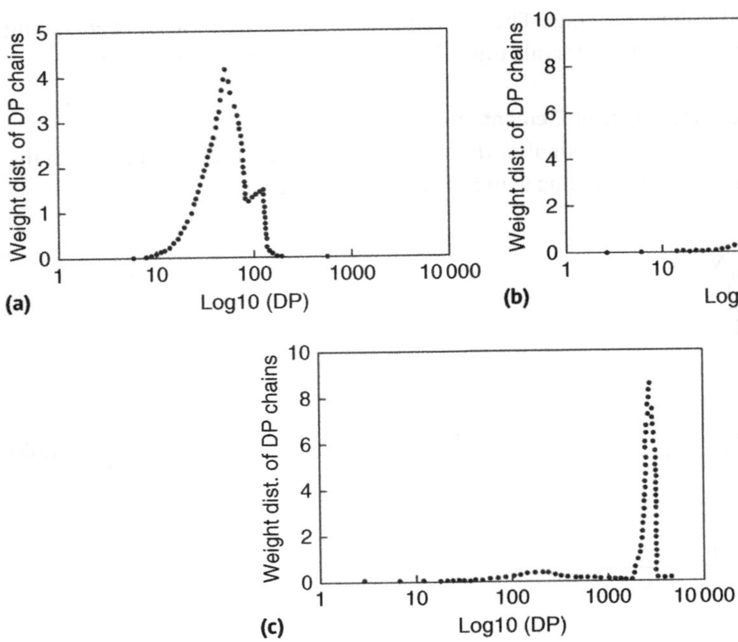

Figure 11.5 Weight fraction DP distribution $P_w\ (DP = n)$ of chains versus $\log_{10}\ (DP = n)$ in emulsion polymerization of BD at 25 °C for (a) $\alpha = 20\%$, (b) $\alpha = 55\%$, and (c) $\alpha = 75\%$ (redrawn from Ref. [28]).

in the micelles because the concentration of terminated chains is relatively low for chain transfer to polymer to occur. As a result, at conversions of <20%, the DP distribution of polymer chains is unimodal. As conversion is increased to 20% (Figure 11.5a), a shoulder appears in the distribution around DP ~ 100, which is attributed to the onset of chain transfer to polymer, branching, and crosslinking. It should be noted that chain termination rate is relatively low in micelles because most micelles contain only one propagating chain.

As conversion is increased to 55% (Figure 11.5b), the polymer concentration is considerably increased in the micelles, and the shoulder becomes a larger second peak compared to the same peak at low conversions. For 55% conversion, the second peak at DP ~ 1200 is attributed to the increased rate of chain transfer to polymer, trifunctional branching, and tetra-functional crosslinking. With a further increase in conversion to 75%, the second peak dominates the chain distribution with peak position increasing to DP ~ 2000. When the BD polymerization is simulated in the absence of tetrafunctional crosslinking (i.e., $R_{ctpdb,gm,lk} = 0$), only a tail, as opposed to a second peak, is observed in the distribution for all conversions. This indicates that chain transfer to C═C bonds of PBD leads to a bimodal DP distribution in BD emulsion polymerization. The frequency of trifunctional branching (3FB) by transfer to polymer C—H bonds and tetrafunctional crosslinking (4FC) by transfer to polymer C═C bonds are shown in Figure 11.6a. The frequency of 3FB and 4FC increased almost linearly with conversion; however, the 4FC frequency was an order of magnitude higher than 3FB. The emergence and growth of a second peak in the DP distribution, as shown in Figure 11.5, is due to the high frequency of crosslinking by 4FC. The polydispersity index (PDI) of the produced polymer chains as a function of conversion is shown in Figure 11.6b. Below 20% conversion, PDI is close to three, but increases to five after 50% conversion, which is attributed to tetrafunctional crosslinking (4FC). Although there is a shift in the slope of PDI at approximately

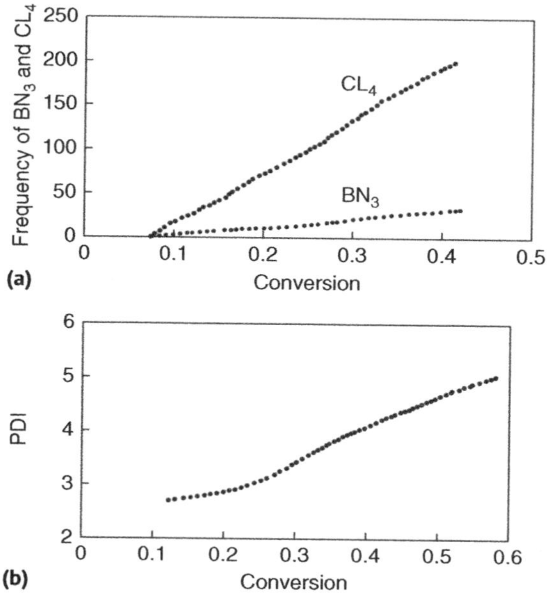

Figure 11.6 (a) Frequency of trifunctional branching (3FB) and tetrafunctional crosslinking (4FC) versus conversion and (b) polydispersity index (PDI) of the polymer chains versus conversion for Monte Carlo simulation of emulsion polymerization of BD at 25 °C (redrawn from Ref. [28]).

20% conversion due to tetrafunctional crosslinking, PDI increases continuously, not abruptly, with conversion.

The results in Figures 11.5 and 11.6 illustrate that the DP distribution in emulsion polymerization of BD, as well as other divinyl monomers, is different from that of bulk polymerization. In bulk polymerization of divinyl monomers, the average DP approaches infinity with increasing transfer to polymer C=C bonds (4FC), which results in microgel formation. Conversely, in emulsion polymerization of divinyl monomers, the average DP increases with increasing 4FC, but it approaches a finite value and the formation of a micellar microgel. This is attributed to confinement of the polymerization reaction to the finite, small, nanoscale size of the micelles with a finite number of chains. Tetrafunctional crosslinking in polymerization of divinyl monomers can be intramolecular or intermolecular. Intramolecular crosslinking leads to formation of loops in polymer chains, whereas intermolecular crosslinking leads to microgel formation. In the above Monte Carlo simulation, all 4FC reactions were assumed to be intermolecular as vinyl bonds on the propagating chains were not distinguished from terminated polymer chains. The simulation can be modified to account for loop formation by intramolecular C=C chain transfer versus microgel formation by intermolecular crosslinking by uniquely defining each propagating chain and each terminated polymer chain in each growing micelle in the emulsion polymerization.

11.10 Summary

This chapter described the method of Monte Carlo for simulation of polymerization reactions. The Monte Carlo method capitalizes on the random nature of the elementary reactions to simulate the polymerization in a microscopic 1×10^6 to 1×10^8 nm^3 volume of the reaction using the following approach. First, the elementary reactions are defined for the polymerization system, and a probability is assigned to each elementary reaction based on its relative reaction rate. Next, a random number generator algorithm is used to generate random numbers that pass the tests of uniformity, interval, and serial correlation. Then, a time interval orders of magnitude smaller than the overall reaction time is chosen. Next, polymerization is initiated by using the random numbers to select one of the elementary reactions in the polymerization system based on the probability assigned to that reaction. This process is continued for all time intervals until the desired reaction time is reached. Then, the averages of the chain distribution and polydispersity index are calculated from the simulated DP distribution. The Monte Carlo method requires the knowledge of elementary reactions and their rates, and initial reaction conditions as well as access to a random number generator to simulate the reaction at a microscopic scale. The Monte Carlo method is especially applicable to simulating chain distribution in polymerizations in confined spaces like emulsion polymerization and propagating chains that carry multiple reactive species (propagating polyradical). This chapter demonstrates through examples the methodology for testing random numbers for uniformity, interval, and serial correlation. It also demonstrates through examples the application of the method of Monte Carlo to simulation of reactive species production by initiator dissociation and PD distribution of polymer chains in emulsion polymerization of divinyl monomers.

11.11 List of Symbols and Abbreviations

a, b: lower and upper bound values of independent variable x

CL: confidence level which is equal to $100 - \alpha\%$

C_m: ratio of rate constants of chain transfer to monomer to propagation

CTA: chain transfer agent

CTA^*: inactive chain transfer agent produced by the reaction of a reactive propagating chain with a chain transfer agent

$[CTA_{gm,l}]$: concentration of chain transfer agent in l^{th} micelle

BR: long branch produced on a terminated polymer chain by reaction and combination of a reactive propagating chain with a pendant double bond on a terminated polymer chain

E: percent error between the estimated and theoretical value of a random variable

EL: percent error limit between the estimated and theoretical value of a random variable

$E\,|h\,(x)|$: expected value of the function $h\,(x)$

$f_{r.n.}\,(x)$: a weighing function for each value of independent variable x used in determination of the expected value of function $f\,(x)$

$f\,(x)$: function f which depends on independent variable x

F: function with a χ^2 distribution

F_{ic}: interval correlation function describing distribution of random numbers in subintervals for testing interval correlation

F_{sc}: serial correlation function describing distribution of random numbers for testing serial correlation

GM: number of growing micelles in the simulated volume

$\overline{h}$: expected value of the function $h\,(x)$

$\hat{h}$: Monte Carlo simulated value of $h\,(x)$

$h\,(\varepsilon_i)$: function $h\,(x)$ evaluated at $x = \varepsilon_i$

$h\,(x)$: function h, dependent on variable x, which is used in estimation of error of a random number generator

I: actual value of an integral or initiator in a polymerization reaction
$\widehat{I}_{m.c.}$: Monte-Carlo estimated value of an integral
$[I]$: initiator concentration
$[I]_0$: initial concentration of initiator
k: number of independent normally distributed random variables
k_{ci}: rate constant for chain initiation reaction between a reactive species and a monomer to generate a reactive propagating chain with one repeat unit
k_{cta}: rate constant for chain transfer to the chain transfer agent
k_{ctm}: rate constant for chain transfer to monomer
k_{ctpha}: rate constant for chain transfer to terminated polymer chains by hydrogen abstraction
k_{ctpdb}: rate constant for chain transfer by reaction and combination with a double bond on a terminated polymer chains
k_d: rate constant for dissociation of the initiator
k_p: rate constant for propagation reaction
k_{tc}: rate constant for termination of two propagating chains by combination to form a terminated polymer chain
k_{td}: rate constant for termination of two propagating chains by disproportionation to form two terminated polymer chains
m, n: number of monomer residues on a polymer chain
M: number of subintervals in an interval of random numbers for testing subinterval correlation or monomer in a reaction
$[M]$: monomer concentration at any time during the reaction
$[M]_0$: initial concentration of monomer in the reaction
$[M_{gm,l}]$: concentration of monomer in l^{th} micelle
$[MTP_{gm,lq}]$: concentration of monomer residues on q^{th} terminated polymer chain in l^{th} micelle
$n_{CTA,gm,lk}$: number of chain transfer agent molecules in l^{th} micelle
$n_{M,gm,lk}$: number of monomer molecules in l^{th} micelle
$n_{MTP,gm,lk}$: number of monomer residues on terminated polymer chains on kth propagating chain in l^{th} micelle
n_{random}: random number in between zero and one generated by a random number generator
$n_{RS,gm,lk}$: number of reactive species on kth propagating chain in l^{th} micelle
$n_{sc,h,jk}$: number of times a random number x_i in subinterval j is followed in h steps by a random number x_{i+h} in subinterval k for testing serial correlation
$n_{si,j}$: number of random numbers in a subinterval j between zero and one
$\overline{n}_{sc,h}$: theoretical number of times a random number x_i in subinterval j is followed in h steps by a random number
x_{i+h} in subinterval k for testing serial correlation
$\overline{n}_{si}$: theoretical number of random numbers in a subinterval j between zero and one
N: number of sampling times
N_{AV}: Avogadro's number
$N_{I,0}$: initial number of initiator molecules in the simulation volume
$N_{M,0}$: initial number of monomer molecules in the simulation volume
$N_{R,0}$: initial number of reactive species in the simulation volume generated by initiator dissociation
$N_{RP,0}$: initial number of reactive propagating chains in the simulation volume
$p_{cta,gm,lk}$: probability of chain transfer to transfer agent of kth propagating chain in l^{th} micelle
p_{ctm}: probability of chain transfer to monomer in a reaction
$p_{ctm,gm,lk}$: probability of chain transfer to monomer of kth propagating chain in l^{th} micelle

$p_{ctpdb,gm,lk}$: probability of chain transfer a double bond on terminated polymer chains for kth propagating chain in l^{th} micelle
$p_{ctpha,gm,lk}$: probability of chain transfer to terminated polymer chains by hydrogen abstraction for kth propagating chain in l^{th} micelle
p_{dis}: probability of initiator dissociation to reactive species
$p_{dis,i}$: probability of initiator dissociation in the time interval Δt_i
p_{nodis}: probability of no initiator dissociation
p_p: probability of propagation of a reactive propagating chain at any time t in the reaction
$p_{p,gm,lk}$: probability of propagation of kth propagating chain in l^{th} micelle
p_{tc}: probability of termination by combination of a reactive propagating chain to form a terminated polymer chain at any time t in the reaction
p_{td}: probability of termination by disproportionation of two reactive propagating chains to form two terminated polymer chains at any time t in the reaction
$p_{td,gm,lk}$: probability of chain termination by disproportionation for kth propagating chain in l^{th} micelle
pdf: probability distribution function
$p(x)$: probability function p which is a function of independent variable x
R: reactive species generated from dissociation of initiator
$R_{cta,gm,lk}$: rate of chain transfer to the transfer agent for kth propagating chain in l^{th} micelle
$R_{ctm,gm,lk}$: rate of chain transfer to monomer for kth propagating chain in l^{th} micelle
$R_{ctpdb,gm,lk}$: rate of chain transfer to a double bond on terminated polymer chains for kth propagating chain in l^{th} micelle
$R_{ctpha,gm,lk}$: rate of chain transfer to terminated polymer chains by hydrogen abstraction for kth propagating chain in l^{th} micelle
R_p: rate of propagation reaction
$R_{p,gm,lk}$: rate of propagation reaction for kth propagating chain in l^{th} micelle
R_{tc}: rate of termination reaction by combination
R_{td}: rate of termination reaction by disproportionation
$R_{td,gm,lk}$: rate of chain termination by disproportionation for kth propagating chain in l^{th} micelle
$R_{T,gm,lk}$: total reaction rate for kth propagating chain in lth micelle which is the sum of rates of propagation, chain transfer to monomer, transfer agent, and polymer, and termination
RP_n: reactive propagating chain with n repeat units
$[RP]$: total concentration of reactive propagating chains
$[RP]_0$: initial concentration of reactive propagating chains in the simulation volume
$[RP]_n$: concentration of reactive propagating chains with n repeat units
$[RS_{gm,lk}]$: concentration of reactive species on kth propagating chain in l^{th} micelle
t: reaction time
$t_{1/2}$: half-life of the initiator
t_{rxn}: reaction time
T_g: polymer glass transition temperature
TP_n: terminated polymer chain with n repeat units
$[TP]$: concentration of terminated polymer chains
V: variance of a random variable with a probability distribution
$V_{gm,l}$: simulation volume of lth micelle
V_{sim}: Volume of the simulation
x_i: i^{th} random number
x_{ij}: random number generated to choose the fate of initiator i in time interval j

$x_{rn,ij}$: random number x generated for testing the ith reactive propagating chain in the time interval j for propagation or termination

z_i: experimental values of the ith independent normally distributed random variables

$\overline{z}$: theoretical expectation of the independent normally distributed random variables

Z_i: i^{th} independent normally distributed random variables

α: significance level or the acceptable probability of being correct about the value of a random variable or reaction conversion

α_i: theoretical expectation of the ith independent normally distributed random variables

$\beta_{CL\%}$: area under a normal distribution corresponding to percent confidence level of $CL\%$

Γ: the gamma function

$\delta t_{cta,gm,lk}$: elapsed time for chain transfer to transfer agent of kth propagating chain in l^{th} micelle

$\delta t_{cta,gm,ilk}$: elapsed time for chain transfer to transfer agent of kth propagating chain in l^{th} micelle in the time interval Δt_i

$\delta t_{ctpdb,gm,lk}$: elapsed time for chain transfer of a double bond on terminated polymer chains for kth propagating chain in l^{th} micelle

$\delta t_{ctm,gm,lk}$: elapsed time for chain transfer to monomer of kth propagating chain in l^{th} micelle

$\delta t_{ctm,gm,ilk}$: elapsed time for chain transfer to monomer of kth propagating chain in l^{th} micelle in the time interval Δt_i

$\delta t_{ctpdb,gm,ilk}$: elapsed time for chain transfer of a double bond on terminated polymer chains for kth propagating chain in l^{th} micelle in the time interval Δt_i

$\delta t_{ctpha,gm,lk}$: elapsed time for chain transfer to terminated polymer chains by hydrogen abstraction for kth propagating chain in l^{th} micelle

$\delta t_{ctpha,gm,ilk}$: elapsed time for chain transfer to terminated polymer chains by hydrogen abstraction for kth propagating chain in l^{th} micelle in the time interval Δt_i

$\delta t_{dis,i}$: elapsed time for dissociation of an initiator molecule in the time interval Δt_i

$\delta t_{dis,ij}$: elapsed time for dissociation of jth initiator molecule in the time interval Δt_i

$\delta t_{p,gm,lk}$: elapsed time for propagation of kth propagating chain in l^{th} micelle

$\delta t_{p,gm,ilk}$: elapsed time for propagation of kth propagating chain in l^{th} micelle in the time interval Δt_i

$\delta t_{td,gm,lk}$: elapsed time for chain termination by disproportionation for kth propagating chain in l^{th} micelle

$\delta t_{td,gm,ilk}$: elapsed time for chain termination by disproportionation for kth propagating chain in l^{th} micelle in the time interval Δt_i

ε_i: i^{th} randomly selected number between zero and one generated by a random number generator

λ: expected value of a random variable with a probability distribution

μ: mean value of a random variable with a probability distribution

μ_i: experimental expectation of the ith independent normally distributed random variables

$\mu_{rn,k}$: kth order moment of the distribution of the generated random numbers

σ: standard deviation from the mean of a random variable with a probability distribution

ρ_{BD}: density of butadiene monomer

ρ_{PBD}: density of polybutadiene

υ: degrees of freedom of the function $F$with a χ^2 distribution

χ^2: a statistical test used to measure how expectation of a random variable compares with its actual or modeled value

11.12 Practice Problems

Problem 11.1 The error limit, EL, which is used to quantify the overall uniformity for a given sample of random numbers is defined by the following relation (Eq. 11.32 in Section 11.6.1):

$$EL \leq \frac{100\,\beta_{CL\%}\sigma}{\sqrt{N}\mu_{rn,1}} = \frac{100\,\beta_{CL\%}V^{1/2}}{N^{1/2}\mu_{rn,1}}$$

where σ and V are standard deviation and variance of the distribution of the sample of random numbers about the mean, respectively. Using the defining equation for V, derive the following relation for EL in terms of the 1st and 2nd order mole-fraction moments of the distribution of numbers, $\mu_{rn,1}$ and $\mu_{rn,2}$ (see Eq. 11.33 in Section 11.6.1):

$$EL \leq \frac{100\,\beta_{CL\%}V^{1/2}}{N^{1/2}\mu_{RN,1}} = \frac{100\,\beta_{CL\%}\left\{\mu_{RN,2} - \mu_{RN,1}^2\right\}^{1/2}}{N^{1/2}\mu_{RN,1}}$$

Problem 11.2 The following 20 numbers are generated by a random number generator program:

$x_1 = 0.45$	0.88	0.24	0.65	0.44	0.21	0.89	0.63	0.59	$x_{10} = 0.68$
$x_{11} = 0.62$	0.91	0.38	0.61	0.28	0.60	0.97	0.55	0.81	$x_{20} = 0.48$

Test the generated random numbers for the following properties:

a. overall uniformity by computing $\mu_{rn,1}$, $\mu_{rn,2}$, $\mu_{rn,3}$, and the percent error limit EL for 99% confidence level.
b. interval correlation with five equal subintervals and confidence level of 99%.
c. one-step (h = 1) serial correlation with five equal subintervals and confidence level of 99%.
d. overall uniformity, interval correlation, and 1-step serial correlation with five equal subintervals and confidence level of 99.9%.

Problem 11.3 The following 100 random numbers are generated by a random number generator program.

$x_1 = 0.32$	0.74	0.14	0.61	0.46	0.70	0.75	0.30	0.81	0.63
$x_{11} = 0.54$	0.09	0.45	0.52	0.56	0.82	0.21	0.95	0.07	0.17
$x_{21} = 0.19$	0.26	0.24	0.60	0.80	0.33	0.35	0.77	0.71	0.87
$x_{31} = 0.96$	0.91	0.03	1.00	0.40	0.51	0.55	0.97	0.53	0.31
$x_{41} = 0.13$	0.43	0.20	0.29	0.49	0.08	0.34	0.78	0.92	0.99
$x_{51} = 0.28$	0.38	0.05	0.27	0.42	0.83	0.64	0.01	0.68	0.12
$x_{61} = 0.04$	0.37	0.25	0.48	0.44	0.79	0.15	0.39	0.76	0.16
$x_{71} = 0.41$	0.86	0.89	0.18	0.85	0.88	0.10	0.62	0.23	0.94
$x_{81} = 0.72$	0.06	0.50	0.00	0.22	0.73	0.66	0.58	0.57	0.11
$x_{91} = 0.59$	0.36	0.47	0.98	0.84	0.02	0.90	0.93	0.67	0.65

Test the generated random numbers for the following properties:

a. overall uniformity of the random numbers by computing $\mu_{rn,1}$, $\mu_{rn,2}$, $\mu_{rn,3}$ and the percent error limit EL for 99% confidence level.

b. interval correlation of the random numbers with five equal subintervals and confidence level of 99%.
c. 1-step (h = 1) serial correlation of the random numbers with five equal subintervals and confidence level of 99%.
d. overall uniformity, interval correlation, and 1-step serial correlation with five equal subintervals and confidence level of 90%.

Problem 11.4 Benzoyl peroxide (BPO) is used as a thermal initiator in radical polymerization [70, 117, 118]. The reaction for the dissociation of BPO is as follows (See Section 1.2.1.1 in Chapter 1):

$$H_5C_6\overset{\overset{O}{\|}}{C}O{-}O\overset{\overset{O}{\|}}{C}C_6H_5 \xrightarrow{k_{ds}} {}^2H_5C_6\overset{\overset{O}{\|}}{C}O^{\bullet} \xrightarrow{k_{di}} {}^2H_5C_6^{\bullet} + {}^2CO_2$$

The weak $-O-O-$ bond of BPO dissociates with rate constant k_{ds} to generate two benzoyloxy radicals. As the benzoyloxy radical is unstable, it further disintegrates into a more stable benzyl radical with a rate constant k_{di} with evolution of carbon dioxide gas [119]. Overall, the dissociation reaction of BPO can be written as follows:

$$I \xrightarrow{k_d} {}^2R + {}^2CO_2$$

where k_d is the dissociation rate constant. In a radical polymerization system, the initial BPO concentration is $[I]_0 = 0.001$ M and the half-life of BPO at 70 °C is $t_{1/2} = 7.3$ h. The probability of dissociation is given by the following relation (see Eq. 11.50 in Section 11.7):

$$p_{dis} = \left|\frac{\Delta [I]}{[I]}\right| = \left[\frac{\ln (2)}{t_{1/2}}\right]\Delta t = k_d \Delta t$$

where the time interval $\Delta t = 10$ s.

a. determine the size of the simulation volume for initial number of initiator molecules $N_{I,0} = 10$, $N_{I,0} = 100$, $N_{I,0} = 1000$, and $N_{I,0} = 10\,000$.
b. use Monte Carlo method to simulate the dissociation of BPO as a function of time with $N_{I,0} = 10$, $N_{I,0} = 100$, $N_{I,0} = 1000$, and $N_{I,0} = 10\,000$.
c. using the rate equation for dissociation, determine the theoretical dissociation profile of BPO as a function of time for $t_{1/2} = 7.3$ h.
d. by comparing the simulated and theoretical dissociation profiles of BPO, determine the minimum $N_{I,0}$ and the corresponding size of the simulation volume that is representative of the actual reaction system.
e. using Monte Carlo simulation with $N_{I,0} = 10\,000$, determine the concentration profile of the produced reactive species as a function of time.

Problem 11.5 A two-initiator thermal initiation system consisting of AIBN and BPO is used for a radical polymerization reaction at 70 °C [68–70, 117, 118]. The two initiators are used at equal concentration of $[I]_{AIBN,0} = [I]_{BPO,0} = 0.0005$ M. The half-lives of AIBN at 70 °C is $t_{1/2,AIBN} = 4.8\ h$ and that of BPO is $t_{1/2,BPO} = 7.3$ h.

a. determine the size of the simulation volume for the initial number of initiator molecules $N_{I,AIBN,0} = 5000$ and $N_{I,BPO,0} = 5000$.
b. using the rate equations for dissociation of AIBN and BPO and their corresponding initial concentrations of $[I]_{AIBN,0} = [I]_{BPO,0} = 0.0005$ M, determine the theoretical concentration profile of the produced reactive species as a function of time.

c. use Monte Carlo method to simulate the number of produced reactive species as a function of time with $N_{I,AIBN,0} = 5000$ and $N_{I,BPO,0} = 5000$. Is the chosen initial number of AIBN and BPO molecules sufficiently large to represent the actual reaction system.

References

1 Zhou, Y.N. and Luo, Z.H., State-of-the-art and progress in method of moments for the model-based reversible-deactivation radical polymerization. *Macromolecular Reaction Engineering*, 2016, **10**(6), 516.

2 Rosa, R.P., Ferreira, F.V., and Lona, L.M.F., Modeling of ring opening polymerization: a short review with insights on how to develop the method of moments. *Chemical Engineering Science*, 2021, **246**, 116934.

3 Mastan, E. and Zhu, S.P., Method of moments: a versatile tool for deterministic modeling of polymerization kinetics. *European Polymer Journal*, 2015, **68**, 139.

4 Bachmann, R., Extension of the method of moments in nonlinear free radical polymerization. *Macromolecular Theory and Simulations*, 2017, **26**(4), 1600097.

5 Wang, J., Wang, T.T., Luo, Z.H., and Zhou, Y.N., Simulation of irreversible and reversible degradation kinetics of linear polymers using sectional moment method. *Chemical Engineering Science*, 2023, **275**, 118711.

6 Manuiko, G.V. et al., Mathematical modeling of 1,3-butadiene polymerization over a neodymium-based catalyst in a batch reactor with account taken of the multisite nature of the catalyst and chain transfer to the polymer. *Theoretical Foundations of Chemical Engineering*, 2010, **44**(2), 139.

7 Zhou, Y.N. and Luo, Z.H., An old kinetic method for a new polymerization mechanism: toward photochemically mediated ATRP. *Aiche Journal*, 2015, **61**(6), 1947.

8 Liu, S. et al., Backbiting and beta-scission reactions in free-radical polymerization of methyl acrylate. *International Journal of Quantum Chemistry*, 2014, **114**(5), 345.

9 Dossi, M., Storti, G., and Moscatelli, D., Relevance of backbiting and beta-scission reactions in the free radical polymerization of acrylonitrile. *Macromolecular Symposia*, 2010, **289**, 119.

10 Nikitin, A.N., Hutchinson, R.A., Buback, M., and Hesse, P., Determination of intramolecular chain transfer and midchain radical propagation rate coefficients for butyl acrylate by pulsed laser polymerization. *Macromolecules*, 2007, **40**(24), 8631.

11 Marien, Y.W. et al., An alternative method to estimate the bulk backbiting rate coefficient in acrylate radical polymerization. *Polymer Chemistry*, 2016, **7**(42), 6521.

12 Nieboer, V., Fanjul-Mosteirín, N., Olsen, P., and Odelius, K., Mastering macromolecular architecture by controlling backbiting kinetics during anionic ring-opening polymerization. *Macromolecules*, 2024, **57**(7), 3397.

13 Klages, H.G.G., Ermis, N., Luinstra, G.A., and Zentel, K.M., Coupling kinetic modelling with SAOS and LAOS rheology of poly(-butyl acrylate). *Macromolecular Rapid Communications*, 2022, **43**(12), 2100620.

14 Drache, M., Hosemann, B., Laba, T., and Beuermann, S., Modeling of branching distributions in butyl acrylate polymerization applying Monte Carlo methods. *Macromolecular Theory and Simulations*, 2015, **24**(4), 301.

15 Özçam, D.D. and Teymour, F., Chain-by-chain Monte Carlo simulation: a novel hybrid method for modeling polymerization. Part I. Linear controlled radical polymerization systems. *Macromolecular Reaction Engineering*, 2017, **11**(1), 1600042.

16 Yaghini, N. and Iedema, P.D., Predicting molecular weight distribution by deterministic modeling and Monte Carlo simulations of radical polymerization with branching and scission allowing for multiradicals and gelation in various reactor configurations. *Chemical Engineering Science*, 2015, **130**, 310.

17 Zarand, S.M.G. et al., Modeling of styrene/butadiene co-polymerization in emulsion with Monte Carlo simulation. *Chemical Engineering & Technology*, 2003, **26**(9), 969.

18 Ghiass, M. et al., Monte Carlo simulation of non-linear free radical polymerization using a percolation kinetic gelation model (I): free radical homopolymerization. *Iranian Polymer Journal*, 2001, **10**(5), 305.

19 Mohammadi, Y. and Jabbari, E., Monte Carlo simulation of degradation of porous poly(lactide) scaffolds. 1. Effect of porosity on pH. *Macromolecular Theory and Simulations*, 2006, **15**(9), 643.

20 Mohammadi, Y. et al., Intelligent machine learning: tailor-making macromolecules. *Polymers (Basel)*, 2019, **11**(4), 579.

21 Mohammadi, Y. et al., Toward olefin multiblock copolymers with tailored properties: a molecular perspective. *Macromolecular Theory and Simulations*, 2021, **30**(3), 2100003.

22 Mohammadi, Y. et al., Intelligent Monte Carlo: a new paradigm for inverse polymerization engineering. *Macromolecular Theory and Simulations*, 2018, **27**(3), 1700106.

23 He, J.P., Zhang, H.D., Chen, J.M., and Yang, Y.L., Monte Carlo simulation of kinetics and chain length distributions in living free-radical polymerization. *Macromolecules*, 1997, **30**(25), 8010.

24 Platkowski, K. and Reichert, K.H., Application of Monte Carlo methods for modelling of polymerization reactions. *Polymer*, 1999, **40**(4), 1057.

25 Beveridge, D.L., *Monte Carlo and molecular dynamics simulations in polymer science*. 1995, Oxford University Press.

26 Vitalis, A. and Pappu, R.V., Methods for Monte Carlo simulations of biomacromolecules. *Annual Reports in Computational Chemistry*, 2009, **5**, 49.

27 Mavrantzas, V.G., Using Monte Carlo to simulate complex polymer systems: recent progress and outlook. *Frontiers in Physics*, 2021, **9**, 661367.

28 Jabbari, E., Monte Carlo simulation of tri-functional branching and tetra-functional crosslinking in emulsion polymerization of butadiene. *Polymer*, 2001, **42**(11), 4873.

29 Knezo, D. and Vagaská, A., Monte Carlo method application and generation of random numbers by usage of numerical methods. *Models and Theories in Social Systems*, 2019, **179**, 197.

30 Mohanty, S., Mohanty, A.K., and Carminati, F., Efficient pseudo-random number generation for Monte-Carlo simulations using graphic processors, in *4th international workshop on advanced computing and analysis techniques in physics research (ACAT 2011)*. 2012, IOP Science, p. 012024.

31 L'Ecuyer, P., Comparison of point sets and sequences for quasi-Monte Carlo and for random number generation, in *Sequences and their applications – SETA 2008*. 2008, Springer.

32 Umeno, K., Chaotic Monte Carlo computation: a dynamical effect of random-number generations. *Japanese Journal of Applied Physics Part 1-Regular Papers Short Notes & Review Papers*, 2000, **39**, 1442.

33 Dang, B.J. et al., Physically transient true random number generators based on paired threshold switches enabling Monte Carlo method applications. *IEEE Electron Device Letters*, 2019, **40**(7), 1096.

34 Ghersi, D., Parakh, A., and Mezei, M., Comparison of a quantum random number generator with pseudorandom number generators for their use in molecular Monte Carlo simulations. *Journal of Computational Chemistry*, 2017, **38**(31), 2713.

35 Mcewan, P. et al., Random number generators in Monte Carlo simulation. *Value in Health*, 2012, **15**(7), A469.

36 Click, T.H., Liu, A.B., and Kaminski, G.A., Quality of random number generators significantly affects results of Monte Carlo simulations for organic and biological systems. *Journal of Computational Chemistry*, 2011, **32**(3), 513.
37 L'Ecuyer, P., Uniform random number generation. *Simulation*, 2006, **13**, 55.
38 L'Ecuyer, P., *Random number generation*. 2012, Springer.
39 Balcerzak, T., Magnetic-properties of a disordered ising lattice with a rectangular probability-distribution of the exchange integrals. *Physical Review B*, 1993, **47**(22), 15294.
40 Lebrun, R., Efficient time/space algorithm to compute rectangular probabilities of multinomial, multivariate hypergeometric and multivariate PLYA distributions. *Statistics and Computing*, 2013, **23**(5), 615.
41 Aroian, L.A., On the probability function of the quotient of sample ranges from a rectangular distribution. *Annals of Mathematical Statistics*, 1953, **24**(4), 680.
42 Gebhardt, F., Generating normally distributed random numbers by inverting normal distribution function. *Mathematics of Computation*, 1964, **18**(86), 302.
43 Wang, L. and Liu, Z.G., Normal distribution random number simulation application, in *Proceedings of the 2016 International Conference on Computer Engineering and Information Systems*. 2016, Atlantis Press, pp. 224–227.
44 Fernández, J.F. and Criado, C., Algorithm for normal random numbers. *Physical Review E*, 1999, **60**(3), 3361.
45 Hull, T.E. and Dobell, A.R., Random number generators. *Siam Review*, 1962, **4**(3), 230.
46 Ahrens, J.H. and Dieter, U., Extensions of Forsythe's method for random sampling from normal distribution. *Mathematics of Computation*, 1973, **27**(124), 927.
47 Snow, R.H., Generator of random numbers satisfying Poisson distribution. *Communications of the ACM*, 1968, **11**(12), 819.
48 Schaffer, H.E., Generator of random numbers satisfying poison distribution. *Communications of the ACM*, 1970, **13**(1), 49.
49 Devroye, L., The computer-generation of Poisson random-variables. *Computing*, 1981, **26**(3), 197.
50 Jodrá, P., Computer generation of random variables with Lindley or Poisson-Lindley distribution via the Lambert function. *Mathematics and Computers in Simulation*, 2010, **81**(4), 851.
51 Lapko, A.V., Lapko, V.A., and Bakhtina, A.V., Comparison of methods for testing the hypothesis of independence of random variables based on a nonparametric classifier and Pearson's chi-squared test. *Optoelectronics Instrumentation and Data Processing*, 2023, **59**(5), 551.
52 Dogan, O., Taspinar, S., and Bera, A.K., A Bayesian robust chi-squared test for testing simple hypotheses. *Journal of Econometrics*, 2021, **222**(2), 933.
53 Dickhaus, T. and Royen, T., A survey on multivariate chi-square distributions and their applications in testing multiple hypotheses. *Statistics*, 2015, **49**(2), 427.
54 Greenwood, P.E. and Nikulin, M.S., *A guide to chi-squared testing*. Wiley Series in Probability and Statistics, Vol. 280. 1996, John Wiley & Sons, 304.
55 Plackett, R.L., Karl Pearson and the chi-squared test. *International Statistical Review*, 1983, **51**(1), 59.
56 Kwak, S.G. and Kim, J.H., Central limit theorem: the cornerstone of modern statistics. *Korean Journal of Anesthesiology*, 2017, **70**(2), 144.
57 Mannalatha, V., Mishra, S., and Pathak, A., A comprehensive review of quantum random number generators: concepts, classification and the origin of randomness. *Quantum Information Processing*, 2023, **22**(12), 439.
58 Crocetti, L. et al., Review of methodologies and metrics for assessing the quality of random number generators. *Electronics*, 2023, **12**(3), 723.

59 Navarro, M.A. et al., A review of the use of quasi-random number generators to initialize the population in meta-heuristic algorithms. *Archives of Computational Methods in Engineering*, 2022, **29**(7), 5149.

60 L'Ecuyer, P., Uniform random number generators: a review, in *Proceedings of the 1997 winter simulation conference*. 1997, IEEE Computer Society.

61 Petrila, I., Manta, V., and Ungureanu, F., Uniformity and correlation test parameters for random numbers generators, in *18th International Conference System Theory, Control and Computing (ICSTTC)*. 2014, IEEE Xplore, pp. 231–236.

62 L'Ecuyer, P., Simard, R., and Wegenkittl, S., Sparse serial tests of uniformity for random number generators. *Siam Journal on Scientific Computing*, 2002, **24**(2), 652.

63 Bernhofen, L.T., Dudewicz, E.J., Levendovszky, J., and vander Meulen, E.C., Ranking of the best random number generators via entropy-uniformity theory. *American Journal of Mathematical and Management Sciences*, 1996, **16**(1&2), 49.

64 Grassberger, P., On correlations in good random number generators. *Physics Letters A*, 1993, **181**(1), 43.

65 Dematteis, A. and Pagnutti, S., Long-range correlation-analysis of the Wichmann-Hill random number generator. *Statistics and Computing*, 1993, **3**(2), 67.

66 Filk, T., Marcu, M., and Fredenhagen, K., Long-range correlations in random number generators and their influence on Monte-Carlo simulations. *Physics Letters B*, 1985, **165**(1-3), 125.

67 Harter, H.L., New table of percentage points of chi-square distribution. *Biometrika*, 1964, **51**(1-2), 231.

68 Li, X.R., Wang, X.L., and Koseki, H., Study on thermal decomposition characteristics of AIBN. *Journal of Hazardous Materials*, 2008, **159**(1), 13.

69 Arutyunyan, R.S., Nalbandyan, D.M., and Beileryan, N.M., Mechanism of dissociation of 2,2'-azobisisobutyronitrile in aqueous emulsions. *Kinetics and Catalysis*, 1985, **26**(5), 906.

70 Ng, S.C. and Chee, K.K., Determination of rate constants of benzoyl peroxide and azobisisobutyronitrile dissociation via polymerization kinetics. *Journal of Polymer Science Part a-Polymer Chemistry*, 1982, **20**(2), 409.

71 Shen, J.C., Yuan, T., Wang, G.B., and Yang, M.L., Modeling and Kinetic-Study on radical polymerization of methyl-methacrylate in bulk. 1. Propagation and termination rate coefficients and initiation efficiency. *Makromolekulare Chemie-Macromolecular Chemistry and Physics*, 1991, **192**(11), 2669.

72 Victoria-Valenzuela, D. et al., Bulk free radical polymerization of methyl methacrylate and vinyl acetate: a comparative study. *Macromolecular Reaction Engineering*, 2016, **10**(6), 577.

73 Jaso, V., Stoiljkovic, D., Radicevic, R., and Bera, O., Kinetic modeling of bulk free-radical polymerization of methyl methacrylate. *Polymer Journal*, 2013, **45**(6), 631.

74 Jaso, V.M., Stoiljkovic, D.M., and Radicevic, R.Z., Effect of temperature on initial stage of methyl methacrylate free radical bulk polymerization. *Hemijska Industrija*, 2010, **64**(6), 503.

75 Barth, J., Buback, M., Russell, G.T., and Smolne, S., Chain-length-dependent termination in radical polymerization of acrylates. *Macromolecular Chemistry and Physics*, 2011, **212**(13), 1366.

76 Nakamura, Y., Lee, R., Coote, M.L., and Yamago, S., Termination mechanism of the radical polymerization of acrylates. *Macromolecular Rapid Communications*, 2016, **37**(6), 506.

77 Nakamura, Y. and Yamago, S., Termination mechanism in the radical polymerization of methyl methacrylate and styrene determined by the reaction of structurally well-defined polymer end radicals. *Macromolecules*, 2015, **48**(18), 6450.

78 Bevington, J.C., Melville, H.W., and Taylor, R.P., The termination reaction in radical polymerizations. 2. Polymerizations of styrene at 60-degrees and of methyl methacrylate at 0 and

60-degrees, and the copolymerization of these monomers at 60-degrees. *Journal of Polymer Science*, 1954, **14**(77), 463.
79 Beuermann, S. and Buback, M., Rate coefficients of free-radical polymerization deduced from pulsed laser experiments. *Progress in Polymer Science*, 2002, **27**(2), 191.
80 Asua, J.M., Emulsion polymerization: from fundamental mechanisms to process developments. *Journal of Polymer Science Part a-Polymer Chemistry*, 2004, **42**(5), 1025.
81 Lovell, P.A. and Schork, F.J., Fundamentals of emulsion polymerization. *Biomacromolecules*, 2020, **21**(11), 4396.
82 Prescott, S.W., Ballard, M.J., and Gilbert, R.G., Average termination rate coefficients in emulsion polymerization: effect of compartmentalization on free-radical lifetimes. *Journal of Polymer Science Part a-Polymer Chemistry*, 2005, **43**(5), 1076.
83 Fitzwater, S., Chang, H.R., Parker, H.Y., and Westmoreland, D.G., Propagating radical termination at high conversion in emulsion polymerization of MMA. Rate coefficient determination from ESR data. *Macromolecules*, 1999, **32**(10), 3183.
84 Marburger, C. and Kreuter, J., Production of nanoparticles: polymerization termination and molecular weights during radical induced emulsifier free emulsion polymerization (heterogeneous polymerization). *Horizons 2000 - Aspects of Colloid and Interface Science at the Turn of the Millenium*, 1998, **109**, 270.
85 Tobita, H., Molecular-weight distribution in free-radical polymerization with chain-length-dependent bimolecular termination. 2. Emulsion polymerization. *Macromolecules*, 1995, **28**(14), 5128.
86 Cutting, G.R. and Tabner, B.J., Radical termination and radical concentrations during the batch emulsion polymerization of methyl-methacrylate studied by electron-spin-resonance spectroscopy. *Macromolecules*, 1993, **26**(5), 951.
87 Urrea-Quintero, J.H., Marino, M., Hernandez, H., and Ochoa, S., Multiscale modeling of a free-radical emulsion polymerization process: numerical approximation by the finite element method. *Computers & Chemical Engineering*, 2020, **140**, 106974.
88 Fellows, C.M., Murison, R.D., and Russell, G.T., Model discrimination of radical desorption kinetics in emulsion polymerisation. *Macromolecular Theory and Simulations*, 2011, **20**(6), 425.
89 Pohn, J. et al., Stable free radical polymerization in emulsion: modeling the thermodynamics of monomer transfer between droplets and particles. *Macromolecular Theory and Simulations*, 2008, **17**(2-3), 73.
90 Dong, Y. and Sundberg, D.C., Radical entry in emulsion polymerization: estimation of the critical length of entry radicals via a simple lattice model. *Macromolecules*, 2002, **35**(21), 8185.
91 Morrison, B.R. et al., Free-radical exit in emulsion polymerization. 2. Model discrimination via experiment. *Journal of Polymer Science Part a-Polymer Chemistry*, 1994, **32**(4), 631.
92 Schneider, H.J. and Wolf, H.J., Radical transport in emulsion polymerization. 1. Model of the reaction-kinetics including chain transfer to the monomer. *Acta Polymerica*, 1983, **34**(7), 413.
93 Meimaroglou, D., Krallis, A., Saliakas, V., and Kiparissides, C., Prediction of the bivariate molecular weight - long chain branching distribution in highly branched polymerization systems using Monte Carlo and sectional grid methods. *Macromolecules*, 2007, **40**(6), 2224.
94 Iedema, P.D., Wulkow, M., and Hoefsloot, H.C.J., Conditional Monte Carlo sampling to find branching architectures of polymers from radical polymerizations with transfer to polymer and recombination termination. *Polymer*, 2007, **48**(6), 1770.
95 Iedema, P.D. and Hoefsloot, H.C.J., Conditional Monte Carlo sampling to find branching architectures of polymers from radical polymerizations with transfer to polymer. *Macromolecules*, 2006, **39**(8), 3081.

96 Iedema, P.D. and Hoefsloot, H.C.J., Computing the trivariate chain length/degree of branching/number of combination points distribution for radical polymerization with transfer to polymer and recombination termination. *Macromolecular Theory and Simulations*, 2005, **14**(8), 505.

97 Iedema, P.D. and Hoefsloot, H.C.J., Role of polyradicals in predicting chain length distribution and gelation for radical polymerization with transfer to polymer. *Macromolecules*, 2004, **37**(26), 10155.

98 Tobita, H., Monte-Carlo simulation of emulsion polymerization - linear, branched, and cross-linked polymers. *Acta Polymerica*, 1995, **46**(3), 185.

99 Tobita, H., Kinetics of long-chain branching in emulsion polymerization. 1. Chain transfer to polymer. *Polymer*, 1994, **35**(14), 3023.

100 Bravo, C. et al., Coordinative chain transfer polymerization of butadiene using nickel(II) allyl systems: a straightforward route to branched polybutadiene. *Macromolecules*, 2023, **56**(20), 8106.

101 Vega, J.R., Estenoz, D.A., Oliva, H.M., and Meira, G.R., Analysis of a styrene-butadiene graft copolymer by size exclusion chromatography II. Determination of the branching exponent with the help of a polymerization model. *International Journal of Polymer Analysis and Characterization*, 2001, **6**(3-4), 339.

102 Fathi, H., Hamielec, A.E., and Davison, E.J., Modelling the anionic solution polymerization of butadiene - the effects of chain termination and long chain branching on molecular weight distribution development. *Polymer Reaction Engineering*, 1996, **4**(4), 243.

103 Basova, R.V., Litvinenko, G.I., and Arestyakubovich, A.A., Evaluation of intensity of chain transfer to polymer and degree of branching of polymers formed in homogeneous polymerization of butadiene under the action of sodium initiators. *Vysokomolekulyarnye Soedineniya Seriya B*, 1988, **30**(8), 586.

104 Mottweiler, R. and Scheinert, W., Molecular-weight distribution and long-chain branching in anionic-polymerization of butadiene. *Angewandte Makromolekulare Chemie*, 1981, **94**, 147.

105 Burnett, G.M. and Cameron, G.G., Branching and crosslinking in styrene-butadiene polymerizations. *Advances in Chemistry Series*, 1973, **128**(128), 102.

106 Cragg, L.H. and Fern, G.R.H., Viscometric detection of branching in polymers. 2. Branching in poly(butadiene-co-styrene) as a function of polymerization temperature. *Journal of Polymer Science*, 1953, **10**(2), 185.

107 Tobisch, S., Mechanism of the chain termination of the allylnickel(II)-catalyzed polymerization of 1,3-butadiene. A density functional investigation for the cationic [Ni-II(RC_3H_4)(*cis*-C_4H_6)L](+) active catalyst. *Macromolecules*, 2003, **36**(16), 6235.

108 Pradel, J.L., Boutevin, B., and Ameduri, B., Controlled radical polymerization of 1,3-butadiene. II. Initiation by hydrogen peroxide and reversible termination by TEMPO. *Journal of Polymer Science Part a-Polymer Chemistry*, 2000, **38**(18), 3293.

109 Wendler, K. and Nghiep, P.H., Chain transfer-reactions in emulsion polymerization of butadiene styrene. *Makromolekulare Chemie-Macromolecular Symposia*, 1990, **31**, 249.

110 Flory, P.J., *Principles of polymer chemistry*. 1953, Cornell University Press, 688.

111 Odian, G., *Principles of polymerization*, 4e. 1970, Wiley-Interscience, 839.

112 Brandrup, J., Immergut, E.H., and Grulke, E.A., eds. *Polymer Handbook*, 4e, Vol. 2. 1998, Wiley.

113 Mansilla, M.A. et al., Evolution of the free volume and glass transition temperature with the degree of cure of polybutadiene rubbers. *Polymer Testing*, 2013, **32**(4), 686.

114 Krishnamoorti, R. et al., Thermodynamic interactions in polybutadiene blends. *Macromolecules*, 1998, **31**(7), 2312.

115 Friedman, E.M., Modality of molecular-weight distributions. *Polymer Engineering and Science*, 1990, **30**(10), 569.

116 Tobita, H., Yamamoto, Y., and Ito, K., Molecular-weight distribution in random cross-linking of polymers – modality of the molecular-weight distribution. *Macromolecular Theory and Simulations*, 1994, **3**(6), 1033.

117 Berger, K.C., Deb, P.C., and Meyerhoff, G., Radical vinyl polymerization – reactions of benzoyl peroxide during bulk-polymerization of styrene with labeled initiator. *Macromolecules*, 1977, **10**(5), 1075.

118 Lee, M.H. et al., Simulation approach to benzoyl peroxide decomposition kinetics by thermal calorimetric technique. *Journal of the Taiwan Institute of Chemical Engineers*, 2014, **45**(1), 115.

119 Bevington, J.C., The nature of the initiation reaction in the polymerization of styrene sensitized by benzoyl peroxide. *Proceedings of the Royal Society of London Series a-Mathematical and Physical Sciences*, 1957, **239**(1218), 420.

Index

d

e

k

l

m

n

o

p

Printed and bound by CPI Group (UK) Ltd, Croydon, CR0 4YY
21/09/2026
14965823-0003